Finite Elements in Water Resources

Proceedings of the 5th International Conference, Burlington, Vermont, U.S.A., June 1984

Editors:
J.P. Laible
C.A. Brebbia
W. Gray
G. Pinder

Springer-Verlag Berlin Heidelberg GmbH

J.P. LAIBLE
C.A. BREBBIA
W. GRAY
G. PINDER
Computational Mechanics Centre
Ashurst Lodge,
Ashurst,
Southampton,
Hampshire SO4 2AA
U.K.

British Library Cataloguing in Publication Data

Finite elements in water resources.
1. Water resources development
Mathematics 2. Finite element method
I. Laible, J.P.
628.1'61'01515353 TC409

ISBN 978-3-662-11746-0 ISBN 978-3-662-11744-6 (eBook)
DOI 10.1007/978-3-662-11744-6

Originally published by Computational Mechanics Centre, Southampton 1984
Softcover reprint of the hardcover 1st edition 1984

CONTENTS

SOLUTION OF THE NAVIER STOKES EQUATIONS

SEDIMENTATION PROCESSES

BOUNDARY ELEMENT METHOD

PREFACE

This book is the edited proceedings of the Fifth International Conference on Finite Elements in Water Resources, held at the University of Vermont, USA in June 1984. This Conference continues the successful series started at Princeton University in 1976, followed by the Conference in Imperial College, London, UK in 1978, the third Conference at the University of Mississippi, USA in 1980 and the fourth at the University of Hannover, Germany in 1982.

The objective of this Conference is to provide engineers and scientists interested in water resources with the state-of-the-art on finite element modelling. The Proceedings review the basic theory and applications of the technique in groundwater and seepage, transport phenomena, viscous flow, river, lake and ocean modelling. The fundamentals of the numerical techniques employed in finite elements are also discussed. Many applications illustrate the versatility and generality of the Finite Element Method for the simulation of a wide range of problems in water resources. More recent schemes, in particular, boundary elements, are also presented, together with a series of advanced numerical techniques.

The Conference has become an internationally accepted forum for the presentation of new developments of finite elements in water resources techniques. Because of this, a large number of abstracts were submitted to the Organizing Committee and it is our only regret that it was impossible to accept all these contributions. The overwhelming response to our Call for Papers has ensured the high quality of these proceedings.

The Organizing Committee gratefully acknowledges the sponsorship given by the following organisations:

National Science Foundation
International Society of Computational Methods in Engineering
International Association of Hydraulic Research
American Society of Civil Engineers
American Institute of Chemical Engineers
American Geophysical Union
Society for Computational Simulation
American Society of Engineering Education
Water Information Center
National Well Water Association

J.P. Laible
C. Brebbia
W. Gray
G.F. Pinder

(Editors).

The papers have been reproduced directly from the original material submitted by the authors who are totally responsible for them.

MATHEMATICS AND NUMERICAL TECHNIQUES

ADJOINT SENSITIVITY THEORY FOR THE FINITE ELEMENT METHOD

J.F. Sykes, Department of Civil Engineering, University of Waterloo
J.L. Wilson, Department of Geoscience, New Mexico Institute of Mining and Technology

INTRODUCTION

Sensitivity analysis is an important aspect of many model studies. The objective is to determine how sensitive model results are to perturbations in the model input or system parameters. The input of concern generally includes boundary conditions, forcing functions, model parameters and for transient analyses, initial conditions. The sensitivity of the model state variables can be investigated or alternatively the sensitivity of a selected performance measure or response function may be of interest. The performance measure which may be written as

$$P = P(\{\phi\}, \{\alpha\}) \tag{1}$$

is a function of the system state $\{\phi\}$ and the system parameters $\{\alpha\}$. For some problems, P may be independent of some or all of the parameters $\{\alpha\}$. In this paper $\{\phi\}$ and $\{\alpha\}$ represent column vectors of finite element nodal state variables and system parameters respectively.

In many analyses, the form of $P(\{\phi\}, \{\alpha\})$ can include

$$P = \sum_i (\phi - \phi')^2 \tag{2a}$$

$$P = \{g\}^T \{\phi\}. \tag{2b}$$

Equation (2a) represents a general optimization function given as the sum of the squares of the difference between predicted and measured (ϕ') state variables. In equation (2b) the performance measure is a linear function of the nodal state variables with $\{g\}^T$ being a row vector of arbitrary weights identifying the state variables of importance. Specific analyses such as groundwater flow may have a performance measure of the magnitude of the Darcy velocity at a selected point, the measure thus being a function of the hydraulic conductivity and the gradient of the state variable of hydraulic head. The performance measure of interest in contaminant transport problems is often mass discharge. Thus, P is a function of both the fluid flux and the species concentration.

For problems in mechanics, the performance measures may include stresses or strains at selected points throughout the domain of interest. Both stress and strain are functions of the state variable of displacement.

The sensitivity of the performance measure to a specific system parameter α_k is determined by differentiating equation (1) giving

$$dP/d\alpha_k = \frac{\partial P(\{\phi\},\{\alpha\})}{\partial \alpha_k} + \frac{\partial P(\{\phi\},\{\alpha\})}{\partial \{\phi\}^T}\{\psi_k\} \tag{3}$$

where the state sensitivity $\{\psi_k\} = \{d\phi/d\alpha_k\}$ is the sensitivity of the system state to the parameter α_k. The quantity $dP/d\alpha_k$ is a marginal sensitivity and gives the change in P resulting from a unit change in α_k. The first term on the right-hand-side of equation (3) represents a "direct effect" indicating the explicit dependence of P on α_k. The last term in equation (3) represents an indirect effect due to the implicit dependence of P on α_k through the state variables (Oblow, 1978).

Direct parameter sampling is commonly used to evaluate the marginal sensitivity of equation (3). In this costly trial-and-error procedure the α_k parameters are perturbed and the system model reanalyzed.

This paper develops an efficient alternative procedure for sensitivity determination. This procedure involves the development of an adjoint problem from the primary system model. For linear problems, the adjoint problem is similar in formulation to the dual problem in linear programming. With the adjoint methodology, the sensitivity of the performance measure to all of the system parameters can be determined with the solution of the primary problem and the adjoint problem.

The adjoint method of sensitivity analysis has been used in petroleum reservoir history matching (Chavent et al, 1975), in electrical engineering problems (Director and Rohrer, 1969) and in nuclear reactor assessments (Oblow, 1978). In comparison, the method has received little attention in fluid flow and contaminant transport analyses. Neuman (1980a, 1980b) used an adjoint methodology as part of an aquifer hydrology parameter estimation routine. Intera (1983) and Sykes et al (1983) applied adjoint sensitivity theory to a two-dimensional steady state finite element groundwater flow model. The resulting model was used as part of the performance assessment package for proposed high level nuclear waste repository sites.

The adjoint method can be applied to both linear (Sykes et al, 1983) and nonlinear analyses (Oblow, 1978; Cacuci, 1981). The adjoint problem can be developed from either the continuum equations of the primary problem or from its discretized equivalent. The discretized approach ensures that the sensitivity of the performance measure is obtained for the system parameters, spatial discretization and temporal discretization used in the primary problem.

ADJOINT FORMULATION FOR DISCRETE EQUATIONS

The finite element (or finite difference) method replaces the partial differential equations, boundary conditions and initial conditions describing a given system with algebraic equations of the generalized form

$$[A(\{\phi\}, \{\alpha\})]\{\phi\} = \{R(\{\phi\}, \{\alpha\})\} \tag{4}$$

Equation (4) represents a nonlinear system with both the banded coefficient matrix [A] and load term {R} being expressed as functions of the discretized state variable $\{\phi\}$. Both [A] and {R} are functions of the system spatial discretization, element interpolation functions, model parameters, boundary conditions and estimates of the state variable. Temporally discretized transient problems can be represented by equation (4). For a time increment, the discretized equations can be written as

$$[b(\{\phi\}, \{\alpha\})^t]\{\phi^{t-1}\} + [a(\{\phi\}, \{\alpha\})^t]\{\phi^t\} = \{r(\{\phi\},\{\alpha\})^t\} \tag{5}$$

where $\{\phi^{t-1}\}$ and $\{\phi^t\}$ are the nodal state variables at times t-1 and t respectively and the coefficient matrices $[a^t]$ and $[b^t]$ and the load vector $\{r^t\}$ are unspecified functions of the system parameters and state variables of the time period t-1 to t. For illustrative purposes, consider the entire model simulation to cover only three time increments. The coefficient matrix [A], the load vector {R} and the state variable $\{\phi\}$ of equation (4) then represent

$$[A(\{\phi\}, \{\alpha\})] = \begin{bmatrix} [a\ ] & 0 & 0 \\ [b^2] & [a^2] & 0 \\ 0 & [b^3] & [a^3] \end{bmatrix} \tag{6a}$$

$$\{R(\{\phi\}, \{\alpha\})\}^T = \{\ -\{\phi^\circ\}^T[b^1]^T + \{r^1\}^T, \{r^2\}^T, \{r^3\}^T\ \} \tag{6b}$$

$$\{\phi\}^T = \{\{\phi^1\}^T, \{\phi^2\}^T, \{\phi^3\}^T\} \tag{6c}$$

where $\{\phi^\circ\}^T$ is the vector of specified state variables at time zero.

The derivative of equation (4) with respect to a parameter α_k leads to the direct state sensitivity equation

$$\frac{d}{d\alpha_k}\left([A(\{\phi\}, \{\alpha\})]\{\phi\} - \{R(\{\phi\}, \{\alpha\})\}\right) =$$

$$\left[[A(\{\phi\}, \{\alpha\})] + \frac{\partial[A(\{\phi\},\{\alpha\})]}{\partial\{\phi\}^T}\{\phi\} - \frac{\partial\{R(\{\phi\},\{\alpha\})\}}{\partial\{\phi\}^T}\right]\{\psi_k\}$$

$$+ \{R_\psi(\{\phi\}, \{\alpha\})\} = 0 \qquad (7)$$

where

$$\{R_\psi(\{\phi\}, \{\alpha\})\} = \frac{\partial[A(\{\phi\},\{\alpha\})]}{\partial\alpha_k}\{\phi\} - \frac{\partial\{R(\{\phi\},\{\alpha\})\}}{\partial\alpha_k} \qquad (8)$$

While equation (4) is nonlinear in $\{\phi\}$, equation (7) is linear in $\{\psi_k\}$. Equation (7) uses the system parameters $\{\alpha\}$ and state variable $\{\phi\}$ developed from the system model equation (4). As a result, the state sensitivities $\{\psi_k\} = d\{\phi\}/d\alpha_k$ are local derivatives and may change with a change in the system parameters. Equation (7) must be formulated and solved for each parameter α_k of interest.

The adjoint problem is developed by multiplying the direct state sensitivity equation (7) by an arbitrary constant $\{\psi^*\}$ and subtracting the result from the marginal performance sensitivity as expressed by equation (3) yielding

$$\frac{dP}{d\alpha_k} = \frac{\partial P(\{\phi\},\{\alpha\})}{\partial\alpha_k} + \frac{\partial P(\{\phi\},\{\alpha\})}{\partial\{\phi\}^T}\{\psi_k\}$$

$$- \{\psi^*\}^T\left[[A(\{\phi\},\{\alpha\})] + \frac{\partial[A(\{\phi\},\{\alpha\})]}{\partial\{\phi\}^T}\{\phi\}\right.$$

$$\left.- \frac{\partial\{R(\{\phi\},\{\alpha\})\}}{\partial\{\phi\}^T}\right]\{\psi_k\} - \{\psi^*\}^T\{R_\psi(\{\phi\},\{\alpha\})\} \qquad (9)$$

Since $\{\psi^*\}$ is arbitrary, terms containing $\{\psi_k\}$ can be eliminated by letting

$$\{\psi_k\}^T\frac{\partial P(\{\phi\},\{\alpha\})}{\partial\{\phi\}} - \{\psi_k\}^T\left[[A(\{\phi\},\{\alpha\})] + \frac{\partial[A(\{\phi\},\{\alpha\}]}{\partial\{\phi\}^T}\{\phi\}\right.$$

$$\left.- \frac{\partial\{R(\{\phi\},\{\alpha\})}{\partial\{\phi\}^T}\right]^T\{\psi^*\} = 0 \qquad (10)$$

The adjoint problem is then defined as

$$\left[[A(\{\phi\},\{\alpha\})] + \frac{\partial [A(\{\phi\},\{\alpha\})]}{\partial\{\phi\}^T} \{\phi\} - \frac{\partial\{R(\{\phi\},\{\alpha\})\}}{\partial\{\phi\}^T} \right]^T \{\psi^*\}$$

$$= \frac{\partial P(\{\phi\},\{\alpha\})}{\partial\{\phi\}} \tag{11}$$

in which $\{\psi^*\}$ represents the nodal values of the adjoint state or importance function. The marginal performance sensitivity then becomes

$$\frac{dP}{d\alpha_k} = \frac{\partial P(\{\phi\},\{\alpha\})}{\partial\alpha_k} - \{\psi^*\}^T \{R_\psi(\{\phi\},\{\alpha\})\} \tag{12}$$

where all terms on the right-hand-side are defined.

The adjoint problem as defined by equation (11) is linear in $\{\psi^*\}$. The terms of the coefficient matrices and load vector are defined with the specification of the primary problem and performance measure respectively. The adjoint problem is a function of the parameters and state variables of the primary problem. Where the primary problem has specified values of the state variable ϕ, the corresponding nodal values of the adjoint state variable ψ^* are specified as zero in the adjoint problem. Neumann boundary conditions in the primary problem result in zero Neumann conditions on the corresponding boundaries of the adjoint problem.

The adjoint problem as compared to the primary problem is reversed in space and time. Transient analyses are solved backwards in time from a final condition of $\{\psi^{*t}\} = 0 : t =$ final to the initial time. Consider the transient problem of equation (6) and assume that both the coefficient matrix [A] and the load vector {R} are independent of $\{\phi\}$. Following Equation (11), the adjoint problem can be developed as

$$\begin{bmatrix} [a^1]^T & [b^2]^T & 0 \\ 0 & [a^2]^T & [b^3]^T \\ 0 & 0 & [a^3]^T \end{bmatrix} \begin{Bmatrix} \{\psi^{*o}\} \\ \{\psi^{*1}\} \\ \{\psi^{*2}\} \end{Bmatrix} = \begin{Bmatrix} \partial P/\partial\{\phi^1\} \\ \partial P/\partial\{\phi^2\} \\ \partial P/\partial\{\phi^3\} \end{Bmatrix} \tag{13}$$

The solution of equation (13) proceeds with the determination in order of $\{\psi^{*2}\}$, $\{\psi^{*1}\}$ and then $\{\psi^{*o}\}$. The time increments are those specified in the formulation of the $[a^t]$ and $[b^t]$ coefficient matrices. For temporal analyses, the primary and adjoint problems are also known as the forward and backward problems respectively.

The calculation of the marginal sensitivities of equation (12) requires the determination of the derivatives in $\{R_\psi\}$ of the coefficient matrix [A] and load vector { R} with respect to a parameter α_k. These derivatives can be determined on an element-by-element basis and then multiplied by the corresponding nodal values of $\{\psi^*\}$ and where required $\{\phi\}$ to yield element marginal sensitivities. These marginal sensitivities can then be added to reflect the zones of the primary problem having the same parameter value α_k.

For linear problems the derivatives in $\{R_\psi\}$ are relatively simple. For example, the vector of marginal sensitivities of P to the initial nodal values $\{\phi^o\}$ of a linear transient problem are from equations (6b), (8) and (12):

$$\frac{dP}{d\{\phi^o\}^T} = \frac{\partial P(\{\alpha\})}{\partial\{\phi^o\}^T} - \{\psi^{*o}\}^T \; [b^1] \tag{14}$$

Normalized sensitivity coefficients for a parameter α_k can be calculated from the marginal sensitivities of P to α_k by

$$S_k = \left(\frac{dP_k}{d\alpha_k}\right)\left(\frac{\alpha_k}{P}\right) = \frac{d(\ell n\ P)}{d(\ell n\ \alpha_k)} \tag{15}$$

These values describe the percentage change of performance measure P to a one percent change of parameter α_k. The net change in the performance measure P is given for an m parameter system by

$$dP = \{dP/d\alpha_k\}^T \{d\alpha\} \tag{16}$$

where $\{d\alpha\}$ is the m x 1 column vector of fractional perturbations of the parameters α_k (k=1,m) and $\{dP/d\alpha_k\}^T$ is the 1 x m row vector of marginal sensitivity coefficients.

PROCEDURE

The procedure for determining the marginal and normalized sensitivities of a performance measure using the adjoint method is as follows:

1. solve the primary problem equation (4) for the state variable $\{\phi\}$

2. select an appropriate performance measure P

3. determine the derivatives $\partial P(\{\phi\}, \{\alpha\})/\partial\{\phi\}^T$, $\partial[A(\{\phi\},\{\alpha\})]/\partial\{\phi\}^T$ and $\partial\{R(\{\phi\},\{\alpha\})\}/\partial\{\phi\}^T$

4. solve the adjoint state problem of equation (11) for the adjoint state variable or importance function $\{\psi^*\}$

5. determine the derivatives $\partial P(\{\phi\},\{\alpha\})/\partial\alpha_k$, $\partial[A(\{\phi\},\{\alpha\})]/\partial\alpha_k$ and $\partial\{R(\{\alpha\},\{\alpha\})\}/\partial\alpha_k$ for the α_k parameters of interest

6. calculate the performance measure marginal sensitivity coefficients of equation (12) and multiply when appropriate by (α_k/P) to yield the normalized sensitivity coefficients.

The majority of the work in the preceding procedure is associated with the determination of the derivatives of steps 3 and 5. For linear finite element problems these derivatives are quite simple and the solution effort is greatly reduced. For example, from equation (11) a linear primary problem having a symmetric coefficient matrix will have an adjoint equation of

$$[A(\{\alpha\})]\{\psi^*\} = \frac{\partial P(\{\phi\},\{\alpha\})}{\partial\{\phi\}} \qquad (17)$$

The primary problem equation (4) and the adjoint equation (17) differ in the load term only. Solution procedures should take advantage of this fact. The element-by-element calculation of the derivatives of steps 3 and 5 can follow a similar procedure to that used in developing the element coefficient matrices of the primary problem. Thus, through careful formulation of computer codes, the calculation of the sensitivity coefficients can be very efficient.

AN EXAMPLE

Following the methodology detailed in the preceding sections, an adjoint sensitivity analysis was developed for a two-dimensional finite element model of groundwater flow in an integrated depth confined aquifer. The governing steady state flow equation of the model is

$$\frac{\partial}{\partial x_i}[bK_{ij}\frac{\partial h}{\partial x_j}] + Q = 0 \qquad i,j = 1,2 \qquad (18)$$

where b is the aquifer thickness; K_{ij} is the hydraulic conductivity tensor; h is the hydraulic head; Q is the aquifer areal recharge or discharge and x_i represent a two-dimensional Cartesian coordinate system. Both prescribed head and prescribed flux boundary conditions are used in the model. Discretization is based on the four noded isoparametric quadrilateral element. The Galerkin finite element equations of the primary flow problem are linear and the coefficient matrix $[A(\{\alpha\})]$ is symmetric.

The selected performance measure, after equation (2b), is

$$P = \{g\}^T\{h\} \qquad (19)$$

where { h} is the vector of unspecified nodal hydraulic heads. The sensitivity of P to the parameters K_{ij}, Q, b and the prescribed heads of the boundary nodes are desired.

The adjoint equations that are formulated from the primary problem correspond to those of equation (18) with the load vector being

$$\partial P(\{h\})/\partial\{h\} = \{g\} \qquad (20)$$

The nodal adjoint state variable $\{\psi^*\}$ can be interpreted as the change in P given a unit volume influx of water at the nodes. The primary and adjoint problems were solved using a Gaussian elimination solver. As the coefficient matrices of both problems are identical, only one upper triangularization is necessary.

The use of the adjoint sensitivity model is illustrated by the analysis of groundwater flow in the Culebra dolomite aquifer of the Rustler formation in New Mexico. The spatial domain of interest includes the Waste Isolation Pilot Plant, a proposed site for the storage of radioactive wastes. The measured potentiometric surface is superimposed on the discretized spatial domain in Figure 1. The areas of recharge/discharge are shown in Figure 2. A distribution of the aquifer hydraulic conductivity that yields a calculated potentiometric surface that matches the measured surface is shown in Figure 3. The hydraulic head was specified for all boundary nodes. The weights { g} in the performance measure of equation (19) were selected as unity at the marked nodes of Figure 4 and zero at all other nodes.

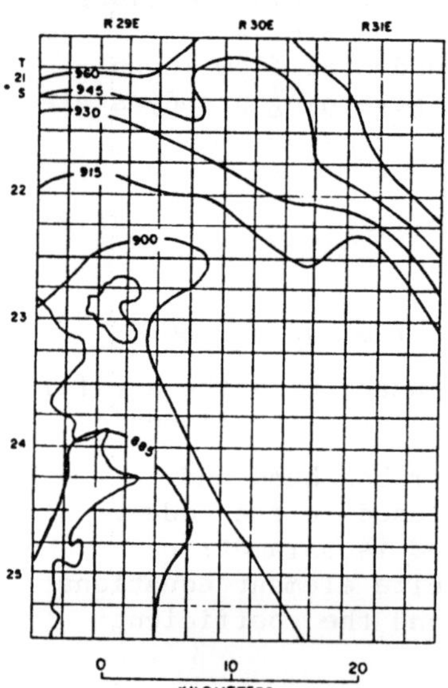

Figure 1: Measured Potentiometric Surface (m) Culebra Dolomite

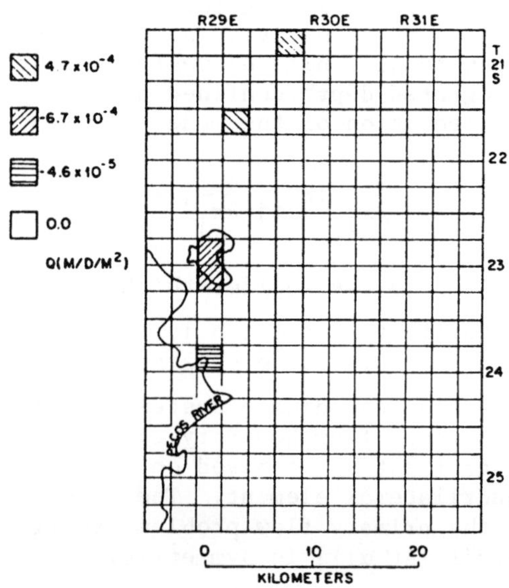

Figure 2: Recharge and Discharge Zones

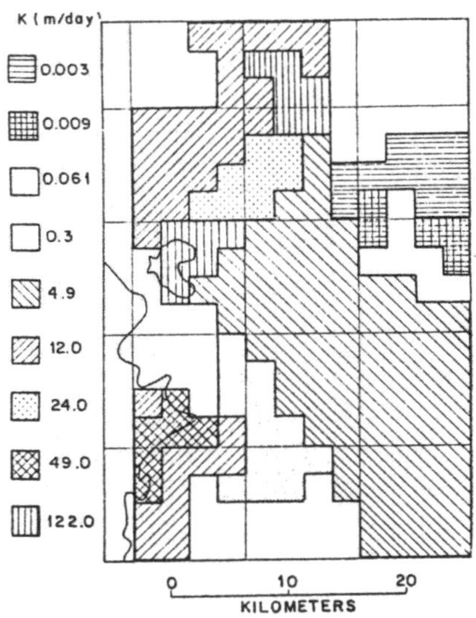

Figure 3: Hydraulic Conductivity Zones

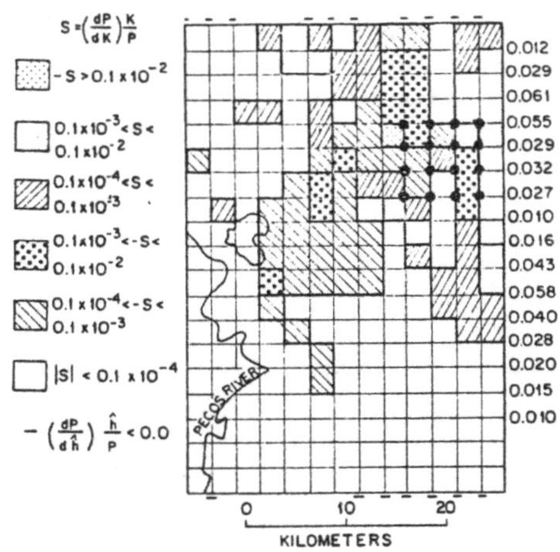

Figure 4: Normalized Sensitivity Coefficients for Element Hydraulic Conductivities and Prescribed Boundary Nodes

The normalized sensitivity coefficients for the performance measure are given in Figures 4 and 5. The values represent the percent change in P for a one percent change in the parameter value. A negative value indicates that a parameter increase/decrease will result in a decrease/increase of P. The sensitivity coefficients for the prescribed boundary heads are shown adjacent to the respective nodes in Figure 4. Sensitivities smaller than 0.01, in absolute value, are represented only by their sign. The selected performance measure is most sensitive to the boundary nodes adjacent to the higher conductivity zones on the right inflow boundary.

The sensitivity coefficients for the element hydraulic conductivities are shown for various orders of magnitude in Figure 4. The sensitivities for the elements downgradient of the region of importance are negative while as expected, the upgradient elements tend to have positive sensitivities. The values are largest in the vicinity of flow discharge and recharge and along the streamlines that pass through the performance measure nodes. The sensitivity coefficients for the aquifer recharge and discharge are shown in Figure 5.

CONCLUSIONS

The determination of the sensitivity of a performance measure to all of the system parameters of a finite element model requires the matrix solution of the primary problem for the state variable and an adjoint problem for the adjoint state variable. Costly trial-and-error sampling with repeated solving of the primary problem can be avoided. The utility of the adjoint method of sensitivity analysis is demonstrated in this study with the analysis of steady state groundwater flow. The method is equally applicable to transient and nonlinear problems.

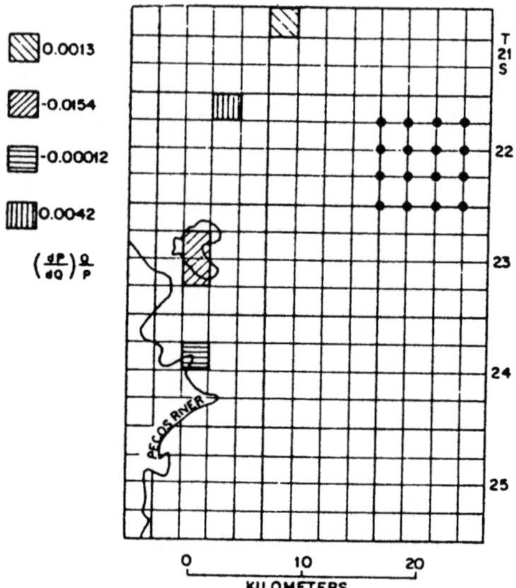

Figure 5: Normalized Sensitivity Coefficients for Recharge/Discharge

REFERENCES

Cacuci, D.G., (1981) "Sensitivity Theory for Nonlinear Systems No. 1: Nonlinear Functional Analysis", App. J. Math Phys., Vol. 22 No. 12, p2794-2802.

Chavent, G., M. Dupuy and P. Lemonnier, (1975) "History Matching by Use of Optimal Control Theory", Soc. Petrol. Eng. J., p74-86, February.

Director, S.W. and R.A. Rohrer, (1969) "The Generalized Adjoint Network and Network Sensitivities", IEEE Trans-Circuit Thy, CT-16, pp. 313-323.

Dogru, A.H. and J.H. Seinfeld, (1981) "Comparison of Sensitivity Coefficient Calculation Methods in Automatic History Matching", Soc. Petrol. Eng. J., p551-557, October.

Intera Environmental Consultants Inc., (1983) "Adjoint Sensitivity Theory for Steady-State Ground-Water Flow", ONWI-515. Battelle Memorial Institute, Columbus, Ohio.

Neuman, S.P., (1980a) "Adjoint-State Finite Element Equations for Parameter Estimation", Proc. of Third Int. Congress on Finite Elements in Water Res., Univ. of Mississippi, p2.66-2.75.

Neuman, S.P., (1980b) 1 "A Statistical Approach to the Inverse Problem of Aquifer Hydrology #3. Improved Solution Method and Added Perspective", WRR, 16(2), p331-346.

Oblow, E.M. (1978) "Sensitivity Theory for General Nonlinear Algebraic Equations with Constraints", Nuclear Sci. Eng., 65, p187-191.

Sykes, J.F., J.L. Wilson and R.W. Andrews (1983) "Adjoint Sensitivity Theory for Steady-State Ground-Water Flow", Submitted to Water Resources Research.

WHY UPWINDING IS REASONABLE

Myron B. Allen

University of Wyoming

INTRODUCTION

Upwind-biased discrete approximations have a distinguished history in numerical fluid mechanics, dating at least to von Neumann and Richtmyer (1950). Lately, however, upwinding has come under fire in water resources engineering. Among the most effective critics of upwind techniques are Gresho and Lee (1980), who take umbrage at the smearing of steep gradients in solutions of partial differential equations. While this viewpoint has cogency, a blanket condemnation of upwinding would be injudicious. There exist fluid flows for which upstream-biased discretizations are not only valid but in fact mathematically more appropriate than central approximations having higher-order accuracy.

Figures 1 and 2 illustrate the source of the controversy. Both plots show numerical solutions to a convection-dominated species transport equation using finite-element collocation. Figure 1, the result of a centered scheme, shows a solution having unrealistic wiggles near the concentration front; Figure 2, from an upwind scheme, exhibits nonphysical smearing. The wiggles in the centered scheme disappear altogether when the spatial step Δx is small enough, whereas the smearing associated with upstream weighting decreases continuously with Δx. Gresho and Lee argue that the wiggles indicate an inappropriate spatial grid and that suppressing them via upwinding eliminates useful symptoms in favor of a less informative flaw, smearing.

Were wiggles the only difficulty with centered schemes, proscribing upwind methods might be in order. However, as we shall see, for certain types of equations centered schemes can fail to converge. This difficulty is not symptomatic of an unsuitable grid; rather, it betrays an inability of centered schemes to impose proper uniqueness criteria. For such equations, upwinding can be reasonable.

SOME UPWINDING TECHNIQUES

Of various discrete methods used in numerical fluid mechanics, finite differences, Galerkin methods, and finite-element collocation have proved to be among the most attractive. A brief review of techniques in each of these

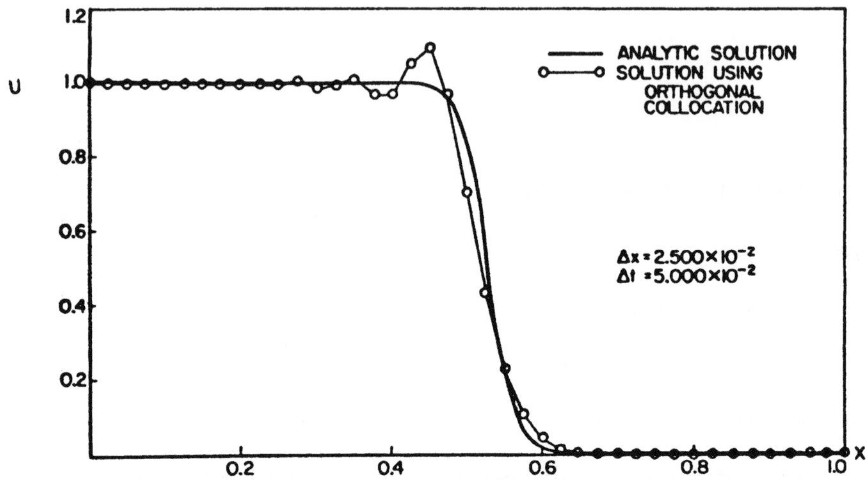

Figure 1. Solution to the convection-diffusion equation using standard collocation on finite elements.

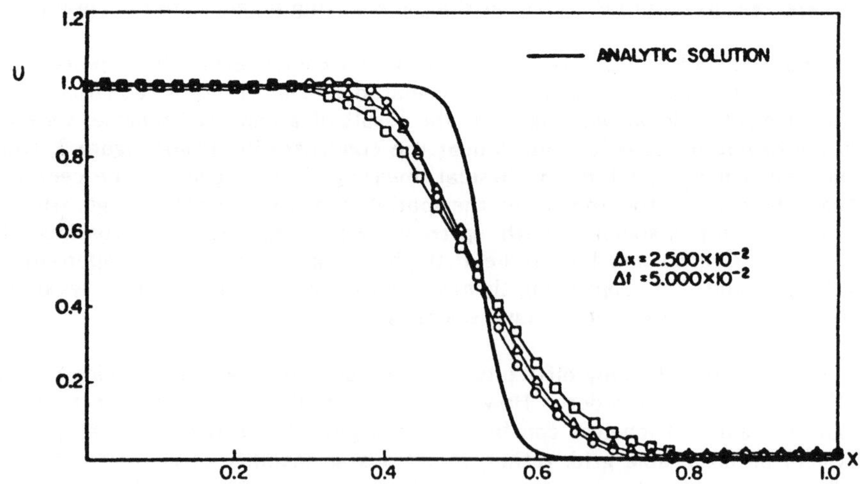

Figure 2. Solutions to the convection-diffusion equation using several choices of upstream collocation points.

make matters concrete, consider as a paradigm of convection-dominated flows the constant-coefficient convection-diffusion problem

$$\begin{gathered}\partial u/\partial t + \partial u/\partial x - Pe^{-1}\partial^2 u/\partial x^2 = 0 \\ u(x,0) = 0, \quad x \in (0,1) \\ u(0,t) = 1 \quad \text{and} \quad \partial u(1,t)/\partial x = 0, \quad t \geq 0 \end{gathered} \tag{1}$$

where x and t are dimensionless space and time coordinates, $u(x,t)$ denotes a normalized concentration, and the Peclet number Pe measures the degree to which convection dominates diffusion.

As commonly implemented, each of the discrete interior methods calls for a partition Δ_N: $0 = x_0 < x_1 < \ldots < x_{N+1} = 1$ of the spatial domain. Suppose for simplicity that Δ_N has uniform mesh Δx. Then the following finite-difference analog of Equation (1) has truncation error $O(\Delta x^2)$:

$$du_i/dt = -(2\Delta x)^{-1}(u_{i+1}-u_{i-1}) + (Pe\ \Delta x^2)^{-1}(u_{i+1}-2u_i+u_{i-1}) \tag{2}$$

where $u_i(t)$ signifies the approximate solution at $x = i\Delta x$ and time t. The problem with Equation (2) is that, unless Δx is sufficiently small, the numerical solution exhibits spurious wiggles near sharp concentration fronts. A desire to avoid these wiggles in favor of smearing prompts many analysts to resort to upwind schemes.

The simplest way to derive an upwind scheme from Equation (2) is to replace the analog of the convective term $\partial u/\partial x$ by a one-sided difference, yielding

$$du_i/dt = -\Delta x^{-1}(u_i-u_{i-1}) + (Pe\ \Delta x^2)^{-1}(u_{i+1}-2u_i+u_{i-1}) \tag{3}$$

The truncation error of Equation (3), $O(\Delta x)$, is larger than that for Equation (2), and writing the lowest error term explicitly shows

$$\begin{aligned} &du_i/dt + \Delta x^{-1}(u_i-u_{i-1}) - (Pe\ \Delta x^2)^{-1}(u_{i+1}-2u_i+u_{i-1}) \\ &\quad = \partial u/\partial t + \partial u/\partial x - (Pe^{-1}+\Delta x/2)\partial^2 u/\partial x^2 + O(\Delta x^2) \end{aligned} \tag{4}$$

where the terms on the right are evaluated at $x = x_i$. It is clear that the upwind difference scheme augments physical diffusion by an amount proportional to Δx. Hence by sacrificing one order of spatial accuracy one can suppress nonphysical wiggles at the cost of numerically induced dissipation.

Upwinding techniques also exist for finite-element Galerkin schemes. One such technique is the use of upstream-biased test functions proposed by Heinrich et al. (1977). Consider the standard Galerkin method applied to Equation (1). This method seeks a trial function

$$\hat{u}(x,t) = u_\partial(x) + \sum_{i=1}^{N} u_i(t)L_i(x)$$

approximating the true solution $u(x,t)$. Here $u_\partial(x)$ is a chapeau function on Δ_N satisfying the boundary conditions and vanishing at each interior node x_i, $i = 1,\ldots,N$, and the functions $L_i(x)$ are elements of the chapeau basis on Δ_N.

To determine the nodal values $u_i(t)$, we force the residual $\partial\hat{u}/\partial t + \partial\hat{u}/\partial x - Pe^{-1}\partial^2\hat{u}/\partial x^2$ to be orthogonal to each basis function $L_j(x)$, $j = 1,\ldots,N$, with respect to the inner product $\langle f,g\rangle = \int_0^1 f(x)g(x)dx$. This requirement leads to a set of N ordinary differential equations for the $u_i(t)$:

$$\sum_{i=1}^{N} (\langle L_i,L_j\rangle\, du_i/dt + \langle L_i',L_j\rangle u_i + Pe^{-1}\langle L_i',L_j'\rangle u_i) = 0, \quad j = 1,\ldots,N \tag{5}$$

Computing the integrals then gives

$$\frac{\Delta x}{6}\frac{d}{dt}(u_{j-1}+4u_j+u_{j+1}) + \frac{1}{2}(u_{j+1}-u_{j-1}) - (\Delta x\, Pe)^{-1}(u_{j+1}-2u_j+u_{j-1}) = 0 \tag{6}$$

Equation (6) is similar to the centered difference formula, Equation (3), the only difference being a peculiar spatial average of time derivatives in Equation (6). This scheme, like Equation (3), produces wiggles when Δx is too large.

The method advanced by Heinrich et al. modifies Equation (5) by replacing the test functions in the convective term with asymmetric functions $L_j^*(x) = L_j(x) + \alpha A_j(x)$, where

$$A_j(x) = \begin{cases} 2[\Delta x^{-2}(x-x_{j-1})^2 - \Delta x^{-1}(x-x_{j-1})], & x \in [x_{j-1},x_j] \\ -2[\Delta x^{-2}(x-x_j)^2 - \Delta x^{-1}(x-x_j)], & x \in [x_j,x_{j+1}] \\ 0, \quad x \notin [x_{j-1},x_{j+1}] \end{cases}$$

and $\alpha > 0$. Thus the integral $\langle L_i',L_j^*\rangle$ appears in Equation (5) instead of $\langle L_i',L_j\rangle$, and the upstream-weighted Galerkin scheme differs from Equation (6) by $\alpha\langle L_i',A_j\rangle$. Simple calculation shows

$$\langle L_i',A_j\rangle = \begin{cases} -1/2, & i = j\pm1 \\ 1, & i = j \\ 0, & |i-j| > 1 \end{cases}$$

so the scheme proposed by Heinrich et al. reduces to

$$\frac{\Delta x}{6}\frac{d}{dt}(u_{j-1}+4u_j+u_{j+1}) + \frac{1}{2}(u_{j+1}-u_{j-1}) - \Delta x^{-1}(Pe^{-1}+\alpha\Delta x/2)(u_{j+1}-2u_j+u_{j-1}) \tag{7}$$

Therefore in a manner analogous to Equation (3), Equation (7) augments physical diffusion by an amount proportional to Δx, and this numerical dissipation mitigates wiggles at the expense of smearing.

Shapiro and Pinder (1981) introduce a related upwinding technique for use with finite-element collocation. Their approach entails the use of Hermite cubic interpolating bases (see Prenter, 1975, Chapter 3), except they perturb the trial function in the convective term by a piecewise quartic biased in the upstream direction. Shapiro and Pinder present a detailed Fourier analysis showing the dissipative effects of their upstream weighting on the propagation of sharp fronts.

There is another upwinding technique for finite-element collocation. Consider the standard implementation, in which the trial function has the form

$$\hat{u}(x,t) = u_{\partial}(x) + \sum_{i=1}^{N} [u_i(t)H_{0i}(x) + u_i'(t)H_{1i}(x)]$$

Here the coefficients $u_i(t)$, $u_i'(t)$ approximate the nodal values $u(x_i,t)$, $\partial u(x_i,t)/\partial x$, respectively, and $\{H_{0i},H_{1i}\}_{i=0}^{N+1}$ is the basis for Hermite cubic interpolation on Δ_N. The standard collocation method, which has truncation error $O(\Delta x^4)$, requires the residual to vanish,

$$\frac{\partial \hat{u}}{\partial t}(\bar{x}_k,t) + \frac{\partial \hat{u}}{\partial x}(\bar{x}_k,t) - Pe^{-1}\frac{\partial^2 \hat{u}}{\partial x^2}(\bar{x}_k,t) = 0 \tag{8}$$

at each of $2N$ collocation points $\bar{x}_k = x_i + \Delta x/2 \pm \Delta x/\sqrt{3}$, $i = 1,\ldots,N$. As Figure 1 shows, unless Δx is sufficiently small the scheme generates wiggles near sharp fronts (Jensen and Finlayson, 1980).

A technique called upstream collocation (Allen and Pinder, 1983) offers a simple remedy to the wiggles, at the usual cost of smearing as shown in Figure 2. To implement the technique, simply shift the collocation points $\bar{x}_k$ in the convective term of Equation (8) to upstream points $\bar{x}_k^* = \bar{x}_k - \zeta\Delta x$. This gives

$$\frac{\partial \hat{u}}{\partial t}(\bar{x}_k,t) + \frac{\partial \hat{u}}{\partial x}(\bar{x}_k^*,t) - Pe^{-1}\frac{\partial^2 \hat{u}}{\partial x^2}(\bar{x}_k,t) = 0 \tag{9}$$

in which the differentiated basis-function values $H_{mi}'(\bar{x}_k^*)$ appear in the convective term instead of $H_{mi}'(\bar{x}_k)$. By Taylor's theorem, the difference between these two values is $-\zeta\Delta x\, H_{mi}''(\bar{x}_k) + (\zeta^2\Delta x^2/2)H_{mi}'''(\bar{x}_k)$. Thus, to within $O(\Delta x^2)$, Equation 9 is equivalent to

$$\frac{\partial \hat{u}}{\partial t}(\bar{x}_k,t) + \frac{\partial \hat{u}}{\partial x}(\bar{x}_k,t) - (Pe^{-1} + \zeta\Delta x)\frac{\partial^2 \hat{u}}{\partial x^2}(\bar{x}_k,t) = 0 \tag{10}$$

It is clear from this equation that numerical diffusion is again the mechanism by which the scheme mitigates wiggles.

Upstream collocation is closely related to an upwind Galerkin scheme introduced by Hughes (1978). This latter scheme involves numerical evaluation of the Galerkin integrals using quadrature points shifted upstream from the Gauss points. In fact, one can show an algebraic correspondence between upstream collocation and a variant of Hughes' method using reduced integration on Hermite trial spaces (Allen, to appear).

A NONLINEAR HYPERBOLIC EQUATION

Numerical dissipation makes upwinding attractive to modelers wishing to avoid wiggles in convection-dominated parabolic flows. It is precisely such motives that provoke justified ire in Gresho and Lee. There is, nevertheless, another motive for using upwind-biased schemes. Many physical systems combine minute dissipation with nonlinearity, obeying governing equations that are effectively hyperbolic. For these systems high-order discrete schemes may be mathematically inappropriate, not because they generate wiggles, but because they fail to converge. Upwinding techniques then provide reasonable alternatives.

The Buckley-Leverett problem furnishes a simple example of a nonlinear hyperbolic equation for which high-order approximations fail. Consider a typical Cauchy problem for this equation:

$$\frac{\partial S}{\partial t} + \frac{\partial}{\partial x} f(S) = 0, \quad x,t > 0$$
$$S(0,t) = 1 - S_{or}, \quad S(x,0) = S_{wr} \tag{11}$$

Here S stands for water saturation; f(S) is a nonconvex, monotonically increasing function giving the flux of S; and S_{or} and S_{wr} are the minimum oil and water saturations, respectively, for the rock-fluid mixture. Equation (11) models immiscible flows in porous media in which capillarity exerts a negligible influence on fluid velocities. A prime feature of Equation (11) is the propagation of a saturation shock through the porous medium. This problem serves as a prototype for many kinds of nonlinear, hyperbolic or nearly hyperbolic systems of flow equations that occur in applications where convective forces are dominant.

It is widely known that spatially centered approximations to Equation (11) can converge to incorrect solutions. Allen and Pinder (1983), for example, examine the finite-element collocation approximation to this problem using both the conventional formulation and the upstream collocation scheme discussed above. As Figures 3 and 4 show, the conventional method predicts a saturation shock that is too slow and too strong, whereas the upstream method gives good approximations to the true shock. No amount of grid refinement can correct the failure of the conventional scheme. The difficulty here is not one of spurious wiggles; it is a deeper problem concerning the suitability of numerical methods from a mathematical standpoint.

Others have reported results similar to those displayed in Figures 3 and 4. Huyakorn and Pinder (1978) use the upstream-weighted Galerkin scheme reviewed above to overcome convergence difficulties with the standard Galerkin method in solving Equation (11). Mercer and Faust (1977) accomplish the same end by adding an adjustable capillary term to the discrete equations. Shapiro and Pinder (1980) use their upstream-weighted collocation method to produce convergent solutions to Equation (11). Indeed, upstream weighting has become standard practice for immiscible flow modeling in the oil industry (Aziz and Settari, 1979, Chapter 5).

UNIQUENESS AND HYPERBOLIC CONSERVATION LAWS

To see why upwind schemes converge for the Buckley-Leverett equation and similar problems, it is useful to review some mathematical facts about Equation (11). This equation is a quasilinear hyperbolic conservation law. When such equations have nonconvex flux functions like f(S), one cannot expect Cauchy problems for the equations to possess classical solutions. Instead, one may have to settle for weak solutions, defined for Equation (11) by the integral criterion

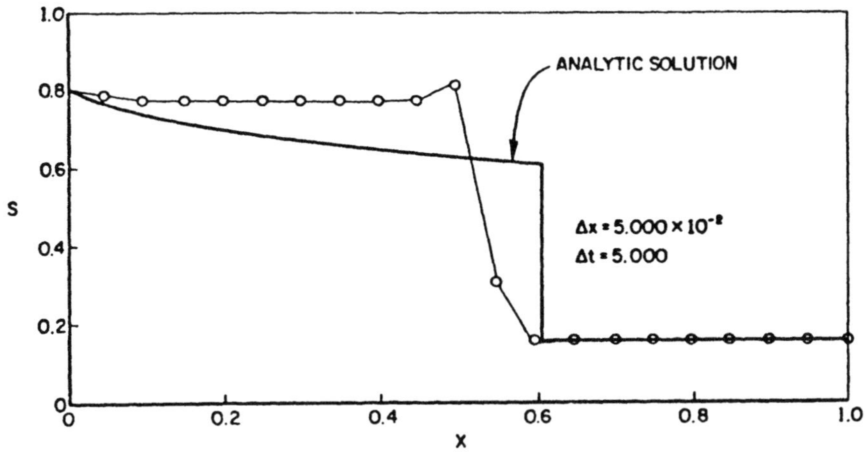

Figure 3. Solution to the Buckley-Leverett problem using standard collocation on finite elements.

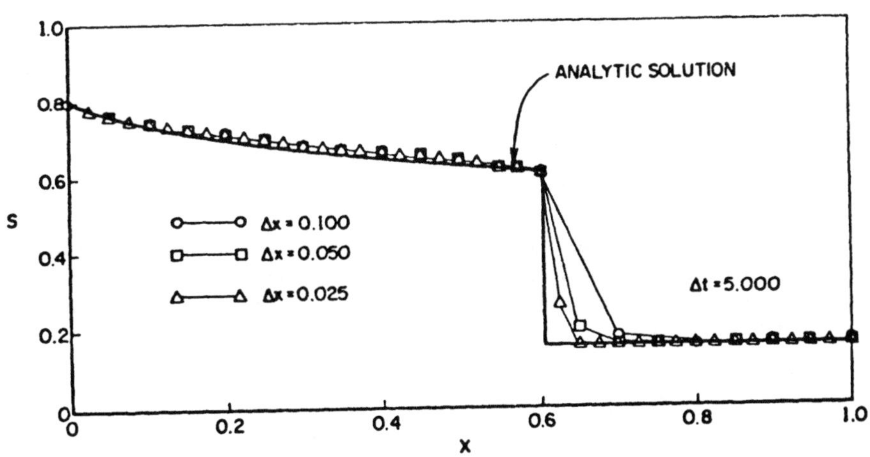

Figure 4. Solutions to the Buckley-Leverett problem using upstream collocation with spatial grids of varying mesh.

$$\int_0^\infty \int_0^\infty \left[S \frac{\partial \phi}{\partial t} + f(S) \frac{\partial \phi}{\partial x} \right] dx\, dt = 0$$

for any C^∞ real-valued function $\phi(x,t)$ with compact support. This criterion reduces to Equation (11) when $S(x,t)$ is continuously differentiable but also admits solutions $S(x,t)$ having shocks.

However, Cauchy problems like Equation (11) may not have unique weak solutions. To guarantee uniqueness for general initial data requires an additional constraint. The correct constraint, or shock condition, requires the weak solution to depend continuously and stably on the initial data. Equivalently, characteristic curves emanating from both sides of a discontinuity must intersect the curve on which the initial data are given. Oleinik (1963) proves a uniqueness condition on weak solutions that is mathematically equivalent to the shock condition but has more immediate implications for discrete approximations. Her criterion essentially states that the solution to the hyperbolic equation must be the limit of solutions, for comparable data, to a parabolic equation differing from the hyperbolic one by a dissipative second-order term of vanishing influence. In gas dynamics, this second-order term is called 'vanishing viscosity'; for the Buckley-Leverett problem the term 'vanishing capillarity' is perhaps more appropriate.

High-order, spatially centered discretizations of the Buckley-Leverett problem, though formally consistent with Equation (11), yield approximate weak solutions that are physically and mathematically incorrect. From a physical standpoint the neglected capillary term in Equation (11), which has the form

$$-\frac{\partial}{\partial x}\left[F(S)\frac{\partial S}{\partial x}\right], \quad F(S) \geq 0 \tag{12}$$

exerts an important influence in a microscopic region of what appears to macroscopic observers as a saturation shock. Thus while the global effects of capillarity may be legitimately neglected in the macroscopic flow equation, some device must remain to guarantee that the solution $S(x,t)$ respect the microscopic physics. Artificial capillarity is such a device.

It is a relatively simple matter to see how various upwinded approximations contribute artificial capillarity. For example, an upstream-weighted difference approximation to Equation (11) that is analogous to Equation (3) yields a flux term that has the form

$$\begin{aligned}\Delta x^{-1}(f_i - f_{i-1}) &= \left.\frac{\partial f}{\partial x}\right|_i - (\Delta x/2)\left.\frac{\partial^2 f}{\partial x^2}\right|_i + O(\Delta x^2)\\ &= \left.\frac{\partial f}{\partial x}\right|_i - (\Delta x/2)\left.\frac{\partial}{\partial x}\left[f'(S)\frac{\partial S}{\partial x}\right]\right|_i + O(\Delta x^2)\end{aligned} \tag{13}$$

Since $f'(S) \geq 0$, it is clear that the lowest error term here mimics the missing physical capillarity of Equation (12), while preserving consistency.

Similarly, the use of upstream-biased test functions in the Galerkin scheme analogous to Equation (7) yields approximations to the flux term of Equation

(11) having an error

$$\langle \partial\hat{f}/\partial x, A_j\rangle = \sum_{i=1}^{N} f_i \alpha \langle L_i', A_j\rangle$$

$$= -(\alpha/2)[f_{j+1} - 2f_j + f_{j-1}] = -(\alpha\Delta x^2/2)\left[\frac{\partial^2 \hat{f}}{\partial x^2}\Big|_j + O(\Delta x^2)\right]$$

$$= -(\alpha\Delta x^2/2)\,\frac{\partial}{\partial x}\left[f'(\hat{S})\,\frac{\partial \hat{S}}{\partial x}\right]_j + O(\Delta x^2)$$

where $f_i = f(\hat{S}_i(t))$ and $\hat{S}$ is the trial function approximating S. Notice that, parallelling Equation (13), the use of asymmetric test functions adds a numerical capillarity that is $O(\Delta x)$ smaller than the approximations to physical terms in Equation (11).

Upstream collocation also adds artificial capillarity to the Buckley-Leverett problem. In this case we project the flux term of Equation (11) to a Hermite interpolating space and collocate the result $\partial\hat{f}/\partial x$ at points $\bar{x}_k^* = \bar{x}_k - \zeta\Delta x$ upstream of the usual collocation points. A Taylor expansion shows

$$\frac{\partial \hat{f}}{\partial x}(\bar{x}_k^*,t) - \frac{\partial \hat{f}}{\partial x}(\bar{x}_k,t) = -\zeta\Delta x\,\frac{\partial^2 \hat{f}}{\partial x^2}(\bar{x}_k,t) + O(\Delta x^2)$$

$$= -\zeta\Delta x\,\frac{\partial}{\partial x}\left[f'(\hat{S}(\bar{x}_k,t))\,\frac{\partial \hat{S}}{\partial x}(\bar{x}_k,t)\right] + O(\Delta x^2)$$

Again, the upwind scheme adds the necessary artificial dissipation in the form of a 'vanishing capillarity.'

Various physics give rise to effectively hyperbolic systems for which uniqueness of weak solutions is an important issue. For such systems formal neglect of dissipation, while arguably valid from an engineering viewpoint, necessitates a device like upwinding to guarantee qualitatively correct numerical solutions. Other examples of interest in water resources engineering include hydraulic jumps (Whitham, 1974, Chapter 13) and wetting fronts in variably saturated soils (Nakano, 1980).

CONCLUSIONS

Upwinding can serve two purposes: it can suppress wiggles or, for certain equations, it can guarantee convergence. As Gresho and Lee observe, the first purpose is largely cosmetic, and the attendant smearing may be a more difficult flaw to recognize than spurious oscillations. However, in the case of conservation laws with discontinuous weak solutions upwinding can be a legitimate practice. The aim in this second case is to formulate consistent approximations that have built-in mechanisms for accommodating the peculiarities of hyperbolic or nearly hyperbolic flows. The lower spatial accuracy inherent in upwind schemes is far preferable to the convergence failures of higher-order schemes. Indeed, in the neighborhood of a discontinuity the very notion of 'order of accuracy' can be problematic.

Despite the mathematical validity of upwinding, the problem of smearing remains. While numerical dissipation vanishes with Δx, in practice Δx never vanishes and may be so large that artificial smearing is unacceptable. One of the most promising remedies to this difficulty is adaptive local grid refinement. Here one uses a spatial grid having smaller elements in portions of the flow field where steep solution gradients drive numerical dissipation. Algorithms combining local grid refinement with upwinding allow both for convergence to correct weak solutions and for the reduction of artificial smearing near sharp fronts. There is no denying the formidable coding difficulties in adaptive local grid refinement; however, progress in this field is encouraging (see Ewing, to appear).

ACKNOWLEDGEMENTS

Partial support for this work came from National Science Foundation grant number NSF-CEE-8111240 and from the Wyoming Water Research Center.

REFERENCES

1. Allen, M.B. (to appear), "How upstream collocation works," Int. J. Num. Meth. Engrg.
2. Allen, M.B., and Pinder, G.F. (1983), "Collocation simulation of multiphase porous-medium flow," Soc. Pet. Eng. J. (February), 135-142.
3. Aziz, K., and Settari, A. (1979), Petroleum Reservoir Simulation, London: Applied Science.
4. Ewing, R. (to appear), "Adaptive mesh refinement in reservoir simulation applications," to be presented at the International Conference on Accuracy Estimates and Adaptive Refinement in Finite Element Computations, Lisbon, Portugal, June, 1984.
5. Gresho, P.M., and Lee, R.L. (1980), "Don't suppress the wiggles--they're telling you something!", in Finite Element Methods for Convection Dominated Flows, ed. by T.J.R. Hughes, New York: American Society of Mechanical Engineers, 37-61.
6. Heinrich, J.C., Huyakorn, P.S., Zienkiewicz, O.C., and Mitchell, A.R. (1977), "An 'upwind' finite element scheme for two-dimensional convective transport equation," Int. J. Num. Meth. Engrg. 11, 131-143.
7. Hughes, T.J.R. (1978), "A simple scheme for developing upwind finite elements," Int. J. Num. Meth. Engrg. 12, 1359-1365.
8. Huyakorn, P.S., and Pinder, G.F. (1978), "A new finite element technique for the solution of two-phase flow through porous media," Adv. Water Resources 1:5, 285-298.
9. Jensen, O.K., and Finlayson, B.A. (1980), " Oscillation limits for weighted residual methods applied to convective diffusion equations," Int. J. Num. Meth. Engrg. 15, 1681-1689.
10. Mercer, J.W., and Faust, C.R. (1977), "The application of finite-element techniques to immiscible flow in porous media," presented at the First International Conference on Finite Elements in Water Resources, Princeton, NJ, July, 1976.
11. Nakano, Y. (1980), "Application of recent results in functional analysis to the problem of wetting fronts," Water Res. Research 16, 314-318.
12. Oleinik, O.A. (1963), "Construction of a generalized solution of the Cauchy problem for a quasi-linear equation of the first order by the introduction of a 'vanishing viscosity'," Am. Math. Soc. Transl. (Ser. 2), 33, 277-283.
13. Prenter, P.M. (1975), Splines and Variational Methods, New York: Wiley.
14. Shapiro, A.M., and Pinder, G.F. (1981), "Analysis of an upstream weighted approximation to the transport equation," J. Comp. Phys. 39,

46-71.

15. Shapiro, A.M., and Pinder, G.F. (1982), "Solution of immiscible displacement in porous media using the collocation finite element method," presented at the Fourth International Conference on Finite Elements in Water Resources, Hannover, FRG, June, 1982.
16. Von Neumann, J., and Richtmyer, R.D. (1950), "A method for the numerical calculations of hydrodynamical shocks," J. Appl. Phys. 21, 232-237.
17. Whitman, G.B. (1974), Linear and Nonlinear Waves, New York: Wiley.

A COUPLED FINITE ELEMENT MODEL OF FLOW IN POROUS LAYERED MEDIA

G. Gambolati[1], A.M.Perdon[1] and G.Ricceri[2]

1- Istituto di Matematica Applicata
2- Istituto di Costruzioni Marittime e Geotecnica
University of Padua, Italy

INTRODUCTION

The set of differential equations describing the behavior of the flow field and the stress field in deforming porous media were originally derived by Biot (1941) for a general three-dimensional system under the assumption of linearity. However coupled models of flow based on the solution of the Biot equations are not frequently used because of their numerical complexity and of the heavy computational burden required by their application to aquifer basins of realistic size.

Soil engineers who are concerned with consolidation problems of small dimension first provided finite element solutions to the coupled equations and used the results to predict the settlement of building areas. Consolidation analyses by finite elements were carried out by Sandhu and Wilson (1969), Christian and Boehmer (1970), Hwang et al. (1971), Desai (1975) , Sandhu (1976), Smith and Hobbs (1976) and Verruijt (1977). Although some authors have alerted to the risk of numerical instabilities (e.g. Booker and Small (1975) and Vermeer and Verruijt (1981)) none has paid adequate attention to the problem of efficiency which is particularly important if natural systems are to be dealt with.

In the present paper an efficient finite element model is developed for the solution of the coupled equations of flow in layered transversally anisotropic soils in the presence of radial symmetry. A fast solver is set up wherein the finite element matrix is stored in compact form and preliminarily reordered with independent row and column permutations. These are intended to minimize the number of elementary operations in the triangular factorization and to mantain the triangular factors as sparse as possible. The efficiency of the model is further increased by symbolically factorizing the f.e. matrix, i.e. by finding in advance the positions of the non-zero coefficients of the triangular factors. If the time step is kept constant the numerical factorization is performed once as well and the backward/forward substitution is the only operation to be repeatedly executed during the entire simulation. In this way a great saving of both computer storage and processing time may be achieved.

The model is first tested on available analytical solutions and then applied to a consolidation problem with real-world data. Future analyses will investigate the applicability range of the traditional flow equation in three-dimensional aquifer systems of realistically large size.

COUPLED MODEL AND FINITE ELEMENT SOLUTION

Under the assumption of linear elastic consolidation the coupled model of flow is given by the following set of equations.

Equilibrium equations of the solid matrix :

$$\frac{\partial \sigma_r'}{\partial r} + \frac{\partial \tau_{rz}}{\partial z} + \frac{\sigma_r' - \sigma_\theta'}{r} = \frac{\partial p}{\partial r}$$

$$\frac{\partial \sigma_z'}{\partial z} + \frac{\partial \tau_{rz}}{\partial r} + \frac{\tau_{rz}}{r} = \frac{\partial p}{\partial z} \qquad (1)$$

$$\frac{\partial \sigma_\theta'}{\partial \theta} = 0$$

where σ_θ', σ_r', σ_z' are the effective inter-granular normal stress, p is the pore pressure (assumed to be positive in compression)and τ_{rz} is the shear stress.

Hook's law for transversally anisotropic soils:

$$\varepsilon_z = \frac{1}{E_2}\left[\sigma_z' - \nu_2\left(\sigma_r' + \sigma_\theta'\right)\right]$$

$$\varepsilon_r = \frac{1}{E_1}\left[\sigma_r' - \nu_1\sigma_\theta' - \nu_2\frac{E_1}{E_2}\sigma_z'\right] \qquad (2)$$

$$\varepsilon_\theta = \frac{1}{E_1}\left[\sigma_\theta' - \nu_1\sigma_r' - \nu_2\frac{E_1}{E_2}\sigma_z'\right]$$

$$\gamma_{rz} = \frac{\tau_{rz}}{G_2}$$

where the meaning of Young's moduli E_1 and E_2, Poisson's ratios ν_1 and ν_2 and shear modulus G_2 is shown in Figure 1 (G_1 is dependent on E_1 and ν_1).

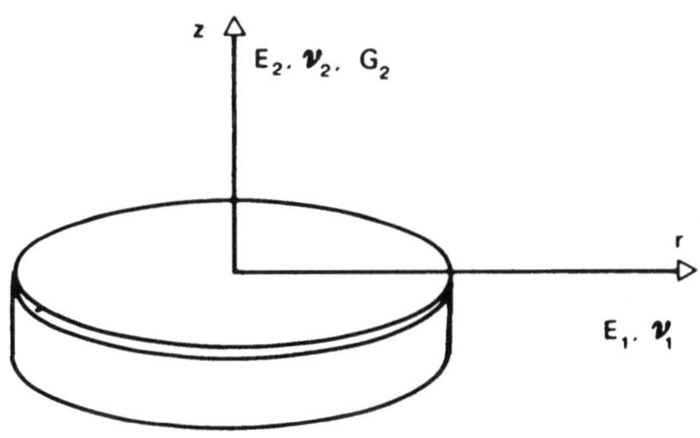

Figure 1 Scheme of a mechanical transversally anisotropic soil.

Continuity equation for water using Darcy's law:

$$\frac{\partial}{\partial r}\left(K_r\frac{\partial p}{\partial r}\right) + \frac{K_r}{r}\frac{\partial p}{\partial r} + \frac{\partial}{\partial z}\left(K_z\frac{\partial p}{\partial z}\right) = \gamma\left(n\beta\frac{\partial p}{\partial t} + \frac{\partial \varepsilon}{\partial t}\right) \qquad (3)$$

where:

$\varepsilon = \partial u_r/\partial r + \partial u_z/\partial z + u_r/r$ is the volume ditation

u_r, u_z are the vertical and horizontal components of the soil displacement;

K_r, K_z are the anisotropic hydraulic conductivities

γ, β are the specific weight and the compressibility of water

n is the porosity of soil

Solutions to eqs.(1) through (3) is obtained after specifying the initial and boundary conditions. Since we are dealing with the variations from a pre-existing state of equilibrium, at the initial time all the variables are zero. Boundary conditions require that u_r and u_z be given on portions of the boundary and σ_r', σ_z' and τ_{rz} be given on the complementary boundary. As far as p is concerned either its value or its normal derivative may be prescribed.

In the finite element solution to the coupled model, the axi-symmetric porous medium is subdivided into annular elements of triangular cross-section. Three unknowns are associated with each single node: the components of the displacement and the pore pressure.

By the aid of relationships (2), a discrete solution to eq.(1) is obtained by applying the "principle of virtual works". This states that for any time at the elastic equilibrium the work done by the internal forces is equal to that of the external forces (i.e. external loads and pore pressure forces which are here regarded as being known) for any virtual displacement satisfying the boundary conditions. The application of the principle of virtual works yields two algebraic equations for each node. A third equation is supplied by the Galerkin integral on eq.(3). Details of the formal numerical approach may be found in Desai (1975), Sandhu (1976), Verruijt (1977) and Zienkiewicz (1977).

Denoting by $\{\delta\}$ and $\{p\}$ the vectors of the nodal displacements and pore pressure, the method outlined above leads to a set of linear differential equations which may be written as:

$$\begin{bmatrix} K & L_1 \\ \phi & H \end{bmatrix} \begin{Bmatrix} \delta \\ p \end{Bmatrix} + \begin{bmatrix} \phi & \phi \\ L_2 & P \end{bmatrix} \begin{Bmatrix} \partial\delta/\partial t \\ \partial p/\partial t \end{Bmatrix} + q_t = 0 \qquad (4)$$

where K and H are the stiffness matrices of the elastic and groundwater problems, respectively, P is the capacity matrix (accounting exclusively for the term n) and L and L are rectangular matrices which "couple" the elastic and the hydraulic problems. The known vector q accounts for applied loading and well pumpage. Eqs.(4) are integrated in time with the Crank-Nicolson (1947) scheme which is second-order accurate and unconditionally stable.

For the resulting algebraic system an especially efficient solver is used. To reduce first the band of the matrices the three nodal unknowns are numbered sequentially. Thus over node i the associated variables have indices 3i, 3i-1 and 3i-2. Next the matrices are stored in vectorial form with the elimination of all zero coefficients. The mesh is then automatically reordered to minimize the number of built-ups, namely the number of elements which are generated in the triangular decomposition. The preliminary optimal reordering is based on the criterion of the minimal row and column length. However to avoid small pivotal elements in the permuted matrix and a consequent substantial loss of accuracy, sub-optimal reorderings are allowed. Using the sub-optimal row and column permutation vectors the finite element matrix is factorized first symbolically and then numerically. Finally the solution is arrived at by backward and forward substitution.

The solver described above proved very efficient, especially because the preliminary reordering and the symbolic factorization need to be performed only once for a given mesh, i.e. for a given problem, before starting the simulations. It has to be noted that a further advantage is obtained if the time step is held constant and the parameter values do not vary in time, as is usually the case, since the only operation to be repeated during the simulation is the backward/forward substitution. The use of a uniform time step is recommended because it leads to a saving of computer time which largely offsets the benefit related to an increasing time step.

PRELIMINARY RISULTS

The coupled model has been tested on two analytical solutions related to the consolidation of an infinite layer of soil overlying a smooth rigid and a rough rigid base and given by Gibson et al. (1970) and by Booker (1974).

A uniform loading is applied to the upper surface. The loaded area is a circle whose radius R is equal the layer thickness b. The loading is applied instantaneously and then held constant. The outer rigid boundary is set at a distance from the symmetry axis equal to 10 times b. Free drainage is allowed at the top of the layer while the base is assumed to be impermeable, rigid, smooth and rough, respectively.

A vertical cross-section of the finite element mesh is shown in Figure 2. Each rectangle is divided into two

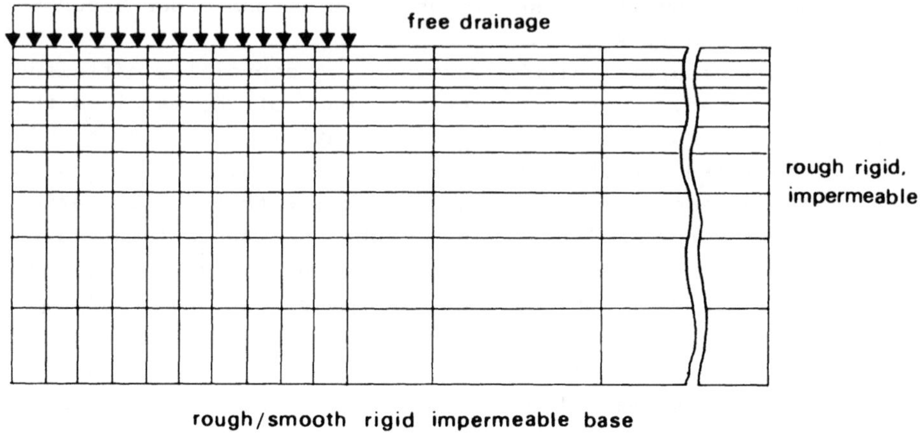

Figure 2. Finite element mesh for the test case.

Table 1 The preliminary optimal reordering of the finite element mesh leads to a reduction of the number of the non-zero coefficients in the triangular factors.

example	mesh size	matrix size	matrix non - zero coefficients	non-zero coefficients of the triangolar factors	
				without reordering	with reordering
test case	160	480	9162	30553	23893
field case	210	630	12186	45451	38859

triangles. The corresponding nodes are 160 and the overall number of nodal unknowns is equal to 160x3 = 480. The non-zero coefficients of the finite element matrix turn out to be 9162, those of the related triangular factors 23893. The latter are the result of a sub-optimal reordering strategy which has proven effective in preserving a good accuracy. The advantage of reordering the mesh is evidenced by Table 1 which also provides the number of coefficients of the triangular factors without the preliminary reordering.

It must be stressed that even a limited reduction of built-ups generated by the factorization turns into the saving of a significant number of numeric operations and hence of computational time. Figure 3 compares the analytical and the finite element results and shows that the model solution is quite satisfactory for engineering purposes. The dimensionless settlement of the centre of the loaded area:

$$\eta = \frac{E u_z}{2(1+\nu) b q}$$

(q being the uniform external loading) is plotted against the time factor:

$$\tau = \frac{E}{1+\nu} \frac{k}{\gamma} \frac{t}{b^2}$$

In Figure 3 Poisson's ratio is set equal to zero.

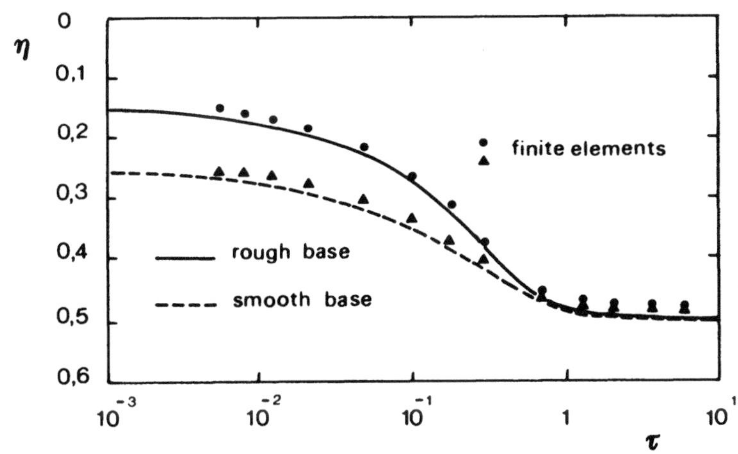

Figure 3 Comparison between analytical and numerical solutions to the coupled model for the test case.

The model has then been applied to analyse the settlement of a multilayer soil overlain by a tank filled with water and located in the Venice Lagoon, Italy. Actually metal tanks represent well enough the theoretical case of uniform vertical loading with negligibile shear stresses.

The tank which has a diameter equal to 8m and is 11m high, was built on reclaimed land between 1914 and 1938. The stratigraphic column down to 30m depth is given in Table 2 which also provides the values of the laboratory vertical compressibilities α of each layer. Assuming the values of Poisson's ratios shown in Table 2, Young's moduli of soil has been derived from the equation :

$$E = \frac{(1-2\nu)(1+\nu)}{\alpha\,(1-\nu)}$$

Table 2 also gives the water content of the clayey soils and the estimated coefficients of vertical permeability. The time behavior of the water raise in the tank is reported in Figure 4a). The experimental settlement of the point located westward on the boundary of the tank is compared in Figure 4b) with the results supplied by the coupled model. The agreement is quite good although it should be mentioned that a larger settlement was recorded on the eastern point of the boundary (actually the thickness of the underlying clay is not uniform and the real case is not a true axi-symmetric case).

Figure 4c) shows the behavior of the incremental pressure head as predicted by the model in the middle of the first clay layer below the center of the tank. These values are well in keeping with the data measured by piezometers situated below similar tanks in the Venice Lagoon.

The numerical performance of the model proved high as may be seen by looking at Table 1 which shows the reduction of the built-ups after the preliminary optimal reordering.

Table 2 Soil structure and average hydro-elastic parameter values used in the finite element analysis of the tank settlement.

Stratigraphic column (depth, m)		Water Content w(%) 20 40 60 80 100	Vertical compress. $\alpha (cm^2/Kg)$	Poisson ratio ν	Young modulus $E(Kg/cm^2)$	Permeabil. k(cm/sec)
0	sand		10^{-2}	0.30	74.3	10^{-2}
5	clay, silty-clay and peat	wp w wl	$3.3 \cdot 10^{-2}$	0.45	7.9	$0.5 \cdot 10^{-6}$
10	sand		$0.22 \cdot 10^{-2}$	0.30	334.0	10^{-2}
	clay		$2.5 \cdot 10^{-2}$	0.45	10.6	$0.5 \cdot 10^{-6}$
15	silty sand		$0.22 \cdot 10^{-2}$	0.30	334.0	10^{-2}
20, 25, 30	clay and silty clay		$2.5 \cdot 10^{-2}$	0.45	10.6	$0.5 \cdot 10^{-6}$

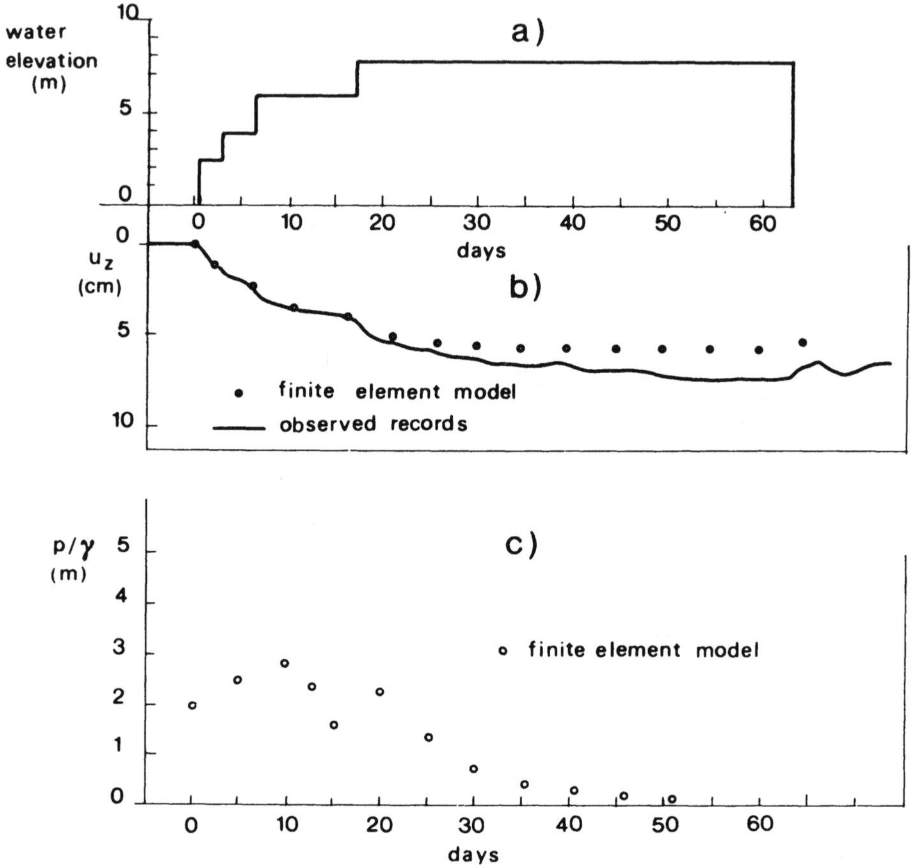

Figure 4 Loading exerted by the tank (expressed as water elevation (m)) a); observed settlement and finite element settlement at the outer radius of the tank b); finite element incremental pressure as predicted by the coupled model at the central point of the upper clayey layer c).

CONCLUSION

An efficient coupled finite element model of flow has been developed. The numerical efficiency stems from a direct solver which stores the f.e. matrix in compact vectorial form and factorizes the matrix symbolically after a preliminary optimal or sub-optimal reordering of the mesh. Reordering and symbolical factorization may be considered as operations related to the definition of the mesh, i.e. to the given problem. If the soil parameters are constant in time, an uniform time step proves particularly effective since also numeric factorization is to be performed only once before starting the simulations.

The model has been validated against available analytical solutions and applied to a field problem where the surface settlement of a multilayer soil loaded by a tank progressively filled with water has been measured experimentally. Experimental and numerical results show a satisfactory agreement.

The coupled model developed in this paper will be applied to large natural aquifer systems to investigate the practical reliability of the traditional flow equation which is universally used in groundwater hydrology.

REFERENCES

Biot M.A.(1941), General theory of three-dimensional consolidation, J. Appl. Phys., 12, 155-164.

Booker H.R.(1974), The consolidation of a finite layer subject to surface loading, Int. J. Solids Struct., 10, 1053-1063.

Booker J.R. and Small J.C.(1975), An investigation of the stability of numerical solutions of Biot's equations of consolidation, Int. J. Solid Struct., 11, 907-917.

Christian J.J. and Boehmer J.W.(1970), Plain strain consoli dation by finite elements, J. Soil Mech. and Found. Div., Proc. ASCE, 96, 1435-1457.

Crank J. and Nicolson P.(1947), A practical method for numerical evaluation of solutions of partial differential equations of the heat conduction type, Proc. Camb. Phil. Soc., 43, 50-67

Desai C.S.(1975), Analysis of consolidation by nu- merical methods, Proc. Symp. on Recent Developments in the Analysis of Soil Behavior and Application to Geotech. Struct., University of New South Wales, Sydney.

Gibson R.E., Schiffman R.L. and Pu S.L.(1970), Plane strain and axially symmetric consolidation of a clay layer on a smooth impervious base, Quart. J. Mech. Appl. Math., 23, 505-519.

Sandhu R.S.(1976), Finite element analysis of soil consoli dation, Geotech. Engng. Rep. No. 6 to National Science Foundation, The Ohio State University.

Sandhu R.S. and Wilson E.L.(1969), Finite element analysis of seepage in elastic media, J. Eng. Mech. Div. Proc. ASCE, 95, 641-652.

Smith I.M. and Hobbs R.(1976), Biot analysis of consolidation beneath embankments, Geotechnique, 26, 149-160.

Vermeer P.A. and Verruijt A.(1981), An accuracy condition for consolidation by finite elements, Int. J. Num. Anal. Methods Geomech., 5, 1-14.

Verruijt A.(1977), Generation and dissipation of pore water pressure, Finite Elements in Geomechanics, G. Gudehus Ed., Wiley, 293-317, London.

Zienkiewicz O.C.(1977), The finite element method, 3-rd edn., McGraw-Hill, New York.

HYBRID METHODS FROM A NEW PERSPECTIVE

Ismael Herrera

Institute of Geophysics, National University of Mexico.
Apdo. Postal 21-524, Deleg. Coyoacán, 04000 México, D.F.

INTRODUCTION

Usually, by a hybrid method it is understood one in which continuity requirements are reduced or eliminated altogether [Babuska, et al., 1977,1978; Pian & Tong, 1969; Tong, 1970] by introducing auxiliary dependent variables. Finite element methods for elliptic equations of order 2m, are said to be nonconforming when in the evaluation of the energy, which involves derivatives of order m, the approximations to derivatives of order m-1 may have simple discontinuities [Nitsche, 1974; Strang, 1972].

In this paper a systematic formulation of hybrid and nonconforming element methods is briefly explained and illustrated by applying it to some simple elliptic operators. The approach here presented is quite general, since it is applicable to any linear operator. It is based on the Algebraic Theory of Boundary Value Problems developed by the author [Herrera, 1980, 1981, 1982, 1984a]and which has just been published in book form [Herrera, 1984b].

The approach here presented apparently supplies a useful tool of analysis which can be used not only to evaluate the error of approximate methods but also gives insight and orientation to develop more efficient ones. In the case of ordinary differential equations it yields systems which permit computing the exact values of the sought solutions and its derivatives at a finite number of nodes. For special choices of the test functions some results obtained by Rose [1964,1975] are also derived in this manner. However, the new approach goes beyond, since it supplies algorithms which permit computing the exact values of the derivatives at prescribed nodes without computing the values of the function there.

FINITE ELEMENT FORMULATION

Generally, boundary value problems can be formulated as follows

[Herrera, 1984b]. Let D be a linear space of functions and D* its dual (space of linear functionals defined on D). Let $P:D \to D^*$ and $B:D \to D^*$ be operators, such that B is a boundary operator for P. Given $U \in D$ and $V \in D$, let $f = PU$ and $g = BV$. Then, $u \in D$ is a solution of the boundary value problem, if and only if,

$$(P+B)u = f+g \tag{1}$$

This equation is equivalent to

$$\langle Pu,v\rangle + \langle Bu,v\rangle = \langle f,v\rangle + \langle g,v\rangle \;\forall\; v \in D \tag{2}$$

To be specific, let

$$\mathcal{L}u = \Delta u \pm u \;;\; \text{in } \Omega \tag{3}$$

or alternatively

$$\mathcal{L}u = \Delta u \quad ;\; \text{in } \Omega \tag{3'}$$

where Ω is an open bounded domain of a two-dimensional Euclidean space satisfying the same restrictions as in [Babuska, et al., 1977]. Define $P:D\to D^*$, $B:D\to D^*$, $f \in D^*$ and $g \in D^*$ by

$$\langle Pu,v\rangle = \int_\Omega v\mathcal{L}u\,dx \quad ; \quad \langle Bu,v\rangle = \int_{\partial\Omega} \frac{\partial v}{\partial n}\gamma_0 u\,dx \tag{4}$$

$$\langle f,v\rangle = \int_\Omega v f_\Omega dx \quad \text{and} \quad \langle g,v\rangle = \int_{\partial\Omega} \frac{\partial v}{\partial n} g_{\partial\Omega} dx \tag{5}$$

trace operator and the Laplacian is understood in the sense of distributions. Then (2), or equivalently (1), is satisfied if and only if

$$\mathcal{L}u = f_\Omega \quad \text{and} \quad \gamma_0 u = g_{\partial\Omega} \tag{6}$$

Let π be a partition of Ω into a finite collection of E subdomains $\Omega_e, 1,2,\ldots,E(\pi)$ [Babuska, et al., 1977, 1978]. We define D_e and $P_e:D_e\to D_e^*$, replacing Ω by Ω_e in previous definitions. In addition, $B_e:D_e\to D_e^*$ is defined replacing $\partial\Omega$ by $\partial\Omega\cap\partial\Omega_e$. Let $\hat{D}$ be the product space $D_1\times D_2\times\ldots\times D_E$. Define $\hat{P}:\hat{D}\to\hat{D}^*$ and $\hat{B}:\hat{D}\to\hat{D}^*$ by

$$\langle\hat{P}\hat{u},\hat{v}\rangle = \sum_{e=1}^{E} \langle P_e u_e, v_e\rangle \;;\; \langle\hat{B}\hat{u},\hat{v}\rangle = \sum_{e=1}^{E} \langle B_e u_e, v_e\rangle \tag{7}$$

Then equations (4) are satisfied by $\hat{P}$ and $\hat{B}$. The operator $\hat{P}+\hat{B}$ is non-symmetric. However, the restriction of the bilinear functional $\langle(\hat{P}+\hat{B})\hat{u},\hat{v}\rangle$ to smooth functions is symmetric. Using this fact, it can be seen that $\hat{P}+\hat{B}-\hat{J}$ is symmetric, where

$$\langle\hat{J}\hat{u},\hat{v}\rangle = \int_\Gamma\{[u]\frac{\partial\dot{v}}{\partial n} - \dot{v}[\frac{\partial u}{\partial n}]\}dx \tag{8}$$

Here, Γ is the union of all the interelement boundaries. More formally:

$$\Gamma = \bigcup_{e\neq f} (\partial\Omega_e \cap \partial\Omega_f) \tag{9}$$

In addition

$$[u] = u_f - u_e \; ; \; [\frac{\partial u}{\partial n}] = \frac{\partial u_f}{\partial n} - \frac{\partial u_e}{\partial n} \qquad (10a)$$

$$\dot{u} = \frac{1}{2}(u_f + u_e) \; ; \; \frac{\partial \dot{u}}{\partial n} = \frac{1}{2}(\frac{\partial u_f}{\partial n} + \frac{\partial u_e}{\partial n}) \qquad (10b)$$

where given any point $x \in \Gamma$, it is assumed that Ω_e and Ω_f are such that $\underset{\sim}{x} \in \partial\Omega_e \cap \partial\Omega_f$. Also, the unit normal vector in (8), must be taken pointing outwards from Ω_e into Ω_f. Then, equation

$$\hat{P}\hat{u} + \hat{B}\hat{u} - \hat{J}\hat{u} = \hat{f} + \hat{g} \qquad (11)$$

or equivalently

$$\langle\hat{P}\hat{u},\hat{v}\rangle + \langle\hat{B}\hat{u},\hat{v}\rangle - \langle\hat{J}\hat{u},\hat{v}\rangle = \langle\hat{f},\hat{v}\rangle + \langle\hat{g},\hat{v}\rangle \;\forall\; \hat{v} \in \hat{D} \qquad (12)$$

if and only if $\hat{u} = \{u_1, u_2, \ldots, u_E\}$, where $u_e \in H^1(\Omega_e)$ are the restrictions to Ω_e of the solution $u \in H^1(\Omega)$ of problem (6).

Observe,

$$\langle\hat{J}^*\hat{u},\hat{v}\rangle = \int_\Gamma \{[v]\frac{\partial \dot{u}}{\partial n} - \dot{u}[\frac{\partial v}{\partial n}]\}dx \qquad (13)$$

by virtue of (8). Given $\hat{u} \in \hat{D}$, the functionals $\hat{J}\hat{u}$ and $\hat{J}^*\hat{u}$ are called the jump and the average values of $\hat{u}$, respectively. Equations (8) and (13) show that this nomenclature is appropriate, since $\hat{J}\hat{u}$ defines uniquely the functions $[u]$ and $[\partial u/\partial n]$ on Γ, while $\hat{J}^*\hat{u}$ defines uniquely the functions $\dot{u}$ and $\partial\dot{u}/\partial n$. Observe that $\hat{J} = \hat{J}_1 + \hat{J}_2$, where

$$\langle\hat{J}_1\hat{u},\hat{v}\rangle = \int_\Gamma [u]\frac{\partial \dot{v}}{\partial n}dx \quad , \quad \langle\hat{J}_2\hat{u},\hat{v}\rangle = -\int_\Gamma \dot{v}[\frac{\partial u}{\partial n}]\,dx \qquad (14)$$

Clearly $\hat{J}_1\hat{u}$, $\hat{J}_2\hat{u}$, $\hat{J}_2^*$ and $\hat{J}_1^*$, characterize $[u]$, $[\frac{\partial u}{\partial n}]$, $\dot{u}$ and $\partial\dot{u}/\partial n$, respectively.

To obtain an approximate solution $\hat{u}' \in \hat{D}$, in the finite element method one replaces (12) by

$$\langle\hat{P}\hat{u}',\hat{\varphi}_\alpha\rangle + \langle\hat{B}\hat{u}',\hat{\varphi}_\alpha\rangle - \langle\hat{J}\hat{u}',\hat{\varphi}_\alpha\rangle = \langle\hat{f},\hat{\varphi}_\alpha\rangle + \langle\hat{g},\hat{\varphi}_\alpha\rangle \;,\; \alpha=1,\ldots,N \qquad (15)$$

where $\{\hat{\varphi}_1, \ldots, \hat{\varphi}_N\} \subset \hat{D}$ is a family of "test functions." Usually, one imposes the additional <u>constraint</u> that $\hat{u}' = \sum_{\beta=1}^{N} a_\beta \hat{\Phi}_\beta$ where $\{\hat{\Phi}_1, \ldots, \hat{\Phi}_N\}$ are the basis functions. However, this latter constraint is alien to the problem, while equation (15) is necessarily satisfied by the solution of the problem. Therefore, it has interest to analyze the restrictions implied by (15), when no assumption is made on the representation of the solution.

To this end, observe that (15) can also be written as

$$\langle\hat{P}\hat{\varphi}_\alpha,\hat{u}'\rangle + \langle\hat{B}^*\hat{u}',\hat{\varphi}_\alpha\rangle - \langle\hat{J}^*\hat{u}',\hat{\varphi}_\alpha\rangle = \langle\hat{f},\hat{\varphi}_\alpha\rangle + \langle\hat{g},\hat{\varphi}_\alpha\rangle \;,\; \alpha=1,\ldots,N \qquad (16)$$

Taking into account (4) and (13), this imposes conditions on the possible values taken by

$$\int_{\Omega} u'\mathcal{L}\varphi_{\alpha}dx \ ; \ \int_{\partial\Omega} \varphi_{\alpha} \frac{\partial u'}{\partial n} dx \ ; \ \int_{\Gamma} \dot{u}[\frac{\partial\varphi_{\alpha}}{\partial n}] dx \text{ and } \int_{\Gamma}[\varphi_{\alpha}]\frac{\partial\dot{u}}{\partial n} dx$$

These in turn, can be interpreted as restrictions on the possible values of u' in the region Ω, $\partial u'/\partial n$ on $\partial\Omega$, u' and $\partial u'/\partial n$ on Γ. The kind of such conditions depends of course on the particular choice of the family of test functions. Thus, for example, in standard finite element formulations $\hat{\varphi}_{\alpha}$ is required to be continous; i.e. $[\hat{\varphi}_{\alpha}] = 0$. Then

$$<\hat{J}_1^*\hat{u}',\hat{\varphi}_{\alpha}> = \int_{\Gamma} [\varphi_{\alpha}] \frac{\partial\dot{u}'}{\partial n} dx = 0 \qquad (17)$$

and no condition is imposed on $\partial\dot{u}'/\partial n$. Similarly, in some hybrid element methods [Babuska, et al., 1978], it is assumed that $\hat{P}\hat{\varphi}_{\alpha} = 0$. Hence $\int_{\Omega} u'\mathcal{L}\varphi_{\alpha}dx = 0$ identically and no restrictions are imposed on the possible values of u' in the region Ω.

Observe that equation (16) is also satisfied by the exact solution $\hat{u}\in\hat{D}$. Therefore, by substraction, one gets

$$<\hat{P}\hat{\varphi}_{\alpha},\hat{u} - \hat{u}'> + <B^*(\hat{u} - \hat{u}'),\hat{\varphi}_{\alpha}> - <\hat{J}^*(\hat{u} - \hat{u}'),\hat{\varphi}_{\alpha}> = 0 \qquad (18)$$

In view of (18), it is clear that the system of equations (16) supplies information about the exact solution $\hat{u}$. Again the kind of information depends on the specific family $\{\hat{\varphi}_1,\ldots,\hat{\varphi}_N\}$, of test functions chosen. In particular, if $\hat{\varphi}_{\alpha}$ are required to be continuous, no information about $\partial u/\partial n = \partial\dot{u}/\partial n$, is gotten. Correspondingly, if $\hat{\varphi}_{\alpha}$ satisfies the differential equation locally, then no information about the function u on Ω is obtained. It has interest to observe, because this may be useful in some applications, that when the normal derivatives are continuous across interelement boundaries (i.e. $[\partial\varphi_{\alpha}/\partial n] = 0$), then no information is obtained about $u = \dot{u}$. These facts may be used to focus the information in some aspects of the solution, depending on the applications. Such procedures are illustrated in the following Section.

EXAMPLES

Let us apply the forgoing discussion for the one dimensional case, taking Ω as the open interval (0,1). Thus, the problems to be considered are

$$\mathcal{L}u = \frac{d^2u}{dx^2} = f_{\Omega}, \quad \text{in } \Omega = (0,1) \qquad (19a)$$

or alternatively

$$\mathcal{L}u = \frac{d^2u}{dx^2} \pm u = f_{\Omega}, \quad \text{in } \Omega = (0,1) \qquad (19'a)$$

subject to

$$u(0) = g_0 \quad \text{and} \quad u(1) = g_1 \qquad (19b)$$

Let the partition contain E subintervals of equal length (x_{k-1},x_k), with $k=1,2,\ldots E$. Thus $h=|x_k-x_{k-1}| = 1/E$. Assume the system of test functions $\{\hat{\varphi}_1,\hat{\varphi}_2,\ldots,\hat{\varphi}_N\}$ satisfies $\mathcal{L}\varphi_{\alpha k}=0$; $k=1,\ldots,E$. Then equation (16) and (18) are

$$\varphi_\alpha \frac{du'}{dx}\Big|_0^1 + \sum_{k=1}^{E-1} \{\dot{u}'[\frac{d\varphi_\alpha}{dx}] - [\varphi_\alpha] \frac{d\dot{u}'}{dx}\}_k = \int_0^1 f_\Omega \varphi_\alpha dx + g_1 \frac{d\varphi_\alpha}{dx}(1) - g_o \frac{d\varphi_\alpha}{dx}(0) \qquad (20)$$

and

$$\varphi_\alpha (\frac{du}{dx} - \frac{du'}{dx})\Big|_0^1 + \sum_{k=1}^{E-1} \{(u - \dot{u}')[\frac{d\varphi_\alpha}{dx}] - [\varphi_\alpha] (\frac{du}{dx} - \frac{d\dot{u}'}{dx})\}_k = 0 \qquad (21)$$

Here $\alpha = 1,\ldots,N$. Also, the dots have been deleted from $\dot{u}_k$ and $(d\dot{u}/dx)_k$, since the exact solution u is smooth. Equations (20) can be interpreted as a system of N equations for the 2E unknows $\dot{u}'_k$, $(d\dot{u}'/dx)_k$, $k=1,\ldots,E-1$, $(du'/dx)_o$ and $(du'/dx)_E$. If $N = 2E$, there is a possibility this system had a unique solution; in such case (21) shows that

$$u_k = \dot{u}'_k \quad ; \quad (du/dx)_k = (d\dot{u}'dx)_k \qquad (22)$$

where $k = 1,\ldots,E-1$, for the first equation while $k = 0,1,\ldots,E$ for the second one. Also, when $k = 0$ or 1, the dot must be deleted.

A first possibility is to take $N = 2E$ with $\{\varphi_1,\ldots,\varphi_{2E}\}$ in such a manner that $\{\varphi_1,\varphi_2\}$ is a system of two linear independent solutions of the homogeneous equation $\mathcal{L}\varphi = 0$ in the interval (x_o,x_1) and zero outside. More generally, the system $\{\varphi_{2j-1},\varphi_{2j}\}$ is a system of two linear independent solutions of the homogeneous equation in the interval (x_{j-1},x_j). Using general results of the Algebraic Theory [Herrera, 1984b], it can be shown that such system is T-complete and the system of equations (20) has a unique solution. Thus, the solution of (20) yields the exact values of u and its derivatives at all the nodes.

If we are not interested on the derivative, it may be convenient to eliminate it from the system of equations. This is achieved by imposing the additional conditions on the test functions $\hat{\varphi}_\alpha$; namely that $\hat{\varphi}_\alpha$ be continuous and $\hat{\varphi}_\alpha(0) = \hat{\varphi}_\alpha(1) = 0$. A complete system satisfying these conditions can be chosen so that, for every $\alpha = 1,\ldots,N$, $\hat{\varphi}_\alpha$ has support on the interval $(x_{\alpha-1},x_{\alpha+1})$ and $N = E-1$. Then the system (20) reduces to

$$\sum_{k=\alpha-1}^{\alpha+1} u_k[\frac{d\varphi_\alpha}{dx}]_k = \int_0^1 f_\Omega \varphi_\Omega dx + g_1 \frac{d\varphi_\alpha}{dx}(1) - g_o \frac{d\varphi_\alpha}{dx}(0) \qquad (23)$$

Taking the test functions as indicated in Table 1, one obtains the following systems of difference equations for the <u>exact</u> values u_k. When $\mathcal{L} = \Delta$, then

TABLE 1. TEST FUNCTIONS WHEN u_k IS SOUGHT $(b = h)$

$\mathcal{L}$	$x_{\alpha-1}<x<x_\alpha$	$x_\alpha < x < x_{\alpha+1}$
Δu	$\varphi_\alpha = x - x_{\alpha-1}$	$\varphi_\alpha = (x_{\alpha+1} - x)$
$\Delta u - u$	$\varphi_\alpha = \sinh(x - x_{\alpha-1})$	$\varphi_\alpha = \sinh(x_{\alpha+1} - x)$
$\Delta u + u$	$\varphi_\alpha = \sin(x - x_{\alpha-1})$	$\varphi_\alpha = \sin(x_{\alpha+1} - x)$

$$\frac{u_{\alpha+1} + u_{\alpha-1} - 2u_\alpha}{h^2} = \frac{1}{h^2}\int_{x_{\alpha-1}}^{x_{\alpha+1}} f_\Omega \varphi_\alpha dx \quad ; \quad \alpha = 2,\ldots,N-1=E-2 \tag{24a}$$

$$\frac{u_2 - 2u_1}{h^2} = \frac{1}{h^2}\int_0^h f_\Omega \varphi_o dx - g_o/h^2 \quad ; \quad \frac{u_N - 2u_{N+1}}{h^2} = \frac{1}{h^2}\int_{1-h}^{1} f_\Omega \varphi_N dx - $$

$$- g_1/h^2 \tag{24b}$$

Notice that these are central differences. When $\mathcal{L} = \Delta - 1$

$$\frac{u_{\alpha+1} + u_{\alpha-1} - 2u_\alpha \cosh b}{b^2} = \frac{1}{b^2}\int_{x_{\alpha-1}}^{x_{\alpha+1}} f_\Omega \varphi_\alpha dx \quad ; \quad \alpha = 2,\ldots,N-1=E-2 \tag{25a}$$

$$\frac{u_2 - 2u_1 \cosh b}{b} = \frac{1}{b^2}\int_0^b f_\Omega \varphi_o dx - g_o/b^2 \quad ; \quad \frac{u_N - 2u_{N+1}}{b^2} =$$

$$= \frac{1}{b^2}\int_{1-b}^{1} f_\Omega \varphi_N dx - g_1/b^2 \tag{25b}$$

where we have written $b = h$, for clarity. Finally, when $\mathcal{L} = \Delta + 1$

$$\frac{u_{\alpha+1} + u_{\alpha-1} - 2u_\alpha \cos b}{b^2} = \frac{1}{b^2}\int_{x_{\alpha-1}}^{x_{\alpha+1}} f_\Omega \varphi_\alpha dx \quad ; \quad \alpha = 2,\ldots,N-1=E-2 \tag{26a}$$

$$\frac{u_2 - 2u_1 \cos b}{h^2} = \frac{1}{b^2}\int_0^b f_\Omega \varphi_o dx - g_o/b^2 \quad ; \quad \frac{u_N - 2u_{N+1}}{b^2} =$$

$$= \frac{1}{b^2}\int_{1-b}^{1} f_\Omega \varphi_N dx - g_1/b^2 \tag{26b}$$

The results just presented, can also be derived applying a method proposed by Rose [1964, 1975] for positive definite self-adjoint operators on the basis of energy considerations. However, the methodology here discussed, is applicable to arbitrary linear operators even if they are non-symmetric, as will be shown elsewhere.

If we were interested in the values $(du/dx)_k$ of the derivative at all the nodes ($k=0,1,\ldots,E$), one can impose the condition that the derivative of the test functions be continuous (i.e. $[d\varphi/dx]=0$). Such families of test functions are given in Table 2. Substitution of them in equation (20) yields the following systems of equations. When $\mathcal{L}=\Delta-1$

$$\left(\frac{du}{dx}\right)_{\alpha+1}+\left(\frac{du}{dx}\right)_{\alpha-1}-2\left(\frac{du}{dx}\right)_{\alpha}\cosh b=\int_{x_{\alpha-1}}^{x_{\alpha+1}} f_{\Omega}\varphi_{\alpha}\,dx\,;\quad \alpha=1,\ldots,E-1 \tag{27a}$$

$$\left(\frac{du}{dx}\right)_{1}-\left(\frac{du}{dx}\right)_{0}\cosh b=\int_{0}^{h} f_{\Omega}\varphi_{o}\,dx+g_{o}\sinh b\,;$$
$$\left(\frac{du}{dx}\right)_{E}-\left(\frac{du}{dx}\right)_{E-1}\cosh b=\int_{1-h}^{1} f_{\Omega}\varphi_{E}\,dx-g_{1}\sinh b \tag{27b}$$

TABLE 2. TEST FUNCTIONS WHEN $(du/dx)_k$ IS SOUGHT ($b=h$)

$\mathcal{L}$	$x_{\alpha-1}<x<x_{\alpha}$	$x_{\alpha}<x<x_{\alpha+1}$
Δu	$\varphi_{\alpha}=1\ ;\ \alpha=1,\ldots,E$	$\varphi_{\alpha}=0$
$\Delta u-u$	$\varphi_{\alpha}=-\cosh(x-x_{\alpha-1}),\ \alpha\neq 0$	$\varphi_{\alpha}=\cosh(x_{\alpha+1}-x)\ ;\ \alpha\neq E$
	$\varphi_{o}=0$	$\varphi_{E}=0$
$\Delta u+u$	$\varphi_{\alpha}=-\cos(x-x_{\alpha-1}),\ \alpha\neq 0$	$\varphi_{\alpha}=\cos(x_{\alpha+1}-x)\ ;\ \alpha\neq E$
	$\varphi_{o}=0$	$\varphi_{E}=0$

When $\mathcal{L}=\Delta+1$ and $2h\neq\pi,3\pi,\ldots$, equations (27) remain valid, if cosh and sinh are replaced by cos and -sin, respectively. When $\mathcal{L}=\Delta u$, the system of test functions given in Table 2, is not complete because it contains only E functions. Is is necessary to take $N=E+1$, with $\varphi_{E+1}=x, 0<x<1$, which clearly has continuous derivatives and is linearly independent of the family given in Table 2. Although the system of difference equations that is obtained for this case is rather trivial, the example has some theoretical interest because it has the peculiarity that it is not possible to construct a function of local support, with continuous derivative and linearly independent of the system $\{\hat{\varphi}_{o},\hat{\varphi}_{1},\ldots,\hat{\varphi}_{E}\}$.

It may be of greater interest to mention that by combining the previous constructions, it is possible to construct finite difference schemes which yields the exact values of the solution at some nodes and the exact values of the derivatives at other ones.

REFERENCES

Babuska, I., Oden, J.T., and Lee, J.K. (1977) Mixed-Hybrid Finite Element Approximations of Second-Order Elliptic Boundary-Value Problems. Computer Methods in Applied Mechanics and Engineering, 11, 175-206.

Babuska, I., Oden, J.T., and Lee, J.K. (1978) Mixed-Hybrid Finite Element Approximation of Second Order-Elliptic Boundary-Value Problems. Par 2. Weak-Hybrid Methods. Computer Methods in Applied Mechanics and Engineering, 14, 1-22.

Herrera, I. (1980) Variational Principles for Problems with Linear Constraints. Prescribed Jumps and Continuation Type Restrictions. Jour. Inst. Maths. and its Applics., 25, 67-96.

Herrera, I. (1981) An Algebraic Theory of Boundary Value Problems. Kinam, 3, 161-230.

Herrera, I. (1982) Boundary Methods for Fluids, in: Finite Elements in Fluids, Vol. IV, R.H. Gallagher, D. Norrie, J.T. Oden, and O.C. Zienkiewics, (eds.), John Wiley & Sons, Ltd., Chapter 19, 403-432.

Herrera, I. (1984a) Trefftz Method, in: Progress in Boundary Element Methods, Vol. 3, C.A. Brebbia, (ed.), Springer-Verlag, Chapter. (in press)

Herrera, I. (1984b) Boundary Methods: An Algebraic Theory, Pitman Publishing Co., London (in press).

Nitsche, J. (1974) Convergence of Nonconforming Methods, in: Mathematical Aspects of Finite Elements in Partial Differential Equations, C. de Boor, (ed.), Academic Press, N.Y., 15-53.

Pian, T.H.H. and Tong, P. (1969) Basis of Finite Element Methods for Solid Continua. Int. J. Numer. Maths. Eng., 1, 3-85.

Rose, M.E. (1964) Finite Difference Schemes for Differential Equations. Mathematics of Computations, 18, 179-195.

Rose, M.E. (1975) Weak-Element Approximations to Elliptic Differential Equations. Numer. Math., 24, 185-204.

Strang, G. (1972) Variational Crimes in the Finite Element Method, in: Mathematical Foundations of the Finite Element Method with Applications to Partial Differential Equations, A.K. Aziz (ed.), Academic Press, N.Y. 689-710.

Tong, P. (1970) New Displacement Hybrid Finite Element Models for Solid Continua. Int. J. Numer. Meths. Eng., 2, 73-85.

NUMERICAL EXPERIMENTS USING 'ADAPTIVE' FINITE ELEMENTS WITH COLLOCATION

M. F. N. Mohsen

Civil Engineering Department
University of Petroleum & Minerals
Dhahran, Saudi Arabia

INTRODUCTION

Partial differential equations with a near-frontal or frontal behaviour occur in a number of practical situations. Notable examples are the diffusion-convection and the Buckley-Leverett equations. The diffusion convection equation is a second order equation that includes a first order term. The dominance of the first order term renders the equation to a hyperbolic form exhibiting a frontal character. The Buckley-Leverett equation is a first order non-linear equation characterised with a frontal behaviour.

The diffusion-convection equation and previous works

The form of diffusion-convection equation used in this work is

$$D \frac{\partial^2 c}{\partial x^2} - V \frac{\partial c}{\partial x} = \frac{\partial c}{\partial t} \qquad [1]$$

where $c(x,t)$ is the concentration of the migrating species $[M/L^3]$
D is the coefficient of diffusion $[L^2/T]$
V is the fluid velocity $[L/T]$
x is the spatial coordinate (L), and
t is time [T].

The initial and boundary conditions under which equation [1] has to be solved are

$$c(x,o) = 0 \qquad [2]$$

$$c(o,t) = c_1 \qquad [3]$$

$$c(\ell,t) = 0 \qquad [4]$$

Equation [1] may be non-dimensionalized to

$$\frac{\partial^2\phi}{\partial X^2} - P\frac{\partial\phi}{\partial X} = \frac{\partial\phi}{\partial T} \qquad [5]$$

where $\phi(X,T) = c(x,t)/c_1$
$X = x/\ell$
$T = Dt/\ell^2$
$P = V\ell/D$, the Peclet number.

The initial condition [2] and the boundary conditions [3] and [4] under this transformation becomes

$$\phi(X,0) = 0 \qquad [6]$$
$$\phi(0,T) = 1 \qquad [7]$$
$$\phi(1,T) = 0 \qquad [8]$$

Attempts to solve equation [5] for large P by any straightforward numerical technique leads to oscillatory results (Gladwell & Wait, 1979). Over the last two decades a number of novel techniques have evolved to circumvent this difficulty. An exhaustive survey cannot be presented here but some key contributions are tabulated below.

	Investigator	Year	Comments
(1)	Peaceman and Rachford	1962	Finite difference
(2)	Price,Cavendish & Varga	1968	Oscillation noted
(3)	Culham and Varga	1969	Classical Galerkin
(4)	Guymon	1970	Transforms $\overline{\phi} = \phi\exp(\frac{P}{2})$ (limited to small P)
(5)	Smith,Faradmy & O'Conner	1973	Galerkin Finite Element
(6)	Christie et al	1976	Galerkin Finite Element (upwind)

A more complete literature review is given by Gladwell & Wait (1979).

In general, two strategies have been taken in attempts to avoid oscillations:

(1) Use of improved spatial approximations
 (a) Higher order elements (C°)
 (b) Higher class of elements (C^1)
(2) Use of asymetric weighting functions (upwinding).

Use of higher-order elements does not result in any striking improvement (Heinrich et al., 1977). The use of higher class of elements yielded better results at the cost of twice as many degrees of freedom. The method of "upwinding" has been very successful in eliminating oscillations but may cost excessive smearing.

In this paper a strategy is used to combine the desirable effects of higher class of elements (C^1) along with some higher order (C^1) elements placed in the vicinity of the front. A collocation finite element method is used for spatial discretization.

The adaptive element method for the diffusion convection equation

In this paper Cubic Hermitian trial functions (Figures 1a,1b, Appendix I) are used every where except in the neighbourhood of the front where quintic Hermitian trial functions are used.

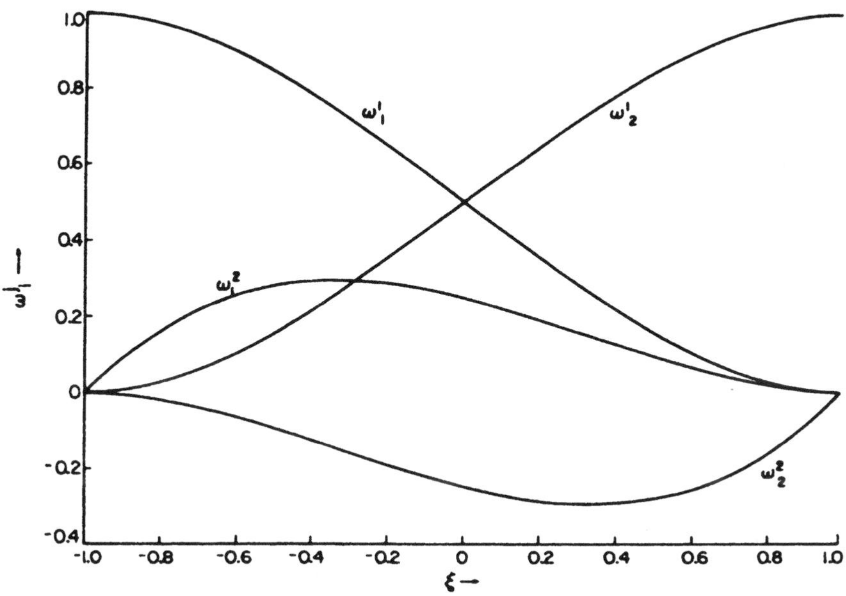

Figure 1a: Cubic Hermite Polynomials ω_i^j [i,j ε 1,2]

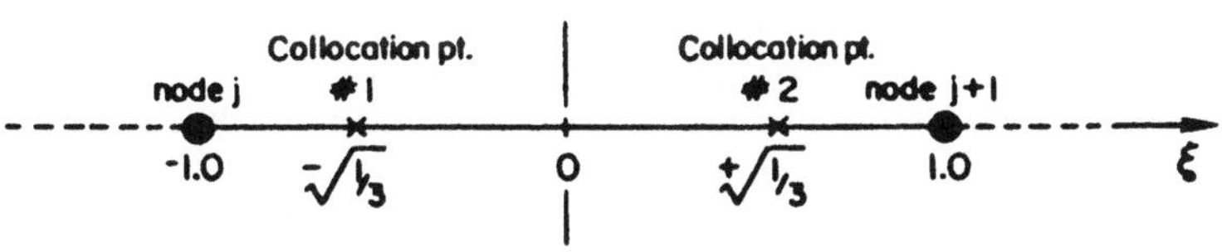

Figure 1b: A typical cubic element showing the locations of nodes and orthogonal collocation points

$$[\xi = \frac{2(x-x_j)}{(x_{j+1}x_j)} - 1]$$

$$\phi(x,t) \simeq \hat{\phi}(x,t) = \sum_{i=1}^{4} N_i\phi_i \qquad [9]$$

$\hat{\phi}$ = approximation of ϕ

ϕ_i = if i is odd value of ϕ at (i+1)/2 th.node
if i is even value of $\frac{\partial\phi}{\partial X}$ at i/2 th.node

N_1, N_2 = ω_2^1, ω_2^2 the Cubic Hermitians for node 1 [Fig.1a]

N_1, N_2 = ω_2^1, ω_2^2 the Cubic Hermitians for node 2 [Fig. 1a]

Substituting approximation [9] into equation [5] and orthogonal collocation at two Gauss points ξ_j per element results

$$\left[\sum_{i=1}^{4}\left(\frac{\partial^2 N_i}{\partial X^2} - P\,\frac{\partial N_i}{\partial X}\right)\phi_i\right]_{\xi_j} = \left[N_i\,\frac{\partial \phi_i}{\partial t}\right]_{\xi_j} \qquad [10]$$

In matrix form this may be written as

$$[B]\,\{u\} = [A]\,\left\{\frac{\partial\phi}{\partial t}\right\} \qquad [11]$$

where

$$\left.\begin{aligned} B_{ij} &= \left[\frac{\partial^2 N_i}{\partial X^2} - P\,\frac{\partial N_i}{\partial X}\right]_{\xi_j} \\ A_{ij} &= [N_i]_{\xi_j} \end{aligned}\right\}\ \begin{aligned} i &\in 1,4 \\ j &\in 1,2 \end{aligned}$$

The global equations are formed by collecting element equation [11] and applying the boundary conditions. The equations were solved explicitly in time.

For quintic elements,one additional node is placed at the centre of the element. The approximating functional form is

$$\phi(X,T) = \hat{\phi}(X,T) = \sum_{i=1}^{6} \bar{N}_i \phi_i$$

where $\bar{N}_1$, $\bar{N}_2$ = $\tilde{\omega}_1^1$, $\tilde{\omega}_1^2$, respectively,the quintic Hermitians for node 1

$\bar{N}_3$, $\bar{N}_4$ = $\tilde{\omega}_2^1$, $\tilde{\omega}_2^2$, respectively,the quintic Hermitians for node 2

$\bar{N}_5$, $\bar{N}_6$ = $\tilde{\omega}_3^1$, $\tilde{\omega}_3^2$, respectively,the quintic Hermitians for node 3 (Figures 2a & 2b)

The location of the Gauss points for cubic elements are shown in Figure 1b and those for quintic in Fig. 2c. For quintic elements,collocations are carried over 4 Gauss points ξ_j instead of two as was for cubic elements. Thus the block matrices now are 4x6 instead of 2x4.

The quintic basis functions were employed only over elements where additional accuracy is warranted. These elements were "reduced" at the element level so that the Global matrix sees the same number of unknowns as if only cubic elements were

Collocation points marked by X

$\xi_1 = -0.86113$

$\xi_2 = -0.33998$

$\xi_3 = +0.33998$

$\xi_4 = +0.86113$

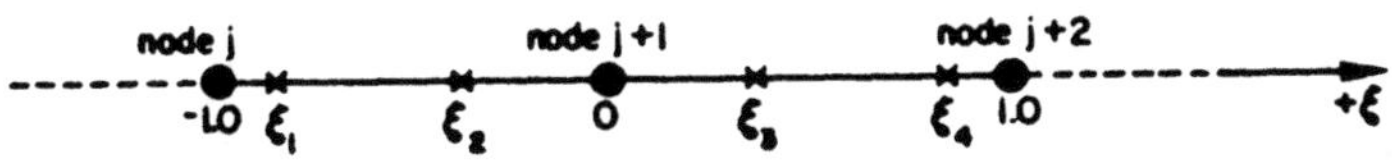

Figure 2c : A typical quintic element showing the locations of nodes and orthogonal collocation points

$$[\xi = \frac{2(x - x_j)}{(x_{j+2} - x_j)} - 1]$$

The "reduction" on condensation of the degrees of freedom associated with the mid-side node was carried out in the following manner.

The element matrix is restructured such that those equations identified with the mid-side node are isolated; that is we partition the 4x6 matrix as follows

$$\begin{bmatrix} [k]_{11} & [k]_{12} \\ [k]_{21} & [k]_{22} \end{bmatrix} \begin{Bmatrix} \{u\}_1 \\ \{u\}_2 \end{Bmatrix} \begin{matrix} = \\ = \end{matrix} \begin{Bmatrix} \{F_1\} \\ \{F_2\} \end{Bmatrix} \qquad [12]$$

where $[k]_{11}$ and $[k]_{21}$ are 2x4 matrices while $[k]_{12}$ and $[k]_{22}$ are 2x2. $\{u\}_1$ and $\{u\}_2$ contains degrees of freedom corresponding to the end and centre nodes respectively. Similarly, $\{F\}_1$ and $\{F\}_2$ contains known information corresponding to end and centre nodes respectively. $\{u\}_2$ is formally solved using the second row of equation [12].

$$\{u\}_2 = [k]_{22}^{-1} \{F_2\} - [k]_{22}^{-1} [k]_{21}\{u\}_1 \qquad [13]$$

Substituting result [13] into upper half of equation [12], we get

$$[k]_{11} \{u\}_1 + [k]_{12} ([k]_{22}^{-1} \{F\}_2 - [k]_{22}^{-1} [k]_{21}\{u\}_1) = \{F\}_1 \qquad [14]$$

Rearranging equation [14], we get

$$[\overline{k}] \{u\}_1 = \{\overline{f}\} \qquad [15]$$

where $[\overline{k}] = ([k]_{11} - [k]_{12} [k]_{22}^{-1} [k]_{21})$

$\{\overline{f}\} = \{F\}_1 - [k]_{12} [k]_{22}^{-1} \{F\}_2$

employed. After each time step, the location of the quintics are changed upon examination of the solution.

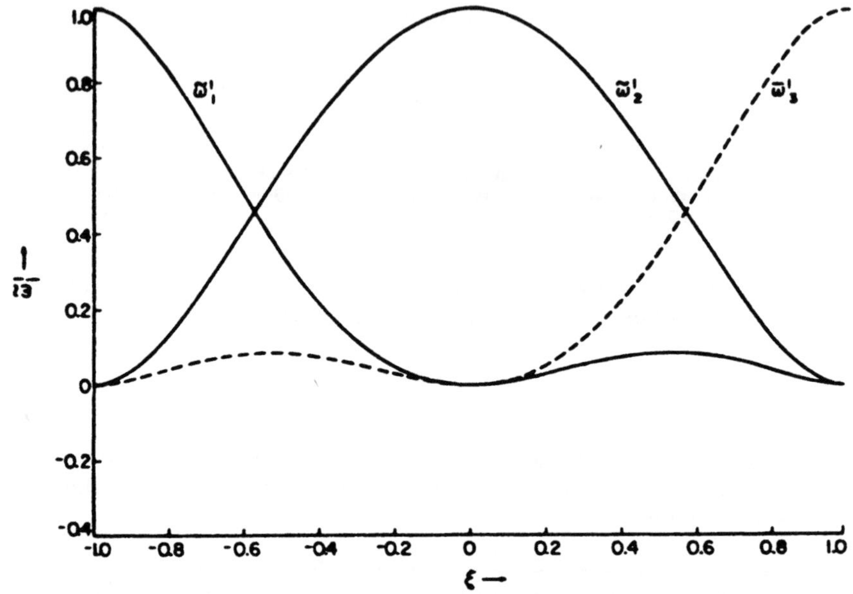

Figure 2a : Quintic Hermite Polynomials $\tilde{\omega}_i^1$ [iε 1,2,3]

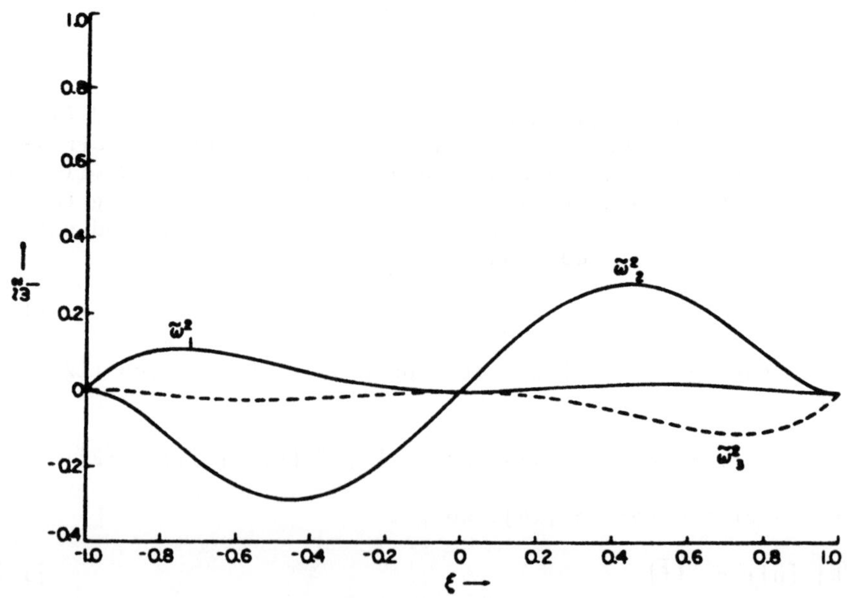

Figure 2b : Quintic Hermite Polymials $\tilde{\omega}_i^2$ [iε 1,2,3]

The "reduced" coefficient matrix $[\bar{k}]$ in [15] is now 2x4 with only the degress of freedom corresponding to the end nodes. Thus it is now compatible with the structure of the cubic elements and may be incorporated directly. The higher order description now has been tailored to the structure of the lower order elements. Once $\{u\}_1$ is determined, the degrees of freedom for the mid side node may be determined using equation [13].

Numerical application

For an error analysis the following values of constants and discretization sizes were used

$P = 50.0$
$\Delta T = 0.000015625$
$\Delta x = .25$ to $.025$

Numerical solutions of the diffusion-convection equation are sually compared against the complimentary error-function-type analytical solution by Grobner & Hofreiter, 1950. There are two known draw backs in the use of this solution. Firstly, the solution is not for a finite domain and secondly, no space-discretized scheme can model the step-like combination of the chosen initial and boundary condition. The result is that such space-discretized schemes starts with some initial mass present over the domain whereas in fact there was none. The error due to the inability to model the initial condition clouds the error of the numerical scheme. The numerical results of this paper are compared with an analytical solution by Mohsen & Pinder,1983which isfor a finite domain and for a polynomially expressed initial condition. The modified initial condition for Cubic Hermitial elements are

$$\phi(X,0) = \begin{cases} 1-3(\frac{X}{\ell_1})^2 + 2(\frac{X}{\ell_1})^3 \;\ldots\ldots & 0 \le X \le \ell_1 \\ 0 \;\ldots\ldots\ldots\ldots\ldots\ldots\ldots & \ell_1 < X \le 1 \end{cases} \qquad [16]$$

where ℓ_1 is the length of the 1st element.

For a quintic Hermitian element, the initial condition is modified to

$$\phi(X,0) = \begin{cases} 1-23(\frac{X}{\ell_1})^2 + 66(\frac{X}{\ell_1})^3 - 68(\frac{X}{\ell_1})^4 + 24(\frac{X}{\ell_1})^5 \ldots & 0 \le X \le \ell_1 \\ 0 \;\ldots\ldots\ldots\ldots\ldots\ldots\ldots\ldots\ldots & \ell_1 < X \le 1 \end{cases} \qquad [17]$$

The analytical solution consistent with these initial conditions is given by

$$\phi(X,T) = \bar{\bar{\phi}}(X) + e^{PX/2} \sum_{n=1}^{\infty} \{ (\frac{2}{\ell} \sum_{q=0}^{P} a_q I_q) - \frac{n\pi}{n^2\pi^2+(P/2)^2} \} e^{-[n^2\pi^2+(P/2)^2]T} \qquad [18]$$

where

$$\bar{\bar{\phi}}(X) = \frac{e^{PX} - e^{P}}{1 - e^{P}} \qquad [19]$$

a_q = coefficients of the polynomials in [16] or [17]

$$I_q = \{e^{-P\ell_1/2} \sum_{r=0}^{q} \frac{(-1)^r q!\ell_1^{q-r}}{\rho^{r+1}(q-r)!} \sin[n\pi\ell_1 - (r+1)\alpha]\}$$

$$- \frac{(-1)^q q!}{\rho^{q+1}} \sin[-(q+1)\alpha] \qquad [20]$$

where

$$\rho = n^2\pi^2 + (P/2)^2 \qquad [21]$$

$$\alpha = \cos^{-1}(-P/2\ell\rho) \qquad [22]$$

Figure 3 shows the results of a typical run as compared to the analytical solution.

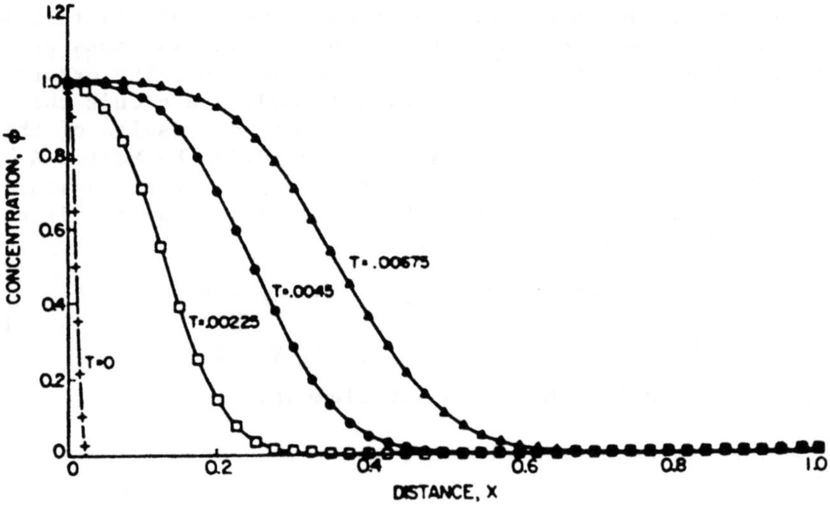

Figure 3 : A typical solution using all cubic element [d.o.f = 80]

For all the results presented in this paper for the diffusion-convection equation only one 'adaptive' element was used near the front, while cubic elements span the rest of the domain. At the end of each time step the 'adaptive' element was relocated at the region of maximum concentration differential. The necessary values at the mid-side node of the newly-converted quintic element is obtained by a cubic interpolation at the old time level results.

The computer program developed was tested for Peclet number P=106 and no oscillations were observed at the nodal values. For the purpose of error analysis a value of Peclet number P=50 was chosen as the analytical solution could not be satisfactorily evaluated beyond this value.

Error analysis

The following two error measures were used in evaluating the results.

$$\text{Sup norm } \ell_\alpha = |\phi_{n_i} - \phi_{a_i}|_{max} \qquad [23]$$

where ϕ_{n_i} = numerical solution at the ith node

ϕ_{a_i} = analytical solution at the ith node

and

$$\ell_2\text{-norm} = \sum_{n=1}^{m} (\phi_{n_i} - \phi_{a_i})^2 \qquad [24]$$

where $m = d/2 + 1$

d = degrees of freedom

Figure 4 shows the comparison of the ℓ_α and ℓ_2 norms using (a) all cubic elements, (b) all quintic elements, and (c) one 'adaptive' element at the front and cubic elements everywhere else.

Each of the above schemes were evaluated for degrees of freedom increasing from 8(Δx = .25) to 80(Δx = .025). In Figure 4, the all-quintic element result shows a better rate of convergence than the all-cubic element result. The 'adaptive' element result is significantly better than the all-cubic result and at higher degrees of freedom becomes somewhat better than the all-quintic result. Recalling that the comparison is based on degrees of freedom, the unexpected rapid rate of convergence may be attributed to the more judicious allocation of compulational effort. At very small meshes, smaller than illustrated in Figure 4, it was observed that the 'adaptive' element scheme does not converge monotonically due to an error dependent upon the time increment size.

The Buckley-Leverett equation

The simultaneous flow of two immiscible, incompressible fluids through a homogeneous porous medium is described by the Buckley-Leverett equation expressed as

$$\frac{\partial S}{\partial t} + q\,\frac{\partial f}{\partial x} = 0 \qquad [25]$$

where

$S = S[x,t]$ = wetting fluid saturation [dimensionless]

q = sum of the interstitial velocities of the two phases [L/T]
f = f[S]=fractional flow rate of the wetting phase [dimensionless]
x = spatial coordinate [L] ($x \in \Omega = [0,1]$)
t = time [T] ($t \in [0,\infty]$)

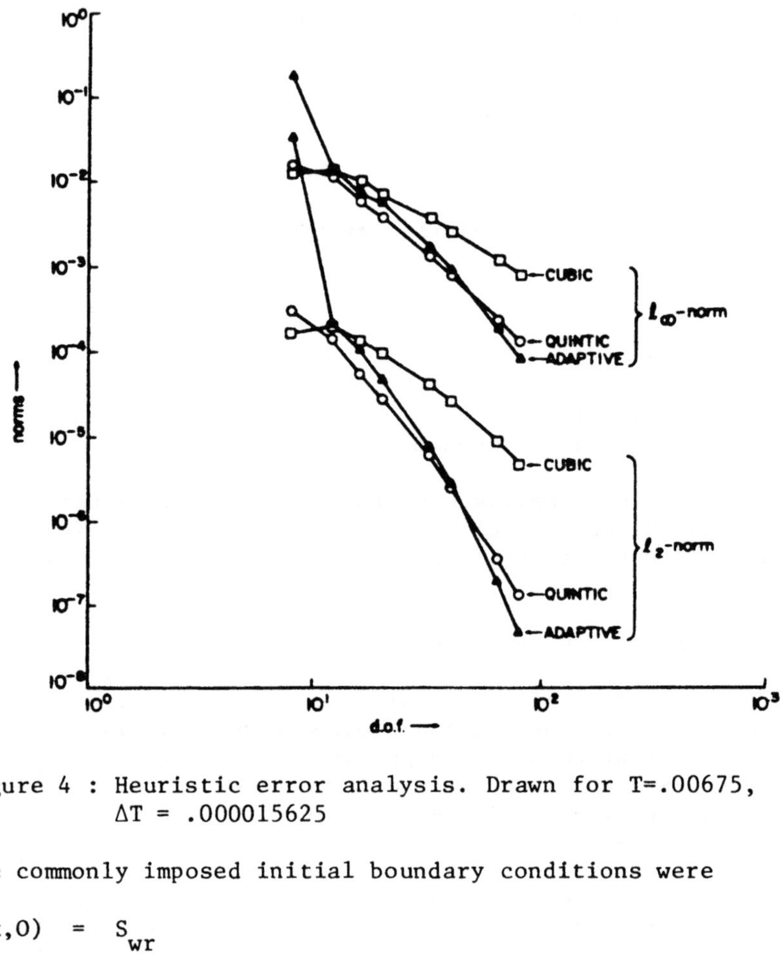

Figure 4 : Heuristic error analysis. Drawn for T=.00675, ΔT = .000015625

The commonly imposed initial boundary conditions were

$$S[x,0] = S_{wr} \qquad [26]$$

$$S[0,t] = 1 - S_{nr} \qquad [27]$$

where S_{wr} = irreducible wetting fluid saturation
S_{nr} = irreducible non-wetting fluid saturation.

The Hermitian approximations for S(x,t), and f(x,t) is given by

$$S(x,t) \simeq \hat{S}(x,t) = \sum_{i=1}^{M} S_i N_i \qquad [28]$$

$$f(x,t) \simeq \hat{f}(x,t) = \sum_{i=1}^{M} f_i N_i \qquad [29]$$

M = 4 for cubic and 6 for quintic elements

where $\hat{S}$ = approximation of S

$\hat{f}$ = approximation of f

$$S_i, f_i = \begin{cases} \text{if } i \text{ is odd, values of S and f respectively at } (i+1)/2 \text{ th.node} \\ \text{if } i \text{ is even, values of } \frac{\partial S}{\partial x} \text{ and } \frac{\partial f}{\partial x} \text{ respectively at } i/2 \text{ th.node} \end{cases}$$

Substituting approximations [28] and [29] into equation [25] and collocation at two points ξ_j per element results

$$[q \frac{\partial \hat{f}}{\partial x} + \frac{\partial \hat{S}}{\partial t}]_{\xi_j} = 0 \qquad [30]$$

It is to be noted that the Buckeley-Leverett equation in some sense is a limiting case of the diffusion convection equation where the second order term has been reduced to zero. As such Buckley-Leverett has a complete frontal behaviour.

It has been shown by Dalen, 1979, Aziz & Settari, 1979, that to have a convergent solution to the Buckley-Leverett problem some form of upstream weighting is required for the convective term. Recently, Allen & Pinder 1983 have shown that orthogonal collocation yields non-convergent solution. However, if the convective terms are evaluated at some upstream points than satisfactory results may be obtained. Persuing these ideas, in the present work, the 'adaptive' element technique is applied to the Buckley-Leverett equation with only the convective term evaluated at some upstream position ξ^*_j while the rest of the quantities are evaluated at the Gauss points ξ_j.Under this procedure equation [30] may be re-written as

$$[q \frac{\partial \hat{f}}{\partial x}]_{\xi^*_j} + [\frac{\partial \hat{S}}{\partial t}]_{\xi_j} = 0 \qquad [31]$$

Equation [31] is integrated in time utilizing a forward difference technique. This will yield

$$[q \frac{\partial \hat{f}}{\partial x}]_{\xi^*_j} + [\frac{\Delta_t \hat{S}}{\Delta t}]_{\xi_j} = 0 \qquad [32]$$

where $\Delta_t \hat{S} = \hat{S}(t + \Delta t) - \hat{S}(t)$

Instead of solving for S, the code was developed to solve for increment $\Delta_t \hat{S}$ The Hermiate representation of $\Delta_t \hat{S}$ may be arrived from [28].

$$\Delta_t \hat{S} = \sum_{i=1}^{M} \Delta_t S_i N_i$$

$$\text{where } \Delta_t S = \begin{cases} \text{if it is odd, value of } \Delta_t \hat{S} \text{ at } (i+1)/2 \text{ th.node} \\ \text{if it is even, value of } \frac{\partial}{\partial x}(\Delta_t \hat{S}) \text{ at } i/2 \text{ th.node} \end{cases}$$

The non-linear aspect of the stated problem was addressed by taking a sufficiently small time step, such that within a given tolerance, the solution of the first two iterations were virtually the same.

Boundary condition

The use of Hermitian interpolation in a first order problem generates the necessity of an additional boundary condition. For n elements equation [32] furnishes 2n equation. However, we have 2(n+1) degrees of freedom for n + 1 nodes. Thus we need 2 additional condition. One is obtained from the imposed condition at x = 0. A computational boundary condition was imposed on the downstream end, $\frac{dS}{dx}\Big|_{x=1.0} = 0.0$. This specification creates no difficulties provided the saturation front does not approach the end of the domain.

Application of adaptive elements to Buckley-Leverett equation and strategy for changing bases

The procedure employed is similar to that in section 'The adaptive element method for the diffusion convection equation', and will not be repeated here. The only new aspect is that two different sets of collocation points are used. This, however, does not change the matrix structure, nor does it effect the node condensation procedure.

Numerical application

For the purpose of numerical application the following sample problem was chosen

$$\lambda_\omega = (S - S_{\omega r})^2 \qquad [33]$$

$$\lambda_n = (1 - S - S_{nr})^2 \qquad [34]$$

The fractional flow f(S), is then defined as

$$f(S) = \begin{cases} \frac{\lambda_\omega}{\lambda_\omega + \lambda_S} & S_{\omega r} \leq S \leq 1 - S_{nr} \\ 1 & 1 - S_{nr} < S < 1.0 \\ 0 & 0 < S < S_{\omega r} \end{cases} \qquad [35]$$

The following data for constants were used

$S_{\omega r} = 0.16$

$S_{nr} = 0.20$

$q = 2.134 \times 10^{-4} [LT^{-1}]$

Figure 5 shows the function λ_ω, λ_n, and f.

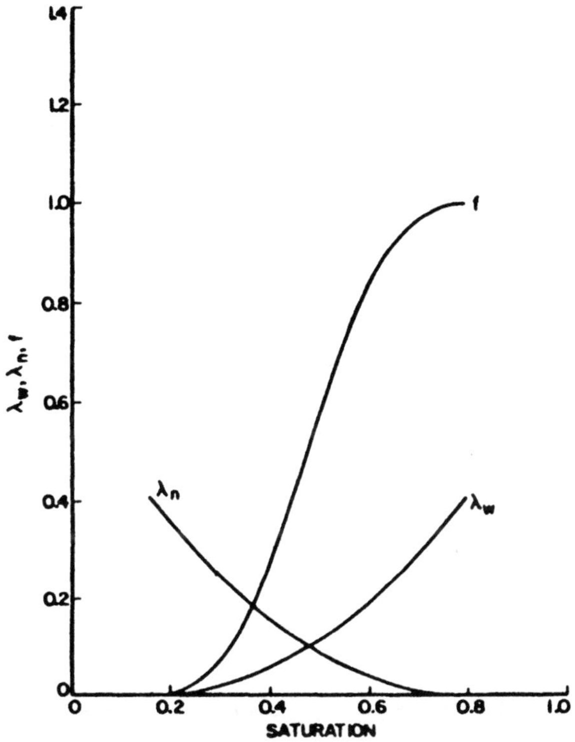

Figure 5 : Fractional flow curves used in the test problem

Discussion of results

The aim of this paper is to investigate the improvements obtained by using 'adaptive' elements at special locations in collocation finite element as opposed to using uniform bases everywhere. In order to compare results the system was first solved using all cubic elements with $\xi_1^* = -0.75$ and $\xi_2^* = +0.25$. The results are shown in Figure 6. Only nodal saturations connected by broken straight lines are shown. In this figure,

(a) The front is correctly located.
(b) The solution upstream of the front is satisfactory.
(c) There are some oscillations immediately downstream of the front.

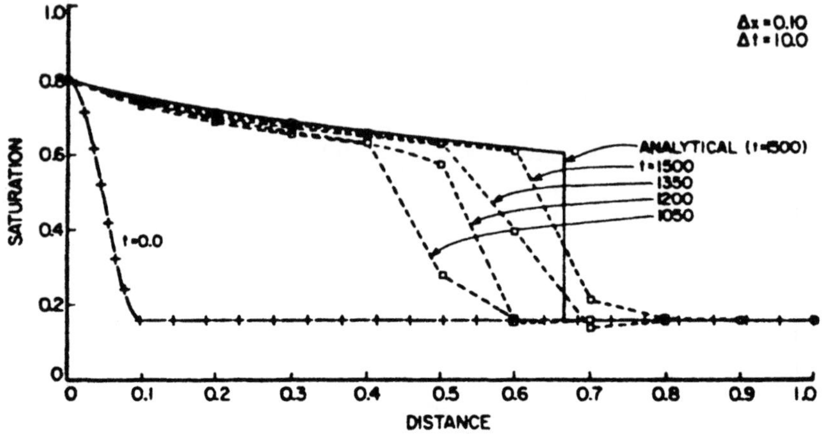

Figure 6 : Solution using all cubic elements (nodal values connected by st. lines)

Figure 7 shows a solution using <u>one</u> 'adaptive' element. The upstream collocation points for the adaptive elements used for all the adaptive element results were

$\xi_1^* = -1.0 \quad \xi_2^* = -0.75 \quad \xi_3^* = 0.25 \quad \& \quad \xi_4^* = 0.50$

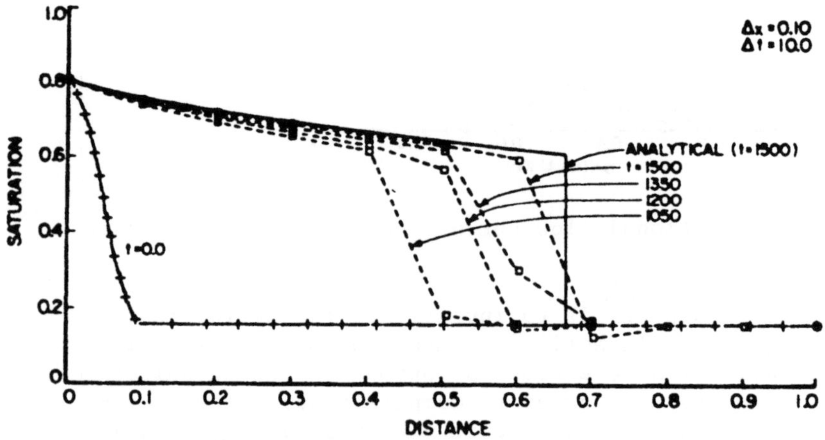

Figure 7 : Solution using 1 'adaptive' element at front and cubic elements on rest of the domain (nodal values connected by st. lines)

In Figure 7, there is a small improvement in the frontal definition but no improvement in the downstream oscillation is seen. Numerical experiments revealed that just one 'adaptive' element was insufficient to model the severe spatial inflection near the front. Figure 8 shows the result obtained with two 'adaptive' elements - one at the largest saturation differential and the other immediately downstream. The 'nodal' oscillations at the downstream of the front have been eliminated at the front.

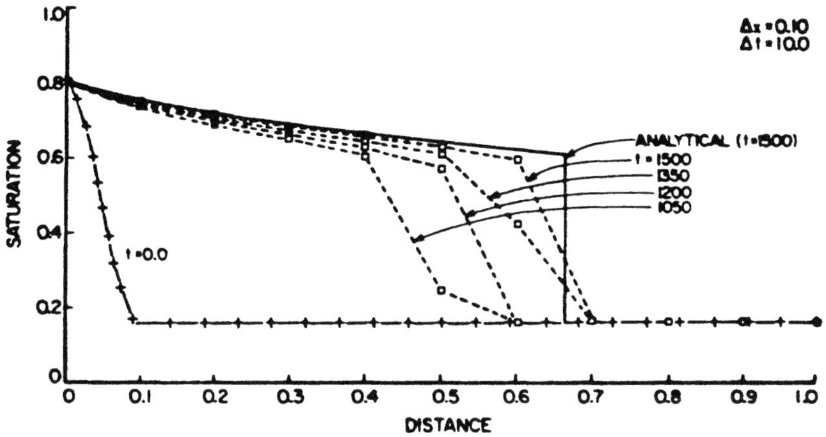

Figure 8 : Solution using 2 adaptive elements on the front and cubic elements over the rest of the domain (nodal values connected by st. lines)

CONCLUSIONS

For the diffusion-convection equation we may conclude that the use of 'adaptive' elements resulted in very accurate results with orthogonal collocation. No 'upstreaming' was necessary at least upto Peclet NO.106. Using one 'adaptive' element near the front and Cubic Hermitian elements everywhere also yielded accuracy comparable to all-quintic Hermitian result. In the case of Buckley-Leverett equation using two 'adaptive' elements with upstream collocation resulted into convergent and non-oscillatory solution.

ACKNOWLEDGEMENT

The author wishes to acknowledge the support provided by the University of Petroleum and Minerals in carrying out this research.

APPENDIX I

Hermitian basis functions

A. Cubic [2 nodes]

$$\omega_1^1 = \frac{1}{4}(2+\xi)(1-\xi)^2$$

$$\omega_1^2 = \frac{1}{4}(1+\xi)(1-\xi)^2(L/2)$$

$$\omega_1^2 = \frac{1}{4}(2-\xi)(1+\xi)^2$$

$$\omega_2^2 = -\frac{1}{4}(1-\xi)(1+\xi)^2(L/2)$$

where $\xi = (\frac{2}{L})x-1 \quad \xi\varepsilon[-1, +1]$

L = length of the element

x = local coordinate $[0,L]$

B. Quintic [with midside node]

$$\tilde{\omega}_1^1 = (\frac{1}{4})\xi^2 (4+3\xi)(\xi-1)^2$$

$$\tilde{\omega}_1^2 = (\frac{1}{4})\xi^2 (1+\xi)(\xi-1)^2(L/2)$$

$$\tilde{\omega}_2^1 = (1+\xi^2)(\xi-1)^2$$

$$\omega_2^2 = \xi(1+\xi^2)(\xi-1)^2(L/2)$$

$$\tilde{\omega}_3^1 = (\frac{1}{4})\xi^2 (4-3\xi)(1+\xi)^2$$

$$\tilde{\omega}_3^2 = (\frac{1}{4})\xi^2 (\xi-1)(1+\xi)^2(L/2)$$

REFERENCES

Allen, M.B., and Pinder, G.F., "Collocation Simulation of Multiphase Porous-Medium Flow", Soc. Pet. Eng. J., Feb.1983, pp.135-142.

Aziz, K. and Settari, A., "Petroleum Reservoir Simulation" Applied Science Publishers, Ltd. London (1979) Sec.5.5.

Christie, I., Griffiths, D.F., Mitchell, A.R. & Zienkiewicz, O.C., "Finite Element Methods for Second Order Differential Equations with Significant First Derivatives", Int.J. Num. Meth. Engng., (1976), 10, pp.1389-1396.

Culham, W.E. & Varga, R.S. "Numerical Methods for Time Dependent, Non-linear Boundary Value Problems", Soc. Pet. Eng. J (1971), pp.374-388.

Dalen, V. "Simplified Finite Element Models for Reservoir Flow Problems", Soc. Pet. Eng. J. (Oct.1979) 333-43.

Gladwell, I. and Wait, R., "A Survey of Numerical Methods for Partial Differential Eqauations", Oxford University Press,

1979, pp.195-211.

Grobner, W. and Hofreiter, N. "Integraltafel", Springer, Vienna, Vol.1, 1949, Vol.2, 1950.

Guymon, G.L., "A Finite Element Solution of the One-Dimensional Diffusion-Convection Equation", Water Resources Res.,(1970) 6, pp 204-210.

Heinrich, J.C., Huyakorn, P.S. & Zienkiewicz, O.C., "An 'Upwind' Finite Element Scheme for Two-Dimensional Convective Transport Equations", Int. J. Num. Meth. Engng., (1977), 11, pp.131-143.

Mohsen, M.F.N. & Pinder, G.F., "Analytical Solution of the Transport Equation Using Polynomial Initial Condition for Verification of Numerical Simulators". To appear in Int. J. of Num. Meth. in Fluids (1983).

Peaceman, D.W. & Rachford, H.H. Jr., "Numerical Calculation of Multi-dimensional Miscible Displacement", Soc. Pet. Eng. J., (1962), 237, pp 327-338.

Price, H.S., Cavendish, J.C., & Varga, R.S., "Numerical Methods of High-Order Accuracy for Diffusion-Convection Equations", Soc. Pet. Eng. J. (1968), 243, pp 293-303.

Smith, I.M., Faraday, R.V. & O'Conner, B.A., "Raleigh-Ritz and Galerkin Finite Elements for Diffusion-Convection Problems" Water Resources Res., (1973), 9, pp.593-606.

DYNAMIC VELOCITY PROFILES ON SIMPLE ELEMENTS - MODEL DESIGN AND TESTING

Daniel R. Lynch and Jean-Louis Hazard

Thayer School of Engineering, Dartmouth College, Hanover, NH 03755 U.S.A.

INTRODUCTION

Enormous progress has been made over the past eight years in the simulation of vertically homogeneous tidal circulation problems. It is in this 2-D, horizontal context that the geometric flexibility of finite elements holds the greatest promise, and yet historically this advantage had been overshadowed by several seemingly insurmountable problems: unexplained instabilities even for small-amplitude motions, uncontrollable parasitic modes in the solutions, and economic disadvantages associated with mass matrix inversion. Fundamental to our present sense of confidence in this arena is a new understanding of the importance of the form of the shallow water equations which is to be implemented on finite elements. Today very efficient explicit and implicit algorithms are available which are free of parasitic gravity waves and which accurately propagate the real, physical waves (Lynch and Gray, 1979; Lynch, 1984). While many detailed and important questions remain, we assert that finite elements have finally become a realistic and competitive alternative in vertically averaged simulation.

As one contemplates adding vertical structure to shallow water simulations, as must ultimately be done, one might reasonably react with extreme reservation if not outright terror in view of the historical 2-D difficulties. However the situation is far from terrible, as the vertical dimension in shallow water is fundamentally different than the horizontal. Gravity waves are not a feature of the vertical structure; essentially the behavior is diffusive (parabolic), although coupled with/ driven by the horizontal hyperbolic structure. Since parabolic problems, at least the simple ones, are quite well understood in the finite element arena, there is at least hope that we may proceed without major roadblocks to full, 3-D simulation which is competitive with finite difference techniques.

In the absence of any evidence to the contrary, then, our hypothesis is that such success can be achieved by starting with the best algorithms for horizontal simulation and appending the vertical diffusion of momentum. In this paper we first review some of the options which might be used in the vertical. Thereafter we explore some algorithm design considerations with respect to space and time discretization. Finally, we present some model tests against analytic solutions, some of which have not appeared in the literature before.

Our entire discussion from this point on is limited to the laterally averaged equations. We view this as a necessary (but not sufficient) prelude to full three-dimensional simulation. Further, the laterally averaged equations have themselves been quite useful in several scientific investigations (e.g., Blumberg, 1977; Hamilton, 1977; Wang and Kravitz, 1980) although no systematic evaluation of finite elements has been carried out in this context. Our interest here is in the testing of the barotropic mode; we defer the simulation of the longer-term dynamics of gravitational circulation and assume the density is either constant (most of the tests) or has a known horizontal distribution. Of course, we restrict our attention to algorithms which do not overtly rely on these various simplifications; otherwise the generalization to the more complete problem is lost at the outset.

GOVERNING EQUATIONS

We employ the standard set of laterally averaged equations, with the usual assumptions (Lynch, 1984). At any point (x,z) the local continuity and momentum equations are

$$\frac{\partial}{\partial x}(BU) + \frac{\partial}{\partial z}(BW) = 0 \tag{1}$$

$$B\frac{DU}{Dt} + Bg\frac{\partial \zeta}{\partial x} - \frac{\partial}{\partial z}\left(BN\frac{\partial U}{\partial z}\right) = -\frac{Bg}{\rho_o} I_x \tag{2}$$

where B = channel width
U,W = laterally averaged velocity components
g = gravity
ρ_o = reference density
N = vertical eddy viscosity
ζ = free surface elevation

(Here we have the z coordinate positive upwards, running from $z = -H$ at the bottom to $z = \zeta$ at the free surface.) As mentioned above, the baroclinic term I_x is retained as an exogenous forcing term, with density structure assumed known:

$$I_x(x,z) = \int_z^{\zeta} \frac{\partial \rho}{\partial x}\, dz' \tag{3}$$

We will also employ the vertically integrated forms of Equations (1) and (2):

$$B \frac{\partial \zeta}{\partial t} + \frac{\partial Q}{\partial x} = 0 \tag{4}$$

$$\frac{\partial Q}{\partial t} + \frac{\partial}{\partial x} \int_{-H}^{\zeta} BUUdz + gA \frac{\partial \zeta}{\partial x} = -\frac{g}{\rho_o} \int_{-H}^{\zeta} BI_x dz + BN \frac{\partial U}{\partial z} \Bigg|_{-H}^{\zeta} \tag{5}$$

in which A = channel cross-sectional area and Q is the cross-section discharge:

$$Q = \int_{-H}^{\zeta} BUdz \tag{6}$$

Equations (4) and (5) are essentially the governing equations for the gravity waves; while Equations (1) and (2) govern the details of the velocity profile.

OPTIONS IN DISCRETIZATION

Although we take as given the use of finite elements for horizontal discretization, one still has considerable freedom in the vertical. For example, several investigators have expanded the vertical structure in terms of the eigenfunctions of the self-adjoint operator $(\partial/\partial z)(N\partial f/\partial z)$, subject to natural boundary conditions at the surface and bottom (Davies, 1980; Humphreys and Pearce, 1981; Laible, 1982). Since such functions constitute an orthogonal basis, the Galerkin procedure yields an uncoupling of the equations for the individual modal amplitudes which are then functions of (x,y,t) only. While this procedure is quite elegant and appealing from a classical viewpoint, one must be prepared to compute the eigenfunctions numerically when N varies with depth; and to recompute them when the density structure changes. In this more general problem setting, then, conventional finite element discretization on local bases becomes attractive. Some results have been reported along these lines by J.D. Wang (1979) and Laible (1982) and we will explore this option with the simplest possible elements: bilinear quadrilaterals in (x,z). As shown in Figure 1, we essentially deploy a string of nodes directly below each horizontal node, in effect creating a very close cousin to a finite difference treatment. In the simplicity of these elements lies their appeal: they have very limited bandwidth; inner products can be evaluated with low level quadrature; integral lumping is feasible and renders the mass matrix diagonal without risking convergence; and graphical output is nearly trivial. For these reasons we have elected to test these elements, with the goal of either disqualifying them or establishing a base of comparison for experiments with higher, more complex elements.

STABILITY

Consider the explicit leapfrog integration of Equations (1-5) with lagged friction on an infinite, uniform rectangular grid. For the finite difference case or the lumped finite element case, the simplest stability analyses yield three stability constraints:

$$\frac{\Delta t}{\Delta x}\sqrt{gH} < 1 \qquad \text{(first Courant condition)} \qquad (7)$$

$$\frac{\Delta t N}{\Delta z^2} < \frac{1}{4} \qquad \text{(diffusion condition)} \qquad (8)$$

$$\frac{\Delta t U}{\Delta x} < 1 \qquad \text{(second Courant condition)} \qquad (9)$$

Suppose we have a wave with period P propagating on this grid, and that the horizontal wavelength is divided into n_x elements:

$$\Delta x = \frac{\sqrt{gh}P}{n_x} \qquad (10)$$

Similarly, let the vertical "wavelength" ($2\sqrt{\pi NP}$ for periodic diffusion) be divided into n_z elements:

$$\Delta z = \frac{2\sqrt{\pi NP}}{n_z} \qquad (11)$$

The three critical time steps from above thus become

$$\Delta t_{C1} = \frac{P}{n_x} \qquad \text{(first Courant condition)} \qquad (12)$$

$$\Delta t_N = \frac{\pi P}{n_z^2} \qquad \text{(diffusion condition)} \qquad (13)$$

$$\Delta t_{C2} = \frac{\sqrt{gh}}{U}\frac{P}{n_x} \qquad \text{(second Courant condition)} \qquad (14)$$

and two useful ratios may be deduced

$$\frac{\Delta t_{C2}}{\Delta t_{C1}} = \frac{\sqrt{gh}}{U} \qquad (15)$$

$$\frac{\Delta t_N}{\Delta t_{C1}} = \frac{\pi n_x}{n_z^2} \qquad (16)$$

Equation (15) is a familiar result - the gravity wave Courant condition is much more restrictive than the advective condition. Equation (16) is more interesting. First, it is independant of physical quantities and depends solely on the relative magnitude of the x and z resolutions. In cases where these are roughly equal, it is clear that the diffusion constraint is binding first and by far. Thus, to achieve reasonably large time steps one must render both the gravity waves and the vertical diffusion implicit. The experience of Blumberg and Mellor (1980), in which vertical diffusion is treated implicitly and gravity waves explicitly in a very complex simulation, appears to corroborate Equation (16).

In many practical cases, however, the horizontal resolution is dictated by the need to resolve detailed geometrical features. As a result, Δx is much smaller than it would need to be in an unbounded medium, and $n_x \gg n_z$. Here we find that the first Courant condition may be the limiting one, and several investigators have achieved large (30 min. for the M2 tide) time steps with implicit gravity waves and explicit diffusion (Hamilton, 1977; Wang and Kravitz, 1980). It is clear from Equation (13) that n_z will be limited to 8 or 9 in this case, which is rather coarse.

Finally, it is instructive to introduce the temporal discretization $n_t = P/\Delta t$, and to reexpress the explicit diffusion constraint Equation (13) as

$$n_t \geq \frac{n_z^2}{\pi} \tag{17}$$

The severity of Equation (17) is clear for reasonable values of $n_z \geq 20$.

TEMPORAL DISCRETIZATION

In light of the above, our strategy is to integrate the gravity waves and the momentum diffusion implicitly. Since the equations are nonlinear, we adopt a three-level "leapfrog" type scheme, with the nonlinear contributions centered but explicit. The gravity wave formulation is based on Kwizak and Robert (1971), as introduced on finite elements by Gray and Lynch (1979). With $\Delta\zeta = \zeta^{k+1} - \zeta^{k-1}$, and k the time level, we have from Equations (4) and (5)

$$B\Delta\zeta + \Delta t \frac{\partial}{\partial x} (\Delta Q + 2Q^{k-1}) = 0 \tag{18}$$

$$\Delta Q = -\Delta t g A^k \frac{\partial}{\partial x} (\Delta\zeta + 2\zeta^{k-1}) - 2\Delta t \left\{ \frac{\partial}{\partial x} \int BUU dz + \frac{g}{\rho_o} \int BI_x dz - BN \frac{\partial U}{\partial z} \bigg|_{-H}^{\zeta} \right\}^k \tag{19}$$

Elimination of ΔQ yields

$$B\Delta\zeta - \Delta t^2 \frac{\partial}{\partial x}\left(gA^k \frac{\partial \Delta\zeta}{\partial x}\right) = 2\Delta t^2 \frac{\partial}{\partial x}\left[gA^k \frac{\partial}{\partial x} \zeta^{k-1} + \left\{ \frac{\partial}{\partial x} \int BUU dz + \frac{g}{\rho_o} \int BI_x dz - BN \frac{\partial U}{\partial z}\bigg|_{-H}^{\zeta} \right\}^k \right] - 2\Delta t \frac{\partial Q^{k-1}}{\partial x} \tag{20}$$

In a given time step, Equation (20) is first used to find $\Delta\zeta$. Then, the velocity profile is obtained from an implicit time-differenced form of Equation (2):

$$B\Delta U - 2\Delta t \frac{\partial}{\partial z} BN \frac{\partial \overline{U}}{\partial z} = -\Delta t Bg \frac{\partial}{\partial x} (\Delta\zeta + 2\zeta^{k-1}) - 2\Delta t B \left\{ U \frac{\partial U}{\partial x} + \frac{g}{\rho_o} I_x \right\}^k \tag{21}$$

where $\overline{U}$ is the implicit version of the momentum diffusion:

$$\overline{U} = \frac{\varepsilon}{2} (U^{k+1} + U^{k-1}) + (1-\varepsilon) U^k \tag{22}$$

In the simplest linear analysis, Equation (21) will be unconditionally stable for $\varepsilon > .5$ and unconditionally unstable for $\varepsilon < .5$. Note that $\Delta\zeta$ appears as a known forcing term in Equation (21) since it is obtained first in any time step. The final calculation in any time step is the calculation of the vertical velocity W, via the continuity equation (1) at time level k+1. We thus have an implicit formulation in which the calculations for ζ^{k+1}, U^{k+1} and W^{k+1} may be made sequentially and without iteration, rather than simultaneously. The vertically integrated momentum equation (19) is never employed directly; the variable Q appearing in Equation (20) is integrated directly over the vertical, as are the advective terms in Equation (20).

ELEMENTS AND QUADRATURE

We employ a standard Galerkin treatment of Equations (20), (21) and (1) on 4-node isoparametric quadrilaterals as in figure 1. Integration on these elements is achieved by quadrature, <u>with the quadrature points at the nodes</u>. Specifically, the integral of any function f(x,y) is obtained as

$\langle f \rangle \simeq \sum_{i=1}^{4} |J|_i f(x_i, z_i)$ where i refers to the local node number. In figure 2 we introduce a local node numbering convention, and in table 1 we display the derivatives of the bases at the nodes. Of course, the bases themselves vanish at 3 of the 4 nodes, which has the effect of reducing the "mass matrix" $\langle \phi_i \phi_j \rangle$ to a diagonal matrix. Table 1 reveals an additional, significant feature: $\partial\phi_i/\partial z$ vanishes at nodes which are not on the same vertical line as node i. As a consequence, the matrix

$\left\langle N \frac{\partial \phi_i}{\partial z} \frac{\partial \phi_j}{\partial z} \right\rangle$ in the implicit momentum equation will be tridiagonal, i.e., the momentum equation at node i will involve only U_i and its two nearest vertical neighbors! This same tridiagonal structure occurs in the matrix

$\left\langle B \frac{\partial \phi_i}{\partial z} \phi_j \right\rangle$ which multiplies the unknown W_i in the continuity equation. As a result, we achieve an implicit model, free of linear stability constraints, which has only tridiagonal matrices. It is readily verified from table 1 that this quadrature exactly integrates the area of an element.

ANALYTIC SOLUTIONS

We have tested our model against analytic solutions to date. Officer (1976) summarizes several steady solutions which are driven by wind, through flow, and density gradient. Additionally, we employ periodic solutions to the linearized equations, constructed as follows.

First, periodic solutions are available (Lynch and Gray, 1978; Prandle and Rahman, 1980) for the barotropic response when the bottom stress may be formulated as $\tau H\overline{U}$, where τ is a constant linear friction parameter. In the steady state, it is easy to show that $\tau=3N/H^2$. In the dynamic case, τ becomes complex, indicating that the bottom stress is out of phase with the vertical average velocity $\overline{U}$. For constant N over the depth, and no slip at the bottom, Lynch (1984) shows that

$$\tau = j\omega\left[\frac{\tanh(\lambda)}{\lambda-\tanh(\lambda)}\right] \tag{23}$$

where $j = \sqrt{-1}$, ω is the radian frequency, and $\lambda = \sqrt{j\omega H^2/N}$. (The low frequency limit is readily verified by a Taylor series expansion.) With this expression for τ, the barotropic response for ζ may be computed from the existing solutions. Subsequently, the velocity profile is obtained from the local barotropic forcing:

$$U = -\frac{g}{j\omega}\frac{\partial\zeta}{\partial x}\left[1 - \frac{\cosh(\lambda z/H)}{\cosh(\lambda)}\right] \tag{24}$$

In all comparison runs the nonlinear terms were retained but suppressed by small-amplitude forcing.

RESULTS

Our first experiments involved time-stepping to steady state with various combinations of sea slope, density gradient, wind stress, and throughflow imposed. Typical results, obtained on a grid with 5 nodes on the length and 5 to 10 nodes on the depth, appear in figure 3. In this and subsequent figures the analytic solution is shown as a line, with computed nodal values plotted as points.

Next, we ran several periodic problems which were forced by specifying $\zeta(t)$ at one end of a channel which was closed at the other. In each of these cases the period was 12.4 hours, and we report results for $\zeta(x)$, $\overline{U}(x)$ (vertically averaged velocity) and/or $U(z)$ at two points in time: when the forcing peaks ($\omega t=\pi/2$) and when it vanishes ($\omega t=0$).

The first such case is a channel 60 km long, with constant depth (10 m) and width (700 m). The forcing amplitude was 0.3 m, and N=10 cm^2/sec. everywhere. There were 7 nodes on the length and 5 on the depth. Under these conditions Δt_N=780 sec. and $\Delta t_{Cl} \approx 1000$ sec. In figures 4a-c we show computed results with ε=.6 and various values of Δt. The smallest value, 620 sec., corresponds to 72 Δt per period. Stable results were obtained with Δt increased by a factor of 6, with an obvious degradation of accuracy - at this point, there are only 12 timesteps per period, and figure 4 clearly shows the growth of this temporal discretization error. Essentially identical results were obtained with ε=.5 and ε=1. These are excellent results, given the coarseness of the grid.

The second periodic test case is a constant width channel with depth which increases as x^2, from H=10 m at the closed end to H=62.5 m at the open end. Channel length is 300 km. The analytic solution for this problem (Lynch and Gray, 1978) assumes that τ is constant, and in light of Equation (23) we let N=N(x) such that N/H^2 is constant. The value N=20 cm^2/sec. was used at the closed end. Computed results with 72 time steps per period were excellent, and appear in figures 5a-d. There were 10 nodes on the horizontal and 5 on the depth.

Finally, we tested a constant depth (10 ft) channel with width which increased linearly with x, from 100 m at the closed end to 250 m at the open end. Overall length was 15×10^4ft. N was 2 cm^2/sec.; the forcing amplitude was .1 ft, 1% of the depth. Results appear in figures 6a-d for 7 nodes on the length, 5 on the depth. The analytic solution for ζ and $\overline{U}$ is from Prandle and Rahman (1980).

Further tests may be found in Hazard (1983).

CONCLUSION

We regard these preliminary tests as very promising, and look forward to several adventures:

- quantification of the importance of the moving boundary condition at the free surface (not accounted for to date),
- incorporation of salt transport and testing of the baroclinic response,
- 3-D simulation with the Implicit Wave Equation formulation for the gravity waves.

ACKNOWLEDGEMENT

This work was supported by the U.S. Office of Naval Research, contract numbers 81-K-0550 and NR-642-001. We are indebted to Charles B. Officer for guidance during the early stage of the work.

REFERENCES

Blumberg, A.F. (1977) Numerical model of estuarine circulation. J. Hyd. Div. ASCE, 103, HY3.

Blumberg, A.F. and G.L. Mellor (1980) A coastal ocean numerical model, Mathematical Modeling of Estuarine Physics, J. Sundermann and K-P. Holz, eds. Springer-Verlag, Berlin, 203-219.

Davies, A.M. (1980) Application of the Galerkin method to the formulation of a three-dimensional non-linear hydrodynamic numerical sea model, Appl. Mathematical Modelling 4, 245-256.

Gray, W.G. and D.R. Lynch (1979) On the control of noise in finite element tidal computations: a semi-implicit approach, Computers and Fluids, 7, 1:47-67.

Hamilton, P. (1977) On the numerical formulation of a time-dependent multi-level model of an estuary, with particular reference to boundary conditions, Estuarine Processes II, M. Wiley, ed., Academic Press, 347-364.

Hazard, J-L. (1983) Finite Element Model of Laterally Averaged Estuarine Circulation, M.Sc. Thesis, Thayer School of Engineering, Dartmouth College, Hanover, NH.

Humphreys, A.C. III and B.R. Pearce (1981) Currents in Penobscot Bay, Maine, Proc. Oceans'81 II, IEEE Pub. #81Ch1685-7, 805-809.

Kwizak, M. and A.J. Robert (1971) A semi-implicit scheme for grid point atmospheric models of the primitive equations, Monthly Weather Review, 99, p. 32ff.

Laible, J.P. (1982) Formulating layered and semi-analytic finite element flow programs with variable eddy viscosity, Finite Elements in Water Res. IV, Holz et al., eds., Springer-Verlag, Berlin, 5-79 to 5-90.

Lynch, D.R. (Jan. 23-26, 1984) Mass Balance in Shallow Water Simulations, Proc. 5th Int'l Symposium on Finite Elements in Flow Problems, Carey and Oden, eds., U. Texas at Austin, 1-5.

Lynch, D.R. (1984) Basic Hydrodynamic Equations for Lakes, Ch. 2 of Physics-Based Modeling of Lakes, Reservoirs, and Impoundments, W.G. Gray, ed., ASCE Report of Engineering Practice, in press.

Lynch, D.R. (1984) Analytic Solutions for Periodic Velocity Profiles, in preparation.

Lynch, D.R. and W.G. Gray (1978) Analytic solutions for computer flow model testing, J. Hydraul. Div., ASCE 104, HY10:1409-1428.

Lynch, D.R. and W.G. Gray (1979) A wave equation model for finite element tidal computations, Computers and Fluids 7, 3:207-228.

Officer, C.B. (1976) Physical Oceanography of Estuaries and Associated Coastal Waters, Wiley.

Prandle, D. and M. Rahman (1980) Tidal response in estuaries, J. Phys. Oceanogr. 10, 1552-1573.

Wang, J.D. (1979) Finite element model of 2-D stratified flow, J. Hydraul. Div. ASCE 105, HY12:1473-1485.

Wang, D-P., and D.W. Kravitz (1980) A semi-implicit two-dimensional model of estuarine circulation, J. Phys. Oceanogr. 10, 441-454.

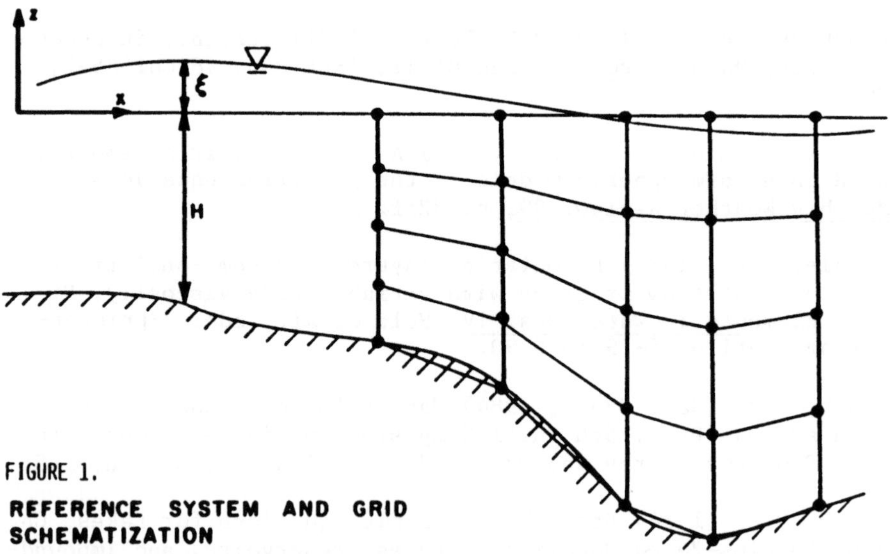

FIGURE 1.

REFERENCE SYSTEM AND GRID SCHEMATIZATION

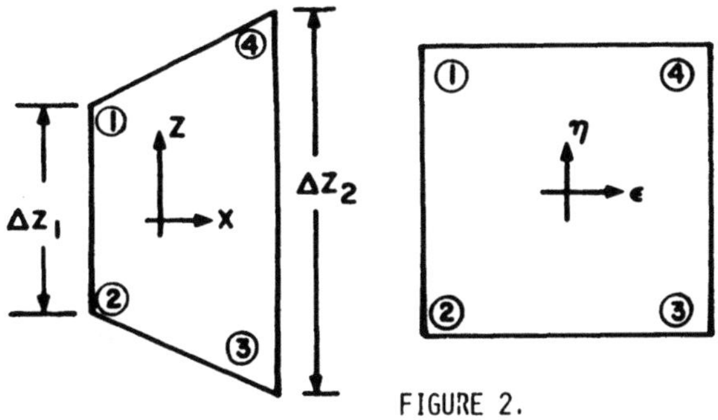

FIGURE 2.

Nodes	1	2	3	4
Coord. (ε,η)	-1,1	-1,-1	1,-1	1,1
$\lvert J \rvert$	$\frac{\Delta x}{4}\Delta Z_1$	$\frac{\Delta x}{4}\Delta Z_1$	$\frac{\Delta x}{4}\Delta Z_2$	$\frac{\Delta x}{4}\Delta Z_2$
$\partial\phi_1/\partial z$	$\frac{1}{\Delta Z_1}$	$\frac{1}{\Delta Z_1}$	0	0
$\partial\phi_2/\partial z$	$\frac{1}{-\Delta Z_1}$	$\frac{1}{-\Delta Z_1}$	0	0
$\partial\phi_3/\partial z$	0	0	$\frac{1}{-\Delta Z_2}$	$\frac{1}{-\Delta Z_2}$
$\partial\phi_4/\partial z$	0	0	$\frac{1}{\Delta Z_2}$	$\frac{1}{\Delta Z_2}$
$\frac{\partial\phi_1}{\partial x}$	$\frac{-(Z_4-Z_2)}{\Delta x\Delta Z_1}$	$\frac{-(Z_3-Z_2)}{\Delta x\Delta Z_1}$	0	$\frac{-1}{\Delta x}$
$\frac{\partial\phi_2}{\partial x}$	$\frac{-(Z_1-Z_4)}{\Delta x\Delta Z_1}$	$\frac{-(Z_1-Z_3)}{\Delta x\Delta Z_1}$	$\frac{-1}{\Delta x}$	0
$\frac{\partial\phi_3}{\partial x}$	0	$\frac{1}{\Delta x}$	$\frac{(Z_4-Z_2)}{\Delta x\Delta Z_2}$	$\frac{(Z_4-Z_1)}{\Delta x\Delta Z_2}$
$\frac{\partial\phi_4}{\partial x}$	$\frac{1}{\Delta x}$	0	$\frac{(Z_2-Z_3)}{\Delta x\Delta Z_2}$	$\frac{(Z_1-Z_3)}{\Delta x\Delta Z_2}$

TABLE 1

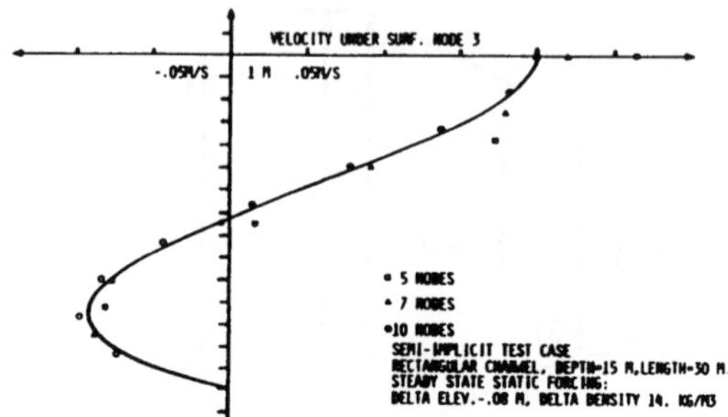

FIGURE 3.

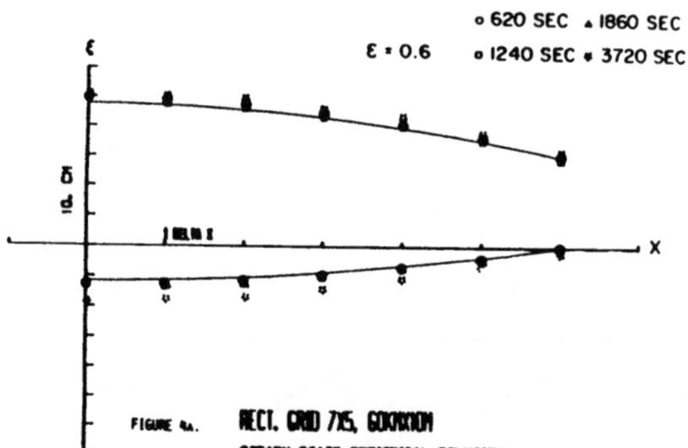

FIGURE 4A. RECT. GRID 7X5, 60KMX10M
STEADY STATE PERIODICAL SOLUTION
SURFACE ELEVATION AS FCQ,FOR VT=0 AND PI/2

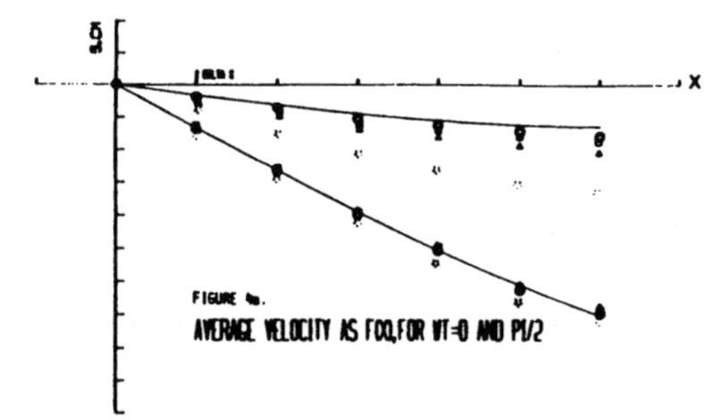

FIGURE 4B.
AVERAGE VELOCITY AS FCQ,FOR VT=0 AND PI/2

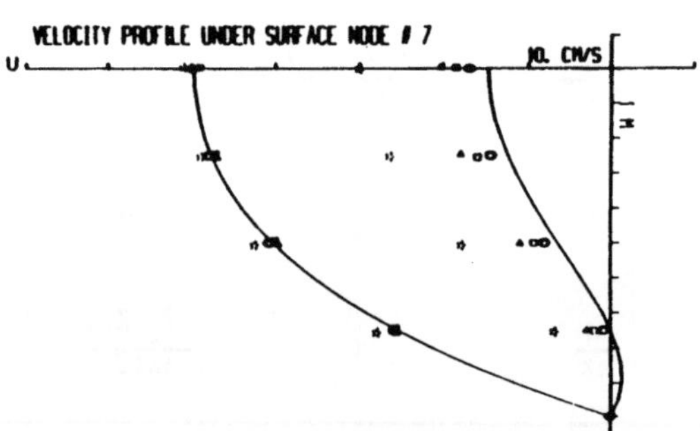

FIGURE 4C.

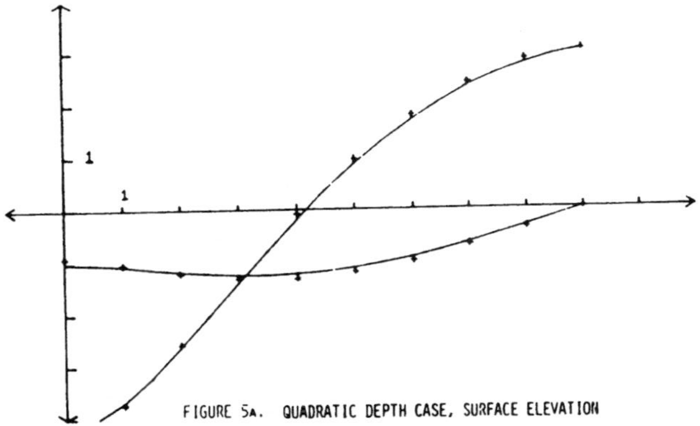

FIGURE 5A. QUADRATIC DEPTH CASE, SURFACE ELEVATION

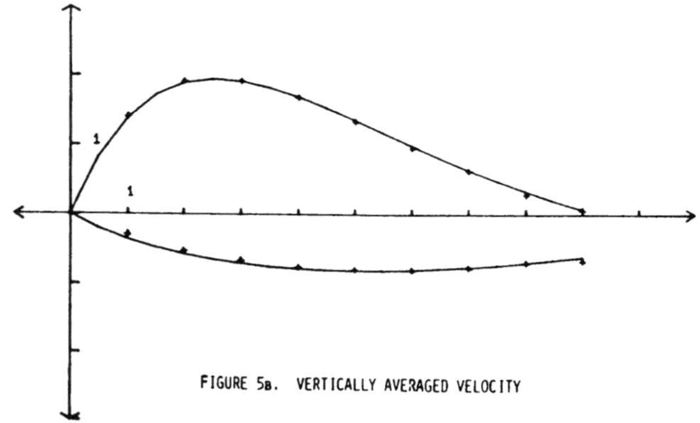

FIGURE 5B. VERTICALLY AVERAGED VELOCITY

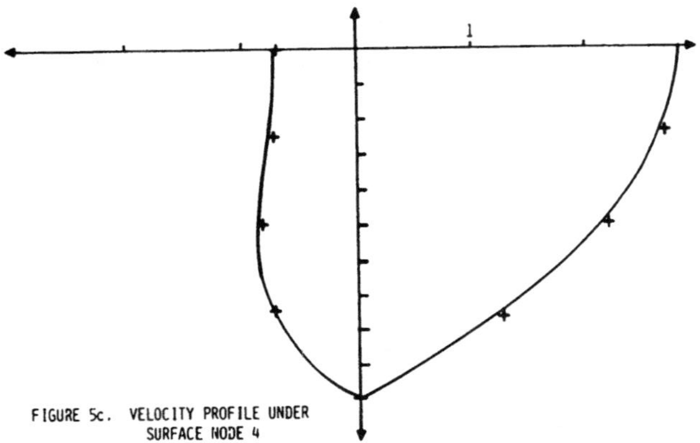

FIGURE 5C. VELOCITY PROFILE UNDER SURFACE NODE 4

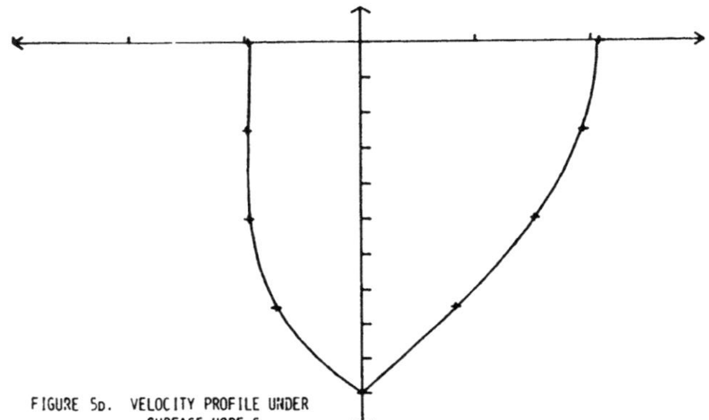

FIGURE 5D. VELOCITY PROFILE UNDER SURFACE NODE 6

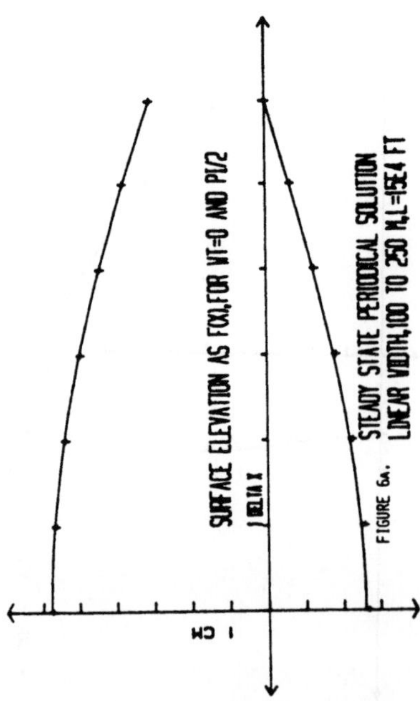

FIGURE 6A.

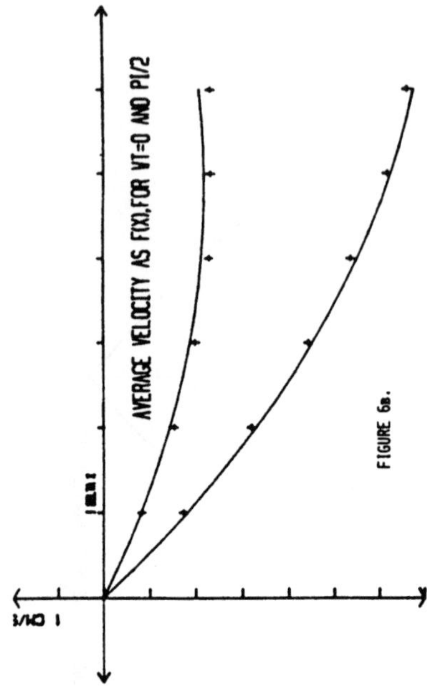

FIGURE 6B.

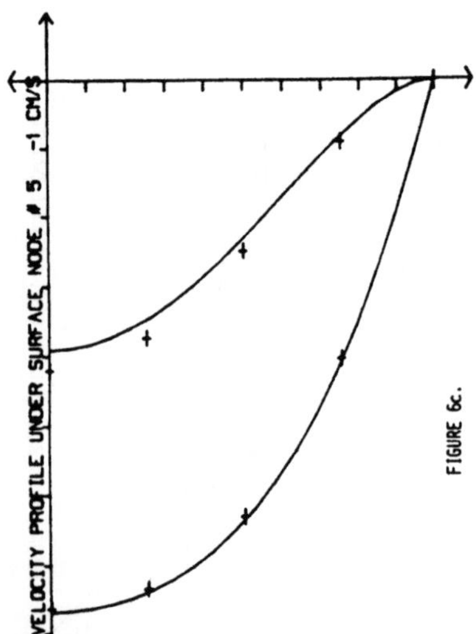

FIGURE 6C.

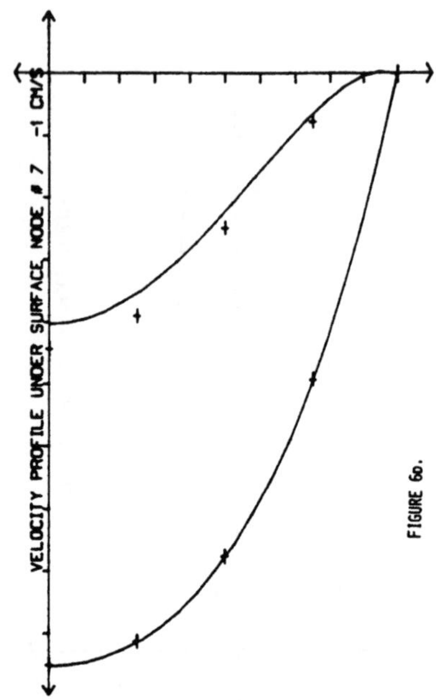

FIGURE 6D.

GROUNDWATER FLOW AND SEEPAGE

FINITE ELEMENTS IN ANALYSIS OF EFFECT OF STOCHASTIC VARIABILITY IN AQUIFER CONDUCTIVITY ON HYDRAULIC HEAD

T. S. Virdee, and N. T. Kottegoda
Department of Civil Engineering, University of Birmingham, Birmingham B15 2TT, England.

ABSTRACT

In a given two-dimensional region, the hydraulic conductivity values are assumed to be log-normally distributed with first-order nearest-neighbour correlation. With a prescribed set of boundary conditions, the appropriate partial differential equations are solved numerically via finite element, finite difference, and random walk methods. This is repeated over a number of simulations of the conductivity values and the head, the variance, and the covariance structure obtained for the elements. The relative merits of the methods in terms of CPU time are examined. The results from the numerical methods are compared with the analogous analytical solution for a statistically homogeneous and isotropic media, and it is shown that the numerical and analytical methods give consistent solutions.

INTRODUCTION

Of all the numerical methods available for the solution of the groundwater flow equation, the finite element method has a particular attraction in certain cases where the aquifer properties are considered to be realizations of a stochastic process. One such case occurs in the optimal sampling strategy of an aquifer -it can be shown that the maximum information per sample, in analysis of aquifer parameters, corresponds to the case where the sampling points are located on the vertices of an isoceles triangle (Dalenius et al. (1961)). Thus, if such a strategy were used in aquifer assessment, isoceles elements could be used, with the inherent ease that the finite element method provides in the direct solution to the problem of head and flow prediction. This is in contrast to the finite difference method, and the random walk simulation technique which generally require the use of rectangular elements. Indeed, one of the major advantages of the finite element method is a more accurate representation of the aquifer material and its boundaries. The fact that application of boundary conditions is easier is an added bonus. The random walk simulation technique is, perhaps, the most realistic simulation method in that one actually realizes the statistical nature in reaching equilibrium in stochastic field. However the computation time required to get an adequate and stable solution, even when one uses variance reduction techniques (see, for example, Hammersley and Handscomb (1964)), is usually prohibitively large.

In reality samples are not usually taken according to the above mentioned optimal strategy so as to exhaust data requirements - they are usually taken according to what one regards as the best regional or immediate benefit/cost ratio, which is not only usually quite subjective, but also perturbs the pattern of sampling, if any. Even in such a case the finite element method provides the natural method for solution of the direct groundwater problem, with a slight increase in the computational time required in dealing with more complex elemental topologies.

In this paper we consider square elements for the purpose of comparison of the computational cost (measured as CPU time on DEC2060) of the above mentioned three numerical methods. The analytical method does not require the extensive use of the computer, and thus the computational cost can be regarded as virtually zero. It should, however, be kept in mind that in certain cases not all of these methods will be suitable for the solution of a specific problem, so that CPU time may not always be the limiting factor, although it may still be an important consideration.

METHODS OF SOLUTION

The equation governing steady-state two-dimensional groundwater flow in an isotropic and heterogeneous confined aquifer is

$$\frac{\partial}{\partial x}\left(K\frac{\partial \phi}{\partial x}\right)+\frac{\partial}{\partial y}\left(K\frac{\partial \phi}{\partial y}\right)=0 \qquad (1)$$

where ϕ is the hydraulic head, K is the hydraulic conductivity, and (x,y) are spatial coordinates. If we denote the probability space parameter by λ, then $K=K(x,y;\lambda)\equiv K(r;\lambda)$, and $\phi=\phi(r;\lambda)$. Equation (1) can be solved in a number of ways, depending upon the nature of K. In this paper we consider an analytical solution , and the above mentioned three numerical methods.

Analytical Solution

It has been shown by Virdee and Kottegoda (1984) that

$$\phi(r) = \langle\phi(r;\lambda)\rangle = -Jr \tag{2}$$

and $$R_{\phi\phi}(r,r_1) \cong J^2\iiiint R_{ff}(r,r_1)dr d\theta dr_1 d\theta_1 \tag{3}$$

where J is the average hydraulic gradient, r and r_1 are two points, and the angles θ and θ_1 are measured from the direction of hydraulic gradient plane. The evaluation of the head covariance $R_{\phi\phi}(r,r_1)$ requires the knowledge of the covariance structure of the logrithm of hydraulic conductivity, $R_{ff}(r,r_1)$. Mizell et al. (1982) consider consider modified Whittle B covariance:

$$R_{ff}(\xi) = \sigma^2[(1+\zeta^2/8)\zeta K_1(\zeta) - \zeta^2 K_0(\zeta)] \tag{4}$$

with $\zeta = 3\pi\xi/16\lambda$, where λ is the integral scale, σ^2 is the variance of the Whittle spectrum, K_1 and K_0 are modified Bessel functions of second kind of first and zeroth order respectively (see Abromowitz and Stegun (1972) for tables), and $\xi = |r-r_1|$. (Whittle B covariance was used because the nearest-neighbour first-order model used in the numerical models for generation of conductivity realizations was found to fit it closely, thus facilitating comparison). Substituting (4) into (3), and averaging over angles, we find that

$$R_{\phi\phi}(\xi) = J^2\sigma_f^{\,2}\lambda^2\pi\left(\frac{8}{3\pi}\right)^2[(2-\zeta^2/2)\zeta K_1(\zeta) + \zeta^2 K_0(\zeta)] \tag{5}$$

Numerical Methods

Conductivity Structure and Its Generation. There are various methods of generating autocorrelated log-normally distributed conductivity field. Although it is recoganized that the turning band method (TBM) introduced by Matheron (1973) (also see Journel and Huijbregts (1978)) is superior to the nearest-

neighbour first-order model in that prescribed field statistics are reproduced more accurately and at a faster rate (Mantoglou and Wilson (1982)), the nearest-neighbour model is used in this paper as it is more basic, and offers the possibility of generating non-stationary field more convinently.
If the region of flow is divided into mxn blocks (10x10 in this this particular case), then

$$f_{i,j} = \alpha_x(f_{i-1,j}+f_{i+1,j})+\alpha_y(f_{i,j-1}+f_{i,j+1})+\varepsilon \qquad (6)$$

where f=log(K), α_x and α_y are autoregressive parameters which describe the degree of interdependence of values of $f_{i,j}$ in x and y directions, and ε is a random number uniformly distributed on the real line interval [-1,+1]. Smith and Freeze (1979) have described the procedure in fair detail, and the reader is referred to their paper for further details.

Finite Element Method. It can be shown that solving (1) is equivalent to finding the extremal value of the integral

$$I(x,y) = \iint_{\Omega} K \operatorname{grad}\phi \, dxdy \qquad (7)$$

The method of discretizing the integral to give extremal value of I has been described adequately by Pinder and Gray (1977). The boundary conditions were as follows. At two opposing sides, the head was supposed to be fixed, and the other two boundaries were assumed to be no flow boundaries. The assignment of the boundaries values was arbitrary, except in that they were not equal. The domain Ω was divided into square elements, and the resulting set of equations for nodal hydraulic head solved by the Gauss-Seidel iteration technique. To obtain correlation function, the deterministic component was removed.

Finite Difference Method. The alternating direction implicit (ADI) method was used to solve (1), in a manner similar to Trescott and Larson (1977). The equations at horizontal traverse (x-direction) is

$$(1/p+2)\phi^{K+1}_{i,j} = (1/p-2)\phi^{k}_{i,j}+\phi^{K+1}_{i-1,j}+\phi^{K+1}_{i+1,j}+\phi^{k}_{i,j-1}+\phi^{k}_{i,j+1}$$
$$+\frac{K_{i,j+1}-K_{i,j}}{4K_{i,j}}(\phi^{k}_{i,j+1}-\phi^{k}_{i,j}) + \frac{K_{i+1,j}-K_{i,j}}{4K_{i,j}}(\phi^{K+1}_{i+1,j}-\phi^{K+1}_{i,j}) \qquad (8)$$

where $\phi^{k+1}_{i,j}$ is head at (k+1)st iteration, and p is the acceleration parameter for convergence. A similar equaiton can be written for the vertical traverse (y direction) and corresponds to head at (k+2)nd iteration. The resulting tri-diagonal matrix has a convinent LU decompostion, and the iteration is repeated until a suitable residual (the difference between values at sucessive iterations) is reached (see discussion).

Random Walk Simulation. As all the numerical methods considered above can be regarded as Monte Carlo simulations, the random walk method, although a Monte Carlo method, is conceptually fundamentally different because of its inherent statistical character even in the solution of deterministic problems. Its essence in the solution of equation (1) is as follows. If e_1, e_2, e_3, and e_4 are four uniformally distributed random numbers on the same interval (eg., (0,1]) then a hypothetical particle is supposed to move in a direction corresponding to one which gives maximum value for

$$K_{i-1,j}/e_1, K_{i+1,j}/e_2, K_{i,j-1}/e_3, K_{i,j+1}/e_4. \qquad (9)$$

If a block has a no flow boundary, then the probability of motion in that direction is set to zero. If a block boundary has a fied head, and the particle moves in that direction, then the value of that head is assigned to the block from which the random walk started. The usual procedure, then is to repeat this process until a desired accuracy is achieved. However if we treat K as being a stochastic variable, then for each realization of K, we perform one complete random walk for the block. This can be done due to the steady state conditions involved. This is equivalent to saying that

$$\lim_{mc\to\infty}\ \lim_{kn\to\infty} \sum_{k=1}^{mc}\sum_{l=1}^{kn} \frac{\phi_{ijkl}}{mc.kn} = \lim_{mc\to\infty} \sum_{k=1}^{mc} \frac{\phi_{ijk}}{mc} \qquad (10)$$

where kn is the number of complete random walks (one from each block), ϕ_{ijkl} is the accumulative head obtained from l^{th} random walk of kth conductivity realization. Despite even this advantage, however, the CPU time required to evaluate (10) for an acceptible error is still large.

RESULTS AND CONCLUSIONS

Figure 1 shows the correlation function of the generated hydraulic conductivity sequence. Also shown is the autocorrelation function given by equation (4).

Figure 1. Conductivity correlation simulation. Symbols represent nearest-neighbour simulation and the solid line represents Whittle B correlation for $\sigma = 1$, and $\lambda = 12$. Total no. of conductivity realizations (mc) is 50.

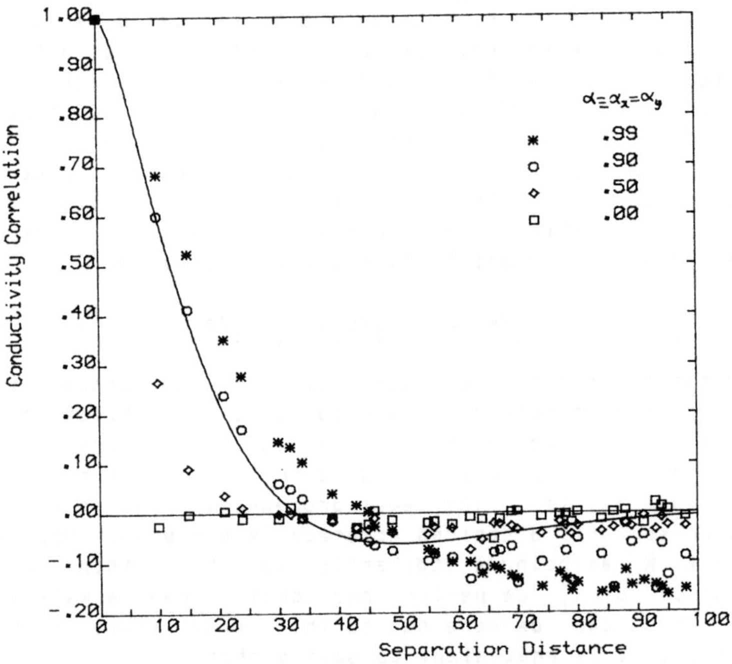

Figure 2 shows the correlation in head after solving equation (1) in various ways. We see that even when conductivity field is not 'strongly' correlated, the head still shows correlation. This is due to the filtering effect of the differential operator in equation (1). However, as correlation in conductivity increases, so does the correlation between hydraulic head. The boundaries of the flow domain appear to have a significant effect, and futher investigations are being made. To obtain correlation function from the random walk technique was not attempted as it takes too long. The results from the finite element and finite difference method were similar, and only the results from the finite element method are shown.

Figure 2. Hydraulic head correlation with same parameters as Fig. 1.

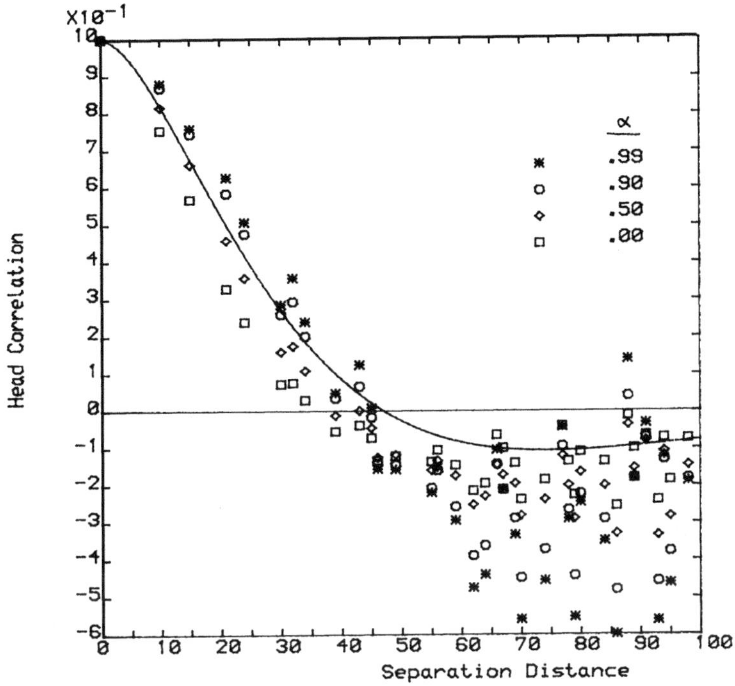

Figure 3 shows the central processor unit (CPU) time to achieve a desired value of root mean average residual squared (RMARS), defined by

$$RMARS^2 = \sum_{k=1}^{mc} \sum_{j=1}^{m} \sum_{i=1}^{n} (mc \times n \times m)^{-1} [\phi_{ijk} - \phi(r)]^2 \qquad (11)$$

where $\phi(r)$ is given by (2). Although this was found to be the satisfactory statistic characterizing the CPU time required to acheive a desired accuracy, other parameters (such as the acceleration factor p in equation 8, and the tolerance in Gauss-Seidel iteration) should be choosen with care, as they could have significantly different values without affacting RMARS appreciably. Figure (3) thus indicates that for the problem studies, the finite element method, and finite difference method do not show significantly different results, although for problems of greater dimension (no. of blocks) and more complex boundaries, the finite element method would be recommended. Random walk procedures should be avoided unless there is no better alternative (which is rare for groundwater problems).

Figure 3. Comparison of computational cost of the three numerical methods (mc=50)

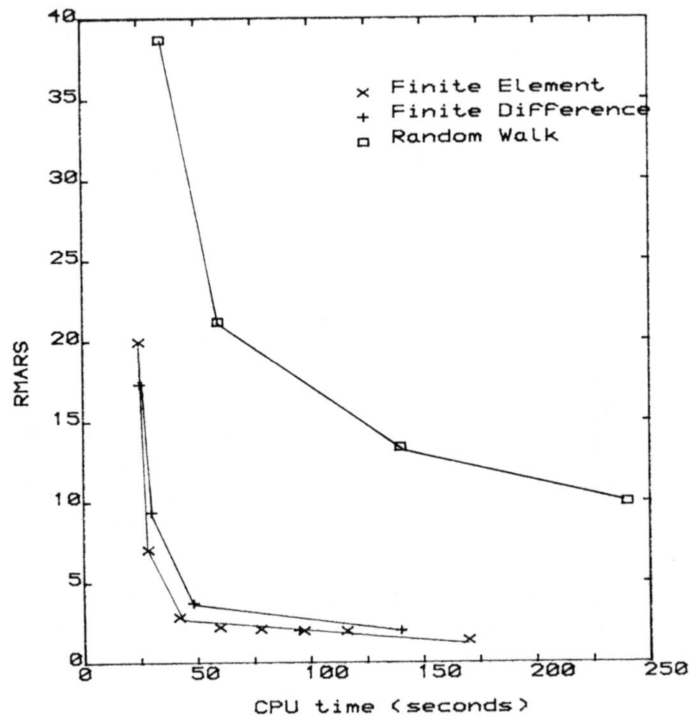

Accknowlegdement

One of the authors (TSV) accknowledges NERC research studentship for financial support. Computer programs for all the above mentioned numerical methods are available from the authors at a small cost.

REFERENCES

Abromowitz, A., and I. A. Stegun (eds.) (1972) Handbook of Mathematical Functions, Dover Publications, Inc., New York.

Dalenius, T., J. Hajek, and S. Zubrzycki (1961) On Plane Sampling and Related Geometrical Problems, Proceedings of the Fourth Berkeley Symposium on Mathematical Statistics and Probability, 1, 125-150.

Hammersley, J. M., and D. C. Handscomb (1964) Monte Carlo Methods, Matheun & Co. Ltd., London.

Journel, A. G., and Ch. J. Huijbregts (1978) Mining Geostatistics, Academic Press, London.

Mantoglou, A., and J. L. Wilson (1982) The Turning Bands Method for Simulation of Random Fields Using Line Generation by a Spectral Method, Water Resour. Res., **18**(5), 1379-1394.

Matheron, G. (1973) The Intrinsic Random Functions and Their Applications, Advan. Appl. Prob., **5**, 438-468.

Mizell, S. A., A. L. Gutjahr, and L. W. Gelhar (1982) Stochastic Analysis of Spatial Variability in Two-Dimensional Steady Groundwater Flow Assuming Stationary and Nonstationaty Heads, Water Resour. Res., **18**(4), 1053-1067.

Pinder G. F., and W. G. Gray (1977) Finite Elements Simulation in Surface and Subsurface Hydrology, Academic Press, New York.

Smith, L., and R. A. Freeze (1979) Stochastic Analysis of Steady State Groundwater in a Bounded Domain. 2. Two-Dimensional Simulation, Water Resour. Res., **15**(6), 1543-1559.

Trescott, P. C., and S. P. Larson, Comparison of Iterative Methods for Solving Two Dimensional Groundwater Flow Equation, Water Resour. Res., **13**(1), 125-136.

Virdee, T. S., and N. T. Kottegoda (1984), Stochastic Analysis of the Effect of Spatial Variability in Conductivity on Groundwater Head, submitted to Water Resour. Res.

LOGARITHMIC INTERPOLATION FOR GROUNDWATER FLOW NEAR WELLS

Xue Yuqun* and Xie Chunhong

Nanjing University
Nanjing 210008, P.R. China

ABSTRACT

A new logarithmic finite element interpolation has been developed to model groundwater flow near wells in two or three dimensional confined flows. An important feature of the new method is its ability to represent the curved outline of the well and to model the logarithmic head distribution in its vicinity. The logarithmic elements are compatible with linear elements and thus, both can be used in the same grid.

INTRODUCTION

In regional finite element models of groundwater flow, wells are usually treated as points. Head distribution near the well is generally described by polynomial interpolation. This makes it difficult to represent the curved outline of the well, which is important for some engineering designs, and the actual near-logarithmic head variation with distance from the well.

Since 1978 several Chinese researchers have studied this problem [5]. In 1979 they developed a new method of simulating wells in two-dimensional flow, both for steady and unsteady flow through porous media. Their method takes into account both the well diameter and the near-logarithmic radial head variations. In 1982 T. M. Tharp developed a method which models the logarithmic head distribution around a well [3]. However, his method is limited to two-dimensional steady state, confined flow, and it cannot be used for unsteady flow. This paper presents an extension of our earlier work to three dimensions.

---

*On leave at the Department of Hydrology and Water Resources, University of Arizona, Tucson, Arizona 85721

THEORY

The basic aquifer equation may be written as

$$\nabla\cdot(K\nabla H) + W = S_s \frac{\partial H}{\partial t} \qquad \text{on } D \tag{1}$$

where H is the hydraulic head, K is the hydraulic conductivity, W is the rate of fluid generation per volume of aquifer (source term), S_s is the specific storage, t is time, and ∇ is the gradient operator. The equation is to be solved for H, subject to the initial and boundary conditions

$$H\ (\underline{x},o) = H_o\ (\underline{x}) \tag{2}$$

$$H\ (\underline{x},t) = \phi\ (\underline{x},t) \qquad \text{on } \Gamma_1 \tag{3}$$

$$(K\nabla H)\cdot\underline{n} = q \qquad \text{on } \Gamma_2 \tag{4}$$

where $\underline{x}$ is a position vector, Γ_1 and Γ_2 are boundaries of D, $\underline{n}$ is a unit vector normal to Γ_2 and pointing outward, and H_o, ϕ, and q are prescribed functions.

The wells in the aquifer can be fully or partially penetrating, and they can be treated as Dirichlet boundaries if the well drawdowns rather than the flow rate are known; if the well withdrawal rates are known, they can be treated as Neumann boundaries with

$$2\ \pi r_{wi} \ell_i\ K\ \frac{\partial H}{\partial r} = Q_i \qquad \text{on } \Gamma_{wi} \tag{5}$$

where Q_i is the flow rate of well i, r_{wi} is the radius of well i, ℓ_i is the length of the screen in well i, and Γ_{wi} is the surface of well i.

In order to construct an interpolation function which is logarithmic in the radial direction, the region surrounding the well is divided into cylinders. The radius of the inner cylinder is the radius of the pumped well (if there are many cylinders, the radius of the smallest inner cylinder is the radius of the well). The region near the cylinders is divided into four triangular prisms, each with one curved surface. The other areas are all divided into the ordinary triangular prisms. The area of transition from cylinders to ordinary triangular prisms is composed of triangular prisms with curved surfaces.

These areas are denoted by D_1, D_2 and D_3 (figure 1). For the purpose of integration, each cylinder (D_1) and each triangular prism with a curved surface (D_2) are turned into a hexahedron, D_{1i} or D_{2i}, by the transformation of the respective variables.

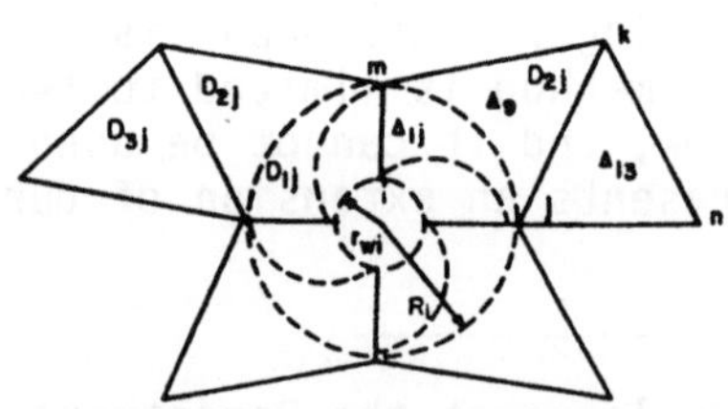

Figure 1. Finite element

In D_{1i}, we make the transformation grids near the pumping well of variables

$$x - \alpha_i = e^p \cos\Theta$$

$$y - \beta_i = e^p \sin\Theta \qquad 0<\Theta<2\pi \tag{6}$$

$$Z = Z$$

where $p = \ln r$, $r = \sqrt{(X-\alpha_i)^2 + (y-\beta_i)^2}$, (α_i, β_i) is the plane coordinate of the center of the well i and $r > 0$ is demanded. The transformation is illustrated in Figure 2, where $P_1=\ln r_{wi}$, $P_2=\ln R$, and R is the radius of the outer cylinder. If necessary, more than two concentric cylinders may be used. At that time, r_{wi} and R will be replaced by the radii of the inner and outer cylinders, respectively.

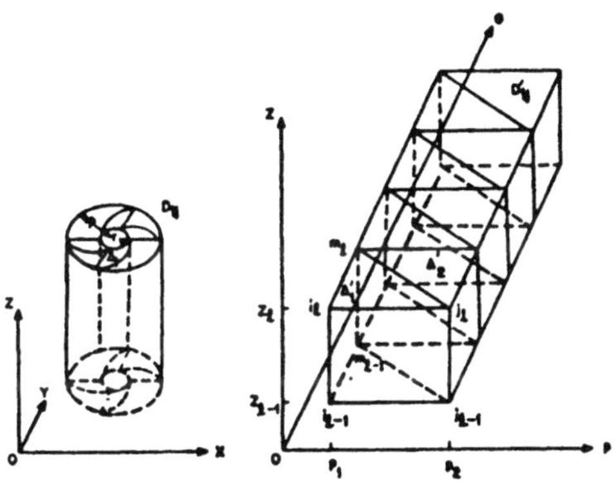

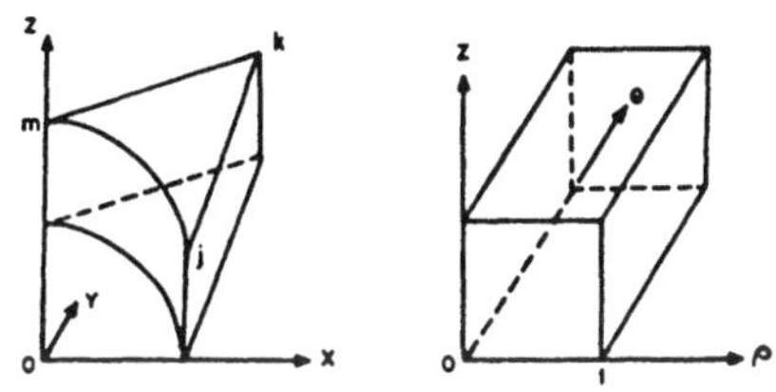

Figure 3. A triangular prism element with curved surface and turning it into a hexahedron

Figure 2. The Clyinder element and turning it into a hexahydron

If we subdivide the D_{1i}' into eight ordinary triangular prisms then each of them corresponds to one of the triangular prisms with curved surfaces in cylinder D_{1i}. In the triangular prism Δ_{1i}' (figure 2), the head at the bottom plane $H_{\ell-1}$, can be taken as a linear function. So the following equation

$$H_{\ell-1} = ap + b\Theta + c \tag{7}$$

becomes

$$H_{\ell-1} = a \ln r + b\Theta + c \tag{8}$$

in the original system (x, y, z). Here a, b and c are parameters. Therefore, $H_{\ell-1}$ is logarithmic in the radial direction, when Θ is fixed. Obviously, we have

$$H_\ell = a'p + b'\Theta + c' = a'\ln r + b'\Theta + c'$$

Here H_ℓ is the head at the top plane of Δ_1', a', b' and c' are parameters.

Because j and m are located on the outer cylinder, we can use the interpolation formula to determine H on Δ_{1i}' as follows

$$H = (1 - \frac{2\Theta}{\pi})H_j + (\frac{p - p_2}{p_2 - p_1} + \frac{2\Theta}{\pi})H_m + \frac{p_2 - p}{p_2 - p_1} H_i$$

On the arc jm , the formula becomes

$$H = (1 - \frac{2\Theta}{\pi})H_j + \frac{2\Theta}{\pi} H_m \tag{9}$$

If the head of apex k, which is located on the triangular prism with curved surface, jkm ($\Delta 9$), is designated by H_k, we make an interpolation between point k (x_k, y_k) and arc jm and obtain

$$H = (1-\rho)H_k + \rho \left[(1 - \frac{2\Theta}{\pi})H_j + \frac{2\Theta}{\pi} H_m \right] \tag{10}$$

In the above triangular prism, jkm, we make the transformation of variables

$$\begin{aligned} X &= (1-\rho)X_k + \rho(\alpha_i + {}_e p_2 \cos \Theta) \\ y &= (1-p)y_k + \rho(\beta_i + {}_e p_2 \sin \Theta) \\ Z &= Z \end{aligned} \tag{11}$$

then jkm is turned into a hexahedron. In the hexahedron plane ρ=0 corresponds to the line $K_\ell K_{\ell-1}$, plane ρ=1 corresponds to the surface jm, plane Θ=0 corresponds to the plane jk, and plane $\Theta = \pi/2$ corresponds to plane mk (figure 3).

Because the Jacobian satisfies

$$J = \begin{vmatrix} \alpha_i + e p_2 \cos\Theta - x_k & \beta_i + e p_2 \sin\Theta - y_k \\ - \rho_e p_2 \sin\Theta & \rho_e p_2 \cos\Theta \end{vmatrix} = \rho Jo$$

only under the circumstance where point k is not located on the four lines which are tangent with the outer cylinder and parallel to the coordinate axes, $J \neq 0$ and the transformation is permitted.

Numerical Approach

Just as O. C. Zienkiewiez [6] points out, the solution to the above-mentioned problem may be transformed into minimizing the functional

$$E(H)=\iiint_D \{\frac{K}{2}[(\frac{\partial H}{\partial x})^2+(\frac{\partial H}{\partial y})^2+(\frac{\partial H}{\partial z})^2] + S_s \frac{\partial H}{\partial t}H-WH\} \, dxdydz-\iint_{\Gamma_2} qHds$$

$$+ \sum_{i=1}^{m_1} \iint_{\Gamma_{wi}} \frac{Q_i H}{2\pi r_{wi} \ell_i} ds \qquad (12)$$

in the set of functions whose value is given by equation (3). According to the discussion above, it is necessary that the functions have a logarithmic singularity near the well and are smooth at all other points. For convenience we assume w=0.

As mentioned above, we adopt three types of subregions for D--D_1, D_2, D_3--to be discretized (equation 12) in space. In D_1 and D_1', each cylinder and hexahedron are further subdivided into eight triangular prisms, but there is no need to know the expression of the curved surface in D_1 during computation. The apexes of the triangular prisms are designated as nodal points. If the head at the three apexes, i, j, and m, at the bottom ℓ=1 of the triangular prism (figure 2) are designated by $H_{i,\ell-1}$, $H_{j,\ell-1}$ and $H_{m,\ell-1}$, then

$$H_{\ell-1} = N_i(p,\Theta)H_{i,\ell-1} + N_j(p,\Theta)H_{j,\ell-1} + N_m(p,\Theta)H_{m,\ell-1} \qquad (13)$$

The head at the top of plane ℓ of Δ_1' is

$$H_\ell = N_i(p,\Theta)H_{i,\ell} + N_j(p,\Theta)H_{j,\ell} + N_m(p,\Theta)H_{m,\ell} \qquad (14)$$

Here N_i, N_j and N_m are prescribed basis functions satisfying $N_i(p_j,\Theta_j) = \delta_{ij}$, δ_{ij} being Kronecker delta (i.e., δ_{ij} = 1, if i=j and δ_{ij} = 0 if i ≠ j), and are defined as

$$N_i = \frac{1}{2\Delta}(a_i + b_i p + c_i\Theta) \qquad N_j = \frac{1}{2\Delta}(a_j + b_j p + c_j\Theta)$$

$$N_m = \frac{1}{2\Delta}(a_m + b_m p + c_m\Theta) \qquad (15)$$

where

$$a_i=p_j\Theta_m-p_m\Theta_j \qquad b_i=\Theta_j-\Theta_m \qquad c_i=p_m-p_j$$

$$a_j=p_m\Theta_i-p_i\Theta_m \qquad b_j=\Theta_m-\Theta_i \qquad c_j=p_i-p_m \qquad (16)$$

$$a_m=p_i\Theta_j-p_j\Theta_i \qquad b_m=\Theta_i-\Theta_j \qquad c_m=p_j-p_i$$

and

$$\Delta = \frac{1}{2}(b_i c_j + b_j c_i)$$

the latter being the area of the triangle ijm.

In order to show the relationship between corresponding H's on plane ℓ and plane ℓ-1 we introduce two new functions,

$$\Psi_{\ell-1}(Z) = \frac{Z_\ell - Z}{Z_\ell - Z_{\ell-1}}, \qquad \Psi_\ell(Z) = \frac{Z - Z_{\ell-1}}{Z_\ell - Z_{\ell-1}} \tag{17}$$

so that

$$\Psi_{\ell-1}(Z_{\ell-1}) = 1 \qquad \Psi_{\ell-1}(Z_\ell) = 0$$

$$\Psi_\ell(Z_{\ell-1}) = 0 \qquad \Psi_\ell(Z_\ell) = 1$$

Hence, in a triangular prism, we have

$$H = H_\ell \Psi_\ell + H_{\ell-1}\Psi_{\ell-1} \tag{18}$$

Substituting Equations (13) and (14) into Equation (18), we obtain

$$H = N_i(p,\Theta)H_i(z,t) + N_j(p,\Theta)H_j(z,t) + N_m(p,\Theta)H_m(z,t) \tag{19}$$

Then we can consider the integral on a typical element, Δ_1'; after transformation the functional on Δ_1' will be

$$E^{\Delta_1'}(H) = \iiint_{\Delta_1'} \{ \frac{K^e}{2} [(\frac{\partial H}{\partial p})^2 + (\frac{\partial H}{\partial \Theta})^2] + \frac{K^e}{2} e^{2p} (\frac{\partial H}{\partial Z})^2$$

$$+ S_s H \frac{\partial H}{\partial t} e^{2p} \} \, dp\, d\Theta\, dZ + \iint_{\Gamma_{wi} \cap \Delta_1'} \frac{Q_i H}{2\pi r_{wi} \ell_i} \, ds \tag{20}$$

where $\Gamma_{wi} \cap \Delta_1'$ stands for the common part between the boundary of Δ_1 and the well wall, Γ_{wi}, e means element, and K^e is the hydraulic conductivity in e.

After substituting Equation (19) into Equation (20) and forming the derivatives with respect to $H_{i,\ell-1}$, $H_{j,\ell-1}$, $H_{m,\ell-1}$, $H_{i,\ell}$, $H_{j,\ell}$ and $H_{m,\ell}$ successively, we obtain the following matrix equation

$$\frac{\partial E}{\partial \underline{H}} = \underline{\underline{d}}\, \underline{H} + \underline{\underline{p}} \frac{d\underline{H}}{dt} + \underline{f} \tag{21}$$

Here $\frac{\partial E}{\partial \underline{H}}$, $\underline{H}$, $\frac{d\underline{H}}{dt}$ and $\underline{f}$ are six-dimensional vectors, $\underline{\underline{d}}$ and $\underline{\underline{p}}$ are 6x6 symmetric matrices whose terms are given by

$$
d_{ij} = \begin{cases} 2A_{ij} + B & \text{if } i \neq j \quad \begin{cases} i = 1,2,3 & j = 1,2,3 \\ i = 4,5,6 & j = 4,5,6 \end{cases} \\ A_{ij} - B & \text{if} \begin{cases} i = 1,2,3 \quad j = 4,5,6 & (i \neq j-3) \\ i = 4,5,6 \quad j = 1,2,3 & (i \neq j+3) \end{cases} \\ A_{ij} - 2B & \text{if } i = j+3 \text{ or } i = j-3 \\ 2A_{ij} + 2B & \text{if } i = j \end{cases} \tag{22}
$$

$$
\rho_{ij} = \begin{cases} \dfrac{S_s \xi \Delta (r_1 + r_2)^2}{144} & \text{if } i \neq j \begin{cases} i=1,2,3 & j=1,2,3 \\ i=4,5,6 & j=4,5,6 \\ i=j+3 & i=j-3 \end{cases} \\ \dfrac{S_s \xi \Delta (r_1 + r_2)^2}{288} & \text{if} \begin{cases} i=1,2,3 \quad j=4,5,6 & (i \neq j-3) \\ i=4,5,6 \quad j=1,2,3 & (i \neq j+3) \end{cases} \\ \dfrac{S_s \xi \Delta (r_1+r_2)^2}{72} & \text{if } i=j \end{cases} \tag{23}
$$

here

$$
A_{ij} = \frac{K^e(b_i b_j + c_i c_j)}{24\Delta} \xi \qquad B_{ij} = \frac{K^e(r_1 + r_2)^2 \Delta}{48\xi}
$$

$\xi = Z_\ell - Z_{\ell-1}$, the height between plane ℓ and ℓ-1.

If i and m are located on Γ_{wi}, $f_{i,\ell-1} = f_{i\ \ell} = \frac{Q}{16}$, $f_{m,\ell-1} = f_{m,\ell} = \frac{Q}{16}$ and $f_{j,\ell-1} = f_{j,\ \ell} = 0$. If Γ_{wi} does not have any common boundary with Δ_1', $\underline{f} = 0$.

Next we integrate the functional (12) on the triangular prism with curved surface D_{2j}. The integral of functional (12) on the triangular prism with curved surface Δ_9 (figure 3) becomes

$$
E^{\Delta 9}(H) = \int_0^{\pi/2} \int_0^1 \int_{Z_{\ell-1}}^{Z_\ell} \{\frac{K^e}{2|J|} (\frac{\partial H}{\partial \rho})^2[(\frac{\partial x}{\partial \Theta}) + (\frac{\partial y}{\partial \Theta})^2] - 2 \frac{\partial H}{\partial \rho} \frac{\partial H}{\partial \Theta}
$$

$$
[\frac{\partial x}{\partial \rho} \frac{\partial x}{\partial \Theta} + \frac{\partial y}{\partial \rho} \frac{\partial y}{\partial \Theta}] + (\frac{\partial H}{\partial \Theta})^2 [(\frac{\partial X}{\partial \rho})^2 + (\frac{\partial y}{\partial \rho})^2]\} d\rho d\Theta dZ
$$

$$
+ \int_0^{\pi/2} \int_0^1 \int_{Z_{\ell-1}}^{Z_\ell} \frac{K^e}{2} |J| (\frac{\partial H}{\partial Z})^2 d\rho d\Theta dZ + \int_0^{\pi/2} \int_0^1 \int_{Z_{\ell-1}}^{Z_\ell} S_s |J| H \frac{\partial H}{\partial t} d\rho d\Theta dZ \tag{24}
$$

Using Equations (18) and (10) we obtain

$$H = (1-\rho)(H_{k,\ell}\Psi_\ell + H_{k,\ell-1}\Psi_{\ell-1}) + \rho[(1-\frac{2\Theta}{\pi})(H_{j,\ell}\Psi_\ell + H_{j,\ell-1}\Psi_{\ell-1}) + \frac{2\Theta}{\pi}(H_{m,\ell}\Psi_\ell + H_{m,\ell-1}\Psi_{\ell-1}) \tag{25}$$

Substituting Equations (25) and (11) into Equation (24) and forming the derivatives with respect to $H_{j,\ell-1}$, $H_{k,\ell-1}$, $H_{m,\ell-1}$, $H_{j,\ell}$, $H_{k,\ell}$ and $H_{m,\ell}$ successively, we obtain

$$\frac{\partial E}{\partial \underline{H}} = \underline{d}\ \underline{H} + \underline{p}\ \frac{d\underline{H}}{dt} \tag{26}$$

Here $\underline{d}$ and $\underline{p}$ are 6x6 symmetric matrices whose terms are given by

$$d_{jk} = \begin{cases} 2C_{jk} + e_{jk} & \begin{cases} j=1,2,3 & k=1,2,3 \\ j=4,5,6 & k=4,5,6 \end{cases} \\ C_{jk} - e_{jk} & \begin{cases} j=1,2,3 & k=4,5,6 \\ j=4,5,6 & k=1,2,3 \end{cases} \end{cases} \tag{27}$$

$$C_{jj} = \int_0^{\pi/2} \frac{K^e}{2|Jo|}\frac{\xi}{6}\left[e^{2p_2}(1-\frac{2\Theta}{\pi})^2 + \frac{4}{\pi} r_{1k}e^{p_2}(1-\frac{2\Theta}{\pi}) + \frac{4}{\pi^2} r_{2k}\right] d\Theta$$

$$C_{jk} = \int_0^{\pi/2} \frac{K^e}{2|Jo|}\frac{\xi}{6}\left[-e^{2p_2}(1-\frac{2\Theta}{\pi}) - \frac{2}{\pi} e^{p_2} r_{1k}\right] d\Theta$$

$$C_{jm} = \int_0^{\pi/2} \frac{K^e}{2|Jo|}\frac{\xi}{6}\left[e^{2p_2}\frac{2\Theta}{\pi}(1-\frac{2\Theta}{\pi}) - \frac{2}{\pi} r_{1k}e^{p_2}(1-\frac{4\Theta}{\pi}) - \frac{4}{\pi^2} r_{2k}\right] d\Theta$$

$$C_{kk} = \int_0^{\pi/2} \frac{K^e}{2|Jo|}\frac{\xi}{6} e^{2p_2}\, d\Theta$$

$$C_{km} = \int_0^{\pi/2} \frac{K^e}{2|Jo|}\frac{\xi}{6}\left[-e^{2p_2}\frac{2\Theta}{\pi} + \frac{2}{\pi} r_{1k}e^{p_2}\right] d\Theta$$

$$C_{mm} = \int_0^{\pi/2} \frac{K^e}{2|Jo|}\frac{\xi}{6}\left[e^{2p_2}(\frac{2\Theta}{\pi})^2 - \frac{4}{\pi} r_{1k}e^{p_2}\frac{2\Theta}{\pi} + \frac{4}{\pi^2} r_{2k}\right] d\Theta$$

$$r_{1k} = -(\alpha_i - X_k)\sin\Theta + (\beta_i - y_k)\cos\Theta$$

$$r_{2k} = (\alpha_i-X_k)^2+(\beta_i-y_k)^2 + 2[(\alpha_i-X_k)\cos\Theta + (\beta_i-y_k)\sin\Theta]e^{p_2}+e^{2p_2}$$

$$e_{jj} = \int_0^{\pi/2} \frac{K^e}{4\xi}|Jo|(1-\frac{2\Theta}{\pi})^2 d\Theta , \qquad e_{jk} = \int_0^{\pi/2} \frac{K^e}{12\xi} |Jo|(1-\frac{2\Theta}{\pi})d\Theta$$

$$e_{jm} = \int_0^{\pi/2} \frac{K^e}{2\pi\xi} |Jo| (1 - \frac{2\Theta}{\pi}) \Theta d\Theta , \quad e_{kk} = \int_0^{\pi/2} \frac{K^e}{12\xi} |Jo| d\Theta$$

$$e_{km} = \int_0^{\pi/2} \frac{K^e}{6\pi\xi} |Jo| \Theta d\Theta , \qquad e_{mm} = \int_0^{\pi/2} \frac{K^e}{\pi^2\xi} |Jo| \Theta^2 d\Theta$$

$$p_{jk} = \begin{cases} 2\, G_{jk} & \begin{cases} j=1,2,3 & k=1,2,3 \\ j=4,5,6 & k=4,5,6 \end{cases} \\ G_{jk} & \begin{cases} j=1,2,3 & k=4,5,6 \\ j=4,5,6 & k=1,2,3 \end{cases} \end{cases} \tag{28}$$

$$G_{jj} = \int_0^{\pi/2} \frac{S_s}{4}|Jo|(1-\frac{2\Theta}{\pi})^2 \frac{\xi}{6} d\Theta , \quad G_{jk} = \int_0^{\pi/2} \frac{S_s}{12}|Jo|(1-\frac{2\Theta}{\pi}) \frac{\xi}{6} d\Theta$$

$$G_{jm} = \int_0^{\pi/2} \frac{S_s}{2\pi}|Jo|\Theta(1-\frac{2\Theta}{\pi}) \frac{\xi}{6} d\Theta , \quad G_{kk} = \int_0^{\pi/2} \frac{S_s}{12}|Jo| \frac{\xi}{6} d\Theta$$

$$G_{km} = \int_0^{\pi/2} \frac{S_S}{6\pi}|Jo|\Theta \frac{\xi}{6} d\Theta , \qquad G_{mm} = \int_0^{\pi/2} \frac{S_S}{4}|Jo|(\frac{2\Theta}{\pi})^2 \frac{\xi}{6} d\Theta$$

Consider the integral of functional (12) on the triangular prism D_{3j} (figure 1). In this case we know that

$$H = N_j(x,y)H_j(z,\ell) + N_n(x,y)H_n(z,t) + N_k(x,y)H_k(z,t) \tag{29}$$

where N_j, N_n and N_k are prescribed basis functions that satisfy $N_j\ (X_n,Y_n) = \delta_{jn}$. On the typical element Δ_{13} we have

$$E^{\Delta 13}(H) = \iiint_{\Delta_{13}} \{ \frac{K^e}{2}[(\frac{\partial H}{\partial x})^2+(\frac{\partial H}{\partial y})^2 + (\frac{\partial H}{\partial z})^2] + S_s H\frac{\partial H}{\partial t}\}dx\, dy\, dz \tag{30}$$

and the derivative of Equation (30) with respect to $H_{j,\ell-1}$, $H_{n,\ell-1}$, $H_{k,\ell-1}$, $H_{j,\ell}$, $H_{n,\ell}$ and $H_{n,\ell}$ can be expressed as

$$\frac{\partial E}{\partial \underline{H}} = \underline{d}\ \underline{H} + \underline{p}\ \frac{d\underline{H}}{dt} \tag{31}$$

where

$$d_{jn} = \begin{cases} 2A_{jn} + B & \text{if } j \neq n \quad \begin{cases} j=1,2,3 & n=1,2,3 \\ j=4,5,6 & n=4,5,6 \end{cases} \\ A_{jn} - B & \text{if} \begin{cases} j=1,2,3 & n=4,5,6 \quad (j \neq n-3) \\ j=4,5,6 & n=1,2,3 \quad (j \neq n+3) \end{cases} \\ A_{jn} - 2B & \text{if } j = n+3 \text{ or } j = n-3 \\ 2A_{jn} + 2B & \text{if } j = n \end{cases}$$

$$A_{jn} = \frac{K^e(b_j b_n + C_j C_n)}{24\Delta}\,\xi \qquad\qquad B = \frac{K^e \Delta}{12\xi}$$

Using the minimum principle we obtain

$$\frac{\partial E}{\partial H_i} = \sum_{j=1}^{m} \frac{\partial E_j^e}{\partial H_i} = 0 \qquad i = 1,2, \ldots, n \tag{33}$$

where m is the total number of elements, n the total number of nodes whose heads are unknown. Substituting Equations (21), (26) and (31) into Equation (33) we obtain

$$\underline{D}\,\underline{H} + \underline{P}\,\frac{\partial \underline{H}}{\partial t} = \underline{F'} \tag{34}$$

where $\underline{D}$ is a symmetric "conductance matrix" of order n whose i, jth term, D_{ij} , is the sum of dij, $\underline{P}$ is a symmetric "capacity matrix" of order n in which the term P_{ij} is the sum of p_{ij} (We can change the matrix $\underline{P}$ to a diagonal matrix as S.P. Neuman mentioned in 1977 [2] so that some abnormal phenomena in water head fluctuations can be avoided) and $\underline{F'}$ is an n-dimensional vector of known term which is given by

$$F_i' = \iint_{\Gamma_2} qH ds - \iint_{\Gamma_{wi} \cap \Delta_{1i}'} \frac{Q_i H}{2\pi r_{wi} \ell_i}\, ds \tag{35}$$

If w ≠ 0, the terms of $\underline{F'}$ are given by

$$F_i' = \iint_{\Gamma_2} qH ds + \iiint_D WH dx dy dz - \iint_{\Gamma_{wi} \cap \Delta_{1i}'} \frac{Q_i H}{2\pi r_{wi} \ell_i}\, ds \tag{36}$$

The last term in Equations (35) and (36) is equal to $\frac{Q_i}{16}$ if the nodes are located on Γ_{wi} or 0 provided Γ_{wi} does not have any common boundary with element e.

If an implicit difference scheme for Equation (34) is used, Equation (34) becomes

$$\underline{\underline{D}}\ \underline{H}^{k+1} + \underline{\underline{P}}\ \frac{\underline{H}^{k+1} - \underline{H}^{k}}{\Delta t} = F' \qquad (37)$$

Where Δt is time increment and k indicates the number of time steps, $\underline{H}^k$ is the n-dimensional vector of nodal head value at time t_k, and $\underline{H}^{k+1}$ is the same vector at time $t_{k+1} = t_k + \Delta t$. Equation (37) can be rewritten as

$$\underline{\underline{G}}\ \underline{H}^{k+1} = \underline{F} \qquad (38)$$

Where $\underline{\underline{G}}$ is an nxn symmetric matrix and $\underline{F}$ is an n-dimensional vector of known term.

Applications

To illustrate the salient features of our finite element model based on logarithmic interpolation, the following two examples are presented. The first example deals with a fully penetrating well in a Quaternary fine sand and silt aquifer and the second with a partially penetrating well in a Quaternary pebble aquifer in China.

Example 1 The aquifer is uniform, isotropic and of infinite lateral extent. Initially, the head distribution in the aqui- is uniform. The hydraulic conductivity and specific storage of the aquifer are 8.01 $^m/day$ and $1.54 \times 10^{-4} m^{-1}$, respectively. At time = 0 the well, whose diameter is 0.2 m., starts pumping at a constant rate Q. The pumping test lasts 97 hours.

To solve this for a maximum time of 97 hours required 14 time time steps, starting with $\Delta t = 6,944 \times 10^{-3}$ day (10 min) and ending with $\Delta t = 1.15$ day (27 hr). If we reduce the value of Δt, the accuracy may be increased. Total execution time on the cyber 175 computer was 3.45 s. Figure 4 shows how our results compare with those in an observation well (r= 117.85m) and with the Theis solution (r=1 and 117.85m). The agreement appears to be good. Our results differ only slightly from those observed and from the Theis solution, provided the computing point is not very close to the pumping well. The differences between our method and the Theis solution decrease as the distance from the computing point to the pumping well increases. Under the same conditions, if we use linear interpolations to compute the ordinary triangular prism elements which contain the same number of nodes (the coordinates of the nodes are the same) and elements, the results are not as accurate. The difference between our method and the Theis solution is far smaller than with linear interpolation particularly near the pumping well.

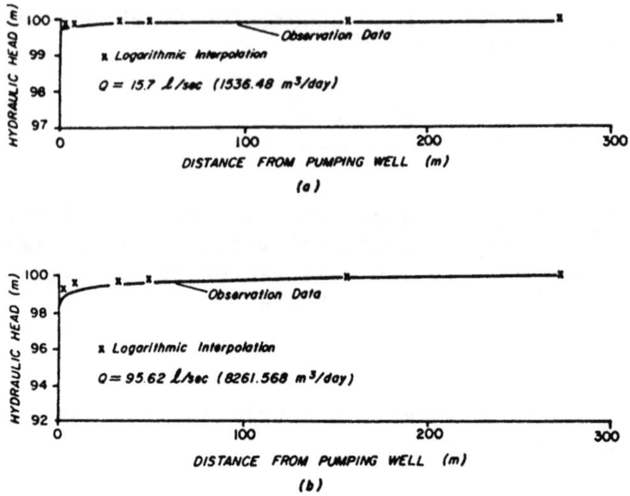

Figure 4. Comparison of results for example 1 with those of Theis solution, observation data and linear interpolation

Example 2 Our second example concerns a partially penetrating well in a nearly uniform, isotropic aquifer which is composed of Quaternary pebble mixed with 30-40% sand and gravel. The bottom confining bed is Tertiary sandstone, and the top confining bed is Quarternary clay and silt. The filter is placed in the upper part of the aquifer.

Because we only obtained steady state data for the pumping test, the problem was solved using a steady state equation. The finite element mesh which we adopted is shown in figure 5. In the vertical direction we subdivided the mesh into four elements, two elements, one above and one below the level of the filter, and two are level with the filter.

The results are shown in figures 6 and 7. They seem to be good, especially far from the pumping well.

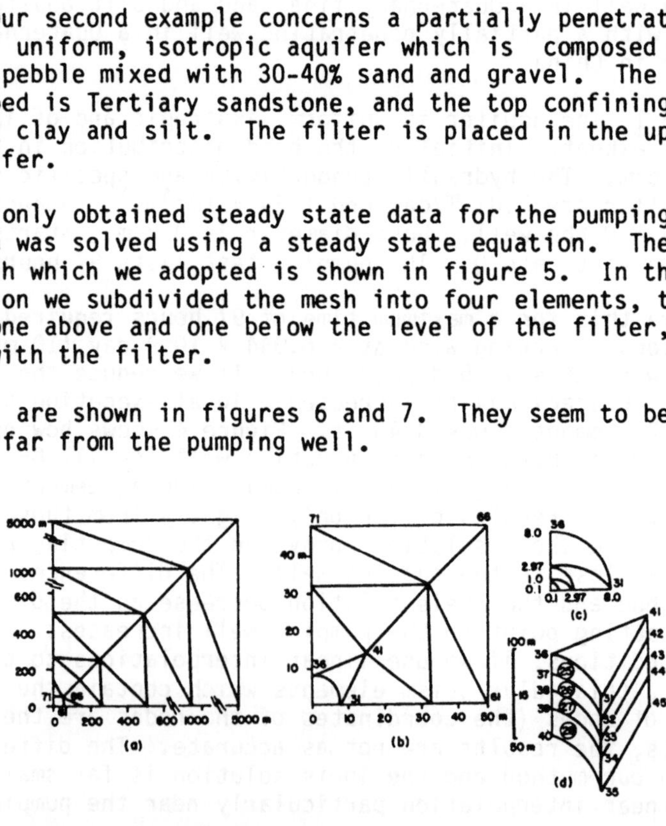

Figure 5. Finite element mesh for example 2. (25)-No. of elements; 45-No. of nodes.

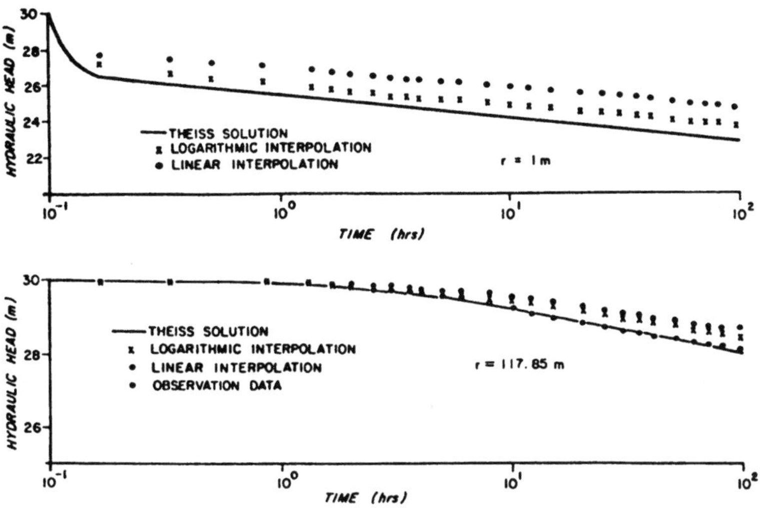

Figure 6. Results of example 2 with 6 (six) observation wells.

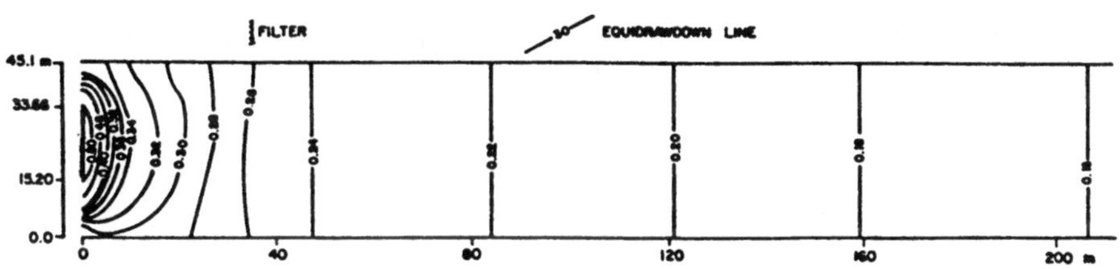

Figure 7. Equidrawdown line for example 2.

SUMMARY

The problem of confined groundwater flow near a well (in two or three dimensions) can be solved efficiently with a numerical method which uses logarithmic interpolation. This method accounts for the diameter of a well and for the near logarithmic head variation in its vicinity. The method is suitable for multiple pumping wells and unconfined aquifers; this will be discussed in another paper. Compared with linear interpolation (trianglular element or triangular prism element) our method results in superior accuracy.

ACKNOWLEDGEMENT

We wish to thank the Department of Hydrology and Water Resources at the University of Arizona for providing computer funds in support of this work, and especially to Professor S.P. Neuman for useful discussions of the subject.

REFERENCES

1. Bear, J. (1972) Dynamics of Fluids in Porous Media. Elsevier, New York.

2. Neuman, S.P., T.N. Narasimhan and P.A. Witherspoon (1977) Application of Mixed Explicit-Implicit Finite Element Method to Nonlinear Diffusion-type Problems, in "Finite Elements in Water Resources". (W.G. Gray, G.F. Pinder and C.A. Brebbia Eds.), Pentech, London.

3. Tharp, T.M. (1982) An Enriched Finite Element for Simulation of Groundwater Flow to a Well or Drain. Journal of Hydrology, 55:237-245.

4. Xue, Yugun and Zhu Xueyu (1979) Dynamics of Groundwater, Geological Press, Beijing.

5. Xue, Yugun and Xie Chunhong (1980) Numerical Method in Hydrogeology. Coal Industry Press, Beijing.

6. Zienkiewicz, O.C. (1971) The Finite Element Method in Engineering Science. McGraw-Hill.

A MASS-CONSERVATIVE PROCEDURE FOR TIME-STEPPING IN MODELS OF UNSATURATED FLOW

P. C. D. Milly

Department of Civil Engineering
Princeton University

INTRODUCTION

In numerical models of moisture in the unsaturated zone, it is frequently desirable to choose the matric (or pressure) head, rather than the moisture content,as the dependent variable. Two important advantages of this choice are that it facilitates modeling of soil systems that are locally saturated and that it provides a state variable that is always continuous in space, regardless of soil inhomogeneities.

Since the relation between moisture content and head is non-linear, however, the use of matric head as the state variable can lead to serious problems in the balancing of computed fluxes with computed changes of storage. The error arises because the storage coefficient (change in storage with respect to state) is itself a function of the state and therefore variable in time. As a result, the specific time level chosen (e.g., beginning of time step, center of time step, end of time step) to evaluate the coefficient is not necessarily representative of the time step as a whole.

The error in mass balance can be made arbitrarily small at the expense of decreasing both the time step and the element size, and hence decreasing the variability of the storage coefficient during the time step within an element. An alternative approach, developed here, evaluates the storage coefficient as an effective, or average, value over the element and over the duration of the time step. It results in arbitrarily small errors of mass balance, at the expense of iteration, during a time step of arbitrary duration with elements of arbitrary size. The latter approach appears to be superior to the former, yielding significantly better balances of mass for a fixed amount of computing effort. It also provides solutions that are closer to the true solutions for a fixed amount of effort.

Due to instabilities that arise from the oscillatory behavior of the standard Galerkin formulation of this problem, aggravated by the inherent nonlinearities, various schemes for diagonalizing, or lumping, the storage matrix have been developed. We consider the proposed time-stepping procedure in conjunction with two such schemes and with the standard approach. Only the lumped schemes lead to useful solution algorithms when the proposed time-stepping procedure is employed.

THEORY

Governing Equations and the Choice of a State Variable

Flow of moisture in an unsaturated porous medium whose air is always at atmospheric pressure is governed by the Richards equation, which is a combination of the mass conservation equation,

$$\frac{\partial\theta}{\partial t} = - \underset{\sim}{\nabla} \cdot \underset{\sim}{q} \tag{1}$$

and the Darcy-Buckingham equation,

$$\underset{\sim}{q} = - K \underset{\sim}{\nabla} (\psi + Z) \tag{2}$$

In equations (1) and (2), θ is volumetric moisture content, t is time, $\underset{\sim}{q}$ is volumetric flux of water, K is unsaturated hydraulic conductivity, ψ is matric head, and Z is elevation. Herein we shall assume unique relations among θ, ψ, and K. Generalizations to accommodate hysteresis and inhomogeneities are not difficult to make.

Multiplication of (1) by an arbitrary function ϕ_i and subsequent integration of the result over a fixed volume V leads eventually to

$$\int_V \frac{\partial\theta}{\partial t} \phi_i \, dV - \int_V \underset{\sim}{q} \cdot \underset{\sim}{\nabla} \phi_i \, dV = -\int_S \underset{\sim}{q} \cdot \underset{\sim}{n} \phi_i \, dS \tag{3}$$

in which S is the surface bounding V and $\underset{\sim}{n}$ is the unit vector normal to S.

If the function ϕ_i in Equation (3) is viewed as a weighting function, then that equation represents a method of weighted residuals. In Galerkin's method, a state variable would be identified and expressed approximately as a linear combination of these weighting functions.

Of the three dependent variables (θ, ψ, K), the first two are both reasonable choices for the state variable. The choice of θ or ψ has various modeling implications, and neither variable is entirely superior to the other. Generally, however, the ψ-based equation has been favored as the more flexible approach, as mentioned in the Introduction. In the ψ-based approach, the storage term is expanded by the chain rule,

$$\frac{\partial\theta}{\partial t} = \frac{d\theta}{d\psi}\frac{\partial\psi}{\partial t} = C\frac{\partial\psi}{\partial t} \tag{4}$$

leading to definition of the moisture capacity C.

The Standard Formulation and Variants

With the head-based approach, the Galerkin formulation is obtained using the approximation

$$\psi(\underset{\sim}{x}, t) \simeq \hat{\psi}(\underset{\sim}{x}, t) \equiv \sum_j \psi_j(t)\, \phi_j(\underset{\sim}{x}) \tag{5}$$

(Henceforth, summations are understood to be made over all nodes, unless otherwise indicated.) The approximated distributions of θ, C, and K should then be determined by $\hat{\psi}$. Throughout this work, we shall assume that these dependent variables are each given by interpolation between their respective nodal values, using the interpolation functions ϕ_i. The nodal values are evaluated using the nodal values of ψ and the appropriate soil physical functions. Thus

$$\theta(\underset{\sim}{x}, t) \simeq \hat{\theta}(\underset{\sim}{x}, t) \equiv \sum_j \theta_j\, \phi_j \tag{6}$$

where

$$\theta_j \equiv \theta(\psi_j) \tag{7}$$

and $\hat{C}$ and $\hat{K}$ are similarly defined. Another, more consistent, approach is to employ $\theta(\hat{\psi})$, $C(\hat{\psi})$, and $K(\hat{\psi})$ as the approximations for θ, C, and K, but this leads to significant additional computational expense when it is necessary to obtain accurate volume integrals of mass in the system. Limitations of space prohibit further discussion of this point here.

Substitution of Equations (4) and (5) into Equation (3) yields

$$\sum_j \frac{d\psi_j}{dt}\int_V \hat{C}\phi_i\,\phi_j\,dV + \sum_j \psi_j \int_V K\,\underset{\sim}{\nabla}\phi_i \cdot \underset{\sim}{\nabla}\phi_j\,dV + \int_V \hat{K}\,\underset{\sim}{\nabla} Z \cdot \underset{\sim}{\nabla}\phi_i\,dV = -\int_S \underset{\sim}{q}\cdot\underset{\sim}{n}\,\phi_i\,dS \tag{8}$$

It is widely known that a diagonalized variant of the storage matrix in Equation (8) is numerically superior to the distributed matrix for the problem of unsaturated flow. In this work, we consider two distinct lumped variants, "L1" and "L2". The first, L1, was employed by Neuman (1973) and by later researchers. It is obtained by replacing the integral in the first term of Equation (8) by the expression

$$\delta_{ij}\int_V \hat{C}\,\phi_i\,dV \qquad \text{"L1"}$$

This is equivalent to shifting the off-diagonal elements of the storage matrix onto the diagonal. The second variant, L2, is a further simplification of L1 that is similar to the storage matrix obtained using the standard finite difference approach. With L2, the aforementioned integral is replaced by

$$\delta_{ij} \; C_i \int_V \phi_i \; dV \qquad \text{"L2"}$$

Numerous finite difference schemes have been employed for the integration of Equation (8) with respect to time. We shall focus here on a common family of schemes. Equation (8) is evaluated at the $t^{k+\alpha}$ time level,

$$t^{k+\alpha} = (1 - \alpha) \; t^k + \alpha t^{k+1} \tag{9}$$

where α lies between zero and unity. All time dependent terms are then evaluated directly by assuming that ψ_j varies linearly with time during an integration interval. The resulting system of equations is

$$\begin{aligned} &(t^{k+1} - t^k)^{-1} \sum_j (\psi_j^{k+1} - \psi_j^k) \int_V \hat{C}^{k+\alpha} \; \phi_i \; \phi_j \; dV \\ &+ \sum_j \left[(1-\alpha)\psi_j^k + \alpha\psi_j^{k+1}\right] \int_V \hat{K}^{k+\alpha} \; \underset{\sim}{\nabla} \; \phi \; \cdot \underset{\sim}{\nabla} \; \phi_j \; dV \\ &+ \int_V \hat{K}^{k+\alpha} \; \underset{\sim}{\nabla} \; Z \cdot \underset{\sim}{\nabla} \; \phi_i \; dV = - \int_S \underset{\sim}{q}^{k+\alpha} \cdot \underset{\sim}{n} \; \phi_i \; dS \end{aligned} \tag{10}$$

where k is a time index.

When one of the mass-lumping schemes is used, the integral of the moisture capacity is replaced by one of the expressions mentioned earlier.

Definition of Mass and Mass Balances

We define $\hat{M}$ as the approximated total mass in the volume V, expressed as an equivalent volume of incompressible water,

$$\hat{M} = \int_V \hat{\theta} \; dV \tag{11}$$

We shall say that a given numerical scheme conserves mass if it guarantees the relation

$$\frac{d\hat{M}}{dt} = - \int_S \underset{\sim}{q} \cdot \underset{\sim}{n} \; dS \tag{12}$$

Using the definitions of $\hat{M}$ and $\hat{\theta}$, we can express this equivalently in time-integrated form as

$$\int_V (\hat{\theta}^{k+1} - \hat{\theta}^k)\, dV = \int_S \left(\int_{t^k}^{t^{k+1}} \underset{\sim}{q}\, dt \right) \cdot \underset{\sim}{n}\, dS \qquad (13)$$

Mass Balance of the Standard Formulation

The behavior of Equation (10) with respect to mass conservation can be examined by summing it over all values of i. With the sum over all i of the interpolation functions identically equal to unity, this yields the simple expression

$$\int_V \hat{C}^{k+\alpha}(\hat{\psi}^{k+1} - \hat{\psi}^k)\, dV = -(t^{k+1} - t^k) \int_S \underset{\sim}{q}^{k+\alpha} \cdot \underset{\sim}{n}\, dS \qquad (14a)$$

This equation is also obtained if the storage matrix is diagonalized according to the procedure L1. If L2 is used, the corresponding relation is

$$\int_V \sum_j C_j^{k+\alpha}(\psi_j^{k+1} - \psi_j^k)\phi_j\, dV = -(t^{k+1} - t^k) \int_S \underset{\sim}{q}^{k+\alpha} \cdot \underset{\sim}{n}\, dS \qquad (14b)$$

(Henceforth the letters a and b, suffixed to equation numbers, are keyed to the storage matrices in the manner employed above.)

If $\hat{C}$ were constant during a time step and piecewise constant in space, and if the net inflow at time $t^{k+\alpha}$ were identical to its time average during the time step, then Equations (13) and (14a) would be exactly equivalent, and hence Equation (10) and the L1 variant could be said to conserve mass. The former condition is never met (except in the trivial case of uniform moisture content) since $\hat{C}$ was prescribed to vary continuously between nodes, with its nodal values determined by the nodal values of ψ. The procedures defined by Equation (10) and its L1 variant therefore do not generally conserve mass, except in the limit as the time step duration and the element size go to zero.

Similarly, Equations (13) and (14b) are equivalent only if the nodal values of C are constants through the duration of the time step. Therefore the L2 variant of Equation (10) does not generally conserve mass, except in the limit as the time step duration goes to zero.

The Mass-Conservative Formulation

Comparison of Equations (13) and (14) suggests modifications of the numerical scheme that will cause it to conserve mass. Suppose we replace $q^{k+\alpha}$ in Equations (10) and (14) by $\bar{q}^k$, where the latter is the average value of q during the time step. Then the right-hand-side of Equation (14) has the desired value Suppose further we replace $\hat{C}^{k+\alpha}$ by $\bar{C}^k$, with the latter to be defined in such a way that

$$\int_V \bar{C}^k (\hat{\psi}^{k+1} - \hat{\psi}^k)\, dV = \int_V (\hat{\theta}^{k+1} - \hat{\theta}^k)\, dV \qquad (15a)$$

or, for L2,

$$\sum_j \bar{C}_j^{\,k}(\psi_j^{\,k+1} - \psi_j^{\,k}) \int_V \phi_j dV = \int_V (\hat{\theta}^{k+1} - \hat{\theta}^k)\, dV \tag{15b}$$

Then the family of schemes defined by Equation (10) and so modified will conserve mass, regardless of the temporal and spatial discretizations.

Equations (15) alone are not sufficient to determine $\bar{C}^k$ uniquely. If we require that the integral equalities hold on an element-by-element basis, and if we specify that $\bar{C}^k$ can be expressed in terms of the interpolation functions within an element, then the conditions on the nodal values of $\bar{C}^k$ are

$$\sum_j \left\{ \Delta\theta_j^{\,k} \int_{V_e} \phi_j \, dV_e - \bar{C}_j^k \sum_m \Delta\psi_m^{\,k} \int_{V_e} \phi_j \phi_m \, dV_e \right\} = 0 \tag{16a}$$

or, for L2,

$$\sum_j (\Delta\theta_j^{\,k} - \bar{C}_j^{\,k}\, \Delta\psi_j^{\,k}) \int_{V_e} \phi_j \, dV_e = 0 \tag{16b}$$

in which $\Delta\theta_j^{\,k}$ and $\Delta\psi_j^{\,k}$ are changes in the nodal values of θ and ψ during the time step. Either of Equations (16) represents L constraints on N values of $\bar{C}_j^k$, where L is the total number of elements and N is the total number of nodes. In the case of (16b), in fact, there is only one independent constraint, since the integral is always non-zero. Both systems are still underdetermined, since L is generally less than N, so further constraints must be introduced.

We propose that Equations (16) be satisfied on a term-by-term basis (i.e., for each j separately). Then the system in (16a) becomes overdetermined and $\bar{C}^k$ must be allowed to have discontinuities across element boundaries. The new description of the $\bar{C}^k$ field is then

$$\bar{C}^k = \sum_{j \subset \Omega_e} \bar{C}^k_{ej}\, \phi_j(\underset{\sim}{x}) \qquad \underset{\sim}{x} \subset V_e \tag{17a}$$

where Ω_e is the set of all nodes associated with element e, and $\bar{C}^k_{ej}$ is the value of $\bar{C}^k$ in element e at node j. This value is then given by

$$\bar{C}^k_{ej} = \frac{\Delta\theta_j^k \int_{V_e} \phi_j \, dV_e}{\sum_{m \subset \Omega_e} \Delta\psi_m^{\,k} \int_{V_e} \phi_j \phi_m \, dV_e} \tag{18a}$$

For L2, $\bar{C}^k$ can be continuous at element boundaries and its nodal values are given by

$$\bar{C}_j^k = \frac{\Delta\theta_j^{\ k}}{\Delta\psi_j^{\ k}} \qquad (18b)$$

APPLICATION

Description of Problem

The behavior of the proposed procedure is illustrated for the problem of vertical infiltration into a soil. The soil properties of Yolo light clay, fitted by Haverkamp et al. (1977), are employed. They are described by the equations

$$\theta = \begin{cases} 0.371\left[1 + \left(\dfrac{\log(-\psi)}{2.26}\right)^4\right]^{-1} + 0.124 & \psi < -1 \text{ cm} \\ 0.495 & \psi \geq -1 \text{ cm} \end{cases} \qquad (19)$$

and

$$K = (1.23 \times 10^{-5} \text{ cm s}^{-1}) \left[1 + \left(\frac{-\psi}{15.3}\right)^{1.77}\right]^{-1} \qquad (20)$$

Linear elements of length 1 cm are employed, with a total depth of 40 cm modeled. The initial condition has ψ equal to -600 cm at all nodes except the surface, where it equals 0 cm. This latter head is maintained as the surface boundary condition, while the bottom boundary condition allows only gravity-induced flow. Total time simulated is 10^4 seconds, during which a little more than 1 cm of water infiltrates.

Comparisons will be made using various numbers of iterations per time step (1,2,4,8) and various time step durations (10^2,10^3,10^4 s). Results discussed here are for the fully implicit case (α=1.0). Results for α=0.5 are similar.

Solution of the Nonlinear Algebraic Equations

The modified forms of Equation (10) described above are systems of nonlinear algebraic equations for the nodal values of ψ. In most cases, they can be solved using an iterative quasi-linearization approach whereby the coefficient matrices are evaluated using the latest estimate of the ψ_j^{k+1}, and new values are then found. The initial estimate for the first iteration is obtained by linear extrapolation from ψ_j^{k-1} and ψ_j^k. This iterative procedure usually converges rapidly when standard time-stepping is employed in conjunction with standard or lumped storage matrices. This is also true for the new time-stepping algorithm used in conjunction with either of the lumping schemes.

For the case where the proposed time-stepping procedure is

employed with the standard distributed storage matrix, the above-mentioned iteration procedure is either non- convergent or very slowly convergent. This is true also of the several other iteration techniques that we tried (the Newton method and variations, and the secant method). Results were thus obtained only for the five other combinations of storage matrices and time-differencing procedures.

Since head changes may be negligible at some nodes during a given time step, the denominators in Equations (18) are sometimes zero and it is then necessary, and justifiable, to replace the nodal value of $\bar{C}^k$ by $C(\psi_j^{k+1})$.

Comparison of Solutions

We introduce two indices that characterize the simulation results. The first is the relative error in cumulative infiltration at the end of the simulation,

$$\varepsilon_1 = \left| \frac{\Delta M - (\Delta M)_o}{(\Delta M)_o} \right| \tag{21}$$

where ΔM is simulated change in storage and $(\Delta M)_o$ is the "true" value, taken to be the value obtained with time step of 10 s and with two iterations. Different values of $(\Delta M)_o$ are found for the different combinations of storage matrices and time-stepping procedures, since there are inherent differences in the spatial discretization error, even though the element size is the same for all combinations.

The second index measures errors of mass balance. It is defined as

$$\varepsilon_2 = \left| \frac{F - \Delta M}{\Delta M} \right| \tag{22}$$

where F is the calculated cumulative influx of water. This is computed for each time step using the modified Equation (10), together with the final estimates of ψ_i, evaluated for values of i corresponding to the boundary nodes.

Theoretically, ε_2 goes to zero for the new time-stepping procedure as the number of iterations increases. It also goes to zero with standard time-stepping applied to L2. In the other cases it cannot go to zero unless the element size is also decreased.

In Figure 1, the errors ε_1 and ε_2 are plotted against a measure of relative computational effort (number of time steps times number of iterations per time step). For clarity of presentation, we include in Figure 1 the results for the L2 lumping procedure only. Of the five combinations of procedures examined, the results, expressed in terms of ε_1 and ε_2, were significantly more sensitive to the choice of the time-stepping procedure than to the choice of a storage matrix.

Since all points for the standard time-stepping scheme fall roughly along the same lines, the errors are relatively insensitive to the tradeoff between iterations and steps. It appears from the ε_1 data that two iterations would be optimal. Lines are drawn in Figure 1 joining the ε_1 and ε_2 points generated by two iterations and the standard time-stepping scheme.

For the new time-stepping procedure, the data are more scattered, reflecting a significant sensitivity to iteration. While two to four iterations provide the most accurate prediction of storage for given effort, the mass balance error decreases markedly with further iteration. Lines have been plotted in Figure 1 joining the points for four iterations. This seems to be a reasonable choice, since it makes ε_1 and ε_2 have the same order of magnitude.

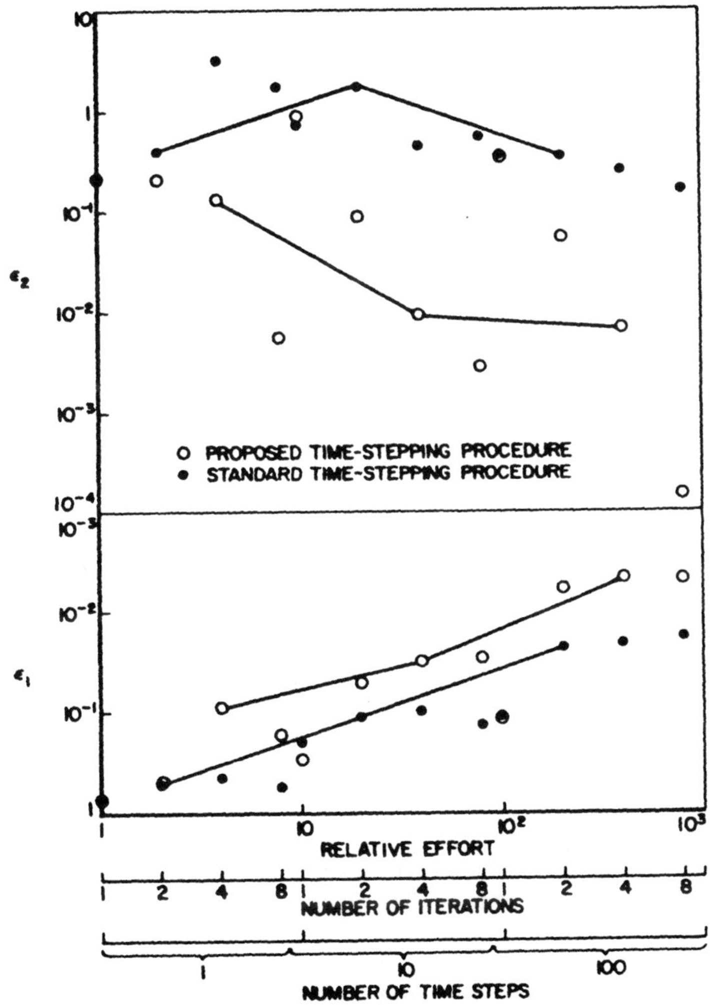

Figure 1. Errors ε_1 and ε_2 obtained using the L2 diagonalization together with standard and mass-conservative time-stepping.

The advantage of the proposed procedure over standard time-stepping for this problem becomes apparent upon comparison of the plotted lines in Figure 1. Errors in storage change are reduced by a factor of three and mass balance errors are reduced by a factor of ten to one hundred. Alternatively, this may be viewed as equivalent reductions in the amount of computational effort required to achieve the desired accuracy.

DISCUSSION

Our results are obviously dependent on the degree of nonlinearity of $\theta(\psi)$. In the present case C ranges from a value of 8×10^{-5} cm^{-1} at the initial head, to a maximum of 2×10^{-3} cm^{-1} at a head of -13 cm, to zero at -1 cm and above. Since such ranges (and even larger ones) are commonplace in soil physics, it is reasonable to expect similar rewards for applying this method in other problems of unsaturated flow.

A notable feature of Figure 1, supported by the unplotted data for the other storage, is that mass balance errors for the standard time-stepping scheme are all greater than 10%, a value that many modelers would agree is above the limit of acceptability. Significant improvement comes only with an increase of one or two orders of magnitude of effort.

The current analysis points to one important advantage of the θ-based Richards equation that is often overlooked. When θ is employed as the state variable in a finite element model, the balance of mass is guaranteed, regardless of element size, time step duration, or number of iterations. Whatever mass is forced into the numerical model is always conserved; only its distribution is affected by discretization and iteration.

SUMMARY

When used in conjunction with the head-based form of the Richards equation, the standard method of discretizing the time domain in a finite element model conserves mass only at the expense of great computational effort. An alternative approach for evaluating the storage coefficient, proposed here, achieves a balance of mass at significantly reduced cost. A spinoff of the proposed procedure is that it has improved convergence properties.

REFERENCES

Haverkamp, R., M. Vauclin, J. Touma, P.J. Wierenga, and G. Vachaud (1977). A comparison of numerical simulation models for one-dimensional infiltration, Soil Sci. Soc. Am. Proc., 41, 2: 285-294.

Neuman, S.P. (1973) Saturated-unsaturated seepage by finite elements, Proc. ASCE, J. Hydraul. Div., 99, HY12:2233-2250.

SATURATED-UNSATURATED TRANSIENT FINITE ELEMENT SEEPAGE MODEL FOR GEOTECHNICAL ENGINEERING

L. Lam and D. G. Fredlund

Department of Civil Engineering, College of Engineering, University of Saskatchewan, Saskatoon, Canada, S7N 0W0

INTRODUCTION

The analysis of seepage through porous media is of interest to many disciplines of engineering. In geotechnical engineering, seepage analysis is required in slope stability analysis, in groundwater contamination control and in the design of various hydraulic structures. Geotechnical engineers have tended to focus their attention on the saturated zone in seepage analysis.

Actually, unsaturated flow plays an important role in many engineering problems. For example, the movement of toxic leachates beneath tailings impoundment and sanitary landfills take place mainly under unsaturated conditions (Neuman, 1973). The transient analysis of seepage through a dam due to fluctuations in reservoir level is also highly influenced by the condition of the unsaturated zone (Freeze, 1971). Furthermore, in groundwater seepage problems involving infiltration and evaporation, realistic solutions can only be obtained if the unsaturated zone is properly taken into account.

The problems of water flow in saturated-unsaturated soils lead to nonlinear partial differential equations that are difficult to solve by graphical or analytical methods. With the development of high speed digital computers, numerical methods are increasingly used in solving the governing differential equation.

A two-dimensional transient finite element seepage analysis model is proposed. Two example problems are presented to demonstrate the capabilities of the model.

BACKGROUND

Some of the earliest theoretical work in the area of flow through unsaturated soils was presented by Richards in 1931.

Richards (1931) recognized that Darcy's law, which was originally proposed for saturated soils can equally be applied to unsaturated soils. The essential difference is that the coefficient of permeability is a constant for saturated soil, while it is a function of suction or moisture content for the unsaturated soil.

In 1937, Casagrande proposed a graphical method for the solution of seepage problems. The flow-net method of solution is practical when the boundary of the region under analysed is clearly defined and the soil medium is homogeneous and isotropic. However, as is often the case, when the region is nonhomogeneous and highly anisotropic, a solution by means of sketching flow nets is not practical. Furthermore, Casagrande (1937) assumed that water flowed only in the saturated zone of the porous media. As a result, the line of seepage was considered to be the uppermost flow line, and there should be no water flow across the line of seepage from the saturated zone to the unsaturated zone.

In 1958, Gardner proposed a general relationship between the coefficient of permeability and the capillary pressure for unsaturated soils. In general, Gardner's equation can be written as:

$$K = \frac{K_s}{1 + a|h|^n} \tag{1}$$

where: K_s = coefficient of permeability at saturation.
h = negative water pressure head (suction).
n = positive dimensionless constant.
a = constant depending on the units of permeability and suction.

Taylor and Brown (1967) presented a solution by finite element in seepage through earth dam with a free surface. They assumed that water flowed in the saturated zone only and the free surface was the uppermost flow line. They stated, "the principal problem is locating the position of the surface that has both zero flow normal to it and prescribed pressure". To locate the position of the free surface, Taylor and Brown (1967) proposed a special trial and error procedure. The procedure was tedious and required a considerable amount of computing time. For the case of transient state, solution by this procedure is difficult.

In 1971, Freeze developed a three-dimensional finite difference model for transient flow through saturated-unsaturated soils. The complete subsurface regime was treated as a unified whole because the flow in the unsaturated zone was integrated with the saturated flow. Consequently, instead of treating the free surface as the uppermost boundary, the ground surface was taken as the uppermost boundary. Furthermore Freeze

(1971) demonstrated that the free surface was not the uppermost flow line. It was stated to simply be a zero pressure isobar in the flow region which delineates the boundary of the saturated and unsaturated zone. Freeze (1971) also showed that the conditions of the unsaturated zone may stronly influence the position of the phreatic line, and failure to include the unsaturated zone can lead to results that are in error.

Most of the research concerning unsaturated flow have been undertaken by soil scientists. Geotechnical engineers have accepted analyses from different backgrounds without retaining a consistent framework for their formulation. Fredlund and Morgenstern (1976) provided the understanding of stress states and the relevant constitutive equations for an unsaturated soil. Furthermore, in 1981, Fredlund presented a transient flow theory through unsaturated soil with a framework most familiar to geotechnical engineers. Consequently, all the necessary physical relations for the formulation of a rigorous transient flow model through saturated-unsaturated soil systems are available.

In 1982, Papagianakis proposed a two-dimensional finite element model for flow in saturated-unsaturated soils. A computer program, SEEP, was developed for steady state flow conditions. Lam (1983) extended the research to a general steady and transient state finite element flow model called TRASEE.

THEORY AND FINITE ELEMENT FORMULATION

The proposed model describes the flow of water in a two-dimensional saturated-unsaturated soil systems. In the unsaturated zone, the air phase is assumed continuous and under atmospheric pressure. Therefore, the model considers only the flow of water. For both saturated and unsaturated soil, the water discharge velocity is assumed to be defined by Darcy's law. In the unsaturated soil, it is assumed that the coefficients of permeability of the medium depends on the negative pressure head. The formulation of the model is based on the Galerkin principle of the weighted-residual method. To integrate with time, the time-centered (Crank-Nicolson) iterative scheme is used.

Governing Differential Equation

The differential equation governing the flow is derived by equating the net flow quantity from a soil element to the time derivative of the constitutive relationship for an unsaturated soil (Fredlund and Morgenstern, 1976).

$$\frac{\partial}{\partial x}\left[k_{xx}\frac{\partial h}{\partial x} + k_{xy}\frac{\partial h}{\partial y}\right] + \frac{\partial}{\partial y}\left[k_{yx}\frac{\partial h}{\partial x} + k_{yy}\frac{\partial h}{\partial y}\right] = \rho_w g m_2^w \frac{\partial h}{\partial t} \qquad (2)$$

where: k_{xx}, k_{xy}, k_{yx}, k_{yy} = permeability tensor of an anisotropic soil
h = total head
m_2^w = slope of the (u_a-u_w) versus Θ_w when $d(\sigma-u_a) = 0$
u_a = pore-air pressure
u_w = pore-water pressure
Θ_w = net inflow or outflow of water for the element
σ = total stress
ρ_w = density of water
g = acceleration of gravity

For steady state conditions, the right-hand side of equation (2) goes to zero. Equation 2 is nonlinear because the values of the coefficient of permeability are not constant but depend upon the unknown pressure head.

Flow Equation for an Element

Using the Galerkin principle of weight-residuals, the flow equation for an element is derived as follows:

$$\int_A \{B\}^T [K] \{B\} dA \{h^n\} + \int_A \{L\}^T \lambda \{L\} \frac{\partial \{h^n\}}{\partial t} dA - \int_S \{L\}^T q ds = 0 \quad (3)$$

where: $\{B\} = \frac{1}{2A} \begin{Bmatrix} y_2 - y_3, & y_3 - y_1, & y_1 - y_2 \\ x_3 - x_2, & x_1 - x_3, & x_2 - x_1 \end{Bmatrix}$, with x_i, y_i the Cartesian coordinates of the nodes of the element.

$[K] = \begin{bmatrix} k_{xx} & k_{xy} \\ k_{yx} & k_{yy} \end{bmatrix}$, with k_{xx}, k_{yy}, k_{xy}, k_{yx} the components of permeability tensor, K

$\{h^n\} = \begin{bmatrix} h_1 \\ h_2 \\ h_3 \end{bmatrix}$, with h_i the total head at the nodes of the element

$\{L\}^T = \{L_1, L_2, L_3\}$, with L_i the area coordinates

$\lambda = \rho_w g m_2^w$
q = flow across the perimeter of the element
A = area of the element
s = perimeter of the element
t = time

Integration Over Time

For transient flow, equation 3 is integrated in time using the time centered, Crank-Nicolson finite difference scheme (Zienkiewicz, 1977). The relationship between nodal heads in two successive time steps of an element can be expressed as follows:

$$\left[\frac{2}{\Delta t}[C] + [D]^{t+\Delta t/2}\right]\{h^n\}_{t+1} = \left[\frac{2}{\Delta t}[C] - [D]^{t+\Delta t/2}\right]\{h^n\}_t + 2\{F\}^{t+\Delta t/2} \quad (4)$$

where: $[D]^{t+\Delta t/2} = \{B\}^T [K]^{t+\Delta t/2} \{B\}$ A, with D being the stiffness matrix evaluated at $t+\Delta t/2$

$$[C] = \frac{\lambda A}{12}\begin{bmatrix} 2 & 1 & 1 \\ 1 & 2 & 1 \\ 1 & 1 & 2 \end{bmatrix}$$, with C being the capacitance matrix

$\{F\}^{t+\Delta t/2}$ = load vector reflecting the boundary conditions evaluated at $t+\Delta t/2$

$\{h^n\}_t$ = known nodal total heads at time, t

$\{h^n\}_{t+1}$ = unknown nodal total heads at time, t+1

Global Head - Flow Equation

Equation 4 can be written in a simpler expression as:

$$[S]\,[h^n]_{t+1} = [Y] \qquad (5)$$

After matrix [S] and [Y] in equation 5 for an element are formed, the next step is to construct the algebraic equations for the whole system from the equations for the individual element. Nodal compatibility is used as the basis for this process (Desai and Abel, 1972). This simple requirement states that all elements adjacent to a particular node must have the same total head at that node. The set of linear simultaneous equations given by equation 5 are solved with respect to nodal total head using Gaussion elimination technique.

COMPUTER IMPLEMENTATION

The formulated finite element solutions are computer implemented in the program TRASEE. The computer facility used is the DIGITAL VAX 11/780 system at the College of Engineering, University of Saskatchewan, Saskatoon.

TRASEE is a two-dimensional finite element computer program that can model both steady state and transient flow conditions. TRASEE can accommodate complex geometries with arbitrary degrees of heterogeneity and anisotropy for up to 12 different soils. The present version of TRASEE has the capability of modelling a discretized soil system of up to 1300 elements and 800 nodes. Anisotropic conditions can also be handled where the direction of the major coefficient of permeability is at any specified angle to the x-axis.

Boundary conditions of either total head or flow can be specified at certain nodes. Flow boundary conditions such as infiltration and evapotranspiration and pumping can be simulated by designating a positive or negative flow at certain nodes. Changing boundary conditions during the transient process can be handled as a prescribed step function. A special procedure is also included for revising the boundary conditions along seepage faces. Furthermore, the program also has the capability to calculate the flux through any sections of the soil system.

For every time step, TRASEE calculates the total head at each node, then calculates element nodal velocities and gradients in the x and y directions. Finally, the flux quantity through different sections are calculated. Auxiliary plotting programs have been written to facilitate in the interpretation and presentation of the computed results.

RESULTS

Two example problems are presented. Example 1 is seepage through a homogeneous dam with a horizontal drain. At time equal to zero, the reservoir level is raised from 4 meters to 10 meters instantaneously. Figure 1a illustrates the rising of the phreatic lines with time. Figures 1b and 1c present the velocity field and the total nodal head values 32 minutes after the raising of the reservoir level. Figures 1d and 1e present the solution at final steady state condition.

Example 2 involves groundwater seepage under a tailings impoundment. Initially the pond is empty, and the soil underneath the pond is in equilibrium with the water table located at 5 meters below ground surface. At a time equal to zero, the pond is assumed to be filled with tailings which set up a water pressure head of 1.0 meter. Figure 2a illustrates the position of the groundwater table due to the influence of the tailings pond with time. Figures 2b to 2d present the velocity field at different times during the transient flow process.

DISCUSSION AND CONCLUSION

The fact that the phreatic line is not a flow line is confirmed by the velocity fields presented in Figures 1b and 1d. In these figures, the average nodal velocity vectors indicate that there is a considerable amount of water flow across the phreatic line. As is shown in Figures 1c and 1e, the equipotential lines extend all the way through the unsaturated zone. This means that there is potential gradient in the unsaturated zone, and there can be water flow in the unsaturated zone. This phenomena agree with the solution by Freeze (1971).

The results have also been compared with Casagrande's classic solutions based on the flow net technique. The solution of the final steady state condition (Figure 1e) compares satisfactorily with the flow net solution. The small differences are due to the fact that the model takes into account water flow in the unsaturated zone, while it is not considered using the flow net technique.

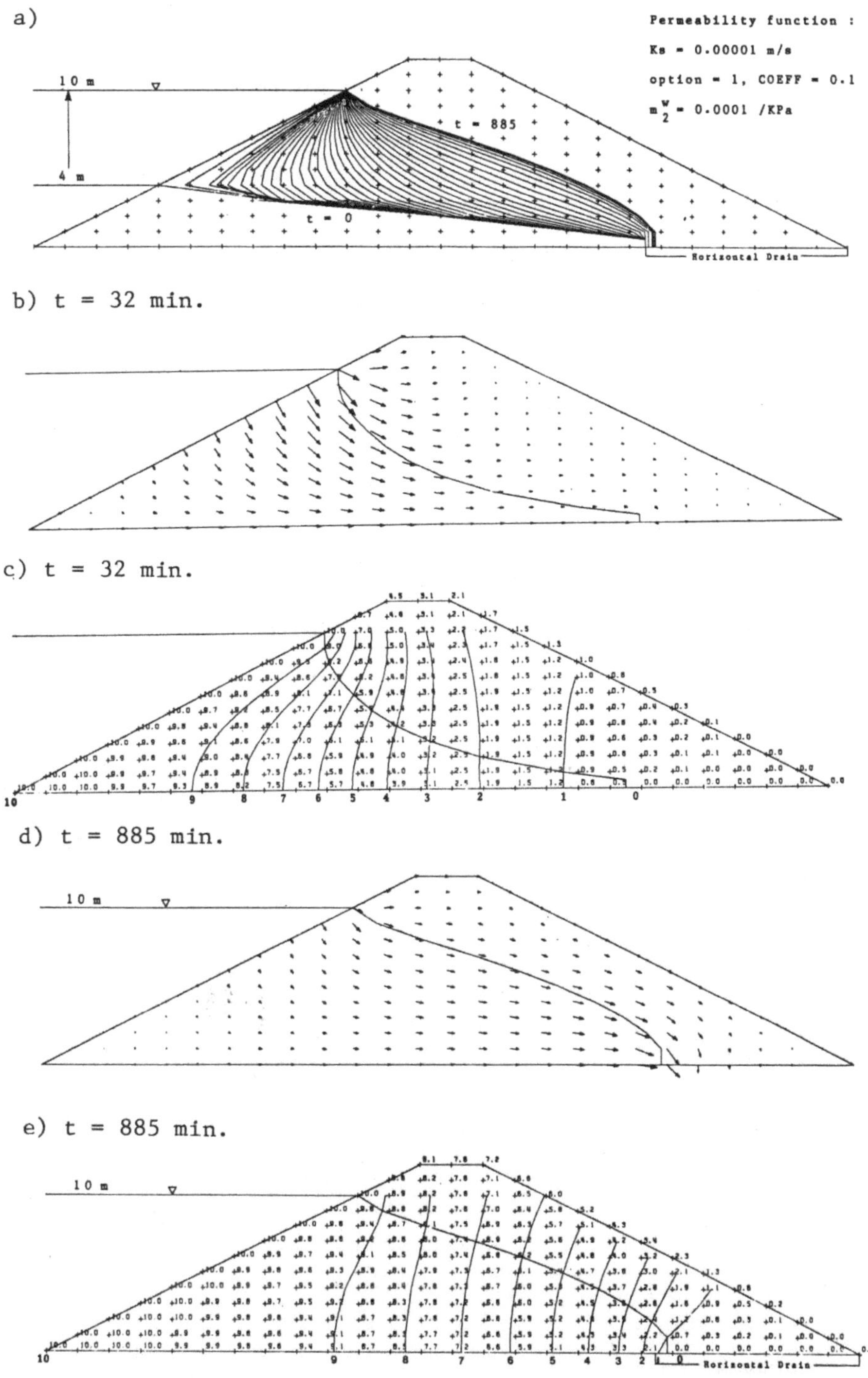

FIGURE 1 SEEPAGE THROUGH HOMOGENEOUS DAM WITH HORIZONTAL DRAIN
a) RISING OF PHREATIC LINES WITH TIME
b) to e) VELOCITY FIELD AND TOTAL HEAD AT DIFFERENT TIME

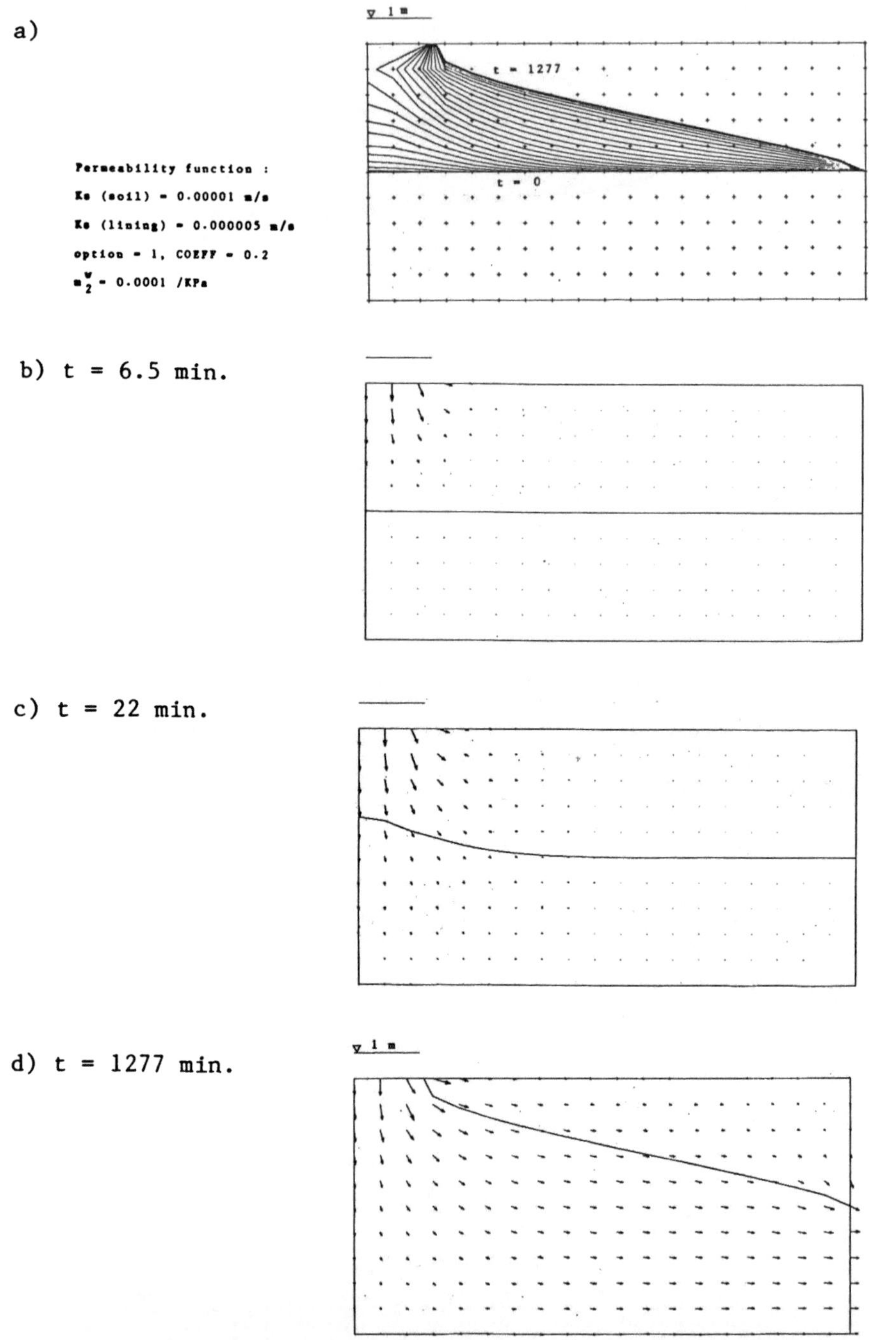

FIGURE 2 GROUNDWATER SEEPAGE UNDER TAILINGS IMPOUNDMENT
a) RISING OF WATER TABLE WITH TIME
b) to d) VELOCITY FIELD SHOWING THE THREE SEEPAGE STAGES

Example 2 is a class of problem to which the conventional "saturated only" methods of analysis cannot be applied. Generally, seepage under an impoundment is considered as taking place in three stages. These three stages can be illustrated by Figures 2b to 2d. Stage 1 (Figure 2b) is the advance of water from the impoundment through the unsaturated zone to the saturated zone which is delineated by the water table. Stage 2 (Figure 2c) is when the water contacts the water table and starts building up a groundwater mound toward the impoundment. Stage 3 (Figure 2d) occurs if the groundwater mound comes in contact with the impoundment. At this stage, the soil underneath the impoundment is saturated, and water flows from the impoundment directly into the saturated zone.

The transient solution of example 2 compares satisfactory with the transient process presented by McWhorter and Nelson (1979). Similar types of analysis can be performed for tailings impoundment design and for the prediction of contaminant migration into the groundwater regime.

The traditional "saturated-only" approach was found to be a special case to the saturated-unsaturated model. The saturated-unsaturated model provides a more realistic representation of the actual situations, and applies to a much wider range of engineering problems.

REFERENCES

Casagrande, A. (1937). Seepage through dams. New England Water Works, Vol. II, No. 2.

Desai, C.S. and J.F., Abel (1972). Introduction to the Finite Element Method, a Numerical Method for Engineering Analysis, Van Nostrand Reinhold Company, N.Y.

Fredlund, D.G. and Morgenstern, N.R., (1976). Constitutive Relations for Volume Change in Unsaturated Soils. Canadian Geotechnical Journal, Vol. 13, No. 3, pp. 261-276.

Fredlund, D.G. (1981). Seepage in Unsaturated Soils. Panel Discussion: Groundwater and Seepage Problems. Tenth International Conference on Soil Mechanics and Foundation Engineering, Stockholm, Sweden.

Freeze, R.A. (1971). Influence of the Unsaturated Flow Domain on Seepage through Earth Dams. Water Resources Research, Vol. 7, No. 4, pp. 929-940.

Gardner, W.R. (1958). Some Steady-state Solutions of the Unsaturated Moisture Flow Equation with Application to Evaporation from a Water Table. Soil Science, Vol. 85, No. 4.

Lam, L. (1983). Saturated-Unsaturated Transient Finite Element Seepage Model. M.Sc. Thesis, University of Saskatchewan, Saskatoon, Saskatchewan, Canada.

McWhorter, D.B. and Nelson, J.O. (1979). Unsaturated Flow Beneath Tailings Impoundments, ASCE, J. Geot. Eng. Dn. GT11, pp. 1317-1334.

Neuman, S.P. (1973). Saturated-Unsaturated Seepage by Finite Elements. ASCE, Hydraulics Division, Vol. 99, HY. 12, pp. 2233-2250.

Papagianakis, A.T. (1982). A Steady State Model for Flow in Saturated-Unsaturated Soils. M.Sc. Thesis, University of Saskatchewan, Saskatoon, Saskatchewan, Canada.

Richards, L.A., (1931). Capillary Conduction of Liquids through Porous Mediums. Physics, Vol. 1.

Taylor. R.L. and Brown, C.B., (1967). Darcy Flow with a Free Surface. ASCE, Hydraulics Division, Vol. 93, p. 25.

Zienkiewicz, O.C. (1971). The Finite Element Method in Engineering Science. McGraw-Hill.

NUMERICAL MODELLING OF GROUNDWATER FLOW IN A LARGE AGRICULTURAL AREA

I. Kazda, M. Kněžek, M. Žemlička

Technical University, Water Research Institute, Czech Hydrometeorological Institute, Prague, ČSSR

The protection and utilization of groundwater represent a significant part of water resources management. The withdrawal of groundwater or its pollution may as a rule endanger the interests of a number of further users, primarily of agriculture and of the population. It is therefore necessary to consider in advance any planned intervention concerning the regime of groundwater flow and to determine its effects. One of the possible means to this end is the use of a suitable numerical model. This report describes a simple model for unconfined groundwater flow in an nonhomogeneous and anisotropic permeable aquifer. The results of testing this model in the analysis of groundwater flow in a large agricultural area are presented.

Assuming that the Dupuit theorem holds valid the unconfined steady state groundwater flow may be described by the equation

$$\frac{\partial}{\partial x}\left(k_x H \frac{\partial h}{\partial x}\right) + \frac{\partial}{\partial y}\left(k_y H \frac{\partial h}{\partial y}\right) = 0 \qquad /1/$$

where h is the piezometric head, H is the height of the water table above the aquitard surface, and

k_x and k_y are coefficients of permeability in the directions of the axes x and y. These axes are co-linear with the axes of anisotropy.

Let the height of the datum plane above the surface of the aquitard be denoted as t_v according to Figure 1. As the expression $H = t_v + h$ holds valid it is possible to introduce a new notation

$$T_x = k_x H = k_x (t_v + h) \qquad /2/$$

$$T_y = k_y H = k_y (t_v + h) \qquad /3/$$

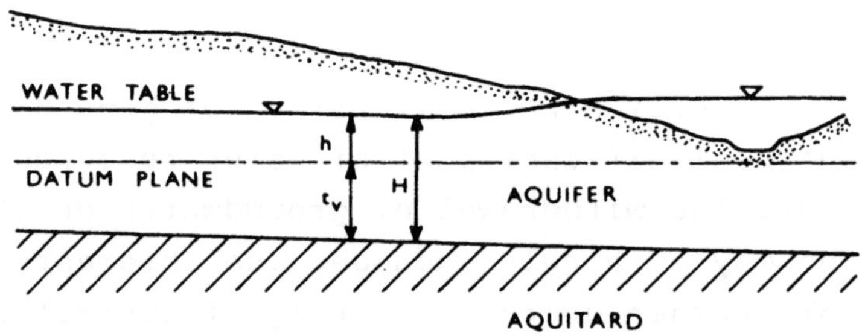

Figure 1 Unconfined groundwater flow

Equation 1 would now read

$$\frac{\partial}{\partial x}\left(T_x \frac{\partial h}{\partial x}\right) + \frac{\partial}{\partial y}\left(T_y \frac{\partial h}{\partial y}\right) = 0 \qquad /4/$$

In the solution of pracical problems h must be found in the given domain G with the boundary $\Gamma = \Gamma_1 + \Gamma_2$ so that it may satisfy both Equation 4 and the boundary conditions

$$h = h_o(x,y) \quad \text{on} \quad \Gamma_1 \qquad /5/$$

and

$$T_x \frac{\partial h}{\partial x} n_x + T_y \frac{\partial h}{\partial y} n_y + q_o = 0 \quad \text{on} \quad \Gamma_2 \qquad /6/$$

where h_o is the piezometric head given on Γ_1, q_o is the given flux and n_x and n_y are the direction cosines of the outward normal to the boundary.

This boundary value problem of the non-linear

partial Equation 4 may be substituted with a variational problem. It is possible to use the well-known functional

$$Fh = \frac{1}{2}\int_G \left\{ T_x\left(\frac{\partial h}{\partial x}\right)^2 + T_y\left(\frac{\partial h}{\partial y}\right)^2 \right\} dG + \int_{\Gamma_2} q_o h d\Gamma \qquad /7/$$

It is advantageous to minimize this functional with the application of the method of successive approximations. At a fixed point $h_1 = h_1(x,y)$ we approximate the functional Fh by the functional Fh_1 for which it holds valid

$$Fh_1 = \frac{1}{2}\int_G \left\{ T_x^{(1)}\left(\frac{\partial h}{\partial x}\right)^2 + T_y^{(1)}\left(\frac{\partial h}{\partial y}\right)^2 \right\} dG +$$

$$+ \int_{\Gamma_2} q_o h \, d\Gamma \qquad /8/$$

and $T_x^{(1)}$ and $T_y^{(1)}$ are determined by the realtions

$$T_x^{(1)} = T_x \, (h_1,x,y) \qquad /9/$$

$$T_y^{(1)} = T_y \, (h_1,x,y) \qquad /10/$$

so that the functional Fh_1 is a quadratic functional and its minimizing function h_2 may be easily found numerically by the finite element method. At the point h_2 it is again possible to substitute the functional Fh with the near functional Fh_2, when $T_x^{(2)}$ and $T_y^{(2)}$ have been determined similarly as $T_x^{(1)}$ and $T_y^{(1)}$. In this way a minimizing sequence h_i x i=1,2,.. arises converging towards the solution of the original boundary value problem. The performed calculations have shown that the convergence is fast and 3-4steps were usually sufficient to obtain the satisfactory accuracy.

For the estimation of the accuracy and of the possibilities of the elaborated numerical model groundwater flow was analyzed in an area situated on the right-hand bank of the river Elbe at a distance of about 40 km from Prague. In this area four observation boreholes are situated, the water table level observations covering several years.

The area belongs to the southern part of the Bohemian Cretaceous Table. Cretaceous rocks are represented by sandy mudstones to siltstones. These rocks form the bedrock of quaternary sediments in which groundwater is concentrated. In comparison with the bedrock this sediments are well permeable.

The chosen domain has a rather extensive area of 28.74 km^2. It is obvious that without very expensive investigations it would be impossible to obtain a sufficient number of data concerning the aquifer characteristics. For this reason a different approach was applied. Several hypotheses were formed concerning the composition of the chosen area and the numerical model was gradually adapted to suit the individual hypotheses. The first task was the delimitation of the domain to be considered and the determination of boundary conditions. This was easiest to do at the part of the boundary formed by the river Elbe /Fig. 2/ where the stable boundary condition was determined by two weir impoundments. The two shorter sides of the domain are formed by the two creeks. It was supposed that their stream beds correspond to the direction of the groundwater flow, the respective parts of the boundary were given as no flow boundaries. For the remaining part of the boundary the 170 m line of equal head was used derived from the observation results concerning the groundwater table in the boreholes.

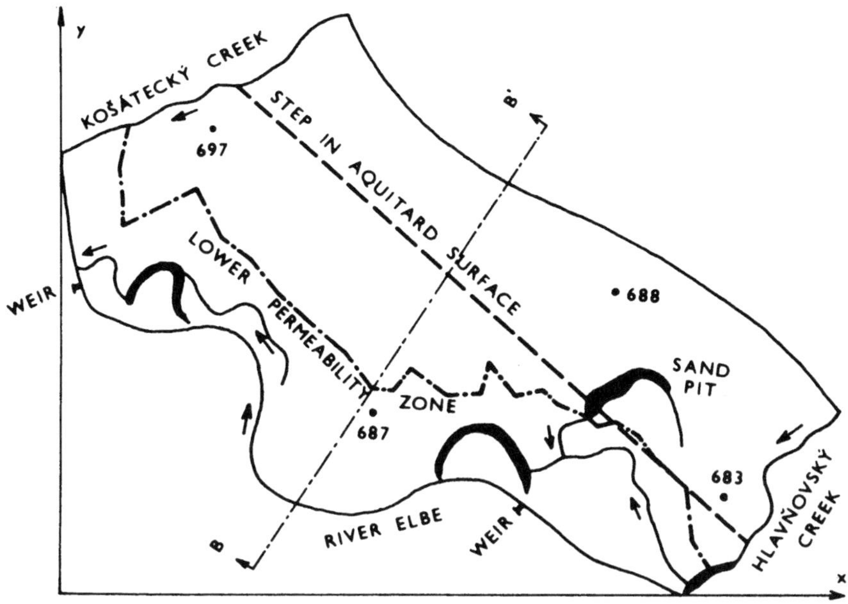

Figure 2 Agricultural area under consideration

Within the domain it was necessary to model two abandoned meanders of the Elbe and an old sand borrow-pit which is at present flooded with water. Another abandoned meander, located in front of the mouth of the Hlavňovský creek joining the Elbe, is connected by means of a channel /marked as channel G/ with the abandoned meander in the vicinity of the weir in the right-hand lower corner of the domain. The small brook running through the old sand borrow-pit joins the channel G.

The first case modelled was the simplest one. The aquitard surface was approximated by a plane with small slopes in both x and y directions. The percolated aquifer was modelled by the homogeneous and isotropic continuum with the permeability coefficient $k_o = 4 \times 10^{-4} \mathrm{m\ s^{-1}}$. The second case was modified in such a way as to simulated the effect of the assumed zone with lower permeability along the Elbe. The domain was divided into two parts differing in the coefficient of permeability. In the zone along the river the permeability coefficient was reduced to $k_1 = 4 \times 10^{-5}$ m s^{-1} and in the remaining part the original value k_o was applied. To verify the effect of the magnitude of k_1 it was necessary to solve the third case in which k_1 was substitued by a greater permeability coefficient $k_2 = 2 \times 10^{-4}$ m s^{-1}.

In the fourth case the aquitard surface was shaped into two terraces. The lower terrace in the vicinity of the Elbe had the same surface as the aquitard in the previous cases. The surface of the upper terrace was made parallel to it, but shifted 5 m upwards. The boundary between the terraces is shown in Fig. 2. Aquifer was modelled in the same way as in the first case.

The last case also assumed the existence of two terraces but similarly to the second case the zone along the river Elbe with the permeability coefficient k_1 was considered.

Table 1 Differences between computed and measured water table level /m/

Borehole	Case				
No.	1	2	3	4	5
697	2.55	5.43	3.10	1.57	4.56
687	-0.64	1.27	-0.40	-0.83	0.88
688	0.02	0.23	0.06	0.09	0.30
683	-0.64	-0.39	-0.56	-0.69	-0.45

In Table 1 are displayed the differences between the calculated and measured water table levels for all five cases. In all cases to be solved the calculated groundwater level in the borehole 697 lay much higher than we should expect in reality.

Either the borehole is locally affected by considerably different permeability of the surrounding soil or the boundary condition given for the part of the boundary the borehole 697 lying close to it is not correctly defined.

In order to better understand the importance of the deviations found at the other boreholes it is necessary to show how in the domain under consideration both the effect of the zone with lower permeability and the step in the aquitard surface will be plainly shown. The zone with lower permeability will cause impoundment of the groundwater in the remaining part of the domain. In Fig. 3 the form of the water table is displayed in section

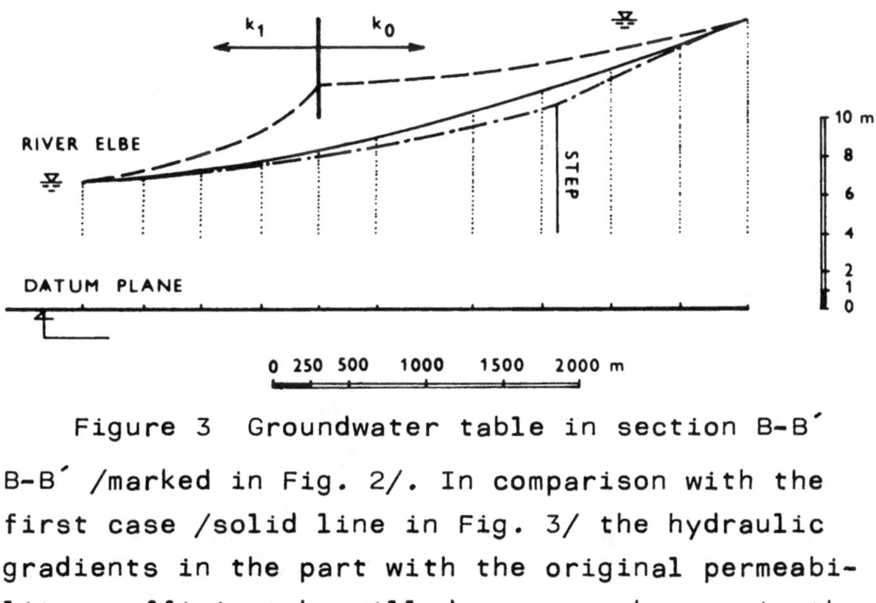

Figure 3 Groundwater table in section B-B´

B-B´ /marked in Fig. 2/. In comparison with the first case /solid line in Fig. 3/ the hydraulic gradients in the part with the original permeability coefficient k_o will decrease, whereas in the zone with k_1 they will increase. The greatest changes will naturally appear at the boundary dividing both parts.

Considerations concerning two terraces in the aquitard lead to a qualitatively different effect.

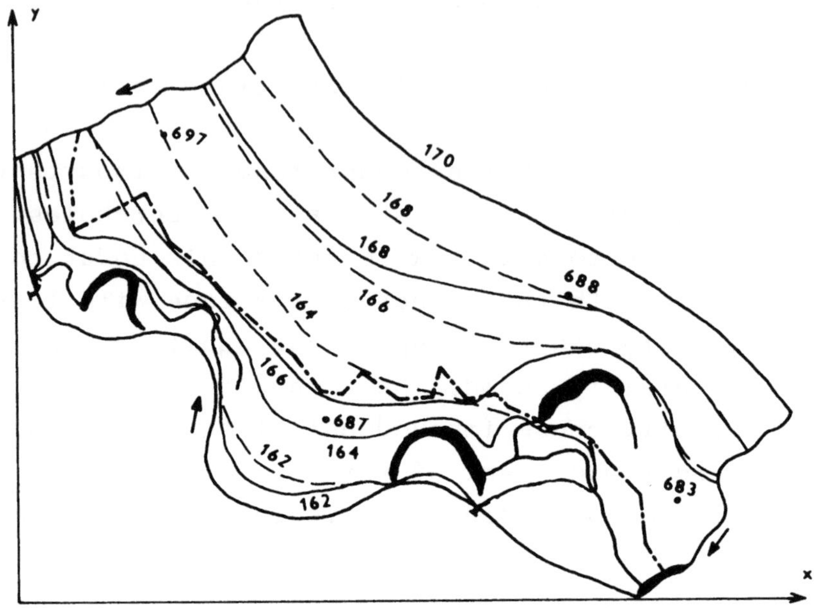

Figure 4 Comparison of cases 2 and 1

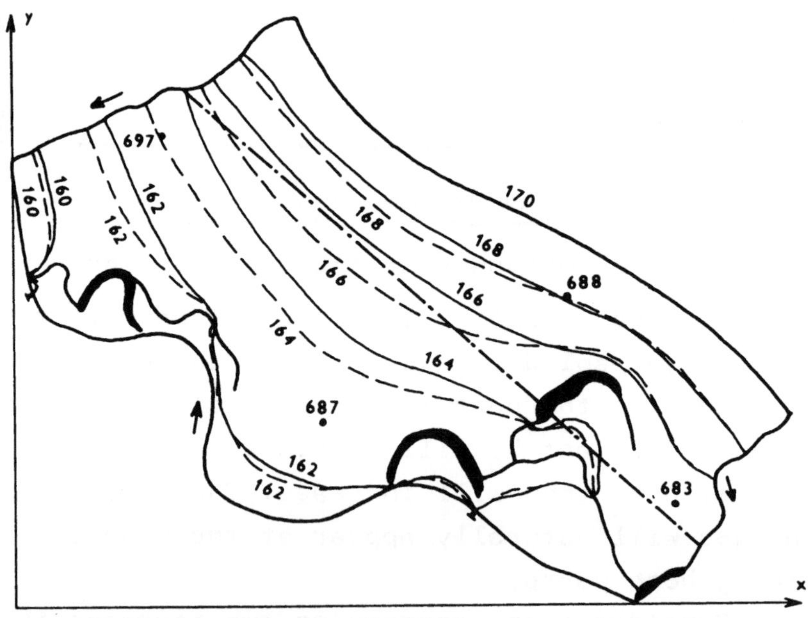

Figure 5 Comparison of cases 4 and 1

In Fig. 3 the water table is displayed for this case as well. Above the step the hydraulic gradients will increase, below it they will tend to decrease.

If the whole domain is taken into account, the situation does not seem to be so clear as in section B-B´ since the boundary conditions remain the same. In Fig. 4 the lines of equal head found for the second case with two zones /solid lines/ and those found for the first case /broken lines/ are compared. In Fig. 5 the same comparison is made for the case with a step and for the first one.

The borehole 687 is situated in the zone with lower permeability quite close to its boundary with other part of the domain. Therefore the water level in this borehole reacts to the magnitude of the permeability coefficient in the zone with lower permeability. Thus in the second case the computed water level is considerably higher than that in the first one and in the third case the level is 0.24 m higher in comparison with the first one. This borehole is situated at a great distance from the step in the aquitard surface so that in the fourth case the calculated level is only 0.19 m lower than that in the first one. In the last case the effect of the zone with k_1 becomes prevalent over the effect of the step and the computed water level is again higher than the observed one.

The borehole 688 lies in the vicinity of the upper part of the domain boundary created by the line of equal head. Therefore the computed level is in all cases affected by the stable boundary condition represented by this line. Out of all boreholes in that designated as 688 there exists a best agreement between the observed water level and those calculated for all cases. This may be proved by the fact that the form of the boundary line of equal head was derived in a reliable way and is in good agreement mainly with cases 1, 3 and 4.

The last borehole 683 is situated closely above the estimated step in the aquitard surface. The difference between the measured and observed water level is favourably affected by the existence of the zone with lower permeability but the step in the aquitard will produce only a very small effect on the calculated level.

On the base of the presented analysis it is possible to draw two main conclusions concerning the individual cases:

- the step in the aquitard surface is of relatively small significance and affects the agreement between measurement and calculation in an unfavourable way,
- the assumption concerning the existence of the zone with lower permeability along the Elbe has proved to be justified but its width in the part between the abandoned meanders is apparently too large.

With regard to the fact that the numerical model was elaborated with a relatively small amount of input data it is possible to consider the obtained results to be satisfactory. Also the method applied to the solution of non-linear flow has proved to be advantageous. With three to four iterations the elapsed time of the CPU of the ICL 4-72 computer amounted from 500 to 650 seconds.

GROUNDWATER DYNAMICS INVESTIGATION OF ENJEBI ISLAND, ENEWETAK ATOLL: AN INTERPRETIVE COMPUTER MODEL SIMULATION

M.E. Herman and S.W. Wheatcraft

Desert Research Institute, Univ. of Nevada System

INTRODUCTION

Development of groundwater resources is often dependent on the presence of potable water supplies. To effectively manage the groundwater resources of small islands, it is necessary to understand the groundwater dynamics of these areas. Using a finite element model, the authors identify factors affecting groundwater flow within atoll island aquifers. Simulated results are compared with field data from Enjebi Island, Enewetak Atoll.

PREVIOUS WORK

Although groundwater researchers have applied the Dupuit assumptions to island subsurface flow problems (Fetter, 1972; Lloyd et al., 1980), the authors of this paper believe these assumptions are not necessarily appropriate for small island groundwater investigations. Previous research by Fang et al. (1972), Buddemeier and Holladay (1977), and Wheatcraft and Buddemeier (1981) directly supports the hypothesis that vertical flow in atoll island aquifers can not be ignored.

On Enjebi Island, Enewetak Atoll (Figure 1), tide lags decrease with increasing well depth (Buddemeier and Holladay, 1977). For the same island, Wheatcraft and Buddemeier (1981) found that tidal efficiencies increase with increasing well depth, and that lags as well as efficiencies are essentially independent of distance from the shoreline (Figure 2). These two sets of observations have led to the hypothesis that tidal signals, and thus groundwater flow, are transmitted vertically from a deeper aquifer to the surficial Enjebi aquifer. This is in contrast to the traditional theory that tidal signals are propagated horizontally through island aquifers.

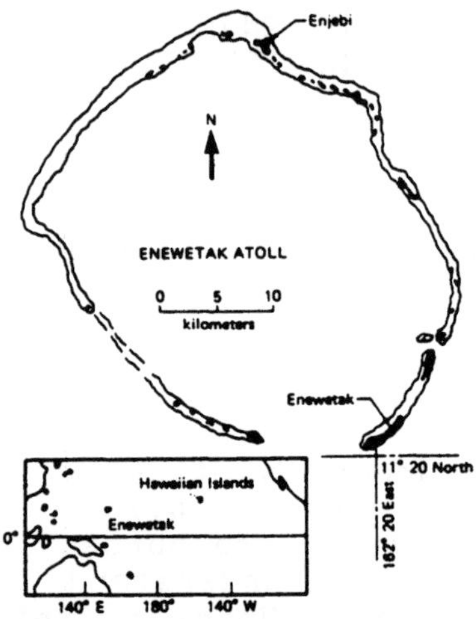

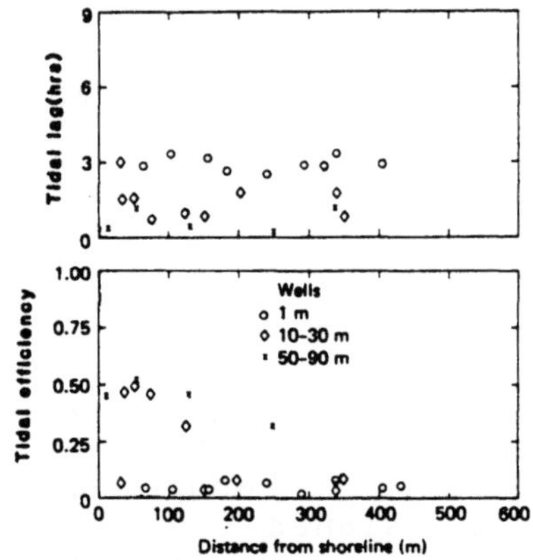

Figure 1. Enewetak Atoll   Figure 2. Enjebi data (modified from Wheatcraft and Buddemeier, 1981)

CURRENT RESEARCH

The computer program FEMWATER (Yeh and Ward, 1980) is chosen to assess the influence of tidal fluctuations upon the groundwater system of Enjebi Island. Several assumptions are made so that this model can be used. First, field data from Enjebi Island demonstrate patterns of tidal response similar to Enewetak Island data, and since Enewetak Island is long and narrow, Enjebi was modeled in the two-dimensional vertical plane. Secondly, it is assumed that density-induced flow is negligible compared to tidally-driven flow.

HYDROGEOLOGY

Enjebi is composed of Holocene carbonate sands and gravels (surficial aquifer) underlain by a consolidated, Pleistocene carbonate base (lower aquifer). A solution unconformity, located at a depth of approximately 15 m(49 ft) below island surface, marks the interface between the surficial aquifer and the lower aquifer (Wheatcraft and Buddemeier, 1981). Tides are semidiurnal, with a maximum range of 1.8 m(5.9 ft) between high and low tide. In addition, ocean tides are in phase with and have the same amplitudes as lagoon tides.

GOVERNING EQUATIONS

FEMWATER employs six governing equations (Yeh, 1982):

fluid continuity:

$$\frac{\partial \rho_f S n_e}{\partial t} + \nabla \cdot \rho_f S n_e \vec{v}_s + \nabla \cdot \rho_f \vec{q}_{fs} = 0 \qquad (1)$$

medium continuity:

$$\nabla \cdot \vec{v}_s = \frac{\partial n_e}{\partial t} + \nabla \cdot n_e \vec{v}_s \qquad (2)$$

fluid motion (Darcy's law):

$$\vec{q}_{fs} = -\bar{\bar{K}} \cdot (\nabla h + \nabla z) \qquad (3)$$

fluid compressibility:

$$\rho_f = \rho_f^o e^{\beta(p - p_o)} \qquad (4)$$

specific moisture capacity:

$$\frac{d\theta}{dh} = S\frac{\partial n_e}{\partial h} + n_e\frac{\partial S}{\partial h} \qquad (5)$$

consolidation:

$$\nabla \cdot \vec{v}_s = \alpha \rho_f g \frac{\partial h}{\partial t} \qquad (6)$$

Equations 1-6 are combined to form one equation that describes transient, simultaneous, saturated-unsaturated groundwater flow in a porous media:

$$\frac{d\theta}{dh} + \theta\beta\rho_f g + \frac{\theta}{n_e}\alpha\rho_f g \frac{\partial h}{\partial t} =$$
$$\nabla \cdot \bar{\bar{K}} \cdot (\nabla h + \nabla z) \qquad (7)$$

Equation 7 is based on the following assumptions: porous material is compressible but individual grains are incompressible; bulk density of the medium is constant; changes in fluid density do not affect work done on the fluid by increases in pressure; pressure changes cause fluid density changes; for the unsaturated zone, changes in effective porosity with respect to pressure head are negligible; horizontal consolidation is negligible; and spatial changes in fluid density are negligible.

INITIAL AND BOUNDARY CONDITIONS

For the initial time step, a static water table of mean sea level was assigned to the surficial aquifer. Aside from a no-flow boundary along the atoll base and a water table boundary within the Holocene aquifer, the only other type of boundary employed for the island simulations is a prescribed pressure-head boundary that varies with each time step according to the tidal change in sea level. The FEMWATER program is modified to specify these changes in prescribed pressure head as a means for reproducing tidal fluctuations on the island boundaries.

PROBLEM FORMULATION

A data file of appropriate Enjebi hydrogeologic characteristics is used as input for the modified FEMWATER program. The atoll cross section (Figure 3), from which a finite element grid is constructed, is derived from a topographic map of Enewetak Atoll and Vicinity (Emery et al., 1954). The vertical boundary XY is chosen arbitrarily in order to limit the grid size.

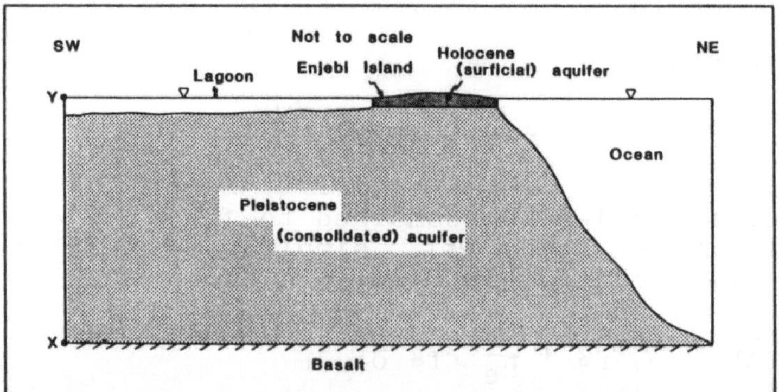

Figure 3. Vertical cross section of Enewetak Atoll

The cross section is divided geologically into three horizontal lithologic units: (1) basalt volcano; (2) consolidated Pleistocene aquifer; and (3) surficial Holocene aquifer. Both of the aquifers are isotropic and homogeneous. An enlarged profile of the surficial aquifer is shown in Figure 4.

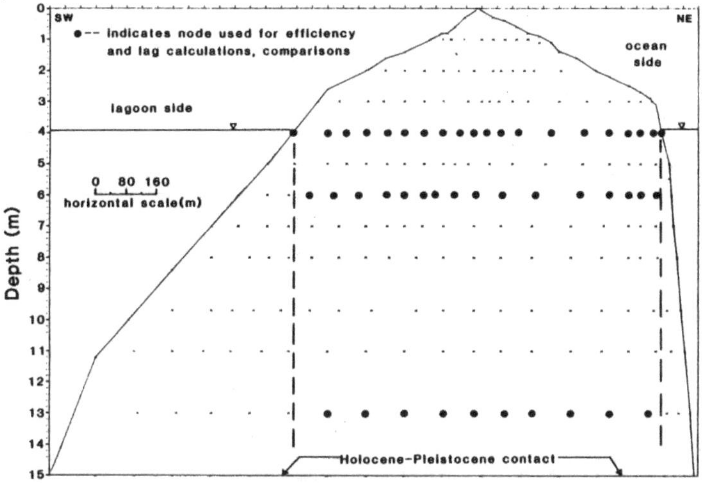

Figure 4. Cross section of surficial aquifer

The lower aquifer remains saturated at all times. Moisture characteristic curves for the upper aquifer are constructed from laboratory analyses of shallow Enjebi soils (Buddemeier, 1983).

Consecutive one-hour time steps are run until pseudo steady-state conditions are achieved, or in other words, when pressure distributions from one time step were approximately the same for a time step exactly 12 hours later. The time-step results include pressure head and total head values for every grid node.

TIDAL EFFICIENCY AND LAG

Tidal efficiencies and average tidal lags for selected nodes are plotted against distance from a shoreline (Figure 5a,b). These computations provide numerical results that can be directly compared with the field data. Horizontal distance from a shoreline is relative to the low tide-shoreline intersection.

Results show that tidal efficiency increases with increasing depth and is essentially independent of distance from a shoreline (Figure 5a), while tidal lag decreases with increasing depth and remains relatively constant with respect to horizontal distance (Figure 5b). These calculations suggest: (1) that vertical groundwater flow within the Holocene aquifer is significant; and (2) that a hydraulic connection between the Holocene and Pleistocene aquifers does exist. The results are also consistent with Enjebi field data and therefore serve to validate the conceptual model.

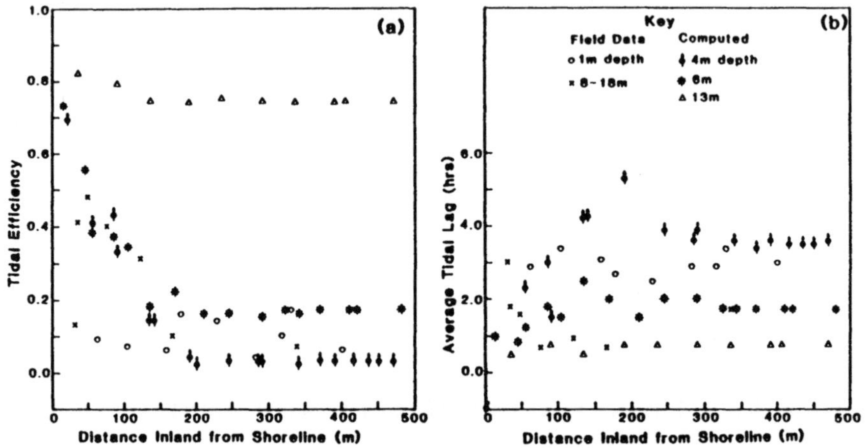

Figure 5. Comparison of computed results vs field data: (a) tidal efficiency; (b) tidal lag

TOTAL HEAD CONTOUR MAPS

Further evidence of significant vertical groundwater flow is obtained by constructing total head contour maps of the surficial aquifer. Only the maps of pertinent time steps appear in this paper (Figures 6-9). Sea level for each time step is also shown.

Steps 0 and 3 represent water levels at mid-tide and high tide, respectively (Figures 6 and 7), and show that groundwater movement at the aquifer margins is directed from the ocean (or lagoon) inward. The horizontal orientation of contour lines within the central portion of the Holocene aquifer demonstrates that groundwater is flowing in a vertical direction.

For the first time step (Figure 6), groundwater flow in the central portion of the aquifer is in an upward direction, reflecting the influential nature of the lower aquifer. As time progresses (Figure 7), the rising tide, via the lower aquifer, causes continued upward groundwater movement. Following mid-tide (Figures 8 and 9), upward groundwater flow is reduced significantly and downward groundwater movement becomes pronounced.

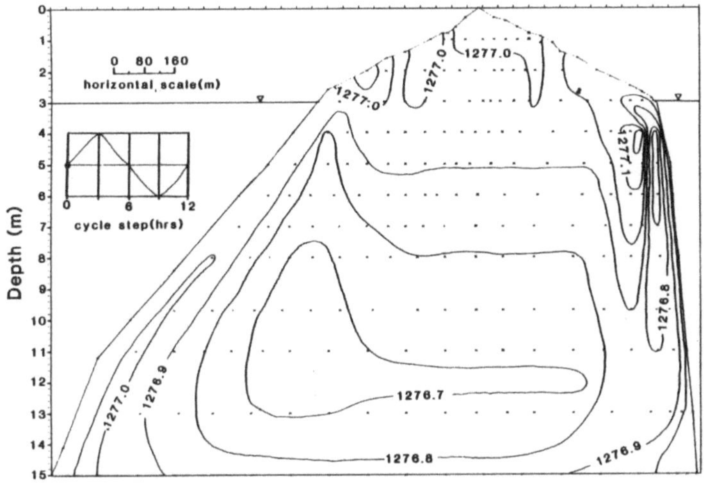

Figure 6. Total head map of surficial aquifer; t = 60 hrs; contour interval = 0.1 m

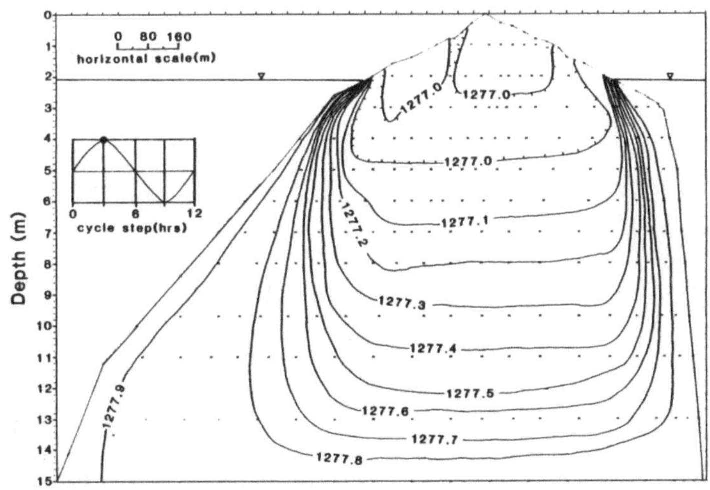

Figure 7. Total head map of surficial aquifer; t = 63 hrs; contour interval = 0.1 m

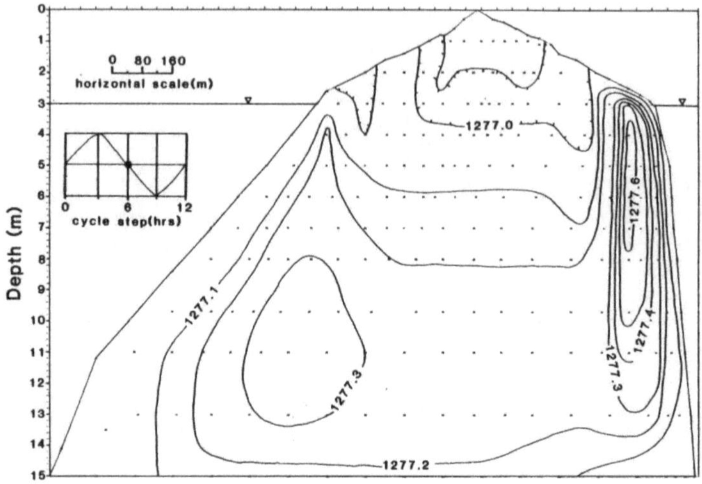

Figure 8. Total head map of surficial aquifer; t = 66 hrs; contour interval = 0.1 m

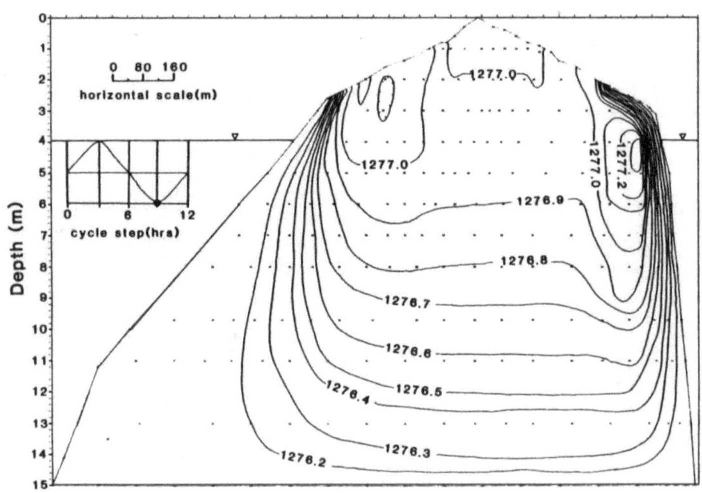

Figure 9. Total head map of surficial aquifer; t = 69 hrs; contour interval = 0.1 m

SUMMARY

A finite element groundwater model is modified to simulate tidal boundary fluctuations for a two-dimensional cross section. The modified computer program is used as an interpretive tool to investigate the groundwater flow system of Enjebi Island, Enewetak Atoll. Tidal lag and efficiency calculations for the simulated results are consistent with similar calculations for Enjebi Island field data.

Total head contour maps of the surficial aquifer, combined with lag and efficiency results, clearly indicate the significance of vertical groundwater flow in small island aquifers.

NOMENCLATURE

g = gravity constant
h = pressure head
$\bar{\bar{K}}$ = hydraulic conductivity tensor
n_e = effective porosity
p = fluid pressure
$\vec{q}_{fs}$ = darcy velocity of fluid relative to solids
S = degree of saturation
t = time
$\vec{v}_s$ = average velocity of solids
z = elevation head
α = coefficient of compressibility of medium
β = coefficient of compressibility of water
θ = Sn_e = moisture content
ρ_f = fluid density
ρ_f^o = fluid density at datum pressure

ACKNOWLEDGEMENTS

The authors wish to acknowledge the logistical support provided by the Water Resources Center, Desert Research Institute and Lawrence Livermore National Laboratory. In particular, a special thanks to Dr. R.W. Buddemeier for his valuable suggestions and advice. Our gratitude is also extended to W.A. McKay for his graphics work.

REFERENCES

Buddemeier, R.W., and G. Holladay (1977) Atoll hydrology: Island groundwater characteristics and their relationship to diagenesis. Proc., Third Int. Coral Reef Symp., Rosenstiel School of Marine and Atmospheric Science, University of Miami, Florida. May, pp. 167-174.

Buddemeier, R.W. (1983) Unpublished laboratory data. Lawrence Livermore National Laboratory, Livermore, California.

Emery, K.O., J.I. Tracy, and H.S. Ladd (1954) Geology of Bikini and nearby atolls. U.S. Geol. Survey Prof. Paper 260-A.

Fang, C.S., S.N. Wang, and W. Harrison (1972) Groundwater flow in a sandy tidal beach, 2. two-dimensional finite element analysis. Water Resour. Res., 8, 1:121-128.

Fetter, C.W. Jr. (1972) Position of the saline water interface beneath oceanic islands. Water Resour. Res., 2, 8:1307-1315.

Lloyd, J.W., J.C. Miles, G.R. Chessman, and S.F. Bugg (1980) A ground water resources study of a Pacific Ocean atoll--Tarawa, Gilbert Islands. Water Resour. Bull., Am. Water Res. Assoc., 16, 4:646-653.

Wheatcraft, S.W. and R.W. Buddemeier (1981) Atoll island hydrology. Ground Water, 19, 3:311-320.

Yeh, G.T., and D.S. Ward (1980) FEMWATER: A finite element model of water flow through saturated-unsaturated porous media. Rep. ORNL-5567. Oak Ridge Nat. Lab., Oak Ridge, Tenn., 153 pp.

Yeh, G.T. (1982) Training course no. 1: The implementation of FEMWATER (ORNL-5567) computer program final report. Rep. ORNL/TM-8327. Oak Ridge Nat. Lab., Oak Ridge, Tenn., 117 pp.

THE IMPACT OF PHASE EXCHANGE TRANSFERS AND UNSTEADY HYDRAULIC CONDITIONS ON SALTWATER INTRUSION FRONTS

Aaron A. Jennings

University of Notre Dame

INTRODUCTION

Saltwater intrusions into coastal aquifers are the result of a rather delicate balance between the seaward flow of "fresh" groundwater and the landward flow of intruding saltwater. When this balance is disrupted, the degree of intrusion can be greatly exaggerated. Because increasing salinity can have profound effects on coastal groundwater supplies, there appears to be a genuine demand for accurate predictions of salinity ahead of the full strength salt front.

The saltwater intrusion problem is best described as a density-driven, convective, dispersive mass transfer. Although considerable success has been reported on the ability of "immiscible fluids" formulations, one must account for the mass transfer across the freshwater/saltwater interface to accurately predict salinity profiles. It is this transfer that supports the constant landward flux of saltwater and attenuates the spacial intensity of the density driving gradient.

Over the last ten years, several finite element formulations have been proposed to solve the "miscible fluids" problems. However, it is still common to observe substantial differences between the predictions of these models and actual field measurements. It is the goal of this paper to discuss some additional physical and mathematical characteristics of the problem that are not commonly considered. The implications of prescribed flux boundary conditions, particularly in combination with transient prescribed potential conditions will be illustrated. The ramifications of phase exchange mass transfers and hydraulic conductivity changes due to successive saltwater floods will also be considered. It is believed that incorporation of these less obvious considerations can improve our ability to predict dispersed saltwater fronts.

FORMULATION DEVELOPMENT

Although the presence of a landward flow of saltwater and a dispersed saltwater interface were postulated in the early works of Drabbe and Badon Ghyben (c. 1889; Cooper, 1964) and Nomitsu, et al. (1927), the first successful analysis of the phenomenon was not developed until 1960. Henry (1960) solved the problem for a confined aquifer using a Fourier-Galerkin series approximation based on Darcy's Law for volumetric flux (q), and steady water (ρ_w) and salt (C) mass balances

$$\vec{q} + \frac{k}{\mu}(\nabla P - \rho\vec{g}) = 0 \tag{1}$$

$$\nabla \cdot \rho_w \vec{q} = 0 \tag{2}$$

$$-\nabla \cdot C\vec{q} + \nabla \cdot D \nabla C = 0 \tag{3}$$

where $\rho = \rho(C)$. However, in addition to steady state, Henry's solution required several simplifying assumptions including a constant dispersion coefficient (i.e., $D \neq D(q)$). Nevertheless, Henry's problem (if not his solution) has become a standard of comparison for nearly all miscible fluids formulations. For a finite element solution of this same problem, see Lee and Cheng (1974). Also, see Pinder and Cooper (1970) for a transient solution based on the method of characteristics.

Segol, et al. (1975, 1976) presented the first major finite element solution of the transient intrusion problem (i.e., $\partial C/\partial t \neq 0$ in Equation (3)), and also incorporated a velocity-dependent dispersion tensor and space-dependent hydraulic conductivity ($k = k(X_i)$). Segol, et al., also pioneered the "three-equation technique" of solving to yield a continuous velocity field. See Huyakorn and Taylor (1976) for a finite element treatment of an alternative "one-equation" technique. Frind (1980) extended the physical description to include both domain (α) and fluid (β) compressibility by posing a transient water mass balance to replace Equation (2).

$$\nabla \cdot \rho\vec{q} + \rho(\alpha + \phi\beta)\frac{\partial P}{\partial t} = 0 \tag{4}$$

Currently, Equations (1), (3), and (4) represent the state of the art in describing the problem. Several improvements have been made on alternative solution strategies and numerical accuracy (see Frind, 1982; Anand and Pandit, 1982; or Volker and Rushton, 1982, etc.) However, to date, little work appears to have been done on alternative physical descriptions.

FORMULATION EXTENSION

For the purpose of this study, the saltwater intrusion formulation has been extended to include mass transfer resistances.

It has commonly been observed that solutions for simple convective, dispersive transport do not always reproduce the results of miscible displacement experiments. Often, breakthrough curves exhibit retarded convective velocities and long "tails" before they converge to full strength.

These effects are illustrated in Figure I. This presents the normalized concentration breakthrough curves generated when saltwater displaces freshwater (or freshwater displaces saltwater) at various constant velocities. These particular data are for a column packed with uniform, non-reactive glass beads, but sand columns yielded similar results. Note that retardation of the front increases as velocity decreases. Less obvious is the fact that divergence from a theoretical one-parameter solution (adjusted for linear retardation) also increases as velocity decreases.

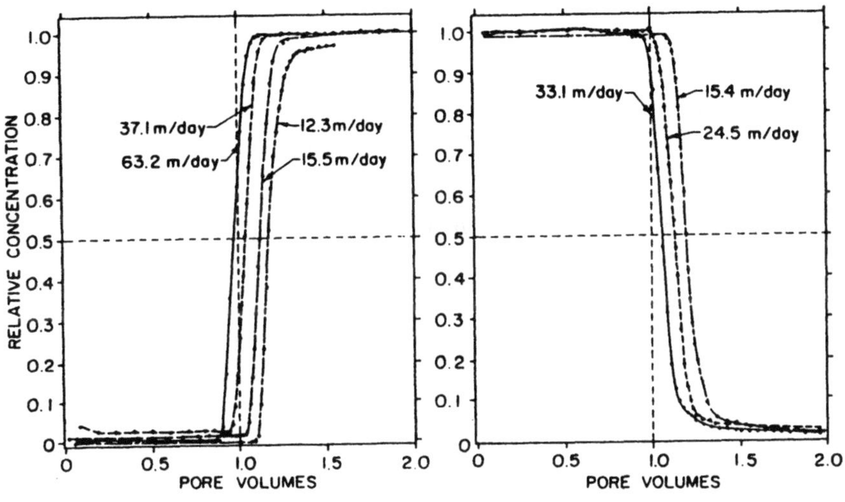

Figure I. Saltwater breakthrough curves for 0.25 - 0.425 mm Diameter Glass Beads at One Meter

Coats and Smith (1964) reasoned that effects of this type were due to rate-limited diffusion into or out of "dead end" pores. van Genuchten and Wierenga (1976) extended this concept to include mass transfers into a fixed film around each soil grain as well as a sorptive transfer to the solid phase. Actually, Nomitsu, et al. (1927) speculated about the effects of salt adsorption on saltwater intrusions, but could measure no significant sorptive transfers for the sands they studied. Since no effect could be detected for the soils used in this study either, the mechanism was not included. However, readers are cautioned that this probably does not generalize well beyond very sandy soils.

Therefore, after Segol, et al. (1975), Frind (1980), and Van Genuchten and Wierenga (1976), the following physical description is proposed for the saltwater intrusion problem.

$$q_i + \frac{kT_{ij}}{\mu}\left[\frac{\partial P}{\partial X_j} + \rho g_j\right] = 0 \qquad (5)$$

$$\rho(\alpha + \phi\beta)\frac{\partial P}{\partial t} + \frac{\partial}{\partial X_i}(\rho q_i) = 0 \qquad (6)$$

$$\frac{\partial}{\partial t}(\phi_m C_m) + \frac{\partial}{\partial t}(\phi_{im} C_{im}) + \frac{\partial}{\partial X_i}(q_i C_m) - \frac{\partial}{\partial X_i}\left(\phi_m D_{ij}\frac{\partial C}{\partial X_j}\right) = 0 \qquad (7)$$

$$\frac{\partial}{\partial t}(\phi_{im} C_{im}) + r_f \phi_m C_m - r_r \phi_{im} C_{im} = 0 \qquad (8)$$

Note that the fluid has been discretized into a mobile and immobile phase ($\phi=\phi_m+\phi_{im}$) with C_m and C_{im} being the salt concentration in each phase. In the above, q_i are the components of the Darcy velocity vector, kT_{ij} is the intrinsic permeability tensor, μ is the dynamic viscosity of the mobile phase, P is the fluid pressure in the mobile phase, g_j are the components of gravity aligned with X_j, ϕ is the total domain porosity, D_{ij} is the dispersion tensor, and r_f and r_r are the forward and reverse film transfer rate constants. The following functional relationships have been presumed.

$$T_{ij}, \alpha, \phi = f(X_i); \qquad k, \mu, \rho = f(C_m)$$

$$\phi_m, \phi_{im} = f(\phi, q_i); \qquad D_{ij} = f(k, T_{ij}, \rho, \mu, q_i)$$

Therefore, Equations (5) through (8) yield a non-linear set of 5 partial differential equations in a two-dimensional space.

The first set of coefficient functions are taken to be prescribed domain properties. The dispersion tensor is computed as per Scheidegger (1961) with the effective longitudinal and traverse components based on intrinsic permeability Reynolds number correlations (Bruch and Street, 1967). Density and viscosity are treated as linear function of C_m. The partitioning of fluid into the mobile and immobile phase currently requires an empirical relationship evaluated over the velocity range to be experienced by each soil. However, recent data have indicated that these approach a constant value (for each soil) at common groundwater velocities.

Special attention is called to the relationship between intrinsic permeability and salt concentration. Figure II illustrates the dramatic permeability changes that can be associated with successive saltwater floods. Both reversible and irreversible changes over several orders of magnitude have been observed (Goldenberg, et al., 1983). These effects have been attributed to precipitation/ dissolution reactions of carbonate minerals and to swelling/ dispersion or flocculation/swelling scenarios for trace clay contents. The results of Figure II were produced by 1% (by weight) of montmorillonite evenly distributed through fine Ottawa sand. Although the mechanism and degree of response are distinctly material-dependent, the phenomenon argues for a careful consideration of the $k=k(C_m)$ possibility, for the consideration of domain compressibility (or swelling) in Equation (6) and possibly for the inclusion of a true sorptive exchange in Equation (7).

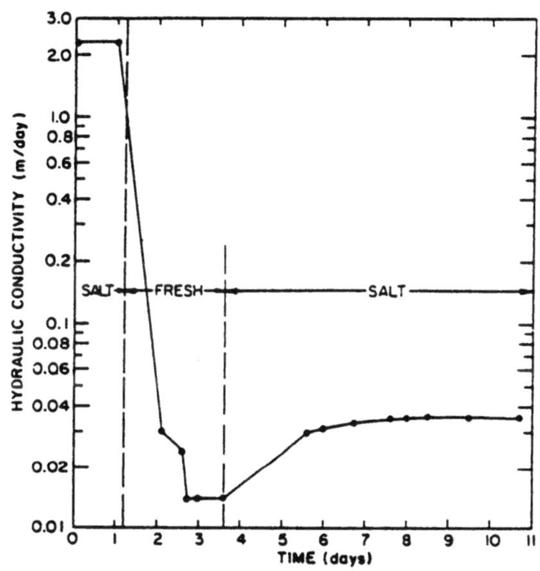

Figure II. Permeability Decline from a Salt/Fresh Water Flood

SOLUTION IMPLEMENTATION

The coupled system of solvent and solute transport equations were solved by the Galerkin finite element method. All spacial integrations were performed by Gaussian quadrature integration, and linear interpolation functions were used for all space-dependent coefficients. Integration over space yields (in matrix form)

$$\mathbf{K}\delta + \mathbf{M}\ \frac{d}{dt}\ \{\delta\} = \mathbf{F} \tag{9}$$

where **K** and **M** denote non-linear coefficient matrices, δ denotes the vectors of unknown nodal coefficients, and **F** denotes a vector of boundary condition contributions. A simple backwards difference time discretization of Equation (9) yields

$$K_t\ \delta_t + (1/\Delta t)M_t\{\delta_t - v\delta_{t-\Delta t}\} = F_t \qquad (10)$$

where **K** and **M** are evaluated for δ_t and F_t is evaluated at time t for all prescribed transient boundary conditions. If $\delta^n{}_t$ is then defined as the nth iterate such that $\delta^0{}_t = \delta^\infty{}_{t-\Delta t}$ and the coefficient functions are always evaluated with the nth iterate, one may seek the solution

$$\delta_t{}^{n+1} = \left[K_t{}^n + (1/\Delta t)M_t{}^n\right]^{-1}\left\{(1/\Delta t)M_t{}^n\ \delta^\infty{}_{t-\Delta t} + F_t\right\} \qquad (11)$$

until the difference between successive iterates satisfies convergence criterion.

This solution was developed for the stylized two-dimensional domain illustrated in Figure III. The initial and boundary conditions have also been inluded on this Figure. The reader should note three significant departures from the classical "Henry's Problem". First, the seaward groundwater flux is generated by a prescribed potential boundary condition. Secondly, the seepage face approach of Frind (1982) has been adopted to eliminate logical inconsistencies implied in Henry's treatment of the seaward boundary. Finally, provisions have been made to allow the seaward boundary potential to vary in time. This has been included to study the effects of tidal oscillations and storm surges.

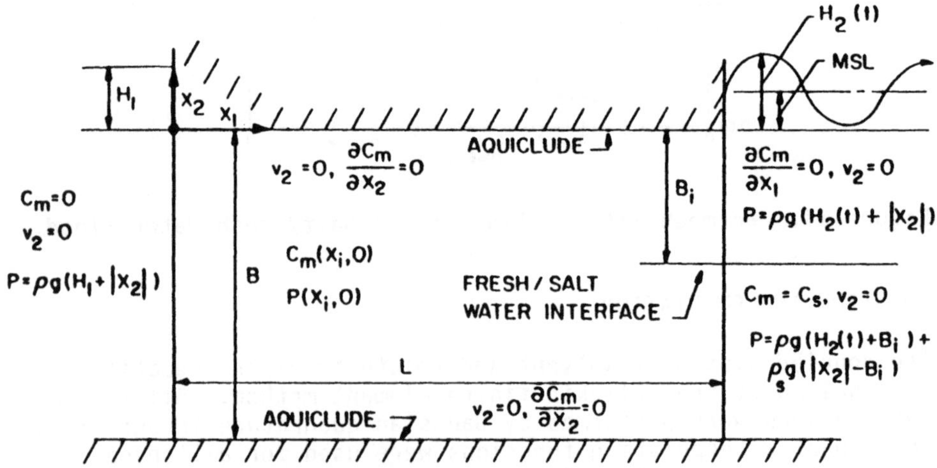

FIGURE III. Domain Definition

DISCUSSION OF RESULTS

To reproduce Henry's problem, the following constant values must be specified for the domain of Figure III.

$\phi = 0.35$, $k = 1.336\text{x}10^{-9}\ \text{m}^2$, $\rho_g = 1000.0\ \text{kg/m}^3$

$\mu = 1.334 \times 10^{-4}\ \text{kg/m}\cdot\text{sec}$, $L = 2.0\ \text{m}$, $B = 1.0\ \text{m}$,

$B_i = 0.25\text{m}$, $C_S = 24.5\ \text{kg/m}^3$, $C_m(X_i, 0) = 0.0$

One must also let α, β, ϕ_{im}, r_r, and r_f equal zero in Equations (5) - (8), let H_2 be a constant valve, and specify the total potential drop ($H_1 - H_2$) to yield a total (positive) freshwater flux of $Q = 6.6\text{x}10^{-5}\ \text{m}^3/\text{sec}$. This may be accomplished by specifying boundary conditions such that (H_1 - MSL) = 0.020 m. However, it should be noted that this will not produce the constant flow field that has commonly been specified on the landward boundary. This prescribed potential will yield a seaward flux at $X_1 = 0$ that varies non-linearly with depth from $6.68\text{x}10^{-5}$ m/sec to $6.41\text{x}10^{-5}$ m/sec at time zero and $6.74\text{x}10^{-5}$ m/sec to $6.35\text{x}10^{-5}$ m/sec at equilibrium. These results are more consistent with the concept of flow towards a partial obstruction.

On Tidal Oscillation

Figure IV presents the solution of Henry's problem based on a "field" scale-up factor of 10 applied to the dispersion correlations of Bruch and Street (1967). The dispersed front (from the 0.1 to 0.9 isochlor) has been indicated as the 0th tidal cycle solution. The 0.5 isochlor essentially reproduces the previously reported solutions of this problem.

The possible effects of tidal oscillations have also been indicated on Figures IV. Tidal effects on dispersed fronts have been observed in field and laboratory (Brown, 1925, Kohout, 1960, Rumer and Harleman, 1963). Using the data of Kohout, Henry (1964) estimated that the flux through the Biscayne aquifer would vary over an order of magnitude from high to low tide. However, recent evaluations of the significance of tides have been more modest (Tyagi, 1971). The result has been their omission from most saltwater intrusion models.

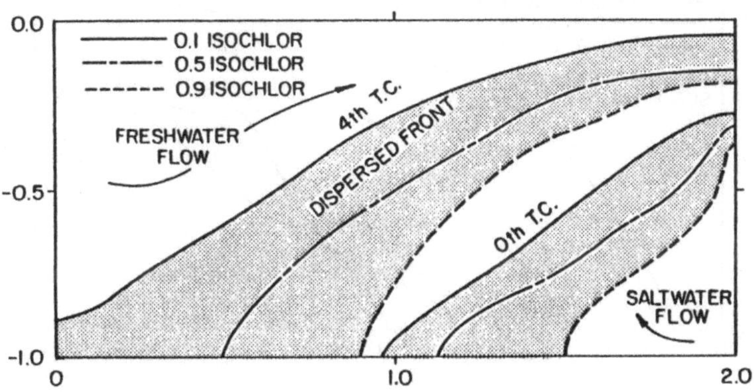

FIGURE IV. Effects of Saltwater Head Oscilations on Henry's Problem

Figure IV presents the result of subjecting Henry's problem to a head oscillation of ±0.001 m (10% of the total specified gradient) on the seaward boundary. For this particular solution, mass transfer resistances were omitted, and oscillations were assumed to be completely damped out at $X_1 = 0$. A very short period was also applied to ease the computational burden. This was based on the observation that the problem responds very rapidly when incompressibility is assumed, and that results did not appear to be sensitive to longer periods.

Figure IV illustrates the irreversible advance of the dispersed front under these conditions. This is accompanied by a permanent decline of the seaward freshwater flux. Prescribing the flux on the landward boundary "desensitizes" the simulation to head oscillations and yields minor fluctuations about the mean sea level solution. When potentials are prescribed, each advance of the front adds new, permanent resistance to the domain and the seaward water flux declines.

For this case, including mass transfer resistances slows but does not prevent the landward advance of the front. Without some attendant change in the landward boundary head, a new equilibrium position will not be established until meters inland of the original domain.

On Dispersion Scale

The consequence of inflating dispersion to help develop a broader dispersed front can be seen by comparing Figures IV and V. The only difference between the 0th T.C. solution of Figure IV and the SPSF=1 solution of Figure V is an additional factor of ten applied to the dispersion correlation. This broadens the front, but the degree of full strength penetration is substantially reduced. For similar results see Volker and Rushton (1982). Therefore, this may not be a desirable technique for "tuning" model predictions.

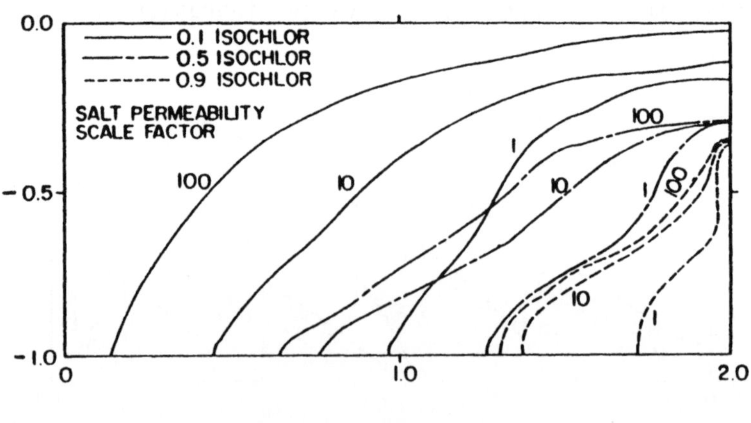

FIGURE V. Effects of Permeability Changes on Henry's Problem

On Permeability Modifications

Figure V also illustrates the potential impact of salt dependent permeability changes. On this figure the profiles resulting from permeability increases of one and two orders of magnitude (taken as linear functions of C_m) have been plotted. Note that the first order of magnitude leads to a substantial increase in the penetration of full strength saltwater. Beyond this, increases only tend to further disperse the front. Although better quantitative information about $k=k(C_m)$ is required, the qualitative behavior shown here could explain anomalies in the Biscayne aquifer data that have previously been attributed to anisotropy.

CONCLUSIONS

Results have been presented to illustrate two mechanisms that tend to displace and broaden the predictions of dispersed saltwater fronts without overinflating the dispersion coefficient. It can also be shown that reversible mass transfers have this same effect when existing fronts are displaced by one-time hydrology transients. It is believed that incorporation of some of these effects can help fill any gap between model predictions and field observations.

ACKNOWLEDGEMENTS

This work was supported in part by grant number CME81-05549 from the National Science Foundation. The assistance of Edward Mehnert is also gratefully acknowledged.

REFERENCES

Anand, S.C. and Pandit, A. (1982) Finite Element Solutions of Coupled Groundwater Flow and Transport Equations ... Proc. 4th Int. Conf. on Finite Elements in Water Resour., 13-3-13-14.

Brown, J.S. (1925) A Study of Coastal Ground Water U.S.G.S. Water Supply Paper 537, 101 p.

Bruch, J.C. and Street, R.L. (1967) Two-Dimensional Dispersion, J. San. Eng. Div., ASCE, 93, SA6:17-39.

Coats, K.H. and Smith, B.D. (1964) Dead-End Pore Volume and Dispersion in Porous Media, Soc. Petrol. Eng. J. 4, 73-84.

Cooper, H.H., Jr. (1964) A Hypothesis Concerning the Dynamic Balance of Fresh Water and Salt Water in a Coastal Aquifer, U.S.G.S. Prof. Paper 1613-C: C1-C12.

Frind, E.O. (1980) Seawater Intrusion in Continuous Coastal Aquifer - Aquitard Systems, Proc. 3rd Int. Conf. on Finite Elements in Water Resour., 2.177-2.198.

Goldenberg, L.C., Magaritz, M. and Mandel, S. (1983) Experimental Investigation on Irreversible Changes of Hydraulic Conductivity ... Water Resour. Res., 19, 1:77-85.

Henry, H.R. (1964) Interfaces Between Salt Water and Fresh Water in Coastal Aquifers, U.S.G.S. Prof. Paper 1613-C:C35-C70.

Henry, H.R. (1960) Salt Intrusion into Coastal Aquifers, Int. Asso. of Sci. Hydrol., 52, 478-487.

Huyakorn, P. and Taylor, C. (1976) Finite Element Models for Coupled Groundwater Flow and Convective Dispersion, Proc. 1st Int. Conf. on Finite Elements in Water Resour., 1:131-151.

Kohout, F.A. (1960) Cyclic Flow of Salt Water in the Biscayne Aquifer of Southeastern Florida, J. Geo. Res., 65, 7:2133-2141.

Lee, C.H. and Cheng, R.T. (1974) On Seawater Encroachment in Coastal Aquifers, Water Resour. Res., 10, 5:1039-1043.

Nomitsu, T., Toyohara, Y. and Kamimoto, R. (1927) On the Contact Surface of Fresh- and Salt-Water Under the Ground ..., Mem. Coll. of Science, Kyoto Univ., Series A, 10, 279-303.

Pinder, G.F. and Cooper, H.H. (1970) A Numerical Technique for Calculating the Transient Position of the Saltwater Front, Water Resour. Res., 6, 3:875-882.

Rumer, R.R. and Harleman, D.R.F. (1963) Intruded Salt-Water Wedge in Porous Media, J. Hyd. Div., ASCE 89, HY6:193-220.

Scheidegger, A.E. (1961) General Theory of Dispersion in Porous Media, J. Geo. Res., 66, 10:3273-3278.

Segol, G. and Pinder, G.F. (1976) Transient Simulation of Salt-water Intrusion in Southern Florida, Water Resour. Res., 12, 1:65-70.

Segol, G., Pinder, G.F., and Gray, W.G. (1975) A Galerkin-Finite Element Technique for Calculating the Transient Position of the Saltwater Front, Water Resour. Res., 11, 2:343-347.

Tyagi, A.K. (1971) Dynamics of the Transition Zone Between Fresh and Salt Waters in Coastal Aquifers, Dissertation, Univ. of Calf., Berkeley, 168p.

van Genuchten, M. TH. and Wierenga, P.J. (1976) Mass Transfer Studies in Sorbing Porous Media, Soil Sci. Soc. Amer. J., 40, 4:473-479.

Volker, R.E. and Rushton, K.R. (1982) An Assessment of the Importance of Some Parameters for Seawater Intrusion ... J. of Hydrol., 56, 239-250.

FINITE ELEMENT SIMULATION OF KINEMATIC SURFACE RUNOFF

George E. Blandford and Michael E. Meadows

Respectively, Assistant Professor, Department of Civil Engineering, University of Kentucky, Lexington, Kentucky 40506-0046; and Associate Professor of Civil Engineering, College of Engineering, University of South Carolina, Columbia, South Carolina 29208.

SUMMARY

A deterministic model to simulate stormwater runoff from impervious surfaces is presented. The surface runoff model is based on a one dimensional, variable width, kinetmatic wave approximation. Resistance to flow is described with the Manning equation for turbulent flow.

A Galerkin finite element formulation utilizing linear, quadratic and cubic Lagrangian elements is presented for the spatial discretization of the kinematic runoff equations. The temporal discretization is based on a linear time variation which includes both explicit and implicit time marching algorithms, depending on the value of the time weighting parameter. The implicit algorithm results in a system of nonlinear equations which are solved using a modified Newton-Raphson iteration scheme. Initial estimates to the solution are generated with a linear finite difference procedure.

A method to estimate a suitable computational time step size based on the Courant condition is presented. Simulation results are compared with the experimental data of Izzard (1946) and conclusions based on these results are presented.

INTRODUCTION

Stormwater simulation models are used to predict watershed hydrologic response to known rainfall patterns at ungauged watersheds and for events not contained in the record at gauged watersheds. These simulations provide basic data for such purposes as hydraulic structure design, impact studies, and the design of land use controls to minimize erosion and nonpoint source pollution. Suitable models must be capable of reliably estimating runoff for spatially variable watershed and rainfall conditions while requiring a minimum of calibration data.

Available models generally are classified as distributed and lumped parameter models. Distributed models, also identified as deterministic models in some references (e.g., Overton and Meadows, 1976), account for spatial variability in watershed characteristics and rainfall pattern. They have a theoretical structure based on physical laws and measures of initial and boundary conditions. When conditions are adequately described, the output from such models should be known with a high degree of certainty. On the other hand, lumped parameter models evaluate the response of an entire watershed as a single hydrologic unit. Model equations generally are descriptive only of a concept of the runoff process and not governing physical laws. As such, model parameters are not always defined as meaningful physical entities, and therefore, require calibration studies to establish suitable values. As a result, deterministic models are better applied in simulations when spatial variability is important.

Most deterministic models have been formulated using the kinematic wave approximation for channel and surface runoff routing. The governing partial differential equations have no general analytical solutions, therefore, model equations are either analytical solutions developed for simplified rainfall patterns and watershed geometries or finite difference solutions. Only recently have finite elements been applied (e.g., Jayawardena and White, 1977; Ross, et al., 1979). The finite element method is attractive since it is easier to conform to the physical geometry of a watershed and to use variable grid spacing and/or higher order elements for regions with rapid variations in flow depth.

The purpose of this paper is to present an implicit finite element formulation of the kinematic wave approximation for routing surface runoff. The uniqueness of the present formulation is that an explicit finite difference algorithm is utilized to predict the solution. The implicit finite element discretization is utilized to iteratively correct the solution via a modified Newton-Raphson procedure which requires no system matrix updates. This solution strategy has been found to be both accurate and efficient (Blandford, et al., 1983). The formulation details and sample results are presented in the following sections.

KINEMATIC WAVE APPROXIMATION

The governing equations for one dimensional kinematic overland flow are continuity and a simplification to the momentum equation obtained by neglecting the dynamic terms. This assumption has been shown to be valid for surface runoff from hillslopes, parking lots, etc., i.e., upland watersheds. Respectively, the two equations are written

$$w\frac{\partial h}{\partial t} + \frac{\partial Q}{\partial x} = wi_e \qquad (1)$$

and $$S_f = S_o \qquad (2)$$

where w is the variable place width, h is the flow depth, Q is the flowrate, i_e is the rainfall excess intensity, S_f is the friction slope, S_o is the slope of the surface runoff plane, x is the distance along the plane and t is the time.

Equation 2 suggests the friction slope can be approximated with a uniform flow formula, e.g., Manning's or Chezy's equation. For this study, Manning's equation for turbulent, uniform flow was used. Equation 2 thus becomes

$$Q = \frac{1.486}{n} \sqrt{S_o} \quad wh^{5/3} \qquad (3)$$

where n is Manning's n value.

FINITE ELEMENT FORMULATION

The finite element spatial discretization of Eq. 1 is obtained using the Galerkin weighted residual method

$$\int_{\Omega^e} \{N\}(w \frac{\partial h}{\partial t} + \frac{\partial Q(h)}{\partial x} - wi_e)d\Omega = \{0\} \qquad (4)$$

where {N} is the column vector of element shape functions, e.g., the cubic Lagrangian element of Figure 1, and Ω^e is the element domain. Substituting the spatial approximations

$$w = \lfloor N \rfloor \{w\}$$

$$\dot{h} \equiv \frac{dh}{dt} = \lfloor N \rfloor \{\dot{h}\}$$

$$Q(h) = \lfloor N \rfloor \{Q(h)\}$$

$$i_e = \lfloor N \rfloor \{i_e\}$$

into Eq. 4 gives

$$\int_{\Omega^e} \{N\}(\lfloor N \rfloor\{w\}) \lfloor N \rfloor \, d\Omega \, \{\dot{h}\} + \int_{\Omega^e} \{N\} \frac{d\lfloor N \rfloor}{dx} d\Omega \, \{Q(h)\} = \int_{\Omega^e} \{N\}(\lfloor N \rfloor\{w\}) \lfloor N \rfloor d\Omega \, \{i_e\} \qquad (5)$$

which is written in general matrix form as

$$[a] \, \{\dot{h}\} + [b]\{Q(h)\} = [a] \, \{\hat{i}_e\} \qquad (6)$$

Due to the inclusion of a variable width in the coefficient matrix [a], the coefficients of [a] are evaluated numerically using Gaussian quadrature. However, [b] does not contain element dependent values; therefore, it was integrated analytically and included as a constant matrix.

Equation 6 shows that the kinematic equations require a temporal discretization in addition to the previously discussed spatial discretization. The time varying response is discretized using a single step numerical time integration scheme. The scheme is based on a linear time variation, as shown in Figure 2. Evaluating Eq. 6 at time $t+\theta\Delta t$ gives

$$[a]\,\{\dot{h}\}_{t+\theta\Delta t} + [b]\,\{Q(h)\}_{t+\theta\Delta t} = \{i_e\}_{t+\theta\Delta t} \quad (7)$$

Using the linear time variation of Figure 2, the various terms in Eq. 7 are approximated as

$$\begin{aligned} \{\dot{h}\}_{t+\theta\Delta t} &= (\{h\}_{t+\Delta t} - \{h\}_t)/\Delta t \\ \{Q(h)\}_{t+\theta\Delta t} &= \theta\{Q(h)\}_{t+\Delta t} + (1-\theta)\{Q(h)\}_t \\ \{\hat{i}_e\}_{t+\Delta t} &= \theta\{\hat{i}_e\}_{t+ t} + (1-\theta)\{\hat{i}_e\}_t \end{aligned} \quad (8)$$

Substituting these relationships into Eq. 7 yields

$$\begin{aligned} [a]\{h\}_{t+\Delta t} = {} & \Delta t\theta\,\{\hat{i}_e\}_{t+\Delta t} + \Delta t(1-\theta)\,\{\hat{i}_e\}_t + [a]\{h\}_t \\ & - \Delta t(1-\theta)\,[b]\,\{Q(h)\}_t + \Delta t\theta\,[b]\,\{Q(h)\}_{t+\Delta t} \end{aligned} \quad (9)$$

Equation 9 is the element level recurrence equation. The time weighting coefficient, θ, can be varied from 0 to 1. Due to the occurrence of $\{Q(h)\}_{t+\Delta t}$ on the right hand side, a nonzero value of θ results in a nonlinear algebraic system of equations. Hughes (1977) has shown that for $1/2 \leq \theta \leq 1$, the time integration scheme is unconditionally stable for nonlinear problems. However, choosing $\theta = 0$ (Euler forward difference) results in a linear system of algebraic equations. Due to the conditional stability associated with $\theta = 0$, the only time weighting coefficients considered were $1/2 \leq \theta \leq 1$.

Once the element matrix equations, Eq. 9, are generated, they are assembled to represent the total solution domain. The assembly procedure is designed to ensure continuity of flow depth, h, and flowrate, Q, at the element junctions. The assembly is executed in a manner consistent with the symmetric profile solution technique of Taylor (1977). Furthermore, the upstream boundary condition on flow depth is included in the resulting equations.

Equation 9 is represented in global assembled form as

$$[A]\{h\}_{t+\Delta t} = \{C\}_t - \Delta t\theta\,[B]\{Q(h)\}_{t+\Delta t} \quad (10)$$

where

$$[A] = \sum_e [a]$$

$$[B] = \sum_e [b]$$

$$\{C\}_t = \sum_e (\Delta t\theta\{\hat{i}_e\}_{t+\Delta t} + \Delta t(1-\theta)\{\hat{i}_e\}_t) + [A]\{h\}_t$$
$$- \Delta t(1-\theta)[B]\{Q(h)\}_t$$

$\sum_e$ represents the summation over the elements. Note that $\{C\}_t$ on the right hand side of Eq. 10 is constant for a particular time step.

NONLINEAR SOLUTION STRATEGY

The implicit form of Eq. 10, $1/2 \leq \theta \leq 1$, requires an initial estimate for the vector $\{h\}_{t+\Delta t}$ and an iteration strategy to generate an equilibrated solution. A modified Newton-Raphson procedure is used.

Consistent with the modified Newton-Raphson solution scheme, Eq. 10 is written in recursive form as

$$[A]\{h\}^{i+1}_{t+\Delta t} + \{C\}_t - \Delta t\theta\ [B]\{Q(h)^{i}_{t+\Delta t}\} \tag{11}$$

where the subscripts i and i+1 represent (successive) estimates of $\{h\}_{t+\Delta t}$. The first estimate for the nodal flow depth values is obtained with an explicit finite difference approximation of Eq. 1 (see grid point 2 in Figure 3). Using forward in time and backward in space finite difference quotients, the following algebraic equation for the flow depths at the advanced time step is obtained

$$h_4 = h_2 + \Delta t\ i_{e_2} - \frac{\Delta t}{w_2}\frac{Q_2 - Q_1}{\Delta x} \tag{12}$$

Using the current time level boundary conditions and the depths and flowrates at the previous time level, this equation is solved for the initial flow depth vector $\{h\}^1_{t+\Delta t}$. These initial flow depths are then used with Manning's equation, Eq. 3, to compute the initial flowrate vector. Equation 11 is then solved for the new vector of flow depths, $\{h\}^2_{t+\Delta t}$. For subsequent iterations, the computed values of $\{h\}_{t+\Delta t}$ are used to update the flowrate vector of Eq. 11.

The iteration procedure is continued until the convergence criterion is satisfied, i.e.,

$$\left| \frac{h^{i+1}_{t+\Delta t} - h^{i}_{t+\Delta t}}{h^{i+1}_{t+\Delta t}} \right| < \varepsilon \tag{13}$$

where ε is the prescribed relative error (0.001) for each nondesignated flow depth node. Computations stop if the maxi-

conditions for an unsteady rainfall. Since a nonequilibrium peak rate cannot be estimated confidently, the velocity is computed for equilibrium conditions for the maximum rainfall intensity. Additionally, the depth for this equilibrium condition is determined.

With the velocity and flow depth approximations calculated, the kinematic flow criteria can be checked, i.e.,

$$Fr = v/\sqrt{gh} \tag{18}$$

$$k = LS_o g/v^2 \tag{19}$$

where Fr is the Froude number, k is the kinematic flow number developed by Woolhiser and Liggett (1967), g is the gravitational acceleration and L is the length of the overland flow plane.

The calculated time increment from Eq. 17 represents the maximum time increment which should be used. However, using $\Delta t = \Delta t_c$ may not produce accurate results with both explicit and implicit schemes Viessman, et al. (1977) state that for the explicit time integration scheme ($\theta = 0$), the best results are obtained with a time increment approximately 20 percent of that defined by the Courant condition. During this study, it was determined that $\Delta t \simeq \Delta t_c$ for the implicit schemes results in sufficiently accurate simulations.

Temporal Rainfall Distribution

Event rainfall data typically is reported at regular time intervals, e.g., every 5 minutes, or at breakpoint intervals. Regardless, during analysis the data is displayed as a hyetograph such as shown by the solid lines in Figure 5. Such a discontinuous loading function can cause convergence problems for numerical solution techniques including the finite element method. Therefore, to eliminate abrupt discontinuities in the rainfall intensity, changes are linearly transitioned over two time steps as shown by the dashed lines.

EXAMPLE SIMULATIONS

To verify the model and to demonstrate its capabilities, simulations were performed and the results compared with published experimental data. Typical results obtained for the Izzard data (Izzard, 1946) are presented.

Izzard's Run No. 136

Run 136 was a two burst event with an intermediate lull period. A steady rainfall intensity of 3.56 inches/hour was applied for ten minutes, followed by a two minute lull. The same intensity then was applied for an additional eleven minutes. The event has a maximum Froude number of 0.55 and a minimum kinematic flow number of 156; therefore, the kinematic runoff model is applicable.

mum number of iterations (25) is exceeded before convergence. A schematic of the iteration procedure is shown in Figure 4.

GENERAL SOLUTION PROCEDURE

In addition to the solution of the recurrence scheme, certain details must be considered when solving an overland flow problem. The following subsections describe the technique used to select the computational time step increment and the assumed temporal rainfall distribution.

<u>Determination of the Time Increment</u>

Selection of a proper computational time increment, Δt, is essential to an efficient, accurate solution. If the time increment is too large, many iterations may be required during each time step to achieve convergence; and an inaccurate solution may be obtained. Overly small time increments require a large number of computations and therefore are costly.

The time steps are chosen to satisfy the Courant condition. This criterion is based on the characteristic roots of the governing partial differential equations to ensure that the node point at the advanced time level falls within the domain of determinacy bounded by the characteristics and the initial conditions. The kinematic overland flow model has only a forward charcteristic, i.e., the wave form (physical disturbance) propagates downstream with no backwater effects. The characteristic root is

$$\frac{dx}{dt} = c \tag{14}$$

where c is kinematic wave celerity and is determined as

$$c = \frac{d(Q/w)}{dt} \tag{15}$$

For Manning's resistance, this becomes

$$c = \frac{5}{3} v \tag{16}$$

where v is the velocity of flow. By combining Eqs. 14 and 16, the Courant condition is determined as

$$\Delta t_c \leq 0.60 \frac{\Delta x}{v} \tag{17}$$

where Δt_c is the maximum time increment and Δx is the smallest distance between consecutive nodes in a finite element mesh.

Equation 17 shows that the maximum velocity must be estimated to obtain apriori estimate of Δt_c. This occurs during equilibrium runoff for a steady rainfall or peak runoff

Using a single cubic Lagrangian finite element to discretize the plane resulted in a $\Delta t_c = 37$ seconds; $\Delta t = 30$ seconds was used. The computed and observed hydrographs are shown in Figure 6 for a variety of time weighting coefficients. The Crank-Nicolson and Galerkin time integration shemes yielded the most accurate solutions. The Crank-Nicolson scheme exhibited a slight oscillation about the equilibrium solution.

The failure of the model to accurately reproduce the recession during the lull and following the termination of rainfall is attributed in part to using only a turbulent flow equation to approximate the friction slope. Previous research (Overton and Meadows, 1976) has indicated flow in the recession limb is better approximated with a liminar resistance law. Basically, for a given depth of flow, a turbulent flow equation predicts higher discharge than does a laminar flow equation.

<u>Izzard's Run No. 138</u>

Run No. 138 consists of two steady intensities; 1.83 in/hr for the first eight minutes and 3.55 in/hr for the next eight minutes. This simulated rainfall event yields approximately the same maximum Froude number and minimum kinematic flow number as Run No. 136. A single cubic Lagrangian element and $\Delta t = 30$ seconds was used.

The computed and observed hydrographs are shown in Figure 7. Again, several weighting coefficients were used. The results show excellent agreement between the computed and observed hydrographs, particularly for the Crank-Nicolson and Galerkin schemes. Note the Crank-Nicolson scheme again oscillated about equilibrium.

CONCLUSIONS

The explicit finite difference-implicit finite element predictor-corrector pair was shown to yield excellent surface runoff results for the unsteady events considered. Results presented by Blandford, et al. (1983) show the accuracy/efficient of the soultion strategy for pervious as well as impervious surface runoff simulations. An improved formulation may be obtained by including laminar and transition resistance laws for the recession limb(s) of the outflow hydrograph.

ACKNOWLEDGEMENTS

The research reported in this paper was supported by the United States Department of the Interior, Office of Surface Mining (Grant No. G5115213) and the Office of Water Research and Technology (Project No. B-069-KY). The authors wish to thank Mr. J. Patrick Reinert and Mr. Nathaniel Peters, II for their assistance with the numerical calculations and Ms. Elaine Powell for typing the paper.

REFERENCES

Blandford, G.E., Peters, N., II and Meadows, M.E. (1983), Development of Models for Simulating Stormwater Runoff from Surface Coal Mined Lands - Finite Element Simulation of Overland Flow, Vol. II, U.S. Department of the Interior, Office of Surface Mining, Report No. G511213, 161 pp.

Hughes, T.J.R. (1977), "Unconditionally Stable Algorithms for Nonlinear Heat Conduction," Comp. Meth. Mech. Engrg., Vol. 10, pp. 135-139.

Izzard, C.F. (1946), "Hydraulics of Runoff from Developed Surfaces," Highway Res. Board, Proceedings 26th Annual Meeting, pp. 129-150.

Jayawardena, A.W. and White, J.K. (1977), "A Finite Element Distributed Catchment Model, I. Analytical Basis," J. of Hydrology, Vol. 42, pp. 231-249.

Overton, D.E. and Meadows, M.E. (1976), Stormwater Modeling, Academic Press, New York.

Ross, B.B., Contractor, D.N. and Shanholtz, V. (1979), "A Finite Element Model of Overland and Channel Flow for Assessing the Hydrologic Impact of Land-Use Change," J. of Hydrology, Vol. 41, pp. 11-30.

Taylor, R.L. (1977), in O.C. Zienkiewicz, The Finite Element Method, Third Edition, McGraw-Hill, New York, Chapter 24.

Viessman, W. Jr., Knapp, J.W., Lewis, G.L. and Harbaugh, T.E. (1977), Introduction to Hydrology, Second Edition, Harper and Row, New York.

Woolhiser, D.A. and Liggett, J.A. (1967), "Unsteady One-Dimensional Flow Over a Plane - The Rising Hydrograph," Water Resources Res., Vol. 3, No. 3, pp. 753-771.

(a) Element

$$N_1 = 1 - 5.5\,\xi + 9\,\xi^2 - 4.5\,\xi^3, \quad \text{Where} \quad \xi = X/L$$

$$dN_1/dX = (-5.5 + 18\,\xi - 13.5\,\xi^2)/L$$

$$N_2 = 9\,\xi - 22.5\,\xi^2 + 13.5\,\xi^3$$

$$dN_2/dX = (9 - 45\,\xi + 40.5\,\xi^2)/L$$

$$N_3 = -4.5\,\xi + 18\,\xi - 13.5\,\xi^3$$

$$dN_3/dX = (-4.5 + 36\,\xi - 40.5\,\xi^2)/L$$

$$N_4 = \xi - 4.5\,\xi^2 + 4.5\,\xi^3$$

$$dN_4/dX = (1 - 9\,\xi + 13.5\,\xi^2)/L$$

(b) Shape Functions and Derivatives

Figure 1 - Cubic Lagrangian Element

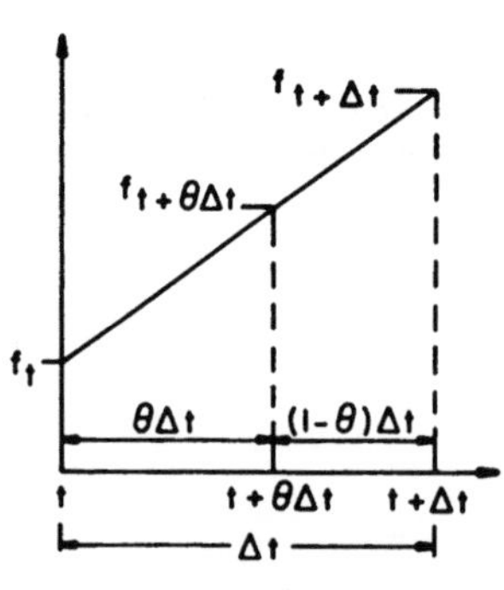

Figure 2 - Linear Time Interpolation

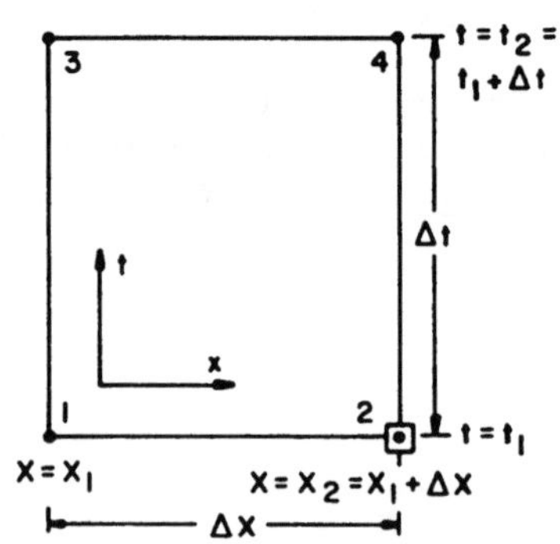

Figure 3 - Finite Difference Grid

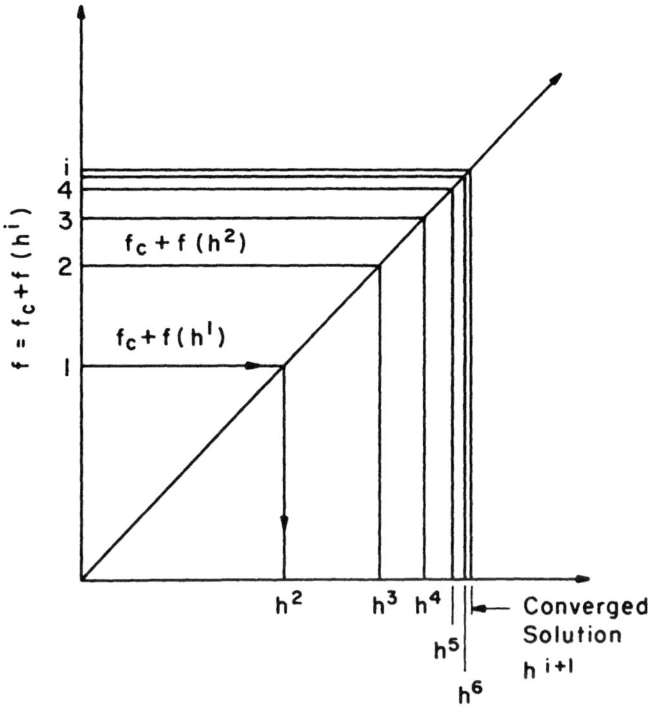

Figure 4 - Modified Newton - Raphson Iteration Schematic

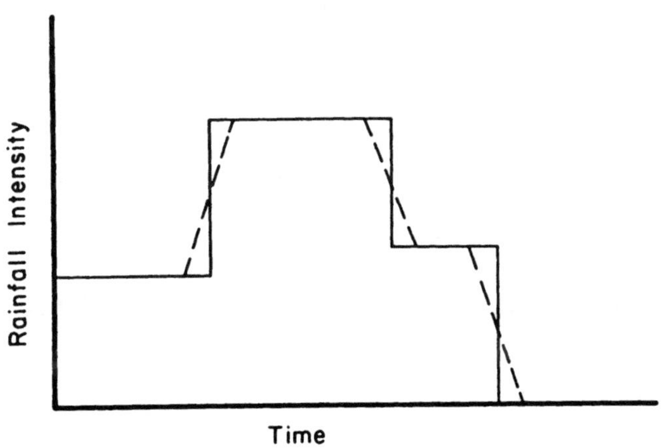

Figure 5 - Rainfall Hyetograph and Assumed Rainfall Distribution

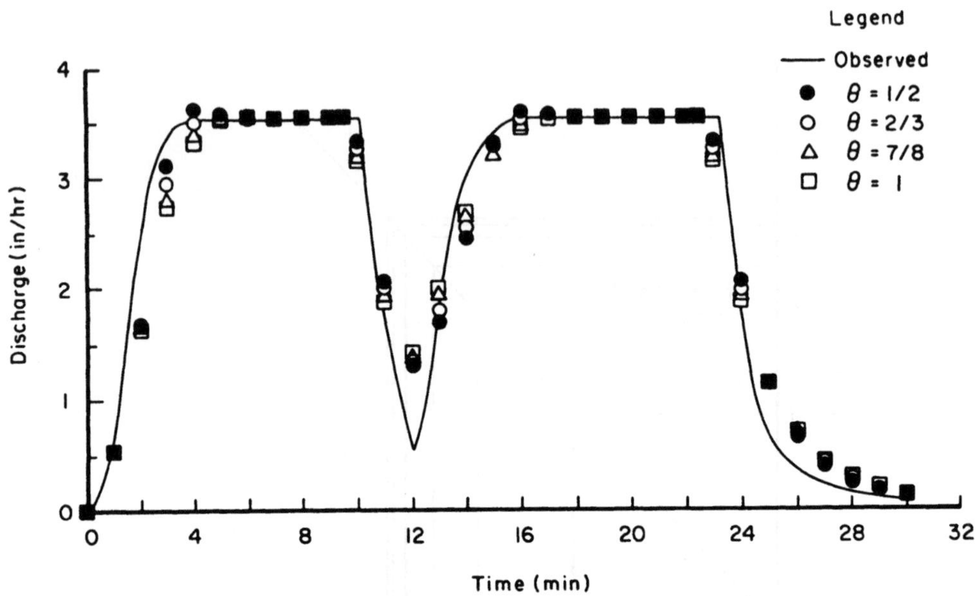

Figure 6 - Outflow Hydrograph for Izzards's Run #136

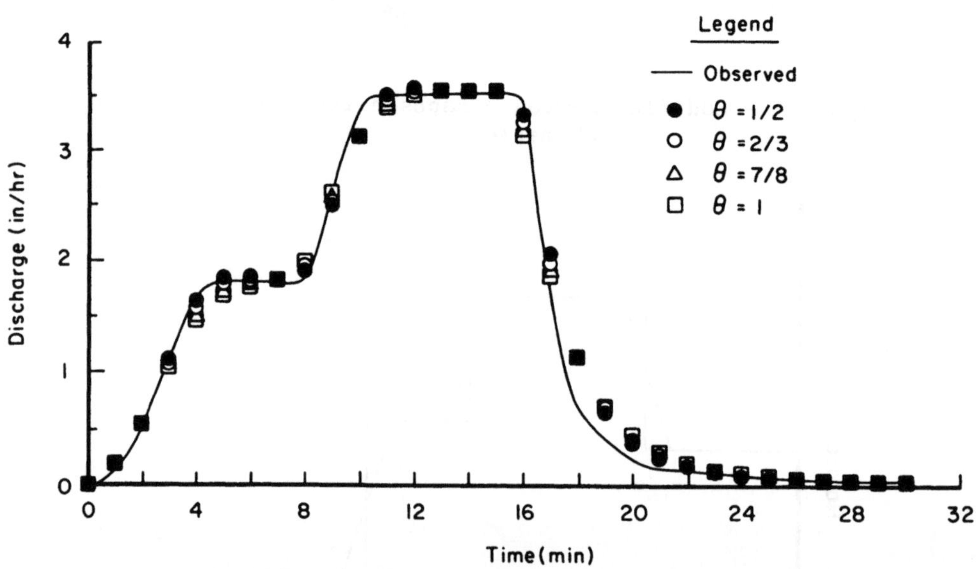

Figure 7 - Outflow Hydrograph for Izzard's Run #138

A FINITE ELEMENT-FINITE DIFFERENCE ALTERNATING DIRECTION ALGORITHM FOR THREE DIMENSIONAL GROUNDWATER TRANSPORT

D. K. Babu
and
G. F. Pinder

Department of Civil Engineering
Princeton University
Princeton, N.J. 08544

While technically feasible, three dimensional finite element groundwater transport simulation has not found widepsread application because of the considerable computational burden inherent in the approach. An operator splitting algorithm which treats the horizontal plane using finite elements in the first step and the vertical dimension using finite differences in the second step provides considerable savings in both computer memory requirements and computational effort. The algorithm reduces the problem of an (N x M) array of nodes from one in which NM equations must be solved simultaneously, to one of solving N equations M times in the first step and M equations N times in the second step. Applications to type problems and field situations indicate that the method is robust and accurate.

INTRODUCTION

The simulation of groundwater transport in three space dimensions using classical numerical approaches constitutes a burdensome computational problem. In this paper, a two-step finite element-finite difference procedure (FEFD) is presented as an efficient and accurate alternative to these standard techniques. The model described herein is based upon a splitting algorithm that combines a finite element formulation in the areal plane with a finite difference formulation in the vertical direction. Heuristic stability analysis indicates the scheme is unconditionally stable.

In the first two sections of this paper, the relevant numerical theory is outlined. In later sections, an example problem involving three dimensional transport is considered along with a heuristic error analysis.

THEORY

Chemical transport through groundwater flows is described by the following equations

$$\nabla\cdot(\bar{\bar{D}}\cdot\nabla c) - \vec{V}\cdot\nabla c + Q_W(c_W-c) = \frac{\partial}{\partial t}[(\theta+S)c] \quad (1)$$

where ∇ is the gradient operator (L^{-1}); $\bar{\bar{D}}$ is the matrix of velocity dependent dispersion coefficients (L^2/T); c is the dimensionless chemical concentration (L^0); θ is the porosity of the medium (L^0); S is the chemical adsorption (L^0); t is the time (T); $\vec{V}$ is the known flow vector (L/T); Q_W is the strength of a source but generally represents a pump in action (1/T); and, c_W is the dimensionless chemical concentration at this source (L^0).

To solve (1), the velocity field must be known. In an earlier paper (Babu et al., 1982), the FEFD algorithm was used to solve the groundwater flow equation. The required velocity field is obtainable, via Darcy's law, from this solution.

The components of $\bar{\bar{D}}$ and $\vec{V}$ are taken as

$$\bar{\bar{D}} \equiv \begin{bmatrix} D_{xx} & D_{xy} & D_{xz} \\ D_{yx} & D_{yy} & D_{yz} \\ D_{zx} & D_{zy} & D_{zz} \end{bmatrix} \quad \text{and} \quad \vec{V} \equiv (V_x, V_y, V_z) \quad (1a)$$

Also

$$D_{xx} = \alpha_L v + (\alpha_L-\alpha_T)V_xV_x/v + D_M$$

$$D_{xy} \equiv (\alpha_L-\alpha_T)V_xV_y/v = D_{yx}, \quad (1b)$$

with similar expressions for other coefficients $D_{yy}, D_{yz}\ldots$.

Here $\alpha_L \equiv$ Longitudinal dispersivity of medium (L)

$\alpha_T \equiv$ Transverse dispersivity of medium (L)

$D_M \equiv$ Molecular Diffusion Coefficient (L^2/T)

$v \equiv (V_x^2 + V_y^2 + V_z^2)^{1/2} \equiv |\vec{V}|$

As in classical finite-element theory, the first step in the formulation is the definition of the trial function for concentration. In this instance the appropriate representation is

$$c \equiv c(x,y,z,t) \sim \hat{c}(x,y,z,t) \equiv \sum_{j=1}^{N} c_j(z,t)\phi_j(x,y) \quad (2)$$

where (x,y) are the coordinates in the areal directions, and z is the vertical coordinate. The basis functions $\phi_j(x,y)$ are identical to those normally employed in two dimensional finite element methods. The Galerkin approximation is now taken as

$$\iint_R \left[\frac{\partial}{\partial t} (\theta c+Sc) - \nabla\cdot(\overline{\overline{D}}\cdot\nabla c) + \vec{V}\cdot\nabla c + Q_W\, c-c_W) \right] \phi_i (x,y)\,dxdy = 0$$

$$i = 1,2,\ldots N \qquad (3)$$

Substitution of (3) into (2), and an integration by parts using Green's theorem in the (x,y) region of R, yields

$$\iint_R \sum_{j=1}^{N} \Big[\frac{\partial}{\partial t} \{(\theta+S) c_j\}\phi_i\phi_j + (c_j)[D_{xx} \frac{\partial\phi_i}{\partial x} \frac{\partial\phi_j}{\partial x}$$

$$+ D_{xy} (\frac{\partial\phi_i}{\partial x} \frac{\partial\phi_j}{\partial y} + \frac{\partial\phi_j}{\partial x} \frac{\partial\phi_i}{\partial y}) + D_{yy} \frac{\partial\phi_i}{\partial y} \frac{\partial\phi_j}{\partial y}$$

$$+ (V_x \frac{\partial\phi_j}{\partial x} + V_y \frac{\partial\phi_j}{\partial y})\phi_i + Q_W\phi_i\phi_j]\Big]dxdy$$

$$= \int_\sigma D_n \frac{\partial\hat{c}}{\partial n} \phi_i d\sigma + \iint_R \sum_{j=1}^{N} [Q_W c_W \phi_i\phi_j]dxdy \qquad (4)$$

$$+ \iint_R \sum_{j=1}^{N} \Big[\frac{\partial}{\partial z} \{(D_{zx} \frac{\partial\phi_j}{\partial x} + D_{yz} \frac{\partial\phi_j}{\partial y})\phi_i c_j\}$$

$$- \{(D_{zx} \frac{\partial\phi_i}{\partial x} + D_{yz} \frac{\partial\phi_i}{\partial y})\phi_j \frac{\partial c_j}{\partial z} \} + \{ \frac{\partial}{\partial z} (D_{zz}\cdot \frac{\partial c_j}{\partial z})$$

$$- V_z \frac{\partial c_j}{\partial z} \}\phi_i\phi_j\Big]dxdy \qquad i = 1,2,\ldots N$$

Here σ is the boundary of the invariant cross section R, n is the outward normal direction on σ, and ($D_n\ \partial\hat{c}/\partial n$) represents the normal component of the dispersive flux across σ. The integrals in (4) may be associated with concentration increases due to the various hydrodynamic and dispersive phenomena in the areal directions, in the vertical directions, and across the boundaries of the flow domain.

A finite difference approximation involving M nodes is next introduced for the vertical components on the right hand side of (4). An approximately horizontal layer of the reservoir is associated with each of the M nodes. A finite element mesh is next placed through the mid sections of each of the M layers.

This finite element mesh, containing the same total of N nodes in each layer, is so structured that corresponding nodes in every layer are located along a single vertical line.

In eq. (4), typical finite difference formulae for the kth layer ($k = 1,2,\ldots M$) may be introduced as follows. Let the vector of nodal concentrations in the kth layer be

$$\{C\}_k \equiv (c_1(z,t),\ c_2(z,t),\ldots c_N(z,t))_k \tag{5a}$$

If the vertical flow component is from layer (k+1) towards layer (k), ($V_z<0$), then

$$\iint_R \frac{\partial}{\partial z}(D_{zx}c_j)_k \phi_i \frac{\partial \phi_j}{\partial x}\,dxdy$$

$$= \left[\iint_R \left(\frac{D_{zx}}{\Delta z}\right)_{k+1} \phi_i \frac{\partial \phi_j}{\partial x}\,dxdy\right](c_j)_{k+1}$$

$$- \left[\iint_R \left(\frac{D_{zx}}{\Delta z}\right)_k \phi_i \frac{\partial \phi_j}{\partial x}\,dxdy\right](c_j)_k \tag{5b}$$

Finally, introducing a central difference scheme for the second order partial z derivatives, and grouping terms of the type (5b), the right hand side of (4) may be written for the kth layer as:

$$L_k\{C\} \equiv [D]_{k-1}\{C\}_{k-1} - ([D]_{k-1} + [D]_{k+1})\{C\}_k + [D]_{k+1}\{C\}_{k+1} \tag{6}$$

In matrix notation, eq. (4) now reads as (subscript k refers to layer k)

$$[A]_k \left\{\frac{\partial C}{\partial t}\right\}_k + [B]_k\ \{C\}_k = \{F\}_k + \{Q_W c_W\}_k + L_k\{C\} \tag{7}$$

where the elements a_{ij}, b_{ij}, f_i and q_i of the matrices [A], [B], {F} and $\{Q_W c_W\}$ are the appropriate integrated values defined in (4), evaluated as shown in (5b), and grouped properly. The formulae (5b) are suitable for handling a layer whose thickness, $(\Delta z)_k$, changes in the (x,y) directions. The same formulae can also accommodate pronounced changes or discontinuities in the medium properties as exhibited in values of D_{xx}, D_{zx}, ... etc.. In such situations, harmonic mean values of the integrands are used to reflect the influence of such violent changes.

As a final step in the solution procedure, the matrices [A], and [D] in (6), are lump-diagonalized by the summations:

$$a_{ii} = \sum_{j=1}^{N} a_{ij}; \quad d_{ii} = \sum_{j=1}^{N} d_{ij}; \quad a_{ij} = d_{ij} = 0, \; i \neq j \tag{8}$$

$$i,j = 1,2,\ldots N$$

COMPUTATIONAL SCHEME

As a result of the procedure in (8), the matrices $[A]_k$ and $[D]_k$ are now in diagonal form. A two-step procedure for solving (7) follows. Introducing the notation,

$$\{C(x,y,z;t)\}^n \equiv \{C(x,y,z;\,n\Delta t)\},$$

equation (7) is expressed in two steps:

Step 1

$$\frac{[A]_k}{\Delta t}(\{C\}_k^* - \{C\}_k^n) + [B]_k\{C\}_k^* = \{F\}_k^* + \{Q_W c_W\}_k^* + L_k(\{C\}^n) \tag{9a}$$

Step 2

$$\frac{[A]_k}{\Delta t}(\{C\}_k^{n+1} - \{C\}_k^n) + [B]_k\{C\}_k^* = \{F\}_k^* + \{Q_W c_W\}^* + L_k(\{C\}^{n+1})$$

$$(k = 1,2,\ldots,M) \tag{9b}$$

Computational effort is greatly reduced if step (9b) is replaced by the difference between (9a) and (9b). Then,

$$\frac{[A]_k}{(\Delta t)}(\{C\}_k^{n+1} - \{C\}_k^n) - L_k(\{C\}^{n+1} - \{C\}^n) = \frac{[A]_k}{(\Delta t)}(\{C\}_k^* - \{C\}_k^n) \tag{9c}$$

Equations (9a) and (9c) constitute, respectively, step 1 and step 2 of the solution mechanism.

COMPUTATIONAL PROCEDURE

In the first step, equation (9a) is solved in each layer k of finite elements, assuming that concentration values $\{C\}^n$ at a previous time level are known for all layers. The result of solving (9a) is the *Intermediate Level Solution* $\{C\}_k^*$. In the second step, eq. (9c) is solved, using the intermediate level values $\{C\}_k^*$, and the previous time level values $\{C\}_k^n$. Remembering that matrices $[A]_k$ and $[D]_k$ are in diagonal form, the set (9c) constitutes a tridiagonal system of equations, for every ith component $(c_i(z,t))_k^{n+1}$ of the vector $\{C\}_k^{n+1}$. Thus the solution of (9c) requires a trivial amount of computational effort.

The procedure thus reduces the three dimensional problem of solving (NxM) equations simultaneously (N equations for each of M layers) - to one of first solving N equations M times, and next solving (with trivial numerical effort) M equations N times.

EXAMPLE PROBLEM

Results of the procedure outlined above were tested against a problem amenable to analytical solution. For this comparison, a steady state, radially convergent flow in a three dimensional spherical annulus was specified (see Fig. 1). The appropriate governing equation in spherical coordinates is

$$\frac{1}{r^2}\frac{\partial}{\partial r}\left(r^2 D_r \frac{\partial c}{\partial r}\right) + V_r \frac{\partial c}{\partial r} = \theta \frac{\partial c}{\partial t}, \quad a < r < b,\ t > 0 \tag{10a}$$

where

$$\begin{aligned}
c &= \text{concentration} && (L^0)\\
V_r &\equiv \text{radial velocity} = \frac{q_o}{r^2} && (L/T)\\
q_o &\equiv \text{volumetric flow rate} && (L^3/T)\\
D_r &\equiv \text{Dispersion coefficient} = \alpha_L V_r && (L^2/T)\\
\alpha_L &\equiv \text{Longitudinal Dispersivity } (L)\\
\alpha_T &\equiv \text{Transverse Dispersivity} \equiv 0 \quad (L)\\
\theta &\equiv \text{porosity of medium } (L^0)
\end{aligned} \tag{10b}$$

Initial and boundary conditions are taken as

$$c = 0,\ t = 0,\ a < r < b;\ c = 1,\ t > 0,\ r = b;\ \frac{\partial c}{\partial r} = 0,$$

$$t > 0,\ r = a \tag{10c}$$

In terms of nondimensional length $\eta \equiv (r/\alpha_L)$, and time $\tau \equiv (q_o t/\theta\alpha_L^3)$, equations (10a) and (10b) reduce to

$$\frac{\partial^2 c}{\partial \eta^2} + \frac{\partial c}{\partial \eta} = \eta^2 \frac{\partial c}{\partial \tau} \qquad \eta_i < \eta < \eta_o,\quad \tau > 0$$

$$\text{along with } c = 0,\ \tau = 0;\ c = 1,\ \eta = \eta_o;\ \text{and } \frac{\partial c}{\partial \eta} = 0,\ \eta = \eta_i \tag{11}$$

Using Laplace Transforms, it is possible to derive analytical solutions to (11) in terms of the Parabolic Cylinder functions. Alternatively, solutions of (11) that possess a high degree of accuracy, also may be generated utilizing finite difference schemes. For the problem considered here, a fully implicit backward finite difference algorithm was employed to solve (11). The finite difference solution thus derived was insensitive to further refinements of mesh and was therefore considered as accurate. This solution was employed as a reference solution in the work that follows.

The domain of analysis is indicated in Fig. (1). It consists of a right prism with vertical sides, a horizontal trapezoidal base, and an inclined trapezoidal top (slope, 1:5). Edges BC, AD, A'D' and B'C' are segments of divergent radii of the sphere centered at 0.

In the finite element formulation of the problem, the prism is divided into five layers in radial directions. The thickness of each layer increases linearly with distance from the origin. A total of 70 nodes and 34 elements (of equal width) constitute the finite element mesh placed in each of the five layers. Fig. (2) shows some body-centered nodes and linear quadrilateral elements in a typical layer. Both length and time variables have been made dimensionless. The following dimensionless parameter values were specified in the finite element equations (1a) and (1b). $\alpha_T \equiv 0$, $S \equiv 0$; $\alpha_L = q_o = \theta \equiv 1.0$. Lastly, the hydraulic head $h(x,y,z)$ was specified as $h = (h_o - 1/r)$, whereafter the computer code generated the flow vector: $\vec{V} \equiv -\nabla(1/r)$.

On the outer lateral face CDD'C', concentration values $C^*(x,y,z;\ t+\Delta t/2)$ computed as solutions to (11) at the 10 nodal locations (Nodes 69 and 70 in each of 5 layers), were specified as time dependent Dirichlet data. On each of the remaining faces, the condition $\partial c/\partial n = 0$ was specified, with the exception of the sloping top on which $\partial c/\partial z = 0$ was specified. Initial values of $c = 0$ were assumed throughout.

The following non-standard features were incorporated into the finite element formulation

1. only linear quadrilateral elements were employed.
2. analytical integration replaced numerical integration in the formulation of the coefficient matrices.
3. element averaged velocities and dispersion coefficients were employed.
4. an upstream weighting procedure was adopted for the convection terms in the transport equation

The solutions obtained using the highly accurate one-dimensional finite difference and three dimensional finite element formulations are presented in Fig. 3. Figure 3a and 3b shows the early-time solutions for each method in layers one and five respectively. Figure 3c shows a late time solution comparison for layer 3.

CONCLUSIONS

The transport equation in three space dimensions can be accurately solved for spherical flow using a FEFD formulation. The FEFD approach, based upon a horizontal-vertical splitting algorithm is a computationally efficient alternative to classical numerical schemes for those analysts who prefer the flexibility of a finite element formulation.

ACKNOWLEDGEMENTS

The authors wish to acknowledge the financial support of the Industrial Support Group and the Department of Energy through grant DE-AC02-83ER60170.

REFERENCES

Babu, D. K., G. F. Pinder and D. K. Sunada, "A Three Dimensional Hybrid Finite Element-Finite Difference Scheme for Groundwater Simulation", Proc. 10th IMACS World Congress on System Simulation and Scientific Computing, p. 292-294, 1982.

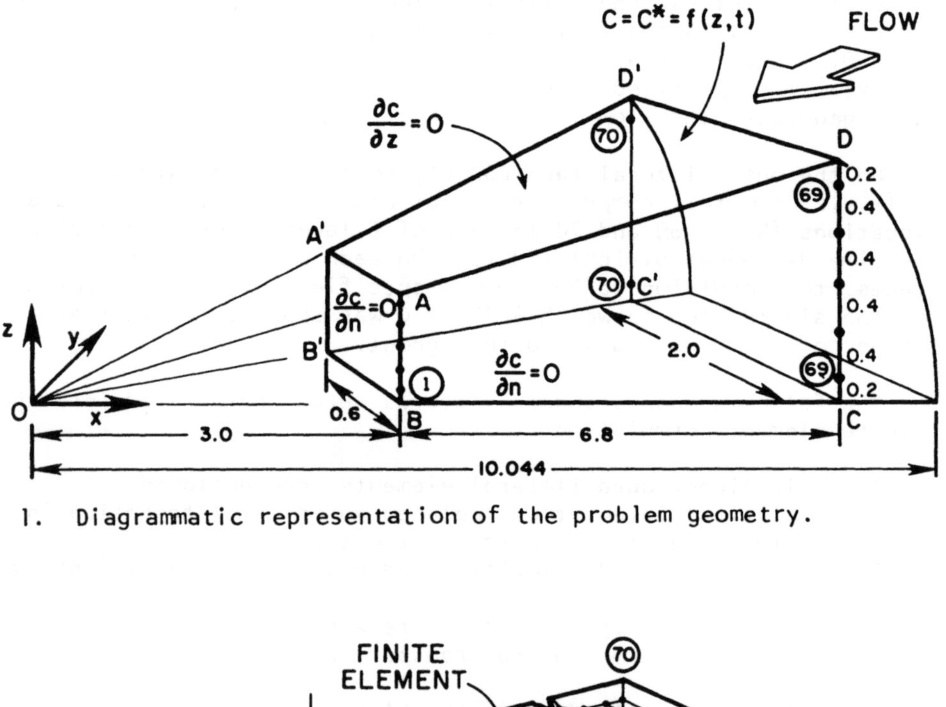

Fig. 1. Diagrammatic representation of the problem geometry.

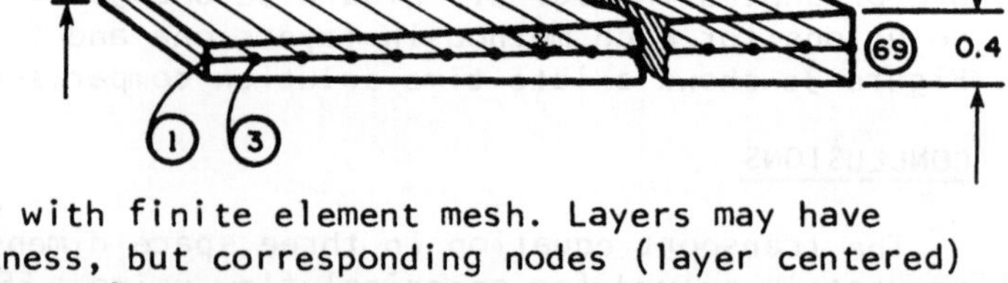

Fig. 2. Typical layer with finite element mesh. Layers may have varying thickness, but corresponding nodes (layer centered) in all layers are located along a vertical line.

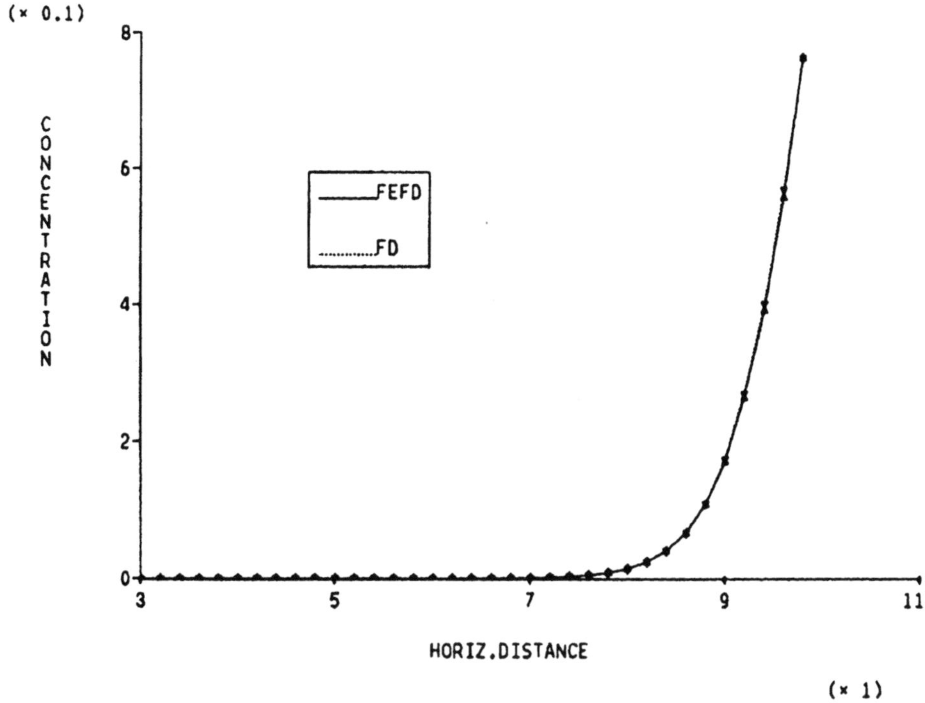

Fig. 3a. Comparison between finite element-finite-difference (FEFD——) and finite difference (FD----) solutions at dimensionless short time τ = 16.0, $\Delta\tau$ = 2.0, for layer 1 (bottom).

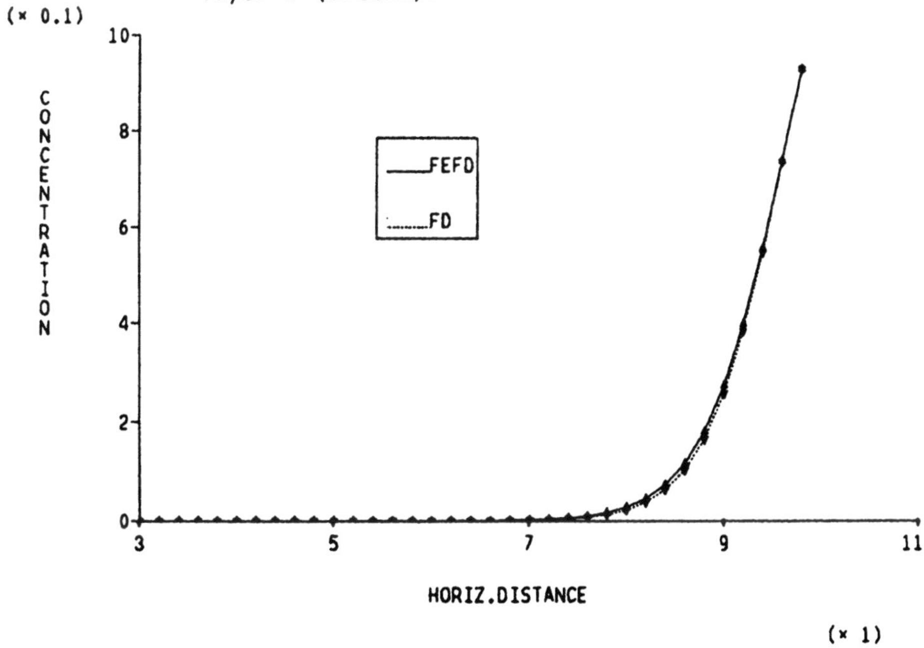

Fig. 3b: Same as Fig. 3a, for τ = 16.0, $\Delta\tau$ = 2.0, for layer 5 (top).

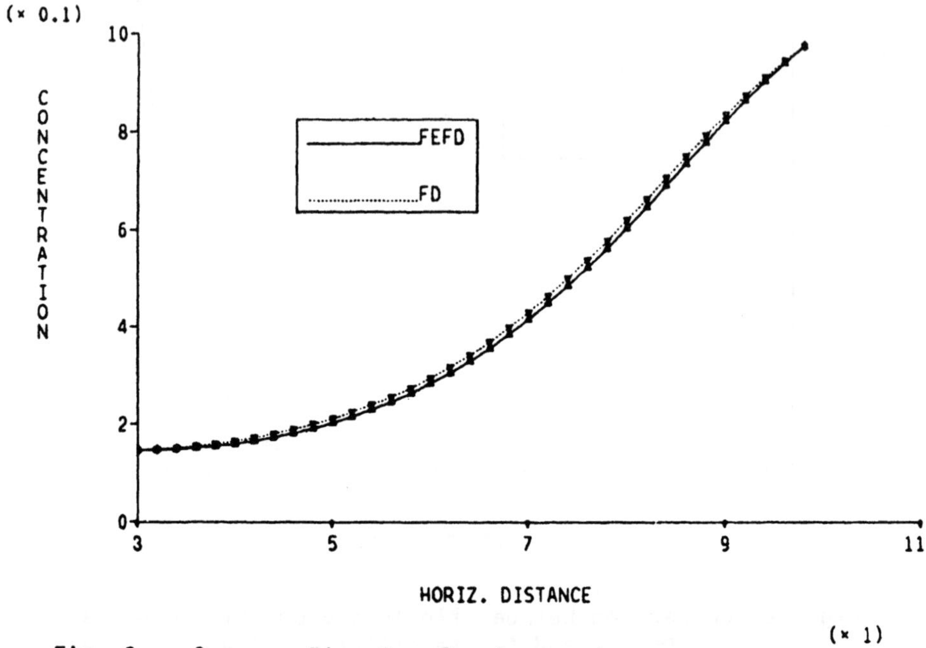

Fig. 3c. Same as Fig. 3a, for large time $\tau = 125.0$, $\Delta\tau = 5.0$, and for middle layer 3.

MODELLING GROUND-WATER FLOW WITH THE GLOBAL AND FINITE ELEMENT METHODS

J.F. Botha and G.N. Bakkes

Institute for Ground-water Studies
University of the Orange Free State, Bloemfontein

INTRODUCTION

The two numerical methods most frequently used to solve differential equations arising in science and engineering are perhaps the finite difference (FDM) and finite element (FEM) methods. The basic philosophy behind both these methods is to replace the true solution by an approximate one constructed from a suitable set of basis functions. In the case of the FEM the basis functions are usually selected from the set of piecewise continuous polynomials, while the FDM is based on interpolation polynomials [Botha and Pinder (1983)]. This reduces the problem to the simple operation of solving a set of equations

$$F(\underline{c}) = 0$$

for the expansion coefficients $\underline{c}$ in the approximating series.

A major disadvantage of both the methods is that, due to the well-known Runge-Meray phenomenon [Stroud (1974)], only low order polynomials can be used. To improve their accuracy, the number of polynomial pieces, i.e. the number of expansion coefficients has to be increased. Such a refinement requires that the domain of interest be discretised anew, a process that can be quite time-consuming, in particular when studying the convergence rate of these methods.

There exist methods whose accuracy can be improved without rediscretising the domain of interest. Such methods use elementary functions that have global support, hence the name global methods.

Global methods have in general some serious limitations, the best known of which is the difficulty to apply them on domains with irregular boundaries. This can, however, be circumvented by using transfinite interpolation [Gordon and Hall (1973)]. Another difficulty is that it is sometimes advantageous to discretise the domain in a number of subdomains. As all the presently known global methods are based on either a variational principle or the method or the method of weighted residuals [Finlayson (1972)], this may prove not to be easy.

The reason for this is not so much connected with the methods themselves, but rather with the restrictions placed on the admissible basis functions. This aspect will be dealt with more fully in the next section where it will be shown how these restrictions can be removed. The paper will then proceed with the global variational method recently introduced by Delves and his co-workers [Delves and Hall (1979), Delves and Freeman (1981)], for elliptic type problems. It will then be shown how the method can be extended to multiple domains and parabolic equations. The method will be finally applied to a few problems from the field of geohydrology and the results compared with those obtained from the FEM.

THE RAYLEIGH-RITZ AND GLOBAL VARIATIONAL METHODS

The Rayleigh-Ritz method (RRM) is, as is well-known, particularly useful for solving elliptic type problems [Oden and Reddy (1976)] such as

$$Au = -\nabla[p\nabla u] + qu = f \qquad 1$$

defined over a closed domain Ω with boundary $\partial\Omega$ where ∇ is the usual gradient operator. Let C_1 and C_2 be two non-overlapping segments of this boundary and assume that Equation 1 is subject to the Dirichlet boundary condition

$$u(\underline{x}) = g(\underline{x}) \qquad \underline{x} \in C_1 \qquad 2$$

and the mixed boundary condition

$$p\nabla u \,.\, \underline{n} = b(\underline{x})u(\underline{x}) + h(\underline{x})\,. \qquad \underline{x} \in C_2 \qquad 3$$

with $\underline{n}$ the unit normal vector on C_2. If the boundary C_1 is sufficiently large, this problem has a unique solution [Delves and Hall (1979)]

$$u(\underline{x}) \in H^1(\Omega) \qquad 4$$

where H^1 denotes the Hilbert space with first derivatives continuous.

To solve this problem by the RRM, one first selects a set of suitable basis functions $\{v_i\}_0^n$, approximates the true solution u by a trial function

$$w = \sum_{i=0}^{n} c_i v_i \qquad 5$$

and then minimizes the functional

$$F[w] = (Aw,w) - 2(f,w)_2$$

where

$$(Aw,w) = \int_\Omega Awwd\Omega \quad \text{and} \quad (f,w)_2 = \int_\Omega fwd\Omega$$

This is equivalent to seeking the function w such that [Oden and Reddy (1976)]

$$(Aw,w) = (f,w)_2 \qquad 6$$

In the case of the operator A defined in Equation 1, application of Green's theorem to the left-hand side of Equation 6 yields

$$\int [p \nabla w \nabla v_i + qwv_i] \, d\Omega = \int_{\Omega} fv_i d\Omega + \int_{\partial\Omega} v_i p \nabla w \cdot \underline{n} ds \qquad 7$$

If ∇w in the line integral on the right-hand side of Equation 7 is known, then this equation represents a system of linear equations for the coefficients c_i of Equation 5.

This is clearly the case for the mixed boundary condition where ∇w can simply be replaced by its equivalent expression from Equation 3. The set of basis functions used with mixed boundary values thus need not satisfy any constraints. This is not possible in the case of a Dirichlet condition.

The method most often employed with a Dirichlet condition is to replace the trial function in Equation 4, by another one of the form

$$w = g + \sum_{i=1}^{n} c_i v_i$$

where the v_i now have to satisfy homogeneous Dirichlet boundary conditions

$$v_i = 0 \text{ on } C_1.$$

The line integral vanishes in this case and Equation 6 reduces again to a system of linear equations with

$$F_i = \int_{\Omega} fv_i \, d\Omega - \int_{\Omega} [p(\nabla g \nabla v_i) + qgv_i] d\Omega$$

This constraint on the basis functions is rather restricting. It prohibits, for example, the use of many useful elementary functions, e.g. Chebyshev polynomials, as basis functions. Moreover, it does not allow the division of the domain into subdomains, because to impose the continuity condtion, Equation 4, would require a non-homogeneous boundary condition. There exists, however, an alternative approach.

It is well-known that Equation 1 is the Euler-Lagrange equation that minimizes the functional in Equation 6. What is less known, is that it also yields a stationary value for the functional

$$F^*[w] = F[w] + \int_{\Omega} \nabla \cdot \underline{G} d\Omega \qquad 8$$

or after applying Gauss' theorem to the integral containing $\underline{G}$

$$F^*[w] = F[w] + \int_{\partial\Omega} \underline{G} \cdot \underline{n} ds \qquad 9$$

where $\underline{G}$ is an arbitrary function of w and its gradient ∇w [Rund (1973)].

Although Equation 1 does not minimize Equation 8 in general, it can be shown [Rund (1973)] that, if $\underline{G}$ is chosen in such a way that it satisfies the transversality conditions [Rund (1973)] associated with Equation 8, then Equation 1 will in fact minimize Equation 8 [Rund (1973), Snyman (1982)]. Hence, if $\underset{\sim}{G}$ is chosen to satisfy the boundary conditions of Equation 1 and at the same time the necessary transversality conditions, Equation 9 will be a new variational functional (in the sense of the RRM) for the solution of Equation 1, but one for which the basis functions need not satisfy any boundary conditions. This method is known as the Global Variational Method (GVM) [Delves and Hall (1979)].

The best way to illustrate the GVM is to apply it to Equation 1. The transversality conditions for Equation 9 is, in general, of the form [Snyman (1982), Bakkes (1984)]

$$(\nabla^* L + D_w \underset{\sim}{G}) \cdot \underset{\sim}{n} = 0$$
$$\nabla^* G_j \cdot \underset{\sim}{n} = 0 \qquad (j = 1,2,3) \qquad 10$$

where

$$L = (p \nabla w \nabla w + qww) - 2fw$$

is the Lagrangian function corresponding to Equation 1 and ∇^* denotes the gradient operator with ∇w as independent variable.

Consider now the function

$$\underset{\sim}{G} = a_1 w^2 + a_2 w$$

defined over the segment C_1, where a_k $(k = 1,2)$ are constants to be determined. The first transversality condition in Equation 10 yields in this case

$$[(2p + a_1)w] \cdot \underset{\sim}{n} = 0$$

and the second

$$(a_1 w + a_2) = 0$$

From these equations follow that

$$a_1 = -2p \text{ and } a_2 = 2pw \qquad 11$$

There will be consistency between Equation 2 and Equation 11 on the boundary C_1 if the second condition in Equation 11 is replaced by

$$a_2 = 2pg$$

The function $\underset{\sim}{G}$ to be used with the Dirichlet boundary condition in Equation 2 is thus completely determined

$$\underset{\sim}{G} = -2pw^2 + 2gpw.$$

Proceeding in a similar fashion, it can be shown [Bakkes (1984)] that the mixed boundary condition on segment C_2 will be satisfied, provided that the function $\underset{\sim}{G}$ is now chosen as

$$\underline{G} = -(bw^2 + 2hw)\underline{n}$$

A variational functional 8, suitable for Equation 1, and for which the basis functions need not satisfy any boundary conditions is thus

$$F^*[w] = \int_{\Omega} [p\nabla w \nabla w + qw^2 - 2fw]\,d\Omega + 2\int_{C_1} (g-w)p\nabla w \cdot \underline{n}ds - 2\int_{C_2} [(b/2)w^2 + hw]\,ds \qquad 12$$

This is a simplified form of the functional originally introduced by Delves and Hall (1979).

THE GLOBAL ELEMENT METHOD (GEM)

The extension of the GVM to domains divided into subdomains, henceforth also referred to as global elements, is straightforward. To show this, consider the domain Ω associated with Equation 1, but now divided into two subdomains Ω_1 and Ω_2 where ($k = 1, 2$ cyclically, i.e. $k + 1 = 1$ if $k + 1 > 2$) as in Figure 1. A solution of Equation 1 can only be accepted on physical grounds if it is continuous across the interface C. This will be true if it satisfies Equation 4, which implies

$$w_1 = w_2 \quad \text{and} \quad \nabla w_1 = \nabla w_2 \quad \text{on } C \qquad 13$$

The functional 12, valid for the k-th global element, can thus be expressed, using Equation 13, as

$$F_k^* = \int_{\Omega} [p(\nabla w_k)^2 + q(w_k)^2 - 2fw_k]\,d\Omega + 2\int_{C_1^k} (g - w_k)p\nabla w_k \cdot \underline{n}ds - 2\int_{C_2^k} [(b/2)(w_k)^2 + hw_k]\,ds - \int_C (w_k - w_{k+1})(p\nabla w_k \cdot \underline{n}_k - p\nabla w_{k+1} \cdot \underline{n}_{k+1})ds$$

where C_m^k denotes that part of the boundary C_m which belongs to the k-th global element. This approach is known as the global element method (GEM) [Delves and Hall (1979), Bakkes (1984)].

The GEM and the GVM was introduced above for elliptic type differential equations. However, it can be easily extended to parabolic problems. Take for example the parabolic problem

$$D_t u(\underline{x},t) = a\nabla^2 u(\underline{x},t) + f \qquad 14$$

where $(\underline{x},t) \in \Omega \otimes [0,T]$ with Ω the same domain as in Equation 1. The functions a and f can in general be functions of $\underline{x}$, t, u and ∇u. To simplify the notation, this dependence of a and f as well as that of u will be suppressed unless required for reasons of clarity.

The boundary conditions will again be taken to be of the form as defined in Equations 2 and 3, while the initial condition is assumed to be given by

$$u(\underline{x},0) = u_0(\underline{x}).$$

By using linear Lagrange polynomials in time and collocation, [Botha (1982)] was able to show that Equation 14 can always be approximated, correct to the second order in time, by the modified Helmholtz equation

$$\nabla^2 u' - m^2 u' = F(\underline{x},t^*) \qquad 15$$

where, with u^n the approximation for $u(\underline{x},t^n)$, at $t = t^n$

$$u' = u^{n+1} - u^n$$

$$F = -(1/\lambda)[(a/2-\lambda)\nabla^2 u^{n+1} + (a/2+\lambda)\nabla^2 u^n + f^{n+1/2}$$

$$m^2 = [1/(\lambda\Delta t)] \qquad (\Delta t = t^{n+1} - t^n)$$

provided that the arbitrary constant λ satisfies the inequality

$$\lambda > \max[a/2]$$

and the boundary conditions are also collocated at $t^{n+1/2}$.

Since Equation 15 is an elliptic equation, it can be solved by either the GVM or GEM.

NUMERICAL EXAMPLES

As an elementary example of the application of the GVM to parabolic problems, consider the very simple ground-water flow equation

$$SD_t u = T\nabla^2 u + Q(\underline{x} - \underline{x}_0) \qquad 16$$

defined over a homogeneous domain of 2 600 by 2 000 m, with S the storage coefficient (= 10^{-3}), u the hydraulic head, T the transmissivity (= 100 m^2) and Q (= - 1 000 m^3/d) the rate at which water is extracted from the single sink situated at the point $\underline{x}_0$ = (1 300, 1 000). For initial and boundary conditions given by

$$u(x,y,0) = 0 \text{ and } u(x,y,t) = 0 \text{ on the boundary,}$$

this problem has the analytical solution [Bear (1979)]

$$u(r,\theta,t) = [Q/(4\pi T)] \int_z^\infty [\exp(-v)/v]\,dv \qquad 17$$

where $z = Sr^2/4Tt$.

Equation 16 contains a logaritmic singularity at the sink which influence the solution adversely [Bakkes and Botha (1980)]. They circumvented this difficulty in the FEM solution of the same problem by subtracting the singularity, using a Green's function approach. This same procedure can also be applied in the global variational and global element methods [Bakkes (1984)].

To investigate the behaviour of the GVM, Equation 16 was solved with the GVM, using a 5 x 5 Chebyshev polynomial approximation and the errors computed from Equation 17. The results are compared with that obtained by Bakkes and Botha (1980) for the finite element solution in Figure 2. The accuracy of the solution is clearly remarkable, particularly if it is kept in mind that the finite element solution was also corrected for the singularity and calculated, using a mesh consisting of 85 quadratic elements (292 nodes).

The final example to be considered here is that of an aquifer, the plan view of which is given in Figure 3. The aquifer is bounded by a series of leaky dykes. The domain of the aquifer can be naturally divided into two subdomains. In element 1 the average transmissivity is 2 000 m^2 and in element 2 1 000 m^2. The storage coefficient is 2×10^{-2} across the aquifer. There are 19 wells present in the aquifer of which nine are used as production wells and the rest for observation. The pumping rates in the nine production wells are constant.

In order to apply the GEM to this irregular domain, transfinite interpolation had to be used. The efficiency of this form of map to transform an irregular domain onto a square one, can be judged from the blended function map of the aquifer given in Figure 4.

The behaviour of the aquifer was simulated, using the GEM with an 8 x 8 Chebyshev approximation on each element. Singularities caused by the production wells was subtracted using the procedure described above. In the case of the finite element simulation, the mesh shown in Figure 5 was used with linear basis functions. The simulation was conducted for the period June 1981 to August 1982.

The observed and computed heads are compared numerically in Table 1 and graphically in Figure 6. Although the drawdowns simulated by the GEM differs from those actually observed, the same applies to the FEM simulation. In fact, the close agreement between the values obtained in both simulations suggests rather that the model was not accurately calibrated. It would thus not be wrong to conclude that the GEM can be used with the same success as the FEM in simulating an actual aquifer.

TABLE 1. Observed and simulated hydraulic heads at the observation wells as for August 1982

Well number	Observed head	GEM	FEM
1	1448,3	1451,6	1450,7
4	1451,3	1450,6	1449,8
5	1454,0	1454,8	1454,2
6	1453,4	1453,9	1453,3
7	1465,0	1461,2	1461,0

CONCLUSION

In conclusion, it can be said that the GEM has some definite properties to recommend it as a method for simulating ground-water flow and related phenomena. The most outstanding of these are perhaps the ease with which the convergence of the method can be studied and singularities handled. This does not imply that the GEM is better than the FEM. In fact, it is doubtful whether the GEM can compete with the FEM in situations where the aquifer is highly inhomogeneous. Nevertheless, as indicated by the present study, there are problems where it may be preferred above the FEM.

REFERENCES

Bakkes, G.N. (1984). Numerical solution of ground-water flow problems by the global element method. Ph.D. thesis submitted at the University of the Orange Free State, Bloemfontein, South Africa.

Bakkes, G.N. and Botha, J.F. (1980). An improved numerical solution for the diffusion equation. Bulletin 11, Institute for Ground-water Studies, University of the Orange Free State, Bloemfontein, South Africa.

Bear, J. (1979). Hydraulics of groundwater. McGraw-Hill International Book Company, New York.

Botha, J.F. (1982). Alternating direction approximation for a parabolic partial differential equation. Internal report, Institute for Ground-water Studies, University of the Orange Free State, Bloemfontein, South Africa.

Delves, L.M. and Freeman, T.L (1981). Analysis of global expansion methods: weakly asymtotically diagonal systems. Academic Press, London.

Delves, L.M. and Hall, C.A. (1979). An implicit matching principle for global element calculations. J. Inst. Maths. Applics., Vol. 23, p. 223-234.

Finlayson, B.A. (1972). The method of weighted residuals and variational principles. Academic Press, New York.

Gordon, W.J. and Hall, C.A. (1973). Transfinite element methods: blending function interpolation over arbitrary curved element domains. Numer. Math., Vol. 21, p. 109-129.

Oden, J.T. and Reddy, J.N. (1976). An introduction to the mathematical theory of finite elements. John Wiley and Sons, New York.

Rund, H. (1973). The Hamilton-Jacobi theory in the calculus of variations. Robert E. Kruger Publishing Company, Huntington, New York.

Snyman, I.M. (1982). Transversality and the complete figure in the second order multiple integral problem in the calculus of variations. Ph.D. thesis, University of South Africa, Pretoria, South Africa.

Stroud, A.H. (1974). Numerical quadrature and the solution of ordinary differential equations. Springer-Verlag, New York.

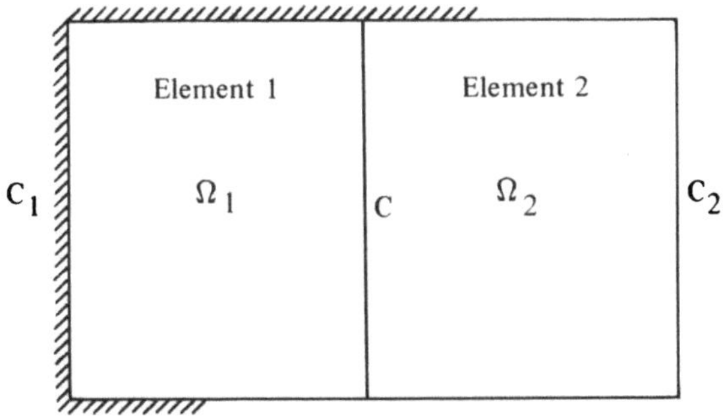

Figure 1. **An example of two global elements.**

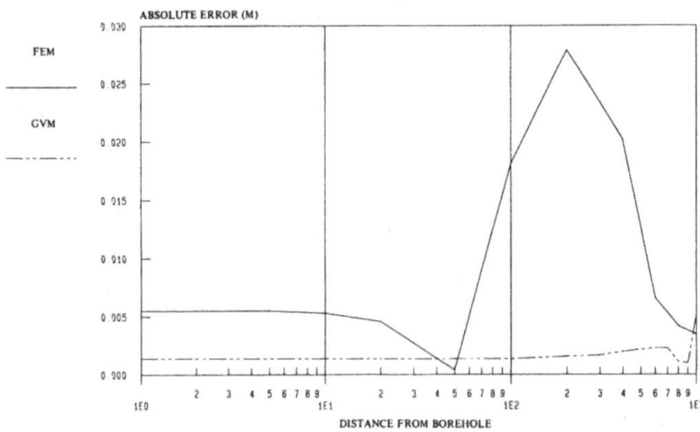

Figure 2. **Maximum absolute errors observed in the GEM and FEM solutions of Equation 16.**

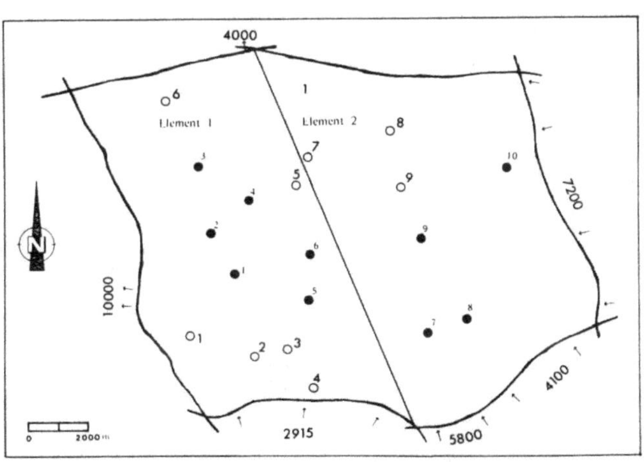

Figure 3. **Plan view of aquifer used in present simulation.**

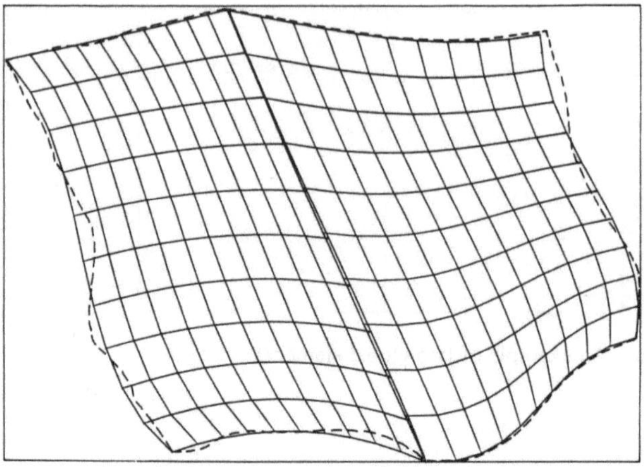

Figure 4. **Blending function map of aquifer in Figure 3.**

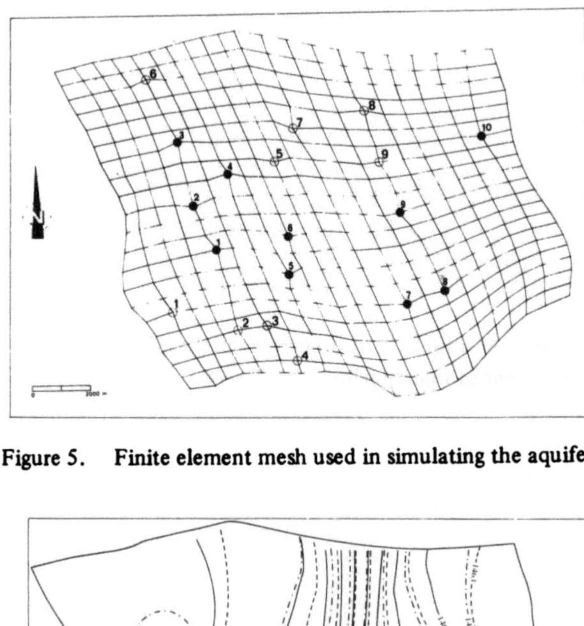

Figure 5. **Finite element mesh used in simulating the aquifer in Figure 3.**

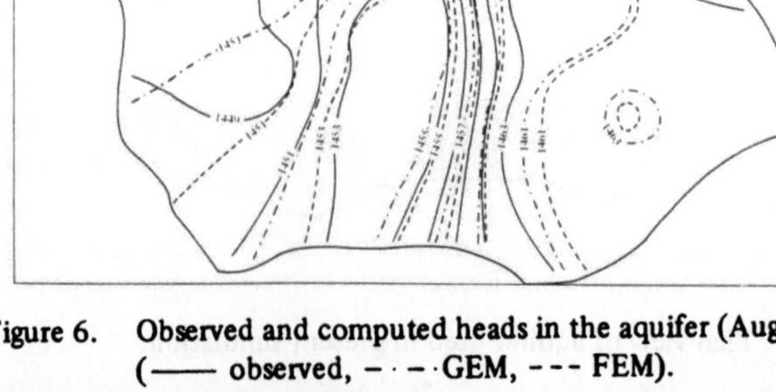

Figure 6. **Observed and computed heads in the aquifer (August 1982). (—— observed, – · – · GEM, - - - FEM).**

SUMMARY

The main objective of the present paper is a study of variational and related methods defined over global regions. It is shown how a variational functional can be constructed for elliptic problems which allows basis functions that do not satisfy any boundary conditions. This functional is then extended to include problems where the domain is divided into a number of subdomains and parabolic problems. A comparison with the finite element method indicates that the method is useful, particularly in those cases where the domain can be divided into a small number of global elements.

TECHNIQUES FOR MAKING FINITE ELEMENTS COMPETITIVE IN MODELING THREE-DIMENSIONAL FLOW AND TRANSPORT

Peter S. Huyakorn, GeoTrans, Inc., Reston, Virginia, U.S.A.

INTRODUCTION

This paper presents the formulation of a three-dimensional finite element model designed to simulate ground-water flow and contaminant transport in complex multi-layered aquifer systems. The model combines the use of two-dimensional basis functions in the x-y plane and one-dimensional basis functions in the z-direction. In addition, coupling of adjacent aquifers and aquitards are performed in a highly efficient manner using one-dimensional analytical solutions. The formulation to be described in the following sections is very general and has several advantages which include great flexibility and highly efficient element matrix computational procedures and equation solution algorithms.

PROBLEM STATEMENT

A typical problem concerned is depicted in Figure 1. The problem is three-dimensional because of hydraulic connections between aquifer and aquitard units. In general, ground-water flow in the aquifers may be described by the following equation:

$$\frac{\partial}{\partial x_i}\left(K_{ij}\frac{\partial h}{\partial x_j}\right) = S_s\frac{\partial h}{\partial t} - q \quad , \quad i,j = 1,2,3 \qquad (1)$$

where K_{ij} is the aquifer hydraulic conductivity tensor, h is the hydraulic head, S_s is the specific storage of the aquifer, t is the time, and q is the volumetric flow rate via sources or sinks per unit volume of the porous medium.

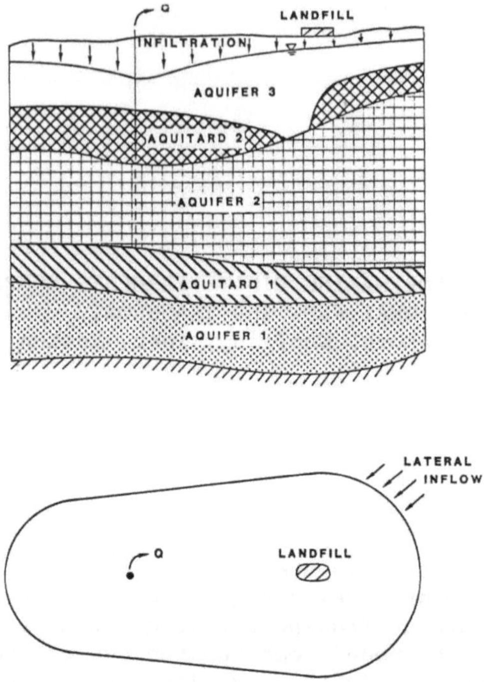

Figure 1. Typical three-dimensional groundwater and contaminant transport problems.

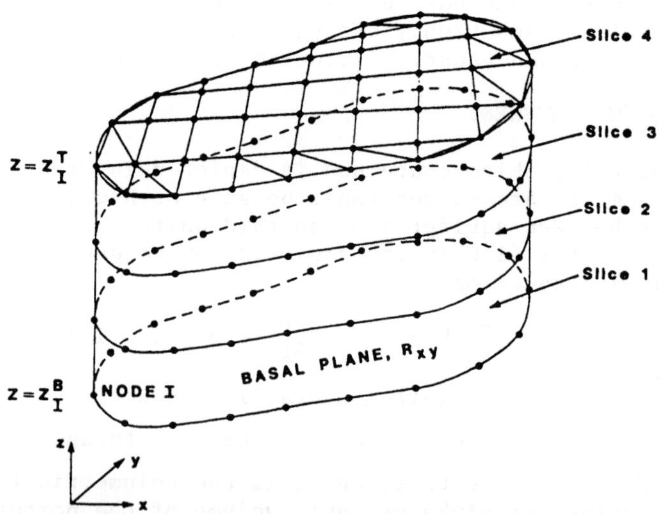

Figure 2. Discretization of an aquifer unit.

Flow in the aquitards may be described approximately by

$$\frac{\partial}{\partial x_3}\left(K' \frac{\partial h'}{\partial x_3}\right) = S_s' \frac{\partial h'}{\partial t} \qquad (2)$$

where the prime is used to denote the aquitard properties. Equations (1) and (2) are coupled via the fluid leakage flux terms at the aquifer-aquitard interfaces where $h = h'$. Transport of a non-conservative species in the aquifer is governed by

$$\frac{\partial}{\partial x_i}\left(D_{ij} \frac{\partial c}{\partial x_j}\right) - V_i \frac{\partial c}{\partial x_i} = \phi\kappa\left(\lambda c + \frac{\partial c}{\partial t}\right) + q\left(c-c^*\right) \qquad (3)$$

where D_{ij} is the hydrodynamic dispersion tensor, c is the solute concentration, V_i is the Darcy velocity vector, ϕ is the effective porosity, λ is the decay coefficient, q is the volumetric fluid injection rate per unit volume, and c^* is the solute concentration in the source fluid.

Transport in the aquitards may be described by

$$\frac{\partial}{\partial x_3}\left(D' \frac{\partial c'}{\partial x_3}\right) = \phi'\kappa'\left(\lambda' c' + \frac{\partial c'}{\partial t}\right) \qquad (4)$$

Equations (3) and (4) are also coupled via solute flux terms at the aquifer-aquitard interfaces, where $c = c'$.

FINITE ELEMENT APPROXIMATION

To illustrate the development of the proposed finite element formulation, it is sufficient to deal with the transport equation. In three dimensional space (x,y,z), the governing equation for transport in an aquifer takes the form

$$\frac{\partial}{\partial x}\left(D_{xx} \frac{\partial c}{\partial x}\right) + \frac{\partial}{\partial y}\left(D_{yy} \frac{\partial c}{\partial y}\right) + \frac{\partial}{\partial z}\left(D_{zz} \frac{\partial c}{\partial z}\right)$$

$$- V_x \frac{\partial c}{\partial x} - V_y \frac{\partial c}{\partial y} - V_z \frac{\partial c}{\partial z} - \phi\kappa \frac{\partial c}{\partial t} = 0 \qquad (5)$$

where, for the sake of simplicity, it is assumed that the terms containing nondiagonal components of the hydrodynamic dispersion tensor, as well as the source and the decay terms are zero. Such an assumption does not lead to any loss of generality because the formulation to be described can be easily extended to include the neglected terms of the transport equation.

To constuct the finite element approximation of equation (5), the three-dimensional aquifer region is divided into a number of slices and each slice is subdivided into a number of elements as shown in Figure 2. Note that the discretization is performed such that each slice has the same projected area (R_{xy}) in the x-y plane. The next step is to represent the function c by a trial function:

$$c \simeq \hat{c}(x,y,z,t) = N_J(x,y)c_J(z,t), \; J = 1,2,\ldots,n_{xy} \qquad (6)$$

where repeated subscripts indicate summation, $N_J(x,y)$ are two-dimensional basis functions in the x-y plane, $c_J(z,t)$ are nodal parameters which are functions of z and t, and n_{xy} is the number of nodes in the basal plane of each slice.

Using the well known Galerkin criterion, the weighted residual approximation of (5) becomes

$$\int_{R_{xy}} N_I \, L\left(\hat{c}\right) dxdy = 0 \quad , \; I = 1,2,\ldots, n_{xy} \tag{7}$$

where $L(\hat{c})$ corresponds to the left hand side of equation (5). Substituting (6) into (7) and performing integration by parts using Green's theorem, one obtains

$$\int_{R_{xy}}\left(D_{xx}\frac{\partial N_I}{\partial x}\frac{\partial N_J}{\partial x} + D_{yy}\frac{\partial N_I}{\partial y}\frac{\partial N_J}{\partial y} + V_x N_I \frac{\partial N_J}{\partial x} + V_y N_I \frac{\partial N_J}{\partial y}\right)dxdy \; c_J$$
$$- \int_{R_{xy}}\left(N_I N_J \, dxdy\right)\left[\frac{\partial}{\partial z}\left(D_{zz}\frac{\partial c_J}{\partial z}\right) - V_z \frac{\partial c_J}{\partial z}\right] + \int_{R_{xy}}\left(\phi\kappa \, N_I N_J dxdy\right)\frac{\partial c_J}{\partial t}$$
$$- \int_{\sigma_{xy}} N_I \left(D_x \frac{\partial \hat{c}}{\partial x} n_x + D_y \frac{\partial \hat{c}}{\partial y} n_y\right) d\sigma = 0$$
$$I = 1,2,\ldots, n_{xy} \tag{8}$$

where σ_{xy} is the boundary of R_{xy}, and n_x and n_y are x and y components of the outward unit vector normal to the boundary σ_{xy}. Equation (8) can be written more concisely as

$$E_{IJ}\, c_J - B_{IJ}\left[\frac{\partial}{\partial z}\left(D_{zz}\frac{\partial c_J}{\partial z}\right) - V_z \frac{\partial c_J}{\partial z}\right]$$
$$+ \tilde{B}_{IJ}\frac{\partial c_J}{\partial t} - f_I^x - f_I^y = 0, \qquad I=1,\, 2,\, \ldots,\, n_{xy} \tag{9}$$

where

$$E_{IJ} = \sum_e \int_{R^e_{xy}} \left(D_{xx}\frac{\partial N_I}{\partial x}\frac{\partial N_J}{\partial x} + D_{yy}\frac{\partial N_I}{\partial y}\frac{\partial N_J}{\partial y} + V_x N_I \frac{\partial N_J}{\partial x} + V_y N_I \frac{\partial N_J}{\partial y}\right) dxdy$$

$$\tilde{B}_{IJ} = \sum_e \int_{R^e_{xy}} N_I N_J dxdy \; ; \quad B_{IJ} = \sum_e \int_{R^e_{xy}} \phi\kappa N_I N_J dxdy$$

$$f_I^x = \sum_e \int_{R^e_{xy}} N_I V_x n_x d\sigma; \quad f_I^y = \sum_e \int_{\sigma^e_{xy}} N_I V_y n_y d\sigma$$

R_{xy} is the element subregion enclosed by boundary σ_{xy}, and Σ_e is the summation over the number elements in the basal plane. Equation (9) is a set of n_{xy} partial differential equations having c_J as the variables that depend on z and t. Therefore, c_J can be represented by trial functions:

$$c_J \approx \hat{c}_J(z,t) = \Phi_J(z) c_{Jj}(t) \quad , \quad j=1,2,\ldots,n_z \qquad (10)$$

where repeated subscripts indicate summation, $\Phi_j(z)$ are one-dimensional basis functions of z, $c_{Jj}(t)$ are nodal values at time t. Note that the double subscripts Jj indicate node J of the j-th slice.

The Galerkin finite element approximation in the z space can be obtained by constructing the following weighted residual integral equations:

$$\int_{z_J^B}^{z_J^T} \Phi_i \left[E_{IJ}\hat{c}_J - B_{IJ} \frac{\partial}{\partial z}\left(D_{zz} \frac{\partial \hat{c}_J}{\partial z}\right) + B_{IJ} V_z \frac{\partial \hat{c}_J}{\partial z} + \tilde{B}_{IJ} \frac{\partial \hat{c}_J}{\partial t} - f_I^x - f_I^y \right] dz = 0, \quad I=1,2,\ldots,n_{xy}, \quad i=1,2,\ldots,n_z \qquad (11)$$

where z_J^B and z_J^T are values of the z-coordinate at node J on the bottom and the top slices. Substituting (10) into (11) and performing integration by parts, one obtains

$$\int_{z_J^B}^{z_J^T} E_{IJ} \Phi_i \Phi_j dz c_{Jj} + B_{IJ} \int_{z_J^B}^{z_J^T} \left(D_{zz} \frac{\partial \Phi_i}{\partial z} \frac{\partial \Phi_j}{\partial z} + V_z \Phi_i \frac{\partial \Phi_j}{\partial z}\right) dz\, c_{Jj}$$
$$+ \left(\int_{z_J^B}^{z_J^T} \tilde{B}_{IJ} \Phi_i \Phi_j dz\right) \frac{dc_{Jj}}{dt} + \int_{z_J^B}^{z_J^T} \Phi_i \left(f_I^x + f_I^y\right) dz$$
$$+ \Phi_i B_{IJ} \left[\left(D_{zz} \frac{\partial \hat{c}_J}{\partial z}\right)^T - \left(D_{zz} \frac{\partial \hat{c}_J}{\partial z}\right)^B\right] = 0$$
$$I = 1,2,\ldots,n_{xy}, \quad i = 1,2,\ldots,n_z \qquad (12)$$

where E_{IJ} and $\tilde{B}_{IJ}$ are kept inside the integrals to allow for the fact that transport parameters of the aquifer concerned may be variable in the z-direction.

For the sake of convenience, it is assumed that one-dimensional basis functions employed are either linear or piecewise constant. Furthermore, the vertical distance between adjacent slices is assumed to be constant to avoid unnecessary complication in the presentation of the integrated equations. In the case where Φ_i and Φ_j are linear, evaluation of the integrals in (12) gives the following equation for node I on a typical internal slice i:

$$\frac{\Delta z}{6}\left[\left(E_{IJ}\right)_{i-\frac{1}{2}}\left(c_{Ji-1}+2c_{Ji}\right)+\left(E_{IJ}\right)_{i+\frac{1}{2}}\left(2c_{Ji}+c_{Ji+1}\right)\right]$$
$$+\frac{B_{IJ}}{\Delta z}\left[\left(D_{zz}\right)_{i-\frac{1}{2}}\left(c_{Ji}-c_{Ji-1}\right)-\left(D_{zz}\right)_{i+\frac{1}{2}}\left(c_{Ji+1}-c_{Ji}\right)\right]$$
$$+\frac{B_{IJ}}{2}\left[\left(V_z\right)_{i-\frac{1}{2}}\left(c_{Ji}-c_{Ji-1}\right)+\left(V_z\right)_{i+\frac{1}{2}}\left(c_{Ji+1}-c_{Ji}\right)\right]$$
$$+\frac{\Delta z}{6}\left[\left(\tilde{B}_{IJ}\right)_{i-\frac{1}{2}}\left(\frac{dc_{Ji-1}}{dt}+2\frac{dc_{Ji}}{dt}\right)\right.$$
$$\left.+\left(\tilde{B}_{IJ}\right)_{i+\frac{1}{2}}\left(\frac{dc_{Ji+1}}{dt}+2\frac{dc_{Ji}}{dt}\right)\right]=0 \qquad (13)$$

where the boundary flux terms has been dropped as it is assumed that node I is an interior node. The equations for boundary nodes can be obtained by including the boundary terms.

Equation (13) can be simplified substantially if the standard matrix lumping procedure is used in the evaluation of matrix elements B_{IJ} and the integral of the product of Φ_i and Φ_j. The simplified equation for node I on an internal slice i takes the form

$$\left(E_{IJ}\right)_i c_{Ji}+\frac{B_{II}}{(\Delta z)^2}\left[\left(D_{zz}\right)_{i-\frac{1}{2}}\left(c_{Ii}-c_{Ii-1}\right)\right.$$
$$\left.-\left(D_{zz}\right)_{i+\frac{1}{2}}\left(c_{Ii+1}-c_{Ii}\right)\right]$$
$$+\frac{B_{II}}{2\Delta z}\left[\left(V_z\right)_{i-\frac{1}{2}}\left(c_{Ii}-c_{Ii-1}\right)+\left(V_z\right)_{i+\frac{1}{2}}\left(c_{Ji+1}-c_{Ji}\right)\right]$$
$$+\tilde{B}_{IJ}\frac{dc_{Ji}}{dt}=0 \qquad (14)$$

In the case where the vertical gradient of the function c is negligible in the aquifer concerned, a constant (unit) one-dimensional basis function may be used to represent the uniform concentration distribution in the z-direction. Consequently, for slice i, which now represents the entire aquifer layer, one obtains

$$\left(\Delta z\right)E_{IJ}c_{Ji}+\left(\Delta z\right)\tilde{B}_{IJ}\frac{dc_{Ji}}{dt}\int_{z_I^B}^{z_I^T}\left(f_I^x+f_I^y\right)dz=0$$
$$I=1,2,\ldots,n_{xy} \qquad (15)$$

where Δz corresponds to the aquifer thickness. It is apparent that the finite element equations (15) can also be derived by direct application of the Galerkin procedure to the two-dimensional transport equation obtainable from equation (5) by setting $\partial c/\partial z$ to zero and integrating over the aquifer thickness.

AQUIFER-AQUITARD COUPLING

The flow and transport processes are in adjacent aquifers and aquitards are coupled by leakage terms which can be evaluated using a numerical solution or a convolution integral approach. In the numerical solution approach, spatial discretizations of the one-dimensional flow and transport equations, (2) and (4), are performed for each nodal column using either the Galerkin procedure or the standard finite difference technique. Although such a numerical approach is very general, it requires a great deal of computer storage and is not as convenient to code as the convolution integral approach, which is based on analytical solutions of the governing equations and needs no spatial discretization of the aquitards.

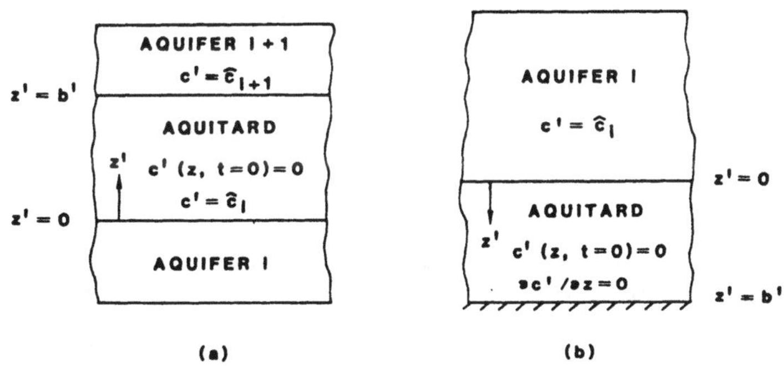

Figure 3. Connections of aquifers and aquitards.

To present a brief outline of the convolution integral approach, we elect to consider once again the solute transport problem and focus on two specific cases illustrated in Figure 3a-3b. In the first case, the aquitard is underlain and overlain by aquifers i and i+1 respectively. The initial and boundary conditions are depicted in Figure 3a. In the second case, the aquitard is underlain and overlain by an impermeable bed rock and aquifer i, respectively. The initial and boundary conditions are depicted in Figure 3b. In both cases the solute flux at the aquifer-aquitard interfaces can be evaluated by using the analytical solutions given by Carslaw and Jaeger (1959) and Duhamel's theorem. For the case shown in Figure 3a, the flux at $z'=0$ is given by

$$q_c'\Big|_{z'=0} = -D' \frac{\partial c'}{\partial z'}\Big|_{z'=0}$$

$$= \frac{D'}{b'}\left(c_i + 2\sum_{n=1}^{\infty}\int_0^t \frac{\partial c_i}{\partial \tau} \exp\left[-\alpha_n (t-\tau)\right] d\tau\right)$$

$$- \frac{D'}{b'}\left(c_{i+1} + 2\sum_{n=1}^{\infty}\cos(n\pi)\int_0^t \frac{\partial c_{i+1}}{\partial \tau} \exp\left[-\alpha_n (t-\tau)\right] d\tau\right) \quad (16)$$

where $\alpha_n = n^2\pi^2 D / (\phi' K' b'^2)$, and D', ϕ', K' and b' are the dispersion coefficient, the porosity, the retardation coefficient and the thickness of the aquitard, respectively. Note that the expression for the flux at $z'=b'$ can be obtained using the left hand side of (16) and interchanging subscripts i and i+1.

For the case shown in Figure 3b, the flux at $z'=0$ is given by

$$q_c' = \frac{2D'}{b'}\sum_{n=0}^{\infty}\int_0^t \frac{\partial c_i}{\partial \tau} \exp\left[-\beta_n (t-\tau)\right] d\tau \quad (17)$$

where $\beta_n = \pi^2(2n+1)^2 D' / (4\phi' K' b'^2)$.

Details on the evaluation of the convolution integrals and the incorporation of the flux terms into the finite element equations can be found in Huyakorn and Pinder (1983). In essence, values of the integrals at the current time level are determined simply by using a recurrence formula that is derived by taking advantage of the special property of the exponential function. To assure fast convergence of the series in equations (16) and (17), an effective thickness b_o' instead of b' in computing β_n. In most cases, it is sufficient to consider only the first three or four leading terms of the series.

ELEMENT MATRIX COMPUTATION

The finite element formulation presented earlier involves coefficient matrices, [E], [B] and $[\widetilde{B}]$ that are contributed by plane elements. In general, these matrices may need to be computed for each time step. To avoid excessive computational cost in element matrix computation, the use of simple rectangular and triangular elements is highly recommended. For such elements, the element matrices can be evaluated in a highly efficient manner using the influence coefficient technique presented by Huyakorn et al. (1984). To illustrate the application of this technique, consider a typical rectangular element with the nodes numbered as 1, 2, 3 and 4 in the counter-clockwise direction. The element matrix $[E]^e$ is first expressed as

$$[E]^e = [E_D]^e + [E_V]^e \tag{18}$$

where $[E_D]^e$ and $[E_V]^e$ are the dispersive and the advective parts. Next, the element matrices $[E_D]^e$ and $[E_V]^e$ are computed from (Huyakorn et al., 1984):

$$[E_D]^e = \langle D_{xx}\rangle \frac{m}{\ell} [E_D^{xx}]^e + \langle D_{yy}\rangle \frac{\ell}{m} [E_D^{yy}]^e \tag{19}$$

and

$$[E_V]^e = \langle V_x\rangle \frac{m}{\ell} ([E_V^x]^e + [E_V^x]_u^e)$$

$$+ \langle V_y\rangle \frac{\ell}{m} ([E_V^y]^e + [E_V^y]_u^e) \tag{20}$$

where $\langle D_{xx}\rangle$, $\langle D_{yy}\rangle$, $\langle V_x\rangle$ and $\langle V_y\rangle$ are centroidal values of D_{xx}, D_{yy}, V_x and V_y, respectively, ℓ and m are the length and the width of the rectangle, and $[E_D^{xx}]^e$, $[E_D^{yy}]^e$, $[E_V^x]^e$, $[E_V^y]^e$, $[E_V^x]_u^e$ and $[E_V^y]_u^e$ are the influence coefficient matrices given in Table 1. Note that $[E_V^x]_u^e$ and $[E_V^y]_u^e$ are the influence coefficient matrices designed to allow for upstream weighting in the x and y directions, respectively. The parameters α_1, α_2, β_1 and β_2 are upstream weighting factors associated with sides 12, 43, 23 and 41 of the rectangular element. Generally, the magnitude of α and β range from 0 to 1. (Zero corresponds to no upstream weighting where one corresponds to full upstream weighting).

SOLUTION PROCEDURE

For convenience in describing the general solution procedure, we refer to Figure 1 in which the aquifers and the aquitards are numbered sequentially from bottom to top. It is assumed for illustrative purposes that aquifers 1 and 3 are modeled using one slice each, and aquifer 2 is modeled using 4 slices.

For each time step, the solution procedure may be outlined as follows:

1. Formulate the finite element matrix equation for aquifer 1 taking into account leakage flux terms contributed by aquitard 1. Perform temporal approximation using either a Crank-Nicolson or a fully-implicit backward difference scheme. Solve the matrix equation using a standard direct solver.

2. Proceed to aquifer 2 and formulate the finite element matrix equation taking into account aquitard leakage fluxes. The flux terms contributed by aquitards 1 and 2 are to be applied to the nodal equations for slices 1 and 4,

Table 1. Influence coefficient matrices for rectangular elements

$$[E_D^{xx}] = \frac{1}{6}\begin{bmatrix} 2 & -2 & -1 & 1 \\ -2 & 2 & 1 & -1 \\ -1 & 1 & 2 & -2 \\ 1 & -1 & -2 & 2 \end{bmatrix} ; \quad [E^{yy}]^e = \frac{1}{6}\begin{bmatrix} 2 & 1 & -1 & -2 \\ 1 & 2 & -2 & -1 \\ -1 & -2 & 2 & 1 \\ -2 & -1 & 1 & 2 \end{bmatrix} ; \quad [M]^e = \frac{1}{9}\begin{bmatrix} 4 & 2 & 1 & 2 \\ 2 & 4 & 2 & 1 \\ 1 & 2 & 4 & 2 \\ 2 & 1 & 2 & 4 \end{bmatrix}$$

$$[E_v^x]^e = \frac{1}{6}\begin{bmatrix} -2 & 2 & 1 & -1 \\ -2 & 2 & 1 & -1 \\ -1 & 1 & 2 & -2 \\ -1 & 1 & 2 & -2 \end{bmatrix} ; \quad [E_v^y]^e = \frac{1}{6}\begin{bmatrix} -2 & -1 & 1 & 2 \\ -1 & -2 & 2 & 1 \\ -1 & -2 & 2 & 1 \\ -2 & -1 & 1 & 2 \end{bmatrix}$$

$$[E_v^x]_u^e = \frac{1}{6}\begin{bmatrix} 2\alpha_1 & -2\alpha_1 & -\alpha_1 & \alpha_1 \\ -2\alpha_1 & 2\alpha_1 & \alpha_1 & -\alpha_1 \\ -\alpha_2 & \alpha_2 & 2\alpha_2 & -2\alpha_2 \\ \alpha_2 & -\alpha_2 & -2\alpha_2 & 2\alpha_2 \end{bmatrix} ; \quad [E_v^y]_u^e = \frac{1}{6}\begin{bmatrix} 2\beta_2 & \beta_2 & -\beta_2 & -2\beta_2 \\ \beta_1 & 2\beta_1 & -2\beta_1 & -\beta_1 \\ -\beta_1 & -2\beta_1 & 2\beta_1 & \beta_1 \\ -2\beta_2 & -\beta_2 & \beta_2 & 2\beta_2 \end{bmatrix}$$

respectively. For this particular aquifer, where a fully three-dimensional finite element approximation has been performed, the time stepping and solution of the matrix equation can be performed using either a two-stage alternating direction implicit method described in Babu and Pinder (1982) or a block successive relaxation method described in Huyakorn et al. (1984).

3. Proceed to aquifer 3 and formulate the finite matrix element equation taking into account the infiltration and the leakage flux contributed by aquitard 2. Solve the matrix equation using the direct solver and a standard Picard iterative procedure to handle nonlinearity due to the water table condition.

COMPUTER CODE DEVELOPMENT AND APPLICATION

A computer code based on the formulation described has been developed. This code provides cost-effective solutions to quasi or fully three-dimensional problems or partially three-dimensional problems that involve both areal and three-dimensional variations of flow and transport variables and aquifer parameters. The code is currently being used to perform ground-water contamination assessment at a site. Results of the simulations will be reported in a future paper.

REFERENCES

Babu, D.K. and G.F. Pinder (1982) A Three Dimensional Hybrid Finite Element-Finite Difference Scheme for Groundwater Simulation, Proc. of 10th IMACS World Congress on System Simulation and Scientific Computation, Montreal, 292-294.

Carslaw, H.S. and J.C. Jaeger (1959) Conduction of Heat in Solids, 2nd ed. Oxford Univ. Press, Oxford, 510 pp.

Huyakorn, P.S. and G.F. Pinder (1983) Computational Methods in Subsurface Flow, Academic Press, New York, 473 pp.

Huyakorn, P.S., Techniques for Making Finite Elements Competitive in the Simulation of Variably-Saturated Porous Media: 2. Two-dimensional Transport Formulation (1984) submitted to Water Resources Research.

Huyakorn, P.S., et al., Techniques for Making Finite Elements Competitive in the Simulation of Variably-Saturated Porous Media: 3. Three-dimensional Formulations (1984) submitted to Water Resources Research.

ADJOINT STATE FINITE ELEMENT ESTIMATION OF AQUIFER PARAMETERS UNDER STEADY-STATE AND TRANSIENT CONDITIONS

Jesus Carrera and Shlomo P. Neuman

Department of Hydrology and Water Resources
University of Arizona, Tucson, Arizona 85721

ABSTRACT

The inverse problem of groundwater hydrology has been recognized as being notoriously difficult to solve. In this paper, a method is presented that overcomes many deficiencies of previous approaches. The method consists of minimizing a statistically-based functional that includes head measurements as well as prior information about the parameters. An adjoint state is computed by solving a partial differential equation that is coupled to the flow equation through the functional to be minimized. This enables one to compute the gradient of the functional with respect to model parameters directly, without the need to compute a sensitivity (or Jacobian) matrix. The latter matrix is required only once for postoptimal sensitivity analysis after the inverse problem has been solved. Any gradient algorithm such as Fletcher-Reeves, Davidon-Fletcher-Powell, or Broyden can then be used. In this paper, the adjoint-state algorithm of Neuman (1980b) is extended to include uncertainty in storage coefficient, sink/source terms, boundary conditions, and initial conditions. The equations are discretized using the finite element method with the added advantage that, since the homogeneous parts of the flow and adjoint-state equations are identical, their solutions are similar. This makes the method both effective and computationally efficient. A case study involving the Colorado River is included.

INTRODUCTION

The estimation of aquifer parameters from head and other related field measurements, usually referred to as "inverse problem", is of paramount importance in groundwater hydrology. Although a considerable effort has been devoted to this problem in the last two decades, a reliable and efficient algorithm for its solution is still lacking. One major difficulty encountered in many published algorithms is lack of stability (i.e. minor variations in measured input values cause severe fluctuations in the computed parameters). As suggested by Neuman (1973, 1980a), the problem of instability can be controlled by introducing prior information about the parameters into the functional which, when minimized, yields the solution to the inverse problem. Another difficulty with many existing algorithms is lack of computational efficiency. This difficulty can be overcome, to a large extent, through the adjoint-state approach of Chavent (1971) which has been extended to the domain of finite elements by Neuman (1980a, 1980b).

In Neuman's papers (1973, 1980a and 1980b), heads and model parameters were characterized statistically by the mean vectors and covariance matrices, and prior information in the functional was weighted by a parameter λ. In his original work on the inverse problem, Neuman (1973) suggested evaluating this parameter by subjective judgement. Later, Neuman (1980a) cast the inverse problem in a statistical framework and recommended evaluating λ by a comparative analysis of the residuals. In the present work, we retain Neuman's characterization of the statistics of heads and prior estimates, but use the maximum likelihood approach to pose the estimation problem. Moreover, we extend his adjoint-state finite element algorithm to account for uncertain initial conditions, boundary conditions, storage coefficients and sink/source terms such as groundwater discharge and recharge. Our paper ends with a transient case study involving a segment of the Colorado River.

PROBLEM DEFINITION

The flow of water through porous media is controlled by the diffusion equation

$$\underline{\nabla} \cdot (\underline{\underline{K}} \cdot \underline{\nabla} h) + q = S_s \frac{\partial h}{\partial t} \qquad \text{on } D \qquad (1)$$

Subject to the initial and boundary conditions

$$h(\underline{x},0) = h_0(\underline{x}) \qquad \text{on } D \qquad (2)$$

$$\underline{\underline{K}} \cdot \underline{\nabla} h \cdot \underline{n} = -\alpha(h-H) + Q \qquad \text{on } \Gamma \qquad (3)$$

Where D is the spatial domain, Γ is its boundary, $\underline{x}$ is the vector of spatial coordinates, t is time, $h(\underline{x},t)$ represents hydraulic head, $\underline{\underline{K}}(\underline{x})$ is the hydraulic conductivity tensor, $q(\underline{x},t)$ is an areally distributed sink/source term, $S_s(\underline{x})$ is specific storage, $\underline{n}$ is a unit outward vector normal to Γ, $H(\underline{x},t)$ is prescribed head at the boundary, $Q(\underline{x},t)$ is prescribed flux at the boundary, and α controls the type of boundary condition: if $\alpha\to\infty$ at a given point $\underline{x} \in \Gamma$, (3) is a Dirichlet condition; if $\alpha=0$, (3) is a Neumann condition; otherwise, (3) is a mixed condition and α represents leakage coefficient. For the case where initial conditions are unknown, we assume that the system is initially at a steady-state. In this case, the unknown initial head $h_0(\underline{x})$ must satisfy the Poisson equation.

$$\underline{\nabla} \cdot (\underline{\underline{K}} \cdot \underline{\nabla} h_0) + q_0 = 0 \qquad \text{on } D \tag{4}$$

subject to the boundary condition

$$\underline{\underline{K}} \cdot \underline{\nabla} h_0 \cdot \underline{n} = - \alpha(h-H_0) + Q_0 \qquad \text{on } \Gamma \tag{5}$$

where q_0, H_0 and Q_0 are the steady-state equivalents of q, H and Q, respectively.

Equations (1) through (5) can be solved by the finite element method. For this, we approximate the hydraulic head by

$$h(\underline{x},t) \simeq \sum_{i=1}^{N} h_i(t)\, \xi_i(\underline{x}) \tag{6}$$

where N is the number of nodes, $h_i(t)$ is head at the ith node, and $\xi_i(\underline{x})$ is the corresponding basis function. In our work we assume (without loss of generality) that the aquifer parameters $\underline{\underline{K}}$, S_s, q, H, Q and α are constant over groups of one or more elements, referred to herein as "zones". For example,

$$\underline{\underline{K}}(\underline{x}) \simeq \sum_{i=1}^{N_k} \underline{\underline{K}}_i \alpha_{ki}(\underline{x}) \tag{7}$$

where N_k is the number of hydraulic conductivity zones, $\underline{\underline{K}}_i$ is the value of the tensor in the ith zone, and $\alpha_{ki}(\underline{x}) = 1$ if $\underline{x}$ is in zone i while $\alpha_{ki}(\underline{x}) = 0$, otherwise. We shall refer to $\underline{\underline{K}}_i$, S_{si}, etc. as "model parameters", and, use $\underline{p_1}$ to denote the vector of zonal hydraulic conductivities (note that each zone may represent up to three components of $\underline{p_1}$ to account for anysotropy), $\underline{p_2}$ the vector of specific storages, etc.

Application of the Galerkin method to (1)-(5) together with a backward finite difference scheme in time leads to the matrix "finite element equation"

$$\underline{\underline{A}}\, \underline{h}^{k+1} + \underline{\underline{D}}\, \frac{\underline{h}^{k+1} - \underline{h}^{k}}{\Delta t^{k+1}} = \underline{Q}^{k} \tag{8}$$

where $\underline{\underline{A}}$ is the conductance (stiffness) matrix, $\underline{\underline{D}}$ is the capacity matrix, $\underline{h}^k$ is the vector of nodal heads at the end of the kth time step, and $\underline{Q}^k$ is the corresponding vector of nodal sink and source terms. The matrices $\underline{\underline{A}}$, $\underline{\underline{D}}$ and vector $\underline{Q}^k$ are functions of the model parameters, and remain unknown as long as the parameters have not been specified.

Let us assume that h has been measured at a discrete set of points x_ℓ in space and t_p in time and let $h^*(\underline{x}_\ell, t_p)$ represent these measurements. Let us further assume that some prior information, independent of h^*, is available about the model parameters. Then, the inverse problem can be stated as finding a particular set of model parameters which, while remaining closest to their prior estimates, lead to an optimum match between measured and computed heads. Maximum likelihood estimation is well suited for this purpose. In our case, a maximum likelihood solution is obtained by minimizing (Carrera, 1984).

$$S = \frac{J_h}{\sigma_h^2} + \sum_{i=1}^{6} \frac{J_i}{\sigma_i^2} + \ell n|\underline{\underline{V}}_h| + \sum_{i=1}^{6} \ell n|\underline{\underline{V}}_i| + N_h \, \ell n \, \sigma_h^2 + \sum_{i=1}^{6} N_i \, \ell n \, \sigma_i^2 + n \, \ell n \, 2\pi \qquad (9)$$

where

$$J_h = (\underline{h}^* - \underline{h})^T \underline{\underline{V}}_h^{-1} (\underline{h}^* - \underline{h})$$

$$J_i = (\underline{p}_i^* - \underline{p}_i)^T \underline{\underline{V}}_i^{-1} (\underline{p}_i^* - \underline{p}_i)$$

are "model performance" and "parameter plausibility" criteria, respectively (i=1 corresponds to hydraulic conductivities, i=2 to specific storage, i=3 to areal recharge, i=4 to head at the boundary, i=5 to prescribed flow at the boundary, and i=6 to leakage at the boundary); $\underline{p}_i^*$ is the vector of prior estimates of model parameters i, $\underline{p}i$; $\sigma_h^2\underline{\underline{V}}_h$ and $\sigma_i^2\underline{\underline{V}}_i$ are the covariance head measurement errors $(\underline{h}^* - \underline{h})$ and prior estimation errors $(\underline{p}_i^* - \underline{p}_i)$ respectively (the decomposition of these into products of a scalar and a matrix allows extra flexibility by treating σ_h and σ_i as unknown parameters which we often find necessary); N_h is the total number of head data; N_i is the number of zones for the ith parameter; and $n = N_h + \sum_{i=1}^{6} N_i$. If $\underline{\underline{V}}_h$ and $\underline{\underline{V}}_i$ are known for all i (as we generally assume is the case) the third, fourth and last term in (9) may be disregarded. If, furthermore, σ_h^2 and σ_i^2 are also known for all i (as is sometimes the case), only the first two terms in (9) involve unknown parameters. Upon multiplying (9) throughout by σ_h^2 and setting $\lambda_i = \frac{\sigma_h^2}{\sigma_i^2}$, the criterion to be minimized may be written as

$$J = J_h + \sum_{i=1}^{6} \lambda_i J_i \tag{10}$$

For convenience we will express J_h as

$$J_h = \sum_k \sum_\ell \sum_m \sum_n \{[h^*(\underline{x}_k,t_m) - h(\underline{x}_k,t_m)] W_{k\ell mn}[h^*(\underline{x}_\ell,t_n) - h(\underline{x}_\ell,t_n)]\} + \sum_k \sum_\ell \{[h_o^*(\underline{x}_k) - h_o(\underline{x}_k)] U_{k\ell}[h_o^*(\underline{x}_\ell) - h_o(\underline{x}_\ell)]\} \tag{11}$$

where $W_{k\ell mn}$ and $U_{k\ell}$ are components of the inverse covariance matrices of head errors under transient and steady state conditions, respectively. The continuous equivalent of (11) is

$$J_h = \int_D \int_T \int_{D'} \int_{T'} \{[h^*(\underline{x}',t') - h(\underline{x}',t')] W(\underline{x},\underline{x}',t,t')[h^*(\underline{x},t) - h(\underline{x},t)]\} dt' d\underline{x}' dt d\underline{x} + \int_D \int_{D'} [h_o^*(\underline{x}') - h_o(\underline{x}')] U(\underline{x}',\underline{x})[h_o^*(\underline{x}) - h_o(\underline{x})] \, d\underline{x}' d\underline{x} \tag{12}$$

where $W(\underline{x}',\underline{x},t',t) = W(\underline{x},\underline{x}',t,t')$ and $U(\underline{x},\underline{x}') = U(\underline{x}',\underline{x})$. Note that Equation (12) reduces to Equation (11) when W and U are defined as summations of appropriate Dirac delta products.

To minimize either S or J, we use various gradient methods, all of which require computing the derivative of the functional with respect to each parameter. Since J_i are simple quadratic forms of $(\underline{p}_i^* - \underline{p}_i)$, their gradients are easily computed. However, computing the gradient of J_h is complicated by the nonlinear relationship between heads and model parameters. A common approach is to iterate by means of the Jacobian or sensitivity matrix, which consists of head derivatives at all measurement points in space and time with respect to each parameter. When the number of measurement points and/or parameters is large, this may require excessive amounts of computer storage and time. The alternative is to work with the adjoint-state, as explained below.

ADJOINT STATE EQUATIONS

Adjoint state equations to compute the gradient of objective functions similar to ours were developed by Chavent (1971). Neuman (1980b) extended Chavent's approach to the domain of finite elements for the case where initial and boundary conditions are known. We now extend Neuman's theory to the more general case where both the initial and boundary conditions are uncertain. As shown in Appendix A, the adjoint state equations for our problem are

$$\underline{\nabla} \cdot (\underline{\underline{K}} \cdot \underline{\nabla}\Psi) = - S_s \frac{\partial \Psi}{\partial t} -2 \int_{D'} \int_{T'} (h^*-h)\, W\, d\underline{x}'\, dt' \quad (13)$$

which must be solved backward in time, starting from the "initial" condition at the finalk time $t=\tau$.

$$\Psi(\tau) = 0 \qquad \text{on } D \quad (14)$$

subject to the boundary condition.

$$\underline{\underline{K}} \cdot \underline{\nabla}\Psi \cdot \underline{n} = - \alpha\Psi \qquad \text{on } \Gamma \quad (15)$$

and

$$\underline{\nabla} \cdot (\underline{\underline{K}} \cdot \underline{\nabla}\Psi_o) = - S_s\Psi(o) - 2 \int_{D'} (h_o^* - h_o)\, U\, dx' \quad (16)$$

which must be solved subject to boundary condition

$$\underline{\underline{K}} \cdot \underline{\nabla}\Psi_o \cdot \underline{n} = - \alpha\Psi_o \qquad \text{on } \Gamma \quad (17)$$

If the functional (Equation 12) is written in its discrete form (Equation 11), the integrals on the right hand side of Equations (13) and (17) must be replaced by summations.

GRADIENT OF J_h

Once the adjoint state has been computed, the derivatives of the functional with respect to the parameters can be easily calculated from Ψ and h. As shown in Appendix B, these derivatives are

$$\frac{\partial J_h}{\partial \underline{\underline{K}}_i} = \int_D \underline{\nabla}\Psi_o \otimes \underline{\nabla}h_o \phi_{ki}\, d\underline{x} + \int_D \int_T \underline{\nabla}\Psi \otimes \underline{\nabla}h\, \phi_{ki}\, dt\, d\underline{x} \quad (18)$$

$$\frac{\partial J_h}{\partial S_{si}} = \int_D \int_T \Psi \frac{\partial h}{\partial t} \phi_{si}\, dt\, d\underline{x} \quad (19)$$

$$\frac{\partial J_h}{\partial q_i} = - \int_D \int_T \Psi\phi_{qi}\, dt\, d\underline{x} \quad (20)$$

$$\frac{\partial J_h}{\partial q_{oi}} = - \int_D \Psi_o\phi_{qi}\, d\underline{x} \quad (21)$$

$$\frac{\partial J_h}{\partial H_i} = - \int_\Gamma \int_T \alpha \Psi \phi_{Hi} \, dt \, d\underline{x} \tag{22}$$

$$\frac{\partial J_h}{\partial H_{oi}} - \int_\Gamma \alpha \Psi_o \phi_{Hi} dt \, d\underline{x} \tag{23}$$

$$\frac{\partial J_h}{\partial Q_i} = - \int_\Gamma \int_T \Psi \phi_{Qi} \, dt \, d\underline{x} \tag{24}$$

$$\frac{\partial J_h}{\partial Q_{oi}} = - \int_\Gamma \Psi_o \phi_{Qi} \, d\underline{x} \tag{25}$$

$$\frac{\partial J_h}{\partial \alpha_i} = \int_\Gamma (h_o - H_o) \Psi_o \phi_{\alpha i} d\underline{x} + \int_\Gamma \int_T (h-H) \Psi \, \phi_{\alpha i} dt \, d\underline{x} \tag{26}$$

The symbol "$\otimes$" denotes tensorial product, and ϕ_{ki}, ϕ_{si}, ϕ_{qi}, ϕ_{Hi}, ϕ_{Qi}, and $\phi_{\alpha i}$ are the ith basis functions of $\underline{\underline{K}}$, S, q, H, Q and α, respectively. At Dirichlet boundaries $\alpha \to \infty$ and the integrals in Equations (22) and (23) are indeterminate. Here one must substitute Equations (15) and (17), which leads to

$$\frac{\partial J_h}{\partial H_i} = \int_\Gamma \int_T \underline{\underline{K}} \cdot \underline{\nabla} \Psi \cdot \underline{n} \, \phi_{Hi} \, dt \, d\underline{x} \tag{27}$$

$$\frac{\partial J_h}{\partial H_{oi}} = \int_\Gamma \underline{\underline{K}} \cdot \underline{\nabla} \Psi_o \cdot \underline{n} \, \phi_{Hi} \, d\underline{x} \tag{28}$$

FINITE ELEMENT DISCRETIZATION OF GRADIENT AND ADJOINT STATE EQUATIONS

The adjoint-state Equations (13) - (15) are discretized following the same methodology we used for the flow Equations (1) - (5). Applying, as we did before, Galerkin's method coupled with a finite difference scheme in time, leads to

$$(\underline{\underline{A}} + \underline{\underline{D}} \,/\, \Delta t^{k+1}) \underline{\Psi}^k = -2 \underline{p}^k + \underline{\underline{D}} \, \underline{\Psi}^{k+1} \,/\, \Delta t^{k+1} \tag{29}$$

and

$$\underline{\underline{A}} \, \underline{\Psi}_o = \underline{\underline{D}} \, \underline{\Psi} \, (o) - 2 \, \underline{p}_o \tag{30}$$

where the components of $\underline{p}^k$ and $\underline{p}_o$, expressed in their discrete form, are

$$p_n^k = \sum_{\ell} \sum_{m} \sum_{j} (h_m^{*k} - h_m{}^k)\, W_{m\ell jk}\, \xi_n(x_\ell) \tag{31}$$

$$p_{on} = \sum_{m} \sum_{\ell} (h_{o_m}^* - h_{o_m})\, U_{m\ell}\, \xi_n(x_\ell) \tag{32}$$

Note that the matrices $\underline{A}$ and $\underline{D}$ of the linear systems (Equations 29 and 30) are the same as those of the finite element flow Equation (8). This results from the flow and adjoint state differential equations having identical homogeneous parts. If similar time steps are used to solve both sets of equations, the same triangular decomposition can be used to solve both the head and adjoint state equations, resulting in considerable savings of computer time

Regarding the gradient equations, it is enough to substitute Equation (6) and its adjoint state equivalent into Equations (18) - (28) to obtain expressions of the form

$$\frac{\partial J_h}{\partial q_i} = - \sum_{n} \sum_{k} \Psi_n{}^k \frac{A^e}{3} \phi_{qi} \left(\frac{\Delta t^k + \Delta t^{k+1}}{2} \right) \tag{33}$$

(for areal recharge constant over time) where the first summation is taken over all nodes n belonging to elements e contained in the ith areal recharge zone, and A^e is the area of element e. Similar expressions are obtained for the remaining parameters (Carrera, 1984).

Once the gradient is known, any gradient minimization algorithm can be used to compute the minimum of Equation (9). We have implemented these equations in a program named INVERT that allows the use of the Broyden, Fletcher-Reeves, or Fletcher Powell-Davidon algorithms, coupled with Newton's method for obtaining the step size to be taken at each iteration.

CASE STUDY

To illustrate some of the capabilities of our method, we shall estimate anisotropic permeabilities in a cross-section perpendicular to the Colorado River. The U.S. Geological Survey, in collaboration with the Bureau of Reclamations, has installed eleven piezometers along eighteen cross-sections normal to the Colorado River downstream of Laguna Dam to monitor irrigation return flows (Loeltz and Leake, 1983). The hydraulic conductivities of the alluvium were estimated by calibrating the response of the aquifer, as measured by the piezometers, to flood events, using a trial-and-error approach.

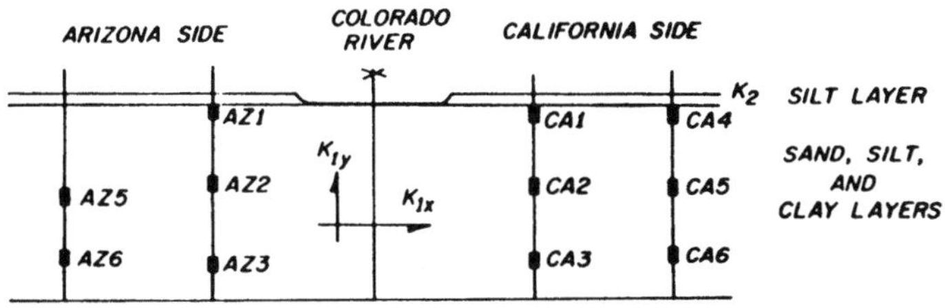

Figure 1: Cross-section used for case study with location of piezometers.

Figure 1 shows cross-section 8 which we have chosen for our case study. The cross-section was discretized into 53 quadrilateral and triangular elements with 65 nodes, and program INVERT was employed to estimate the horizontal and vertical hydraulic conductivities of the alluvium, as well as the isotropic hydraulic conductivity of the overlying silt layer. The specific storage values were assumed fixed for both layers, and no prior information regarding hydraulic conductivities was used. The results of the automatic calibration are illustrated in Figure 2, which shows an excellent agreement between measured and computed heads in all eleven piezometers.

ACKNOWLEDGEMENTS

The senior author was partially supported by the Harshbarger Fellowship. We also wish to thank the U.S. Geological Survey and Mr. R. D. MacNish, District Chief, and Mr. S. A. Leake for providing the data for the case study.

APPENDIX A: DERIVATION OF ADJOINT-STATE EQUATIONS

Let us consider the functional J_h, as defined by Equation (12). Its first variation is

$$\delta J_h = - \int_D \int_{D'} 2[h_o^*(\underline{x}') - h_o(\underline{x}')]U(\underline{x}',\underline{x}) \; \delta h_o(\underline{x}) \; d\underline{x}' \; d\underline{x}$$

$$- \int_D \int_T \int_{D'} \int_{T'} 2[h^*(\underline{x}',t') - h^*(\underline{x}',t')] \; W \; (\underline{x}',\underline{x},t',t)$$

$$\delta h(\underline{x},t) \; dt' \; d\underline{x}' \; dt \; d\underline{x} \tag{A1}$$

Taking the first variations of state Equations (1) through (5) yields

$$\underline{\nabla}\cdot[\underline{\underline{K}}\cdot\underline{\nabla}(\delta h) + \delta\underline{\underline{K}}\cdot\underline{\nabla}h] + \delta q - \delta S_s \frac{\partial h}{\partial t} - S_s \frac{\partial(\delta h)}{\partial t} = 0 \tag{A2}$$

$$\delta h(\underline{x},o) = \delta h_o(\underline{x}) \tag{A3}$$

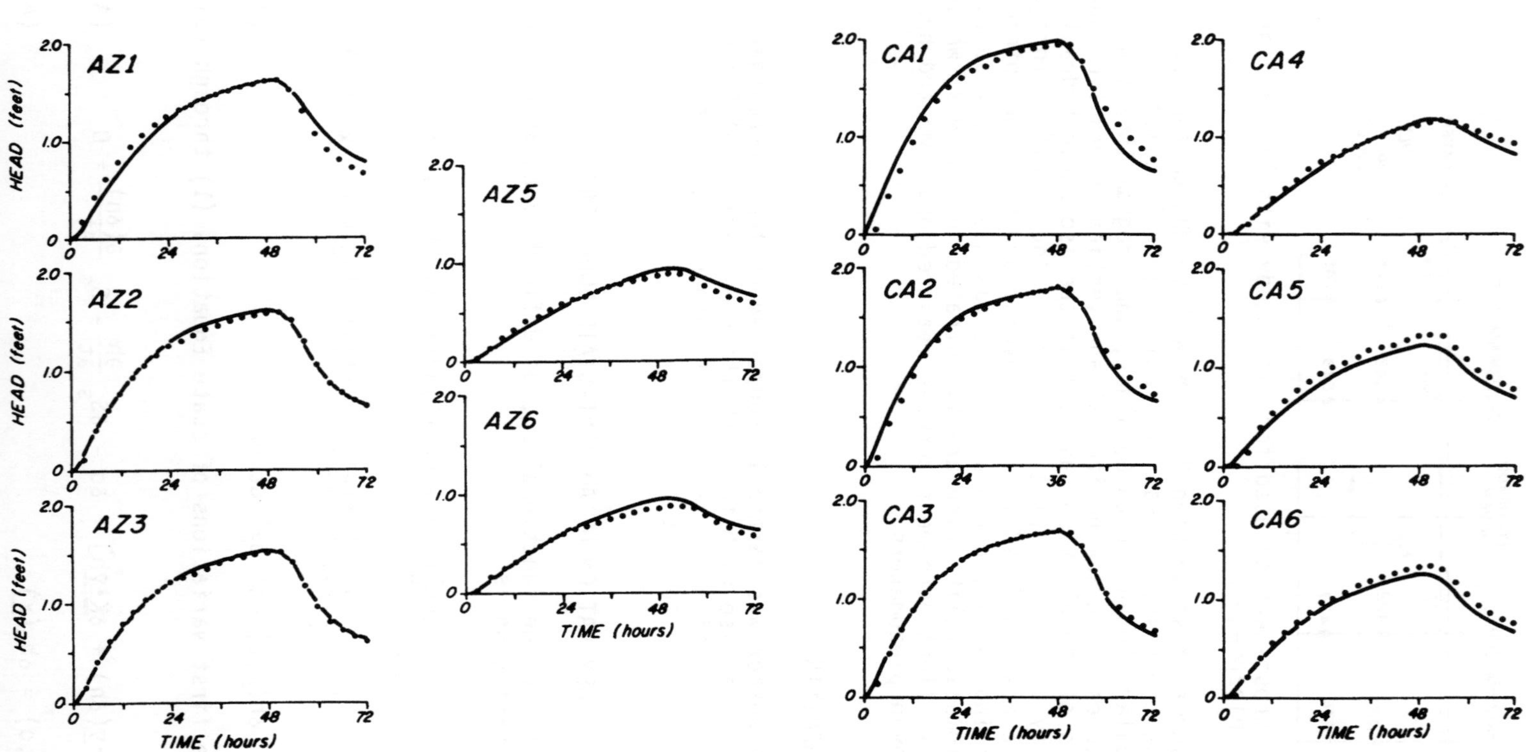

Figure 2: Measured and computed heads at all piezometers in cross-section 8. Continuous lines represent computed heads and dotted lines represent measured heads.

$$[\underline{\underline{K}}\cdot\underline{\nabla}(\delta h) + \delta\underline{\underline{K}}\cdot\underline{\nabla}h]\cdot\underline{n} = -\delta\alpha(h-H)-\alpha(\delta h-\delta H) + \delta Q \qquad (A4)$$

$$\underline{\nabla}\cdot[\underline{\underline{K}}\cdot\nabla(\delta h_o) + \delta\underline{\underline{K}}\cdot\underline{\nabla}h_o] + \delta q_o = 0 \qquad (A5)$$

$$[\underline{\underline{K}}\cdot\underline{\nabla}(\delta h_o) + \delta\underline{\underline{K}}\cdot\nabla h_o]\cdot\underline{n} = -\delta\alpha(h_o-H_o) + \delta Q_o \qquad (A6)$$

Let Ψ and Ψ_o be differentiable functions satisfying $\Psi(\underline{x}',\underline{x},t',t)$ and $\Psi_o(\underline{x},\underline{x}') = \Psi_o(\underline{x}',\underline{x})$. Multiplying (A2) by Ψ and integrating over D and T, multiplying (A4) by Ψ_o and integrating over D, adding up both equations, applying Green's first identity, substituting (A3) and (A5) into the resulting expression, subtracting from (A1), and rearranging leads to

$$\delta J_h = \int_\Gamma [\delta\alpha(h_o-H_o)-\alpha\delta H_o-\delta Q_o]\Psi_o d\underline{x} + \int_D (\underline{\nabla}\Psi_o\cdot\delta\underline{\underline{K}}\cdot\underline{\nabla}h_o-\Psi_o\delta q_o)d\underline{x} +$$

$$+ \int_\Gamma\int_T [\delta\alpha(h-H) - \alpha\delta H - \delta Q]\Psi\, dt\, d\underline{x} +$$

$$\int_D\int_T [\underline{\nabla}\Psi\cdot\delta\underline{\underline{K}}\cdot\underline{\nabla}h - (\delta\Delta q-\delta S_s \frac{\partial h}{\partial t})\Psi]\, dt\, d\underline{x} + F \qquad (A7)$$

where F is

$$F = \int_\Gamma \alpha\Psi_o\delta h_o\, d\underline{x} + \int_D \underline{\nabla}\Psi_o\cdot\underline{\underline{K}}\cdot\underline{\nabla}(\delta h_o)\, d\underline{x} + \int_\Gamma\int_T \alpha\Psi\delta h\, dt\, d\underline{x}$$

$$- \int_D\int_T [\underline{\nabla}\Psi\cdot\underline{\underline{K}}\cdot\underline{\nabla}(\delta h) + \Psi S_s \frac{\partial(\delta h)}{\partial t}]dt d\underline{x} - 2\int_D\int_{D'} (h_o-h_o^*)U\delta h_o d\underline{x}' d\underline{x} -$$

$$-2 \int_D\int_T\int_{D'}\int_{T'} (h^*-h)\, W\, \delta h\, dt' d\underline{x}' dt d\underline{x} \qquad (A8)$$

It is convenient to choose Ψ and Ψ_o so as to insure that F=0. To achieve this, we apply Green's first identity twice to the second integrand and the first term of the fourth integrand, and integrate by parts the second term of the fourth integrand. Upon choosing $\Psi(\tau)=0$, where τ is the final time, and recognizing that $\delta h(o) = \delta h_o$ and that the variations δh and δh_o are arbitrary, it is obvious that the integrands of the different integrals over D, Γ, DT, and ΓT must vanish independently for F to be zero. This immediately results in the adjoint-state equations (13) through (17).

APPENDIX B: GRADIENT OF J_h

In order to compute the derivatives of Equation (12) with respect to the model parameters, let us rewrite the functional in the form

$$J_h = \int_D g(\underline{x})d\underline{x} = \int_D [g_1(\underline{x}) + \int_T g_2(\underline{x},t)dt]\, d\underline{x} \quad \text{(B1)}$$

where

$$g_1(\underline{x}) = \int_{D'} [h_o^*(\underline{x}')-h_o(\underline{x}')]U(\underline{x}',\underline{x})[h_o(\underline{x})-h_o^*(\underline{x})]\, d\underline{x}' \quad \text{(B2)}$$

$$g_2(\underline{x},t) = \int_{D'}\int_{T'} [h^*(\underline{x}',t')-h(\underline{x}',t')]\, W\, (\underline{x}',\underline{x},t',t)$$

$$[h^*(\underline{x},t)-h(\underline{x},t)]dt'dx' \quad \text{(B3)}$$

The first variation of (B1) can be expressed in terms of the first variations of the parameters as

$$\delta J_h = \int_\Gamma (\frac{\partial g}{\partial \alpha}\,\delta\alpha + \frac{\partial g}{\partial H_o}\,\delta H_o + \frac{\partial g}{\partial H}\,\delta H + \frac{\partial g}{\partial Q}\,\delta Q + \frac{\delta g}{\delta Q_o}\,\delta Q_o)\, d\underline{x} +$$

$$\int_D (\frac{\partial g}{\partial \underline{\underline{K}}}\,\delta\underline{\underline{K}} + \frac{\partial g}{\partial S_s}\,\delta S_s + \frac{\partial g}{\partial q}\,\delta q + \frac{\partial g}{\partial q_o}\,\delta q_o)\, d\underline{x} \quad \text{(B4)}$$

Comparison of Equation (B4) with Equation (A7) gives the variational derivative

$$\frac{\partial g}{\partial \alpha}\Big|_\Gamma = \Psi_o(h_o-H_o) + \int_\Gamma \Psi(h-H)\, dt \quad \text{(B5)}$$

$$\frac{\partial g}{\partial H_o}\Big|_\Gamma = -\alpha\Psi_o \quad \text{(B6)}$$

$$\frac{\partial g}{\partial H}\Big|_\Gamma = -\alpha \int_T \Psi dt \quad \text{(B7)}$$

and similar expressions for $\frac{\partial g}{\partial Q}$, $\frac{\partial g}{\partial Q_o}$, $\frac{\partial g}{\partial q}$, $\frac{\partial g}{\partial q_o}$, $\frac{\partial g}{\partial S_s}$ and $\frac{\partial g}{\partial \underline{\underline{K}}}$. In order to obtain the derivatives of the functional with respect to the model parameters, it is enough to discretize each parameter according to equations of the form of Equation (7) and integrate, which immediately leads to Equations (18) through (26).

REFERENCES

1. Carrera, J. (1984) Estimation of Aquifer Parameters Under Transient and Steady-State Conditions. Ph.D. Dissertation, Department of Hydrology, University of Arizona, Tucson, Arizona.

2. Chavent, G. (1971) Analyse Fonctionelle et Identification de Coefficients Repartis dans les Equations aux Derivees Partielles. These de Docteur es Sciences, University of Paris, Paris VI.

3. Loetz, O.J. and S.A. Leake (1983) A Method for Estimating Groundwater Return Flow to the Lower Colorado River in the Yuma Area, Arizona and California. U.S. Geological Survey Water Resources Investigations Report 83-4220, 86 pp.

4. Neuman, S.P. (1980) Adjoint-State Finite Element Equations for Parameter Estimation, Finite Elements in Water Resources. Proceedings of Third International Conference on Finite Elements in Water Resources. (Ed. by S.W. Wang, C.V. Alonso, C.A. Brebbia, W.G. Gray, and G.F. Pinder) University of Mississippi, University, Mississippi.

5. Neuman, S.P. (1980) A Statistical Approach to the Inverse Problem of Aquifer Hydrology. 3. Improved Solution Method and Added Perspective. Water Resources Research, 16(2), 331-346.

6. Neuman, S.P. (1973) Calibration of Distributed Parameter Groundwater Flow Models Viewed as a Multiple-Objective Decision Process Under Uncertainty. Water Resources Research, 9(4), 1006-1020.

COMPUTATION OF POROUS MEDIA NATURAL CONVECTION FLOW AND PHASE CHANGE

Kevin O'Neill and Mary R. Albert

U.S. Army Cold Regions Research and Engineering Laboratory
Hanover, NH, USA

INTRODUCTION

The computation of multi-dimensional phase change in porous media has been a long standing interest of researchers in hydrology, geophysics, and geochemistry. Chemical and mechanical engineers as well have had to concern themselves with the challenges of simulating the complexities of coupled heat and fluid flow in chemical processes and heat exchangers. More recently interest in this domain has developed strongly among civil, petroleum, and geotechnical engineers. This has been stimulated by many developments, including an increased recognition of the continual cost of winter related soil problems, increased exploration and exploitation of northern regions, use of combustion in reservoir engineering, and increased use of ground freezing techniques for mining and construction purposes.

In response to this interest numerous computational techniques have been developed. Typical among these are programs which include latent heat effects by embedding extra weight in the heat capacity in a phase change vicinity (Comini et al, 1974; O'Neill, 1983). In this way phase change may be calculated as a conduction problem with a highly nonlinear coefficient. However in many instances this is not the optimal strategy. For example, artificial gravel islands are sometimes used for shallow water offshore oil drilling. These islands are intentionally frozen to serve as a firm base for heavy equipment and to resist the attack of natural forces. It is important to be able to determine the degree and course of freezing in such situations, which are characterized by coarse grained material and relatively large characteristic lengths. These features produce essentially isothermal freezing, with a sharp discontinuity between the frozen and unfrozen zones, and also foster the development of natural convection fluid flows which may affect the course of freezing. In such cases it makes sense to acknowledge the discontinuity between phases, that is, to treat the matter as a moving boundary problem. While moving boundary phase change calculation has been attempted for conduction only cases on a fixed mesh (eg Yoo and Rubinsky, 1983), the more complex cases of practical interest warrant consideration of a moving boundary - moving mesh approach.

In the moving boundary - moving mesh approach, a moving boundary retains an unambiguous relation to the rest of the mesh, which continually deforms to accomodate changes in phase morphology. A numerical boundary always lies on a phase boundary, and fluid and thermal boundary conditions (including latent heat conditions) may be expressed there clearly and accurately.

The work reported here builds upon earlier results referenced in context below, in which 2-D finite element programs have been developed using linear triangles. A transfinite mapping system of mesh generation is used to control interior mesh deformation as phase change proceeds, and for the first time a solution for coupled fluid flow and heat convection are incorporated in the program. We emphasize that the system operates by including mesh movement directly in the governing equation, so that dependent variable values are solved for directly on a moving node system. In other words, we do not proceed by solving repeatedly on a static mesh, and then interpolating or extrapolating to new positions. The method employed here allows a numerical boundary to follow the phase boundary in a continuous fashion, and facilitates a formulation which conserves mass and energy across the phase discontinuity.

The solution for fluid flow is developed from a stream function formulation, enhanced by a rational smoothing of calculated flow rates through a secondary Galerkin type integral statement. The technique produces C^0 continuous flow fields from piecewise constant derivatives over linear elements. While analogous procedures have been suggested elsewhere to improve global mass balance (Yeh, 1981), we test our formulation here against an analytical solution to determine its local accuracy in defining solution gradients. This is particularly of interest for future applications, in which water temperatures are likely to be near the density inversion point, causing complex reversals of flow within relatively small domains. Indications are given of the positive stability implications of this "reintegration" technique. Ultimately simulations are shown involving natural convection phase change with a linear density fluid, incorporating all aspects of the system for determining heat and fluid flow with mesh deformation.

GOVERNING EQUATIONS

In this moving boundary - moving mesh approach to phase change computation, temperature equations are solved separately within each phase, and the two solutions are linked by a latent heat condition attached to the phase change interface. The temperature equation in the frozen zone is

$$\frac{\partial \theta}{\partial \tau} - \nabla \cdot K^f \nabla \theta = 0 \tag{1}$$

and in the unfrozen zone is

$$C^u_{pm} \frac{\partial \theta}{\partial \tau} + C_r \mathbf{V} \cdot \nabla \theta - \nabla^2 \theta = 0 \tag{2}$$

These and all equations which follow are written in dimensionless quantities. The scaling system and definition of all symbols are contained in the appendix below. We note that a system has been chosen so that both the Stefan and Rayleigh numbers appear. Superscripts f and u denote frozen and unfrozen, respectively.

On the interface between the two phases we apply a condition stating that the rate of normal motion of the phase front due to phase change is proportional to the jump in flux of sensible heat across the interface.

$$\frac{1}{St} \frac{dS}{d\tau} = \left(K \frac{\partial \theta}{\partial N}\right)^f - \left(\frac{\partial \theta}{\partial N}\right)^u \tag{3}$$

Flow solution in the unfrozen zone begins with Darcy's law.

$$\mathbf{V} = -R_a \left(\nabla P + R\hat{e}_y\right) \tag{4}$$

The dimensionless variables in this expression of the law have been defined in such a way that the Rayleigh number appears prior to further differentiation. This makes sense when nonlinear density fluids are considered. The X coordinate is horizontal and the Y coordinate increases in an upward direction. One could proceed to solve the flow field by using equation (4) to produce a pressure based formulation. However here we will proceed in terms of the stream function, in part because it is desirable to examine natural convection patterns in terms of streamlines. Thus the X and Y direction flow rates are expressed respectively as

$$U = -\frac{\partial \psi}{\partial Y}, \quad V = \frac{\partial \psi}{\partial X} \tag{5}$$

and the curl of equation (4) becomes

$$\nabla^2\psi = -R_a\frac{\partial R}{\partial X} \tag{6}$$

At this point we have implicitly enforced the assumption that the flow field is divergence free, despite the fact that the fluid density is considered to vary. This assumption is commonly applied in the context of the Boussinesq approximations. However we note here that this equation (and the specific definition of Ra used) does not assume that fluid density is a linear function of temperature. Thus, together with appropriate boundary and initial conditions, equations (1), (2), (3), and (6) would allow us to solve for natural convection phase change for water near the density inversion, given the known density function R.

For the sake of this preliminary investigation we specialize equation (6) for a linear density fluid as

$$\nabla^2\psi = \frac{R_a}{\theta_w}\frac{\partial \theta}{\partial X} \tag{7}$$

NUMERICAL FORMULATION

Formulations for solving transport equations on a moving and deforming finite element mesh both with and without phase change (O'Neill and Lynch, 1979; Lynch and O'Neill, 1981; O'Neill, 1981; Lynch, 1982; Albert, 1983; Albert and O'Neill, 1983a and 1983b) have been presented elsewhere, and we will not dwell on the details here. One apprehends the basic idea and its relation to the flow field formulation by considering the governing temperature equations in a coordinate system attached to the mesh as it deforms. In computational practice this coordinate system may be identified locally with the normalized coordinates within each element. Partial derivatives with respect to time in equations (1) and (2) may be reexpressed holding the mesh coordinate constant, yielding

$$\left(\frac{\partial \theta}{\partial \tau}\right)_{x_o} - \frac{d\mathbf{X}}{d\tau}\cdot\nabla\theta - \nabla\cdot K^f\nabla\theta = 0 \tag{8}$$

and

$$\left(\frac{\partial \theta}{\partial \tau}\right)_{x_o} + \mathbf{U}\cdot\nabla\theta - \nabla^2\theta = 0$$

$$U = C_r\mathbf{V} - C^u_{pm}\frac{d\mathbf{X}}{d\tau} \tag{9}$$

Here the time derivatives express change in temperature following a point on the mesh. We emphasize that this pertains to all points on the mesh, over element interiors as well as at node points. The mesh velocity now appears in the temperature equations in terms expressing apparent convection due to the mesh coordinate motion.

In finite element terms the dependent variables will be expressed over linear triangular elements as

$$\theta = \theta_j(\tau)\ \phi_j(\mathbf{X}); \qquad \psi = \psi_j(\tau)\phi_j(\mathbf{X}) \tag{10}$$

While the spatial expressions for the basis functions are in general functions of time on a moving mesh, those functions are invariant with respect to the mesh coordinates. Thus

$$\left(\frac{\partial\theta}{\partial\tau}\right)_{x_o} = \frac{d\theta_j}{d\tau}\ \phi_j \tag{11}$$

The mesh velocity itself may be expressed over the whole domain in terms node velocities as

$$\frac{d\mathbf{X}}{d\tau} = \frac{d\mathbf{X}_j}{d\tau}\ \phi_j \tag{12}$$

Thus all terms may be evaluated in the context of usual Galerkin finite element procedures. It should be emphasized that the solutions (coefficient values) obtained pertain directly to moving nodes. Inclusion of the mesh motion directly in the governing equation means that one does not solve on an essentially static system, and then interpolate values to a new, relocated set of nodes. Rather, as long as the mesh motion is specified in a suitable way, one need only apply procedures along the lines of those used normally for convective - dispersive problems. Transformation of the space derivative terms into the moving coordinate system is carried out automatically when one evaluates the element Jacobians, given node locations, while calculating the weighted residual integrals.

All of this presupposes that one includes an algorithm which actually accomplishes continual and nonpathological redistribution of the mesh, including its boundaries. This can be accomplished in a number of ways, with alternative characteristics and limiting considerations, as explained elsewhere (eg Lynch, 1982; Albert and O'Neill, 1983). In brief, the system employed here uses an integral statement of equation (3) to move nodes on the phase change boundary. A transfinite mapping system of mesh generation then develops the interior mesh from the new boundary configuration. This interior mesh control system was selected because it was simple and efficient, involving in the end interpolation between boundary points as opposed to solution of a system of equations.

DETERMINATION OF THE FLOW FIELD; STABILITY IMPLICATIONS

At this point, we draw attention to the fact that the mesh coordinate velocity is C^0continuous throughout the domain, being interpolated exactly over element interiors (equation (12)). However if the fluid flow rate is based simply on derivatives of the stream function solution in accordance with equations (5) and (10), it will be constant over each element and discontinuous between elements. This risks poor resolution of the flow field for transport calculation purposes, and in any case introduces a disparity in the levels of resolution of the mesh and physical velocities. Indications of the stability implications of a discontinuous flow field are also given below.

To achieve an accurate and very efficient formulation which determines a C^0 continuous flow field, we expand on a method suggested by Yeh (1981). Working from a pressure based porous medium flow solution from linear elements, he showed that one could achieve flow continuity and thereby dramatically improve global mass balance in his example problems. In what follows, his method is applied in a stream function flow formulation. Investigations are carried out to determine whether the virtues of the system in terms of global mass balance correspond to a comparable improvement of local accurary.

To proceed, we equate a finite element interpolation of the flow rate with a derivative of the finite element stream function solution.

$$V = V_j \phi_j = \psi_j \frac{\partial \phi_j}{\partial X} \tag{13}$$

The development of the equations needed to obtain U is completely analogous. Multiplying this equation by each basis function in turn and then integrating over the domain produces a Galerkin type formulation which may be solved for nodal values of V.

$$V_j \int_A \phi_i \phi_j dA = \psi_j \int_A \phi_i \frac{\partial \phi_j}{\partial X} dA \tag{14}$$

In this work, each of the governing equations was solved separately, beginning within each time step with equation (7). Coupling of the equations was then accomplished through iteration. Thus at the end of each time step (iteration) we achieve the nodal fluid flow rate values for next use by inserting the most current stream function solution on the right hand side of equation (14). That right hand side is quite simply evaluated in the course of other procedures. The mass matrix on the left hand side is also easily obtained in the course of other matrix assemblies, and greatest economy is achieved if one lumps all its elements onto the diagonal. In the computational experiments performed, it was found that lumping resulted in no discernible loss of accuracy, and avoided the necessity for solution of (14) as a matrix equation. Thus given a current stream function solution, one can immediately specify the values of V_j, and interpolate it continuously throughout using the linear triangular basis functions.

In the face of highly variable flow fields, the increased resolution of flow rates as evidenced below seems intuitively desirable. However we note that at any point these enhanced values are obtained from a weighted integration of values from nearby elements. At the same time, the numerical statement of the transport equations themselves are integral statements. Hence one can ask whether the enhanced flow representation should have any desirable impact in the transport computations, which also integrate flux expressions around each node in any case.

To address this, one does well to examine the effects on stability of alternative formulations. Each individual governing equation treated here was solved implicitly using simple two level finite difference schemes in time which are ordinarily considered stable in themselves. However the traditional stability criteria applied to each individual equation cannot illuminate the effects of interactions between the equations. The effect of flow rate in the transport equations is just such an interaction. Efforts to analyze this are still underway, however some guiding relations may be displayed.

One may generalize an analysis reported by Connor and Christodoulou (1981) to obtain expressions for the eigenvalues of a multi-dimensional finite element system of equations corresponding to either of equation (8) or (9), in the general form

$$\lambda^2 = \frac{\left(m/\Delta t - (1-\varepsilon)(a_s + e)\right)^2 + (1-\varepsilon)^2\, a_{ss}^2}{\left(m/\Delta t + \varepsilon(a_s + e)\right)^2 + \varepsilon^2 a_{ss}^2} \tag{15}$$

The quantities m and e result from the mass and stiffness matrices respectively. Due to the positive definiteness of those matrices, both quantities will be positive. The quantities a_s and a_{ss} derive from symmetric and skew symmetric contributions from the convection term. The epsilon parameter controls the degree of implicitness in the numerical solution, ie, it is equal to 0.5 for a centered-in-time scheme, and 1.0 for a fully implicit scheme. If the magnitude of the eigenvalue is less than one, we expect stability.

It can be shown (O'Neill, 1984) that when the flow field is C^0 continuous and when optimal boundary conditions are applied, a_s will be zero. One easily sees that this implies unconditional stability when epsilon is greater than 0.5, with conditional stability for smaller values. Other favorable boundary conditions will produce positive values of a_s, which can be seen to enhance stability further. However a discontinuous velocity field will produce arbitrary values of a_s, depending upon the particular flow field one happens to encounter. Examination of equation (15) shows that negative values increase the value of the numerator, and hence threaten stability.

Rigorous testing of this analysis awaits the completion of work underway. For the present, the local quality of gradient values obtained from procedures along the lines of equation (13) is examined in some test cases. More complex simulations of natural convection phase change which utilize the method are also carried out successfully.

COMPUTED RESULTS: CONTINUOUS GRADIENTS

As a test case for the calculation of continuous gradients, the heat equation was solved over the rectangle depicted in Figure 1. That is, an equation of the form of (1) was solved, with K^f scaled to 1.0, using linear triangular elements. A coarse mesh was used deliberately to push the method. Initially the dependent variable value was assigned the value 1.0, the X and Y axes were considered to be zero flux boundaries, and solution values on the other boundaries were held at zero. At a dimensionless time of 0.08 we obtain the numerical solution displayed is Figure 2. The accuracy of the solution away from the boundaries was similar to that shown in the figure.

Unimproved solution gradient values along the axes were obtained by differentiating the numerical solution, and these are compared to analytical values in Figure 3. Clearly these piecewise constant values are not very accurate pointwise, although the figure suggests that they may not be so bad in an

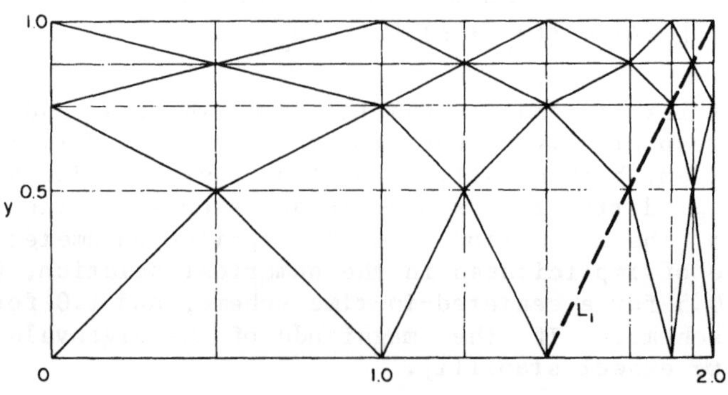

Figure 1. Mesh in dimensionless space for gradient computation test runs.

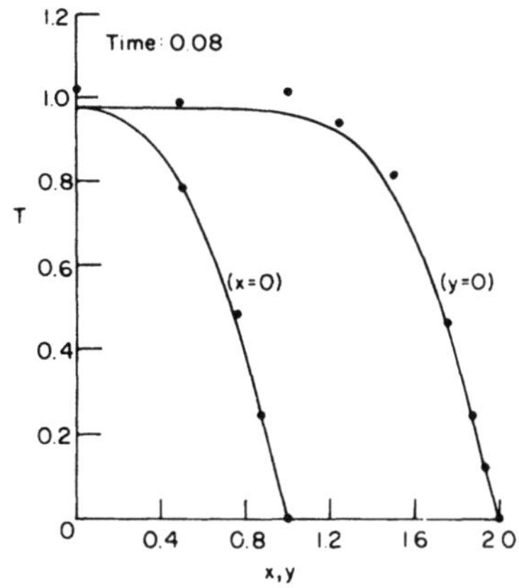

Figure 2. Numerical (dots) versus analytical temperature values along the axes in Figure 1.

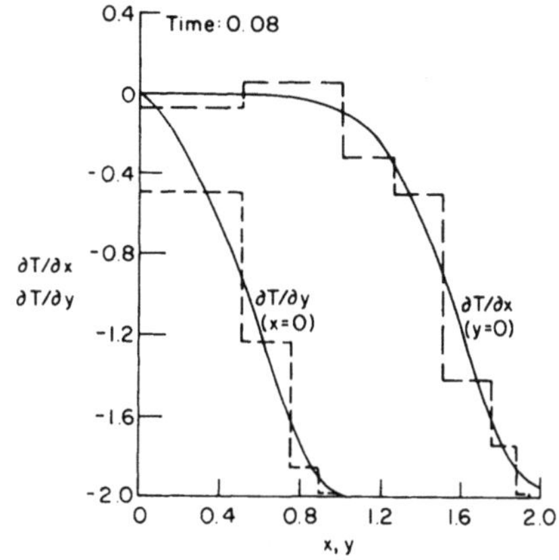

Figure 3. Comparison of numerical (dashed) gradient values with analytical curves along the axes in Figure 1.

integral sense. In Figure 4 analytical values are compared to those obtained numerically from the same solution which produced Figure 3, after the additional step described above (ie, from solution of a system analogous to equation (14)). Gradient values vary linearly between the nodes, and are obviously a much more accurate representation pointwise. In performing the integration like that on the right hand side of equation (14), one obtains nodal gradient values which are effectively averaged from nearby elements. The accuracy of the results attests to the fact that the piecewise linear gradient system was obtained from a rational, error minimization formulation using the numerical results, not from an arbitrary averaging. While the initial solution of the stream function equation attempts to conserve mass, differentiation of it (Figure 3) degrades its virtues. In effect, we have performed a modified reintegration of this differentiation, obtaining continuous gradient values that better reflect the mass conserving virtues of the original numerical solution.

The virtues of reintegration (integration smoothing) are more dramatically illustrated if one considers a situation in which the derivatives of the original solution are more ambiguous in their relation to a locale to be examined. In the previous case, it was clear that one should select for comparison the derivatives of the original solution from elements with sides along the axes. This obscured the fact that the original solution had produced gradients in other elements near the boundary, ie with one node touching the boundary, which do not appear in Figure 3. The reintegration procedure brings into play gradients in all elements contiguous to a locale of interest. This is emphasized by an examination of results near

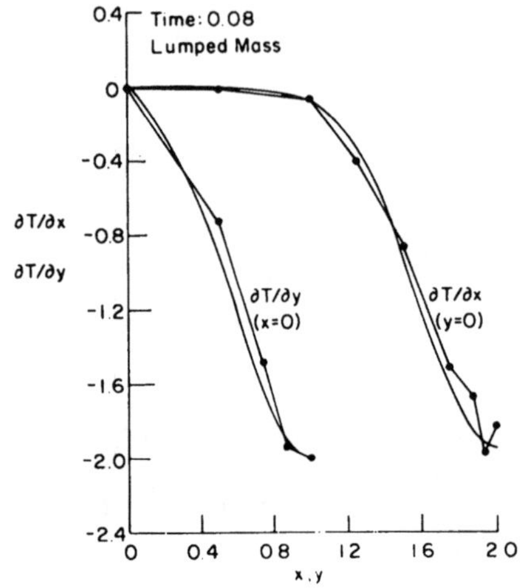

Figure 4. Reintegrated gradient values for the same case as in Figure 3.

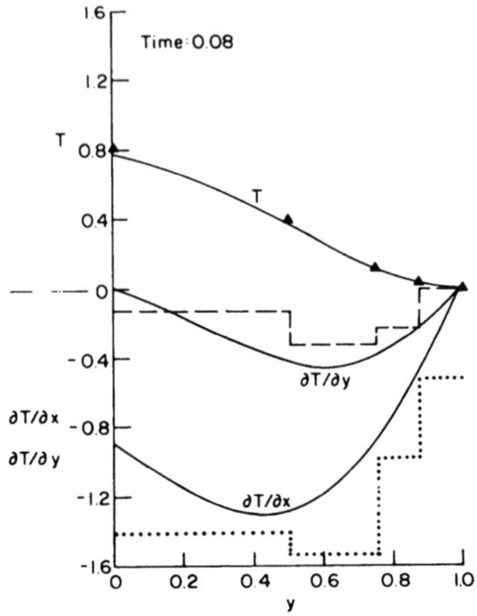

Figure 5. Numerical temperature and gradient values compared to analytical curves along line L1 in Figure 1. Gradients were obtained from elements below the line.

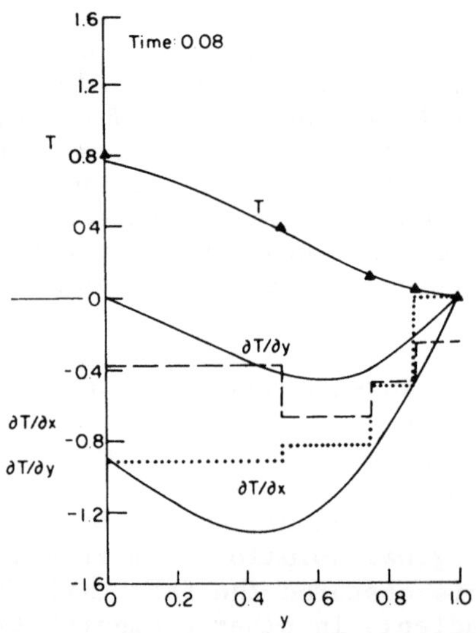

Figure 6. Same as Figure 5, except gradients were obtained from elements above line L1.

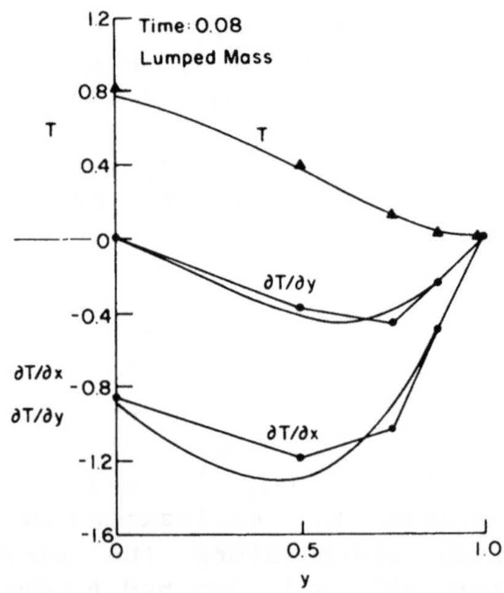

Figure 7. Same as for Figures 5 and 6, except that C^0 continuous gradient values are shown.

the dashed line L1 in Figure 1. Gradient values from the original numerical solution are fundamentally ambiguous along that line, because they suffer a jump as one approaches the line from either side. Figure 5 shows original derivative values near the line from elements below, while Figure 6 shows derivatives from above it. Clearly the values are quite poor, overall, although it appears that a combination of the two would produce a superior approximation. The reintegration method accomplishes this combination and also accounts for gradients in elements touching the line but not along an entire side (Figure 7).

COMPUTED RESULTS: NATURAL CONVECTION

Figures 8 through 14 show results from simulations of heat flow with phase change in the presence of natural convection, using the method outlined above for obtaining the integration smoothed liquid flow rates from the stream function solution. A linear density fluid was assumed and the entire system was formulated on a moving deforming mesh. That is, equations (3), (7), (8), and (9) were solved, with flow rates determined from (13) and an analogous equation for U. Parameters were selected which were deemed reasonable for a porous medium, in particular

$$C_r = 1.5 \qquad St = 0.5$$

$$C^u_{pm} = 1.33 \qquad Ra = 100 \tag{16}$$

$$K^f_h = 1.5 \qquad \theta_w = 1.0$$

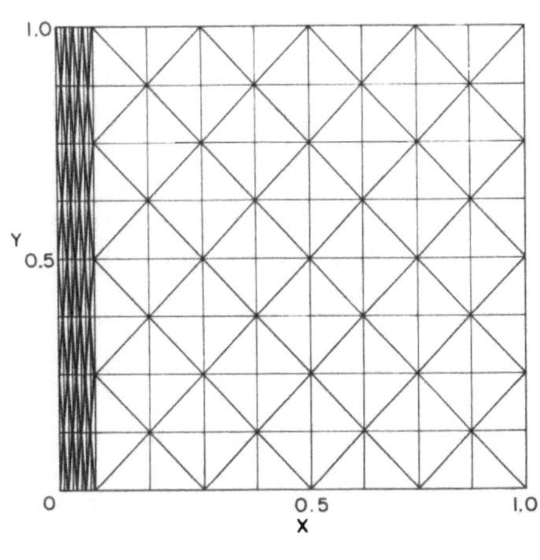

Figure 8. Mesh in dimensionless space used in natural convection simulations. Dense mesh on the left covers the frozen zone.

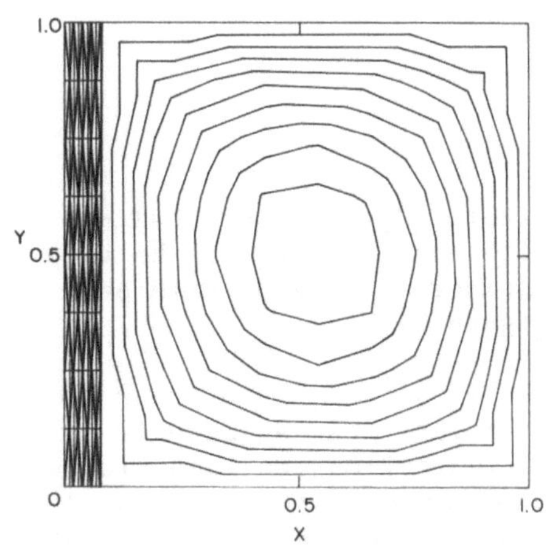

Figure 9. Streamline pattern with equal increments between the streamlines, in early time.

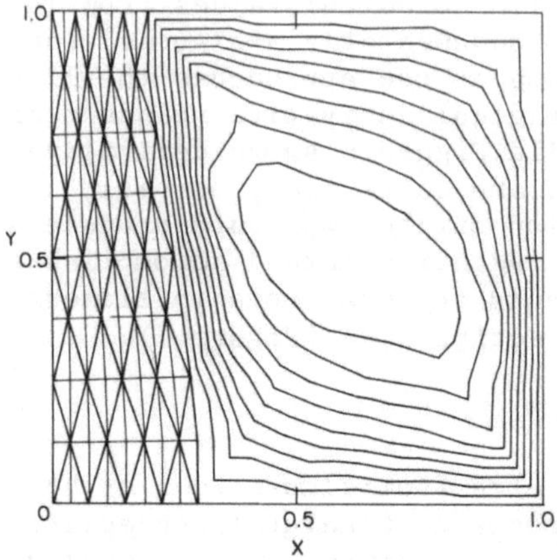

Figure 10. Streamline pattern and frozen zone mesh at an intermediate time.

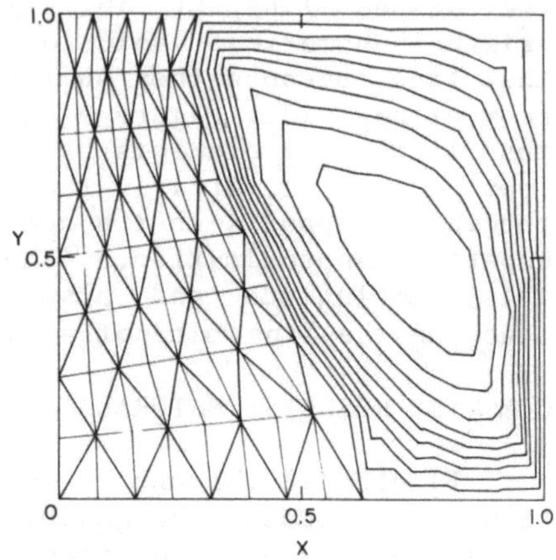

Figure 11. Streamline pattern and frozen zone mesh at steady state.

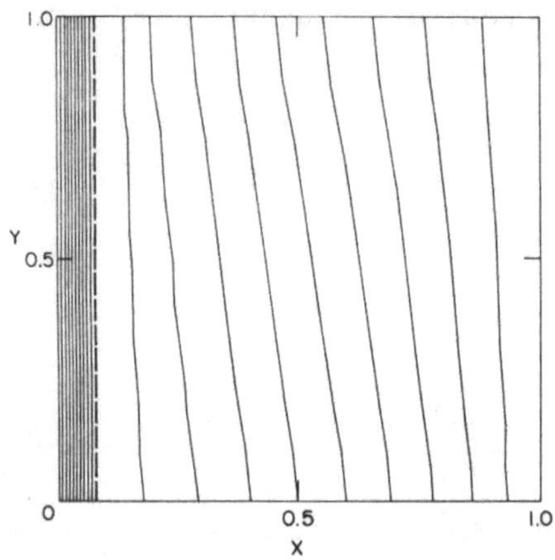

Figure 12. Isotherm pattern, with equal increments between isotherms, at same time as Figure 9.

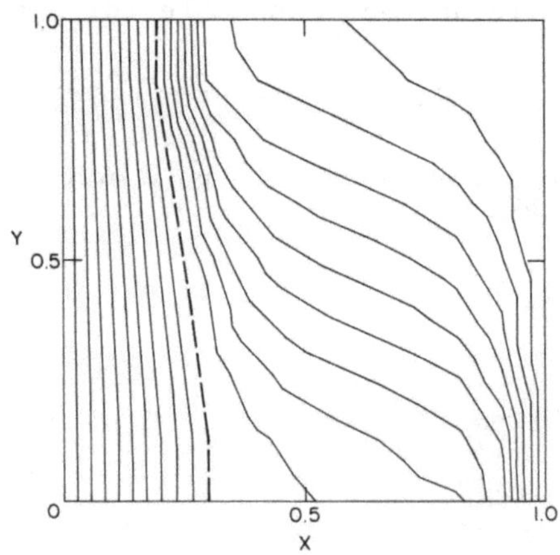

Figure 13. Isotherm pattern at same time as Figure 10.

The initial mesh is shown in Figure 8. The top and bottom of the region are insulated while the left boundary is kept cold at a temperature of −1, with the right boundary kept warm at +1. All boundaries are impermeable. The densely discretized region on the left represents an assumed initial frozen zone, in which the mesh is simply a compacted version of that in the unfrozen zone. The initial temperature was assigned to be +1 over the unfrozen zone, with a linear variation of temperature across the frozen zone from −1 to 0.

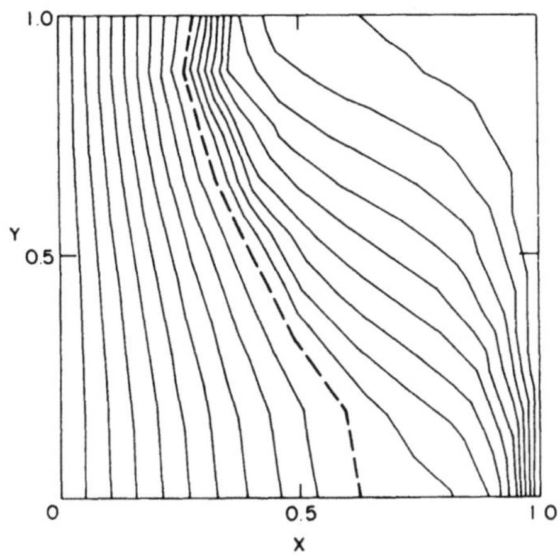

Figure 14. Isotherm pattern at steady state.

Figures 9 through 11 show the evolution of the streamline pattern as freezing progresses. Falling fluid near the freezing front and rising fluid near the warm boundary cause a convection pattern which clearly influences the morphology of the frozen zone. This in turn influences the pattern of convection, until a steady state is achieved as shown in figure 11. The mesh is shown in the frozen zone in these figures, and it is clear that the boundary and mesh steering system function smoothly.

Figures 12 through 14 show isotherm patterns at the same time steps as in the previous three figures. At first the strength of convection is moderate, but it intensifies as the freezing front approaches the warm boundary, as evidenced by increased distortion of the isotherms. This is because both the freezing front and the warm boundary are at constant temperatures, and the narrowing unfrozen zone induces larger density gradients.

SUMMARY AND CONCLUSIONS

These modeling exercizes illustrate the use of a moving boundary - moving mesh system for attacking two-dimensional phase change problems with convection. The system employed was reasonably efficient and produced credible sequences of flow pattern, temperature change, and phase morphology. A rational system for integration smoothing of solution gradients was testd against analytical solutions and it performed very well in terms of local accuracy and efficiency. Use of this system for liquid flow rate determination from the numerical stream function values allowed specification in the temperature equations of a smooth and presumably accurate C^0 continuous liquid flow field. This was possible despite the irregularities and angles in the original streamline solution, which was piecewise linear. These results pave the way for application to more complex phase change situations in which convection is important, in particular, those in which the density inversion of water produces more complicated flow patterns. Computations have already been carried out showing credible configurations of secondary circulation due to the inversion, when phase change is not present. Lab data for cases with freezing are now being developed which will ultimately be used to verify all aspects of the program.

REFERENCES

Albert, M.R. (1983). Modeling two-dimensional freezing using transfinite mappings and a moving mesh finite element system, M.S. Thesis, Thayer School of Engineering, Dartmouth College.

Albert, M.R. and K. O'Neill (1983). The use of transfinite mappings with finite elements on a moving mesh for two-dimensional phase change, in Adaptive Computational Methods for Partial Differential Equations, Babuska et al (eds), SIAM, Philadelphia.

Albert, M.R. and K. O'Neill (1983). 2-D transient freezing in a pipe with turbulent flow, using a continually deforming mesh with finite elements, Proc. 3rd Int'l Conf. Num. Meth. Thermal Prob., Lewis et al (eds), Pineridge Press, 102-111.

Comini, G., S. DelGuidice, R.W. Lewis, and O.C. Zienkiewicz (1974). Finite element solution of nonlinear heat conduction problems with special reference to phase change, Int'l J. Num. Meth. Eng., 8: 613-624.

Connor, J.J. and G.C. Christodoulou (1981). Stability of finite element models for convection - diffusion, Interdisciplinary Finite Element Analysis, Abel et al (eds), Cornell University, 633-666.

Lynch, D.R. (1982). A unified approach to simulation on deforming elements with application to phase change problems, J. Comp. Phys., 47: 387-411.

Lynch, D.R. and K. O'Neill (1981). Continuously deforming finite elements for solution of parabolic problems with and without phase change, Int'l J. Num. Meth. Eng., 17: 81-96.

O'Neill, K. (1981). Highly efficient, oscillation free solution of the transport equation over long times and large spaces, Wat. Resour. Res., 17: 1665-1675.

O'Neill, K. (1983). Fixed mesh finite element solution for Cartesian two-dimensional phase change, ASME J. Energy Resour. Techn., 105: 436-441.

O'Neill, K. (1984). Stability analysis of finite element flow formulations in fixed and moving mesh transport computation, in preparation.

O'Neill, K. and D.R. Lynch (1979). A finite element solution for porous medium freezing, using Hermite basis functions and a continuously deforming coordinate system, Proc. 1st Int'l Conf. Num. Meth. Thermal Prob., Lewis and Morgan (eds), Pineridge Press, 548-559.

Yoo, J. and B. Rubinsky (1983). Numerical computation using finite elements for the moving interface in heat transfer problems with phase transformation, Num. Heat. Transf., 6: 209-222.

Yeh, G-T (1981). On the computation of Darcian velocity and mass balance in the finite element modeling of groundwater flow, Wat. Resour. Res., 17: 1529-1534.

NOMENCLATURE

c_{pm} = volumetric porous medium heat capacity

$C^u_{pm} = c^u_{pm}/c^f_{pm}$

$C_r = (\rho c)_\ell/c^u_{pm}$

$\frac{dS}{d\tau}$ = dimensionless rate of motion of phase interface normal to the frozen zone

g = acceleration of gravity

H = reference length

$K^f = k^f_h/k^u_h$

k_h = thermal conductivity

k_d = permeability

L = latent heat of fusion per unit volume frozen material

N = dimensionless coordinate normal to the phase interface, increasing outwards from the frozen zone

p = pressure

$P = \frac{p + \rho_0 g y}{(\rho_0 - \rho_1)gH}$

$Ra = \frac{k_d H(\rho_0 - \rho_1)g}{\mu\kappa^u}$ = Rayleigh number

$R = \frac{\rho - \rho_0}{\rho_0 - \rho_1}$

$St = \frac{c^f_{pm}(T_p - T_c)}{L}$ = Stefan number

t = time

T = temperature

T_p = phase change temperature

T_c = reference cold (boundary) temperature

T_w = reference warm (boundary) temperature

U = dimensionless X direction liquid flow rate

$\mathbf{v}$ = liquid flow rate

$\mathbf{V}$ = dimensionless liquid flow rate

V = dimensionless Y direction flow rate

x = horizontal space coordinate

x_0 = reference space coordinate, attached to mesh

X = x/H

$\mathbf{X}_j$ = nodal value of (X,Y)

y = vertical space coordinate, positive upwards

Y = y/H

$\theta = \frac{T - T_p}{T_p - T_c}$

$\theta_w = \frac{T_w - T_p}{T_p - T_c}$

$\kappa^u = k^u_h / c^u_{pm}$

μ = dynamic molecular viscosity of water

ρ = liquid density

ρ_0 = maximum liquid density considered

ρ_1 = minimum liquid density considered

$(\rho c)_\ell$ = volumetric heat capacity of liquid

$$\tau = \frac{k_h^{\mu}}{H^2 \ c_{pm}^f} \ t$$

ϕ_j = basis function associated with jth node

ψ = dimensionless stream function

FINITE ELEMENT APPROXIMATION OF NONLINEAR VARIATIONAL INEQUALITIES ARISING IN POROUS MEDIA.

Muhammad Aslam Noor
Mathematics Department, College of Science, King Saud University, Riyadh, Saudi Arabia.

ABSTRACT

In this paper, we are concerned with the approximate solution by the finite element technique of a fairly large class of nonlinear variational inequalities encountered in the study of a model of groudwater flow in a partially saturated porous media. Using the smooth perturbation(regularization) of variational inequalities and piece-wise linear conforming elements, we show that the error for the approximate solution of nonlinear variational inequalities is of order h in the energy norm. In fact, our estimates improve all of the previous known results for elliptic variational inequalities. We also discuss special cases, which can be obtained from our general result.

1. INTRODUCTION

The development of the finite element method has had a unifying effect on solving the variational inequalities numerically, as the method is general, being applicable to both linear and nonlinear problems and accommodates changing in geometry and boundary conditions. The mathematical theory of finite element analysis for linear variational inequalities has been extensively studied and may be considered as a classical part of the numerical analysis for partial differential equations, see Glowinski, Lions and Tremolieres[1981] and Ciarlet [1978]. Recently considerable efforts have been made in order to give a complete mathematical theory for finite element technique in solving variational inequalities containing the terms defining the frictional forces, see Noor [1984], Oden and Pires [1982], and Glowinski et al [1981]. It is known that the functionals representing

the frictional forces are non-differentiable, so they represent a major difficulty in applying the finite element technique effectively coupled with the minimization of the associated potential energy functional. To overcome this difficulty, one usually uses the perturbation technique for the non-differentiable functionals. Such type of perturbation techniques of non-differentiable is known as the regularization method and has been used in nonlinear operator theory and optimization by Brezis [1973]. Campos, Oden and Kikuchi [1982], Glowinski et al [1981] and Noor [1984] used this regularization method as a basis for obtaining the numerical solution of variational inequalities.
We, in this paper, use this regularization of variational inequalities coupled with piecewise linear conforming elements to obtain the error estimates for the approximate solution of nonlinear variational inequalities, which are of order h in the energy norm. Section 2 is devoted to a variational inequality formulation. A perturbed variational principle is introduced in Section 3 and its main properties are discussed. In Section 4, we derive the main results of this paper. Due to the space limitation, we only give few details of the results obtained. The complete results and proofs will appear elsewhere.

2. VARIATIONAL INEQUALITIES FORMULATION.

The general problem of finding the discharge u of the the groundwater flow through a partially saturated media from an aquifer beneath a river into the river channel, see Noor [1984] for mathematical formulation of this model, can be characterized by a class of highly nonlinear variational inequalities of the type:

$$\langle Tu, v-u\rangle + j(v)-j(u) \geq \langle f, v-u\rangle, \quad \text{for all } v\varepsilon K \tag{1}$$

Here

$$\langle Tu, v\rangle = \int_\Omega K(\rho)\nabla u\nabla v \, dxdy, \tag{2}$$

$K(\rho)$ is the pressure-dependent soil conductivity.

$$j(v) = \int_\Omega \{v-(\tfrac{1}{2}h^2-\tfrac{1}{2}\tfrac{1}{b}(T^2-h^2)(x-b)\}^+ dxdy \tag{3}$$

$$\langle f, v\rangle = \int_\Omega v dxdy$$

$H = \{v\varepsilon H^1(D),\ v = 0 \text{ a.e. on } D\}$, a Hilbert Space,

$K = \{v\varepsilon H;\ v \geq 0 \text{ on } D\}$, a closed convex set in H.

D is the whole domain, Ω is the unknown domain, T is the depth of the draining and h is the depth of the impressible layer. Here and throughout this paper, we emply notations and conventions commonly used in the study of partial differential equations in Sobolev spaces, see Ciarlet [1978] and Kikuchi and Oden [1979].

It has been shown by Noor [1984] that under certain conditions on the nonlinear term $K(\rho)$, the operator $T : H \longrightarrow H'$, the dual space of H, defined by (2) is strongly monotone and Lipschitz continuous, that is, there exist constants $\rho > 0$, $\mu > 0$ such that

$$\langle Tu-Tv, u-v\rangle \geq \rho\|u-v\|^2, \quad \text{for all } u,v\in H \tag{4}$$

and

$$\|Tu-Tv\| \leq \mu\|u-v\|, \quad \text{for all } u,v\in H \tag{5}$$

and the functional $j: H \longrightarrow \mathbb{R}$ is convex and lower semi-continuous, then there exists a unique solution $u\in K$ satisfying (1). It is important to note that the functional j(.) arises due to the resistivity of the partially saturated porous media and represent the frictional forces obeying some kind of Coulomb's law of friction.

We also note that if $j(v)=0$, that is in the absence of the frictional forces, we have the problem of finding $u\in K$ such that

$$\langle Tu, v-u\rangle \geq \langle f, v-u\rangle, \quad \text{for all } v\in K,$$

for which the error estimates have been derived by Noor [1981] for the finite element approximation of the solution.
The case, when $K(\rho)$ is independent of the soil conductivity pressure, has been considered by Oden and Kikuchi [1980].

3. A PERTURBED VARIATIONAL PRINCIPLE

It is clear that the operator T, defined by (2) is symmetric, so it is possible to consider the functional

$$I[v] = \tfrac{1}{2}\langle Tv, v\rangle + j(v) - \langle f, v\rangle, \tag{6}$$

associated with (1), which is known as the potential energy functional. It has been shown by Noor [1975, 1982] that the minimum of I[v] on a convex set K in H can be characterized by a class of nonlinear variational inequalities of type (1). Furthermore, we note that the functional I[v] is convex and lower semi-continuous, hence there exists a unique minimizer of

$I[v]$, see Ekland and Temam [1976]. Thus we conclude from the above observations that the function $u \in K$, which minimizes $I[v]$, is also solution of (1) and the converse is also true. It is clear that the functional $j: H \longrightarrow \mathbb{R}$ is non-differentiable, which represent a main difficulty in using the finite element technique effectively. It has been shown by Oden and Pires[1982], and Glowinski et al [1981] that the direct approximation of (1) by finite elements through the minimization of the associated potential energy functional yields a discrete system, which is not a popular method for solving numerically the nonlinear variational inequalities.
Following the idea of Campos, Oden and Kikuchi [1982] , and Glowinski et al [1981], we consider a F-differentiable perturbation $j_\varepsilon(.)$ of the non-differentiable functional $j(.)$, which is a function of a positive real parameter ε, that approximates $j(.)$ arbitrarily closely as $\varepsilon \longrightarrow 0$. Such type of perturbation is known as the regularization(smooth) technique, which has been used to study non-differentiable functionals arising in the nonlinear operator theory and optimization. To be more specific, we introduce the functional $\phi_\varepsilon : H \longrightarrow H'(D)$ defined by

$$\phi_\varepsilon(x) = \begin{cases} v - \tfrac{1}{2}\varepsilon, & \text{for } |v| > \varepsilon \\ \tfrac{1}{2}\tfrac{1}{\varepsilon}\, v.v, & \text{for } |v| \leq \varepsilon \end{cases} \qquad (7)$$

for given $\varepsilon > 0$. It is obvious that $v \longrightarrow \phi_\varepsilon(v)$ is convex and $\phi_\varepsilon(u+r\theta v) \in C^1[0,1]$ for any $u, v \in H$. We now define the approximation of $j(.)$ as follows:

$$j_\varepsilon(v) = \int g\phi_\varepsilon(v)ds \qquad (8)$$

The following result of Campos et al[1982] shows that $j(v)$ can be approximated by $j_\varepsilon(v)$.

<u>Lemma 1.</u>

If $g \in L^\infty(D)$, $g \geq 0$ a.e., then for all $v \in H$,

$$|j(v) - j_\varepsilon(v)| \leq (2\, \|g\|_{0,\infty,D}\ \text{me s}(D)) \qquad (9)$$

We now define the regularized (perturbed) potential energy functional as:

$$I_\varepsilon : H \longrightarrow \mathbb{R}\,,$$

$$I_\varepsilon[v] = \tfrac{1}{2}\langle Tv, v\rangle - \langle f, v\rangle + j_\varepsilon(v)$$

The functional $I_\varepsilon[v]$ is obviously strictly convex, coercive and F-differentiable on all of H. In this case, it can be shown that $u^\varepsilon \varepsilon H$ can be characterized by the variational equation

$$\langle Tu^\varepsilon, v\rangle + \langle j'_\varepsilon(u^\varepsilon), v\rangle = \langle f, v\rangle, \quad \text{for all } v\varepsilon H, \tag{10}$$

where $j'_\varepsilon(u^\varepsilon)$ is the Frechet differential of $j_\varepsilon(u)$.

Now the question arises whether the solution of (10) converges to the exact solution u of (1) as $\varepsilon \longrightarrow 0$? The answer to this question is in affirmative and this is the main motivation of our next result, which is a generalization of a result of Campos et al [1982], and also provides us an estimate of the rate of convergence.

<u>Theorem 1:</u>
Let $u\varepsilon K$ be the solution of (1) and u^ε, the solution of (10) for fixed $\varepsilon > 0$. Then

$$\|u - u^\varepsilon\|_1 \leq C\sqrt{\varepsilon} \tag{11}$$

where C is a constant independent of ε.

<u>Proof:</u>
Taking $v = u^\varepsilon$ in (1) and $v = u^\varepsilon - u$ in (10) and adding these inequalities, we obtain

$$\langle Tu^\varepsilon - Tu, u^\varepsilon - u\rangle \leq j(u^\varepsilon) - j(u) - \langle j'_\varepsilon(u^\varepsilon), u^\varepsilon - u\rangle \tag{12}$$

Since $j_\varepsilon(u)$ is convex and Frechet differentiable, so by a result of Noor [1975], we have

$$j_\varepsilon(u) - j_\varepsilon(u^\varepsilon) \geq \langle j'_\varepsilon(u^\varepsilon), u^\varepsilon - u\rangle \tag{13}$$

From (4), (12) and (13), we get

$$\rho\|u - u^\varepsilon\| \leq |j(u^\varepsilon) - j_\varepsilon(u^\varepsilon)| + |j_\varepsilon(u) - j(u)|,$$

from which, it follows by making use of lemma 1,

$$\|u - u^\varepsilon\|_1 \leq C\sqrt{\varepsilon},$$

the required result.

In order to derive the error estimates, we assume the following:

(A) = { $u\varepsilon K$ satisfying (1) is also in H^2 }.

4. FINITE ELEMENT APPROXIMATION AND ERROR ESTIMATES

Let $\{T_h\}_{h>0}$ be a regular family, see Ciarlet [1978], of triangulation of D and define

$$S_h = \{v_h;\ v_h \varepsilon C^o(D),\ v_h /\varepsilon P_1 \quad \text{for all } T\varepsilon T_h\},$$

where P_1 is the set of all polynomials on $\mathbb{R}^2$ of degree less and equal to one. Clearly S_h is the finite dimensional subspace of H. The finite dimensional convex set K_h is defined as:

$$K_h = \{v_h : v_h \varepsilon S_h;\ v_h \geq 0 \text{ at every vertex of } T_h\}$$

It is clear that $K_h \not\subset K$. It is enough to consider the case when K_h and K are in H.

The finite dimensional approximation of (1) is:
Find $u_h \varepsilon K_h$ such that

$$\langle Tu_h, v_h - u_h\rangle + j(v_h) - j(u_h) \geq \langle f, v_h - u_h\rangle, \text{ for all } v_h \varepsilon K_h. \quad (14)$$

Similarly, the approximation of the regularized problem (10) consists of finding $u_h^\varepsilon \varepsilon S_h$ such that

$$\langle Tu_h^\varepsilon, v_h\rangle + \langle j_\varepsilon'(u_h^\varepsilon), v_h\rangle = \langle f, v_h\rangle, \quad \text{for all} v_h \varepsilon S_h \quad (15)$$

It is obvious that the relations (14) and (15) are similar to (1) and (10), and all the properties of those forms are carried from the continuos problems to its finite dimensiona approximate problems. We may conclude that all the results concerning existence and uniqueness of solution to the finite dimensional problems follows with similar conclusions.
Following essentially the same technique of theorem 1, we have the following estimate for the finite dimensional case:

$$\| u_h - u_h^\varepsilon \|_1 \leq C\sqrt{\varepsilon}, \quad (16)$$

where C is a constant independent of h and ε, and u_h u_h^ε are the solutions of (14) and (15) respectively.

We now derive the general estimate for the error $u - u_h$.

Theorem 2:
Let u and u_h be the solutions of (1) and (14) respestively, then

$$<Tu-Tu_h,u-u_h> \leq <Tu-Tu_h,u-v_h>+<Tu-f,v-u_h> + <Tu-f,v_h-u>+j(v)-j(u)+j(v_h)-j(u_h)$$
for all $v\varepsilon K$ and $v_h\varepsilon K_h$ (17)

Proof:
Adding (1) and (14), we obtain for all $v\varepsilon K$ and $v_h\varepsilon K_h$,

$$<Tu,v-u>+<Tu_h,v_h-u_h>+j(v)-j(u)+j(v_h)-j(u_h) \geq <f,v-u> +<f,v_h-u_h>.$$

Subtracting from both sides the term $<Tu,v_h>+<Tu_h,v>$, and rearranging the terms, we obtain the required result (17).

We now state and prove the main result of this paper.

Theorem 3:
Let f, $Tu\varepsilon L_2(D)$, and the angle condition be satisfied. If u and u_h are solution of (1) and (14) respectively, hypothesis (A) and $K_h \subset K$, then

$$\|u-u_h\|_1 = O(h).$$

Proof:
Since $K_h \subset K$, so by taking $v=u_h$ in equality (17), we have

$$<Tu-Tu_h,u-u_h> \leq <Tu-Tu_h,u-v_h>+<Tu-f,v_h-u>+j(v_h)-j(u) \qquad (18)$$

Let I_h be the operator of S_h-interpolation. Then, since $D \subset \mathbb{R}^2$, we have $H^2 \subset C^0(D)$, and $u\varepsilon H^2$, $u \geq 0$ on D imply $I_h u\varepsilon K$. Taking $v_h=I_h u$ in (18), we obtain

$$<Tu-Tu_h,u-u_h> \leq <Tu-Tu_h,u-I_h u> +<Tu-f,I_h u-u> j(I_h u)-j(u) \qquad (19)$$

Now using the strongly monotonicity, Lipschitz continuity of T and applying the Young's inequality

$$ab \leq \varepsilon a^2 + \tfrac{1}{4}\tfrac{1}{\varepsilon} b^2, \qquad \text{for all } a,b\varepsilon\mathbb{R}, \text{ and } \varepsilon>0.,$$

we have

$$\|u-u_h\|_1^2 \leq \frac{\beta^2}{\rho^2}\|u-I_hu\|_1^2 + \frac{1}{2\rho}\{ \|Tu-f\|_{L_2(D)} \|u-I_hu\|_{L_2(D)} + |j(I_hu)-j(u)|\} \quad (20)$$

Since the angle condition holds and $u\varepsilon H^2$, it is known see Ciarlet[1978], that

$$\|I_hu-u\|_{r,D} \leq Ch^{2-r}\|u\|_{2,D}, \quad r=0,1 \quad (21)$$

In addition, with mild restriction on D and the space S_h, we have the estimate of the following type , to hold on traces, see Campos et al [1982],

$$|j(I_hu)-j(u)| \leq C_2\| g\|_{\frac{1}{2}}h^2\|u\|_{2,D} \quad (21)$$

Thus from (20)—(21), we obtain

$$\|u-u_h\|_1 = O(h), \quad (22)$$

which is the required result.

Special Cases:
i. Note that for $j(v)=0$, that is inthe absence of the frictional forces, the results of theorems 2 and 3 are exactly the same as proved by Noor [1980].
ll. Furthermore, for the case $K = H$, the results obtained above are identical to the ones derived by Campos et al [1982].

Use of (16), (22) and the triangle inequality gives the final error estimate for the approximate problem (15), which we state in the following:

Theorem 4:
Under the assumptions of theorem 2 and 3, we have

$$\|u-u_h^\varepsilon\|_1 = O(h+\sqrt{\varepsilon}),$$

where u and u_h^ε are the solutions of (1) and (15) respectively.

5. CONCLUSION:

In this paper, we have applied the perturbation technique of non-differentiable functionals to improve the previous know results for the approximate solution of the variational inequalities. It would be interesting to see whether this technique can be applied to other type of complicated variational inequalities with success or not arising in finite elasticity and other branches of engimeering sciences. A complete study of the boundary value problems arising on the formulation of flow in the partially saturated porous media with friction is an interesting and very general problem both in the fluid dynamics and mathematical points of view.

6. ACKNOWLEDGEMENT:

The author would like to thank the Dean, College of Science, Chairman, Department of Mathematics, King Saud University, Riyadh, for providing the excellent research facilities.

REFERENCES

Brezis, H. (1973) Operateurs maximaux monotones et semigroups de contractions dans les espaces d'Hilbert, North-Holland, New York.

Ciarlet, P. (1978) The finite element method for elliptic problems, North-Holland, New York.

Campos, L.T., Oden, T. and Kikuchi, N. (1982) A numerical analysis of contact problems with friction in elastostatics, Comp. Meth. in Appl. Mech. Eng. 34, 821-845.

Duvaut, G and Lions, J. (1972) Les inequations en mechanique et en physique, Dunod, Paris.

Ekland, I. and Temam, R. (1976) Convex analysis and variational problems, North-Holland, New York.

Glowinski, R., Lions, J. and Tremolieres, R. (1981) Numerical analysis of variational inequalities, North-Holland, New York.

Noor, M. A. (1975) On Variational Inequalities, Ph.D. Thesis, Brunel University, U. K.

Noor, M. A. (1980) Finite element error analysis of nonlinear variational inequalities, Zent fur Math. Mech. (ZAMM), 60, 631-632.

Noor, M.A. (1982) Mildly nonlinear variational inequalities, Math. Rev. Anal. Num. Th. Approx. 47, 99-110.

Noor, M. A. (1984) Variational inequalities in partially saturated porous medis, Formulation and error analysis, to appear.

Kikuchi, N. and Oden, T. (1979) Contact problems in elasticity, TICOM Rep. 79-8, The university of texas, Austin, U.S.A.

Oden, T. and Kikuchi, N(1980) Theory of variational inequalities with applications to problems of flow through porous media, Int. J. Eng. Sc. 18,1173-1284.

Pires, E.B. and Oden, T.(1981-82) Error estimates for the approximation of a class of variational inequalities arising in unilateral problems with friction, Num. Funct. Anal. and Opt. 4, 397-412.

FINITE ELEMENT COMPUTATION OF UNSATURATED AND SATURATED GROUNDWATER FLOW IN STRATIFIED AQUIFERS

Bruno Herrling, Heinz M. Leismann

Institute of Hydromechanics, University of Karlsruhe, W.Germany

INTRODUCTION

The computation of both unsaturated and saturated groundwater flow together in one model permits simulation of the water movement associated with contaminant transport from the surface to the groundwater, and further into the saturated zone. Other advantages of such a model, which are of special interest in this paper, are: (1) it allows for a relatively simple handling of the free surface in two-dimensional modelling in a vertical plane, and (2) it can be used to simulate artificial groundwater recharge in regions where the groundwater table is far from the surface and where the underground is stratified.

Three-dimensional flow modelling (e.g. Segol (1977)) requires extensive effort, even with today's efficient computers. For this reason, engineers attempt to solve both saturated and unsaturated flow problems, as in our case, by vertical one-dimensional flow simulation or by two-dimensional modelling in a vertical plane.

In the unsaturated zone the hydraulic conductivity and the saturation with water depend considerably on the pressure head, the dependent variable. The functional relation is described by measured or estimated functions. Generally, to compute unsaturated flow problems several iterations are necessary within each time step. Further, the discretization in time and space must be fine because of the rapid change of conductivity and saturation near a moisture front. Much effort was spent on setting up a program requireing a minimum of iterations and computer time. The most useful algorithms were determined through the use of a straight forward program for one-dimensional flow. The knowledge was then transfered to the program for the two-dimensional computation, as will be explained in the subsequent section.

The results of several one- and two-dimensional simulations show that the programs work well. A practical application, the computation of seepage flow during storm tides into a polder which was built to counteract floods in the Hamburg harbor, is presented.

COMPUTATION OF VERTICAL PLANE UNSATURATED AND SATURATED FLOW

Basic equation

The basic equation describing the time-dependent, two-dimensional vertical flow of water in an unsaturated and saturated aquifer is (see e.g. Neuman (1973))

$$(C + \frac{\theta}{\phi} S_s) \frac{\partial \psi}{\partial t} - (K^r K^s_{ij} \psi,_j + K^r K^s_{i3}),_i - Q = 0 \qquad (1)$$

summing over i,j = 1,3. The symbol $,_i$ denotes partial differentiation with respect to the Cartesian coordinates x_i. The dependent variable is the pressure head ψ. The other symbols denote the following quantities: K^r, the relative hydraulic conductivity ($0 \leq K^r \leq 1$); K^s_{ij}, the conductivity tensor at saturation; θ, the volumetric water content; ϕ, the porosity; S_s, the specific storage; $C = \partial\theta/\partial\psi$; t, the time; and Q, a source or sink term defined at nodes.

Besides initial conditions, there have to be specified boundary conditions. These are prescribed hydrographs of pressure head $\bar{\psi}$

$$\psi - \bar{\psi} = 0 \qquad \text{on } S_1 \qquad (2)$$

and of specific discharge q normal to be boundary

$$q_i n_i + \bar{q} = - (K^r K^s_{ij} \psi,_j + K^r K^s_{i3}) n_i + \bar{q} = 0 \quad \text{on } S_2. \qquad (3)$$

$S = S_1 + S_2$ defines the total boundary, n_i the unit vector normal to the boundary and q_i the specific discharge.

Finite element formulation

Eq. 1 is solved using the Galerkin method in space and the finite difference method in time. Simple triangular finite elements have been chosen with linear shape functions for ψ (not in the first term of Eq. 1, see below) and K^r. On the level of an element this means

$$\psi = N_E \psi_E \qquad (4)$$

$$K^r = N_E K^r_E \qquad (5)$$

with E = 1,2,3 and N_E represents the three shape functions within an element. As in the following, the repeated indices signify summation (Einstein's summation convention) for the time being.

After transforming the second term in Eq. 1, and after having applied Green's theorem and implemented the boundary condition (3), the following integration is carried out on the element level:

$$\int_{A_\Delta} N_{(F)}\ \delta_{FE}\ \bar{C}_{(E)}\ \frac{\partial \psi_E}{\partial t}\ dA = E_{(F)}\ \delta_{FE}\ \bar{C}_{(E)}\ \frac{\partial \psi_E}{\partial t} = B_{FE}\ \frac{\partial \psi_E}{\partial t} \qquad (6)$$

$$\int_{A_\Delta} N_{F,i}\ N_G\ N_{E,j}\ K_G^r\ K_{ij}^s \psi_E\ dA = F_{Fi}\ E_G\ F_{Ej}\ K_G^r\ K_{ij}^s\ \psi_E = D_{FE}\ \psi_E \qquad (7)$$

$$\int_{A_\Delta} N_{F,i}\ N_G\ K_G^r\ K_{i3}^s\ dA = F_{Fi}\ E_G\ K_G^r\ K_{i3}^s = G_F \qquad (8)$$

with E,F,G = 1,2,3 ,

$$\int_{A_\Delta} N_F\ dA = E_F \text{ and } \int_{A_\Delta} N_{F,i}\ N_G\ N_{E,j}\ dA = F_{Fi}\ E_G\ F_{Ej}\ . \qquad (9)$$

N_F describes the Galerkin weighting function and A_Δ the area of an element. K^s_{ij} is specified as a constant tensor within each element. δ_{FE} signifies the Kronecker-delta. The indices in parentheses denote that no summation is allowed, for clearness the components of matrix B_{FE} are presented:

$$B_{FE} = \begin{bmatrix} E_1\ \bar{C}_1 & 0 & 0 \\ 0 & E_2\ \bar{C}_2 & 0 \\ 0 & 0 & E_3\ \bar{C}_3 \end{bmatrix} \qquad (10)$$

The formulation of Eq. 6 includes that for the term $C \cdot \partial\psi/\partial t = \partial\Theta/\partial\psi \cdot \partial\psi/\partial t = \partial\Theta/\partial t$, resulting from the primary continuity equation, no shape function in space, neither for C nor for $\partial\psi/\partial t$ is used. Other authors often call this procedure "lumping of an (element) matrix". The time derivative is only considered node by node. This procedure was chosen to obtain a more expeditious convergence of the solution and to avoid wiggles as is similarly described by Neuman et al. (1975).

For simplicity the term $\Theta \cdot S_s/\phi$ (Eq. 1), which is unimportant in an unsaturated flow, is handled in the same manner as C and can be combined to yield

$$\bar{C} = C + \Theta \cdot S_s/\phi \qquad (11)$$

as used in Eq. 6. S_s is described as a constant within each element. Therefore in a saturated flow region the formulation in Eq. 6 is also correct.

After summation over all elements, which means integration on the whole computational area, and after addition of the boundary integrals (H_L) and consideration of the nodal sink or source flux (Q_L) Eq. 12 results:

$$B_{LK} \cdot \partial\psi_K/\partial t + D_{LK}\,\psi_K = -\,G_L + H_L + Q_L \qquad (12)$$

with L, K = 1,2,...,N (N = number of nodes) and

$$H_F = \int_{S_\Delta} N_F\,\bar{q}\,dS \qquad (F = 1,2)$$

For the time integration a fully implicit backward difference scheme is used. When there are both saturated and unsaturated regions within one model, the Crank Nicolson time integration scheme, generally better for unsaturated flows, does not work. The reason is that the basic equation becomes nearly elliptical in the saturated zone because C is zero and $\Theta \cdot S_s/\phi$ is extremely small.

For the accelerated convergence of the computation, the determination of the parameters C and K^r within the time step is extremely important. In a fully implicit backward scheme the differential equation is solved at the new time. But the evaluation of the parameters C and K^r using the unknown ψ^1 at the end of a time step (superscript 0 indicates values at the beginning of a time step) is not recommendable. Extensive test computations have indicated that the number of iterations can be halved with only a minimal loss in accuracy when

$$K^r = K^r\left(\frac{\psi^0 + \psi^1}{2}\right) \qquad (13)$$

as proposed by Neuman et al. (1975). Further, instead of the use of the best estimation of C resulting from the observance of continuity

$$C = \frac{\Theta^1 - \Theta^0}{\psi^1 - \psi^0} \quad , \qquad (14)$$

a more feasible and nearly equivalent method

$$C = C\left(\frac{\psi^0 + \psi^1}{2}\right) \qquad (15)$$

was examined. Using $C = C(\psi^1)$ or $C = C(\psi^0)$ or averaging them results in a poorer approximation.

In an attempt to accelerate the computation, tests have been made applying only $K^r = K^r(\psi^0)$ and $C = C(\psi^0)$, which means that no iterations are necessary (see Finlayson (1977)). When very small time steps are used (smaller than normally chosen) this can be a useful tool e.g. when calibrating a model (see below). The error is acceptable when one considers the inaccuracy of the soil parameters. There is, however, a great risk: Since the number of iterations does not warn the engineer, as it normally does, when the time step or the element network in space is too coarse, the numerical solution can turn out to be very inaccurate without any possibility of identification. On the other hand, however, the possibility of computing without any iteration as opposed to three to six or more iterations normally required, makes this scheme economically attractive, even when the time step has to be reduced.

Another important advance to save computer time is to consider only the modifications of those element matrices in the global equation set whose ψ-values are in the range that K^r, Θ and C change with the new ψ^1:

$$\psi_{min} \leq \psi^1 \leq 0 \qquad \text{and} \qquad |\psi^1 - \psi^{old}| > \varepsilon \tag{16}$$

with ψ^{old} as the last ψ where the element matrix have been altered and ε a prescribed limit. The matrices do not change when an element remains either saturated ($\psi^1 > 0$) or nearly drained ($\psi^1 < \psi_{min}$), so that these above-mentioned parameters remain constant. The time saved by not having to set up the entire global equation set is particularly significant.

NUMERICAL RESULTS

One-dimensional computations

The findings reported in the preceding section have been obtained by comparing the numerical results of various one-dimensional computations with an analytical solution published by Philip (1969). In Fig. 1, good agreement is shown between analytical and numerical solutions using the suggested algorithms for both a fine and a coarse discretization. The one-dimensional column has a height of 2m; the node distance amounts to 1 cm and 5 cm, respectively; the time steps are in the first case 10 s. In the second case, the time steps are 100 s after 8 smaller ones. In each time step 3 or 6 iterations were necessary for the fine or the coarse discretization, respectively. The soil parameters apply the fine sands and have been taken over from Haverkamp et al. (1977). The results demonstrate that a coarse discretization essentially describes the solution with less effort.

To prepare the computations of practical case studies, a seepage flow through a layered soil column has been investigated. The results presented in Figs. 2 and 3 describe a situation, in which a sample of 5 m height (including a 1 m layer of lower conductivity in the middle) was filled and drained by increasing and decreasing the water inflow at the top. At the top and bottom $K^s = 10^{-4}$ m/s and $\phi = 0.07$ have been chosen; in the middle, $K^s = 10^{-5}$ m/s and $\phi = 0.035$. The node distance amounts to 5 cm and the time step is constant at 20 s. In each time step 3 iterations were necessary. The dimensionless parameter curves from Haverkamp et al. (1977) have been chosen. For the soil in the middle of the column the curves were shifted 10 cm in the direction of the negativ ψ-axis. In Fig. 2 starting with a steady state initial condition the infiltration at the top is suddenly modified from 10^{-6} m/s to $2 \cdot 10^{-5}$ m/s. The boundary condition at the bottom is constant, $\bar{\psi} = 0.5$ m.

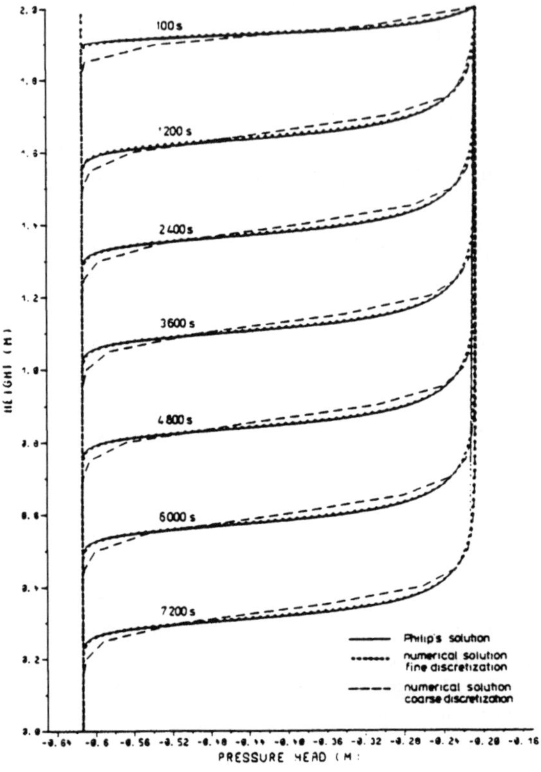

Fig. 1: Comparison of numerical solution with Philip's analytical solution.

The left side of Fig. 2 shows the development of the preasure head profile and the right side that of the saturation profile at time intervals of 1000 s. Fig. 3 demonstrates the reverse situation: The inflow at the top suddenly decreases from 2 • 10^{-5} m/s to 10^{-6} m/s, the boundary condition at the bottom remains constant as before. The profiles of pressure head and saturation are shown at intervals of 1000 s each. The horizontal dashed line marks the layer with low conductivity; the vertical dashed line in the pressure head profiles describes 95% saturation.

The results demonstrate that the program is a useful tool to investigate seepage problems in stratified aquifers with a deep groundwater table as indicated. Further, some general observations can be made: (1) It takes much more time to reach a steady state situation when the infiltration decreases than when it increases. (2) When there is an unsaturated zone beneath a layer of lower conductivity with a saturated condition on top, then the passage from the saturated to the unsaturated zone takes place in the region with low permeability.

<u>Two-dimensional infiltration in a stratified unsaturated aquifer</u>

As a first test principle, the time dependent situation of an artificial groundwater recharge from a ditch being introduced into a stratified aquifer with a groundwater table deep under the surface (Herrling (1983)) is investigated. Previously this problem has been approached in a different manner, in which a horizontal computation on top of a leakage aquitard with moving boundaries was used (Herrling (1982)).

The computation is performed only for the right part of the symmetrical flow field (Fig. 4). Besides the closed boundary S_2 with $\overline{q}$ = 0, a boundary S_1 with $\overline{\psi}$ = 0.5 m of width 1 m is

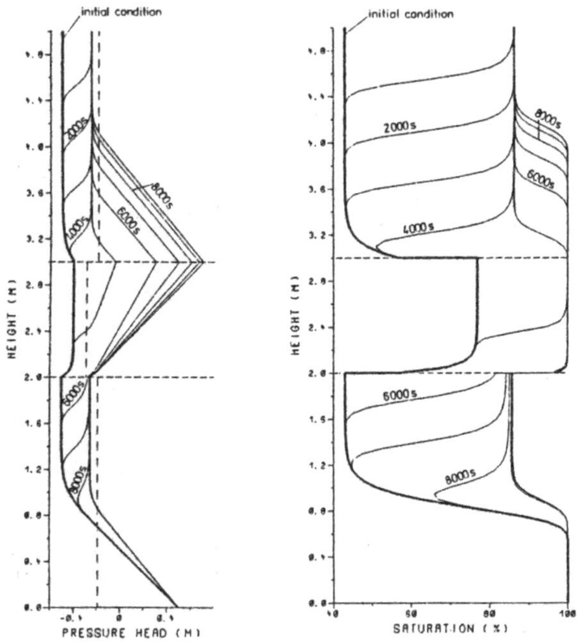

Figure 2: Profiles of pressure head and saturation at intervals of 1000 s after a suddenly increased infiltration.

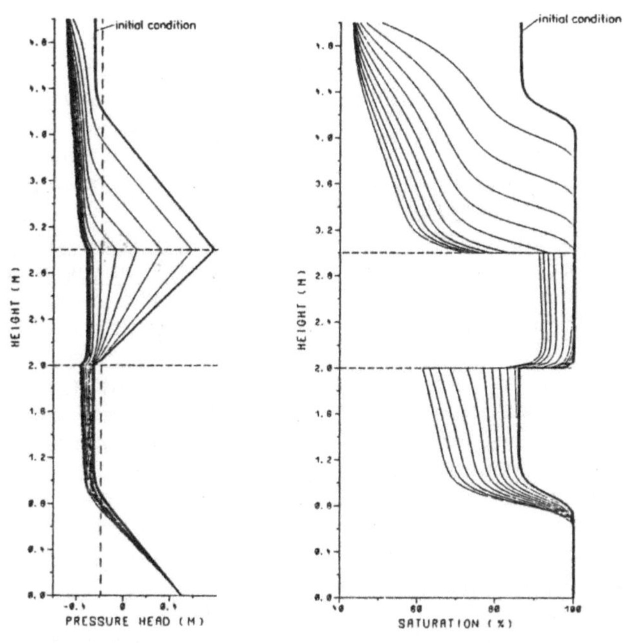

Figure 3: Profiles of pressure head and saturation at intervals of 1000 s after a suddenly decreased infiltration.

prescribed on top (flow field shown is 15 m times 9 m); at the right side, the boundary condition is a hydrostatic $\overline{\Psi}$ with a groundwater level 2 m above the bottom. There are three isotropic layers marked by horizontal lines in Fig. 4; from the top down $K^s = 10^{-4}$ m/s, $\phi = 0{,}15$; $K^s = 5 \cdot 10^{-6}$ m/s, $\phi = 0.1$; $K^s = 10^{-3}$ m/s, $\phi = 0.2$. For the soil parameters the dimensionless parameter function published by Haverkamp et al. (1977) have been taken. 841 nodes and 1550 elements have been used with the smallest vertical node distance of 0.10 m in the middle layer. Employing time steps of 60 s, 2 or 3 iterations were necessary. Starting with a hydrostatic initial without any infiltration condition Fig. 4 shows the flow situation and the zero pressure line (dashed line) after 105 min, 300 min, 720 min, the last being nearly steady state. The results give a good description in what happens in a stratified aquifer.

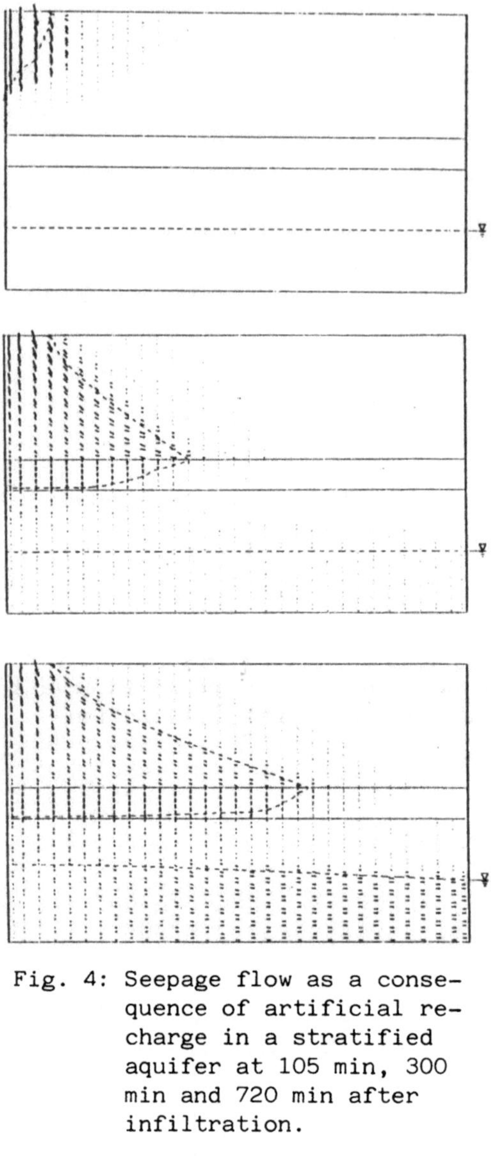

Fig. 4: Seepage flow as a consequence of artificial recharge in a stratified aquifer at 105 min, 300 min and 720 min after infiltration.

<u>Seepage flow in high water polders during sturm tides</u>

After the very momentous storm tide in January of 1976, during which the quays of Hamburg harbor were flooded to depths exceeding 0.75 m, much effort was expended to prevent a recurrence by the construction of polders. The protection consists mainly of sheet pilings surrounding the endangered areas. Up to now no numerical investigation has been carried out to examine in what way future extreme high waters might damage these areas by a seepage flow coming from beneath. In a case study this flow situation was investigated using the described algorithms.

The model area (Fig. 5) situated in the Hamburg harbor was selected because data of a groundwater gauge and of underground analysis from several borings are available. From the preliminary model calibration hydraulic conductivities (m/s) shown in Fig. 5 result; the porosity ranges from 0.05 to 0.15. The left and bottom boundaries are closed. At the right there is a boundary with prescribed $\overline{\Psi}$ values (tide in the Elbe river). Constant time steps of 5 min were used. The calibration was done applying no iteration (see above) after it had been shown that the error in the numerical solution without iteration was very small for this case.

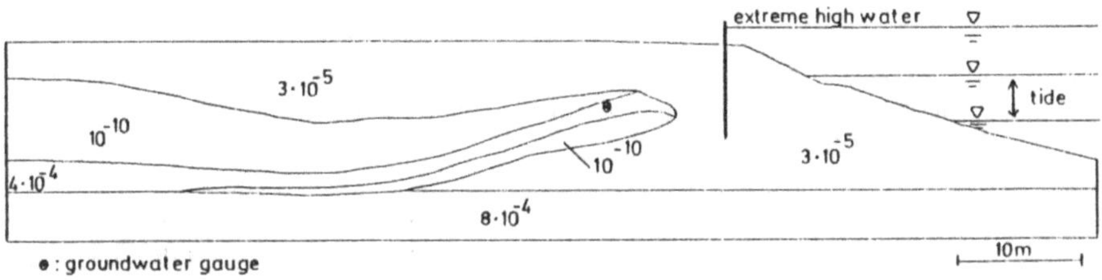

Figure 5: Model area for a seepage flow computation for a polder in Hamburg harbor (K^S in m/s).

The preliminary results of the calibration are demonstrated in Fig. 6. Studying a storm tide under conditions as defined by the local authorities no seepage water emerges within the polder in the case under consideration. But another underground situation could give different results. The program is a useful tool to investigate such flow conditions.

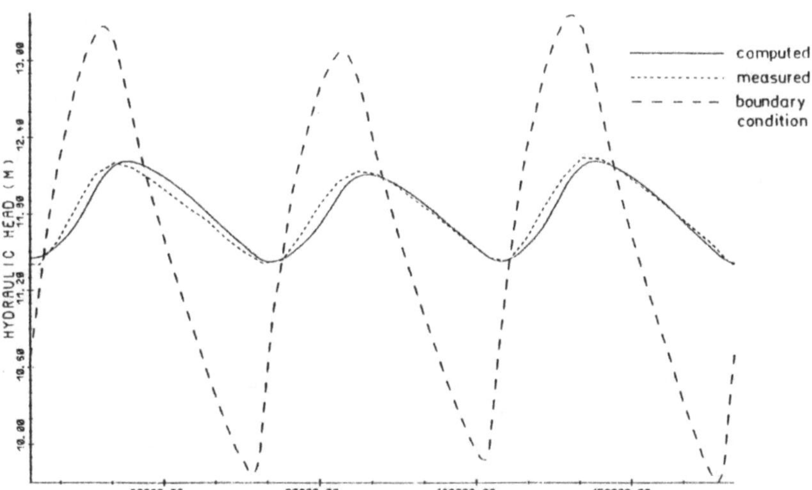

Figure 6: Preliminary results of a model calibration.

ACKNOWLEDGEMENTS

The investigation was supported by the IHP (International Hydrological Programme), German Committee (Grant No. IHP/Th/44). The authors express their appreciation for their subsidy. Further, they thank the consulting office "IGB (Ingenieurbüro für Grundbau und Bodenmechanik)" (Dr. K. Günther) for providing them with data from a polder in the Hamburg harbor.

REFERENCES

Finlayson, B.A. (1977) Water Movement in Desiccated Soils. In: Finite Elements in Water Resources, ed. by Gray et al., Pentech Press: 3.91-3.106.

Haverkamp, R., Vauclin, M., Touma, J., Wievenga, P.J., Vachaud, G. (1977) A Comparison of Numerical Simulation Models for One-Dimensional Infiltration. Soil Sci. Soc. Am. S., Vol. 41: 285-294.

Herrling, B. (1982) Finite Element Computations of Horizontal Groundwater Flow with Moving Boundaries. In: Finite Elements in Water Resources, ed. by K.P. Holz et al., Springer: 10.25 - 10.39.

Herrling, B. (1983) Artificial Groundwater Recharge Using High-Waters of Streams. In: Groundwater in Water Resources Planning, IAHS Publ. No. 142, 795-804.

Neuman, S.P. (1973) Saturated-Unsaturated Seepage by Finite Elements. J. Hydr. Div. Amer. Soc. Civil Eng., Vol. 99 (HY 22): 2233-2250.

Neuman, S.P., Feddes, R.A., Bresler, E. (1975) Finite Element Analysis of Two-Dimensional flow in Soils Considering Water Uptake by Roots: I. Theory. Soil Sci. Soc. Amer. Proc. Vol. 39: 224-237.

Philip, J.F. (1969) Theory of Infiltration. In: Advances in Hydroscience, ed. by Ven Te Chow, Acadèmic Press, New York: 216-296.

Segol, G. (1977) A Three-Dimensional Galerkin-Finite Element Model for the Analysis of Contaminant Transport in Saturated-Unsaturated Porous Media. In: Finite Elements in Water Resources, ed. Gray et al., Pentech Press: 2.123-2.144

SECOND ORDER EFFECTS OF UNCERTAIN TRANSMISSIVITIES ON PREDICTIONS OF PIEZOMETRIC HEADS

Lloyd R. Townley

Centre for Water Research
The University of Western Australia

ABSTRACT

Finite element techniques are frequently used for deterministic analysis of steady state two-dimensional aquifer flow. When the spatial distribution of transmissivities is uncertain, however, the usual solution for piezometric heads is only correct to first order. Using a Taylor series expansion, it is shown that the mean distribution of heads consists of the usual solution plus a second order correction. Numerical results are presented which show agreement between Monte Carlo solutions and the second order Taylor series solutions for expected heads. An efficient procedure is described which avoids the explicit evaluation of the extremely large second derivative matrix.

INTRODUCTION

There has been considerable interest in the recent hydrologic literature in the role of parameter uncertainty in groundwater flow modelling. Studies in one, two and three dimensions have considered the natural variability of permeabilities using theories of continuous spatial stochastic processes. Different analytical approaches include studies in the frequency domain (e.g. Mizell et al., 1982) and in the spatial domain (e.g. Dagan, 1982).

The purpose of this paper is to present another approach, in which concepts of uncertainty are introduced after the formulation of a classical finite element model. Whereas much of the literature has focussed on estimating standard deviations of predicted heads (i.e. a measure of uncertainty in our predictions), this paper concentrates on a second order effect which modifies the predicted distribution of heads itself. Papers by Sagar (1978) and Dettinger and Wilson (1981) cover the theoretical aspects in more detail than is possible here.

MATHEMATICAL FORMULATION

We restrict our attention in this paper to steady state flow in aquifers of large areal extent. Flow systems of this type are well described by the two-dimensional hydraulic equation:

$$\frac{\partial}{\partial x}\left(T\frac{\partial\phi}{\partial x}\right) + \frac{\partial}{\partial z}\left(T\frac{\partial\phi}{\partial z}\right) + Q = 0 \tag{1}$$

in which ϕ is the depth-averaged piezometric head, (x,z) is a horizontal coordinate system, T is the aquifer transmissivity (assumed to be isotropic in this work), and Q represents sources and sinks, such as pumping from wells, evapotranspiration and recharge. In practical applications, with complex aquifer geometries and distributions of T and Q, it is common to discretize (1) using finite element techniques. The problem reduces to solving:

$$\underline{K}\,\underline{h} = \underline{f} \tag{2}$$

where $\underline{h}$ is an (nxl) vector of nodal heads, $\underline{K}$ is an (nxn) symmetric positive-definite matrix which depends on the grid geometry and the spatial distribution of T, and $\underline{f}$ is an (nxl) flux vector which depends on both Q and boundary fluxes at the horizontal extremes of the aquifer. Fixed head boundary conditions are included in (2) by adding a large number N to the diagonal of $\underline{K}$, and the product of N with the known boundary heads to $\underline{f}$.

In this paper, all numerical results are obtained using a simple linear triangular finite element model, however similar results can be expected with any numerical discretization technique. The matrix equation (2) represents a class of algebraic aquifer models, and each discretization technique is simply a set of rules for defining the structure of $\underline{K}$ and $\underline{f}$. Mainly to facilitate contouring of transmissivities, we assume that the spatial distribution of T is parameterized by an (nxl) vector $\underline{t}$ of nodal values of T; other parameterizations, such as defining element values of T, or constant values of T over zones of elements, simply change the dimension of $\underline{t}$ and the rules for constructing $\underline{K}$ from $\underline{t}$.

In situations where the distribution of T is known with certainty, (2) provides the deterministic solution for the distribution of piezometric heads. Alternatively, we can assume that $\underline{t}$ is an uncertain vector with mean $\underline{m}_t$ and covariance $\underline{P}_{tt}$; these two quantities then provide an approximate description of the likely range of values of T. It follows that the distribution of heads is also uncertain, and we therefore seek a best estimate (in some sense) of $\underline{h}$.

An approximate method for obtaining the expected value and the covariance structure of $\underline{h}$ can be devised by expanding $\underline{h}$ in a Taylor series about the expected value $\underline{m}_t$ of the uncertain model parameters. Dettinger and Wilson (1981) adopted a notation developed by Vetter (1973) for the manipulation of derivatives of matrices with respect to other matrices, and a similar notation is adopted here. A summary of the notation is given in Appendix A.

The first three terms in the expansion of $\underline{h}$ and $\underline{m}_t$ can be written as:

$$\underline{h}(\underline{t}) = \underline{h}(\underline{m}_t) + (D_{\underline{t}} \cdot \underline{h})(\underline{t} - \underline{m}_t) + \frac{1}{2}(D_{\underline{t}} \cdot D_{\underline{t}} \cdot \underline{h})(\underline{t} - \underline{m}_t)^{*2} \qquad (3)$$

where $D_{\underline{t}} \cdot \underline{h}$ is an nxn first derivative or sensitivity matrix, $D_{\underline{t}} \cdot D_{\underline{t}} \cdot \underline{h}$ is an nxn^2 second derivative matrix, and $(\underline{t} - \underline{m}_t)^{*2}$ is an n^2x1 vector defined in Appendix A. To first order, the expected heads are obtained from the first two terms of (3), namely:

$$E(\underline{h}(\underline{t})) = \underline{m}_h \overset{1}{=} \underline{h}(\underline{m}_t) \qquad (4)$$

where $\overset{1}{=}$ implies equality to first order. Equivalently, the first order mean satisfies (2):

$$\underline{K}\ \underline{m}_h \overset{1}{=} \underline{f} \qquad (5)$$

and is identical to the usual deterministic solution at $\underline{t} = \underline{m}_t$. Premultiplying (3) by $\underline{K}$, substituting (5), and taking expectations, gives the equation for the second order mean:

$$\underline{K}\ \underline{m}_h \overset{2}{=} \underline{f} + \Delta\underline{f} \qquad (6)$$

where:

$$\Delta\underline{f} = \frac{1}{2}\underline{K}\ (D_{\underline{t}} \cdot D_{\underline{t}} \cdot \underline{h})(cs\underline{P}_{tt}) \qquad (7)$$

and $cs\underline{P}_{tt}$ is the (n^2x1) column string obtained by forming all the columns of $\underline{P}_{tt}$ in a single vector.

At first sight, the evaluation of $\Delta\underline{f}$ would seem to require the explicit evaluation of an extremely large (and in general full) second derivative matrix, with an effort equivalent to n^2 solutions of (2). If we ignore the known sparse structure of $\underline{K}$, the necessary matrix multiplications would then require n^4+n^3 scalar multiplications. Fortunately, however, $\Delta\underline{f}$ can be evaluated efficiently with much less computational effort.

Taking the derivative of (2) with respect to the transpose $\underline{t}'$ of $\underline{t}$, gives in Vetter notation (see Appendix A):

$$\underline{K}(D_{\underline{t}} \cdot \underline{h}) + (D_{\underline{t}} \cdot \underline{K})(\underline{I} * \underline{h}) = \underline{0} \tag{8}$$

since $\underline{f}$ is independent of $\underline{t}$. We note that $D_{\underline{t}} \cdot \underline{K}$ can be visualized as a row of n vectors of the form $(d\underline{K}/dt_i)\underline{h}$. If the i'th node of a finite element grid is connected to six other nodes, for example, $d\underline{K}/dt_i$ has only 7x7 non-zero terms, and its product with $\underline{h}$ has only seven non-zero terms. The use of:

$$\underline{K} \frac{d\underline{h}}{dt_i} = \left(- \frac{d\underline{K}}{dt_i} \underline{h}\right) \tag{9}$$

is a convenient method for calculating $D_{\underline{t}} \cdot \underline{h}$ directly.

Since $\underline{K}$ is linear in $\underline{t}$, further differentiation of (8) gives:

$$\underline{K}(D_{\underline{t}} \cdot D_{\underline{t}} \cdot \underline{h}) + (D_{\underline{t}} \cdot \underline{K})(\underline{I} * D_{\underline{t}} \cdot \underline{h}) + (D_{\underline{t}} \cdot \underline{K})\underline{E}^{nxn}_{nxn}(D_{\underline{t}} \cdot \underline{h} * \underline{I}) = \underline{0} \tag{10}$$

where $\underline{E}^{nxn}_{nxn}$ is a permutation matrix (Vetter, 1973). Utilizing the symmetry of $\underline{P}_{tt}$, the correction term (7) becomes:

$$\Delta\underline{f} = - (D_{\underline{t}} \cdot \underline{K})(\underline{I} * D_{\underline{t}} \cdot \underline{h})(cs\underline{P}_{tt}) \tag{11}$$

where $(D_{\underline{t}} \cdot \underline{K})(\underline{I} * D_{\underline{t}} \cdot \underline{h})$ can be visualized as a row of n matrices of the form $(d\underline{K}/dt_i)(D_{\underline{t}} \cdot \underline{h})$. Continuing our example, each such matrix would have only 7x7 non-zero terms in known locations. Multiplication by the appropriate terms in $cs\underline{P}_{tt}$ leads immediately to $\Delta\underline{f}$.

Sometimes, when dealing with uncertain aquifer properties, it is convenient to utilize the transformation $Y = \log_{10}T$. We then define a vector of model parameters, $\underline{y}$, containing log-transmissivities at the nodes of a finite element grid. In this case, $\underline{K}$ is no longer linear in the model parameters $\underline{y}$, and the results given above must be modified.

Firstly we note that:

$$\underline{K}(D_{\underline{y}} \cdot \underline{h}) = - (D_{\underline{y}} \cdot \underline{K})(\underline{I} * \underline{h}) \tag{12}$$

where: $$D_{\underline{y}} \cdot \underline{K} = (D_{\underline{y}} \cdot \underline{t}' * \underline{I})(\underline{I} * D_{\underline{t}}\underline{K}) \tag{13}$$

is a row of n matrices of the form $(\log_e 10)t_i(d\underline{K}/dt_i)$. Equation (10) is replaced by:

$$\begin{aligned} &\underline{K}(D_{\underline{y}} \cdot D_{\underline{y}} \cdot \underline{h}) + (D_{\underline{y}} \cdot \underline{K})(\underline{I} * D_{\underline{y}} \cdot \underline{h}) + (D_{\underline{y}} \cdot \underline{K})\underline{E}^{nxn}_{nxn}(D_{\underline{y}} \cdot \underline{h} * \underline{I}) \\ &\quad + (D_{\underline{y}} \cdot D_{\underline{y}} \cdot \underline{K})(\underline{I} * \underline{I} * \underline{h}) = \underline{0} \end{aligned} \tag{14}$$

where $D_{\underline{y}} \cdot D_{\underline{y}} \cdot \underline{K}$ is a row of n^2 (nxn) matrices, of which only n are non-zero. For i=1,...,n, the non-zero matrices are in positions given by j = n(i-1)+i, and have the form

$(\log_e 10)^2 t_i (d\underline{K}/dt_i)$. These are multiplied by $\underline{h}$ and scaled by terms from the diagonal of $\underline{P}_{yy}$, but most of these calculations are already required to form the right hand side of (12). Finally, for completeness, the second order correction term becomes:

$$\Delta\underline{f} = -(D_{\underset{\sim}{y}} \cdot \underline{K})(\underline{I} * D_{\underset{\sim}{y}} \cdot \underline{h})(cs\underline{P}_{yy})$$
$$- \frac{1}{2}(D_{\underset{\sim}{y}} \cdot D_{\underset{\sim}{y}} \cdot \underline{K})(\underline{I} * \underline{I} * \underline{h})(cs\underline{P}_{yy}) \quad (15)$$

EXAMPLES OF SECOND ORDER EFFECTS

In order to demonstrate the validity of the second order correction, we present three examples. In each case we choose an aquifer geometry and boundary conditions, as well as a mean and covariance structure for the field of $Y = \log_{10}T$. Assuming Y is normally distributed, we generate 500 realizations of $\underset{\sim}{y}$ using the turning bands method (Mantoglou and Wilson, 1982) to preserve the assumed spatial covariance structure. After solving for heads using (2), we calculate the mean of the 500 sets of nodal heads, and this Monte Carlo mean is compared with the second order mean (6,15).

In all the examples we choose an anisotropic exponential correlation structure defined by:

$$\rho(\Delta x, \Delta z) = \exp\{-((\Delta x/\ell_x)^2 + (\Delta z/\ell_z)^2)^{0.5}\} \quad (16)$$

where $\Delta x, \Delta z$ are the x,z separations between two points. We assume that the mean m_y, standard deviation σ_y, and correlation function $\rho(\Delta x, \Delta z)$ are constant in space, i.e. that the field of Y is statistically stationary. The covariance matrix $\underline{P}_{yy}$ contains terms of the form:

$$P_{yy}(i,j) = \sigma_y^2 \, \rho(x_i - x_j, z_i - z_j) \quad (17)$$

where (x_i, z_i) are the coordinates of the i'th node. The vector $\underline{m}_y$ contains m_y in all rows.

Example 1

First we consider a one-dimensional example similar to that used for Monte Carlo studies by Smith and Freeze (1979a). The aquifer has dimensions $L_x = 100$, $L_z = 1$ and is discretized as shown in Figure 1. Fixed head boundaries are assumed at both ends of the domain, with $\phi = 100$ at $x = 0$ and $\phi = 0$ at $x = 100$. The assumed statistical properties of the distribution of Y are $m_y = 1$, $\sigma_y = 1$ and $\ell_x = 40$. To ensure one-dimensionality in the Monte Carlo simulations, identical values of Y are assigned to each triplet of nodes; for the same reason in the second order calculations, the ratio ℓ_z/ℓ_x is set equal to 10^6 when $\underline{P}_{yy}$ is filled.

The results of this simple example are presented in Table 1. The deterministic solution, with constant Y throughout the aquifer domain, is a linear variation between the boundary values and is identical to the first order mean of the predicted heads (5). The mean of 500 Monte Carlo experiments deviates from the linear solution, however, by being lower for two thirds of the length of the domain and higher for the final third. The second order mean (6,15) is basically consistent with the Monte Carlo result, except that the solution is symmetric about x = 50. It is interesting to note that the correction vector $\Delta\underline{f}$ also exhibits this symmetry, but because of the asymmetry of the finite element grid, about lines of constant x, it is necessary to sum the terms in $\Delta\underline{f}$ at a given value of x to see it. We note that the terms in $\Delta\underline{f}$ can be interpreted as sources or sinks which move the second order mean from the first order solution.

Table 1 Comparison of Monte Carlo and first and second order means for Example 1

x	1st order mean	Monte Carlo mean	2nd order mean	$\sum_1^3 \Delta\underline{f}^{(1)}$
0	100	100	100	-0.16(2)
10	90	86.7	84.8	-3.63
20	80	76.0	73.2	-2.54
30	70	66.7	64.1	-1.61
40	60	57.9	56.7	-0.78
50	50	48.5	50.0	0.0
60	40	38.8	43.3	0.78
70	30	30.6	35.9	1.61
80	20	22.1	26.8	2.54
90	10	12.5	15.2	3.63
100	0	0	0	1.03(2)

Notes: (1) Sum of three terms in $\Delta\underline{f}$ at given x
(2) These have no effect at fixed head boundaries

The physical interpretation of the second order mean solution is difficult to grasp. Numerical experiments show that the difference between the first and second order means increases with σ_y, but decreases as ℓ_x both vanishes (spatially uncorrelated Y) and goes to infinity (perfectly correlated Y).

When $\ell_x/L_x \sim 0.5$, as in this example, simulated realizations of y are likely to have some nodal values greater than m_y, and some less. The quantity $(y - m_y)$ can be expected to have only one or two zero crossings along the length of the domain. The calculated second order mean implies a dominant effect of those realizations which have $Y > m_y$ in the middle of the domain and $Y < m_y$ near the ends. This dominant effect is probably due to the non-linearity of the logarithmic transformation.

Example 2

Moving to a truly two-dimensional example, we consider another problem studied by Smith and Freeze (1979b, Figure 7) using Monte Carlo techniques. The rectangular domain has dimensions $L_x = 200$, $L_z = 100$, and is discretized into an 11 x 9 99-node grid with 160 elements. The grid can be visualized as 8 rows of 20 elements, with diagonals sloping from top left to bottom right, as in the top row of Figure 1. We assume fixed head boundaries along the top of the domain ($\phi = 1.0$) and along three element sides in the lower right corner ($\phi = 0.0$); all other boundaries are assumed to be impermeable. For the statistical properties of the Y field, we assume $m_Y = -2$, $\sigma_Y = 0.43$, $\ell_x = 17.6$ and $\ell_z = 16.9$; these correlation lengths are chosen to be equal to the integral scales calculated by Smith and Freeze (1979b) from realizations of a nearest-neighbour correlation model.

Figure 2 compares the expected head field obtained by Monte Carlo simulations with the first order (5) and second order (6,15) means. Clearly the second order mean is closer to the Monte Carlo mean over most of the domain. Figure 3 shows the spatial distribution of $\Delta \underline{f}$; the obvious features along fixed head boundaries have no effect on the second order mean, however positive values along the bottom boundary and negative values over most of the domain are responsible for moving the second order mean towards the Monte Carlo mean. Oscillatory features in the plotted contours are the result of linear interpolation over triangular elements, rather than unstable behaviour.

This two-dimensional example can be viewed as a distortion of Example 1. Moving along a curved streamline from $\phi = 1$ towards $\phi = 0$, the second order mean is less than the first order mean, just as in Table 1. In fact there is one node (at $x = 180$, $z = 0$) where the second order mean is less than the first order mean, and it seems likely that with a finer mesh, more nodes would exhibit this similarity with the one-dimensional example. The effect of the smaller ratio of ℓ_x/L_x in this example is not immediately apparent.

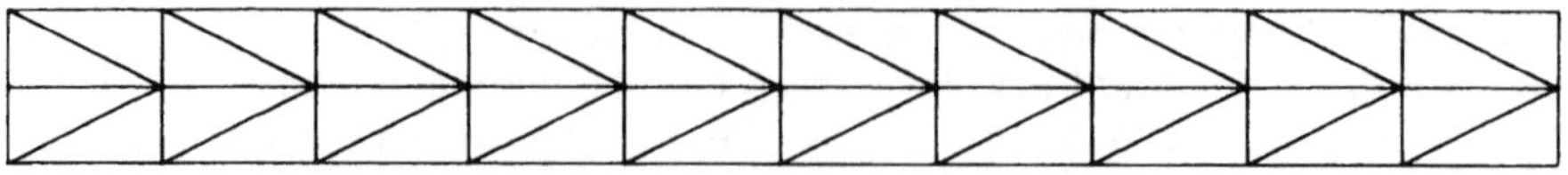

Figure 1 Grid for Example 1 (not to scale)

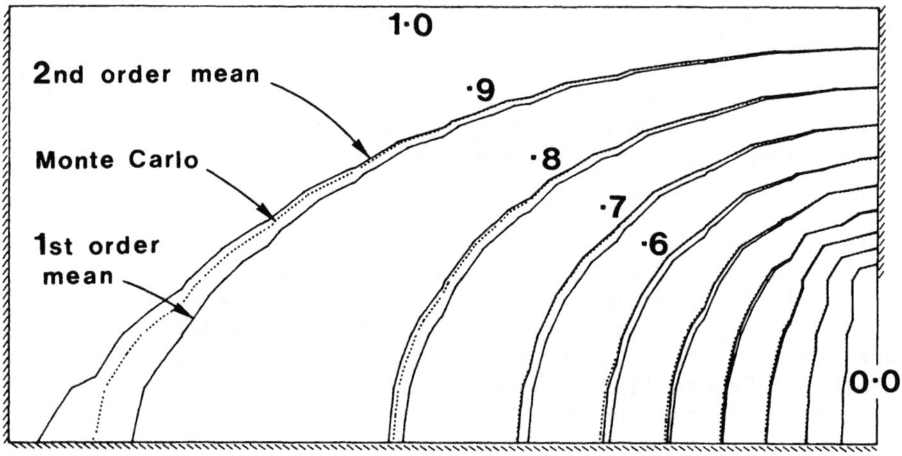

Figure 2 Comparison of mean heads for Example 2

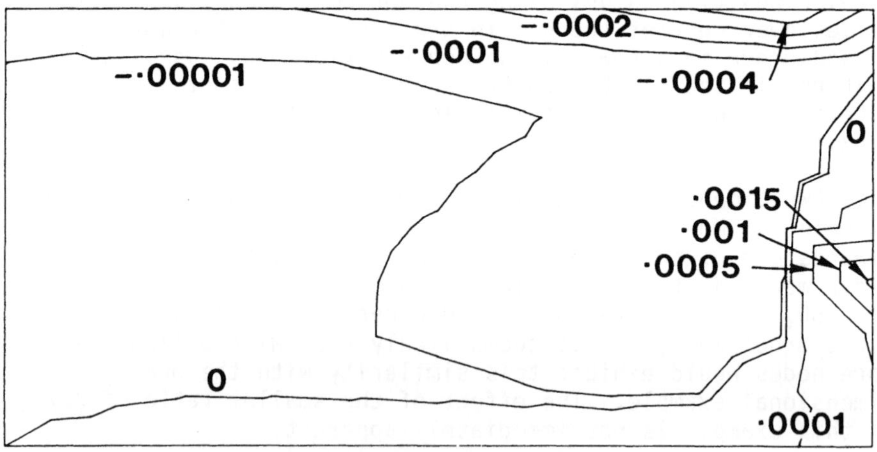

Figure 3 Distribution of $\Delta \underline{f}$ for Example 2

Example 3

Finally we consider a more interesting problem which includes the effect of pumping wells. In this case the pattern of the 99-node grid is as in Example 2, but the dimensions of the domain are scaled to $L_x = 100$ and $L_z = 80$. The boundary at $x = 0$ has known head $\phi = 0$, and wells at locations (30,50) and (80,60) are pumped at the rates of -5 and -4, respectively. For transmissivities, we assume $m_y = 1$, $\sigma_y = 0.2$ and $\ell_x = \ell_z = 40$.

Figure 4 shows the results of 500 Monte Carlo experiments, as well as the first and second order means. Again the second order mean is clearly a better approximation. The main features of the distribution of $\Delta \underline{f}$ in Figure 5 are effective additional withdrawals of -0.282 and -0.229 at the two wells. This phenomenon is clearly due to the high sensitivity of heads at these locations to the distribution of transmissivities. In fact it can be shown that the highest sensitivities are to values of Y at locations between the fixed head boundary and the well at (30,50).

It is interesting to note that large values of Y would tend to reduce the observed drawdown at wells, hence this example seems to behave differently from the first two. If as in Example 1, however, the second order solution (with $\ell_x/L_x \sim 0.5$) is dominated by realizations with large Y in the centre of the domain and small Y near the periphery, then the results are consistent. This follows from the observation that transmissivities between $x = 0$ and (30,50) would be small in the dominant realizations, thus causing increased drawdown at the wells.

THE COVARIANCE OF PREDICTED HEADS

Although the main issue in this paper is the second order effect on the mean predictions, many of the Monte Carlo studies in the literature (e.g. Delhomme, 1979; Smith and Freeze, 1979a,b; Clifton and Neuman, 1982) have concentrated on the standard deviations of predicted heads due to parameter uncertainty. Sagar (1978) and Dettinger and Wilson (1981) developed expressions for the covariance of heads at the nodes of finite difference and finite element grids. Numerical examples of these techniques are presented by Townley (1983) and Townley and Wilson (1983, 1984). Here, we provide a comparison between Monte Carlo and first order estimates of the standard deviations of predicted heads, σ_h.

To first order, the covariance of $\underline{h}$ due to uncertain $\underline{y}$ is given by:

$$\underline{P}_{hh} = E((\underline{h} - \underline{m}_h)(\underline{h} - \underline{m}_h)') = (D_{\underline{y}} \cdot \underline{h})\underline{P}_{yy}(D_{\underline{y}} \cdot \underline{h})' \tag{18}$$

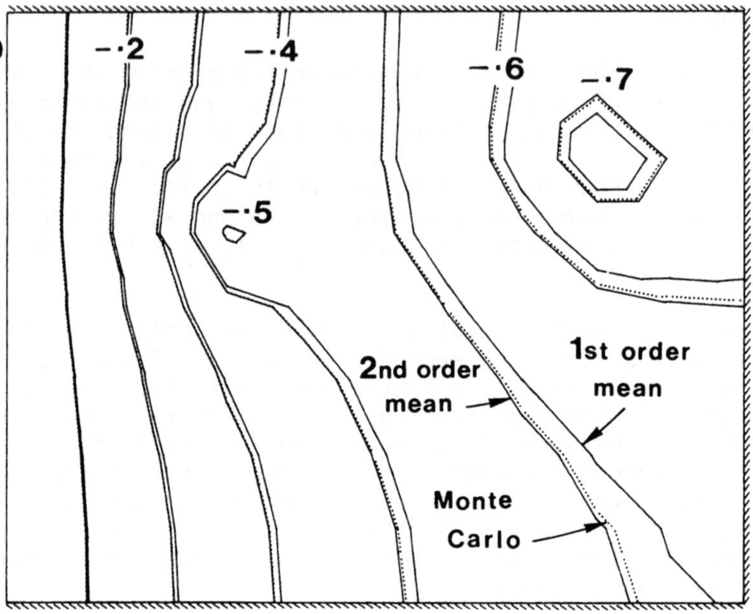

Figure 4 Comparison of mean heads for Example 3

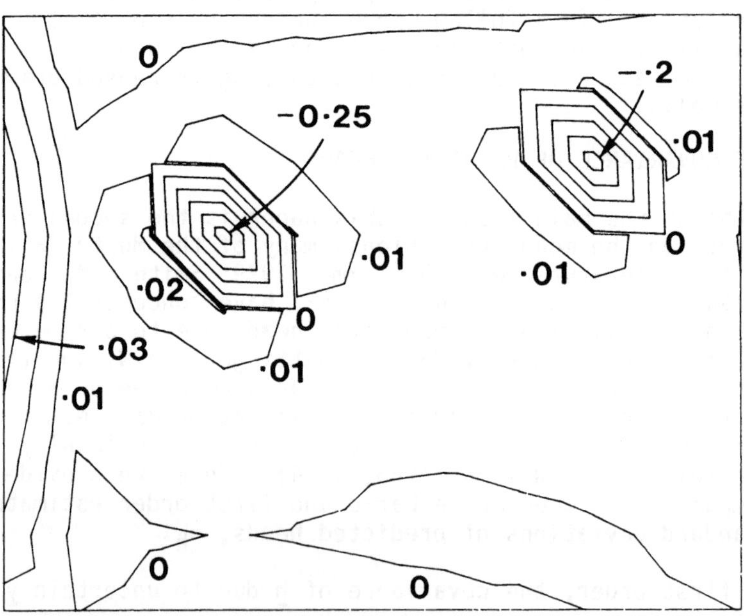

Figure 5 Distribution of $\Delta \underline{f}$ for Example 3

Pre- and post-multiplying by $\underline{K}$, and substituting (12), gives:

$$\underline{K}\,\underline{P}_{hh}\underline{K} = ((D_{\underline{y}}{}'\underline{K})(\underline{I} * \underline{h}))\underline{P}_{yy}((D_{\underline{y}}{}'\underline{K})(\underline{I} * \underline{h}))' \qquad (19)$$

which can be solved efficiently using equation solving techniques (Townley, 1983). We note that the right hand side of (19) is easily formed using sparse matrix routines and does not require the explicit evaluation of $D_{\underline{y}}{}'\underline{h}$. Computational savings are significant when $\underline{y}$ and $\underline{h}$ are of similar dimensions.

Turning to the examples, we compare Monte Carlo standard deviations, calculated relative to the Monte Carlo mean at each node, with theoretical first order standard deviations, whose squares appear on the diagonal of $\underline{P}_{hh}$. In Example 1, the first order σ_h varies parabolically from 0 at x = 0 and x = 100, to 48.6 at x = 50; the Monte Carlo σ_h is almost symmetric and peaks at 29.8 at x = 50. The discrepancy between these solutions indicates a weakness of the first order method for σ_y as large as 1; a standard deviation of 48.6 seems extremely unlikely when the mean head is 50. Comparisons of σ_h for Examples 2 and 3 are made in Figures 6 and 7. In both cases the agreement is qualitatively very good. The difference in peak σ_h in Example 3 is probably due to the longer correlation length which increases the prediction variance.

DISCUSSION

The results of three examples demonstrate good agreement between the theoretical second order mean (6) and the mean computed from a large number of Monte Carlo experiments. First order estimates of the standard deviations of predicted heads are also at least qualitatively in agreement with Monte Carlo results. Given the degree of similarity in the results, these direct numerical techniques have significant advantages over Monte Carlo simulations. Apart from reproducibility (i.e. the direct methods are not dependent on random number generators and a finite number of repetitions), the first and second order methods are also much less expensive. Whereas Examples 2 and 3 took 54 minutes of DEC-10 CPU time to generate 500 $\underline{y}$ fields, and 2.75 minutes to calculate 500 corresponding $\underline{h}$ fields, calculation of the second order mean (6,15) took 16 seconds, and calculation of $\underline{P}_{hh}$ (19) required only 8 seconds.

The choice of a stationary random field for Y can easily be improved by conditioning an estimated field on direct measurements of Y (e.g. by kriging) and also on measurements of heads (e.g. by weighted least squares estimation). The methods described in this paper can be expected to work equally well when $\underline{P}_{yy}$ is the estimation covariance matrix from one of these methods. Monte Carlo methods in these cases require more

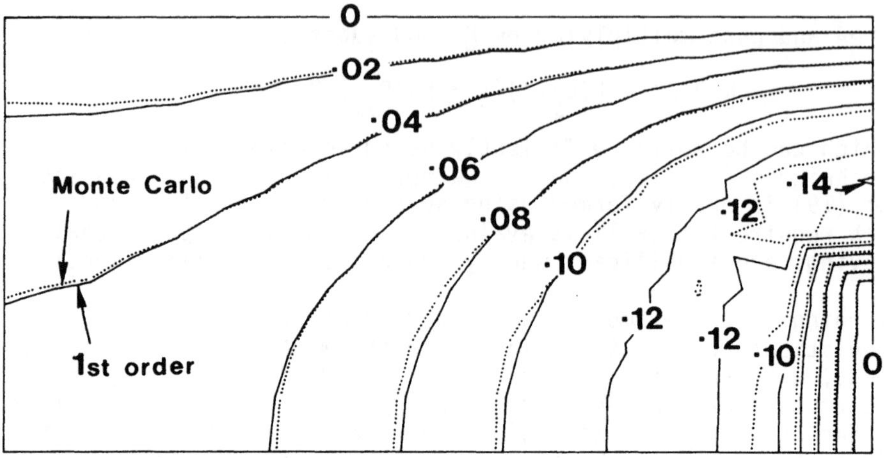

Figure 6   Comparison of σ_h for Example 2

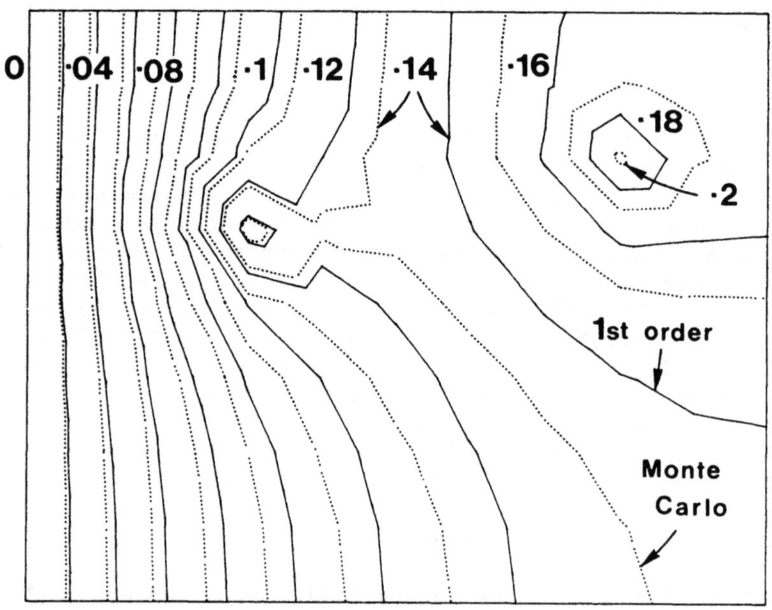

Figure 7   Comparison of σ_h for Example 3

expensive generation techniques (see, e.g., Clifton and Neuman, 1982, (43)) than the turning bands method. The first and second order methods can also be extended to include uncertain storage coefficients, boundary values and sources or sinks.

APPENDIX A

Vetter's (1973) matrix calculus represents the derivatives of matrices with respect to other matrices as two-dimensional arrays. Consider the matrices $\underline{U}(pxq)$, $\underline{V}(qxr)$, $\underline{X}(sxt)$ and $\underline{u}(ux1)$. The Kronecker product $(\underline{V} * \underline{X})$ has dimensions (qs x rt) and is a (qxr) array of (sxt) submatrices of the form $V_{ij}\underline{X}$. The notation $\underline{u}^{*2}$ represents $\underline{u}*\underline{u}$. The derivative $D_{\underline{X}}\underline{V} = d\underline{V}/d\underline{X}$ has dimensions (sq x tr) and is an (sxt) array of (qxr) submatrices of the form $d\underline{V}/dX_{ij}$. The derivative $D_{\underline{u}'}\underline{u}$ is the (uxu) identity matrix $\underline{I}_u$. Finally, the derivative

$$D_{\underline{u}'}(\underline{U}\underline{V}) = (D_{\underline{u}'}\underline{U})(\underline{I}_u * \underline{V}) + \underline{U}(D_{\underline{u}'}\underline{V}) \qquad (A1)$$

has dimensions (p x ur). Other results needed to derive (10) and (13) are available in the literature.

ACKNOWLEDGEMENT

The author is grateful for the provision of low cost computing services by the Western Australian Regional Computing Centre.

REFERENCES

Clifton, P.M. and S.P. Neuman (1982) Effects of kriging and inverse modeling on conditional simulation of the Avra Valley aquifer in southern Arizona, Water Resources Research, 18(4):1215-1234.

Dagan, G. (1982) Stochastic modeling of groundwater flow by unconditional and conditional probabilities. 1. Conditional simulation and the direct problem, Water Resources Research, 18(4):813-833.

Delhomme, J.P. (1979) Spatial variability and uncertainty in groundwater flow parameters: A geostatistical approach, Water Resources Research, 15(2):269-280.

Dettinger, M.D. and J.L. Wilson (1981) First order analysis of uncertainty in numerical models of groundwater flow. Part 1. Mathematical development, Water Resources Research, 17(1):149-161.

Mantoglou, A. and J.L. Wilson (1982) The turning bands method for simulation of random fields using line generation by a spectral method, Water Resources Research, 18(5):1379-94.

Mizell, S.A., A.L. Gutjahr and L.W. Gelhar (1982) Stochastic analysis of spatial variability in two-dimensional steady groundwater flow assuming stationary and nonstationary heads, Water Resources Research, 18(4):1053-1067.

Sagar, B. (1978) Galerkin finite element procedure for analyzing flow through random media, Water Resources Research, 14(6):1035-1044.

Smith, L. and R.A. Freeze (1979a) Stochastic analysis of steady state groundwater flow in a bounded domain. 1. One-dimensional simulations, Water Resources Research, 15(3):521-528.

Smith, L. and R.A. Freeze (1979b) Stochastic analysis of steady state groundwater flow in a bounded domain. 2. Two-dimensional simulations, Water Resources Research, 15(6):1543-1559.

Townley, L.R. (1983) Numerical models of groundwater flow: prediction and parameter estimation in the presence of uncertainty, Ph.D thesis, Department of Civil Engineering, Massachusetts Institute of Technology.

Townley, L.R. and J.L. Wilson (1983) Conditional second moment analysis of groundwater flow: the cumulative effects of transmissivity and head measurements, Papers of the International Conference on Groundwater and Man, Sydney, 5-9 December 1983, Vol. 1, Australian Government Publishing Service.

Townley, L.R. and J.L. Wilson (1984) First order effects of parameter uncertainty on predictions using numerical aquifer flow models, submitted to Water Resources Research.

Vetter, W.J. (1973) Matrix calculus operations and Taylor expansions, SIAM Review, 15(2):352-369.

ADJUSTING THE NODAL POINT DISTRIBUTION IN DOMAIN GROUNDWATER FLOW NUMERICAL MODELS

T. V. Hromadka II and T. J. Durbin

U.S. Geological Survey

ABSTRACT

A procedure for anticipating the relative error which results from a domain model of groundwater flow is presented. The procedure uses the complex variable boundary element method to develop steady-state solutions within the groundwater basin (or portions of the basin). These steady-state solutions are used for comparison with the domain model approximation, and a relative error distribution is determined. The nodal point distribution is adjusted to increase the nodal density in regions of large error, and likewise decrease the nodal density in regions of small error. The resulting nodal distribution is found to produce associated transient problem solutions comparable to approximations generated by increasing the total number of nodal points.

INTRODUCTION

A popular method for approximating groundwater flow effects is by means of numerical modeling. Generally, domain methods such as finite element and finite difference are used, although collocation methods and boundary integral equation methods have also been employed. In the domain methods the problem domain is discretized by nodal points into control volumes or finite elements. The choice as to nodal point placement is usually based on the judgment and experience of the hydrologist. Generally, the nodal point density is increased in regions where the state variable (e.g. water surface) is anticipated to vary rapidly with respect to either space or time. Additional placement of nodal points is governed by the interface between dissimilar materials or boundary condition specifications (e.g. aquifer fractures, groundwater wells, etc.). A complete presentation of such domain numerical methods is given in Pinder and Gray (1977).

In this paper, the main objective is to report on a procedure for identifying regions within the problem domain where the nodal point density needs to be increased in order to increase the numerical accuracy. The basis of the procedure is to examine the accuracy of the numerical model in predicting steady-state conditions where various boundary value problem conditions are considered. In order to examine the steady-state predicted values, the CVBEM (complex variable boundary element method) (Hromadka and Guymon, 1984a; Hromadka, 1984) is used to develop nodal point approximation values and estimates of nodal point relative error. Nodal points are then added (or removed when possible) in regions where the domain model estimates of the steady-state values differ significantly from the CVBEM predicted values. In this fashion, the conduction process modeling error due to choice of discretization is reduced.

Only the two-dimensional discretization of the groundwater basin is considered where all vertical direction effects are averaged by integration. The procedures can be extended to the three-dimensional case by examining several two-dimensional cross sections of the problem domain (such as vertical and horizontal slices of the domain).

The paper is divided into four parts. The first part presents the development for approximation of the transient groundwater flow regime. The NDI (nodal domain integration technique) (Hromadka, et al., 1981) is used to develop the domain numerical solution. An advantage of using the NDI approach is that the Galerkin finite element method, the subdomain integration method, and an integrated finite difference method can be represented by a single computer code. The second part of the paper presents a brief development of the CVBEM for use in determining a highly accurate solution of the two-dimensional Laplace equation. Because the CVBEM affords an immediate and exact evaluation of approximation error and results in an approximation function which exactly solves the Laplace equation, the numerical technique can be used to determine (to a high level of accuracy) an approximation of a Laplace equation boundary value problem. Part three presents the error evaluation technique (Hromadka and Guymon, 1984b) used with the CVBEM. Finally, part four presents two illustrations of the presented approach in locating domain method nodal points.

DOMAIN NUMERICAL MODEL DEVELOPMENT

Two-dimensional models of groundwater basins have been extensively reported in the literature. Generally, either the finite element or finite difference method is used to develop a matrix system of nodal point values as functions of the basin geometry, flow parameters, and boundary conditions. The general equation solved is to the form (for a confined aquifer)

$$\frac{\partial}{\partial x}\left(K_x \frac{\partial \phi}{\partial x}\right) + \frac{\partial}{\partial y}\left(K_y \frac{\partial \phi}{\partial y}\right) = S \frac{\partial \phi}{\partial t} \tag{1}$$

where

ϕ = total energy head with respect to a reference elevation

k_x, k_y = x- and y-direction hydraulic conductivities

S = specific storage

For homogeneous, isotropic aquifers (1) can be rewritten as

$$\frac{\partial^2 \phi}{\partial x^2} + \frac{\partial^2 \phi}{\partial y^2} = \frac{S}{bK} \frac{\partial \phi}{\partial t} \tag{2}$$

Application of the domain numerical method results in a matrix system

$$[K] \underset{\sim}{\phi} + [C] \underset{\sim}{\dot{\phi}} = \underset{\sim}{F} \tag{3}$$

where [k] is a symmetrical banded matrix representing the soil-water flowrates from the nodal point control volumes; [C] is a symmetrical banded matrix representing the capacitance of the nodal point control volumes; $\underset{\sim}{F}$ is a vector of specified nodal point values and flux boundary conditions (with [K] and [C] appropriately modified); and $\underset{\sim}{\phi}$ and $\underset{\sim}{\dot{\phi}}$ are the vectors of nodal point values and their time derivatives. Hromadka, et al. (1981) show that an infinity of domain methods can be described by (3) when written in the form

$$[K] \underset{\sim}{\phi} + [C(\eta)]\underset{\sim}{\dot{\phi}} = \underset{\sim}{F} \tag{4}$$

where η = 2, 22/7, ∞ results in the Galerkin analog, subdomain integration, and an integrated finite difference formulation, respectively.

In this paper, only errors in approximating the flowrates are considered. That is, it is assumed that the integration approximation scheme for the state variable in each nodal point control volume is assumed adequate (i.e., an appropriate η exists in (4)), and the specified flow parameters and boundary conditions are assumed adequately defined. To evaluate the numerical errors resulting from the [K] matrix, a steady-state problem is solved of the form

$$[K] \underset{\sim}{\phi} = \underset{\sim}{\hat{F}} \tag{5}$$

where $\underset{\sim}{\hat{F}}$ is a vector representing the boundary conditions for a selected steady-state scenario. Usually, several groundwater basin scenarios are considered resulting in several approximations from (5) which can be examined for numerical error development. However, to evaluate the error in (5), the $\tilde{\phi}$ vector needs to be compared to the correct solution vector $\tilde{\phi}^*$. Because an analytic solution for the problem approximated by (5) is seldom available, the CVBEM is used to develop another approximation vector $\tilde{\phi}'$ and a corresponding relative error distribution. The $\tilde{\phi}'$ values represent a highly accurate estimate of the exact solution values, $\tilde{\phi}$. The $\tilde{\phi}'$ vector is then used for comparison purposes with the domain model solution of $\tilde{\phi}$ in order to locate regions where the domain method approximation deviates substantially from the CVBEM approximation values.

CVBEM DEVELOPMENT

Hromadka and Guymon (1984a) present a detailed development of the CVBEM. A comprehensive presentation of the method is given in Hromadka (1984). A feature available with the CVBEM is the generation of a relative error measure which can be used to match the known boundary condition values of the problem. Consequently, the method can be used to develop a highly accurate approximation function for the Laplace equation and yet provide a descriptive relative error distribution for analysis purposes. Because the main objective of this paper is to analyze the numerical error in solving (5), it is noted that the Laplace equation is solved throughout the problem domain (if homogeneous) or in connected subregions (if inhomogeneous). Many anisotropic effects can be accommodated by the usual rescaling procedures or by subdividing the total domain into easier-to-handle subproblems. The CVBEM is then applied to the problem domain(s) as discussed in the following.

Let Ω be a simply connected domain with boundary Γ where Γ is a simple closed contour (Figure 1). Discretize Γ by m nodal points into m boundary elements such that a node is placed at every angle point on Γ (Figure 2). Each boundary element is defined by

$$\Gamma_j = \{z: z = z(s) \text{ where } z(s) = z_j + (z_{j+1} - z_j)s,\ 0 \le s \le 1\},\ j \ne m \quad (6)$$

with the exception that on the last element,

$$\Gamma_m = \{z: z = z(s) \text{ where } z(s) = z_m + (z_1 - z_m)s,\ 0 \le s \le 1\}$$

Then

$$\Gamma = \bigcup_{j=1}^{m} \Gamma_j \quad (7)$$

Let each Γ_j be discretized by (k+1) evenly spaced nodes ($k \ge 1$) such that Γ_j is subdivided into k equilength segments (Figure 3). Then Γ_j is said to be a (k+1)-node element. From

Figure 3, each Γ_j has an associated nodal coordinate system such that $z_{j,1} = z_j$ and $z_{j,k+1} = z_{j+1} = z_{j+1,1}$.

On each Γ_j, define a local coordinate system by

$$\begin{aligned} \zeta_j(s) &= z_{j,1} + (z_{j,k+1} - z_{j,1})s, \quad 0 \le s \le 1 \\ &= z_j + (z_{j+1} - z_j)s \end{aligned} \tag{8}$$

where $d\zeta_j = (z_{j,k+1} - z_{j,1})ds$.

On each (k+1)-node element Γ_j, a set of order k polynomial basis functions are uniquely defined by

$$N_{j,i}^{k}(s) = a_{j,i,0} + a_{j,i,1}\, s + \cdots + a_{j,i,k}\, s^k \tag{9}$$

where $i = 1,2,\cdots,(k+1)$ and $0 \le s \le 1$, and where

$$N_{j,k}^{k}\left(\frac{z_{j,n} - z_{j,1}}{z_{j,k+1} - z_{j,1}}\right) = \begin{cases} 1, & n = i \\ 0, & n \neq i \end{cases} \tag{10}$$

The basis functions are further defined to have the property that for $\zeta \varepsilon \Gamma$

$$N_{j,i}^{k}\left(\frac{\zeta - z_{j,1}}{z_{j,k+1} - z_{j,1}}\right) = \begin{cases} N_{j,i}^{k}\left(\dfrac{\zeta - z_{j,1}}{z_{j,k+1} - z_{j,1}}\right), & \zeta \varepsilon \Gamma_j \\ 0, & \zeta \notin \Gamma_j \end{cases} \tag{11}$$

Let $\omega(z)$ be analytic on $\Omega \cup \Gamma$. That is, let $\omega(z)$ be the solution (unknown) to the steady-state boundary condition problem being used to study (5). At each nodal point on Γ, define a specified nodal value by (Figure 3)

$$\bar{\omega}_{j,i} = \omega(z_{j,i}) \tag{12}$$

where from Figure 3, $\bar{\omega}_{j,1} = \bar{\omega}_j = \bar{\omega}_{j-1,k+1}$.

Using (11) and (12), an order k global trial function is defined by

$$G^k(\zeta) = \sum_j G^k(\zeta_j(s)) = \sum_j \sum_{i=1}^{k} \bar{\omega}_{j,i} \, N_{j,i}^{k}\left(\frac{\zeta - z_j}{z_{j+1} - z_j}\right) \tag{13}$$

From (13), the global trial function is continuous on Γ. An H_k approximation function $\hat{\omega}_k(z)$ (Hromadka, 1984) is defined by the Cauchy integral

$$\hat{\omega}_k(z) = \frac{1}{2\pi i}\int_\Gamma \frac{G^k(\zeta)\, d\zeta}{\zeta - z}, \quad z\in\Omega, \quad z\notin\Gamma \tag{14}$$

Because the derivative of $\hat{\omega}_k(z)$ exists for all $z\in\Omega$, then $\hat{\omega}_k(z)$ is analytic in Ω and exactly solves the Laplace equation in Ω.

Expanding (14) and using (7) gives

$$\int_\Gamma \frac{G^k(\zeta)\, d\zeta}{\zeta - z} = \sum_{j=1}^{m} \int_{\Gamma_j} \frac{G^k(\zeta)\, d\zeta}{\zeta - z} \tag{15}$$

Integrating on boundary element j gives (Hromadka, 1984)

$$\int_{\Gamma_j} \frac{G^k(\zeta)\, d\zeta}{\zeta - z} = R_j^{k-1}(z) + \sum_{i=1}^{k} \bar{\omega}_{j,i} N_{j,i}^{k}(\gamma_j) \ln\left(\frac{z - z_{j+1}}{z - z_j}\right) \tag{16}$$

where $R_j^{k-1}(z)$ is an order (k-1) complex polynomial resulting from the circuit around point z (Figure 4) and γ_j is equal to $(z - z_j)/(z_{j+1} - z_j)$.

Thus, the CVBEM results in the approximation function

$$\hat{\omega}_k(z) = \frac{1}{2\pi i} \sum_j \left(R_j^{k-1}(z) + \sum_i \bar{\omega}_{j,i} \, N_{j,i}^{k}(\gamma_j) \ln\left(\frac{z - z_{j+1}}{z - z_j}\right)\right) \tag{17}$$

or in a simpler form (Hromadka, 1984)

$$\hat{\omega}_k(z) = R^k(z) + \frac{1}{2\pi i} \sum_j \ln(z - z_j) \sum_i T_i^k \tag{18}$$

where $T_i^k = \bar{\omega}_{j-1,i} \, N_{j-1,i}^{k}(\gamma_{j-1}) - \bar{\omega}_{j,i} \, N_{j,i}^{k}(\gamma_j)$, and $R^k(z)$ follows from (17).

The approximation function of (18) exactly satisfies the governing flow equation in the problem domain Ω for the approximated boundary conditions on the problem boundary, Γ. Because $\hat{\omega}_k(z)$ is analytic on Ω, then the maximum relative error of $|\omega(z) - \hat{\omega}_k(z)|$ must occur on Γ. Consequently, the total approximation error can be simply evaluated on Γ with the corresponding errors in the interior of Ω being less in magnitude. Because the boundary conditions used to evaluate (18) are known continuously on Γ, then $\hat{\omega}_k(z)$ can be determined within arbitrary accuracy by the addition of nodal points on Γ due to (without proof)

$$2\pi i \lim_{\max|\Gamma_j|\to 0} \hat{\omega}_k(z) = \int_\Gamma \frac{\lim\limits_{\max|\Gamma_j|\to 0} G^k(\zeta)\, d\zeta}{\zeta - z} = \int_\Gamma \frac{\omega(\zeta)\, d\zeta}{\zeta - z} = 2\pi i\, \omega(z) \quad (19)$$

CVBEM ERROR REDUCTION TECHNIQUE

The previous discussion shows that the resulting CVBEM approximation function $\hat{\omega}(z)$ is analytic in the problem domain. Therefore, both the real equipotential approximation function $\hat{\phi}$ and the imaginary approximation $\hat{\psi}$ function satisfy the Laplace equation $\nabla^2\hat{\phi} = \nabla^2\hat{\psi} = 0$ for $z \in \Omega$, and $\hat{\omega} = \hat{\phi} + i\hat{\psi}$.

Therefore, the only approximation remaining is the continuous matching of the boundary conditions on the problem boundary, Γ.

This error of approximation is reflected in the difference between $\hat{\omega}(z)$ and $\omega(z)$ on Γ for the conjugate function specified. For example, if ϕ is known on element Γ_1, then the error used is $e_\phi = \phi - \hat{\phi}$ for $z \in \Gamma_1$. Likewise, if ψ is known on element Γ_2 then the error used is $e_\psi = \psi - \hat{\psi}$. This relative error can be determined continuously on Γ and an appropriate plot of the error in matching the boundary constructed. Hromadka and Guymon (1984b) use this error distribution on Γ to locate additional collocation points on Γ in order to develop a subsequent refined CVBEM approximation function. In this fashion, a highly accurate approximation of the boundary conditions is achieved on Γ while still retaining an exact solution of the governing partial differential equation (Laplace equation) in Ω.

After developing a $\hat{\omega}(z)$ which has an acceptable error distribution on Γ, the approximation function is then used as a substitute for the exact solution of the boundary value problem, $\omega(z)$.

From the maximum modulus theorem, the maximum value of $M = |\omega(z) - \hat{\omega}(z)|$ occurs for some point $z^* \varepsilon \Gamma$. Thus, the CVBEM approximation values for points in the interior of Ω (such as nodal points specified for the discretization of Ω for the NDI model) are guaranteed to have an approximation error less than M.

It is stressed that the principal step in the proposed procedure is the development of an accurate CVBEM approximation function. Generally, two to four attempts are required to develop a $\hat{\omega}(z)$ which has associated e_ϕ and e_ψ magnitudes which can be considered negligible.

APPLICATION

After developing the CVBEM $\hat{\omega}(z)$ function, the accuracy of the NDI domain model can be tested in the solution of the steady-state problems. Using identical boundary conditions on Γ, the CVBEM and the NDI models are compared as to the discrepancy in nodal values for all nodes located in the interior of Ω. The variation in modeling estimates determines a distribution of the approximation error throughout Ω. At regions of large error in Ω, additional nodal points are added to the domain discretization. Continuing in this fashion, the approximation error due to the nodal point distribution in Ω is reduced.

In the following application problems, the domain model is initially applied to a uniform distribution of nodal points in Ω. That is, the density of nodal points ρ_N (number of nodes per unit area) is a constant. After developing a $\hat{\omega}(z)$ function, the error distribution E(z) between the CVBEM and NDI models is determined in Ω. The next step is to redistribute the NDI interior nodes such that the nodal density ρ_N variation is related to an error density distribution ρ_E which is defined by

$$\rho_E(z_j) = \int_{\Omega_j} |E|dA \Big/ \int_{\Omega} |E|dA \tag{20}$$

where z_j is a nodal point coordinate; Ω_j is the domain control volume associated to nodal point z_j; and $|E|$ is the magnitude of the error between the CVBEM and NDI models. The $\rho_E(z_j)$ values may be estimated as simple point values. In the applications, $\rho_N(z_j)$ is reset by letting

$$\rho_N = \rho_E \text{ (total number of nodes).} \quad (21)$$

It is noted that equations (20) and (21) are arbitrary and other relations could be used. It is also noted that in the applications presented, the total number of nodes is assumed fixed, and only a redistribution of nodes is performed. Generally, the analyst would simply add nodes to regions in Ω where the error distribution is large.

Application I

A confined groundwater aquifer of thickness 30.5 m and with flow parameters K = 30.5 m/day and S = 0.01 has a centrally located well. The problem domain is a 360 m radius field with the well located at the centroid (Figure 5). The problem is to calculate the drawdown within the confined aquifer due to a constant pumping of Q = 1.7 m^3/min at the well.

In order to numerically model the problem, the domain has to be discretized by nodal points. Figure 5 shows a uniform nodal point distribution. Assuming a typical drawdown at the well and a uniform constant head boundary condition along the problem domain, the CVBEM is now used to estimate the drawdown values throughout the domain. Comparing the CVBEM values to estimates of drawdown (steady-state conditions) obtained from the domain model of (5) indicates a nearly uniform increase in relative error as the distance to the well decreases (Figure 6).

The problem domain is rediscretized by the nodal densities shown in Figure 7. Figure 8 shows the estimated relative error for steady-state conditions using the selected nodal point densities of Figure 7.

In order to check the performance of the two discretizations in the transient problem, the drawdown can be checked along a radial (for a short simulation time) by using the well-known Theis solution. Figure 9 shows the plots of drawdown obtained by both discretizations along with the drawdown predicted by the Theis analytic solution. (In the domain model, the Crank-Nicolson time advancement algorithm is used with a timestep of 0.1 days.) From the Theis solution, the adjusted nodal point distribution provides a significant reduction in error.

In this application problem, the nodal point density is determined to be increased near the vicinity of the well. The motivation for adjusting the nodal point distribution is to reduce the relative error obtained in solving a steady-state problem which approximates the range of conditions that are anticipated for the associated transient problem. The key to this approach is determining the steady-state problem approximation error in order to evaluate the nodal point densities. Because the CVBEM provides a highly accurate approximation for the Laplace equation, it can be used for determining the relative error by comparison with the results generated by the domain model.

Application II

The above procedure can be applied to improving the nodal point distribution in steady-state problems which are governed by the Laplace equation. The domain problem shown in Figure 10 is initially discretized using the shown nodal point distribution. Of concern is the pressure distribution along the specified pressure surface. Figure 11 shows the relative error (percent) in comparison to the CVBEM predicted values. Using the error distribution as a guide, the 40 nodal points were redistributed to be concentrated more in regions of large relative error. Resolving the domain method steady-state problem produced a significant decrease in relative error along the pressure surface. In comparison, increasing the original 40-node distribution uniformly to 120 nodes resulted in a relative error along the pressure surface of comparable magnitude to the redistributed 40-node model.

The adjusted 40-node distribution and the 120-node models were used to examine several boundary conditions of various water depths upstream and downstream of the dam. In all test trials, both models produced comparable results which indicates that the adjusted nodal point distribution is not significantly affected by changes in magnitude of the problem boundary conditions.

CONCLUSIONS

In using domain models for estimating groundwater flow effects, the choice as to the nodal point distribution is usually left to the experience of the hydrologist. Typically, nodal densities are increased in regions where it is anticipated that there may be a large departure from the assumed trial function distribution within a nodal point's associated control volume.

In this paper, a procedure to examine the nodal point distribution is proposed. Using steady-state conditions in portions of the problem domain, the CVBEM is used to generate a solution of the Laplace equation for comparison with the analogous domain model approximation. Based on the accompanying relative error, the domain model nodal distribution is adjusted to have an increased density in regions of large error between the models.

The adjusted nodal distribution is found to significantly reduce modeling error for the subsequent transient problems, and is not significantly affected by changes in the boundary conditions.

The two applications problems illustrate the proposed procedure in studying the nodal point distributions for two classic problems. For large-scale problems where available computer memory becomes an issue, the procedure can be used to aid in optimizing the nodal point distribution throughout the problem domain. In this way, nodal point densities can be increased in regions of large anticipated relative error and, perhaps more important, decreased in regions of small relative error with the end result of keeping the computer memory requirements at a minimum.

REFERENCES

Hromadka II, T., Guymon, G., and Pardoen, G. (1981) Nodal Domain Integration Model of Unsaturated Two-Dimensional Soil-Water Flow: Development. Water Resources Research, (17), 1425-1430.

Hromadka II, T. (1984) The Complex Variable Boundary Element Method. Springer-Verlag publishers (in press).

Hromadka II, T., and Guymon, G. (1984a) The Complex Variable Boundary Element Method: Development. Int. Jour. Num. Methods in Eng. (in press).

Hromadka II, T., and Guymon, G. (1984b) An Alogrithm to Reduce Approximation Error from the CVBEM Applied to Soil Freezing. Numerical Heat Transfer (in press).

Pinder, G., and Gray, W. (1977) Finite Element Simulation in Surface and Subsurface Hydrology. Academic Press.

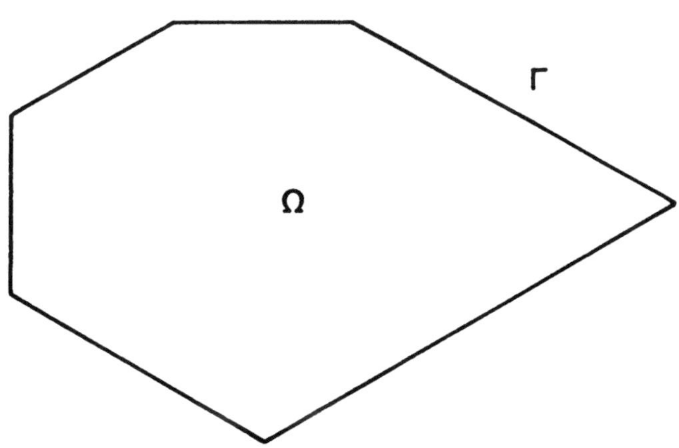

Figure 1. **Simply Connected Domain Ω with Simple Closed Contour Boundary Γ**

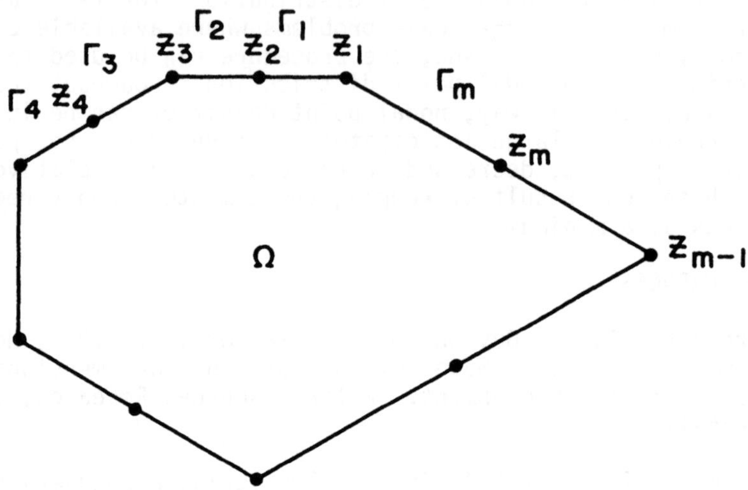

Figure 2. Γ Discretized Into m Boundary Elements

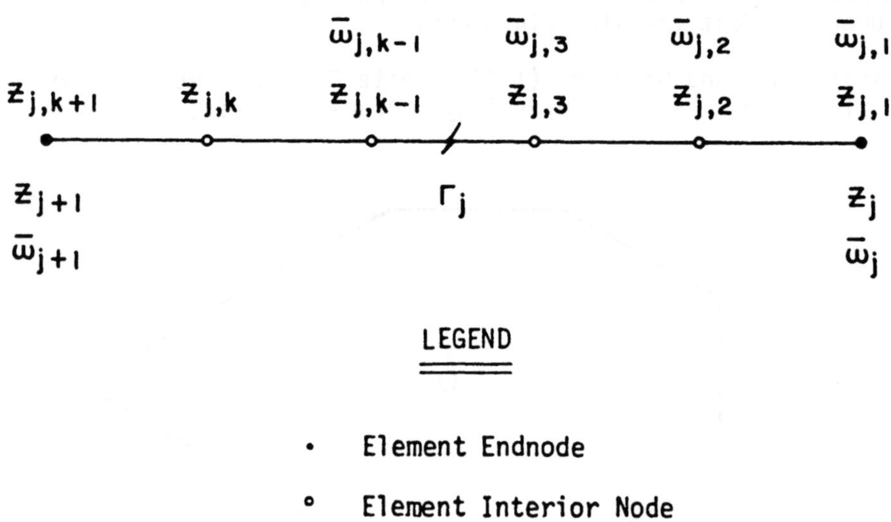

Figure 3. (k+1)-Node Boundary Element Γ_j Nodal Definitions

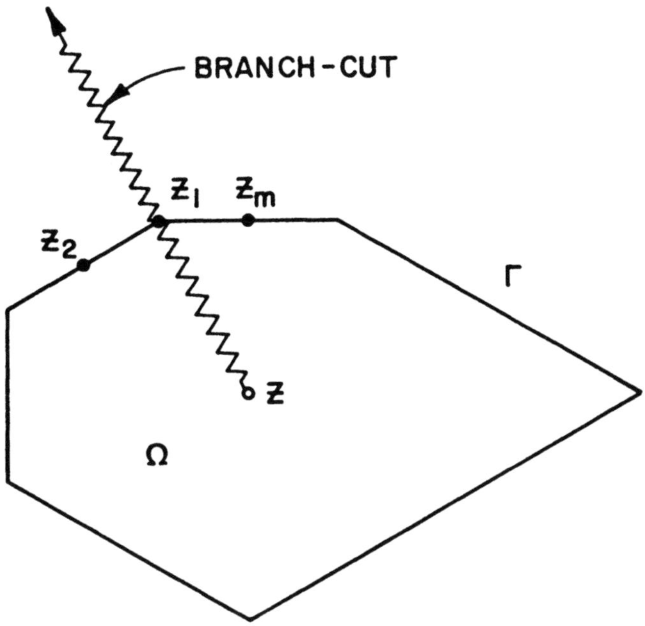

Figure 4. Branch-Cut of $\ln(z - \zeta)$ function, $\zeta \in \Gamma$

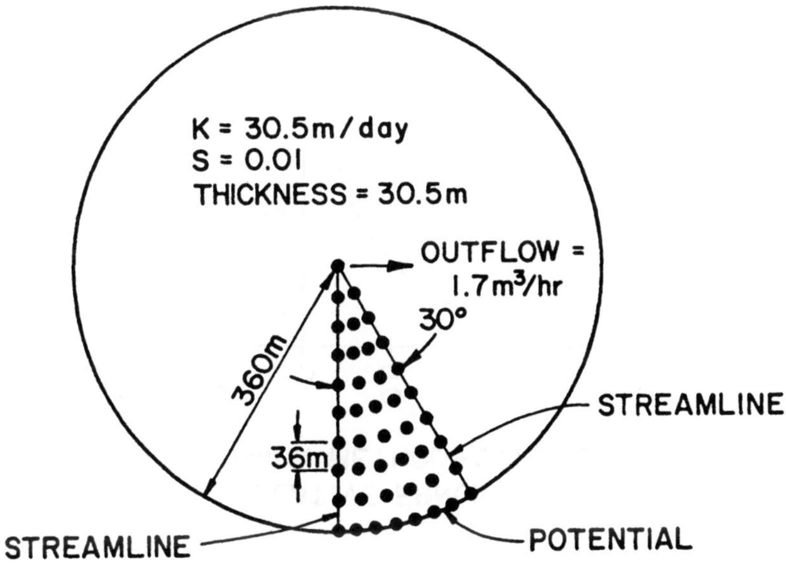

Figure 5. Problem Domain Definition and Uniform Nodal Point Distribution

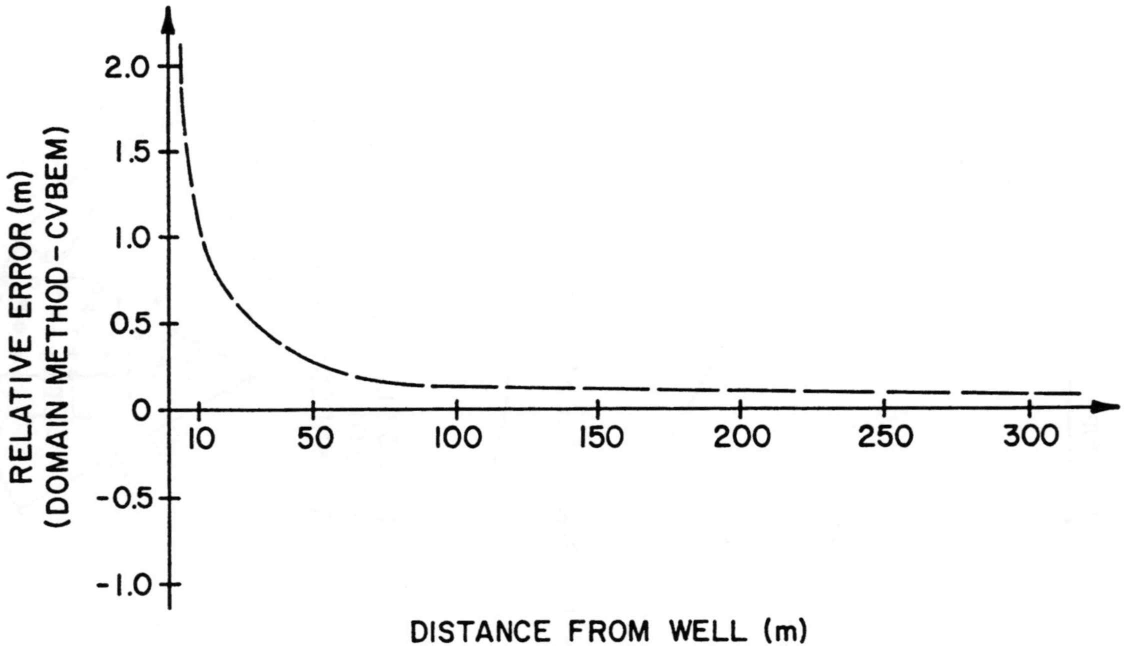

Figure 6. **Relative Error** Using Uniform Radial Nodal Point Density

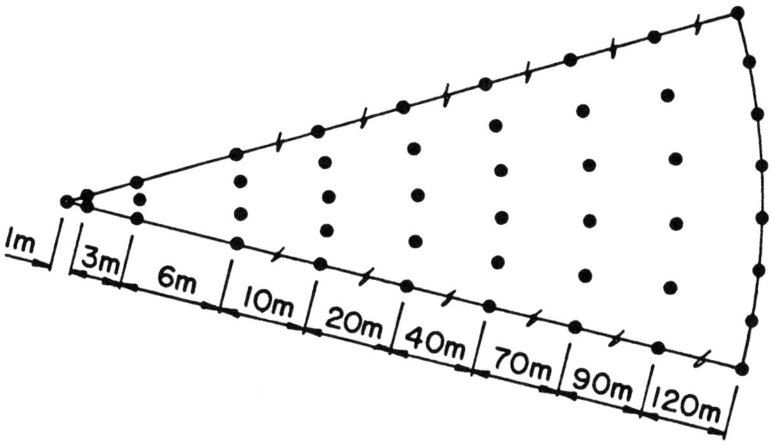

Figure 7. Adjusted Nodal Point Distribution

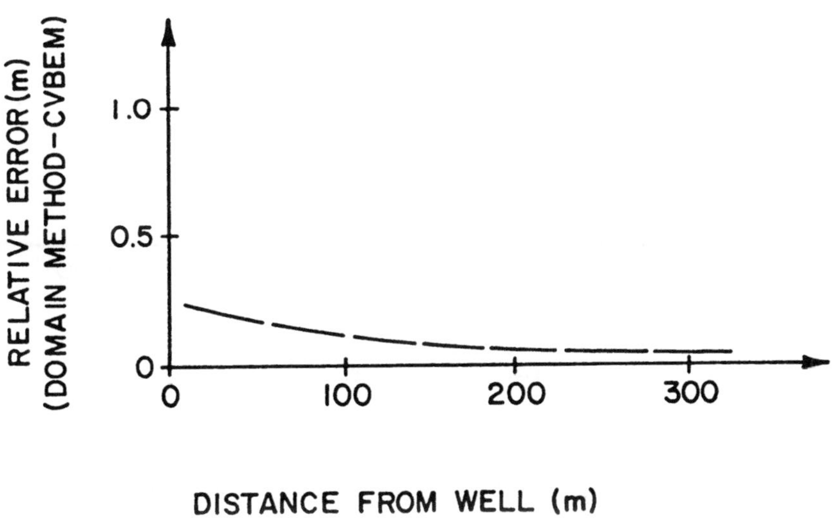

Figure 8. Relative Error Using Adjusted Nodal Point Distribution

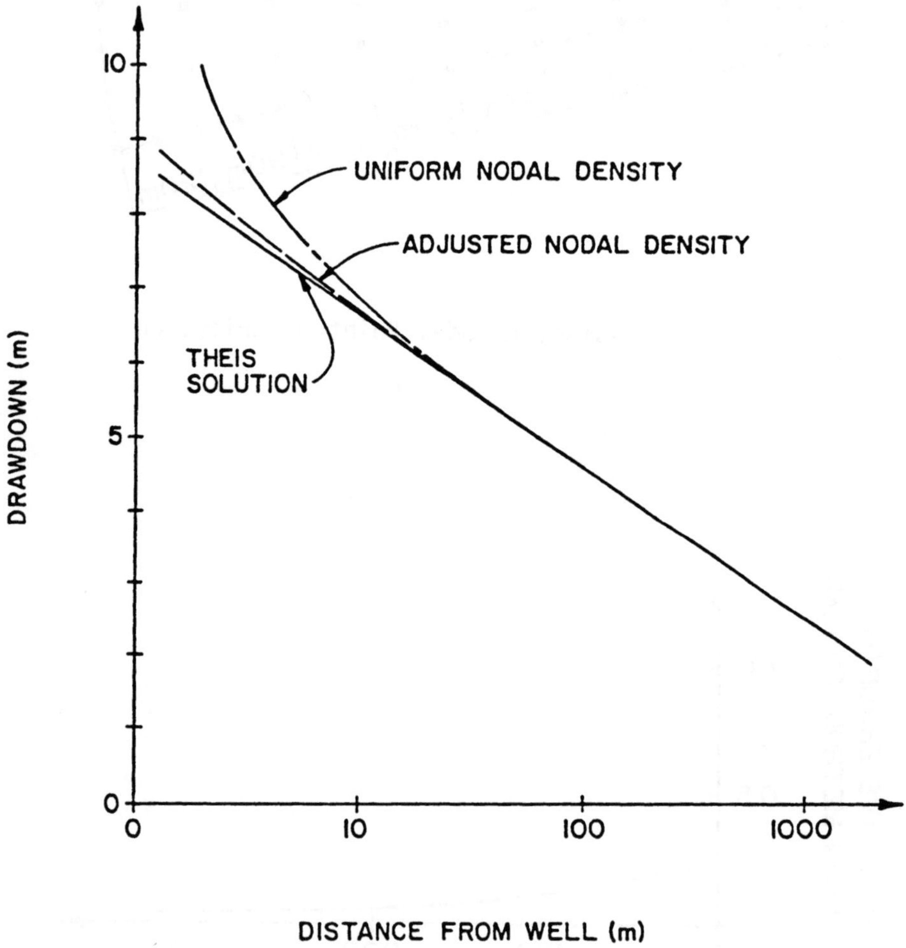

Fig. 9. Radial Drawdown at time = 0.5 days

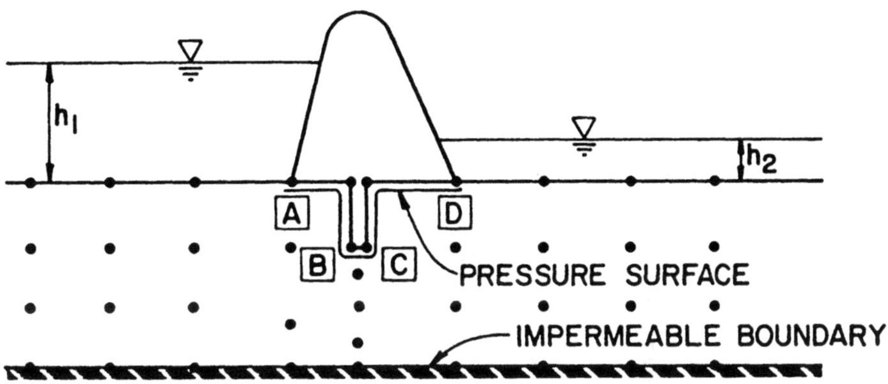

Figure 10. Flowfield Approximated Geometry

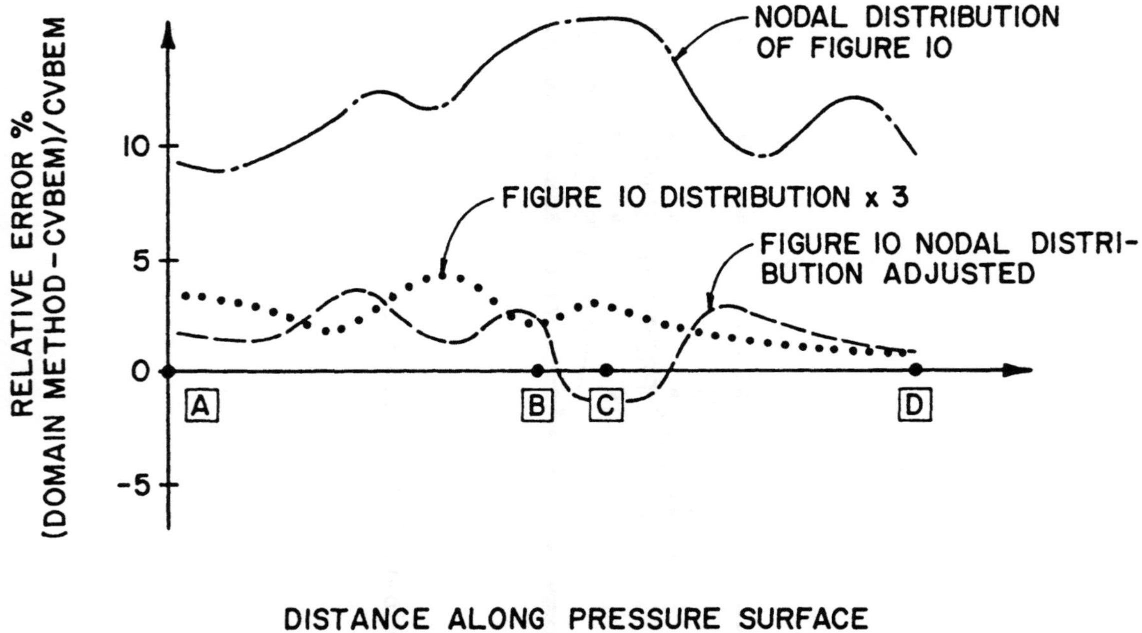

Figure 11. Relative Error Plots for the Flowfield Problem of Figure 10.

TRANSPORT PHENOMENA

SOLUTION OF CONTAMINANT TRANSPORT EQUATIONS USING AN ORTHOGONAL UPSTREAM WEIGHTING FINITE ELEMENT SCHEME

G. T. Yeh

Environmental Sciences Division, Oak Ridge National Laboratory, Oak Ridge, Tennessee 37830, USA.

ABSTRACT

An orthogonal upstream weighting finite element scheme is presented for approximating the contaminant transport equation, that will result in a matrix amenable to the pointwise solution strategies. Orthogonal upstream weighting functions are derived for line, quadrilateral, and triangular elements. Numerical experiments indicate that the direct elimination solutions of the proposed scheme yield comparable results as those obtained by the Galerkin and non-orthogonal upstream schemes. When the successive over-relaxation is applied to solving the matrix equation, the former generates convergent solutions for all cases. In contrast, the latter result in convergent solutions only for dispersion dominant transport, but produce divergent computations for advection dominant transport.

INTRODUCTION

The Galerkin finite element method has been extensively used to solve groundwater flow and contaminant transport problems (Pinder and Gray 1977). Two problems arise: (1) oscillatory, unstable solutions prevail for advection dominated transport, (2) poor matrices that yield divergent solutions with pointwise iterations are encountered for many situations. The oscillatory behavior has recently been overcome by using an upstream weighting function (Huyakorn and Nilkuha 1979). However, poor matrices still persist for contaminant transport in many ocassions. If iterative methods are applied to solve such matrix equations, nonconvergence may occur, thus necessitating the use of direct elimination methods for the solution. When large-field problems are to be dealt with, employing direct band-matrix solver becomes impractical because of excessive requirement on CPU memory for the coefficient matrix.

In order to make the applications of finite element models to large-field problems feasible, means of reducing the CPU memory must be found. Out-of-core solvers or the "frontal methods" can be used to greatly reduce the CPU memory. However, out-of-core solvers are in general extremely slow and may increase CPU time at least one order of magnitude (Segol 1976, Putney and Broyd 1983). The programming of "frontal methods" involves considerable "housekeeping" and is not a task to be undertaken by an inexperienced programmer (Livesley 1983). Iteration methods such as alternating direction implicit procedure (Peaceman and Rachford 1955, Bredehoeft and Pinder 1970), line or block iterations (Young 1954, Freeze 1971), and the strong implicit procedure (Stone 1968, Trescott and Larson 1977) have been commonly used in finite difference models to reduce CPU memory for large two-dimensional or three-dimensional problems. They are seldom employed in finite element models because the matrix entries are not known untill assembling procedures are done. Perhaps the simplest and easiest iteration methods that can be incorporated in finite element models that can drastically reduce the CPU memory without complicating programming efforts and without increasing CPU time significantly are the pointwise iteration methods such as Gauss-Seidel or successive over-relaxation (SOR) iterations (Yeh and Huff 1984). Unfortunately, these methods may fail on matrices resulting from the Galerkin approximation of contaminant transport equation (Zienkiewicz 1977).

An orthogonal upstream weighting (OUW) finite element scheme, extending the orthogonal weighting (OW) presented earlier for solving groundwater flow equation (Yeh 1983) and incorporating the non-orthogonal upstream weighting (NUW) given elsewhere (Huyakorn and Nilkuha 1979), is developed to yield a matrix amenable to pointwise iteration strategies and to smooth the oscillatory behavior. The CPU time for pointwise iteration solutions (PIS) of the OUW finite element scheme is comparable to that for direct elimination solutions (DES). It can be expected that as the bandwidth of the coefficient matrix increases, the CPU time for PIS will be less than that for DES.

The proposed OUW is tested with two simple examples for which analytical solutions are available so that numerical results can be checked. With the new finite element scheme, the system of algebraic equations can be solved one by one with Gauss-Seidel or SOR techniques or by region by region (Yeh et al 1984) to greatly reduce the CPU memory without much increase in CPU time and complication in programming. For example, the CPU memory for the coefficient matrix can be reduced from more than 16 megabytes to 0.8 megabytes for a 100 X 100 node discretization in two-dimensional cases. The CPU time with PIS is comparable with that with DES for all cases in two test examples.

DERIVATION OF ORTHOGONAL UPSTREAM WEIGHTING FUNCTIONS

The governing equation for contaminant transport through aquifers over a horizontal plane x-y can be obtained by vertically integrating the three-dimensional equation (Bear 1977). A general form of the vertically integrated transport equation that considers advection, dispersion, source/sink, decay, degradation, sorption, and the effects of consolidation and leaky beds was given elsewhere (Yeh and Huff 1984). To illustrate the application of OUW finite element scheme, we consider the following simple form of transport equation:

$$\frac{\partial C}{\partial t} + \nabla_H \cdot (\vec{V}C) = \nabla_H \cdot (\overline{\overline{D}} \cdot \nabla_H C) \tag{1}$$

where ∇_H is the del operator over the horizontal coordinate (x,y); C is the concentration; t is the time; $\vec{V}$ is the velocity; and $\overline{\overline{D}}$ is the dispersion coefficient tensor.

Let the trial solution of C be expressed in the form of

$$C \simeq \hat{C} = \sum^{n} C_j(t) N_j(x,y) \tag{2}$$

where C_j is the value of C at node j, N_j is the basis function of node j, and n is the total number of nodes in the region.

It is proposed that the velocity term of the transient transport equation are weighted using "orthogonal upstream" weighting functions to be developed here, and all other terms are weighted using the orthogonal weighting functions given previously (Yeh 1983). Let $U_i(x,y)$ and $W_i(x,y)$ be two sets of orthogonal upstream and orthogonal weighitng functions, respectively, with i = 1, 2, ..., n, then applying the principle of weighted residual, we obtain:

$$[A]\{dC/dt\} + [B]\{C\} = \{Q\} \tag{3}$$

where {C} is the column vector representing the values of C at n nodes; {dC/dt} is the derivatives of C with respect to time at n nodes; the matrices [A] and [B] and the column vector {Q} are defined as:

$$A_{ij} = \sum_{e \in M_e} \int_{R_e} W_\alpha^e N_\beta^e \, dR \tag{4}$$

$$B_{ij} = \sum_{e \in M_e} \int_{R_e} [(\nabla_H W_\alpha^e) \cdot \overline{\overline{D}} \cdot (\nabla_H N_\beta^e) - (\nabla_H U_\alpha^e) \cdot \vec{V} N_\beta^e] \, dR \tag{5}$$

and

$$Q_i = \sum_{e \in K_e} \int_{B_e} \vec{n} \cdot (W_\alpha^e \overline{\overline{D}} \cdot \nabla_H C - U_\alpha^e \vec{V} C) \, dR \tag{6}$$

in which the superscript e denotes the element; $\alpha,\beta = 1,2$ for linear line element, or $\alpha,\beta = 1,2,3$ for linear triangular element, or $\alpha,\beta = 1,2,3,4$ for linear quadrilateral element; R_e is the region of element e; M_e is the set of elements that have a local side α-β coinciding with the global side i-j; W_α^e and N_β^e are the α-th and β-th orthogonal weighting and basis functions of element e; U_α^e is the α-th orthogonal upstream weighting function of element e; B_e is the global boundary of element e; K_e is the set of boundary segments that have a local node α coinciding with the global node i; and $\vec{n}$ is the outward unit vector normal to the boundary.

Performing time integration of Equation (3) using finite difference, we obtain

$$[T]\{C\}_{t+\Delta t} = \{R\} \tag{7}$$

where $\{C\}_{t+\Delta t}$ is the value of {C} evaluated at time t + Δt with Δt being the time step size, and the coefficient matrix [T] and load vector {R} are given as:

$$[T] = [A]/\Delta t + \theta[B] \tag{8}$$

and

$$\{R\} = ([A]/\Delta t - (1-\theta)[B])\{C\}_t + \{Q\} \tag{9}$$

in which $\{C\}_t$ is the value of {C} evaluated at time t, and θ is the time weighting factor with θ = 0.5 for Crank-Nicolson central difference and θ = 1.0 for backward difference.

The reduction of the partial differential equation [Eq. (1)] to the matrix equation [Eq. (7)] simplifies to the evaluation of integrals on the right hand side of the Equations (4) through (6) for every element. The major task remains is the derivation of appropriate expression for the orthogonal upstream weighting functions U_α^e's. This will be carried out for both one- and two-dimensional elements.

Weighting Functions for One-dimensional Elements

The basis functions for nodes 1 and 2 of a linear line element written in terms of local coordinate ξ are (Pinder and Gray 1977)

$$N_1^e = (1 - \xi)/2 \quad \text{and} \quad N_2^e = (1 + \xi)/2 \tag{10}$$

If U_1^e and U_2^e are the orthogonal upstream weighting (OUW) functions for nodes 1 and 2, they are set to satisfy the following compatible conditions:

$$U_1^e = 1, \quad U_2^e = 0 \quad \text{at} \quad \xi = -1 \tag{11}$$

and

$$U_1^e = 0, \quad U_2^e = 1 \quad \text{at} \quad \xi = 1 \tag{12}$$

In addition, it is proposed here that U_1^e and U_2^e be defined to satisfy the following orthogonal conditions:

$$\int_{-1}^{1} U_1^e N_1^e \, d\xi \neq 0, \qquad \int_{-1}^{1} U_2^e N_1^e \, d\xi = 0 \tag{13}$$

and

$$\int_{-1}^{1} U_1^e N_2^e \, d\xi = 0, \qquad \int_{-1}^{1} U_2^e N_2^e \, d\xi \neq 0 \tag{14}$$

Since OUW functions are to satisfy four conditions [Eqs. (11) through (14)], polynominals of third degree can be used. Thus, let

$$U_i^e = a_i + b_i\xi + c_i\xi^2 + d_i\xi^3 \tag{15}$$

where i = 1 or 2. Substituting Eq. (15) into Eqs. (11) through (14), we obtain:

$$U_1^e = [(-2+3\gamma_1) + (6-15\gamma_1)\xi + (6-3\gamma_1)\xi^2 + (-10+15\gamma_1)\xi^3]/8 \tag{16}$$

$$U_2^e = [(-2+3\gamma_2) - (6-15\gamma_2)\xi + (6-3\gamma_2)\xi^2 - (-10+15\gamma_2)\xi^2]/8 \tag{17}$$

where γ_i's are the undetermined coefficients. These coefficients are chosen to satisfy the conditions of total unit weight and upstream weighting

$$\int_{-1}^{1} U_1^e \, d\xi + \int_{-1}^{1} U_2^e \, d\xi = 2 \quad \text{and} \quad \int_{-1}^{1} U_1^e \, d\xi = (1 - \alpha) \tag{18}$$

where α is a weighting factor, ranging from - 1.0 to 1.0 corresponding to the Peclet number of $-\infty$ to ∞. To establish numerical stability with minimum loss of accuracy, the values of α must be carefully chosen. The optimum value of α was given theoretically by Christie et at (1976) as function of the Peclet number for steady state transport. When α is equal to zero, U_i^e's become the orthogonal weighting functions W_i^e's (Yeh 1983).

Substituting Eqs. (16) and (17) into (18) to obtain α, we have following orthogonal upstream weighting functions for the line element e:

$$U_1^e = [(2-3\alpha) - (12-15\alpha)\xi + 3\alpha\xi^2 + (10-15\alpha)\xi^3]/4 \tag{19}$$

and

$$U_2^e = [(2+3\alpha) + (12+15\alpha)\xi - 3\alpha\xi^2 - (10+15\alpha)\xi^3]/4 \tag{20}$$

Equations (19) and (20) written in length coordinate (L_1,L_2) are

$$U_i^e = L_i + L_iL_j(L_i - L_j)(10 - 15\alpha) - 3\alpha L_iL_j \tag{21}$$

in which (i,j) = (1,2) or (2,1), and L_1 and L_2 represent $(x_2 - x)/(x_2 - x_1)$ and $(x - x_1)/(x_2 - x_1)$, respectively.

<u>Weighting Functions for Linear Quadrilateral Elements</u>

Consider a quadrilateral element shown in Figure 1. Let (ξ,η) be a local isoparametric coordinate. Since each element belongs to the Lagrangian family, its weighting functions at four nodes can be obtained by taking appropriate products of functions in each coordinate. Thus, referring to Figure 1, the OUW functions for nodes 1, 2, 3, and 4 of a typical element e are given as follows:

$$U_1^e = F_1(\xi,\alpha_1)F_1(\eta,\beta_2), \quad U_2^e = F_2(\xi,\alpha_1)F_1(\eta,\beta_1) \tag{22}$$

and

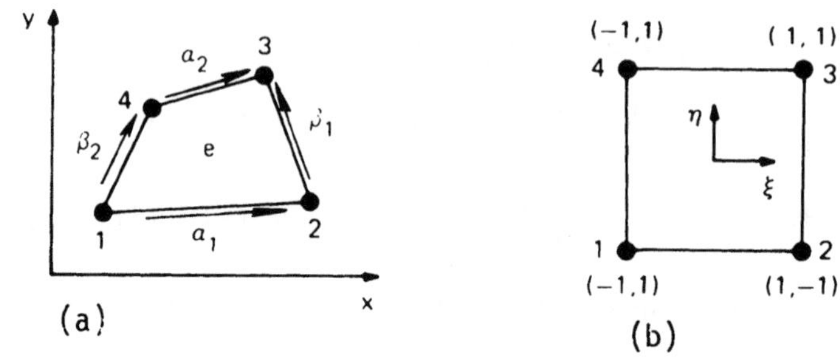

Figure 1. (a) Quadrilateral Element e in Global coordinate and (b) Quadrilateral Element e in Local Coordinate.

$$U_3^e = F_2(\xi,\alpha_2)F_2(\eta,\beta_1), \quad U_4^e = F_1(\xi,\alpha_2)F_2(\eta,\beta_2) \tag{23}$$

where

$$F_1(x,a) = [(2-3a) - (12-15a)x + 3ax^2 + (10-15a)x^3]/4 \tag{24}$$

and

$$F_2(x,a) = [(2+3a) + (12+15a)x - 3ax^2 - (10+15a)x^3]/4 \tag{25}$$

Weighting Functions for Linear Triangular Elements

Consider a typical triangular element (Fig. 2). Let (L_1,L_2,L_3)

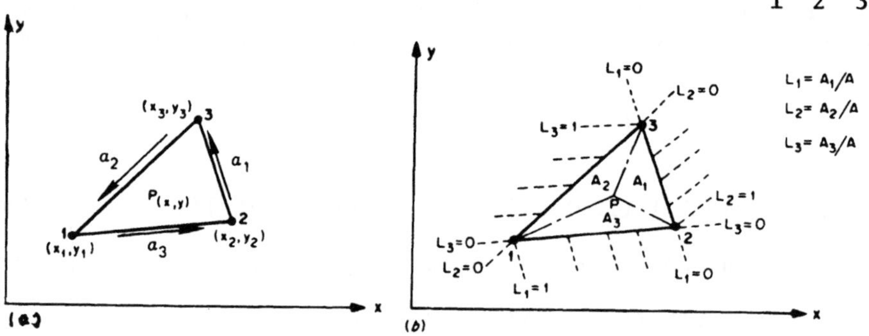

Figure 2. (a) A Typical Triangular Element e in Global Coordinate and (b) Local Coordinate of the Triangular Element e.

be the area coordinate. Let U_i^e denote the OUW functions for nodes i of the triangular element e. Clearly, U_i^e must satisfy the following compatible conditions:

$$U_i^e = 1 \quad \text{at} \quad L_i = 1, \quad L_j = 0, \quad \text{and} \quad L_k = 0 \tag{26}$$

and

$$U_i^e = 0 \quad \text{at} \quad L_i = 0 \tag{27}$$

In order to ensure the compatibility along each common side of quadrilateral and triangular elements when both types of elements are used, it is required that U_i^e should be reducible to Eq. (21) for each of the three sides of element e, i.e.

$$U_i^e = L_i + L_iL_j(L_i-L_j)(10-15\alpha_k) - 3\alpha_kL_iL_j \quad \text{at} \quad L_k = 0 \tag{28}$$

where i, j, k are the permutation index set and α_k is the weighting factor for the side i-j. Another requirement of the weighting functions is that the sum of weighting functions is unity:

$$\int_A U_1^e \, dA + \int_A U_2^e \, dA + \int_A U_3^e \, dA = A \tag{29}$$

in which A is the area of the triangular element e. In addition orthogonal conditions can be stated as follows:

$$\int_A U_i^eN_j^e \, dA = 0 \quad \text{if } j \neq i, \quad \int_A U_i^eN_j^e \, dA \neq 0 \quad \text{if} \quad j = i \tag{30}$$

in which $N_j^e = L_j$ is the basis function for node j (Huebner 1975). It can be verified that the following expression for the weighting function U_i^e satisfy conditions specified by Eqs. (26) through (30):

$$\begin{aligned} U_i^e = {} & L_i + 10L_i^2(L_j+L_k) - 10L_i(L_j^2+L_k^2) - 35L_iL_jL_k + 105L_i^2L_jL_k \\ & - 15L_i^2(\alpha_kL_j-\alpha_jL_k) + 15L_i(\alpha_kL_j^2-\alpha_jL_k^2) - 3L_i(\alpha_kL_j-\alpha_jL_k) + \\ & L_iL_jL_k[84(\alpha_j-\alpha_k)L_i + (21\alpha_k-63\alpha_j)L_j + (63\alpha_k-21\alpha_j)L_k] \end{aligned} \tag{31}$$

Final expressions of U_1^e, U_2^e, and U_3^e and their derivatives with respect to the global coordinate (x,y) can be obtained from Equation (31) and the realtionships between (L_1,L_2,L_3) and (x,y).

To summarize, we have derived orthogonal upstream weighting (OUW) functions for linear line elements in Eqs. (19) and (20), for linear quadrilateral elements in Eqs. (22) and (23), and for linear triangular elements in Eq. (31).

RESULTS AND DISCUSSIONS

Four finite element schemes are included for numerical experiments. These are the orthogonal upstream weighting (OUW) given here, the orthogonal weighting (OW) presented earlier for solving groundwater flow equations (Yeh 1983), the non-orthogonal upstream weighting (NUW) proposed elsewhere (Huyakorn and Nilkuha 1979), and the standard Galerkin weighting (GW) given in any text book on finite element methods. For OUW and NUW simulation the optimum weighting factors determined by Christie et al's formula (1976) are chosen. Both direct elimination solution (DES) and pointwise iteration solutions (PIS) strategies are employed to solve the matrix equations resulting from any of the four schemes. For DES, the compact Crout algorithsm is used (Zienkiewicz 1977). For PIS, the successive over-relaxation (SOR) with an relaxation parameter of 1.8 is chosen (Smith 1965).

To test and compare numerical performance of DES and SOR of all four schemes, we consider two simple problems for which analytical solutions are available. These are: (1) one-dimensional transient transport under constant upstream concentration and (2) two-dimensional transient transport from an upstream strip sources.

One-Dimensional Problem

Initially the concentration over the region $0 \leq x \leq L$ is assumed to be zero everywhere. Boundary conditions are given as: $C = 1$ at $x = 0$ and $\partial C/\partial x = 0$ at $x = L$. The velocity is assumed constant and equal to 1.0. A constant rectangular element size of 0.5 X 0.5 is used. To identify different cases solved, we define the mesh Peclet Number and Courant Number as $P_e = V\Delta x/D$ and $C_r = V\Delta t/\Delta x$. A constant $C_r = 0.4$ (i.e. a constant time step size of 0.2) is used throughout the simulation. Four cases with $P_e = 0.05$, 1.0, 50.0, and ∞ are solved. They represent cases of highly dispersion dominant, dispersion-advection equally dominant, highly advection dominant, and pure advection transport, respectively. These Peclet numbers are made up by varying the dispersion coefficient from 10.0 to 5.0, 0.01, and 0.0. To ensure that the boundary condition at $x = L$ is satisfied, the length L is set to 50, 15, 15, and 10, respectively corresponding to $P_e = 0.05$, 1.0, 50.0, and ∞.

Numerical results were obtained with two computer codes: AQUITRAN (Yeh and Francis 1984) and FEMA (Yeh and Huff 1984). The structure of these two general purpose computer packages are almost identical, except that the former ultilizes the OUW and OW scheme and the latter employs the NUW and GW schemes. Both code include DES and PIS strategies for solving matrix equations. The comparison on CPU time should truly reflect the efficience of the schemes rather than the skill of the programmer since both codes are programmed by the same author.

Figures 3(a-d) show the concentration profiles at $t = 4.0$. For $P_e = 0.05$, all four schemes produce almost identical results and are very accurate compared with the analytical solution. The SOR computations of all four schemes generate nearly the same results as their respective DES calculations. In fact, computer outputs indicated that the SOR solutions are identical to the DES solutions to the fourth digit. For $P_e = 1.0$, DES of all four schemes yield very close solutions to the analytical results. However, only OUW and OW schemes produce convergent SOR solutions, both NUW and GW result in divergent SOR simulations. As the Peclet number is increased to 50.0, OW and GW exhibit better accuracy as compared to their corresponding upstream schemes OUW and NUW, respectively. Nevertheless, these results can be thought comparable considering none of the scheme yields accurate solutions. To obtain reasonably accurate solutions, one has to decrease the mesh size and time step (Huyakorn and Nilkuha 1979), which is not the purpose of this paper. The point to note is that for this highly advection dominant transport only the OUW leads to convergent SOR solutions, all other three schemes fail if SOR

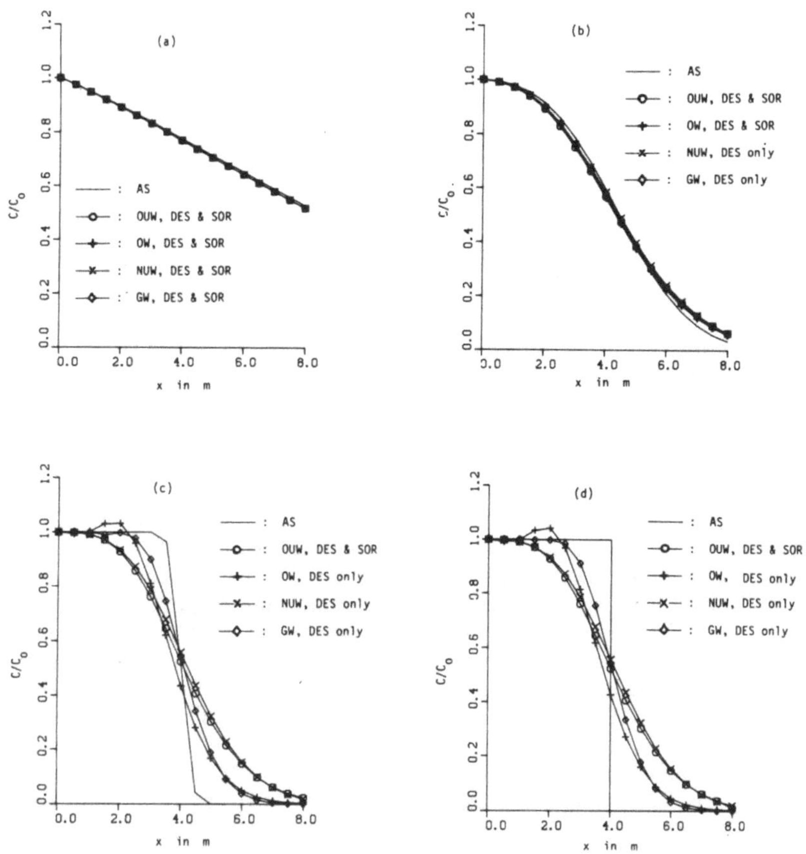

Figure 3. Concentration profiles at t = 4.0 for (a) $P_e = 0.05$, (b) $P_e = 1.0$, (c) $P_e = 50.0$, and (d) $P_e = \infty$

method is applied to solving the matrix equation. To test if further increasing in Peclet number would render OUW with divergent SOR computations, the case of pure advection is simulated. It is seen from Figure 3(d), that the OUW still leads to convergent SOR computations while other schemes fail.

Next, we examine the efficience of DES and SOR simulations of the four schemes in terms of computational time. Table 1 shows the comparison of the CPU time for this simple one-dimensional problem. It is seen that OUW and OW are approximately a factor of two slower than the NUW and GW. This results from the fact that a 3 X 3 Gaussian integration has been used to compute the element matrices in OUW and OW schemes while a 2 X 2 Gaussian integration is used in NUW and GW schemes. Had the same number of Gaussian points been used, the CPU time would have been comparable. It is also noted that SOR simulations, if converge, take about

Table 1. Comparison of CPU time required by DES and SOR simulations of OUW, OW, NUW, and GW schemes for the one-dimensional problem.

P_e	OUW		OW		NUW		GW	
	DES	SOR	DES	SOR	DES	SOR	DES	SOR
0.05	27.08	39.59	22.05	34.45	12.19	23.53	11.27	22.66
1.00	8.80	16.15	7.30	15.18	4.36	*	4.09	*
50.0	8.69	20.01	7.29	*	4.29	*	4.02	*
∞	6.19	13.95	5.21	*	3.24	*	3.08	*

* Divergent Computations

double amount of time compared to the DES computations for all four schemes. Thus, the CPU time with SOR is within the same order of magnitude as with DES even for this simple one-dimensional problem.

Two-Dimensional Problem

In this case, we consider a common problem of two-dimensional transport from a strip source at the upstream end. The flow is assumed to be uni-directional along the x-direction with a velocity of 1.0. The medium is homogeneous and isotropic. The region to be considered is 5 (in y-direction) X 10 (in x-direction). The initial condition is zero concentration everywhere in the region. The boundary conditions are: $C = 1.0$ at $x = 0$ and $2 \leqslant y \leqslant 3$; $C = 0.0$ at $x = 0$ and $y < 2$ or $y > 3$; zero fluxes at $y = 0$ and $y = 5$; and $\partial C/\partial x = 0$ at $x = 10$. For numerical simulations, the region is discretized with 10 X 20 rectangular elements resulting in 231 nodes. Figures 4(a-c) illustrate concentration profiles along $y = 2.5$ at $t = 4.0$ for $P_e = 0.05$, 1.0, and 50.0, respectively. It is seen that the main conclusions drawn from the one-dimensional analysis carry into this two-dimensional example. These are: (1) for dispersion dominant transport, all four schemes produce about the same solutions, all schemes are amenable to SOR computations; (2) for dispersion-advection equally dominant transport, all four schemes generate similar results, but only OUW and OW are suscetible to SOR simulations, and (3) for advection dominant transport, non-upstream schemes exhibit better accuracy than the corresponding upstream schemes, but all results are comparable, only the OUW finite element can be solved with SOR iterations.

Table 2 compares the CPU time as simulated by DES and SOR of the finite element schemes. It is seen that for any of the four schemes, the SOR computations take less CPU time than the DES calculations. This is because the bandwidth of the coefficient matrix is increased moderately even for this small two-dimensional problem. The CPU time is proportional to $N \cdot M^2$ with DES (Pinder and Gray 1977) but is proportional to $9N \cdot K$ with SOR, where N is the number of equations, M is the half-bandwidth minus one, 9 denotes the nine non-zero entries for every row of the coefficient matrix, and K is the number of iterations required for a convergent solution with SOR. Thus, it can be expected that as

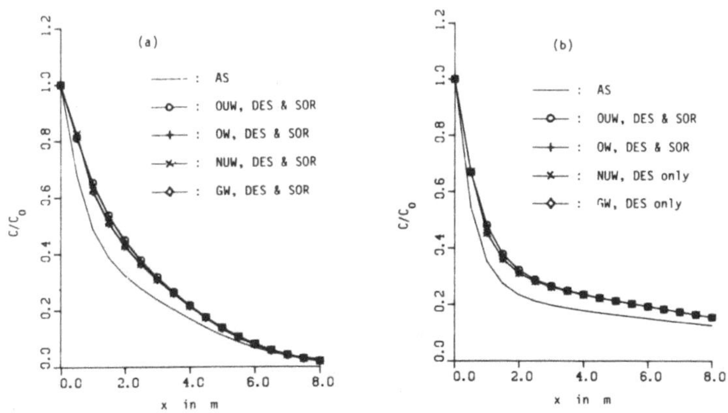

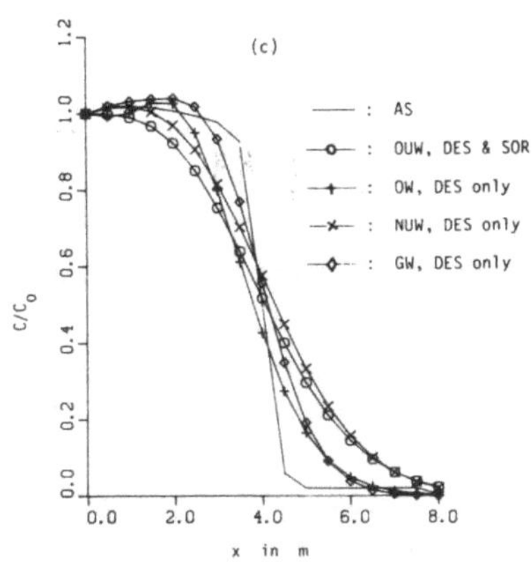

Figure 4. Concentration profiles along y = 3.5 at t = 4.0 for (a) P_e = 0.05, (b) P_e = 1.0, and (c) P_e = 50.0.

Table 2. Comparison of CPU time required by DES and SOR simulations of OUW, OW, NUW, and GW schemes for the two-dimensional problem.

P_e	OUW		OW		NUW		GW	
	DES	SOR	DES	SOR	DES	SOR	DES	SOR
0.05	110.46	61.67	100.52	51.67	26.01	26.05	24.12	24.10
1.00	110.22	75.95	100.23	67.02	25.98	*	24.09	*
50.0	110.35	100.35	100.72	*	25.99	*	24.14	*

* Divergent Computations

the bandwidth of the matrix increases, the saving in CPU time with SOR over that with DES can be substantial.

ACKNOWLEDGEMENT

Research sponsored by the Office of Health and Environmental Research, U. S. Department of Energy under Contract W-7405-eng-26 with Union Carbide Corporation.

REFERENCES

Bear, J. (1977) Hydraulics of Groundwater, McGraw-Hill Book Company, New York, 569 pp.

Bredehoeft, J. D. and G. F. Pinder. (1970) Digital analysis of area flow in multi-aquifer systems: a quasi three-dimensional model, Water Resour. Res. 6, 883-888.

Christie, L., D. F. Griffiths, A. R. Mitchell, and O. C. Zienkiewicz. (1976) Finite element methods for second order differential equations with significant first order derivatives, Int. J. Numer. Meth. Engrg. 10, 1389-1396.

Freeze, R. A. (1971) Three-dimensional transient, saturated-unsaturated flow in groundwater basin, Water Resour. Res. 7, 347-366.

Huebner, K. H. (1975) The Finite Element Methods for Engineers, John Wiley & Sons, Inc., New York, 500 pp.

Huyakorn, S. P. and K. Nilkuha. (1979) Solution of transient transport equation using an upstream finite element scheme, Appl. Math. Modelling 3, 7-17.

Livesley, R. K. (1983) Finite Elements: An Introduction for Engineers, Cambridge University Press, Cambridge, UK, 199 pp.

Peaceman, D. W. and H. H. Rachford, Jr. (1955) The numerical solution of parabolic and elliptic differential equations, J. Soc. Ind. Appl. Math. 3, 28-41.

Pinder, G. F. and W. G. Gray. (1977) Finite Element Simulations in Surface and Subsurface Hydrology, Academic Press, New York, 295 pp.

Putney, J. M. and T. W. Broyd. (1983) The Identification, Acquisition and Assessment of Detailed Geospheric Flow and Migration Computer Programs with Particular Reference to Variably Saturated Media, Stage I, Atkin Research and Development, Epsom, Surrey, England.

Segol, G. (1976) A three-dimensional Galerkin finite element model for the analysis of contaminant transport in saturated-unsaturated porous media, Finite Elements in Water Resources (ed. Gray, et. al), 2.123-2.144, Pentech Press, London.

Smith, G. D. (1965) Numerical Solution of Partial Differential Equations, Oxford University Press, London, 179 pp.

Stone, H. L. (1968) Iterative solution of implicit approximations of multi-dimensional partial differential equations, J. Soc. Ind. Appl. Math. 5, 530-558.

Trescott, P. C. and S. P. Larson. (1977) Comparison of iterative methods of solving two-dimensional groundwater flow equations, Water Resour. Res. 13, 125-136.

Yeh, G. T. (1983) Solution of groundwater flow equations using an orthogonal finite element scheme, Int. AMSE Conference "Modeling and Simulations", Nice, France, Sept. 12-14, 4, 329-351, AMSE Press, 69190, Tassin, France.

Yeh, G. T., J. C. Hwang, and W. C. Cho. (1984) Subregion iteration of finite element method for large-field problems of groundwater solute transport (submitted to J. Hydraulic Division, ASCE).

Yeh, G. T. and D. D. Huff. (1984) FEMA: A Finite Element Model of Material Transport through Aquifers, ORNL- , Oak Ridge National Laboratory, Oak Ridge, TN. (in prep.).

Yeh, G. T. and C. W. Francis. (1984) AQUITRAN: An Orthogonal Upstream Finite Element Approach to Modeling AQUIfer Contaminant TRANsport, ORNL- , Oak Ridge National Laboratory, Oak Ridge, TN. (in preparation).

Young, D. M. (1954) Iterative methods for solving partial differential equations of the elliptic type, Trans. Ameri. Math. Soc. 76, 92-111.

Zienkiewicz, O. C. (1977) The Finite Element Method, Third Edition, McGraw-Hill Book Company (UK) Limited, London, 787 pp.

ADAPTIVE COLLOCATION METHOD FOR THE TRANSPORT PROBLEM INDUCED BY IRREGULAR WELL PATTERNS

Vietchau Nguyen and Georges Abi-Ghanem, EWA, Inc.,
12800 Industrial Park Boulevard, Minneapolis, Minnesota 55441, USA

I. Problem Statement

In this communication, we present a numerical method of solution to the transport problem associated with a multiple source/sink flow field. Consider the miscible displacement equation:

$$(1)\quad \frac{\partial c}{\partial t} + \frac{\partial}{\partial x}(v_x c) + \frac{\partial}{\partial y}(v_y c) = \nabla\cdot(D(x,y)\nabla c), \nabla = \left(\frac{\partial}{\partial x},\frac{\partial}{\partial y}\right),$$

where c is the concentration, D is the variable dispersion coefficient for applications under study, and the flow field is defined by

$$(2)\quad \underset{\sim}{v} = \begin{bmatrix} v_x \\ v_y \end{bmatrix} = \frac{1}{2\pi\phi h}\sum_{m=1}^{N} \frac{q_m}{(x-x_m)^2 + (y-y_m)^2}\begin{bmatrix} x-x_m \\ y-y_m \end{bmatrix} \qquad ,$$

where is the porosity, h is the thickness of the flow field, and q_m is the pumping rate at well (x_m, y_m), $m = 1,\dots,N$. Equation (2) is obtained by first solving the Laplace equation for the hydraulic head, then applying Darcy,s law and the principle of superposition to the N-well system. Note that equation (1) becomes advection-dominated in the neighborhood of the (x_m, y_m)-well, and consequently leads to numerical instability of the solution.

II. Review Of Alternating Direction Collocation (ADC) Over Rectangles (Ref.2,3,4)

Consider the simple parabolic equation on a rectangular domain:

$$(3)\quad \frac{\partial u}{\partial t} - \nabla^2 u = f(t,x,y,\nabla u), \quad \nabla^2 = \frac{\partial^2}{\partial x^2} + \frac{\partial^2}{\partial y^2},$$

Upon using the Crank-Nicolson discretization in time, we have

$$(4)\quad \frac{u^{n+1} - u^n}{\Delta t} - \nabla^2\left(\frac{u^{n+1} + u^n}{2}\right) = f^{n+1/2}$$

where the superscript n denotes the time level. Let

$$(5)\quad u \approx \hat{u} = \sum_{p,q} \xi_{pq}(t)\, \alpha_p(x)\, \beta_q(y)$$

Substitute (5) into (4) yields (after collocation) :

$$(6)\quad \left\{ \sum_{p,q} \left[\frac{\xi_{pq}^{n+1} - \xi_{pq}^{n}}{\Delta t} \alpha_p \beta_q - (\alpha_p'' \beta_q + \alpha_p \beta_q'') \left(\frac{\xi_{pq}^{n+1} + \xi_{pq}^{n}}{2} \right) \right] = f^{n+1/2} \right\} \Bigg|_{\substack{x_i \\ y_j}}, \qquad i = 1,\dots,N_x; j=1,\dots,N_y$$

$$(7)\quad \underset{\approx}{C}_x = [\alpha_p | x_i], \qquad \underset{\approx}{A}_x = [\alpha_p'' | x_i],$$

$$\underset{\approx}{C}_y = [\beta_q | y_j], \qquad \underset{\approx}{A}_y = [\beta_q'' | y_j],$$

and the tensor product $\otimes$ by

$$\underset{\approx}{C}_y \otimes \underset{\approx}{A}_x = [c_{qj}\, [\underset{\approx}{A}_x]] .$$

Equation (6) may now be written in tensor product $\otimes$ form as

$$(8)\quad (\underset{\approx}{C}_y \otimes \underset{\approx}{C}_x) \frac{\underset{\sim}{\xi}^{n+1} - \underset{\sim}{\xi}^{n}}{\Delta t} - (\underset{\approx}{C}_y \otimes \underset{\approx}{A}_x + \underset{\approx}{A}_y \otimes \underset{\approx}{C}_x) \left(\frac{\underset{\sim}{\xi}^{n+1} + \underset{\sim}{\xi}^{n}}{2}\right) = \underset{\sim}{f}$$

Adding $\frac{\Delta t^2}{4} (\underset{\approx}{A}_y \otimes \underset{\approx}{A}_x) (\underset{\sim}{\xi}^{n+1} - \underset{\sim}{\xi}^{n})$, and rearranging terms to obtain

$$(9)\quad \left[(\underset{\approx}{C}_y \otimes \underset{\approx}{I}) - \frac{\Delta t}{2} (\underset{\approx}{A}_y \otimes \underset{\approx}{I})\right]\left[(\underset{\approx}{I} \otimes \underset{\approx}{C}_x) - \frac{\Delta t}{2} (\underset{\approx}{I} \otimes \underset{\approx}{A}_x)\right] \underset{\sim}{\xi}^{n+1}$$

$$= \left[(\underset{\approx}{C}_y \otimes \underset{\approx}{I}) + \frac{\Delta t}{2} (\underset{\approx}{A}_y \otimes \underset{\approx}{I})\right]\left[(\underset{\approx}{I} \otimes \underset{\approx}{C}_x) + \frac{\Delta t}{2} (\underset{\approx}{I} \otimes \underset{\approx}{A}_x)\right] \underset{\sim}{\xi}^{n} + \underset{\sim}{F} .$$

Equation (9) is solved in two successive steps i.e., in x-direction then in y-direction, with an appropriate numbering scheme for the unknowns.

When the transport equation (9) has constant coefficients, the resultant ξ^{n+1} is the same, whether x-direction, or y-direction starting first.

For the case of variable D(x,y), replacing D of Equation (1) with a constant λ, $\lambda_{optimal} = D_{max}/2$ (Ref. 3); and this Laplace-modification scheme in two consecutive iteration step k and k+1 may be written as:

$$(10) \quad \frac{c^{n+1,k+1} - c^n}{\Delta t} - \lambda\nabla^2 \frac{(c^{n+1,k+1} + c^n)}{2} = f^{n+1/2}$$

$$+ (\nabla . D(x,y)\nabla - \lambda\nabla^2) \frac{(c^{n+1,k} + c^n)}{2} .$$

Similar to Equation (9), a temporal perturbation term may be added to (10) to allow for matrix factorization. Equation (10) is iteratively solved for the change in c which reduces the number of arithmetic operation per time step and the relative size of the round-off error (IADC).

The described method of solution could not overcome the hyperbolic nature of Equation (1) near pumping wells. The concentration spatial and temporal distribution has been found unstable, given smooth initial and boundary conditions, and constant pumping rates.

III. Adapative Iterative Alternating Direction Collocation (AIADC) Method

Let us write Equation (1) as

$$(11) \quad \frac{\partial c}{\partial t} + \underset{\sim}{v} . \nabla c - D\nabla^2 c = g(t,x,y,c\nabla . \underset{\sim}{v},\nabla c)$$

The Peclet number $P_e = \|v\| L/D$, where L is a characterstic spatial scale. Define the advection-dominated zone $\mathfrak{D}$ to be the set of (x,y) where $P_e > 100$. $\mathfrak{D}$ contains all the pumping wells.

$$(12) \quad \underset{\approx}{E}_x = [\alpha'_p | _{xi}] \text{ and } \underset{\approx}{E}_y = [\beta'_q | _{yj}] .$$

Rewrite (11) in the tensor-product form using (5):

$$(13) \quad (\underset{\approx}{C}_y \otimes \underset{\approx}{C}_x) \frac{\underset{\sim}{\xi}^{n+1} - \underset{\sim}{\xi}^n}{\Delta t} - (\underset{\approx}{C}_y \otimes v_x \underset{\approx}{E}_x + \underset{\approx}{C}_y \otimes D\underset{\approx}{A}_x + v_y \underset{\approx}{E}_y \otimes \underset{\approx}{C}_x + D\underset{\approx}{A}_y \otimes \underset{\approx}{C}_x)$$

$$\left(\frac{\underset{\sim}{\xi}^{n+1} + \underset{\sim}{\xi}^n}{2}\right) = \underset{\sim}{g}$$

Adding $\frac{\Delta t^2}{4}((v_y \underset{\approx}{E}_y + D\underset{\approx}{A}_y) \otimes (v_x \underset{\approx}{E}_x + D\underset{\approx}{A}_x))(\underset{\sim}{\xi}^{n+1} - \underset{\sim}{\xi}^{n})$ to obtain the

factorized equation

$$(14)\quad \left[\underset{\approx}{C}_y \otimes \underset{\approx}{I} - \frac{\Delta t}{2}(v_y \underset{\approx}{E}_y + D\underset{\approx}{A}_y) \otimes \underset{\approx}{I}\right]\left[\underset{\approx}{I} \otimes \underset{\approx}{C}_x - \frac{\Delta t}{2}\underset{\approx}{I} \otimes (v_x \underset{\approx}{E}_x + D\underset{\approx}{A}_x)\right]\underset{\sim}{\xi}^{n+1}$$

$$= \left[\underset{\approx}{C}_y \otimes \underset{\approx}{I} + \frac{\Delta t}{2}(v_y \underset{\approx}{E}_y + D\underset{\approx}{A}_y) \otimes \underset{\approx}{I}\right]\left[\underset{\approx}{I} \otimes \underset{\approx}{C}_x + \frac{\Delta t}{2}\underset{\approx}{I} \otimes (v_x \underset{\approx}{E}_x + D\underset{\approx}{A}_x)\right]\underset{\sim}{\xi}^{n} + \underset{\sim}{G},$$

in which v_x, v_y and D are functions of x and y.

Adaptive IADC criterion:

When $c(x,y,t)$ changes from initial condition $c(x,y,o)$ in $\mathfrak{D}$, every product terms with v_x, v_y, $\frac{\partial v_x}{\partial x}$, $\frac{\partial v_y}{\partial y}$ in Equation (14) will be upstream collocated in each direction.

See Reference 1 for an introduction to upstream collocation. Numerical experiments show that the alternating-direction upstream collocation damps out oscillations in the concentration fronts away and toward pumping wells. However, in many occasions the computed centers of mass location of the concentration fronts is overestimated as compared to the purely-advective streamtube model using Equation 2. To improve the front tracking results, a Laplace modified scheme, similar to Equation 10, was developed for (14) which assumes the following form for

$\underset{\sim}{\delta}^{n+1} = \underset{\sim}{\xi}^{n+1} - \underset{\sim}{\xi}^{n}$:

$$(15)\quad \left[(\underset{\approx}{C}_y \otimes \underset{\approx}{I}) - \frac{\Delta t}{2}(v_y \underset{\approx}{E}_y + \lambda \underset{\approx}{A}_y) \otimes \underset{\approx}{I}\right]\left[(\underset{\approx}{I} \otimes \underset{\approx}{C}_x) - \frac{\Delta t}{2}\underset{\approx}{I} \otimes (v_x \underset{\approx}{E}_x + \lambda \underset{\approx}{A}_x)\right]\underset{\sim}{\delta}^{n+1,k+1} = \Delta t\left[\underset{\approx}{C}_y \otimes (v_x \underset{\approx}{E}_x + \lambda \underset{\approx}{A}_x) + (v_y \underset{\approx}{E}_y + \lambda \underset{\approx}{A}_y) \otimes \underset{\approx}{C}_x\right]\underset{\sim}{\xi}^{n} + \underset{\sim}{G}^{n+1/2} + \underset{\sim}{H}^{n+1,k}$$

where $\underset{\sim}{H}^{n+1,k}$ is a vector function of $\nabla . D\nabla \underset{\sim}{\xi}^{n+1,k}$ and $\lambda \nabla^2 \underset{\sim}{\xi}^{n}$. Superscript k denotes the iteration level for an alternating-direction estimated-solution sweep.

The Laplace modified scheme when applied to the advection related term yield divergent iteration process, especially when the adaptive IADC criterion is satisfied. A formal analysis of the preceding method of solution will be presented in a future communication.

IV. Application

The preceding AIADC method is now applied to a field problem of aquifer restoration. A set of three purge wells were used to remove a contaminant plume at an industrial site. Initial and boundary conditions, grid dimension, and pumping strategy are shown in Figures 1,2, and 3. Dispersion varies between 0.6 and 2.0m^2/day. Concentration distribution is normalized against 139 ppm. After 90 days, the accumulative contaminant removal curve is presented in Figure 4.

Other applications of this method to the uranium mining problem (Ref. 4 and 5), and to the secondary oil recovery operation will be communicated elsewhere.

V. References

1. M.B. Allen "How Upstream Collocation Works", Intl. J. Num. Meth. Eng., to appear.

2. M.A. Celia and G.F. Pinder (1982) "Transport Simulation Using Three Dimensional Alternating Collocation", in Finite Elements in Water Resources IV, K.P. Holtz et al ed., Hanover, Germany, 14.9 -14.9.

3. J. Douglas and T. Dupont (1971) "Alternating-Direction Galerkin Methods on Rectangles", Symp. on the Numerical Solution of PDEs-II, B. Hubbard ed., Academic, New York, 133-213.

4. V.V. Nguyen (1983) "Collocation Simulator for a System of Coupled Transport Operators", in IMACS Trans. on Scientific Computation Vol. III, W.F. Ames ed., North-Holland, Amsterdam, The Netherlands.

5. V.V. Nguyen et al. (1983) "Numerical Simulation of Uranium In-Situ Mining", Chemical Engng. Sci., 38(11): 1885-1862.

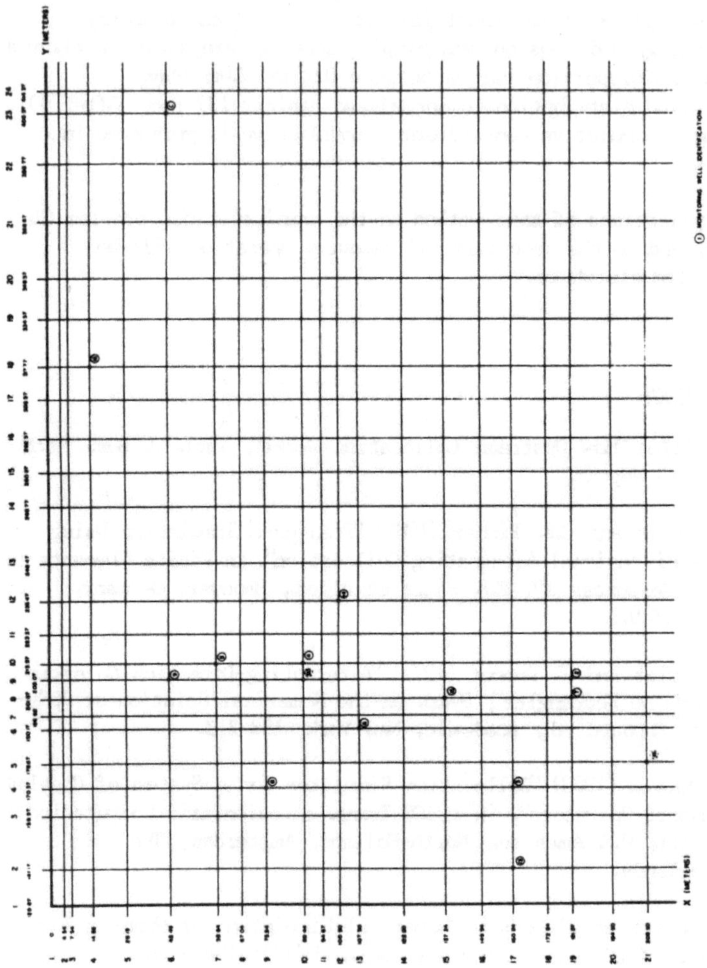

Figure 1. Grid Configuration

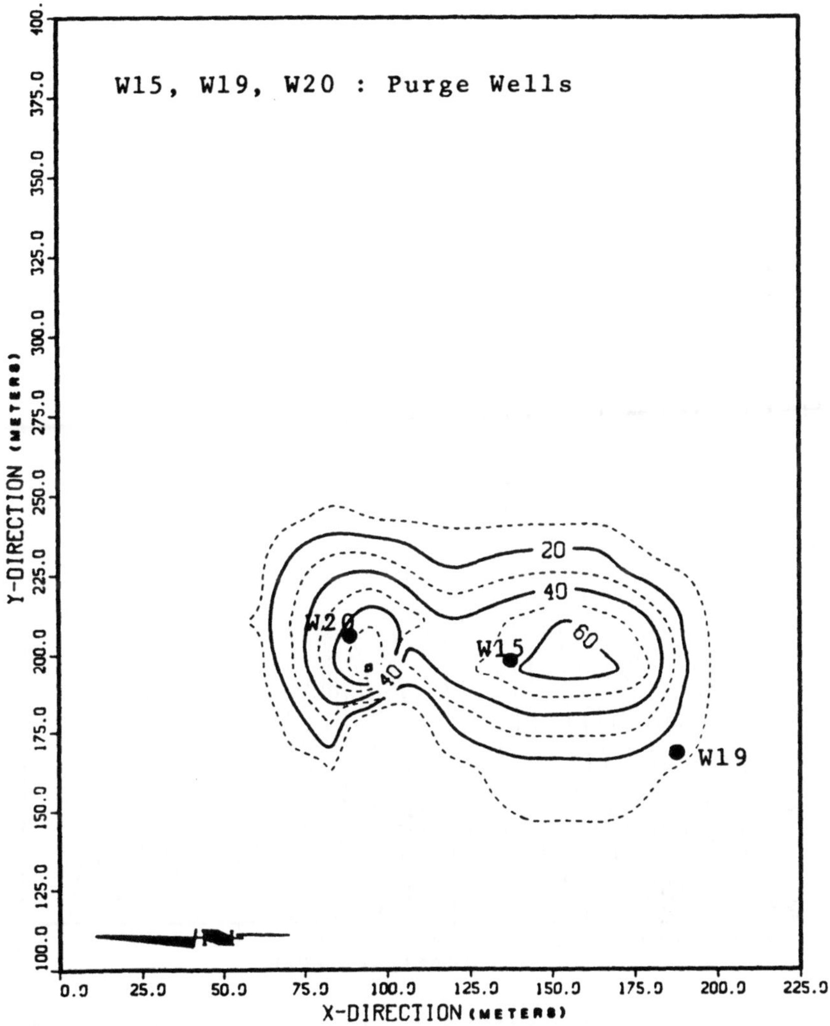

Figure 2. Initial Condition of Contaminant

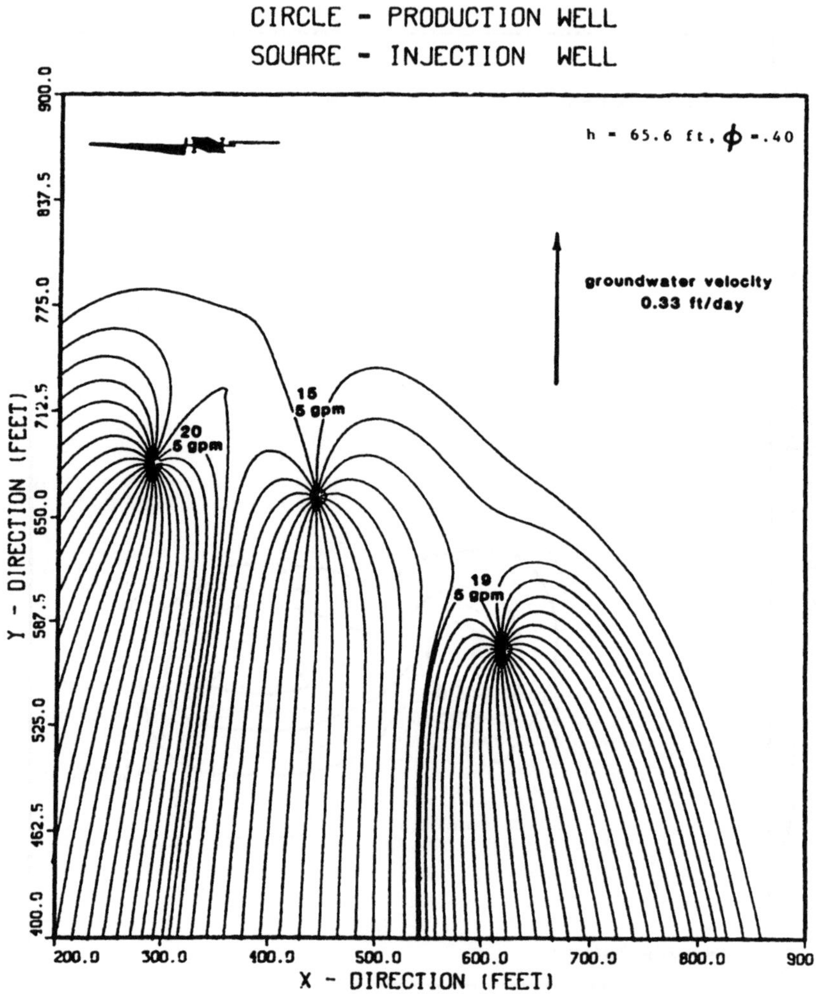

Figure 3. Streamline Pattern For Purge Wells

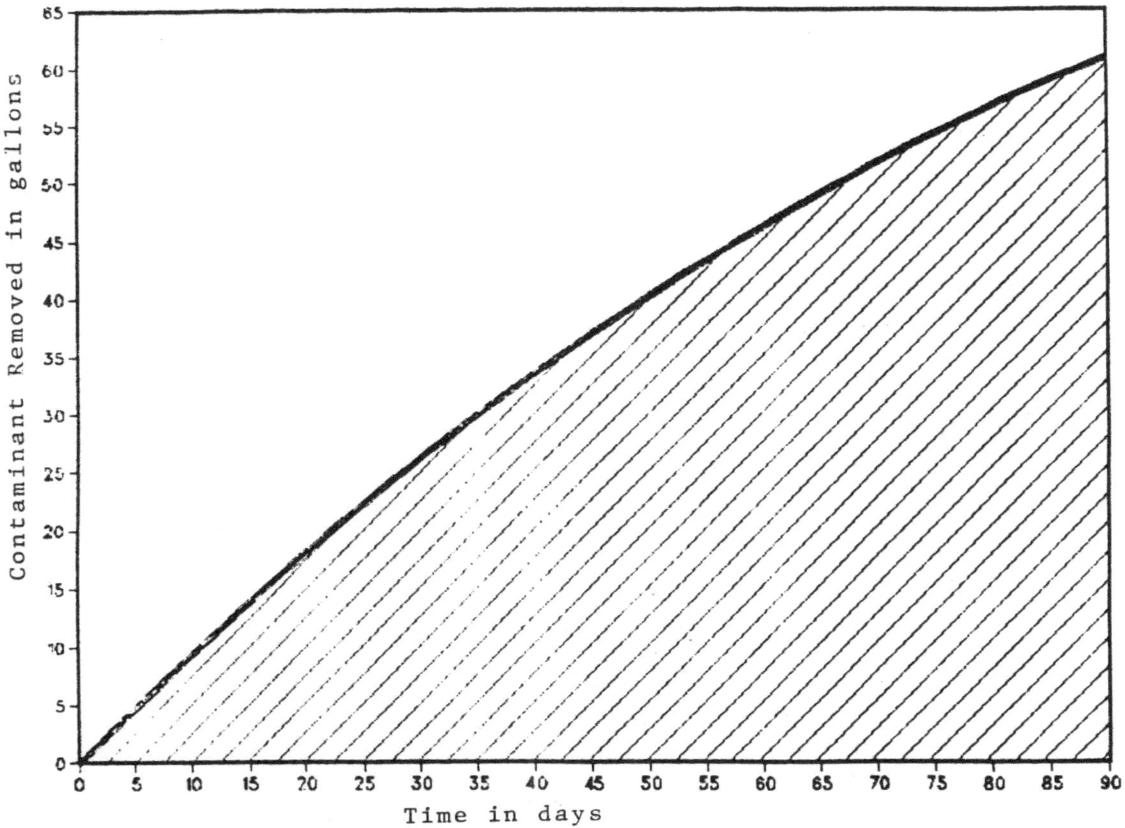

Figure 4. Accumulative Removal Record

FINITE ELEMENT ANALYSIS OF MULTICOMPONENT CONTAMINANT TRANSPORT INCLUDING PRECIPITATION-DISSOLUTION REACTIONS

David J. Kirkner, Howard W. Reeves, and Aaron A. Jennings

University of Notre Dame, Notre Dame, IN 46556

INTRODUCTION

The determination of the fate of contaminants that enter groundwater systems pose difficult mathematical problems for the modeler. This is particularly true when complex multi component interactions occur among mobile and immobile species. The types of chemical reactions that can occur include; acid-base, adsorption-desorption, complex formation, ion exchange, oxidation-reduction and precipitation-dissolution. This paper will focus on the analysis of precipitation-dissolution reactions.

In multi-component transport systems, competition for sorptive sites can have unexpected results. Preferential ion-exchange or surface complexation can produce a concentration "spike" within the domain that exceeds the solute's source strength. This can occur whenever a rapidly moving component occupies surface sites and then is displaced by competition with some less mobile species. The concentration "spike" that develops in the neighborhood of the competition front will continue to grow as the solution penetrates further into the domain. Eventually this can grow until the solution's solubility constraints are violated. At this point a precipitate should form, and then dissolve once the concentration spike has passed. Therefore, to be successful, the mass transport model must be capable of accommodating both precipitation and dissolution reactions even for cases where the source strengths satisfy all solubility constraints.

In this paper, different formulations of the equations governing multicomponent mass transport with solution phase complexation, competitive adsorption and precipitation-dissolution are developed and their relative merits discussed. A finite element method of solution of one of the formulations is presented along with a sample simulation.

MODEL PROBLEM

To avoid the complexity that often accompanies generality, the following model problem is considered. Let c_1, c_2 be the concentrations (mass per unit volume of fluid) of two metals, c_3 the concentration of a nonadsorbing ligand and c_{13} the concentration of a soluble complex formed by the reaction of constituents one and three. A mass balance on each species in solution yields

$$Lc_1 + \partial c_1/\partial t + \partial s_1/\partial t + r_1 = 0 \tag{1}$$

$$Lc_2 + \partial c_2/\partial t + \partial s_2/\partial t + \partial p/\partial t = 0 \tag{2}$$

$$Lc_3 + \partial c_3/\partial t + \partial p/\partial t + r_3 = 0 \tag{3}$$

$$Lc_{13} + \partial c_{13}/\partial t + r_{13} = 0 \quad , \tag{4}$$

where the convection-diffusion operator in one dimension is

$$L(\cdot) = - \frac{1}{P_e} \partial^2(\cdot)/\partial x^2 + \partial(\cdot)/\partial x, \tag{5}$$

and P_e is the Peclét number. In the above system c_1 and c_2 are presumed to engage in competitive adsorption, thus s_1, s_2 are the concentrations of the sorbed forms of c_1, c_2 respectively. Also, r_1, r_3, r_{13} are the rates of accumulation of c_1, c_3, c_{13} due to reaction in solution. It is also presumed that a heterogeneous precipitation-dissolution reaction occurs between species two and three, and p is the concentration of the precipitate.

Each reaction occuring in the system (1) - (4) will be discussed separately.

<u>Solution Phase Complexation.</u> The reaction which forms the soluble complex is

$$\hat{c}_1 + \hat{c}_3 \underset{k_r}{\overset{k_f}{\rightleftarrows}} \hat{c}_{13} \tag{6}$$

where (^) denotes moles of the indicated species, thus the symbol $\hat{c}_1$, for example, means one mole of species one. As noted above, the concentration of species one is $c_1 \equiv [\hat{c}_1]$. For ease in presentation, all stoichiometric coefficients are taken as one, and order corresponds to stoichiometry.

The reaction rates for (6) are

$$r_1 = r_3 = -r_{13} = k_r\, c_{13} - k_f\, c_1\, c_3 \tag{7}$$

For the analysis herein, the local equilibrium assumption is made. This means that $k_r \to \infty$ and $k_f \to \infty$ at a fixed ratio

$K_{13} = k_f/k_r$ while the rates r_1, r_3, r_{13} are assumed to be bounded and not identically zero. The condition for this is

$$c_{13} - K_{13}\, c_1\, c_3 = 0. \tag{8}$$

This is simply the law of mass action for (6).

Sorption. For the system (1) - (4), the sorptive mechanism is assumed to be surface complexation (adsorption-desorption). Ion exchange leads to a similar mathematical system. The reactions are

$$\hat{c}_1 + \hat{x} \rightleftarrows \hat{c}_1\text{-}\hat{x} \equiv \hat{s}_1 \quad ; \qquad K_1 \tag{9}$$

$$\hat{c}_2 + \hat{x} \rightleftarrows \hat{c}_2\text{-}\hat{x} \equiv \hat{s}_2 \quad ; \qquad K_2 \tag{10}$$

where $\hat{x}$ indicates an immobile surface site. It is further assumed that the total concentration of immobile surface sites (free plus occupied) is constant, i.e.,

$$x + s_1 + s_2 = x_T \ . \tag{11}$$

Again invoking local equilibrium, writing the law of mass action for (9) and (10) and using (11) leads to the familiar competitive Langmuir isotherms,

$$s_1 = x_T\, K_1\, c_1/(1 + K_1\, c_1 + K_2\, c_2) \tag{12}$$

$$s_2 = x_T\, K_2\, c_2/(1 + K_1\, c_1 + K_2\, c_2) \tag{13}$$

Precipitation-dissolution. The following reaction

$$\hat{p} \equiv \hat{c}_2\text{-}\hat{c}_3 \rightleftarrows \hat{c}_2 + \hat{c}_3 \ , \tag{14}$$

occurs to enforce the constraint,

$$c_2\, c_3 \leq K_{23} \ , \tag{15}$$

where K_{23} is the solubility product. In essence, (15) is an inequality constraint on the system (1) - (4), with the solid present only when (15) is satisfied but with $p = 0$ if $c_2\, c_3 < K_{23}$.

PROBLEM FORMULATION

The rates r_1, r_3, r_{13} may be eliminated as unknowns as follows. Add (1) to (4) and (3) to (4), utilizing (7) to yield,

$$L(c_1 + c_{13}) + \partial(c_1 + c_{13})/\partial t + \partial s_1/\partial t = 0 \tag{16}$$

$$L(c_3 + c_{13}) + \partial(c_3 + c_{13})/\partial t + \partial p/\partial t = 0. \tag{17}$$

The number of unknowns in the system is now seven; c_1, c_2, c_3, c_{13}, s_1, s_2, p and there are seven equations; (2), (16), (17), (18), (12), (13), (15). If $c_2\, c_3 < K_{23}$ there is one less equation, (15), and one less unknown, p. This approach to problem formulation has recently been taken by Miller and Benson (1983). Therein a finite difference approximation, in space and time, was used for equations (2) and (17). The total number of nonlinear algebraic equations to be solved every time step is then seven times the number of grid points, minus the number of specified degrees of freedom. An alternate formulation is obtained by introducing the total soluble concentration of each component as a primary unknown, i.e. let

$$u_1 = c_1 + c_{13} , \tag{18}$$

$$u_2 \equiv c_2 , \tag{19}$$

$$u_3 = c_3 + c_{13} . \tag{20}$$

Equations (2), (16) and (17) then become,

$$Lu_1 + \partial u_1/\partial t + \partial s_1/\partial t = 0 , \tag{21}$$

$$Lu_2 + \partial u_2/\partial t + \partial s_2/\partial t + \partial p/\partial t = 0 , \tag{22}$$

$$Lu_3 + \partial u_3/\partial t + \partial p/\partial t = 0 . \tag{23}$$

This has simply added three additional equations and three additional unknowns to the system. Note however that (8), the law of mass action for the solution phase reaction, may be substituted into (18) through (20) and that system inverted to yield,

$$c_i = \tilde{c}_i(u_1, u_3) , \quad i = 1,3 . \tag{24}$$

In (24), the distinction between a function, $\tilde{c}_i$, and its value, c_i, is made. The fact that an equation like (24) may be written even for a more general system is guaranteed by the implicit function theorem. Equation (24), in the general case, may be implicit, but since a numerical solution is anticipated this is acceptble.

Equation (24) can now be substituted into (12), (13) and (15) to yield,

$$s_i = \tilde{s}_i(u_1, u_2, u_3) \qquad i = 1,2 \tag{25}$$

$$\tilde{c}_2\, \tilde{c}_3 = K_{23} \tag{26}$$

Equation (25) is now substituted into (21) and (22) to yield

$$Lu_1 + \partial u_1/\partial t + \partial \tilde{s}_1/\partial t = 0 \tag{27}$$

$$Lu_2 + \partial u_2/\partial t + \partial \tilde{s}_2/\partial t + \partial p/\partial t = 0 \tag{28}$$

In essence the number of primary unknowns has been reduced to four; u_1, u_2, u_3 and p with four equations; (27), (28), (23) and (26). For larger systems the reduction in primary unknowns, from the formulation discussed above, is even more dramatic.

The fundamental difference between the two types of solids present in the system should perhaps be noted here. The rate of accumulation of a particular component due to sorption can always be expressed as a function of the total soluble concentration of all components engaged in competition for surface sites. Thus, in effect, the sorption term can be considered a nonlinear source/sink term. The precipitation term, however is clearly not a function of the solution phase concentrations, in essence because p does not appear in the "constraint" equation (26). Therefore in the second formulation the solid accumulated due to precipitation is retained as a primary unknown.

Note that the total number of unknowns in the system has not actually been reduced; the secondary unknowns c_{13}, s_1, s_2 still must be evaluated during the course of the solution process. However, the solution of (24), (25) and (8) at discrete points in space and time is certainly less effort than expanding the total system to include these variables as primary unknowns as is done in the first formulation discussed.

In equations (28) and (23) the ratio of the rate of accumulation of u_2 to u_3 due to precipitation is fixed by the stoichiometry of the reaction (14). It is this fact which differentiates the system considered here from the classic problem of a system subjected to an inequality constraint, where this ratio is determined by the orientation of the "constraint surface"; Dorny (1975). This fact also allows another formulation to be developed where the precipitate p is removed as a primary unknown. Simply subtract (23) from (28) to yield

$$L(u_2 - u_3) + \partial(u_2 - u_3)/\partial t + \partial s_2/\partial t = 0 \tag{29}$$

The primary unknowns are now three; u_1, u_2, u_3 with three equations: (26), (27) and (29). In the case where the stoichiometric coefficients are not one, a linear combination of (23) and (28) will still eliminate p. Again p must still be calculated during the solution process, but this is easily done using (23) or (28) once the primary unknowns are available.

If there is no sorbed form of u_2 then (29) indicates that the difference, $u_2 - u_3$ is transported conservatively and this quantity can be introduced as a primary unknown. This type of problem has been recently discussed by Rubin (1983). This approach cannot be taken with sorption, however, since $\tilde{s}_2$ is not a function of the difference $u_2 - u_3$.

It would seem that this last formulation is the most attractive having the fewest number of primary unknowns. However there are other considerations, such as; ease in programming, efficiency, and ease in generalizing to a larger system. Space does not allow a complete comparison of these alternatives here. Rather, a numerical procedure and sample problem will be presented for one formulation; the primary system governed by (27), (28), (23) and (26), which retains p as an unknown.

NUMERICAL SOLUTION

The unknown functions in (27), (28), (23) and(26) are interpolated as follows

$$u_i(x,t) = \underset{\sim}{\phi}(x) \cdot \underset{\sim}{u}_i \ , \tag{30}$$

$$s_i(x,t) = \underset{\sim}{\phi}(x) \cdot \underset{\sim}{s}_i \ , \tag{31}$$

$$p(x,t) = \underset{\sim}{\Psi}(x) \cdot \underset{\sim}{p} \ , \tag{32}$$

where, for example, $\underset{\sim}{u}_i$ is a vector whose components are concentrations, u_i at the nodal points, i.e.,

$$\underset{\sim}{u}_i = (u_{i1}, u_{i2}, \ldots, u_{iN}) \tag{33}$$

In the example to follow, the vector of shape functions, ϕ are the the "hat" functions, Strang and Fix (1973). Since p may exhibit discontinuities it should not be interpolated with C° shape functions. Here a vector of Dirac delta functions centered at the nodal points is used. Thus

$$p(x,t) = \begin{cases} p_i \text{ if } x = x_i \ (i^{th} \text{ nodal point}) \\ 0 \text{ if } x \neq x_i \ . \end{cases} \tag{34}$$

Employing a standard weighted residual approach then requires that the solubility constraint be satisfied only at the nodal points and precipitate is only accumulated at the nodal points. The weighted residual formulation reduces (27), (28), (23) and (26) to a set of coupled ordinary differential and algebraic equations. The ordinary differential equations are converted to algebraic equations by a finite difference approximation. Herein the backward difference scheme is used. The resulting set of nonlinear algebraic equations are solved iteratively using Newton's method and the solution is marched through time. The constraint, (26), is presumed not to be in effect initially, and it is checked at each node every iteration and is either "turned-on" or "turned-off" appropriately.

RESULTS

A simulation was performed for the model problem introduced above and the results are summarized in Figures 1 through 6. In order to isolate the effects of precipitation as much as possible, aqueous complexation was not considered. Thus u_1, u_2, u_3 become simply c_1, c_2, c_3. The source strength of all solutes were fixed at the upstream boundary and taken equal to c_0. The gradients of all solutes were set to zero at the downstream boundary, which was located a sufficient distance beyond the column length that the model simulates a semi infinite domain. Zero initial conditions were specified for c_1, c_2 and p while c_3 was given a constant initial concentration equal to its source strength. The physical and chemical constants used were: $P_e = 40$, $K_1 = 10.0$, $K_2 = 1.0$, $x_T = 5.0$. To show the effect of precipitation the simulation was performed with $K_{23} = 1.2$ and $K_{23} = \infty$ (i.e., no precipitation) and the results compared.

Figures 1 and 4 show the concentration profiles for the adsorbed solids s_1 and s_2, without and with precipitation respectively. While precipitation has very little effect on the results of sorption the driving force for the precipitation is evident. A chromotographic effect is seen as the arrival of the c_1 front displaces c_2 from the surface and thus back into solution. It is the greater affinity of the c_1 for adsorption which causes the concentration of c_2 to rise above its source strength. Figures 2 and 5 show the soluble concentration profiles at 5.0 pore volumes, without and with precipitation. Without precipitation c_3 is conservative and since the initial condition equals the source strength, c_3 remains constant. The strong chromatographic effect is clearly displayed in Figure 2.

Concentration profiles for the simulation with precipitation are given in Figure 5. The concentration of c_3 is no longer constant. There is a zone where the concentration is elevated above its source strength due to the addition of c_3 to the aqueous phase as the solid dissolves. Beyond this zone c_3 is below its source strength due to precipitation. The c_2 profile is also altered considerably from Figure 2. It also exhibits two zones; the first shows a small increase over the source strength due to dissolution and the second higher plateau due to competition. The concentration in this "competition zone" is not as high as in Figure 2 because c_2 is removed from solution by precipitation. This two stage behavior is effected by the relative source strengths of the solutes, the initial conditions, adsorption, as well as the solubility product, K_{23}.

The concentration of the precipitate, p, is also given in Figure 5. As discussed in the previous section the solubility constraint was imposed pointwise at the nodes, and thus precipitate was only accumulated at the nodes. For graphical purposes, an average concentration was computed at each node and these values then interpolated. This was done in order to show the relative amount and position of the precipitate to the aqueous components. Breakthrough curves for the simulations without and with precipitation are presented in Figures 3 and 6 respectively. Figure 6 shows clearly the lowering of the c_2 peak. A second zone where c_2 is greater than the source strength due to dissolution is exhibited from approximately six to ten pore volumes. The precipitation and dissolution zones are also clearly displayed in the c_3 curve. For comparative purposes the precipitate breakthrough is also given.

CONCLUSION

Several formulations of the equations governing multicomponent mass transport with chemical interactions including precipitation-dissolution reactions were presented. A finite element solution algorithm was presented with a sample simulation. The simulation was for a system where the concentration of one component rises significantly above its source strength due to competitive adsorption with a less mobile solute. With the imposition of a solubility constraint this increase of soluble concentration above the source strength was significantly reduced. This simulation serves as an example of the importance of modeling precipitation-dissolution even in cases where there is no solid initially present and the source strengths satisfy are all solubility constraints.

ACKNOLWEDGEMENT

The work presented in this paper was supported by grant DE-AC02-79EV10253, from the Ecological Research Division, Office of Energy Research, Department of Energy, to the University of Notre Dame.

REFERENCES

Dorny, C.N. (1975) A Vector Space Approach to Models and Optimization, John Wiley and Sons, New York.

Miller, C.W. and Benson, L.V. (1983) Simulation of Solute Transport in a Chemically Reactive Heterogeneous System: Model Development and Application, Water Resources Research, 19, 2:381-391.

Rubin, J. (1983) Transport of Reacting Solutes in Porous Media: Relation Between Mathematical Nature of Problem Formulation and Chemical Nature of Reactions, Water Resources Research, 19, 5:1231-1252.

Strang, G. and Fix, G. (1973) An Analysis of the Finite Element Method, Prentice-Hall, Englewood Cliffs, New Jersey.

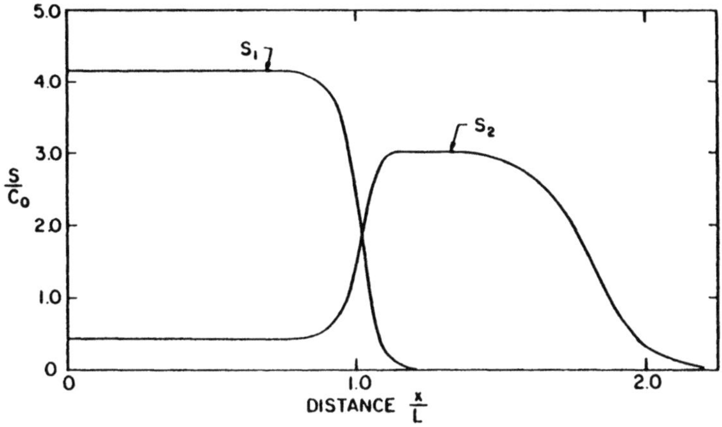

Figure 1. Sorbed Form Profiles without Precipitation

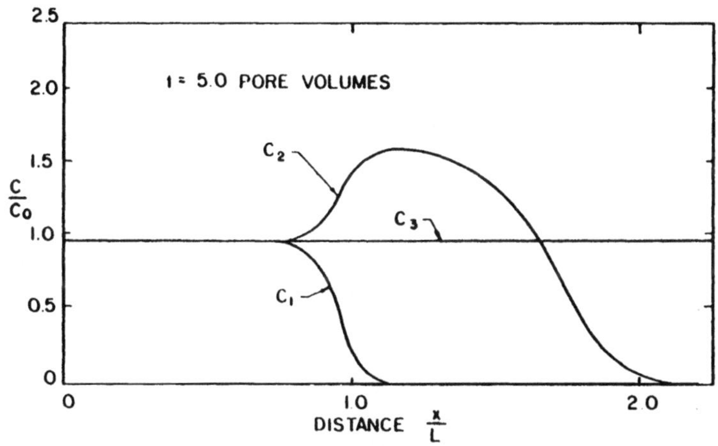

Figure 2. Solute Profiles without Precipitation

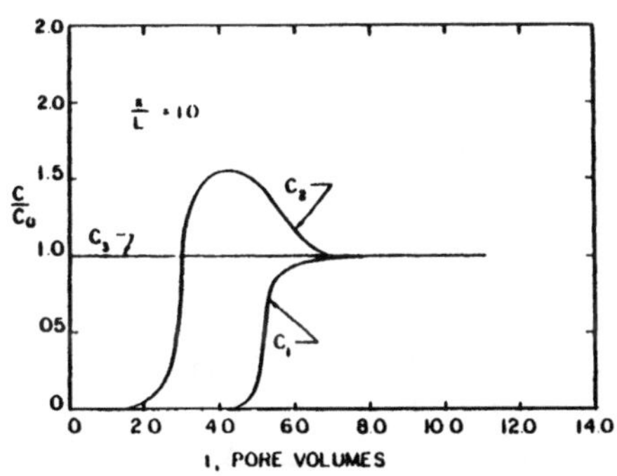

Figure 3. Breakthrough Curves without Precipitation

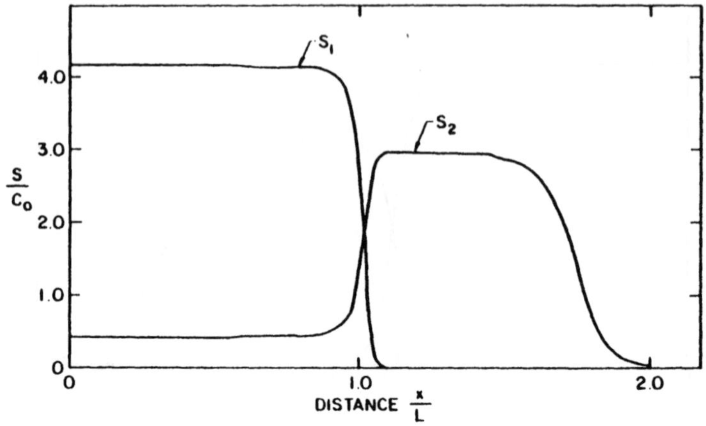

Figure 4. Sorbed Form Profiles with Precipitation

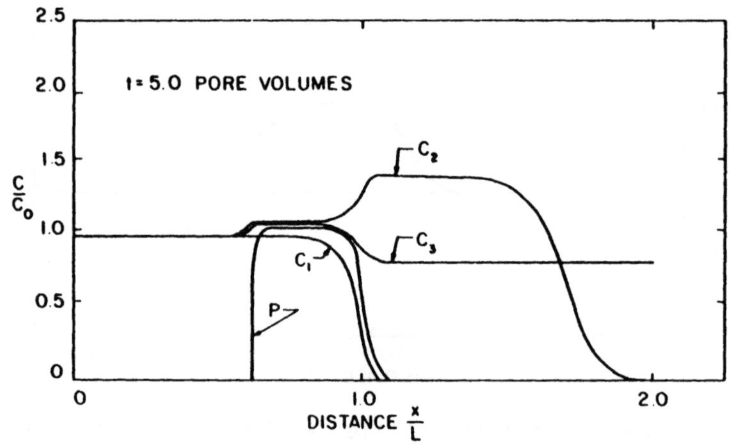

Figure 5. Solute Profiles with Precipitation

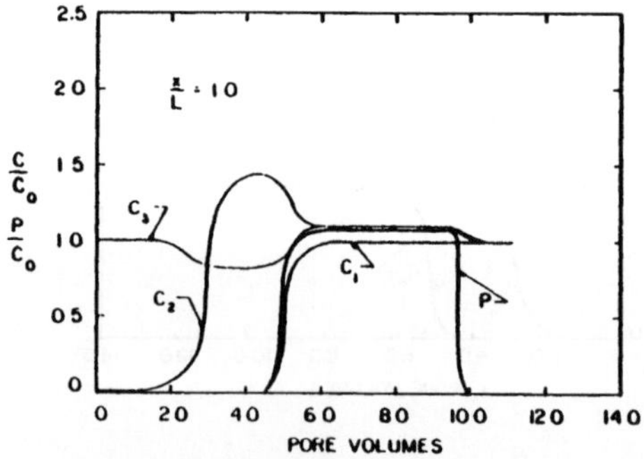

Figure 6. Breakthrough Curves with Precipitation

GROUNDWATER FLOW AND MASS TRANSPORT BY FINITE ELEMENTS - A PARAMETRIC STUDY

A. Pandit and S. C. Anand

Assistant Professor, Florida Institute of Technology and Professor, Clemson University.

INTRODUCTION

The subject of saltwater intrusion in coastal aquifers has been receiving wide attention for the past few years. Henry (1960) was the first investigator to obtain a steady state solution for the coupled flow and mass transport equations in an idealized, rectangular, confined aquifer by means of a Fourier-Galerkin double series expansion. His results confirmed the existence of saltwater circulation in coastal aquifers which had been observed earlier by Cooper (1959). Lee and Cheng (1974) utilized the Galerkin finite element technique to obtain solutions for the steady state, two dimensional saltwater encroachment problem. The governing partial differential equations, in their model, were written in terms of the stream function. Anand et al. (1980), using a similar model, investigated the saltwater intrusion problem for a wide, realistic, range of field parameters. In all these models the coefficient of hydrodynamic dispersion was considered to be a constant.

This paper is concerned with an investigation of Henry's problem utilizing a Hydraulic Head Model similar to the h-c model presented by Huyakorn and Taylor (1976). A detailed parametric study is conducted to determine the influence of various nondimensional field parameters on saltwater encroachment. In this study the coefficient of hydrodynamic dispersion is considered to be a function of soil dispersivity and interstitial pore velocity as proposed by Bear (1960).

MATHEMATICAL FORMULATION

The mathematical model is formulated in terms of a flow equation and a transport equation which are coupled by the velocity and density terms. The equations may be written as

$$\frac{\partial}{\partial x}\left[K_{xx}\frac{\partial h}{\partial x} + \frac{\partial}{\partial y}\left[K_{yy}\left(\frac{\partial h}{\partial y} + \varepsilon\frac{C}{C_1}\right)\right] = 0 \qquad (1)$$

and $$\frac{\partial}{\partial x_j}\left[D_{ij}\ \frac{\partial C}{\partial x_j}\right] - V_1\frac{\partial C}{\partial x_1} = 0 \qquad (2)$$

in which K_{xx} and K_{yy} = hydraulic conductivity tensors, h = hydraulic head, ε = normalized density difference between contaminant and fresh water, C_1 = contaminant concentration at the source, V_i = interstitial pore velocities, and C is concentration. D_{ij} in Equation (2) is the dispersion tensor, and can be expressed as

$$D_{xx} = \alpha_L\ (V_x^2\ /\ |V|) + \alpha_T\ (V_y^2\ /\ |V|) + D_d\tau_{yy}, \qquad (3)$$

$$D_{yy} = \alpha_T\ (V_x^2\ /\ |V|) + \alpha_L\ (V_y^2\ /\ |V|) + D_d\tau_{yy}, \qquad (4)$$

and $$D_{xy} = D_{yx} = (\alpha_L - \alpha_T)\left(V_xV_y\right)\ /\ |V| \qquad (5)$$

in which α_L and α_T = longitudinal and transverse dispersivities, D_d = molecular diffusion coefficient, τ_{xx} and τ_{yy} = principal components of the tortuosity tensor, and n_j = 1 for j = 2 (vertical direction). Introducing the nondimensional variables

$$x' = x/d,\ y' = y/d,\ h' = h/d,\ C' = C/C_1\ ,$$

$$K'_{yy} = K_{yy}/K_{xx},\ K'_{xx} = K_{xx}/K_{xx} = 1,\ V'_x = V_x/U, \qquad (6)$$

and $$V'_y = V_y/U,\ \alpha'_T = \alpha_T/\alpha_L,\ \alpha'_L = \alpha_L/\alpha_L = 1,\ d_L = d/\alpha_L$$

in which U and d = characteristic velocity and depth of aquifer, respectively. Equations (1) and (2) can be nondimensionalized (neglecting molecular diffusion) by substituting Equations (3) to (6) which after rearrangement, and neglecting primes for convenience yield

$$K_{xx}\frac{\partial^2 h}{\partial x^2} + K_{yy}\frac{\partial^2 h}{\partial y^2} = -K_{yy}\,\varepsilon\,\frac{\partial C}{\partial y} \qquad (7)$$

$$\left(\frac{V_x^2}{V} + \alpha_T\frac{V_y^2}{|V|}\right)\frac{\partial^2 C}{\partial x^2} + \left(\alpha_T\frac{V_x^2}{|V|} + \frac{V_y^2}{|V|}\right)\frac{\partial^2 C}{\partial y^2} + (1-\alpha_T)\frac{V_xV_y}{|V|}\frac{\partial^2 C}{\partial x\partial y}$$

$$+ (1-\alpha_T)\frac{V_xV_y}{|V|}\ \frac{\partial^2 C}{\partial y\partial x} - \left(V_x\frac{\partial C}{\partial x} + V_y\frac{\partial C}{\partial y}\right)d_L = 0, \qquad (8)$$

respectively.

Finite Element Technique
Using a triangular element with no internal or midside nodes and assuming the shape functions to be linear, Equations (7) and (8) can be written symbolically in the matrix form as

$$[K]\ \{h\} = \{P\} \tag{9}$$

$$[R]\ \{C\} = \{Q\}, \tag{10}$$

respectively, where [K] and [R] are coefficient matrices and {P} and {Q} are force vectors containing the boundary fluxes. Expressions for these matrices are given by Pandit (1982).

PROBLEM DEFINITION

The saltwater encroachment problem, along with the necessary boundary conditions, is shown in Fig. 1. Freshwater flow is from left to right, and is horizontal at the Freshwater Face. the nondimensional length to depth ratio of the aquifer is a variable.

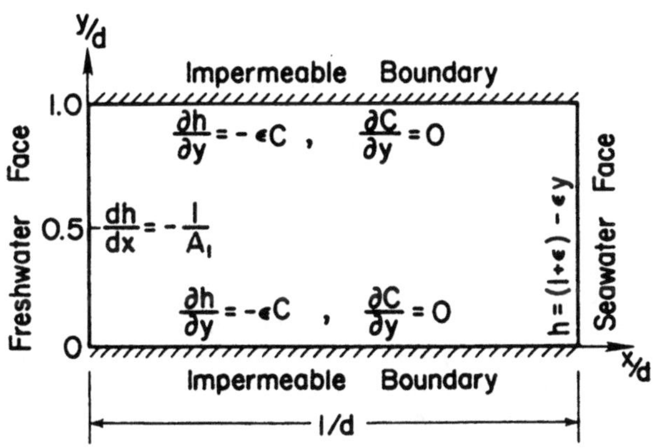

Fig. 1. Boundary for Hydraulic Head Model

The boundary conditions for the flow and transport equations may be given by

$$\frac{\partial h}{\partial x}(o, y, t) = -\frac{1}{A_1} \qquad C(1, y, t) = 0$$

$$\frac{\partial h}{\partial y}(x, 1, t) = -\varepsilon C \qquad \text{and} \qquad \frac{\partial C}{\partial y}(x, o, t) = 0$$

$$h(1, y, t) = (1 + \varepsilon) - \varepsilon y \qquad C(0, y, t) = 1$$

$$\frac{\partial h}{\partial y}(x, o, t) = -\varepsilon C \qquad \frac{\partial C}{\partial y}(x, 1, t) = 0$$

RANGES OF VALUES OF VARIOUS FIELD PARAMETERS

An inspection of the governing equations and the boundary conditions indicates that the saltwater transport depends upon four nondimensional parameters, namely, α_T, K_{yy}, d_L, and A_1 which are defined in Equations (6) and (11). The longitudinal dispersity α_L (L) in the field is typically much larger than its values obtained from laboratory experiments and is reported to be in the range of 0.1 m to 100 m. In view of the fact that the depth, d, of an aquifer may vary anywhere from 1 m to 500 m, it is not unreasonable to assign a range of 0.01 to 20 for d_L keeping in mind that larger aquifers should also have larger dispersivities. Bounds to 1.0 for the nondimensional parameters α_T and K_{yy} are considered quite reasonable. The gradient of the piezometric surface can vary from 0.1% to 0.5% for most realistic situations. This establishes bounds from 10000 to 200 for the parameter A_1.

RESULTS AND DISCUSSION

Saltwater encroachment episodes are simulated for five different values of d_L; 0.01, 0.1, 1.0, 10 and 20 and three values of the nondimensional parameters α_T and K_{yy}; 0.1, 0.5 and 1.0. In addition, three values of A_1 which are 200, 1000, and 10000 are considered. As noted by Anand et al. (1980) the proper selection of the region size (length to depth ratio of the aquifer) is extremely important in order to satisfy the boundary condition for the flow equation at the Freshwater Face. The preliminary study was conducted for a region with a length to depth ratio of 10:1. However, it was observed that for some combinations of the nondimensional parameters the saltwater encroachment was close to the Freshwater Face and the requirement that the flow entering the region should be uniform and horizontal, i.e., $V_x = 1$ and $V_y = 0$, was either not satisfied or only partially satisfied. It has been shown by Anand et al. (1980) that one way to obtain the correct values of concentration is to extend the length to depth ratio of the region. Therefore, if the boundary condition was not fully satisfied, the problem was solved again with a region size of 20:1 or 30:1.

Comparison of Velocities at the Freshwater Face

The influence of the region size can be illustrated by comparing nodal velocities calculated at the Freshwater Face utilizing the three region sizes. The magnitudes of the velocity vectors V_x and V_y at nodes 1 to 6, which lie on the Freshwater Face, are shown in Tables 1 and 2 for two typical cases.

It can be observed from Table 1 that for $d_L = 10$, $\alpha_T = 1.0$, $K_{yy} = 1.0$, and $A_1 = 10{,}000$ the boundary conditions $V_x = 1$ and $V_y = 0$ at the Freshwater Face are not satisfied when a region size of 10:1 is utilized. However, when the region is extended to 20:1 the boundary conditions are fully satisfied.

Table 1. Velocity Vectors, V_x and V_y, at the Freshwater Face Using Various Region Sized for $d_L = 10$, $\alpha_T = 1.0$, $K_{yy} = 1$, and $A_1 = 10{,}000$

Nodal Point Number	V_x			V_y		
	Region Size					
	10:1	20:1	30:1	10:1	20:1	30:1
1	3.77	1.00	--	2.36	0.00	--
2	3.35	1.00	--	4.53	0.00	--
3	3.24	1.00	--	8.29	0.00	--
4	2.34	1.00	--	11.31	0.00	--
5	-1.50	1.00	--	10.50	0.00	--
6	-6.58	1.00	--	8.18	0.00	--

Table 2. Velocity Vectors, V_x and V_y, at the Freshwater Face Using Various Region Sizes for $d_L = 10$, $\alpha_T = 0.5$, $K_{yy} = 1$, and $A_1 = 10{,}000$

Nodal Point Number	V_x			V_y		
	Region Size					
	10:1	20:1	30:1	10:1	20:1	30:1
1	4.37	1.19	1.00	2.73	0.16	0.00
2	4.01	1.18	1.00	5.35	0.36	0.00
3	4.04	1.14	1.00	10.10	0.67	0.00
4	2.99	1.03	1.00	13.99	0.87	0.00
5	-2.11	.80	1.00	13.18	0.75	0.00
6	-9.17	.56	1.00	10.85	0.51	0.00

Similarly it is evident from Table 2 that there is a gradual improvement in the values of the boundary velocity vectors V_x and V_y with an increase in region size when $d_L = 10$, $\alpha_T = 0.5$, $Kyy = 1$, and $A_1 = 10{,}000$. A region size of 30:1 is required before the boundary conditions are fully satisfied.

Isochlors obtained with various regions sizes It has already been mentioned that if the selected region size is not long enough, the flow boundary condition at the Freshwater Face is not met. The importance of selecting the proper region size can be further understood by observing the Isochlors shown in Figs. 2a and b. These Figures show 0.1 and 0.5 Isochlors for region sizes 10:1, 20:1 and 30:1, for a typical case in which $d_L = 10$, $\alpha_T = 0.5$, $K_{yy} = 1$ and $A_1 = 10{,}000$. It is readily observed that a substantial change occurs in the location of both the 0.1 and 0.5 Isochlors when the length to depth ratio of the region is increased from 10:1 to 20:1. The toe of the 0.1 Isochlor advances from a nondimensional distance of 10 from the Seawater Face to a distance of 16.6. A similar advance occurs in the location of the toe of the 0.5 Isochlor, and its position changes from a distance of 6.7 to 8.4 with respect to the ocean boundary. However, only a slight shift in the respective locations of the 0.1 and 0.5 Isochlors is observed when the region size is increased from 20:1 to 30:1. It was observed that a substantial change occurred in the position of the Isochlors for all cases in which A_1 = 10,000 when a longer region size was used.

Effect of K_{yy}

In the results obtained during the course of this analyses, it was observed that the variation in the value of K_{yy} produced only negligible changes in the location or shape of the Isochlors. This is illustrated for some typical cases in Figs. 3a and b. For the cases shown in Fig 3a, d_L = 1, α_T = 0.5, A_1 = 1,000, while the value of K_{yy} is varied from 0.1 to 1.0. It is evident from this figure that the location as well as the shape of the 0.1 or 0.5 Isochlor does not change significantly. A similar phenomenon is observed in Fig. 3b.

Effect of d_L

The effect of the value of d_L on the saltwater encroachment phenomenon for some typical cases can be observed in Fig. 4. For these cases, K_{yy} = 1, α_T = 0.5, and A_1 = 1,000 while the value of d_L varies from 0.01 to 20. It should be noted that all the Isochlors are obtained using a region of size 10:1 consequently, the velocity boundary conditions are not fully satisfied for the cases where d_L is equal to 0.01 or 0.1.

It may be observed from Fig. 4 that as the value of d_L increases from 0.01 to 0.1, the shape of the 0.5 Isochlor remains essentially vertical and that its location, in a nondimentional sense, gets closer to the seawater front. On the other hand, for d_L values of 1.0 and higher the 0.5 Isochlor begins to take an S-form and the encroachment of the saltwater increases at the bottom of the aquifer and diminishes at the top.

An explanation for these phenomena may be found by examining Equations (3) to (5). In these equations, the coefficient of hydrodynamic dispersion is considered to be a function of the longitudinal and transverse dispersivities as well as the interstitial pore velocity. Since $d_L = d/\alpha_L$, it is evident that for a fixed depth d, the value of d_L increases with a decrease in the coefficient of longitudinal dispersivity, α_L. Therefore, with an increase in the value of d_L from 0.01 to 0.1, α_L becomes correspondingly smaller, and as a result, the value of the coefficient of hydrodynamic dispersion coefficient also decreases (from Eqs. (3) and (5)) leading to lesser encroachment in the aquifer. However, when the value of d_L is further increased to 10 or higher, the dispersion effects, due to the reduction in the value of α_L, become negligible and the hydraulic effects governing the cyclic flow, as described by

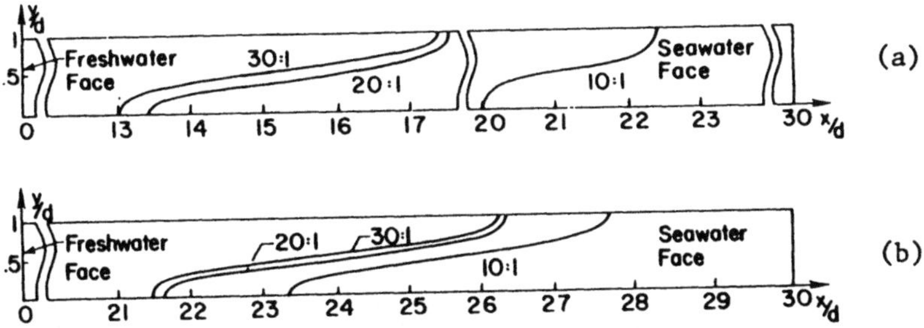

Fig. 2. $d_L = 1$, $\alpha_T = 0.5$, $K_{yy} = 1$, $A_1 = 10{,}000$ Using Various Region Sizes (a) 0.1 Isochlors (b) 0.5 Isochlors

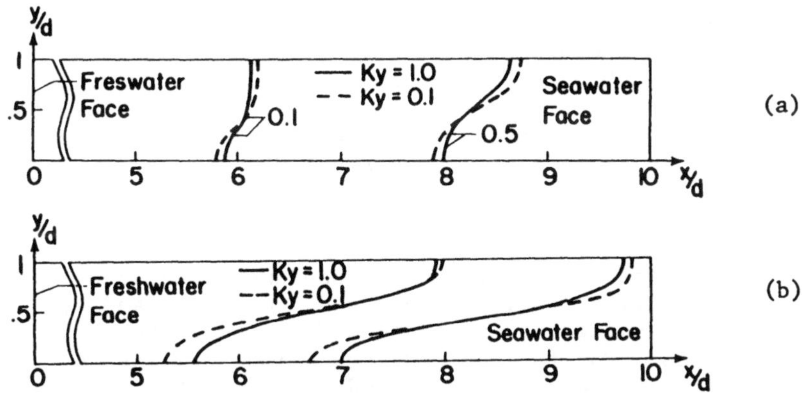

Fig. 3. $d_L = 1$, $\alpha_T = 0.5$, and $A_1 = 1000$, (a) 0.1 Isochlors, (b) 0.5 Isochlors

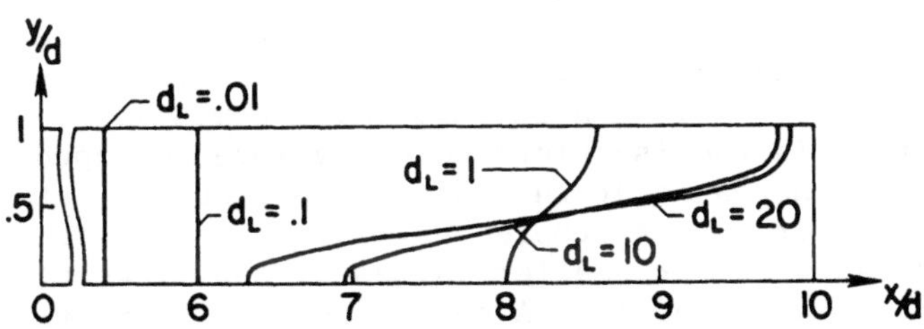

Fig. 4. Velocity Vectors for $\alpha_T = 0.5$, $K_{yy}=1$, and $A_1 = 1000$

Cooper (1959), become prominent. These effects are illustrated in Figs. 5a and 5b. It may be observed from these figures that the velocity flow field is rotational in character, in that the seawater traverses away from the ocean at the bottom of the aquifer and then circulates back into the ocean. This results in the S-shaped form of the 0.5 Isochlor, and the relatively large encroachment at the bottom of the aquifer, for high values of d_L.

Effect of α_T

The effect of the nondimensional parameter α_T on the phenomenon of saltwater encroachment can be studied by observing Figs. 6a and b. For the three cases shown in Fig. 6a the value of $Kyy = 1$, $d_L = 1$ and $A_1 = 1000$ while the value of α_T is equal to 0.1, 0.5 or 1.0. It is observed that for larger values of A_1 the saltwater encroaches farther into the aquifer. This is especially true at the bottom of the aquifer.

Effect of A_1

To illustrate the effect of A_1 on the encroachment process, plots of 0.5 Isochlors are shown for two typical cases in Figs. 7a and b for $\alpha_T = 0.5$, $K_{yy} = 1.0$ and $d_L = 1.0$ and 10.0, respectively. It is observed that for larger values of A_1, the saltwater, encroaches farther into the aquifer. This is especially true at the bottom of the aquifer.

CONCLUSIONS

The flow and transport equations are nondimensionalized leading to four nondimensional parameters d_L, α_T, K_{yy}, and A_1. Their effect on saltwater encroachment is studied for a wide range of values. For the boundary conditions utilized by Henry, results of the parametric study indicate that for this particular problem the ratio of the principal hydraulic conductivity tensors, K_{yy}/K_{xx}, does not greatly affect the extent of saltwater encroachment. Similarly, the ratio of the transverse to logitudinal dispersivity also does not exert a major influence on the encroachment. However, the piezometric gradient and the ratio of the depth of the aquifer logitudinal dispersivity do influence the extent to which saltwater will encroach into the freshwater. It is observed that for values of $d_L = 1$ or larger, cyclic flow in the aquifer becomes more prominent resulting in the Isochlors becoming more S-shaped, i.e., encroachment increases at the bottom and decreases at the top of the aquifer. Furthermore, saltwater encroaches farther into the aquifer for smaller values of the piezometric gradient or larger values of A_1. A comparison of the nodal velocities at the Freshwater Face further reinforces an observation made by Anand et al. (1980) that for this problem the selection of a proper region size is extremely important in simulating correct saltwater concentrations.

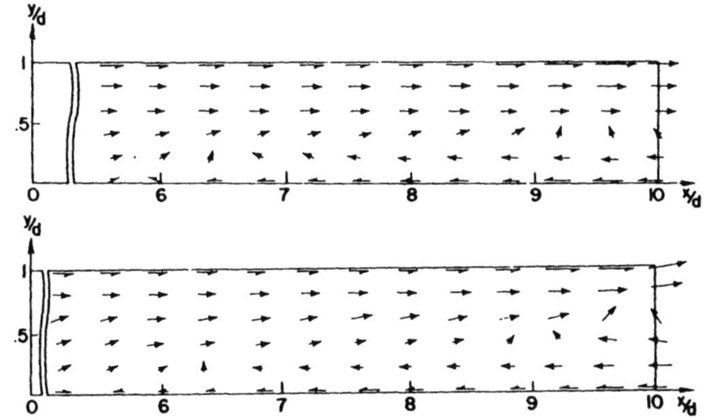

Fig. 5, Velocity Vectors for $\alpha_T = 0.5$, $K_{yy} = 1$, and $A_1 = 1000$. (a) $d_L = 1$, (b) $d_L = 10$.

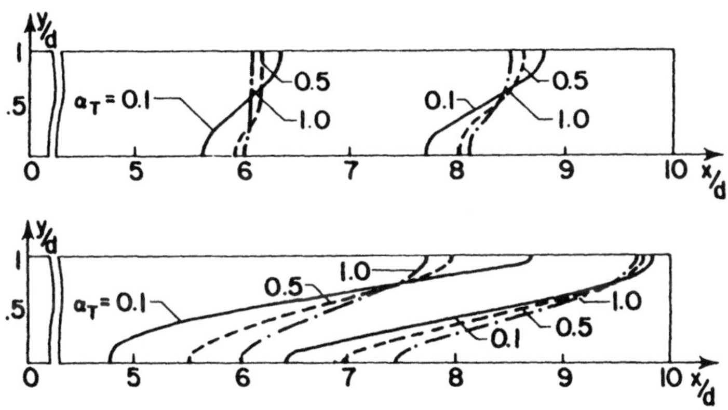

Fig. 6. 0.1 and 0.5 Isochlors for Kyy = 1, $A_1 = 1000$. (a) $d_L = 1.0$, (b) $d_L = 10$.

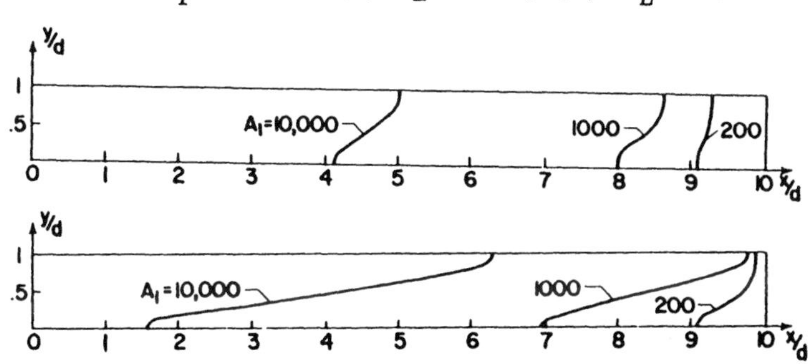

Fig. 7. Isochlors for $d_L = 10$, $\alpha_T = 0.5$, and $K_{yy} = 1$ (a) 0.1 Isochlors, (b) 0.5 Isochlors.

REFERENCES

Anand, S. C., Pandit, A. and B. L. Sill (1980). Some Considerations in Finite Element Solutions to Coupled Groundwater Flow and Convection-Dispersion Equations. Proceedings, Third Int. Conference on Finite Elements in Water Resources, Univ. of Mississippi, Miss. 1,3.98-3.107.

Bear, J. (1961). On the Tensor Form of Dispersion. Jour. Geophys. Res., 66,1185-1197.

Cooper, H. H., Jr. (1959). A Hypothesis Concerning the Dynamic Balance of Freshwater and Saltwater in Coastal Aquifer. Jour. Geophys. Res., 64, 461-467.

Henry, H. R. (1960). Saltwater Intrusion in Coastal Aquifers. Int. Assoc. Sci. Hydrol. Publ., 52,478-487.

Huyakorn, P. and C. Taylor (1976). Finite Element Models for Coupled Groundwater Flow and Convective Dispersion. Proceedings First Int. Conference on Finite Elements in Water Resources. Princeton University, Princeton, 1,131-151.

Lee, C. and R. T. Cheng (1974). On Seawater Encroachment in Coastal Aquifers Water Resour. Res., 10,1039-1043.

Pandit, A. (1982). Numerical Simulation of Contaminant Transport Problems in Groundwater Using the Finite Element Method. Ph.D. Diss., Dept. of Civil Engrg., Clemson University, Clemson, S.C.

SIMULATION OF MASS TRANSPORT IN ROTATING FLOW USING THE FINITE ELEMENT METHOD

V. Alavian[1], S. M. Broeren[1], and D. W. Bintz[2]

[1]Department of Civil Engineering
University of Illinois
Urbana, Illinois 61801 USA

[2]St. Anthony Falls Hydraulics Laboratory
Minneapolis, Minnesota 55414 USA

ABSTRACT

This paper presents simulation of streamline patterns and mass transport within a rotating (circulating) region of flow approximated by a square on the side of a channel. The steady rotating flow is assumed two-dimensional (depth-averaged) and is generated and maintained by a uniform main flow at the open boundary. The equation governing mixing and transport of a finite quantity of conservative tracer instantaneously introduced at a location in the flow field has been numerically solved. The numerical scheme is based on the finite element approximation of the governing differential equation and uses the method of weighted residuals. The flow field is represented by triangular elements, and linear basis functions are used in the interpolation scheme. The unsteady term in the mass transport equation is approximated by finite differencing in full-forward time steps. The diffusion coefficient has been represented as a tensor, the components of which are functions of diffusion coefficients along and normal to the streamlines.

INTRODUCTION

Transport of substances in water is an integral part of the overall environmental quality management. The fate of effluents and sediments entering streams and rivers depends on their physical, chemical, and biological nature as well as the hydraulic characteristics of the body of water. Not all mechanisms responsible for mixing (dilution) and transport of effluents in the water environment are adequately understood and modeled. Among them are mass and momentum exchange between the main flow and a slowly rotating flow adjacent to it. Depending on the size, geometry, and zone of influence; induced rotating (circulating) motions can significantly affect transport and deposition of sediments, mixing and transport of contaminants, and dilution of heated effluents from cooling facilities.

Fischer [1967,1973] and others have shown that in rivers and streams turbulent diffusion resulting from velocity fluctuations along with the differential advection are the basic mechanisms responsible for longitudinal dispersion. Furthermore, Fischer [1966,1967] has specifically shown that in natural streams the transverse variation of velocity across the stream is the dominant factor in estimating the longitudinal dispersion. Channel boundaries, particularly the banks, in natural streams are often irregular with regions of flow separation and circulation. Presence of such regions significantly alters the transverse velocity profile. Day [1974], Pedersen [1977], and Valentine and Wood [1977] have speculated that the interaction between the main flow and the circulating regions can significantly affect the dispersion characteristics of natural streams. In addition, depending on their size and geometry, the circulating regions can serve as temporary storage and a slow release source of dispersing substances. This process can cause concentration distributions markedly different from those predicted by simple advection-diffusion models presently available. A depth-averaged numerical model for circulating flow has been developed by Ponce and Yabusaki [1981]. They state that at the present no comprehensive analysis of the various mechanisms leading to generation of circulation has been attempted using numerical models. However, they do not address the mass exchange between the circulating flow and the main flow, nor do they address mixing and transport of a solute released in the rotating flow.

Westrich [1976] developed a one-dimensional model predicting the mixing process of a conservative tracer mass between a main flow and an adjacent rotating region. He verified the analytical results by physical modeling and found an exponential decrease of the tracer concentration with time for steady flows. His approach is limited in that concentration is assumed constant in the rotating region. This is not the case in practical situations. Tsai and Holley [1979] conducted an analytical and experimental study of longitudinal mixing process in prismatic channels with storage in separation zones. They determined a dimensionless mass exchange coefficient for separation zones at the bottom of a channel. Their values were in some cases up to five times higher than those obtained by direct measurement suggesting a possible problem in formulation of the exchange mechanism. McGuirk and Rodi [1979] have studied square cavity circulation driven by a main stream flow. Their depth-averaged finite difference flow model rigorously computes the velocity and eddy viscosity distribution from the basic conservation and momentum equations and the energy transport equation. The mass transport model is detailed to reveal the concentration distribution within the cavity as well as the mass exchange with the main stream. The numerical solution of the flow field parameters requires substantial computational time and the assignment of values to seven empirical coefficients. The authors do not discuss the effects of numerical diffusion in their finite difference approximation scheme.

The work reported here was undertaken to improve understanding and simulation of mixing and transport of substances in the environment. The objective of this phase of the study was to apply the method of Finite Elements to simulate mass transport in rotating flow and numerically generate concentration patterns. Numerical experiments have focused on illustrating the role of diffusion coefficient and the evolution of concentration contours.

GOVERNING EQUATIONS

The basic flow situation under study is illustrated in Figure 1. The main flow passing the expansion area develops a free shear zone through which energy is transferred from the main flow to the cavity area, setting it into motion. A dividing streamline is formed between the edge of the circulating area and the opposite wall. There is no net flow across this streamline, but mass and momentum are transported in and out of the cavity area. The problem is simplified by assuming that the flow field can be represented by a depth-averaged mathematical model -- a common approach in

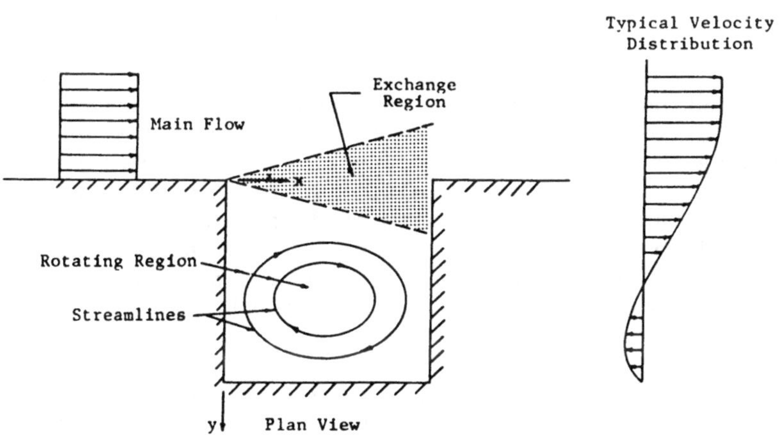

Figure 1. Idealized Sketch of Rotating Region Driven by Uniform Flow

study of this type of flow. Casting the problem in two-dimensions has the added advantage of reducing the computational costs without great loss of generality.

Generation of a Rotating Flow Field

The rotating flow field is obtained by solving the vorticity transport equation governing two-dimensional motion of an incompressible viscous fluid. The rotating motion is generated and maintained by a known uniform flow. The dimensionless form of the vorticity transport equation can be written as

$$\frac{\partial \omega}{\partial t} - \nabla \cdot (\bar{U}\omega) = \frac{1}{R_e} \nabla^2 \omega \tag{1}$$

where ω is the vorticity vector; $\bar{U}$ is the velocity vector; R_e is the Reynolds number based on uniform flow velocity and the size of the rotating region; and t is time. Derivation of equation (1) is given in Alavian et al. [1983] as well as other references. Equation (1) along with its boundary conditions is solved repeatedly until a steady-state solution for the vorticity is obtained. Once the vorticity is known, the stream function and subsequently components of the velocity vector can be computed.

Mass Transport Model

Mixing and transport of conservative pollutants discharged into a moving body of water are caused by the combined effects of advection and diffusion (molecular and turbulent). For non-conservative substances, the process also includes gain or loss of mass due to chemical, biological, and mechanical (e.g., settling or adsorption). Spacial averaging of flow properties introduces the concept of dispersion. Dispersion is a mechanical process caused by the spacial variation in the mean velocity distribution and is considered a major factor in spreading and dilution of pollutants in streams. In shear induced rotating flows, the direction of motion continuously changes. This coupled with the asymmetry of the flow makes it impossible to define spatially averaged conditions for the flow. Subsequently, unlike relatively straight flow, spreading and mixing in rotating flow cannot be formulated through the mechanism of dispersion. Diffusion is considered the mechanism for mixing in rotating flow for the purposes of this study. Some refer to the process as two-dimensional dispersion due to the depth averaging of flow properties.

The unsteady turbulent mass transport equation for a conservative substance released in a two-dimensional (depth-averaged) flow is

$$\frac{\partial c}{\partial t} + u \frac{\partial c}{\partial y} + v \frac{\partial c}{\partial y} = \frac{\partial}{\partial x} [E_{xx} \frac{\partial c}{\partial x}] + \frac{\partial}{\partial y} [E_{yy} \frac{\partial c}{\partial y}] +$$

$$+ \frac{\partial}{\partial x} [E_{xy} \frac{\partial c}{\partial y}] + \frac{\partial}{\partial y} [E_{yx} \frac{\partial c}{\partial x}] \qquad (2)$$

where c is the time-dependent, depth-averaged concentration; E is the diffusion tensor; u and v are components of the mean velocity field in x and y directions, respectively; and t is time. The diffusion tensor is of the second order with components defined as:

$$E = \begin{vmatrix} E_{xx} & E_{xy} \\ E_{yx} & E_{yy} \end{vmatrix} \qquad (3)$$

where E_{xx} and E_{yy} are components along the x and y coordinate axes (see Figure 1), respectively; and E_{xy} and E_{yx} are the cross product terms and account for the interaction between E_{xx} and E_{yy} and are considered equal. It is difficult to define diffusion coefficient along cartesian coordinate axes in rotating flow. Therefore, diffusion is defined along the streamlines, E_L, and transverse to them, E_T. Now E_{xx}, E_{yy}, and E_{xy} can be written in terms of the longitudinal and transverse diffusion coefficients much the same way discussed by Scheidegger [1961] and Bear [1972] for flow in porous media:

$$E_{xx} = E_L \frac{u^2}{|U|^2} + E_T \frac{v^2}{|U|^2} \;;$$

$$E_{yy} = E_T \frac{u^2}{|U|^2} + E_L \frac{v^2}{|U|^2} \;; \text{ and} \qquad (4)$$

$$E_{xy} = E_{yx} = (E_L - E_T) \frac{uv}{|U|^2}$$

where $|U|$ is the magnitude of the velocity vector at each point within the flow field; and u and v are the components of that vector. Equation (4) when evaluated using the spatially averaged one-dimensional flow properties of a straight stream reduces to longitudinal disperison coefficient.

Finite Element Approximation

The solution to equation (2) has been approximated using triangular elements and linear basis functions as the first attempt to simulate mass transport. The linear basis functions are identically the local area coordinates and can be integrated explicitly in the approximation procedure. The integrals can also be found in Pinder and Gray [1977]. Figure 2 shows the flow field discretized by triangular elements.

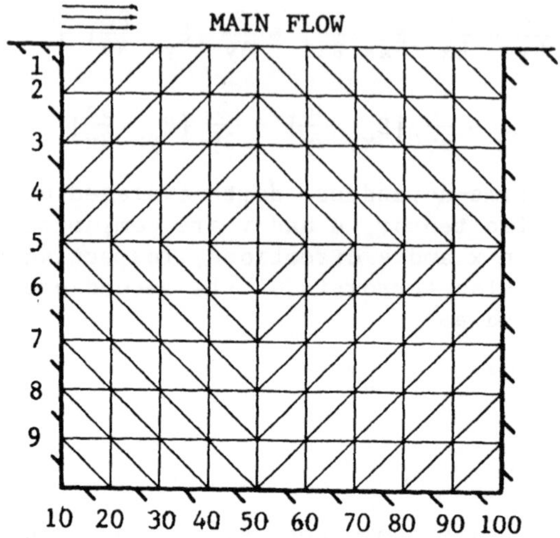

Figure 2. Discretized Flow Field

Linear basis functions, $\phi(x,y)$, are used to approximate the spacial distribution of the concentration, the velocity components and the diffusion tensor components viz:

$$
\begin{aligned}
c(x,y,t) &\approx \sum_{j=1}^{3} a_j(t)\, \phi_j(x,y) \; ; \\
u(x,y) &\approx \sum_{k=1}^{3} u_k\, \phi_k(x,y) \; ; \\
v(x,y) &\approx \sum_{k=1}^{3} v_k\, \phi_k(x,y) \; ; \\
E_{xx}(x,y) &\approx \sum_{m=1}^{3} E_{xxm}\, \phi_m(x,y) \; .
\end{aligned} \tag{5}
$$

where $a_j(t)$, u_k, and v_k are the nodal values of concentration and velocity components in the x and y directions, respectively. The nodal values of the diffusion tensor components E_{xxm}, E_{yym}, and E_{xym}=E_{yxm} are evaluated using equation (4). The last equation above is written similarly for E_{yym} and E_{xym}.

Substitution of equation (5) into equation (2), generates the residual error, R(x,y). Following the Galerkin method, the integral of the weighted residual error over the element area is forced to zero. Thus with the above substitutions and applying Greens theorem with zero mass flux across boundaries, equation (2) integrated over an element area becomes

$$\sum_{j=1}^{3} \int_A \left\{ \frac{\partial a_j}{\partial t} \phi_j \phi_i + [\sum_{k=1}^{3} u_k \phi_k] a_j \frac{\partial \phi_j}{\partial x} \phi_i + [\sum_{k=1}^{3} v_k \phi_k] a_j \frac{\partial \phi_j}{\partial y} \phi_i \right.$$

$$+ [\sum_{m=1}^{3} E_{xxm} \phi_m] a_j \frac{\partial \phi_j}{\partial x} \frac{\partial \phi_i}{\partial x} + [\sum_{m=1}^{3} E_{yym} \phi_m] a_j \frac{\partial \phi_j}{\partial y} \frac{\partial \phi_i}{\partial y}$$

$$\left. + 2[\sum_{m=1}^{3} E_{xym} \phi_m] a_j \frac{\partial \phi_j}{\partial x} \frac{\partial \phi_i}{\partial y} \right\} dA = \int_A R(x,y)\, \phi_i \, dA = 0 \quad (i=1,3) \,. \qquad (6)$$

The time derivative of a_j in the above equation is approximated in a full-forward finite difference manner.

Since equation (6) holds for every element, a global equation can be assembled by summing this equation over the number of elements in the flow field. Concentration gradients at the boundaries are assumed zero. The resulting system of simultaneous equations with the appropriate boundary conditions was solved on the University of Illinois Cyber 175 computer.

SIMULATION RESULTS AND DISCUSSION

Calculations were carried out for a shear induced rotating flow confined to a square region. Results are presented for various formulations of diffusion with a constant flow geometry and uniform driving flow. Exchange of tracer mass with the main flow is being modeled now and is not addressed in this paper. Equation (1) was solved using a standard five-point finite difference approximation routine for a Reynolds number of 10,000. The calculated streamline pattern is shown in Figure 3. This pattern was then used in solving equation (6) for mixing and transport of a known mass instantaneously released at a node point. The initial concentration at the input node was calculated by spreading the input mass over the elements sharing the input node using the linear basis functions.

Concentration contours at 90% and 50% levels of the current maximum value found in the flow field are shown in Figures 4-6 at two and six time units after release of mass. Figure 4 shows advection and diffusion of mass for a scalar diffusion coefficient of 0.6. Figures 5 and 6 show the contours for a tensor diffusion coefficient where the transverse diffusion coefficient, (E_T), is 0.5 and 0.1 times the streamwise diffusion coefficient, (E_L=0.6), respectively. The difference in diffusion coefficient formulation has a distinctive influence on the shape and position of the contours. The spreading of mass across streamlines is greatest for the scalar diffusion coefficient (same spreading rate in all directions). Concentration plots show elongation of the contours with a decrease in the ratio of transverse to longitudinal diffusion coefficient. Limiting the transverse mixing confines the mass to the curving streamline pattern. Concentration patterns become more markedly different as time progresses.

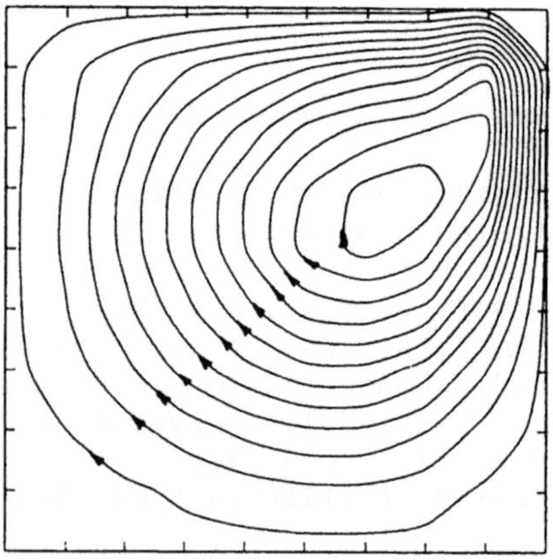

Figure 3. Streamlines

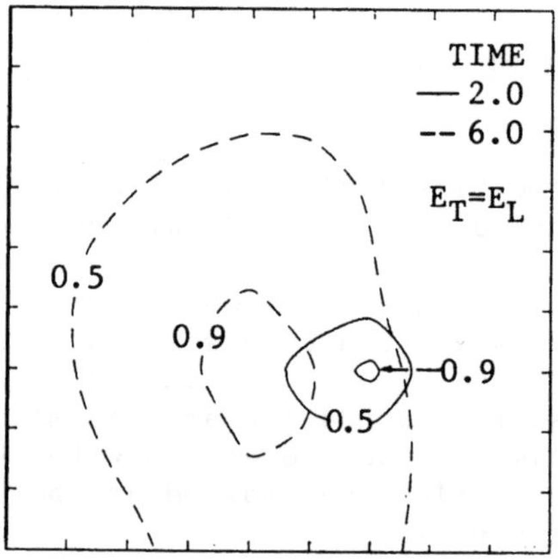

Figure 4. Normalized Concentration Contours

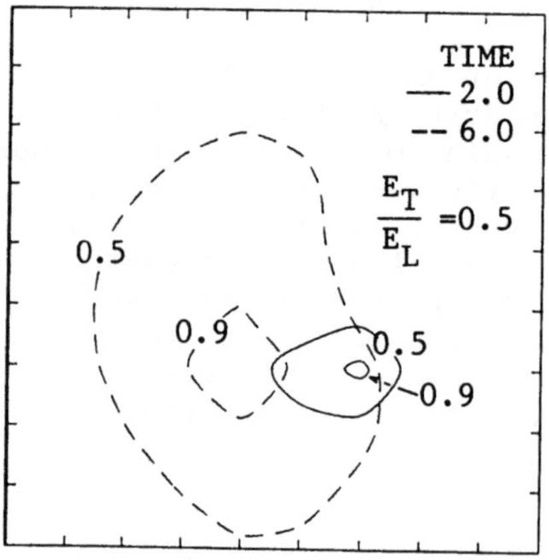

Figure 5. Normalized Concentration Contours

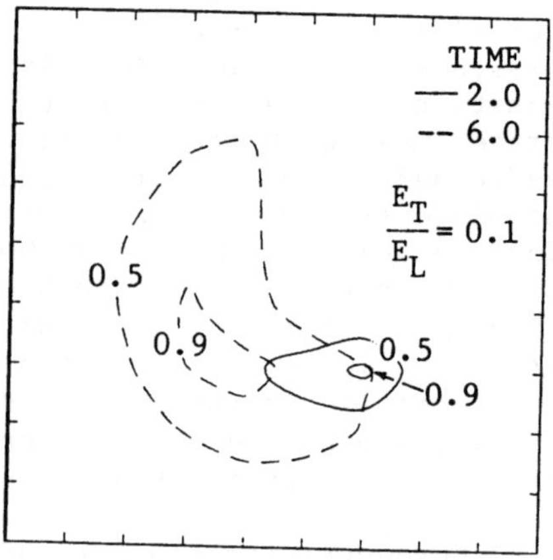

Figure 6. Normalized Concentration Contours

The formulation of the diffusion term does not radically influence the degree of dilution as measured by the peak concentration, differences diminish in time. Figure 7 shows plot of peak concentration decay with time for the three simulations. There is a sharp drop in the peak value early in the computations followed by an exponential decay. Sharp initial decrease of the peak concentration seems to be due to the combined effects of initially large concentration gradient and averaging associated with the numerical procedure.

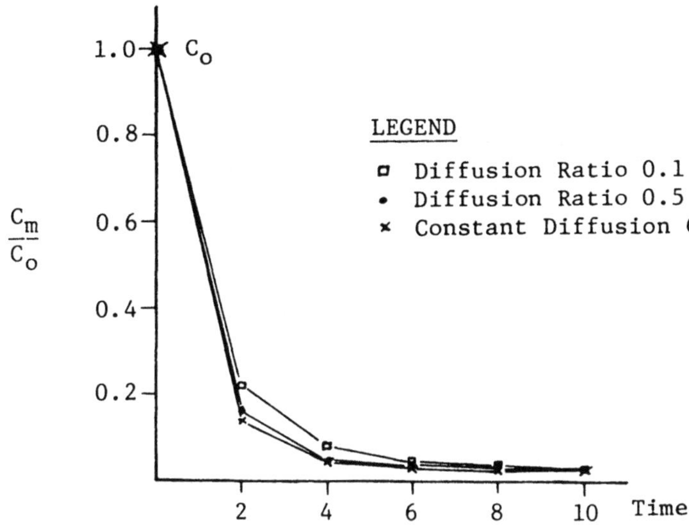

Figure 7. Normalized Maximum Concentration Versus Time

CONCLUSIONS

The finite element method has been successfully used to approximate the two-dimensional equation for transport of mass in a steady rotating flow. The model computes concentration levels at desired time intervals for a specified flow pattern. Diffusion has been formulated as a tensor the components of which, weighted by local velocity components, control the streamwise and transverse spreading of mass. Plans are being made to compare numerical model simulations to mass transport in a rotating flow observed in a laboratory model. This will help gain a better understanding of the diffusion process in such flows as well as help estimate the streamwise and transverse diffusion coefficients for more realistic simulations.

ACKNOWLEDGEMENTS

The study reported here was supported in part by a grant from the United States Department of the Interior as authorized under the Water Resources Research and Development Act of 1978. The project was administered by the University of Illinois Water Resources Center.

REFERENCES

1. Alavian, V., S. M. Broeren, and D. W. Bintz. (1983) Modeling Streamlines and Mass Transport in Circulating Flow. Water Resources Center Report No. 182, University of Illinois, Urbana-Champaign.
2. Bear, J. (1972) Dynamics of Fluids in Porous Media. New York: Am. Elsevier Publishing Co., 764 p.
3. Day, T. (1974) Dispersion in Natural Channels. Thesis presented to the University of Canterbury, New Zealand in partial fulfillment of the requirements for the degree of Doctor of Philosophy.
4. Fischer, H. B. (1966) A Note on the One-Dimensional Dispersion Model. International Journal of Air and Water Pollution, 10, 6/7:443-452.
5. Fischer, H. B. (1967) The Mechanics of Dispersion in Natural Streams. Journal of the Hydraulics Division, ASCE, 93, HY6:187-216, Nov.
6. Fischer, H. B. (1973) Longitudinal Dispersion and Turbulent Mixing in Open Channel Flow. Annual Review of Fluid Mechanics, 5, 8036:59-78.
7. McGuirk, J., and Rodi, W. (1979) Calculation of Unsteady Mass Exchange Between a Main Stream and a Dead Water Zone. Proceedings of International Conference on Hydraulic Engineering in Water Resources Development and Management, 3, XVIII Congress of IAHR, Italy.
8. Pederson, F. B. (1977) Prediction of Longitudinal Dispersion in Natural Streams. Institute of Hydrodynamics and Hydraulic Engineering, Technical University of Denmark, Series Paper No. 14:69, Feb.
9. Pinder, G. F., and Gray, W. G. (1977) Finite Element Simulation in Surface Sand Subsurface Hydrology. New York: Academic Press.
10. Ponce, V. M., and Yabusaki, S. B. (1981) Modeling Circulation in Depth-Averaged Flow. Journal of the Hydraulics Division, ASCE, 107, HY11:1501-1518, Nov.
11. Scheidegger, A. E. (1961) General Theory of Dispersion in Porous Media. Journal Geophysics Research, 66, 10: 3273-3278.
12. Tsai, Y. G., and Holley, E. R. (1979) Temporal and Spatial Moments for Longitudinal Mixing in Prismatic Channels with Storage in Separation Zones. Hydraulic Engineering Series No. 35, University of Illinois, Champaign-Urbana.
13. Valentine, E. M., and Wood, I. R. (1977) Longitudinal Dispersion with Dead Zones. Journal of the Hydraulics Division, ASCE., 103, HY9:975-990, Sept.
14. Westrich, B. (1976) Simulation of Mass Exchange in Dead Zones for Steady and Unsteady Flow Conditions. International Symposium on Unsteady Flow in Open Channels, Newcastle upon-tyne, England.

WASTEWATER UTILIZATION IN THE PROBLEM OF WATER RESOURCES MANAGEMENT

F. Cavalieri
A. Russo Spena

Tecnosistem S.p.A., Napoli, Italy
University of Sassari, Institute of Agricultural Hydraulics, Sassari, Italy

INTRODUCTION

Management of water for different purposes needs thorough investigations in order to evaluate the level of quality under which a given water resource cannot be directly utilized. In presence of limited water resources, we have to take into consideration the possibility of recovery of waters of lower quality than admissible by means of proper purification treatments of wastewater.
The utilization of wastewaters for irrigation is related to the consideration that agricultural uses require lower - and therefore cheaper - levels of water quality and, on the other hand, fertilizer contained in it are useful to vegetation.
However, wastewater utilization can produce many kinds of problems both for plants and for soil and can lead to contamination of vegetation and to modification of cultural techniques. Nevertheless, these harmful effects can be reduced and controlled by means of proper methods regarding the purification treatments of wastewater, the choice of volumes and intervals of irrigation.
In any case, the utilization of wastewater resources must be supported by a profound theoretical knowledge of the water flow and mass transport phenomena in the soil. It is the purpose of this paper to present numerical schemes for solution of the governing one-dimensional flow and mass transport equations and to discuss the results of some example.

THE FLOW EQUATION

We consider the isothermal unsteady flow of a newtonian, homogeneous liquid in a rigid and unsaturated porous medium, which may consist of one or more horizontal homogeneous layers. The one-dimensional flow of water along the z axis, which is vertical and taken as being positive towards the bottom, is described by the equation:

$$L(h) = c(h)\frac{\partial h}{\partial t} - \frac{\partial}{\partial z}\left[k(h)\left(\frac{\partial h}{\partial z} - 1\right)\right] = 0 \qquad (1)$$

where: c is the specific moisture capacity (cm^{-1})
h is the pressure head (cm)
k is the hydraulic conductivity $(cm \cdot min^{-1})$
z is the vertical coordinate (cm)
t is the time (min)

The coefficient c in (1) is given by:

$$c = \frac{d\vartheta}{dh} \qquad (2)$$

where ϑ is the volumetric moisture content $(cm^3 \cdot cm^{-3})$

Eq. (1) must be supplemented by appropriate initial and boundary conditions.
As hysteresis is not considered, ϑ is a single valued function of h and so the initial conditions are:

$$h(z,0) = h_o(z) \qquad t = 0 \qquad (3a)$$
$$\vartheta(z,0) = \vartheta_o^o(z) \qquad t = 0 \qquad (3b)$$

The boundary conditions at the surface of the soil are represented in relations 4a - 4b, while the boundary conditions at the bottom of the soil are represented in relations 4e - 4f; conditions 4c - 4 d are valid for the entire soil profile:

$$\left| k(h)\left(\frac{\partial h}{\partial z} - 1\right)\right|_{z=0} \leq I_p(t) \qquad z = 0,\ t \geq 0 \qquad (4a)$$

$$\left| k(h)\left(\frac{\partial h}{\partial z} - 1 \right) \right|_{z=0} \leq E_p(t) \qquad z = 0,\ t \geq 0 \quad (4b)$$

$$\vartheta \leq \vartheta_s \qquad t \geq 0 \quad (4c)$$

$$h \leq h_c \qquad t \geq 0 \quad (4d)$$

$$h(z,t) = \bar{h}(t) \qquad z = z_{max},\ t \geq 0 \quad (4e)$$

$$\left| k(h)\left(\frac{\partial h}{\partial z} - 1 \right) \right|_{z=z_{max}} = 0 \qquad z = z_{max},\ t \geq 0 \quad (4f)$$

Along soil-air interface, in the absence of ponding, the soil can gain water by infiltration or lose water by evaporation. The exact boundary conditions at the soil surface cannot be predicted "a priori". We assume the filtration rate at the soil surface depending only on the external conditions and equal to the potential rate of infiltration I_p, as long ϑ is less than ϑ_s (in saturation); we assume, however, the filtration rate at the soil surface depending only on the soil characteristics when $\vartheta = \vartheta_s$. In the same way, we assume the actual evaporation flux equal to the potential flux E_p, as long as the pressure h is less than its critical value h_c; in the opposite way, the actual evaporation flux is determined by the condition $h = h_c$.

At the bottom of the soil, h is equal to a prescribed pressure $\bar{h}$, while the water flux through the bottom surface is assumed null.

Eq. 1, together with the initial and boundary conditions, is solved by Galërkin method in conjunction with a finite element space discretization scheme. In using the finite element method, the flow region is subdivided into one-dimensional elements, whose corner points are designated as nodal points. Also, let each element be associated with a set of local coordinate functions $\emptyset_i$, named shape functions, which vary linearly between 0 and 1 inside the element and vanish outside the element. Then, according to the Galërkin method, an approximate solution u, at any given time t, is obtained in the form:

$$h(z,t) \simeq u(z,t) = \sum_{1}^{N}{}_{i}\ \emptyset_i(z)\ a_i(t) \qquad (5)$$

where N is the total number of nodes and a_i are the values of u at nodes i. The nodal values a_i are determined so as to satisfy the initial and boundary conditions of the problem together with the orthogonality requirements expressed by:

$$\int_0^{z_{max}} \emptyset_i \, L(u) \, dz = 0 \qquad (1 \le i \le N) \tag{6}$$

By integrating eq. 6 by parts, we obtain the set of ordinary non linear differential equations:

$$\underline{\underline{C}} \frac{d\underline{a}}{dt} + \underline{\underline{K}} \, \underline{a} = \underline{F} \tag{7}$$

where:

$$C_{ij} = \int_0^{z_{max}} c \, \emptyset_j \emptyset_i \, dz \tag{8a}$$

$$K_{ik} = \int_0^{z_{max}} k \frac{d\emptyset_j}{dz} \frac{d\emptyset_i}{dz} \, dz \tag{8b}$$

$$F_i = \emptyset_i \left(I_p - E_p \right)_{z=0} + \int_0^{z_{max}} k \frac{d\emptyset_i}{dz} \, dz \tag{8c}$$

To integrate eq. 7, the time domain is discretized into a sequence of finite intervals and the time derivatives are replaced by finite differences. Good results can be obtained by employing the time-centered scheme:

$$\underline{\underline{P}}_{t+\frac{1}{2}\Delta t} \; \underline{a}_{t+\Delta t} = \underline{Q}_{t+\frac{1}{2}\Delta t} \; \underline{a}_t + \underline{F}_{t+\frac{1}{2}\Delta t} \tag{9}$$

where:

$$\underline{\underline{P}} = \tfrac{1}{2} \underline{\underline{K}} + \frac{1}{\Delta t} \underline{\underline{C}} \tag{10a}$$

$$\underline{\underline{Q}} = - \tfrac{1}{2} \underline{\underline{K}} + \frac{1}{\Delta t} \underline{\underline{C}} \tag{10b}$$

To be able to solve (7), we need estimates of the coefficients k and c in (9) at the half-time level. Because these coefficients depend on h, it is necessary to have an estimate of the pressure head distribution h at the half-time level. For each new time step this distribution is obtained through linear extrapolation from the old distributions as follows:

$$u_{t + \frac{1}{2} \Delta t} = u_t + \frac{1}{2} \frac{\Delta t_n}{\Delta t_o} (u_t - u_{t - \Delta t}) \tag{11a}$$

where Δt_o e Δt_n are the old and new increments. At each iteration the most recent values are used to obtain an improved estimate of the nodal values at instant $t + \frac{1}{2} \Delta t$, by means of equation :

$$u_{t + \frac{1}{2} \Delta t} = \frac{1}{2} (u_{t + \Delta t} + u_t) \tag{11b}$$

The iterative process continues until a satisfactory degree of convergence is obtained. The criterium of convergence, in its general form, is given by:

$$\frac{\left\| \underline{a}^{p+1} - \underline{a}^{p} \right\|}{\left\| \underline{a}^{p+1} \right\|} < \mu \tag{12}$$

where:
p is the iteration index and μ the selected relative error.

THE TRANSPORT EQUATION

In the preceding section we have examined one-dimensional water flow through unsaturated soil without reference to solutes. In actual fact, water present in the soil is never chemically pure. As it moves through the profile, soil water carries its solute load in its convective stream, leaving some of it behind to the extent that the component salts are adsorbed, taken up by plants or precipitated whenever the concentration exceeds their solubility. The movement of solutes in the soil water occurs as a result of molecular diffusion and mass

transport in the soil-water phase. Because of the general acceptance of the chromatography theory and its applicability to a soil-water system, this approach has been used to describe the vertical movement and distribution of the solutes.
The partial differential equation for one-dimensional solute transport is taken as:

$$L(C) = \varrho \frac{\partial S}{\partial t} + \frac{\partial (\vartheta C)}{\partial t} - \frac{\partial}{\partial z}\left(\vartheta D \frac{\partial C}{\partial z}\right) + \frac{\partial (vC)}{\partial z} + Q = 0 \qquad (13)$$

where: ϱ is the soil bulk density ($g.cm^{-3}$)
S is the amount of solute adsorbed by soil ($g.g^{-1}$)
C is the solute concentration in soil solution ($g.cm^{-3}$)
D is the solute dispersion coefficient ($cm^2.min^{-1}$)
v is the soil water flux ($cm.\ min^{-1}$)
Q is the irreversible solute sink ($g.cm^{-3}.min^{-1}$)

The velocity v is given by Darcy's law:

$$v = -\ k(h)\left(\frac{\partial h}{\partial z} - 1\right) \qquad (14)$$

The initial and boundary conditions used in conjunction with the solute transport equations, assuming that a solute solution of concentration C_s is applied at the soil surface for a time T, are given by:

$$C(z,0) = C_o(z) \qquad t = 0, z \geq 0 \qquad (15a)$$

$$C = C_s \qquad t \leq T, z = 0 \qquad (15b)$$

$$vC_s = -\vartheta D \frac{\partial C}{\partial z} + vC \qquad t \leq T, z = 0 \qquad (15c)$$

$$C = 0 \qquad t > T, z = 0 \qquad (15d)$$

$$0 = -\vartheta D \frac{\partial C}{\partial z} + vC \qquad t > T, z = 0 \qquad (15e)$$

$$0 = -\vartheta D \frac{\partial C}{\partial z} + vC \qquad t \geq 0, z = z_{max} \qquad (15f)$$

The mechanism used to describe the reversible solute adsorption term in (13) is a linearized equilibrium adsorption:

$$S = k_a C \tag{16}$$

where k_a is usually referred to as the solute distribution coefficient ($cm^3.g^{-1}$).

The dispersion coefficient D represents the effect of both molecular diffusion and mechanical dispersion and is assumed to be adequately defined by:

$$D = a + b \frac{|v|}{\vartheta} \tag{17a}$$

or:

$$D = a + b \log \frac{|v|}{\vartheta} \tag{17b}$$

where a is the molecular diffusion coefficent and b is the dispersivity.

The sink term Q represents irreversible processes, as chemical transformation or precipitation, and is assumed as:

$$Q = k_c \vartheta C \tag{18}$$

when k_c is the rate coefficient (min^{-1}) for irreversible solute removal from the soil solution.

From (13), the transport of solute is dependent on ϑ and v, both of which are variable during transient, unsaturated soil water movement. Therefore, the water flow equation is solved prior to the solute transport equation. If the time step Δt is small enough, functions $\vartheta(z,t)$ and $v(z,t)$ can be respectively substituted by functions $\vartheta_m(z)$ and $v_m(z)$ as:

$$\vartheta(z,t) \simeq \vartheta_m(z) = \tfrac{1}{2} \left[\vartheta(z,t) + \vartheta(z,t+\Delta t) \right] \tag{19a}$$

$$v(z,t) \simeq v_m(z) = \tfrac{1}{2} \left[v(z,t) + v(z,t+\Delta t) \right] \tag{19b}$$

Considering (16), (18) and (19), the solute transport equation can be written as:

$$\frac{\partial C}{\partial t} - \bar{D} \frac{\partial^2 C}{\partial z^2} + \bar{A} \frac{\partial C}{\partial z} + \left(\frac{1}{\vartheta_m} \frac{\partial v_m}{\partial z} + k_c \right) C = 0 \tag{20}$$

where:

$$R = 1 + \frac{\varrho k_a}{\vartheta_m} \qquad (21a)$$

$$\bar{D} = D/R \qquad (21b)$$

$$A = \frac{1}{\vartheta_m}\left[v_m - D\frac{\partial \vartheta_m}{\partial z} - \vartheta_m \frac{\partial D}{\partial z}\right] \qquad (21c)$$

$$\bar{A} = A/R \qquad (21d)$$

Application of the finite element method to the solution of the transport equation leads to:

$$\underline{\underline{Z}}\,\frac{d\underline{x}}{dt} + \underline{\underline{M}}\,\underline{x} = \underline{R} \qquad (22)$$

where $\underline{x}$ is the column matrix of the nodal values of concentration and:

$$Z_{ij} = \int_o^{z_{max}} \emptyset_j \emptyset_i \, dz \qquad (23a)$$

$$M_{ij} = \int_o^{z_{max}} \left(\bar{D}\,\frac{d\emptyset_j}{dz} + \bar{A}\emptyset_j \quad \frac{d\emptyset_i}{dz}\right)dz \qquad (23b)$$

$$R_i = \emptyset_i \left(\frac{vC_s}{\vartheta_m}\right)_{z=0} + x_i(t)\int_o^{z_{max}} k_c \, \emptyset_i \, dz \qquad (23c)$$

Terms $\frac{\partial v_m}{\partial z}$ and $\frac{\partial D}{\partial z}$ vanish as velocity v is constant for each element.

Applying the time-centered scheme to (22), we finally obtain:

$$\underline{\underline{A}}_{t+\Delta t}\,\underline{x}_{t+\Delta t} = \underline{\underline{B}}_t\,\underline{x}_t + \underline{R}_t \qquad (24)$$

where:

$$\underline{\underline{A}} = \frac{1}{2}\,\underline{\underline{M}} + \frac{1}{\Delta t}\,\underline{\underline{Z}} \tag{25a}$$

$$\underline{\underline{B}} = -\frac{1}{2}\,\underline{\underline{M}} + \frac{1}{\Delta t}\,\underline{\underline{Z}} \tag{25b}$$

DISCUSSION OF RESULTS

For the purpose of model simulation, solute and water transport were calculated for a single - layered loam soil, during water infiltration and redistribution. The $\vartheta(h)$ and $k(h)$ relationships used in this study were those proposed by Gardner:

$$\vartheta(h) = \frac{\vartheta_s}{1 + (-h/p)^q} \tag{26a}$$

$$k(h) = k_s\, e^{\alpha h} \tag{26b}$$

where ϑ_s and k_s are the soil water content and hydraulic conductivity at saturation, respectively. The values chosen for ϑ_s, k_s and the arbitrary parameters p, q, α were 0.4, 1.0, 300, 1.0, 0.03 for the loam layer, respectively. The dispersion coefficient D was considered as a linear function of the pore-water velocity v/ϑ, according to (17a), with coefficients a and b equal to 0.00041 and 1.026, respectively. The value of soil bulk density ϱ was assumed equal to 1.4. The depth of the loam layer was taken equal to 3 m.

The initial soil water content ϑ_o was defined by the hydrostatic distribution of the pressure head, while the solute concentration C_o was assumed to be zero. The boundary conditions were:

$I_p(t) = 0{,}008$ cm/min	$z = 0$	$0 \le t \le 10^h$	(27a)
$I_p(t) = 0$	$z = 0$	$t > 10^h$	(27b)
$E_p(t) = 0$	$z = 0$	$0 \le t \le 10^h$	(27c)
$E_p(t) = 0{,}00016$ cm/min	$z = 0$	$t > 10^h$	(27d)
$C_s = 100$ g/cm^3	$z = 0$	$0 \le t \le 4^h$	(27e)
$C_s = 0$	$z = 0$	$t > 4^h$	(27f)

Two study cases were considered, the first with a non reactive solute ($k_a = 0$) and the second with a reactive solute ($k_a = 0.2$).

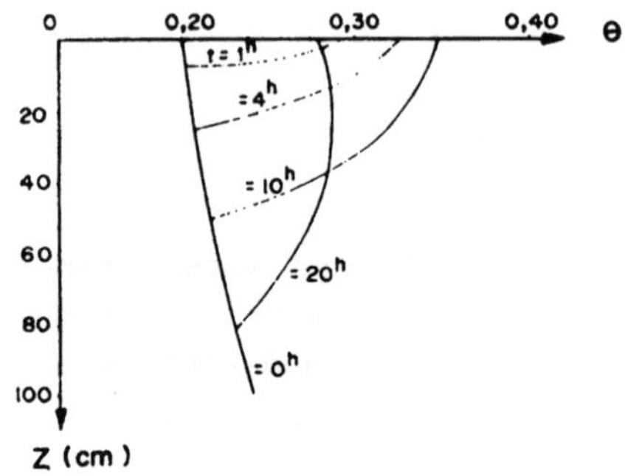

Fig. 1. Water content distribution.

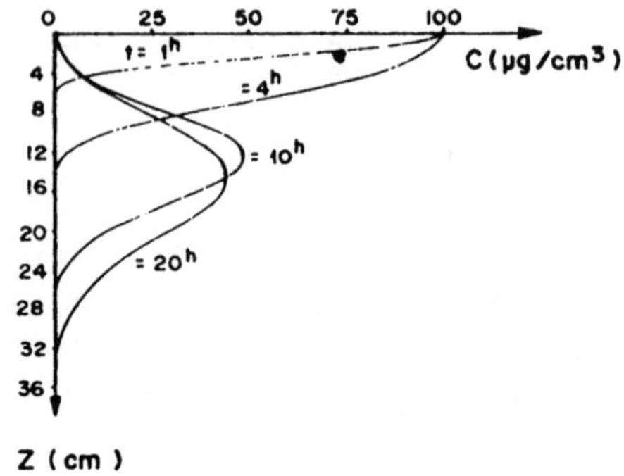

Fig. 2. Concentration distribution for a nonreactive solute.

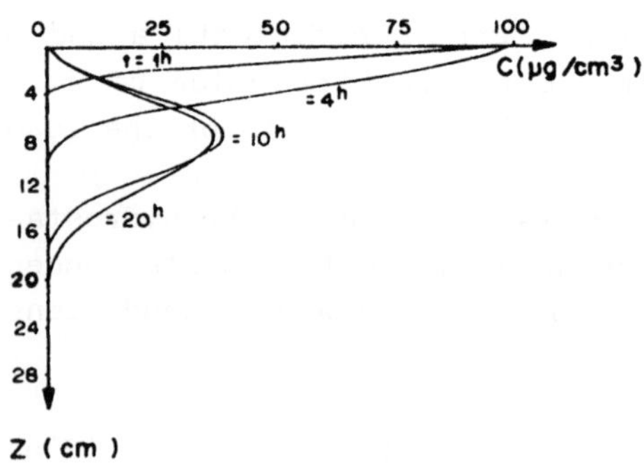

Fig. 3. Concentration distribution for a reactive solute.

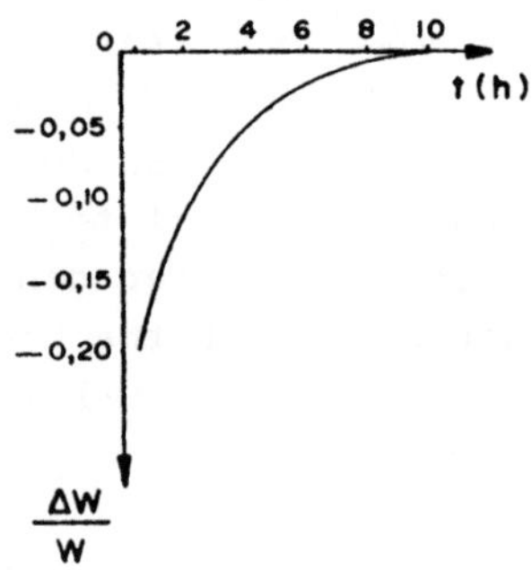

Fig. 4. Relative errors for water volumes.

Water content distributions in the loam soil profile are shown in fig. 1. The wetting front associated with the infiltrating water is gradually increasing at the soil surface up to the value $\vartheta = 0.34$, while zones of increasing depht are interested by wetting process. When the inflow stops, the redistribution process starts: the water content at the soil surface and in the upper layers becomes a decreasing function of time, while new flow spaces are gained at lower layers at the expense of the water drain.

Figs. 2 and 3 show concentration distributions for a non reactive and a reactive solute, respectively. As expected, the linear adsorption in the loam layer resulted both in the retardation of solute movement and in the decrease of solute concentration (about 45% of solute was adsorbed).

Both finite element equations (7) and (22) are first-order approximations and, depending upon a particular choice of Δz and Δt, the convergence of the numerical solution may or may not be guaranted. The best values of Δz and Δt for a convergent solution were determined with short trial runs. A value of 1 cm for Δz gave satisfactory results; however, Δt varied from 0.1 to 2 min for different cases. Calculations were carried out using a PDP 11 computer. Fig. 4 shows differences $\Delta W/W = (W_0 - W)/W$ vs. time between the inflow water volume W_0 and the corresponding water volume W in the soil. The profile of the curve clearly indicates that ΔW is equal to zero after 10 hours and is practically negligible after 4 hours. Initial error is due to the circumstance that, after 1 hour of simulation, the ϑ curve is very sharp and, therefore, small changes of Δz cause large variations of ϑ.

REFERENCES

Carravetta, R., and Santini, A.

1977. Mathematical model for analysis of water infiltration and redistribution in soils. Symposium on "Hydrodynamic diffusion and dispersion in porous media", Pavia, Italy, April 20 - 22.

Cavalieri, F., and Russo Spena, A.

1983. Analysis of the one-dimensional flow filtration and mass transport: the chromatography of a single solute. International Symposium on "Computational hydraulics in agricultural engineering", Wroclaw, Poland, June 21 - 24.

Connor, J. J., and Brebbia, C. A.
1976. Finite Element techniques for fluid flow. Newnes-Buttherworths, London.

Gardner, W. R.
1958. Some steady-state solutions of the unsaturated moisture flow with applications to evaporation from a water table. Soil Sci. 85: 228 - 232.

Giura, R.
1980. Modelli matematici per lo studio degli acquiferi in sistemi permeabili per porosità. Atti del IV Congresso internazionale sulle acque sotterranee, Acireale, Italy, February 17 - 21.

Hillel, D.
1980. Application of soil physics. Academic Press, New York.

Kirda, C., Nielsen, D. R., and Biggar, J.W.
1973. Simultaneous transport of chloride and water during infiltration. Soil Sci., Soc. Am. Proc. 37: 339 - 345.

Neuman, S. P., Feddes, R. A., and Bresler, E.
1975. Finite element analysis of two-dimensional flow in soils considering water uptake by roots. Soil Sci. Soc. Am. Proc. 39: 224 - 237.

Russo Spena, A., and Cavalieri, F.
1981. Finite element analysis of one dimensional flow filtration and redistribution in soils considering water uptake by roots. STUDI SASSARESI, sez. III - Annali della Facoltà di Agraria dell'Università di Sassari, vol. XXVIII - XXIX: 336 - 343.

Santini, A.
1979. Bilancio idrico del sistema suolo-vegetazione: impiego di un modello matematico di simulazione. Rivista di Ingegneria Idraulica, X: 63 - 76.

Santini, A.
1980. Rialimentazione naturale degli acquiferi. Atti del IV Congresso internazionale sulle acque sotterranee, Acireale, Italy, February 17 - 21.

Selim, H. M., Mansell, R. S., and Elzeftawy, A.
1976. Distribution of 2,4 - D and water in soil during infiltration and redistribution. Soil Sci. 121: 176 - 183.

Selim, H. M.
1978. Transport of reactive solute during transient unsaturated water flow in multilayered soils. Soil Sci. 126: 127 - 135.

Van Geunchten, M. Th.
1980. A comparison of numerical solutions of the one-dimensional unsaturated - saturated flow and mass transport equations. In S.Y. Wang et al. (eds.) Finite elements in water resources, pp. 3.49 - 3.66. Rose Printing Company, Inc., Tallahassee, Florida.

Zienkiewicz, O. C.
1977. The finite element method. Mc. Graw-Hill, London.

THE 2-D, UNSTEADY, TRANSPORT EQUATION SOLVED BY THE COMBINED USE OF THE FINITE ELEMENT METHOD AND THE METHOD OF CHARACTERISTICS

António M. Baptista
E. Eric Adams
Keith D. Stolzenbach

Department of Civil Engineering
Massachusetts Institute of Technology

ABSTRACT

The 2-D transport equation is split into separate convection and dispersion equations. The first is solved by a backwards method of characteristics, and the second by a quadratic Galerkin FEM. Accuracy is briefly discussed, both theoretically and through solution of test problems.

1. INTRODUCTION

Our approach splits the transport equation into a pure-convection equation and a pure-dispersion equation; the first is solved by a backwards method of characteristics, and the second by a quadratic Galerkin FEM. This approach avoids the simultaneous treatment of hyperbolic and parabolic terms which, for convection-dominated problems, is the cause of spurious oscillations (wiggles) in centered FEM (e.g., LEIMKUHLER 1974) and excessive numerical damping in "upwind" FEM (e.g. HEINRICH et al 1977, HUGHES and BROOKS 1979).

The splitting approach (often associated with Eulerian-Lagrangian methods) has been used in FD (HOLLY and PREISSMAN 1977, GLASS and RODI 1981, HOLLY and POLATERA 1984, CHENG *et al* 1984), but has seen less use in FE; major works include NEWMAN 1981 and NEWMAN and SOREK 1983, who treated the 1-D transport equation, and BENQUE *et al* 1980, who treated the 2-D Navier-Stokes equations.

2. GOVERNING EQUATION

We consider depth-averaged transport of a passive scalar in

domain Ω, with boundaries $\partial\Omega_1$ and $\partial\Omega_2$. For simplicity we neglect sinks, decay, and vertical transport through bottom and surface boundaries, and assume that principal axes of dispersion coincide with x and y axes. Then

$$\frac{\partial c}{\partial t} + u_i \frac{\partial c}{\partial x_i} = \frac{1}{h} \frac{\partial}{\partial x_i} \left(K^i h \frac{\partial c}{\partial x_i} \right) + p \quad \{i = 1,2\} \tag{1}$$

where $c(x,y,t)$ is a depth-averaged concentration (unknown), $u_i(x,y,t)$ are depth-averaged velocity components, $h(x,y,t)$ is the flow depth, $K^i(x,y,t)$ are dispersion coefficients, and $p(x,y,t)$ is a source term. (Note: $x_1 \equiv x$; $x_2 \equiv y$, $u_1 \equiv u$; $u_2 \equiv v$.) Initial and boundary conditions are

$$c(x,y,t) = \bar{c}_o(x,y) \qquad \text{at} \quad t = 0, \text{ in } \Omega \tag{2}$$

$$c(x,y,t) = \bar{c}(x,y,t) \qquad \text{at} \quad t > 0, \text{ on } \partial\Omega_1 \tag{3}$$

$$q(x,y,t) = \bar{q}(x,y,t) \qquad \text{at} \quad t > 0, \text{ on } \partial\Omega_2 \tag{4}$$

where bars denote known quantities and q denotes normal flux.

3. DECOMPOSITION OF THE TRANSPORT PROBLEM

Equation 1 is written in time-discrete form

$$\frac{c^n - c^{n-1}}{\Delta t} + \left\{u_i \frac{\partial c}{\partial x_i}\right\}_{n-1} = \left\{\frac{1}{h} \frac{\partial}{\partial x_i} \left(K^i h \frac{\partial c}{\partial x_i}\right) + p\right\}_n \tag{5}$$

where Δt is time-step and n, n-1 are current and previous time. Defining an auxiliary variable c^f, and making use of linearity, we decompose Equation 5, without any additional error (see BAPTISTA 1984), into two sub-equations:

$$\frac{c^f - c^{n-1}}{\Delta t} + \left\{u_i \frac{\partial c}{\partial x_i}\right\}_{n-1} = 0 \qquad \text{convection equation} \tag{6}$$

$$\frac{c^n - c^f}{\Delta t} = \left\{\frac{1}{h} \frac{\partial}{\partial x_i} \left(K^i h \frac{\partial c}{\partial x_i}\right) + p\right\}_n \qquad \text{dispersion equation} \tag{7}$$

Note that Equations 6 and 7 are purely hyperbolic and purely parabolic, respectively. c^f could be interpreted as the concentration that would be obtained at time n, if only convection took place between n-1 and n. However, c^f can not be rigidly identified with time n, as it becomes an initial condition for the dispersion equation. A similar decomposition applies to the boundary conditions.

4. SOLUTION OF THE DISPERSION EQUATION

The dispersion equation is solved by a FEM, based on a

quadratic weighted residual Galerkin formulation. The resulting set of ordinary differential equations (in time) is discretized using an implicit backward Euler's scheme. As the formulation is quite conventional, we omit the details (to be included in BAPTISTA 1984) and present only the resulting system of algebraic equations:

$$(\underline{\underline{M}} + \Delta t\ \underline{\underline{A}}^n)\ \underline{\alpha}^n = \underline{\underline{M}}\ \alpha^f + \Delta t\ \underline{P}^n \tag{8}$$

$$\underline{\underline{M}} = \sum_{\substack{\text{all}\\ \text{elements}}} \underline{\underline{M}}^e = \sum_{\substack{\text{all}\\ \text{elements}}} \iint_{\Omega_e} \underline{\phi}^T \underline{\phi}\ dA \tag{9}$$

$$\underline{\underline{A}} = \sum_{\substack{\text{all}\\ \text{elements}}} \underline{\underline{A}}^e = \sum_{\substack{\text{all}\\ \text{elements}}} \iint_{\Omega_e} \frac{\partial}{\partial x_i}\left(\frac{\underline{\phi}^T}{h}\right) K^i h\ \frac{\partial \underline{\phi}}{\partial x_i}\ dA \tag{10}$$

$$\underline{P} = \sum_{\substack{\text{all}\\ \text{elements}}} \underline{P}^e = \sum_{\substack{\text{all}\\ \text{elements}}} \left\{\iint_{\Omega_e} \underline{\phi}^T p dA - \int_{\partial\Omega_{e_2}} \underline{\phi}^T \frac{\bar{q}}{h}\ ds\right\} \tag{11}$$

where $\underline{\alpha}$ is the column vector of nodal concentrations, ϕ is the row vector of interpolation functions and n, f denote current time, after and before dispersion. Boundary conditions on $\partial\Omega_1$ must be added to Equation 8 to complete the problem.

5. SOLUTION OF THE CONVECTION EQUATION

Equation 6 implies that concentration remains constant along "characteristic lines". Let (P,n) denote the position of a particle at node P and time n (Figure 1a). This particle was previously at position (P',n-1). Tracking the characteristic line backwards, we can find P'. Now, by Equation 6, $c(P,n) \equiv c(P',n-1)$, which may be obtained by interpolation. (The same interpolation functions used for the dispersion equation must be used, for consistency.) The procedure is then repeated for the remaining nodes. If an inflow boundary is crossed during back-tracking (e.g., node Q, Figure 1a), concentration is imposed directly from the boundary condition, i.e., $c(Q,n) \equiv \bar{c}(Q',n-\beta)$. In outflow or closed boundaries back-tracking is performed within the domain, and boundary conditions are not required.

Back-tracking of a particle is governed by

$$\frac{dx_j}{dt} = -u_j(x,y,t) \qquad j = 1,2 \tag{12}$$

with initial conditions $x = \bar{x}_n$ at time n. In FE, u_j is given explicitly only at nodes; however, interpolation functions allow definition of velocity over the whole domain. Also, most FE circulation models only compute velocity at fixed times, requiring time-interpolation. Exceptions include frequency-domain approaches that explictly establish the time dependence of the velocity field. Our element-by-element tracking algorithm accounts for element-based spatial interpolation of velocity, and has flexibility to accomodate time-interpolation, when required.

Consider a particle at position (P, n) in Figure 1b where P coincides with a node; at time n-1 this particle was at P'''', which we want to find. We first follow the particle backwards along its characteristic line, until position (P', $n-\beta_1$). As only element K_1 is involved,

$$u_j(x,y,t) = \sum_{i=1}^{m} \psi_i(x,y;\, K_1)\, u_{ij}(t;\, K_1) \quad j = 1,2 \qquad (13)$$

where $\psi_i(x,y;\, K_1)$ are interpolation functions over K_1, associated with the finite element approximation of the flow field; $u_{ij}(t;K_1)$ are the nodal velocities in the j-direction and m is the number of nodes of the element. We assume that nodal velocities are known at all times between n and n-1, either directly or by interpolation.

With velocities u_j from Equation 13, (P', $n-\beta_1$) is computed from Equation 12 using a 4th order Runge-Kutta scheme. A constant time step δt^{K_1} is selected to limit truncation error to a prescribed value. Once (P', $n-\beta_1$) is found, velocity functions u_j are replaced by equivalent ones over element K_2, and the particle is tracked through this element to (P'', $n-\beta_2$). δt^{K_2} may differ from δt^{K_1} to account for differences in flow. The element-by-element tracking is continued until time n-1 is reached, or a physical boundary is crossed (whichever happens first).

6. COMMENTS ON STABILITY AND ACCURACY

The method has no stability limitations on the time step, Δt. For the convection equation, unconditional stability is assured by the fact that the concentration at the foot (P', say) of each characteristic line is always found by interpolation (over the element that contains P'). For the dispersion equation, unconditional stability is assured by the implicit time-discretization scheme that was adopted.

Formal accuracy analysis is difficult. Errors may arise in the splitting technique, the solution of the dispersion equation and the solution of the convection equation.

The error associated with the splitting technique results exclusively from the time-discrete form adopted for the transport equation and is $O(\Delta t)$. The dispersion equation is already in a time-discrete form. Errors inherent to the solution of this equation, therefore, are associated with spatial approximations and, based on the accuracy of equivalent FD techniques, are estimated to be $O(\Delta x^4)$.

Errors with the convection equation may come from back-tracking or from interpolation required to find the concentration at the foot of the characteristic line. For spatially uniform flows, back-tracking errors are $O(\delta t^4)$; for more complex flows, additional error is associated with spatial interpolation for velocities (Equation 13), though this has generally been found to be small. Errors due to interpolation of concentrations are more severe. These errors were formally investigated for the case of pure convection of a harmonic wave (BAPTISTA 1984). The analysis shows no error for integer values of the Courant number; for non-integer values the method introduces numerical damping which depends on the number of computational points per wavelength, N, the fractional part of the Courant number, α, and the number of time steps (Figure 2).

7. TEST PROBLEMS

To illustrate actual accuracy several test problems are presented from BAPTISTA 1984. The problems were solved with 2-D grids consisting of triangular elements; the elements have 6 nodes, and are right isosceles, with length of the small legs $\ell = 2\Delta x$ (Δx=nodal spacing). Independent parameters include the Peclet number $Pe = u\ell/D$, the Courant number $Cu = u\Delta t/\ell$, and the number of elements N over which the source is distributed in the flow direction. (N=source length/ℓ).

7.1 Instantaneous source in a uniform flow

$$\frac{\partial c}{\partial t} + u \frac{\partial c}{\partial x} = D \frac{\partial^2 c}{\partial x^2} \tag{14}$$

$$\text{with } c(x,t) = \begin{cases} 0.5\left[1 + \cos \dfrac{\pi(x-x_o)}{N\Delta x}\right] & \text{for } t=0,\ |x-x_o|<N\Delta x \\ 0 & \text{for } t=0, \text{ elsewhere} \\ & \text{for } t>0,\ x = \pm\infty \end{cases} \tag{15}$$

where x_o is the coordinate of the source centroid. Results are shown in Figures 3 and 4. Mass and phase (measured by the position of the centroid) are nearly exact for all runs. Providing that the source is distributed over sufficient elements (e.g., $N \geq 7$), numerical diffusion is negligible, and significant numerical damping occurs only for high Pe. (Note that results can be improved by increasing N slightly, as suggested in Figure 3.)

The numerical solution is "wiggles"-free, regardless of the specific characteristics of the run. However, negative concentrations (quickly damped in space) may appear on either edge of the distribution; their maximum value tends to increase with decreasing N, and increasing Pe. For the same total time, accuracy increases (numerical damping and maximum negative concentrations are reduced) as Cu increases (i.e., as Δt increases and hence the number of concentration interpolations decreases).

7.2 Continuous source in a uniform flow

$$\frac{\partial c}{\partial t} + u\frac{\partial c}{\partial x} = D\frac{\partial^2 c}{\partial x^2} + p \tag{16}$$

with $c(x,t) = 0$ for $t=0$ and $x = \pm\infty$ (17)

$$p(x,t) = \begin{cases} M\left[1 + \cos\frac{\pi(x-x_o)}{N\Delta x}\right] & \text{at } |x-x_o|<N\Delta x \\ 0 & \text{elsewhere} \end{cases} \tag{18}$$

Results are shown in Figure 5. Excellent accuracy is obtained up to moderately high Pe. For high Pe, a slight overshoot occurs in the downstream edge of the concentration distribution. Again, results can be improved using slightly higher N. As in §7.1 mass and phase are nearly exact.

7.3 Instantaneous source in a rigid body rotation (pure convection)

$$\frac{\partial c}{\partial t} + u\frac{\partial c}{\partial x} + v\frac{\partial c}{\partial y} = 0 \tag{19}$$

$$\text{with } c(x,y,t) = \begin{cases} 0.5\left[1 + \cos\frac{\pi r}{N\Delta x}\right] & \text{for } t=0,\ r\le N\Delta x \\ 0 & \text{for } t=0,\ \text{elsewhere} \\ & \text{for } t>0,\ x = \pm\infty \end{cases} \tag{20}$$

where $r = [(x-x_o)^2 + (y-y_o)^2]^{1/2}$, x_o, y_o are source coordinates, $u=-Wy$, $v=Wx$ and W = angular frequency.

Results are shown in Figure 6. After 5 tours, the "cosine-hill" distribution has neither collapsed nor distorted excessively, even though a small number of elements (N=3) was used to discretize the source. Mass and phase are well reproduced, and no wiggles are present. The discussion of accuracy (and its dependence on N, Pe and Cu) given in §7.1 is valid here, suggesting the method's adequacy in 2-D.

8. CONCLUDING REMARKS

The method has a number of merits, including unconditional stability, and good to excellent accuracy, even for convection-dominated problems. Goals of future research include comparison (in terms of accuracy and computational efficiency) with available upwind FEM and Eulerian-Lagrangian FDM, plus applications to prototype domains.

9. ACKNOWLEDGEMENTS

This study was funded by Sea Grant and by the MIT Energy Laboratory Electric Utility Program. While participating in the study, the first author was on leave from Laboratorió Nacional de Engenharia Civil, Portugal, being partially sponsored by INVOTAN and Luso-American Cultural Commission.

BIBLIOGRAPHY

BAPTISTA, A.M. (1984) Eulerian-Lagrangian Analysis of transport in shallow waters. S.M. thesis, M.I.T.

BENQUE, J.P., et al (1980) A finite element method for Navier-Stokes equations. 3rd Conf. on Fin. Elem. in Flow Problems.

CHENG, R.T., V. CASULLI and S. MILFORD (1984) Eulerian lagragian solution of the convection-diffusion equation in natural coordinates. (to be published).

GLASS, J. and W. RODI (1982) A higher order numerical scheme for scalar transport. Comp. Meth. in Appl. Mech. and Eng., 31, pp. 337-358.

HEINRICH, J.C., et al (1977) An upwind finite element scheme for the two-dimensional convective transport equation. Int. J. Num. Meth. Eng., 11, pp. 131-143.

HOLLY, F.M. and J.M. POLATERA (1984) Dispersion simulation in 2-D tidal flow, to appear in Jour. Hydr. Engrg., ASCE.

HOLLY, F.M., Jr. and A. PREISSMANN (1977) Accurate calculation of transport in two dimensions. Jour. Hydr. Div., ASCE, 103, pp. 1259-1278.

HUGHES, T.J. and A. BROOKS (1979) A multidimensional upwind scheme with no crosswind diffusion in Finite Elements for Convection Dominated Flows, (Ed. T.J. Hughes), AMD ASME, 34, pp. 19-35.

LEIMKUHLER, W.F. (1974) A two-dimensional element dispersion model. Civil Engineer thesis, M.I.T.

NEUMAN, S.P. (1981) An Eulerian-Lagrangian scheme for the dispersion-convection equation using conjugate space-time grids. Jour. Comp. Physics, 41, pp. 270-294.

NEUMAN, S.P. and S. SOREK (1982) Eulerian-Lagrangian Methods for Advection-Dispersion in Finite Elements in Water Resources, (Ed. K.P. Mole et al), Vol. 4, pp. 14.41-14.68.

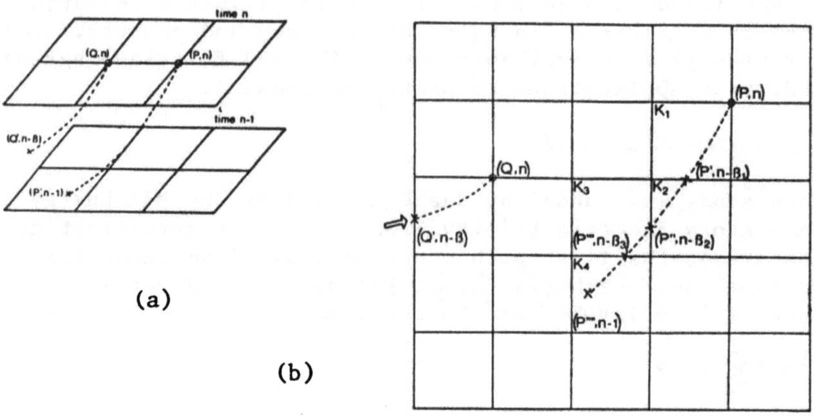

Figure 1 - Illustrative sketch for the solution of convection.

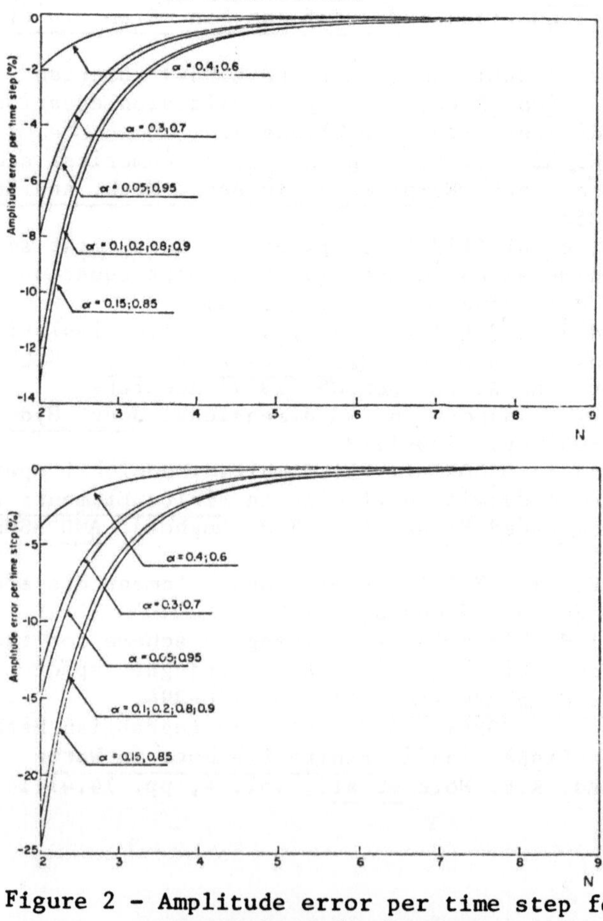

(a)

(b)

Figure 2 - Amplitude error per time step for the convection of a harmonic wave : (a) 1-D and (b) 2-D, triangular elements.

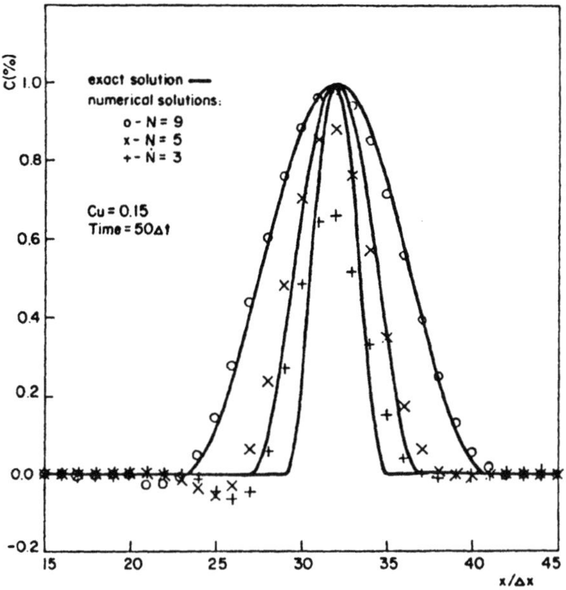

Figure 3 - Instantaneous source in a uniform flow. Concentration profiles for different values of N, for pure convection.

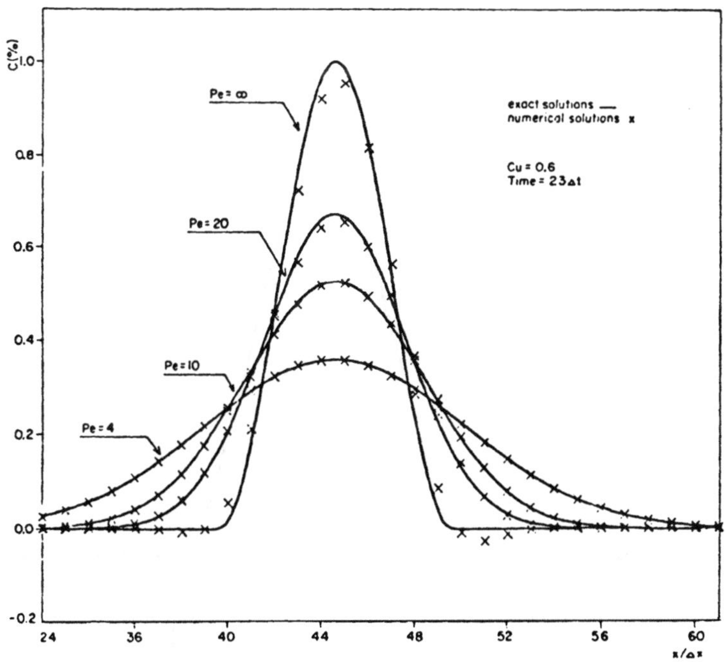

Figure 4 - Instantaneous source in a uniform flow. Concentration profiles for different Pe (N=5).

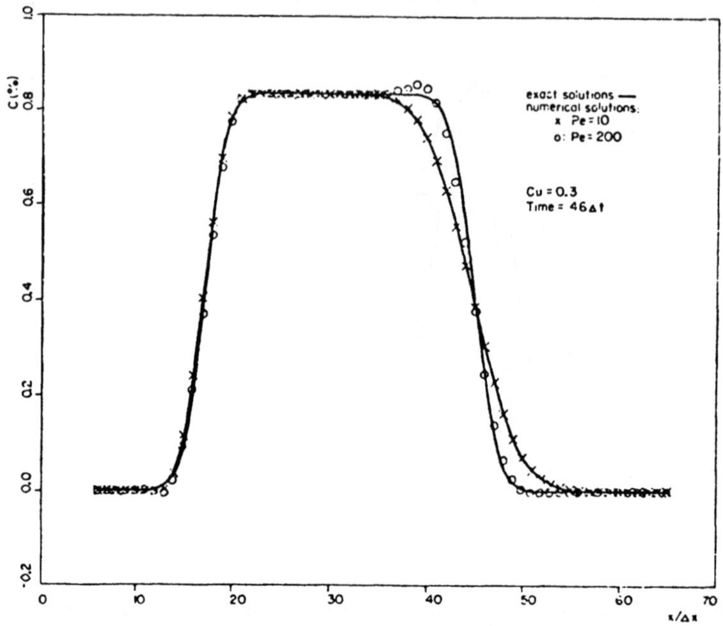

Figure 5 - Continuous source in a uniform flow. Concentration profiles for different Pe (N=5).

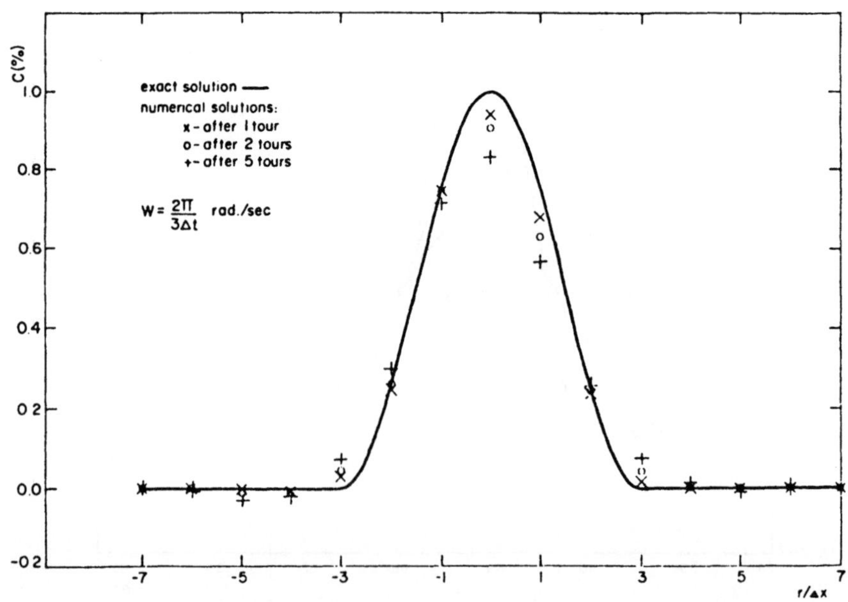

Figure 6 - Instantaneous source in rigid body rotation. Radial concentration profiles (N=3, pure convection).

THE PRINCIPAL DIRECTION TECHNIQUE FOR ADVECTIVE-DISPERSIVE TRANSPORT SIMULATION IN THREE DIMENSIONS

E. O. Frind

Department of Earth Sciences, University of Waterloo, Waterloo, Ontario, Canada

ABSTRACT

The Principal Direction Technique for advection-dispersion problems is extended to three dimensions. The technique combines the versatility of Galerkin finite elements in handling non-rectangular domains with the efficiency and conceptual simplicity of an alternating direction time-stepping scheme. Various time step splitting options are explored. The resulting algorithm appears to have both the accuracy and the efficiency necessary for routine application to 3-D problems of practical significance.

INTRODUCTION

Numerical models for the simulation of groundwater contamination have now become well established in hydrogeologic practice. Used correctly, such models can help the hydrogeologist in the evaluation of contamination scenarios and in the design of disposal facilities.

In order to provide a valid and credible base, a simulation model should be capable of representing all physical mechanisms relevant to the system under investigation. One aspect that is often neglected is the inherently three-dimensional nature of dispersion in natural groundwater systems. Reasons for this neglect are the high computing cost associated with many 3-D models, and the often-mentioned lack of 3-D data. Hence most simulations of contaminant transport today are performed in two dimensions.

It is now recognized however that even a low-order transverse dispersive mechanism, acting over a long distance, can substantially affect the configuration of a plume. This transverse spreading process can only be represented correctly if the third spatial dimension is included. Accurate and efficient 3-D simulation models are therefore needed.

The finite element technique has become, because of its versatility, a favoured simulation method for natural groundwater systems. Unfortunately in its 3-D mode (Frind and Verge, 1978) the method tends to be extremely demanding in terms of computing resources. By contrast, the alternating direction finite difference technique (Peaceman and Rachford, 1955) has what would appear to be optimum efficiency for multi-dimensional problems, but in its basic form is restricted to rectangular domains. Early versions of an alternating direction Galerkin method (Douglas and Dupont, 1971) are similarly restricted to simple geometries.

Recent efforts to combine the advantages of a finite element spatial approximation with an alternating direction time-stepping scheme have led to the development of the Principal Direction Technique (Frind and Pinder, 1982; Frind, 1982) for two-dimensional domains. A more general Alternating Direction Galerkin Technique (Daus and Frind, 1984) that is radically different from the Douglas-Dupont method bearing the same name has also been proposed. Both techniques can handle non-rectangular domains.

Unlike the 2-D alternating direction technique, which naturally yields an optimum time stepping scheme that is unconditionally stable and gives second-order accuracy, the 3-D version is not straightforward and various options are possible. A number of approaches based on finite differences are discussed by Peaceman (1977), who recommends the methods of Brian (1961) and Douglas (1962) for their accuracy. More recently, the collocation method has also been successfully employed in a 3-D alternating direction framework (Celia and Pinder, 1982).

The objective here is to extend the principal direction solution for the advection-dispersion problem to three dimensions and to explore possible time-stepping options with a view toward achieving the accuracy and simplicity needed for practical applications.

THEORETICAL DEVELOPMENT

We will pursue the same strategy as was used previously in the development of the 2-D Principal Direction Technique (Frind and Pinder, 1982; Frind, 1982). The multi-dimensional problem is first spatially decoupled by devising an alternating direction time-stepping mechanism in which only one of the spatial directions is represented implicitly at each step. The spatial decoupling eliminates the large coefficient matrices obtained in direct solutions. Galerkin finite element theory is then applied to the spatially decoupled equations, thus maintaining the capability to represent geometrically irregular domains.

The solution can be formulated in the principal directions of transport, namely the direction of flow and the two directions orthogonal to the flow. The advantages of this approach are the restriction of the advective component to only one of the decoupled equations, and the elimination of cross-derivative terms.

Governing Equation

We will consider the three-dimensional domain $0 \le x \le A$, $0 \le y \le B$, $0 \le z \le C$, where x, y, and z define an orthogonal (but not necessarily Cartesian) coordinate system (Figure 1). The x-coordinate follows the flow lines. The governing equation for advective-dispersive transport for this system is:

$$\frac{\partial}{\partial x}\left(D_x \frac{\partial u}{\partial x}\right) + \frac{\partial}{\partial y}\left(D_y \frac{\partial u}{\partial y}\right) + \frac{\partial}{\partial z}\left(D_z \frac{\partial u}{\partial z}\right) - v \frac{\partial u}{\partial x} - \frac{\partial u}{\partial t} = 0 \qquad (1)$$

where t is time, $u = u(x,y,z,t)$ is the relative concentration, v is the fluid velocity, and D_x, D_y, D_z are the dispersion coefficients in the respective directions. The coefficients can be spatially variable, and the dispersion coefficients can be of the velocity-dependent type $D = \alpha v$, with α being the dispersivity.

The boundary conditions for the entry boundary $x = 0$ are either of the Dirichlet type

$$u = u_o \qquad (2)$$

or the Cauchy type

$$vu - D_x \frac{\partial u}{\partial x} = v_o\, u_o \qquad (3)$$

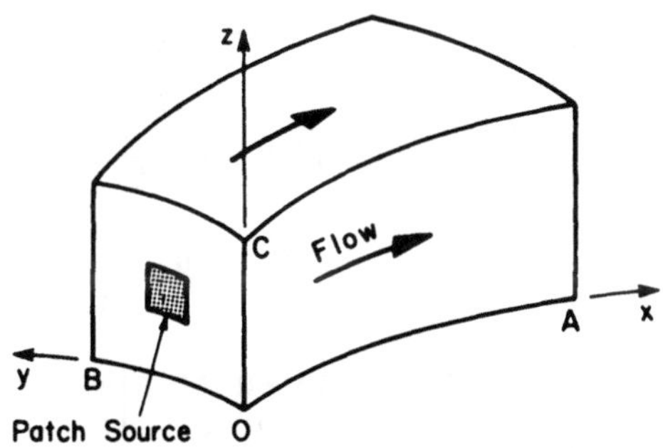

Figure 1. Three-dimensional domain.

where u_o is the specified source concentration, and v_o is the influent velocity. For the remaining boundaries, the boundary condition can be of the Neumann type

$$\frac{\partial u}{\partial \underset{\sim}{n}} = 0 \tag{4}$$

with $\underset{\sim}{n}$ designating the normal to the boundary.

Alternating Direction Time-Stepping Schemes

Equation (1) can be written in the form:

$$(L_x + L_y + L_z)u = \frac{u^{n+1} - u^n}{\Delta t} \tag{5}$$

where L_x, L_y and L_z represent the differential operators, and the time derivative has been replaced by a finite difference approximation, with n indicating the time level and $\Delta t = t^{n+1} - t^n$.

In order to spatially decouple equation (5), it must be split into steps in such a way that at each step only one of the spatial derivatives is represented implicitly. This will leave, at each step, the remaining two spatial derivatives to be represented explicitly. The natural symmetry of the 2-D alternating direction scheme, where each step contains one implicit and one explicit term, is thus not possible in the 3-D scheme.

Peaceman (1977) suggests splitting the time-stepping equation (5) into three steps. The splitting can be done in a variety of ways. Brian (1961) uses a four-step algorithm covering two spatial steps in time. We will here investigate some possible approaches.

Peaceman's Scheme. This basic scheme can be written as:

Step 1 $$L_x u^{n+\frac{1}{3}} + L_y u^n + L_z u^n = \frac{u^{n+\frac{1}{3}} - u^n}{\Delta t/3} \tag{6a}$$

Step 2 $$L_x u^{n+\frac{1}{3}} + L_y u^{n+\frac{2}{3}} + L_z u^n = \frac{u^{n+\frac{2}{3}} - u^{n+\frac{1}{3}}}{\Delta t/3} \tag{6b}$$

Step 3 $$L_x u^{n+\frac{1}{3}} + L_y u^{n+\frac{2}{3}} + L_z u^{n+1} = \frac{u^{n+1} - u^{n+\frac{2}{3}}}{\Delta t/3} \tag{6c}$$

where the implicit terms are located at time $n+\frac{1}{3}$, $n+\frac{2}{3}$, and n+1 respectively for Steps 1, 2, and 3. In Step 1, both explicit terms are lagged $\Delta t/3$ behind the implicit step, while in Steps 2 and 3 one term is lagged $\Delta t/3$ and the other $2\Delta t/3$. On average, the spatial derivatives are centered back of the temporal derivative and the scheme is therefore explicit in nature. According to Peaceman, the scheme is conditionally stable.

Explicit 1 Scheme. The Peaceman Scheme can be slightly modified as follows:

$$\underline{\text{Step 1}} \quad L_x u^{n+\frac{1}{3}} + L_y u^{n-\frac{1}{3}} + L_z u^n = \frac{u^{n+\frac{1}{3}} - u^n}{\Delta t/3} \tag{7a}$$

$$\underline{\text{Step 2}} \quad L_x u^{n+\frac{1}{3}} + L_y u^{n+\frac{2}{3}} + L_z u^n = \frac{u^{n+\frac{2}{3}} - u^{n+\frac{1}{3}}}{\Delta t/3} \tag{7b}$$

$$\underline{\text{Step 3}} \quad L_x u^{n+\frac{1}{3}} + L_y u^{n+\frac{2}{3}} + L_z u^{n+1} = \frac{u^{n+1} - u^{n+\frac{2}{3}}}{\Delta t/3} \tag{7c}$$

where the three steps are now consistent with respect to the placing of the explicit terms. In each step, the three spatial derivatives are centered $\Delta t/6$ back of the center of the temporal derivative, and on the average over the three steps, the spatial derivatives are centered at time $n+\frac{1}{3}$ while the temporal derivative is centered at time $n+\frac{1}{2}$. This scheme is therefore also explicit in nature.

Explicit 2 Scheme. Another option is to write the split equations as:

$$\underline{\text{Step 1}} \quad L_x u^{n+\frac{1}{3}} + L_y u^n + L_z u^n = \frac{u^{n+\frac{1}{3}} - u^n}{\Delta t/3} \tag{8a}$$

$$\underline{\text{Step 2}} \quad L_x u^{n+\frac{1}{3}} + L_y u^{n+\frac{2}{3}} + L_z u^{n+\frac{1}{3}} = \frac{u^{n+\frac{2}{3}} - u^{n+\frac{1}{3}}}{\Delta t/3} \tag{8b}$$

$$\underline{\text{Step 3}} \quad L_x u^{n+\frac{2}{3}} + L_y u^{n+\frac{2}{3}} + L_z u^{n+1} = \frac{u^{n+1} - u^{n+\frac{2}{3}}}{\Delta t/3} \tag{8c}$$

where both of the explicit terms at each step are now lagged by $\Delta t/3$. Because the most recent results would be used in each of the partial solutions, this scheme is expected to be more accurate than the previous schemes. Each equation now has its spatial derivatives located only $\Delta t/18$ behind the temporal derivative, while the average over the three equations shows the left-hand side centered at $n+\frac{4}{9}$ and the right-hand side at $n+\frac{1}{2}$, resulting once again in an explicit scheme.

Centered Scheme. A fully centered (Crank-Nicolson equivalent) scheme can be obtained by writing:

$$\underline{\text{Step 1}} \quad L_x u^{n+\frac{1}{2}} + L_y u^n + L_z u^n = \frac{u^{n+\frac{1}{3}} - u^n}{\Delta t/3} \tag{9a}$$

$$\underline{\text{Step 2}}\quad L_x u^{n+\frac{1}{3}} + L_y u^{n+\frac{5}{6}} + L_z u^{n+\frac{1}{3}} = \frac{u^{n+\frac{2}{3}} - u^{n+\frac{1}{3}}}{\Delta t/3} \tag{9b}$$

$$\underline{\text{Step 3}}\quad L_x u^{n+\frac{2}{3}} + L_y u^{n+\frac{2}{3}} + L_z u^{n+\frac{7}{6}} = \frac{u^{n+1} - u^{n+\frac{2}{3}}}{\Delta t/3} \tag{9c}$$

We now have, in each equation, the spatial derivatives together centered in time exactly at the point where the temporal derivative is centered ($n+\frac{1}{6}$ for Step 1, $n+\frac{1}{2}$ for Step 2, $n+\frac{5}{6}$ for Step 3), and on the average over the three equations, each of the derivatives is located exactly at time $n+\frac{1}{2}$. This three-step scheme is therefore equivalent to a Crank-Nicolson scheme and it should be second-order accurate in time.

Equations (9) can be transformed into a more convenient form by substituting the linear interpolations:

$$u^{n+\frac{1}{2}} = \frac{3}{2} u^{n+\frac{1}{3}} - \frac{1}{2} u^{n}$$

$$u^{n+\frac{5}{6}} = \frac{3}{2} u^{n+\frac{2}{3}} - \frac{1}{2} u^{n+\frac{1}{3}}$$

$$u^{n+\frac{7}{6}} = \frac{3}{2} u^{n+1} - \frac{1}{2} u^{n+\frac{2}{3}}$$

which results in:

<u>Step 1</u>

$$\frac{3}{2} L_x u^{n+\frac{1}{3}} - \frac{1}{2} L_x u^{n} + L_y u^{n} + L_z u^{n} = \frac{u^{n+\frac{1}{3}} - u^{n}}{\Delta t/3} \tag{10a}$$

<u>Step 2</u>

$$L_x u^{n+\frac{1}{3}} + \frac{3}{2} L_y u^{n+\frac{2}{3}} - \frac{1}{2} L_y u^{n+\frac{1}{3}} + L_z u^{n+\frac{1}{3}} = \frac{u^{n+\frac{2}{3}} - u^{n+\frac{1}{3}}}{\Delta t/3} \tag{10b}$$

<u>Step 3</u>

$$L_x u^{n+\frac{2}{3}} + L_y u^{n+\frac{2}{3}} + \frac{3}{2} L_z u^{n+1} - \frac{1}{2} L_z u^{n+\frac{2}{3}} = \frac{u^{n+1} - u^{n+\frac{2}{3}}}{\Delta t/3} \tag{10c}$$

The implicit terms are now located at time $n+\frac{1}{3}$, $n+\frac{2}{3}$, and $n+1$, and time centering is achieved by weighting these terms with their equivalents at the explicit time level. This scheme also uses the most recent information at each step.

<u>*Brian's Scheme.*</u> This scheme (Brian, 1961) is of the form:

Step 1.1

$$L_x \, {}^{*}u^{n+\frac{1}{2}} + L_y \, u^n + L_z \, u^n = \frac{{}^{*}u^{n+\frac{1}{2}} - u^n}{\Delta t/2} \quad (11a)$$

Step 1.2

$$L_x \, {}^{*}u^{n+\frac{1}{2}} + L_y \, {}^{**}u^{n+\frac{1}{2}} + L_z \, u^n = \frac{{}^{**}u^{n+\frac{1}{2}} - u^n}{\Delta t/2} \quad (11b)$$

Step 1.3

$$L_x \, {}^{*}u^{n+\frac{1}{2}} + L_y \, {}^{**}u^{n+\frac{1}{2}} + L_z \, {}^{***}u^{n+\frac{1}{2}} = \frac{{}^{***}u^{n+\frac{1}{2}} - u^n}{\Delta t/2} \quad (11c)$$

Step 2

$$L_x \, {}^{*}u^{n+\frac{1}{2}} + L_y \, {}^{**}u^{n+\frac{1}{2}} + L_z \, {}^{***}u^{n+\frac{1}{2}} = \frac{u^{n+1} - u^n}{\Delta t} \quad (11d)$$

where the * symbol designates successive approximations of u at the $n+\frac{1}{2}$ time level. The scheme thus consists of three iterations of an implicit-explicit step to the $n+\frac{1}{2}$ time level, followed by a fully explicit step to the n+1 time level. According to Brian (1961), it is equivalent to the Crank-Nicolson scheme and unconditionally stable.

Galerkin Finite Element Formulation

We will use the Centered Scheme, Equations 10, as a basis for developing the finite element solution. The development for other time-stepping schemes follows the same procedure.

The domain of Figure 1 can be subdivided into block elements that have quadrilateral (but not necessarily rectangular) sides and that join at the nodes (Figure 2). In formulating the finite element solution, it is important to maintain compatibility with the alternating direction time-stepping

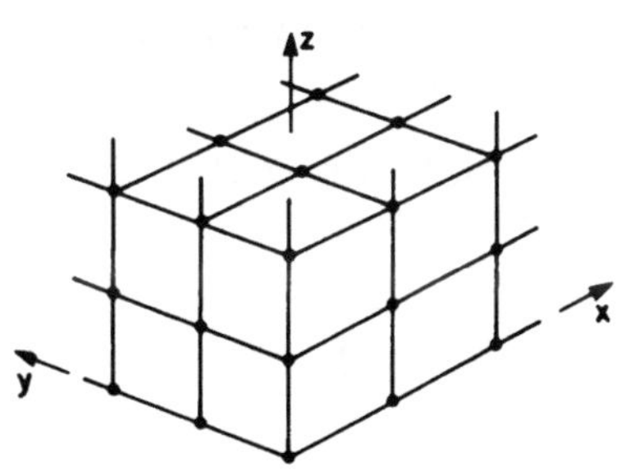

Figure 2. Finite element grid system.

scheme which provides for the decoupling of the spatial coordinates. This compatibility can be achieved through the choice of suitable basis functions.

We define a set of basis functions associated with a node j, for each partial time step, as follows:

Step 1

$$w_j(x) = \begin{cases} \dfrac{x - x_{j-1}}{x_j - x_{j-1}} & x_{j-1} \le x \le x_j \\[2ex] \dfrac{x_{j+1} - x}{x_{j+1} - x_j} & x_j \le x \le x_{j+1} \end{cases} \quad \begin{matrix} -b_j \le y \le b_j \\ -c_j \le z \le c_j \end{matrix} \tag{12a}$$

Step 2

$$w_j(y) = \begin{cases} \dfrac{y - y_{j-1}}{y_j - y_{j-1}} & y_{j-1} \le y \le y_j \\[2ex] \dfrac{y_{j+1} - y}{y_{j+1} - y_j} & y_j \le y \le y_{j+1} \end{cases} \quad \begin{matrix} -a_j \le x \le a_j \\ -c_j \le z \le c_j \end{matrix} \tag{12b}$$

Step 3

$$w_j(z) = \begin{cases} \dfrac{z - z_{j-1}}{z_j - z_{j-1}} & z_{j-1} \le z \le z_j \\[2ex] \dfrac{z_{j+1} - z}{z_{j+1} - z_j} & z_j \le z \le z_{j+1} \end{cases} \quad \begin{matrix} -a_j \le x \le a_j \\ -b_j \le y \le b_j \end{matrix} \tag{12c}$$

where the nodal index j is understood to increase in the implicit (connected) direction always, that is, in the x-direction for Step 1, the y-direction for Step 2, and the z-direction for Step 3, and $\pm a_j$, $\pm b_j$, $\pm c_j$ represent the midpoints between node j and its neighbouring nodes on either side in the two unconnected directions.

Corresponding uni-directional interpolation functions that can be used as trial solutions are:

Step 1 $$u \simeq \hat{u} = \sum_{j=1}^{N} u_j(t)\, w_j(x) \tag{13a}$$

Step 2 $$u \simeq \hat{u} = \sum_{j=1}^{N} u_j(t)\, w_j(y) \tag{13b}$$

Step 3 $$u \simeq \hat{u} = \sum_{j=1}^{N} u_j(t)\, w_j(z) \qquad (13c)$$

where N is the number of nodes, and $u_j(t)$ are nodal values of u.

The Galerkin equations are obtained by substituting the trial solutions (13) into the appropriate alternating direction equations (10) and globally minimizing the residual by setting it orthogonal to nodal weighting functions. The decoupling of the spatial coordinates is maintained by the choice of basis functions (12) as weighting functions for each of the three steps. The following Galerkin equations are obtained:

Step 1

$$\int_x \int_y \int_z \Big\{\frac{3}{2}\Big[\frac{\partial}{\partial x}\Big(D_x \frac{\partial u}{\partial x}\Big) - v\frac{\partial u}{\partial x}\Big]^{n+\frac{1}{3}} - \frac{1}{2}\Big[\frac{\partial}{\partial x}\Big(D_x \frac{\partial u}{\partial x}\Big) - v\frac{\partial u}{\partial x}\Big]^{n}$$

$$+ \Big[\frac{\partial}{\partial y}\Big(D_y \frac{\partial u}{\partial y}\Big)\Big]^{n} + \Big[\frac{\partial}{\partial z}\Big(D_z \frac{\partial u}{\partial z}\Big)\Big]^{n}$$

$$- \frac{u^{n+\frac{1}{3}} - u^{n}}{\Delta t/3}\Big\}\, w_i(x)\, dz\, dy\, dx = 0 \qquad i=1,\ldots,N \qquad (14a)$$

Step 2

$$\int_y \int_z \int_x \Big\{\Big[\frac{\partial}{\partial x}\Big(D_x \frac{\partial u}{\partial x}\Big) - v\frac{\partial u}{\partial x}\Big]^{n+\frac{1}{3}} + \frac{3}{2}\Big[\frac{\partial}{\partial y}\Big(D_y \frac{\partial u}{\partial y}\Big)\Big]^{n+\frac{2}{3}}$$

$$- \frac{1}{2}\Big[\frac{\partial}{\partial y}\Big(D_y \frac{\partial u}{\partial y}\Big)\Big]^{n+\frac{1}{3}} + \Big[\frac{\partial}{\partial z}\Big(D_z \frac{\partial u}{\partial z}\Big)\Big]^{n+\frac{1}{3}}$$

$$- \frac{u^{n+\frac{2}{3}} - u^{n+\frac{1}{3}}}{\Delta t/3}\Big\}\, w_i(y)\, dx\, dz\, dy = 0 \qquad i=1,\ldots,N \qquad (14b)$$

Step 3

$$\int_z \int_x \int_y \Big\{\Big[\frac{\partial}{\partial x}\Big(D_x \frac{\partial u}{\partial x}\Big) - v\frac{\partial u}{\partial x}\Big]^{n+\frac{2}{3}} + \Big[\frac{\partial}{\partial y}\Big(D_y \frac{\partial u}{\partial y}\Big)\Big]^{n+\frac{2}{3}}$$

$$+ \frac{3}{2}\Big[\frac{\partial}{\partial z}\Big(D_z \frac{\partial u}{\partial z}\Big)\Big]^{n+1} - \frac{1}{2}\Big[\frac{\partial}{\partial z}\Big(D_z \frac{\partial u}{\partial z}\Big)\Big]^{n+\frac{2}{3}}$$

$$- \frac{u^{n+1} - u^{n+\frac{2}{3}}}{\Delta t/3}\} \, w_i(z)\, dy\, dx\, dz = 0 \qquad i=1,\ldots,N \qquad (14c)$$

where the ^ symbol has been omitted for convenience. These equations are integrated first in the two explicit directions, that is, in the y- and z-directions for Step 1, the x- and z-directions for Step 2, and the x- and y-directions for Step 3. Integration by parts is then applied in the implicit direction to eliminate the second-derivative terms. Because of the choice of boundary conditions, the resulting natural boundary terms are non-zero only at the entry boundary x = 0. Next, the interpolation functions (13) are substituted into the resulting integral equations. With respect to the time derivative term, there is a choice of either the consistent or the lumped formulation (Frind and Pinder, 1982). Although the consistent formulation has been shown to give better accuracy (Daus et al, 1984), we will here choose the lumped formulation for simplicity.

Assuming a uniform node spacing 2a, 2b, and 2c in the x-, y-, and z-directions respectively, we obtain the equations:

<u>Step 1</u>

$$\sum_{j=1}^{N} u_j^{n+\frac{1}{3}} \frac{3}{2} \int_x [2b\; 2c(D_x \frac{\partial w_i}{\partial x} \frac{\partial w_j}{\partial x} + v \frac{\partial w_j}{\partial x} w_i(x))]dx$$

$$- 2b\; 2c[D_x \frac{\partial u}{\partial x} w_i(x)]_{x=0}$$

$$- \sum_{j=1}^{N} u_j^{n} \frac{1}{2} \int_x [2b\; 2c(D_x \frac{\partial w_i}{\partial x} \frac{\partial w_j}{\partial x} + v \frac{\partial w_j}{\partial x} w_i(x))]dx$$

$$- 2a\; 2c\{[D_y \frac{\partial u}{\partial y}]_b - [D_y \frac{\partial u}{\partial y}]_{-b}\}^n$$

$$- 2a\; 2b\{[D_z \frac{\partial u}{\partial z}]_c - [D_z \frac{\partial u}{\partial z}]_{-c}\}^n$$

$$+ \sum_{j=1}^{N} \frac{u_j^{n+\frac{1}{3}} - u_j^{n}}{\Delta t/3} \int_x 2b\; 2c\; w_i(x)\, dx = 0 \qquad i=1,\ldots,N \qquad (15a)$$

Step 2

$$- 2b\ 2c\{[D_x \frac{\partial u}{\partial x}]_a - [D_x \frac{\partial u}{\partial x}]_{-a} - [vu]_a + [vu]_{-a}\}^{n+\frac{1}{3}}$$

$$+ \sum_{j=1}^{n} u_j^{n+\frac{2}{3}} \frac{3}{2} \int_y 2a\ 2c\ D_y \frac{\partial w_i}{\partial y} \frac{\partial w_j}{\partial y} dy$$

$$- \sum_{j=1}^{N} u_j^{n+\frac{1}{3}} \frac{1}{2} \int_y 2a\ 2c\ D_y \frac{\partial w_i}{\partial y} \frac{\partial w_j}{\partial y} dy$$

$$- 2a\ 2b\{[D_z \frac{\partial u}{\partial z}]_c - [D_z \frac{\partial u}{\partial z}]_{-c}\}^{n+\frac{1}{3}}$$

$$+ \sum_{j=1}^{N} \frac{u_j^{n+\frac{2}{3}} - u_j^{n+\frac{1}{3}}}{\Delta t/3} \int_y 2a\ 2c\ w_i(y)dy = 0 \qquad i=1,\ldots,N \qquad (15b)$$

Step 3

$$- 2b\ 2c\{[D_x \frac{\partial u}{\partial x}]_a - [D_x \frac{\partial u}{\partial x}]_{-a} - [vu]_a + [vu]_{-a}\}^{n+\frac{2}{3}}$$

$$- 2a\ 2c\{[D_y \frac{\partial u}{\partial y}]_b - [D_y \frac{\partial u}{\partial y}]_{-b}\}^{n+\frac{2}{3}}$$

$$+ \sum_{j=1}^{N} u_j^{n+1} \frac{3}{2} \int_z 2a\ 2b\ D_z \frac{\partial w_i}{\partial z} \frac{\partial w_j}{\partial z} dz$$

$$- \sum_{j=1}^{N} u_j^{n+\frac{2}{3}} \frac{1}{2} \int_z 2a\ 2b\ D_z \frac{\partial w_i}{\partial z} \frac{\partial w_j}{\partial z} dz$$

$$+ \sum_{j=1}^{N} \frac{u_j^{n+1} - u_j^{n+\frac{2}{3}}}{\Delta t/3} \int_z 2a\ 2b\ w_i(z)dz = 0 \qquad i=1,\ldots,N \qquad (15c)$$

The explicit directions in each of these equations are represented by means of the mass flux at the midpoint between the neighbouring nodes, lagged in time by $\Delta t/3$. Integration in the connected direction yields the final algebraic equations which can be written in matrix form as:

$$(\tfrac{3}{2}\,[R_x] + [R_t])\ \{u\}^{n+\frac{1}{3}} = (\tfrac{1}{2}\,[R_x] + [R_t])\ \{u\}^{n}$$

$$+ \{F_y\}^{n} + \{F_z\}^{n} + \{F_o\} \qquad \text{(16a)}$$

$$(\tfrac{3}{2}\,[R_y] + [R_t])\ \{u\}^{n+\frac{2}{3}} = (\tfrac{1}{2}\,[R_y] + [R_t])\ \{u\}^{n+\frac{1}{3}}$$

$$+ \{F_x\}^{n+\frac{1}{3}} + \{F_z\}^{n+\frac{1}{3}} \qquad \text{(16b)}$$

$$(\tfrac{3}{2}\,[R_z] + [R_t])\ \{u\}^{n+1} = (\tfrac{1}{2}\,[R_z] + [R_t])\ \{u\}^{n+\frac{2}{3}}$$

$$+ \{F_x\}^{n+\frac{2}{3}} + \{F_y\}^{n+\frac{2}{3}} \qquad \text{(16c)}$$

where the column vector $\{u\}$ contains the nodal values in the implicit direction, $[R_x]$, $[R_y]$ and $[R_z]$ are tridiagonal matrices, $[R_t]$ is a diagonal matrix (tridiagonal in the case of a consistent time derivative representation), $\{F_x\}$, $\{F_y\}$, and $\{F_z\}$ contain the mass flux terms for the explicit directions, and $\{F_o\}$ contains the boundary term arising at the entry boundary. Similar matrix equations can be written, by inspection of Equations (16), for the other time stepping schemes listed in the preceding section.

For the case of a uniformly-spaced rectangular grid, typical rows in matrices $[R_x]$, $[R_y]$ and $[R_z]$ corresponding to interior nodes take the form:

$$R_{x_i} = 2bc\left[\left(-\frac{D_x}{a} - v\right)_{i-1} \quad \left(\frac{2D_x}{a}\right)_i \quad \left(-\frac{D_x}{a} + v\right)_{i+1}\right] \qquad \text{(17a)}$$

$$R_{y_i} = 2ac\left[\left(-\frac{D_y}{b}\right)_{i-1} \quad \left(\frac{2D_y}{b}\right)_i \quad \left(-\frac{D_y}{b}\right)_{i+1}\right] \qquad \text{(17b)}$$

$$R_{z_i} = 2ab\left[\left(-\frac{D_z}{c}\right)_{i-1} \quad \left(\frac{2D_z}{c}\right)_i \quad \left(-\frac{D_z}{c}\right)_{i+1}\right] \qquad \text{(17c)}$$

where the nodal index i increases in the connected direction always. A corresponding term on the diagonal of $[R_t]$ is:

$$R_{t_i} = \frac{24abc}{\Delta t} \qquad \text{(18)}$$

Similarly, the corresponding terms in the transverse flux vectors $\{F_x\}$, $\{F_y\}$ and $\{F_z\}$, written in the form of vector products, are:

$$F_{x_i} = 2bc[(\frac{D_x}{a} + v)_{i-1} \quad (-\frac{2D_x}{a})_i \quad (\frac{D_x}{a} - v)_{i+1}] \begin{Bmatrix} u_{i-1} \\ u_i \\ u_{i+1} \end{Bmatrix} \tag{19a}$$

$$F_{y_i} = 2ac[\ (\frac{D_y}{b})_{i-1} \quad (-\frac{2D_y}{b})_i \quad (\frac{D_y}{b})_{i+1}] \begin{Bmatrix} u_{i-1} \\ u_i \\ u_{i+1} \end{Bmatrix} \tag{19b}$$

$$F_{z_i} = 2ab[\ (\frac{D_z}{c})_{i-1} \quad (-\frac{2D_z}{c})_i \quad (\frac{D_z}{c})_{i+1}] \begin{Bmatrix} u_{i-1} \\ u_i \\ u_{i+1} \end{Bmatrix} \tag{19c}$$

where the values of u are taken at the time levels indicated in eqn. (16). Similar coefficients are obtained for a deformed grid.

ACCURACY AND STABILITY

In the PD technique, numerical oscillations due to the advective component in the x-direction can be controlled by satisfying the Courant and Peclet criteria (Frind and Pinder, 1982):

$$C = \frac{v\ \Delta t}{\Delta x} \leq 1 \tag{20}$$

$$P = \frac{v\ \Delta x}{D_x} \leq 2 \tag{21}$$

The effective grid spacing can be described by the parameters:

$$\rho_x = \frac{D_x\ \Delta t}{\Delta x^2} \tag{22a}$$

$$\rho_y = \frac{D_y\ \Delta t}{\Delta y^2} \tag{22b}$$

$$\rho_z = \frac{D_z\ \Delta t}{\Delta z^2} \tag{22c}$$

Direct-solution finite element techniques require the spacing parameters (22) to be approximately equal for best accuracy. For the 2-D PD technique on the other hand, Frind (1982) obtained accurate results even with greatly differing spacing parameters. For the 3-D case, these criteria remain to be investigated.

For stability, Peaceman's Scheme requires $\rho_x \leq 3/2$ (Peaceman, 1977, p. 60). Similarly, the Explicit 2 Scheme requires $\rho_x \leq 9/4$ while the Centered Scheme requires $\rho_x \leq \sqrt{27/8}$. These conditions are automatically satisfied (provided $v > 0$) by the Courant and Peclet criteria (20) and (21). Brian's Scheme is reported to be unconditionally stable (Brian, 1961).

APPLICATION AND COMPARISON

We will use a simple problem to test the performance of the 3-D PD solution under the various time-stepping schemes. The flow system (Figure 3) consists of a rectangular block $0 \leq x \leq 60$, $0 \leq y \leq 20$, $0 \leq z \leq 20$ with unidirectional flow and a Dirichlet boundary condition at the inflow boundary. The physical parameters are:

$v = 0.1$

$D_x = 0.1$

$D_y = D_z = 0.025$

The discretization parameters are:

$\Delta x = 2$ $\rho_x = 0.5$

$\Delta y = 1$ $\rho_y = 0.5$

$\Delta z = 1$ $\rho_z = 0.5$

$\Delta t = 20$ $C = 1$

$P = 2$

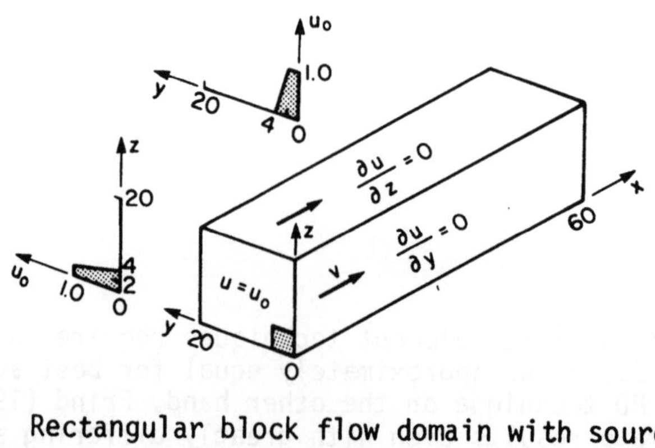

Figure 3. Rectangular block flow domain with source function.

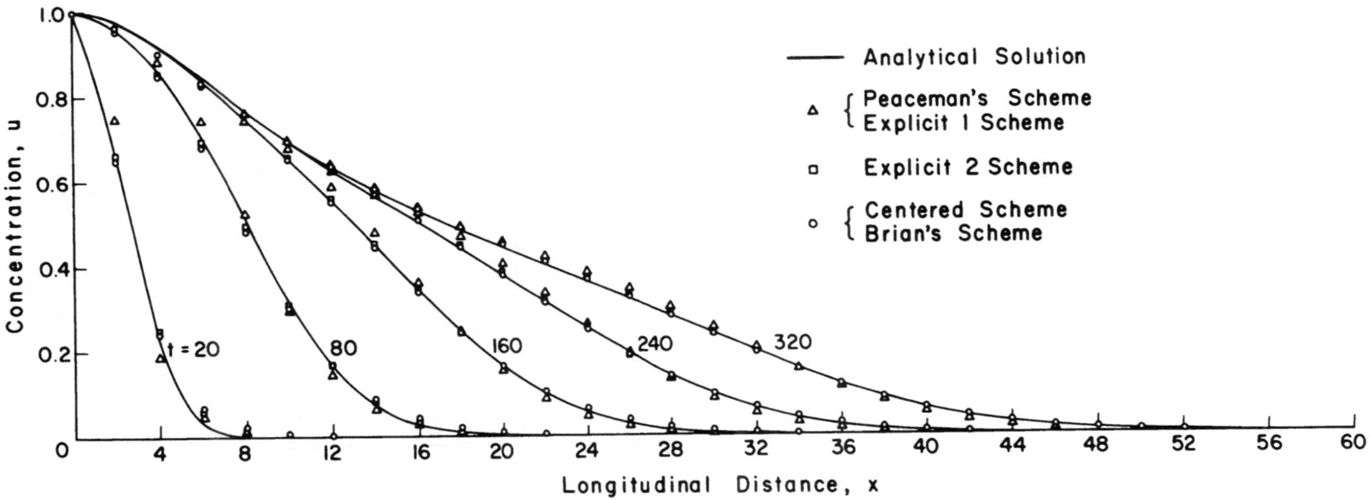

Figure 4. Longitudinal profiles at y = z = 0; for 20, 80, 160, 240, and 320 time units.

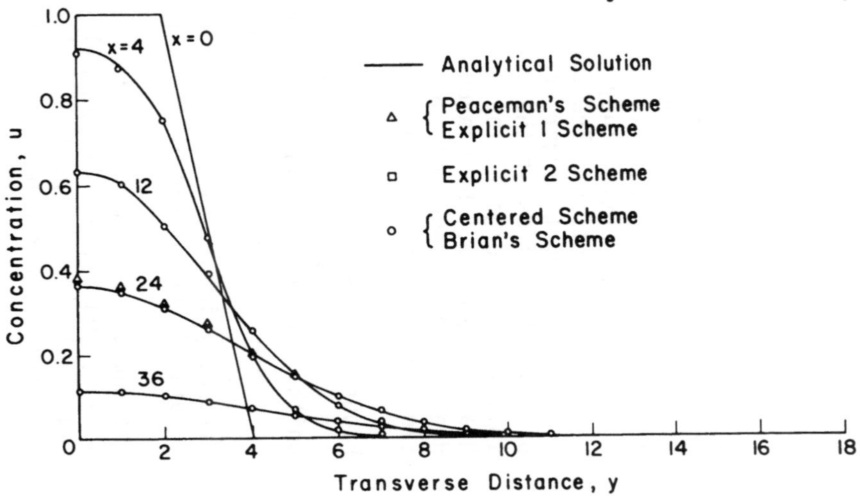

Figure 5. Transverse profiles at z = 0; x = 4, 12, 24 and 36; for 320 time units.

These parameters satisfy the conditions for non-oscillatory behaviour as well as all stability conditions. The grid consists of 31x21x21 = 13,671 nodes and 12,000 block elements. The source function for the numerical model is shown in Figure 3.

The analytical solution for this problem is based on a solution for the 2-D semi-infinite domain developed by Cleary (1978). This solution was extended to three dimensions by Edward Sudicky of the University of Waterloo. The first-type patch source boundary condition for the analytical solution is $u_0 = 1$ for $x = 0$, $y \leq 3$, $z \leq 3$.

The numerical results for Peaceman's Scheme and the Explicit 1 Scheme were found to be identical (to five decimals), as expected because of their similarity. Somewhat less expected, the three-step Centered Scheme and the four-step Brian's Scheme were found to give converging results, with agreement to about two decimals at the end of the first time step t = 20, increasing to five decimals at t = 320.

Figure 4 shows the results of the simulation in terms of profiles along the flow direction at y = z = 0 and at various points in time. Only three distinct sets of profiles are plotted. (Where the numerical solutions coincide, only one symbol is shown.)

The profiles indicate excellent reproduction of the solution for both the Centered Scheme (also Brian's Scheme) and the Explicit 2 Scheme. At the leading toe of the profile, for later times, both Schemes show essentially zero error, while in the middle portion they deviate only slightly from the analytical solution. The Explicit 1 (also Peaceman's) Scheme produces somewhat larger errors in the central portion of the profile as well as at the leading toe. All of the numerical solutions converge to the same steady-state configuration. Near the source, the steady-state solution is slightly in error due to the presence of the abrupt first-type boundary condition; however this error vanishes within a short distance.

Figure 5 shows the transverse profiles in the y-direction at z = 0 and various points on the x-axis, at a time of 320 units. The source configuration for the numerical model is also shown. The agreement between the numerical and analytical solutions is seen to be excellent. (The numerical results are virtually identical except at x = 24.) The maximum error occurs at the center of the plume, x = 0.

The comparison shows that the Explicit 2 Scheme, the Centered Scheme, and Brian's Scheme all yield excellent results, while Peaceman's Scheme and the Explicit 1 Scheme are somewhat less accurate. The reason for the superiority of Explicit 2 over

Explicit 1 would seem to lie in the use of the most recent information at each time step in Explicit 2. The Centered Scheme also uses the most recent information and it has an additional advantage in the exact centering of the spatial derivatives. This latter advantage however seems to be translated into only a minimal improvement in accuracy compared with Explicit 2.

The common factor in both the Centered Scheme and Brian's Scheme that explains their convergence to an identical result is the location of the spatial derivatives at precisely the half time step. The fact that the former is only conditionally stable while the latter is unconditionally stable appears to have no bearing on the result.

CPU times, on an IBM Model 4341 machine under CMS with FORTRAN G, were 11.8 seconds per simulation time step for each of the three-step schemes and 13.3 seconds per time step for Brian's Scheme. These times are very reasonable for a 12,000 element 3-D grid.

COMPUTATIONAL EFFORT

In a conventional 3-D finite element transport model employing a direct Gaussian-type solver, the operations count for solving the matrix equations corresponding to a grid having $n \times m \times \ell$ nodes (where $n \geq m \geq \ell$), is of the order $n(m\ell)^3$ for decomposition and $2n(m\ell)^2$ for forward and backward solution. The decomposition is normally performed only when the time step changes, provided the problem is linear. The three-step PD model which uses a Thomas-type solver has an operations count of approximately 18nm for the three sets of matrix solutions required per full time step on the same grid. The relative computational effort for the matrix solution stage is therefore

$$\frac{PD}{FE} = \frac{18}{m\ell(\frac{m\ell}{k} + 2)} \tag{23}$$

where k is the number of time steps executed in the FE scheme with one matrix decomposition. This comparison neglects the computational effort for the matrix generation stage, which varies linearly with the grid size in both techniques.

Because the cost of the PD solution increases linearly with the grid size while the cost of the FE solution increases at least with the square of the grid cross-section, the cost advantage of PD mounts very rapidly as a grid becomes larger. For example, a grid having 11x11x11 nodes with 16 time steps per matrix decomposition will give a relative operations count of PD/FE = 0.016, while the 31x21x21 grid used in the accuracy comparison will give PD/FE = 0.0014. Because the latter grid is of a size that is relevant to practical problems, a two- to three-order-of-magnitude cost advantage will not be unusual.

Another important factor is the array storage used for execution. A conventional 3-D FE model has storage requirements of the order of $2n(m\ell)^2$, while the PD model has requirements of about $3nm\ell$ with the Explicit 1 Scheme, $2nm\ell$ with the Explicit 2 and the Centered Schemes, and $4nm\ell$ with Brian's Scheme. This means that, given a fixed machine capacity, the PD model can accommodate a much more detailed grid than the FE model. Also, because solution roundoff error is smaller as a result of the reduced size of the coefficient matrices, the PD model can usually operate at a lower level of machine precision. These favourable operating characteristics suggest that problems of practical significance can be handled with a moderately-sized computer.

CONCLUSION

The Principal Direction Technique would appear to be an attractive and much-needed tool for the solution of advective-dispersive transport problems in three dimensions. The technique, like its 2-D counterpart, combines a Galerkin finite element spatial approximation with an alternating direction time-stepping framework, thereby achieving a high degree of overall efficiency as well as the ability to handle non-rectangular domains.

Although the 3-D version lacks the natural symmetry of the 2-D scheme, suitable time step splitting schemes can be found. Either Brian's Scheme, which uses a four-step algorithm to achieve a two-way time step split, or a simpler three-step scheme can deliver results having acceptable accuracy.

Compared with conventional finite elements, the most outstanding characteristic of PD is its substantial cost advantage. It is hoped that this advantage will be sufficient to bring 3-D transport modeling into the realm of practicality.

ACKNOWLEDGEMENTS

The author is grateful to G. F. Pinder for many helpful suggestions during the initial stages of this study, which had its genesis at Princeton University. Thanks also go to E. A. Sudicky for the development of the analytical solution. The work was supported by the Natural Sciences and Engineering Research Council of Canada, Grant A8368.

REFERENCES

Brian, P. L. T. (1961), A Finite-Difference Method of High-Order Accuracy for the Solution of Three-Dimensional Transient Heat Conduction Problems. A.I.Ch.E. Journal, 7(3): 367-370.

Celia, M. A. and G. F. Pinder (1982), Transport Simulation using Three-Dimensional Alternating Direction Collocation. Proceedings, Fourth International Conference on Finite Elements in Water Resources, Hannover, Germany. Springer Verlag, New York, 14: 9-19.

Cleary, R. W. (1978). Analytical Models for Groundwater Pollution and Hydrology. Report 78-WR-15, Department of Civil Engineering, Princeton University, Princeton, N.J.

Daus, A. D. and E. O. Frind (1984). An Alternating Direction Galerkin Technique for Simulation of Contaminant Transport in Complex Groundwater Systems. Water Resources Research, to appear.

Daus, A. D., E. O. Frind and E. A. Sudicky (1984). Comparative Error Analysis in Finite Element Formulations of the Advection-Dispersion Equation. Advances in Water Resources, to appear.

Douglas, J. (1962). Alternating Direction Methods for Three Space Variables. Numerische Mathematik, 4: 41-63.

Douglas, J. and T. Dupont (1971). Numerical Solution of Partial Differential Equations - II. Ed. B. Hubbard, Academic Press, New York, 133-214.

Frind, E. O. and M. J. Verge (1978). Three-Dimensional Modeling of Groundwater Flow Systems. Water Resources Research, 14(5): 844-856.

Frind, E. O. (1982). The Principal Direction Technique: A New Approach to Groundwater Contaminant Transport Modeling. Proceedings, Fourth International Conference on Finite Elements in Water Resources, Hannover, Germany. Springer Verlag, New York, 13: 25-42.

Frind, E. O. and G. F. Pinder (1982). Proceedings, Tenth World Congress of the International Association for Mathematics and Computers in Simulation (IMACS), Concordia University, Montreal.

Peaceman, D. W. (1977). Fundamentals of Numerical Reservoir Simulation, Elsevier, New York.

Peaceman, D. W. and H. H. Rachford (1955). The Numerical Solution of Parabolic and Elliptic Differential Equations. SIAM (Soc. Ind. Appl. Math.) Journal, 3: 28-41.

OCEAN LAKE AND RIVER MODELING

CIRCULATION OVER CONTINENTAL SHELVES AND SHELF-BREAK REGIONS

Francisco E. Werner

School of Oceanography, University of Washington, Seattle, WA 98195

INTRODUCTION

This paper presents a discussion of the wind driven circulation over a continental shelf and shelf-break as predicted by a finite element model. Upwelling over continental shelves, near the coast, is a well documented phenomenon, having received much attention over the past decade (e.g., Allen, 1980). At the same time, a secondary upwelling region, occurring over the continental shelf-break has been discussed both by means of observations (e.g., Petrie, 1983) and theoretically via analytical models (e.g., Johnson and Manja, 1979). The dynamics of this latter type of upwelling is still not well understood.

The shelf break may be defined roughly as a narrow (2-10 km wide) bathymetric feature just seaward of the continental shelf where the seafloor exhibits a rapid increase in slope. It is here where the relatively flat (about a 120 meter drop in 25 km) continental shelf and the steeper (on the order of a 200 meter drop in 1 km) continental slope are connected. Typically the break is located 10-100 km from the coast at a depth ranging between 120 and 200 meters. The sharp change in bathymetry has been invoked in the existing modelling efforts as the cause for the secondary circulation pattern (upwelling and alongshelf jets) observed along the NW African coast and the edge of the European continental shelf.

Two reasons why there has been no clear resolution of the circulation over the break are as follows. Firstly, there is the difficulty in measuring (actually inferring from isopleth displacements) velocity fields occurring at such small scales; and secondly, the analytical and numerical approaches to the problem have required approximations near the break (relaxation of no-slip conditions, coarse meshes, etc.) that have rendered the solutions ambiguous. It should be kept in mind that the vertical velocities under consideration are 100 times smaller than the horizontal velocities which themselves are, in many cases, close to the noise level in the experiment. The finite element method, due to its versatility in providing adequate resolution to problems involving processes at different spatial scales and its flexibility in conforming to irregular boundaries, is a natural approach to solving this problem.

GOVERNING EQUATIONS AND SOLUTION METHOD

The time-dependent, non-linear, unstratified Navier-Stokes equations in a cartesian coordinate system are used. A constant Coriolis parameter, i.e., an f-plane, has been assumed as well. They are:

$$\frac{\partial u}{\partial t} + u\frac{\partial u}{\partial x} + w\frac{\partial u}{\partial z} - fv = -\frac{1}{\rho_o}\frac{\partial p}{\partial x} + A_H\frac{\partial^2 u}{\partial x^2} + A_V\frac{\partial^2 u}{\partial z^2} \qquad (1)$$

$$\frac{\partial v}{\partial t} + u\frac{\partial v}{\partial x} + w\frac{\partial v}{\partial z} + fu = A_H\frac{\partial^2 v}{\partial x^2} + A_V\frac{\partial^2 v}{\partial z^2} \qquad (2)$$

$$\frac{\partial w}{\partial t} + u\frac{\partial w}{\partial x} + w\frac{\partial w}{\partial z} = -\frac{1}{\rho_o}\frac{\partial p}{\partial z} + A_H\frac{\partial^2 w}{\partial x^2} + A_V\frac{\partial^2 w}{\partial z^2} \qquad (3)$$

together with the continuity equation,

$$\frac{\partial u}{\partial x} + \frac{\partial w}{\partial z} = 0 \quad .$$

The velocity components (u,v,w) are in the (x,y,z) directions respectively, where x is positive onshore (eastward), y is positive alongshore (northward) and z positive upward. The shelf considered is in the northern hemisphere, on the western side of a continent at a latitude near 45°N. There are no alongshore variations considered in our solution and hence the y-dependence of the equations has been removed. A_H and A_V are the horizontal and vertical eddy diffusivities respectively and are constant with values $A_H = 10^6$ cm^2/sec and $A_V = 50$ cm^2/sec for the results shown herein. The hydrostatic component of pressure has been removed from the equations leaving only the perturbation pressure. There are no buoyancy terms as the fluid is homogeneous and consequently the perturbation density identically zero. The background density is 1.0 gm/cc.

The boundary conditions are: (i) no-slip conditions (u,v,w)=0 on solid walls; (ii) no horizontal gradients in velocity at the open (seaward) boundary, i.e., $(\partial u/\partial x, \partial v/\partial x, \partial w/\partial x) = 0$; and (iii) a rigid lid at the surface whereby $w = 0$, $\partial u/\partial z = \tau_x/(\rho_o A_V)$ and $\partial v/\partial z = \tau_y/(\rho_o A_V)$. At t=0 all velocities are zero as well as the pressure. Throughout the computations, the pressure is held fixed at an arbitrary node since the pressure cannot be determined aside from an arbitrary constant.

The equations are discretized using the Galerkin variant of the method of weighted residuals. The dependent variables, velocity and pressure, are approximated using mixed interpolation:

$$u \sim Q_i u_i, \quad v \sim Q_i v_i, \quad w \sim Q_i w_i, \quad p \sim L_i p_i$$

where Q_i are 9-node biquadratic (Lagrange) basis functions and L_i are 4-node bilinear (C°) basis functions. An implicit method is adopted to discretize the time dependence. The terms in the equations are weighted between $t+\Delta t$ and $t+\Delta t/2$ as follows:

$$w' \frac{\partial u_i^{j+1}}{\partial t} + (1-w') \frac{\partial u_i^j}{\partial t} = \frac{u_i^{j+1} - u_i^j}{\Delta t}$$

where $0.5 < w' < 1$ ($w' = 0.55$ in these simulations). The resulting system of equations is solved using a standard Newton-Raphson technique, iterating only once per time step. The algorithm was throughly tested with the classical spin-up problem (Werner and Walters, 1984).

RESULTS

Flat shelf

As a first approximation a flat shelf (horizontal scale of 50 km and vertical scale of 200 meters) was considered, yielding results shown in Figures 1(a,b) and 2. Dimensional values of the parameters are as follows. The wind stress was increased 0.10 dynes/cm^2 per 3 hour time step until it reached a value of 1.0 dynes/cm^2 (a wind speed of about 7 m/sec) after which it was held constant for the rest of the computation. The wind is entirely southward and parallel to the coast. A southward wind stress accelerates the water column southward, but more importantly, it will cause an offshore flow in a surface Ekman layer. The offshore mass flux is replaced initially by a subsurface return flow which is uniformly distributed with depth (Figure 1a). The offshore flow causes a drop in sea level at the coast creating an offshore gradient in sea surface height (pressure) which in turn is geostrophically balanced by the Coriolis term and consequently accelerates the v-velocity. The geostrophic (interior) v-velocity, which is independent of depth, induces the formation of a bottom Ekman layer shifting the form of the mass balance from a uniformly distributed onshore flow to an onshore flow contained in the bottom Ekman layer (Figures 1b and 2). The thickness of these layers is on the order of 20 to 50 meters. The sidewall layer in which the upwelling takes place is thicker, on the order of 5 to 15 km. The results agree well with analytic solutions (see pages 189 ff. in Pedlosky, 1979). The mesh (not shown) had 961 nodes and 225 elements. In anticipation of this circulation pattern, the mesh was refined in all these regions. The seaward boundary was refined to avoid oscillations found in problems with open boundaries (Gresho and Lee, 1980).

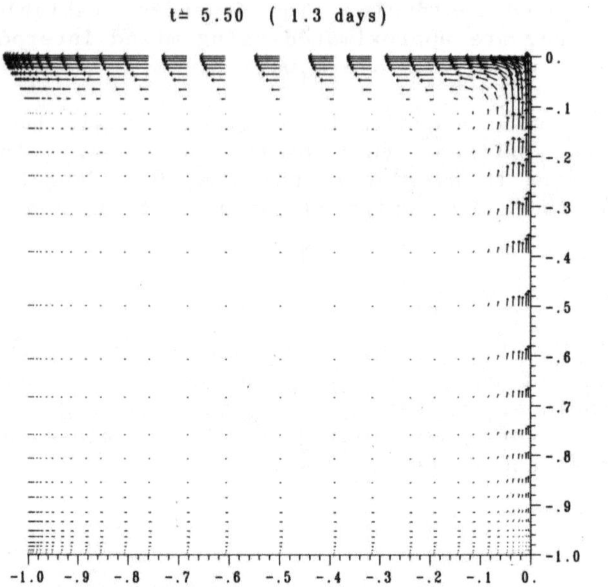

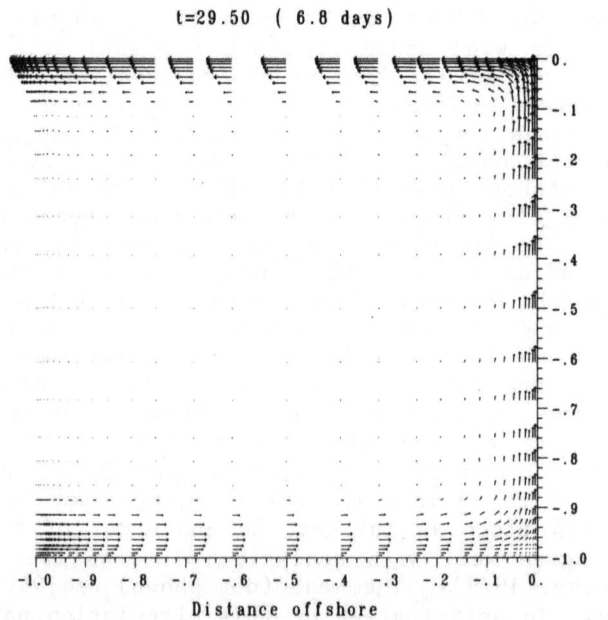

Figure 1a (top): cross-shelf circulation for flat shelf at an early stage; 1b (bottom) after the development of the bottom Ekman layer. Vertical scale of figures is 200 m; horizontal scale is 50 km.

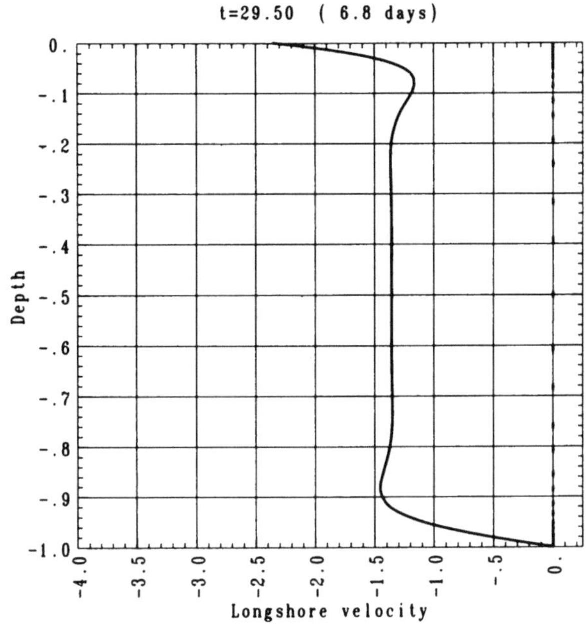

Figure 2. Alongshore (v) velocity profile near x = -0.3 for the same time as Figure 1b. Nodal locations are on v=0 axis. The velocity scale is 20 cm/sec.

Realistic topography

A closer representation to a continental shelf is found in Figures 3(a,b) which show the mesh used in the calculations that follow (1407 nodes and 330 elements). The coast is modelled with a vertical wall at a depth of 50 meters. The continental slope is represented by a vertical wall at a distance of 45 km offshore. In reality, for the topography shown (representative of the shelf off the state of Washington) the continental slope drops approximately 180 meters over a horizontal distance of nearly 1 km. However, when this slope was included in the mesh, the cross-shelf circulation exhibited oscillations in areas where the vertical sides of the elements were tilted from the vertical (results not shown). Distorted elements in the solution of the spin-up of a rotating cylinder showed no particular problems. The difficulty might lie in the difference of the aspect ratios. The terms in Equation (3) are smaller than those in Equations (1) and (2) by a factor of the order of the square of the aspect ratio, making the integration of the equations and subsequent matrix solution more "difficult" for the computer to handle accurately. Also, the element distortion leads to very obtuse and/or acute angles; a characteristic known to cause noise in some finite element schemes. This result is currently under investigation as it might prove to be a serious drawback to the finite element

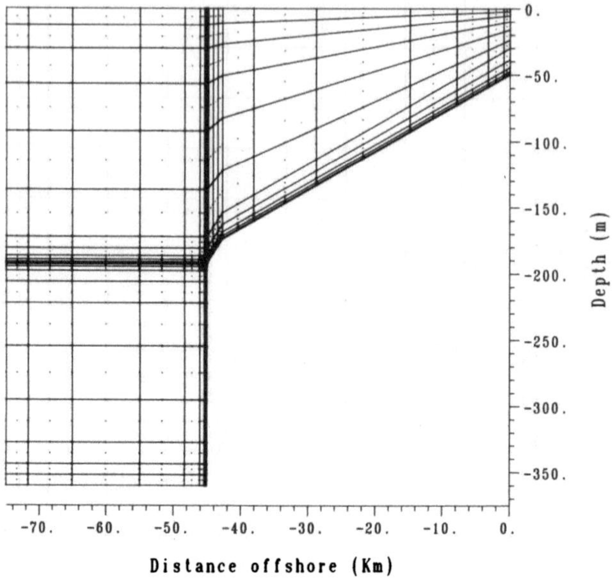

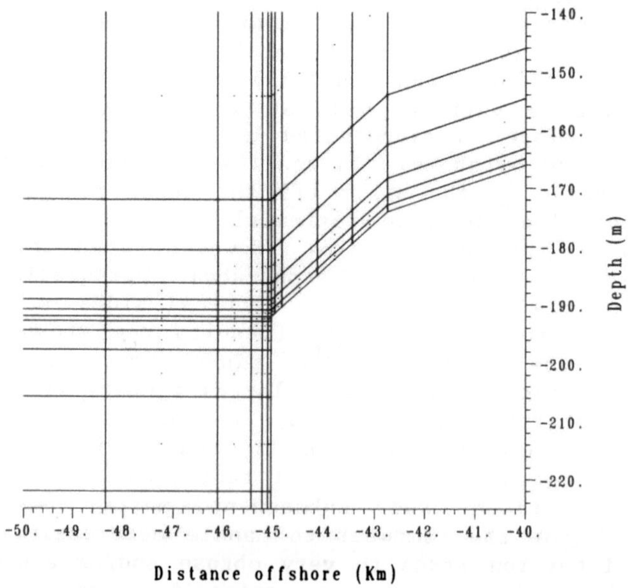

Figure 3a (top): mesh representing realistic topography; 3b (bottom) enlarged view of shelf-break region.

modelling of certain geophysical flows if it cannot be overcome. Presently those results are inconclusive.

Figure 4a shows the resulting cross-shelf flow after 6 days. The surface offshore flow, as in the flat bottom case, is balanced by a return flow in the bottom layer after the alongshore flow developed. The additional feature is the upwelling and recirculation zone just seaward of the shelf break (Figures 4a and b; also see Lill, 1979). The same pattern, although contaminated by noise, also appeared when the real continental slope was included. The magnitude of the vertical velocity is increased by a factor of 50 for plotting purposes. A one to one scaling shows very little vertical structure. Had the real distortion in the plots of 200 to 1 been used, the flow in the bottom Ekman layer on the shelf would be parallel to the bottom as expected, but the shelf break recirculation feature would overwhelm the picture. The maximum w-velocity near the break is about 0.1 cm/sec (a value of 0.2 cm/sec was inferred off the shelf of Nova Scotia by Petrie, 1983). The maximum magnitudes of the surface u-velocity were on the order of 8-10 cm/sec.

DISCUSSION

We have shown that the form of the finite element method used herein yields acceptable results for the barotropic wind driven shelf circulation problem. Upwelling was observed near the break as suggested by previous field and mathematical efforts. In addition, the recirculation feature indicates a complicated structure in the secondary upwelling region. It has also become apparent that the structure of the horizontal Ekman layer over the shelf and the structure of the vertical boundary layer over the slope might be different, namely, 20 to 50 meters thick over the shelf vs order 1 km on the slope.

An instructive case to consider in attempting to understand the nature of the currents at the shelf-break is given by Stewartson (1957). In it, the adjustment processes between two areas of a rotating fluid possessing different flow characteristics are explained. Specifically, he considered two parallel plates rotating uniformly about a common axis. Each plate possesses an inner disk that can spin independently with the constraint that the differential rotation rates are kept small. Further, the inner disks (both of equal radii) have an equal rotational speed that is slightly in excess of that of the larger plates. Stewartson finds that vertical free shear layers, connecting the top and bottom disks, are formed at the point of discontinuity in the angular velocity. They are much thicker than the $O(E^{1/2})$ Ekman layers on the horizontal surfaces of the disks and, are themselves composed of two layers of differing thicknesses. (The Ekman number (E) is a small number defined as $\nu/(\Omega d^2)$ where ν is the kinematic viscosity, Ω the angular velocity and d is the appropriate length scale.)

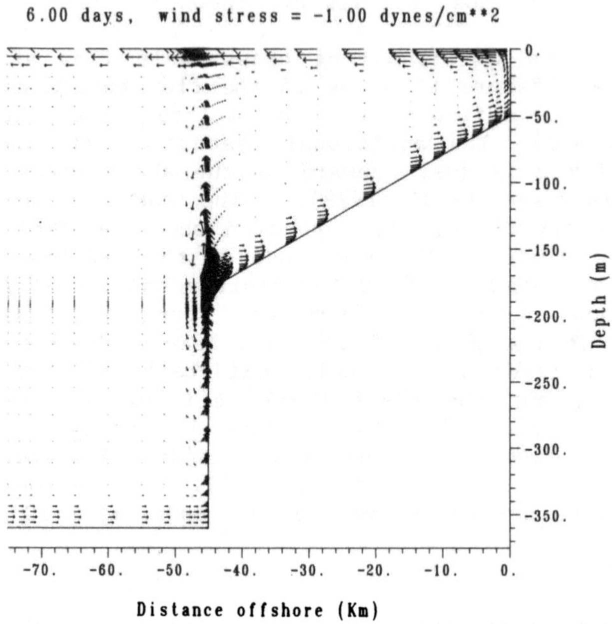

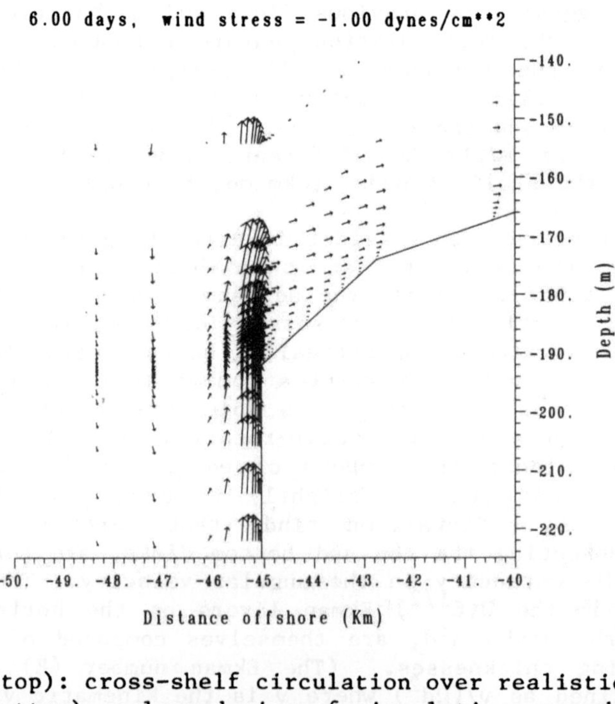

Figure 4a (top): cross-shelf circulation over realistic topography; 4b (bottom) enlarged view of circulation near shelf-break.

One layer, of $O(E^{1/4})$, adjusts the discontinuity in the azimuthal velocity while the other, of $O(E^{1/3})$, satisfies mass transport requirements.

An analogy between Stewartson's findings and the results of the shelf circulation model can be drawn by considering the alongshore and the cross-shelf velocities, counterparts to the azimuthal and meridional velocities. Figure 5 shows the profile of the surface alongshore velocity as a function of offshore distance. The velocity fields on the shelf and in the deeper ocean are clearly different. Over the shelf, away from the coastal layer, the v-velocity decreases in magnitude offshore in response to the shelf's increase in depth. Over the deeper ocean, the topographic effect is lessened, yielding velocities which are smaller in magnitude than the ones on the shelf, and with very little shear in x. In the oceanic case, rather than the forcing at the top or bottom causing the discontinuity in alongshore velocity (or the azimuthal velocity in Stewartson's case), it is a topographic effect whereby shallower waters respond differently than deeper waters when subjected to the same conditions. The subsurface onshore mass flux in the Ekman layers on the shelf and the deeper ocean, although of the same order, is also different. On the shelf, due to higher v-velocities, the bottom transports are higher than in the deeper ocean (Figures 4a and 5). Moreover, the onshore mass transport of the deeper ocean occurs in a layer on the continental slope which is much thicker than the Ekman layer on the horizontal shelf. At the shelf-break, just as in Stewartson's boundary between the inner and outer disks, the flow fields on the shelf and the deeper ocean need to adjust. The adjustment takes place seaward of the shelf-break between x = -45 km and x = -50 km. Upwelling and a recirculation feature, as well as a higher shear zone in the alongshore velocity, are exhibited as part of the adjustment process (Figures 4 and 5). To some extent, the recirculation zone may be due to a non-divergent surface offshore flow over the break set up in response to a curl-free wind stress. The water upwelled at the break reaching the bottom of the surface layer, cannot penetrate the upper layers (or flow horizontally), turning downward and recirculating. The structural details of the shear layer, upwelling and recirculation are not yet clear and are presently being investigated.

ACKNOWLEDGMENTS

This work was supported by the Department of Energy under contract DE-AT06-76EV71025=15. The calculations were performed using the facilities of the National Magnetic Fusion Energy Computer Center at the Lawrence Livermore Laboratories.

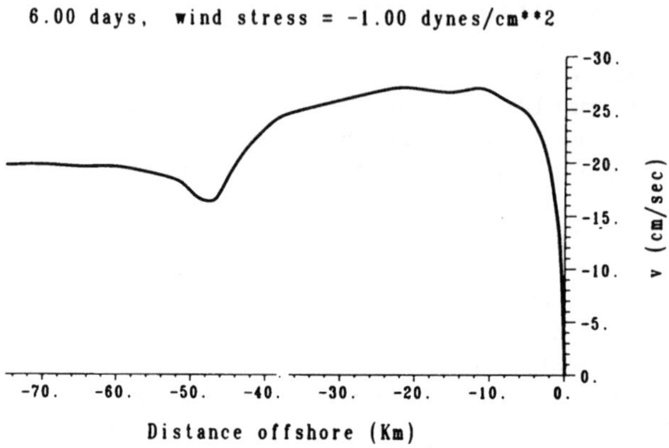

Figure 5. Surface alongshore velocity as a function of x. The shelf break is at x = -45 km.

REFERENCES

Allen, J.S. (1980) Models of wind-driven currents on the continental shelf. Ann. Rev. Fluid Mech., 12: 389-433.

Gresho, P. and R.L. Lee (1980) Don't supress the wiggles-they're telling you something! Computers and Fluids, 9: 223-253.

Johnson, J.A. and B.A. Manja (1979) Shear layers above a break in bottom topography. Geophys. Astrophys. Fluid Dynamics., 14: 45-60.

Lill, C.C. (1979) Upwelling over the shelf break. Journ. Phys. Oceanogr., 9: 1044-1047.

Pedlosky, J. (1979) Geophysical Fluid Dynamics, Springer Verlag, 624 pp.

Petrie, B.D. (1983) Current response at the shelf break to transient wind forcing. Journ. Geophys. Research, 88: 9657-9578.

Stewartson, K. (1957): On almost rigid rotations. Journ. Fluid Mech., 3: 17-26.

Werner, F.E. and R.A. Walters (1984) Finite element solutions of the spin-up of a rotating fluid. Proceed. Fifth Int'l. Symp. on Finite Elem. and Flow Probs., 155-159.

FINITE ELEMENT ANALYSIS OF COMBINED DIFFRACTION - REFRACTION

A. Nachbin and L.C. Wrobel
Civil Engineering Department
COPPE/Federal University of Rio de Janeiro
Rio de Janeiro - Brazil

INTRODUCTION

The problem of combined diffraction - refraction of surface waves propagating in a fluid of constant depth which encounter an area of varying bathymetry and obstacles has been studied by many different numerical methods. The problem is essentially three-dimensional and, as such, can be analysed by the boundary element method (Garrison (1978)), the finite element method (Zienkiewicz et al. (1978)) or the hybrid element method (Yue et al. (1978)). All these three-dimensional models, however, require huge storage and computer time mainly when the incident wave length becomes short.

Berkhoff (1972) developed a general approximated theory in which the variation of the velocity potential along the depth is uncoupled and the problem is then reduced to two dimensions only. The resulting equation, which is usually referred to as the mild-slope equation, is valid for both short and long waves. It reduces to the Helmholtz equation in deep or constant depth waters and to the linear long wave equation in shallow waters.

Solution of this equation is usually carried out by dividing the inner region of analysis, which contains all the relevant features such as obstacles, varying depth, etc., into finite elements and approximating the influence of the far region by one of the following methods: boundary elements, infinite elements, radiation dampers or analytical series expansions.

All the above methods have certain drawbacks, what makes the choice difficult. For instance, radiation dampers generally require the truncation of the finite element mesh at a large distance, although accurate second-order dampers have recently been derived by Bando et al. (1982). Boundary elements and analytical series expansions have widely been used but they destroy the bandedness of the resulting system of equations. Infinite elements generally add a certain number of degrees of freedom to the system and depend on a decay parameter which is rather arbitrary.

In this paper, we employ an efficient parametric infinite element geometrically defined by six reference points but having three nodal points only (Nachbin and Wrobel (1984)). A simple formula for computing the decay parameter is derived from wave energy considerations. Finally, some numerical results are presented and compared to published two - and three - dimensional solutions.

Mathematical formulation

The propagation of periodic, small amplitude surface gravity waves over a slowly-varying depth seabed can be described by the following equation (Berkhoff (1972)):

$$\nabla \cdot (p \nabla \phi) + q \phi = 0 \tag{1}$$

with

$$\nabla = \left(\frac{\partial}{\partial x} , \frac{\partial}{\partial y} \right)$$

$$p = c\, cg = \frac{1}{2} \left(\frac{\omega}{k} \right)^2 \left(1 + \frac{2kh}{\sin(2kh)} \right)$$

$$q = \frac{\omega^2 cg}{c} = \frac{\omega^2}{2} \left(1 + \frac{2kh}{\sin(2kh)} \right)$$

$$\omega^2 = kg \tanh(kh)$$

In the above, $\phi(x,y)$ is the reduced velocity potential, ω the wave angular frequency, $h(x,y)$ the water depth, c the phase velocity, cg the group velocity, k the wave number and g is the gravitational constant.

We assume that at solid boundaries the normal velocity is zero, implying in total reflection of the wave. Furthermore, at infinity, we have the Sommerfeld radiation condition

$$\lim_{r \to \infty} r^{1/2} \left(\frac{\partial \phi}{\partial r} - \frac{i\omega}{c} \phi \right) = 0 \tag{2}$$

which applies only to the diffracted waves and guarantees that they are outgoing.

Dividing the problem domain into an inner region Ω_E of constant depth, where the reduced potential can be divided into an incident (ϕ^I) and a diffracted (ϕ^D) parts, a functional for the problem can be written in the form

$$\Pi = \iint_{\Omega_I} \frac{1}{2} \left[c\, cg\, (\nabla\phi)^2 - \frac{\omega^2 cg}{c} \phi^2 \right] dxdy \; +$$

$$\iint_{\Omega_E} \frac{1}{2} \left[c\, cg\, (\nabla\phi^D)^2 - \frac{\omega^2 cg}{c} (\phi^D)^2 \right] dxdy \; +$$

$$\int_{\Gamma} c\, cg \left[\frac{\partial\phi^I}{\partial x} \phi^D\, dy - \frac{\partial\phi^I}{\partial y} \phi^D\, dx \right] \tag{3}$$

Note that the line integral along the interface Γ provides for the compatibility of fluxes between the regions Ω_I and Ω_E. There is no integral along the boundary at infinity since, in addition to condition (2), we have that both ϕ^D and $\partial\phi^D/\partial n$ tend to zero there.

The inner region Ω_I is discretized with standard quadratic isoparametric finite elements and, to account for the water depth variation, a different wave number is computed at each integration point.

The outer region Ω_E is modelled by using three-noded periodic infinite elements (Nachbin and Wrobel (1984)). These elements are geometrically defined by six reference points in two layers, as if they were standard six-noded Serendipity finite elements. There is no need to place reference points at a large distance. The element shape functions are given as products of the form

$$M_j(s,\eta) = N_j\,(\eta)\;\; e^{-s/L}\;\; e^{iks} \qquad j = 1,2,3 \tag{4}$$

which satisfy the Sommerfeld radiation condition. In expression (4), N_j are quadratic Serendipity polinomials and s is a coordinate in the same direction as the natural coordinate ξ, and related to it by

$$s = T(\xi + 1) \tag{5}$$

where $T = T(\eta)$ is the Jacobian $ds/d\xi$.

The decay length L in expression (4) is calculated based on the same energy criterium suggested by Buragohain and Agrawal (1982). However, while they ended up with an equation dependent on a parameter which was fixed by trial and error, we derived for the actual case the relation

$$L = \frac{2\,(R_2 - R_1)}{\ell n \left(1 + \dfrac{R_2 - R_1}{R_1}\right)} \tag{6}$$

in which R_1 and R_2 are the radius of the two layers of reference points.

NUMERICAL RESULTS

Diffraction and refraction of waves by a parabolic shoal, surmounted by a circular island

This problem has been analysed by several different researchers, but the published results are not all related to the same data. The data employed here is the same as in Zienkiewicz et al. (1978), and can be seen in figure 1 where the finite element mesh employed is also displayed.

A very good agreement was obtained for all frequencies analysed as can be seen in figures 2 and 3, where the numerical results are compared to the analytical solution given by Homma (1950). It is noted that the wave number was calculated through the shallow water formula $k=\omega/(gh)^{1/2}$ rather than the dispersion relation. This is because Homma's solution is specific for long waves, as pointed out by Zienkiewicz and Bettess (1982).

Elliptic island with circular base

In the second example, we study the elliptic island shown in figure 4. Since the slope of the shoal is quite large, the slow

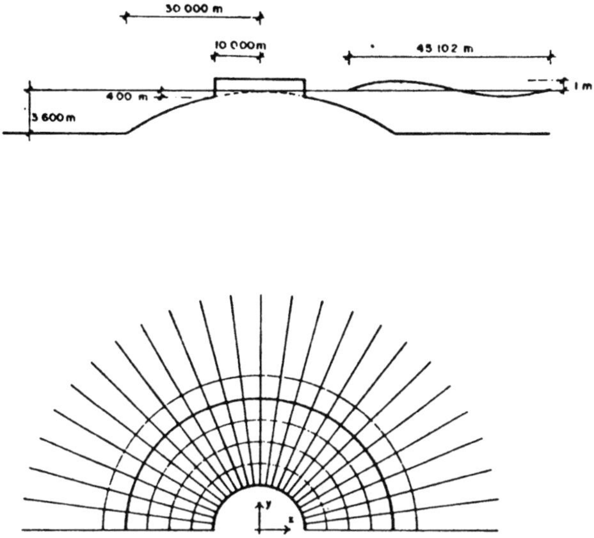

Figure 1 - Geometry and Finite Element Mesh for
Parabolic Shoal.

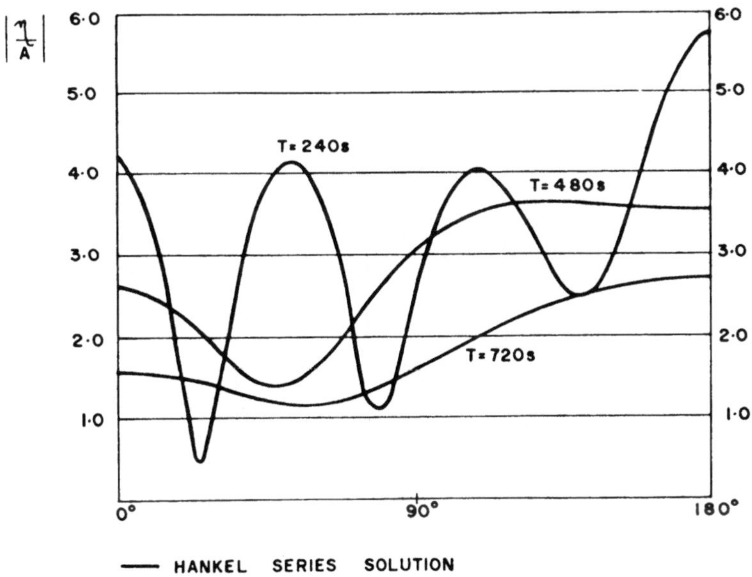

Figure 2 - Analytical Solution for Parabolic
Shoal.

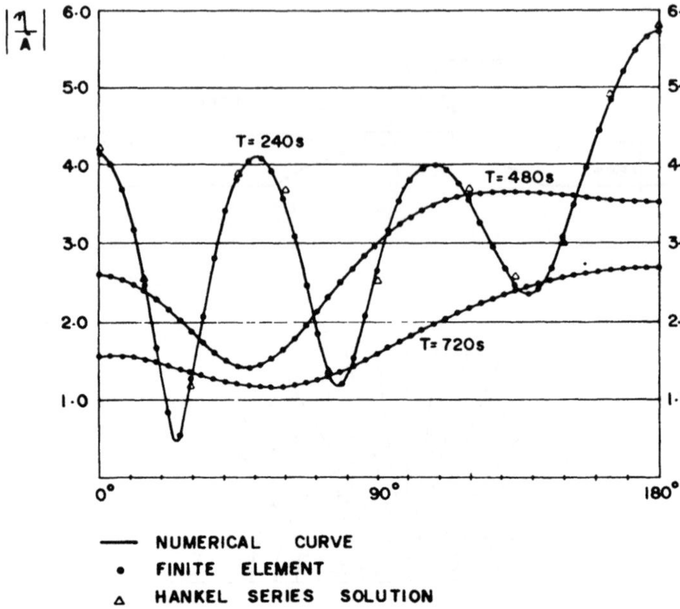

Figure 3 - Comparison between Numerical and Analytical Solutions.

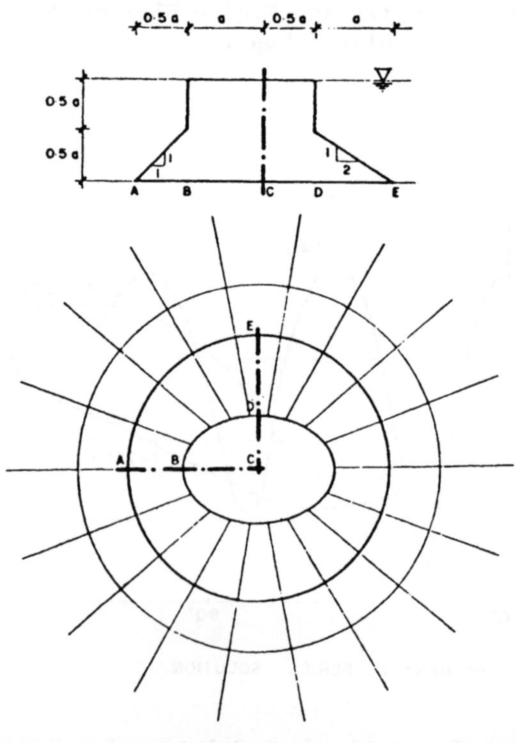

Figure 4 - Geometry and Finite Element Mesh for Elliptic Island.

bathymetry variations assumed for deriving the mild-slope equation are violated. The problem is essentially three-dimensional and, as such, was studied by Yue et al. (1978). However, Houston (1981) and Tsay and Liu (1983) obtained accurate solutions using the mild-slope equation and the hybrid element method.

Two different meshes were used to test the convergence of the present solution. In the first only one layer of finite elements was employed in the varying depth region, with a total of 144 complex equations. The second mesh consists of three layers of elements in the same region, now taking advantage of the symmetry of the problem, with 135 complex equations. It can be seen in figure 5 that the results are virtually the same for both discretizations. The meshes adopted are very coarse in comparison with Houston's (1200 elements) and Tsay and Liu's (720 elements for half domain).

Two incident wave directions were considered. The results for $\alpha = 90^o$ were more accurate than for $\alpha = 180^o$. From figure 4, we can notice that for $\alpha = 90^o$ the wave encounters a 1 : 2 slope while for $\alpha = 180^o$ the slope turns out to be 1 : 1; thus, in the second case, the mild-slope theory is more strongly violated, and the results are consistent.

A parameter λ, which ranges from 0 to 1, is used to determine the cross-section of the island under the elliptic cylinder, in the following equation

$$\frac{x^2}{(\lambda(A - R) + R)^2} + \frac{y^2}{(\lambda(B - R) + R)^2} = 1$$

in which A and B are the major and minor axes of the ellipse, respectivelly, and R is the radius of the circular base.

CONCLUSIONS

In this paper, a solution of the mild-slope equation using finite and infinite elements has been presented. The infinite element employed does not have some of the problems of previous periodic infinite elements: since it is three-noded, it does not increase the final system of equations; its decay parameter is easily estimated in terms of geometrical data only; and its geometry is defined as a standard finite element, what facilitates the mesh generation.

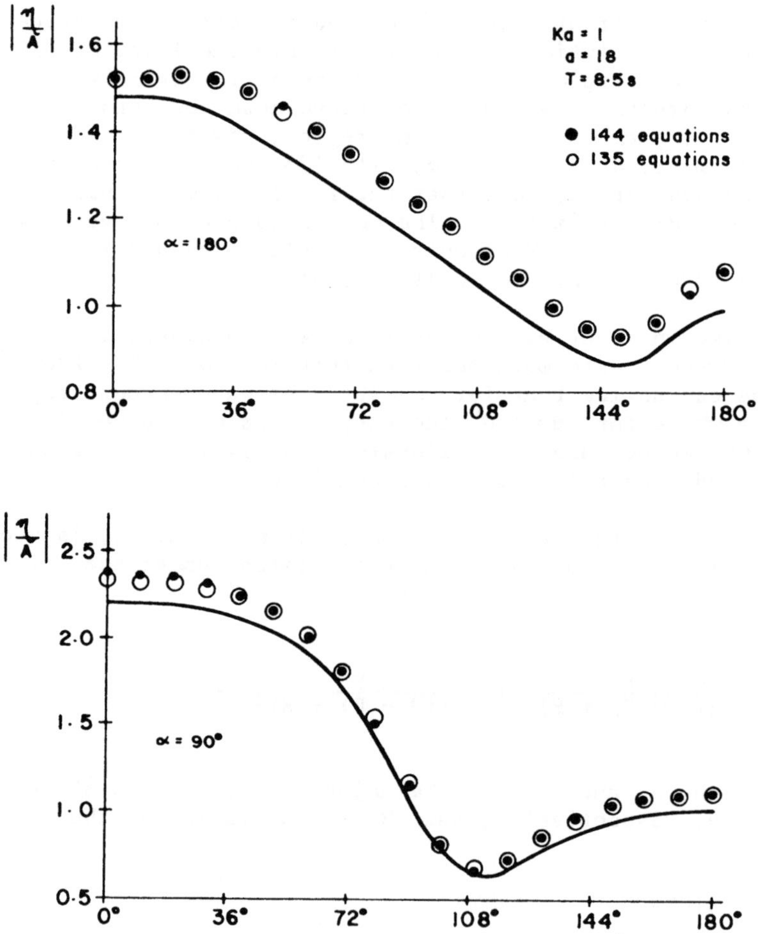

Figure 5 - Numerical Results for the Wave Elevation
a) $\alpha = 180^o$; b) $\alpha = 90^o$

REFERENCES

Bando, K. , Bettess, P. and Emson, C. (1982), The Effectiveness of Dampers for the Analysis of Exterior Scalar Waves Diffraction by Cylinders and Ellipsoids, University College of Swansea Report C/R/430/82, Inst. Num. Meth. Engng.

Berkhoff, J.C.W. (1972) Computation of Combined Refraction - Diffraction, 13th Int. Conf. Coastal Engineering, Vancouver.

Buragohain , D.N. and Agrawal, B.L. (1981) Hydrodynamic Forces on Large Offshore Structures under Ground Excitation, in Numerical Methods for Coupled Problems, E. Hinton et al. (eds.), Pineridge Press, Swansea.

Garrison, C.J. (1978) Hydrodynamic Loading of Large Offshore Structures. Three - Dimensional Source Distribution Methods, in Offshore Engineering, O.C. Zienkiewicz et al. (eds.), Wiley, Chichester.

Homma, S. (1950) On the Behaviour of Seismic Sea Waves Round Circular Island, Geophys. Mag., XXI, 199-208.

Houston, J.R. (1981) Combined Refraction and Diffraction of Short Waves Using the Finite Element Method, Appl. Ocean Research, 3, 163-170.

Nachbin, A. and Wrobel, L.C. (1984) An Efficient Infinite Element for Fluid - Structure Interaction, in Numerical Methods for Transient and Coupled Problems, R.W. Lewis et al. (eds.) , Pineridge Press, Swansea.

Tsay, T.-K. and Liu, P.L.-F. (1983) A Finite Element Model for Wave Diffraction and Refraction, Appl. Ocean Research, 5, 30-37.

Yue, D.K.P., Chen, H.S. and Mei, C.C. (1978) A Hybrid Element Method for Diffraction of Water Waves by Three-Dimensional Bodies, Int. J. Num. Meth. Engng, 12, 245-266.

Zienkiewicz, O.C., Bettess, P. and Kelly, D.W. (1978) The Finite Element Method for Determining Fluid Loadings on Rigid Structures. Two- and Three- Dimensional Formulations, in Numerical Methods in Offshore Engineering, O.C. Zienkiewicz et al. (eds.), Wiley, Chichester.

Zienkiewicz, O.C. and Bettess, P. (1982) Letter to the Editor, Appl. Ocean Research, 4, 124.

LINEAR SHORT WAVE MODELING FOR HARBOURS

W. Rottmann-Söde and W. Zielke

Institute of Fluid Mechanics
University of Hannover
Fed.Republic of Germany

INTRODUCTION

With respect to harbours and coastal areas of irregular shape and variable water depth, it is essential to be able to evaluate the combined effects of diffraction, refraction and reflection. This paper contains the latest results of a current research project dealing with linear short wave calculations considering totally and partially reflecting as well as nonreflecting boundaries, i.e. coastlines, breakwaters, islands and open boundaries. Wedges (singularities) are excluded for the time being.
The developed model is partly based on prior studies by BERKHOFF (1976), and CHEN and MEI (1974). It includes, however, the development of more realistic boundary conditions, such as partial reflection, as well as a more accurate and economical description of the radiation condition at the open boundary. The formulation which has been used previously by the authors (1983) will not be elaborated upon here, because further investigations proved that the combining of the analytical solution in the infinite domain with the source distribution along the open boundary is unable to give proper description of the radiating waves. Therefore, a method will be presented which describes the open boundary at infinity and permits the derivation, with the aid of functional analysis, of a formulation for a finite boundary.

MATHEMATICAL FORMULATION OF THE BOUNDARY-VALUE-PROBLEM AND LINEAR TIME HARMONIC WAVES

Assuming a moderate bottom-slope, BERKHOFF (1976) derived the diffraction-refraction-equation for the complex wave potential

$$\nabla(cc_g\ \nabla\Phi(x,y)) + \omega^2\frac{c_g}{c}\ \Phi(x,y) = 0 \qquad (1)$$

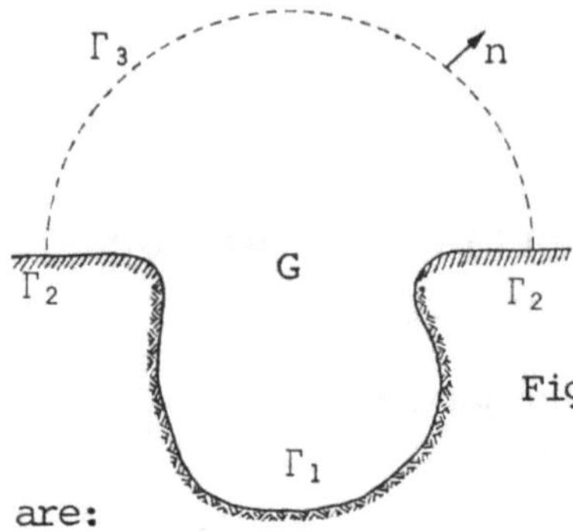

Figure 1.: $\Gamma = \Gamma_1+\Gamma_2+\Gamma_3$

The notations are:

$\Phi(x,y)$ = two-dimensional complex wave potential function

$c(x,y)$ = phase velocity ($= \omega/k$)

$c_g(x,y)$ = group velocity ($= 0.5 + kh/\sinh(2kh)$)

$k(x,y)$ = wave number defined by the relation $\omega^2 = gk\tanh(kh)$

g = acceleration due to gravity

$h(x,y)$ = water depth

x,y = horizontal coordinates of space

∇ = $\left(\frac{\partial}{\partial x},\frac{\partial}{\partial y}\right)$ gradient

This basic equation is of elliptic type and, therefore, requires strict boundary conditions for the entire boundary Γ of the solution domain G. From the engineering point of view the most important conditions are:

a) full reflection at a fixed boundary Γ_1
b) partial reflection at a fixed boundary Γ_2
c) a radiation condition at an open boundary Γ_3 for a given incident wave field.

These boundary conditions are described mathematically by the following expressions:

a) $$\frac{\partial\Phi}{\partial n} = 0 \qquad \text{at } \Gamma_1 \qquad (2)$$

b) $$\frac{\partial\Phi}{\partial n} + a\Phi = 0 \qquad \text{at } \Gamma_2 \qquad (3)$$

where $a = a_1 + ia_2$ is a complex constant, which is related to the commonly used reflection coefficient $\rho = Re^{i\gamma}$ in which R is the amplitude reduction factor, and γ is the phase shift. The relation between the coefficients a and ρ can be written as
(α = direction of the wave)

$$a_1 = k\cos\alpha \frac{2R\sin\gamma}{1 + R^2 + 2R\cos\gamma}, \text{ and}$$

$$a_2 = k\cos\alpha \frac{R^2 - 1}{1 + R^2 + 2R\cos\gamma} .$$

In the special case that $\gamma = 0$, i.e. $\rho = R$, the equations are reduced to

$$a_1 = 0 \quad \text{and} \quad a_2 = k\cos\alpha \frac{R - 1}{R + 1} \qquad \text{(see figure 2).}$$

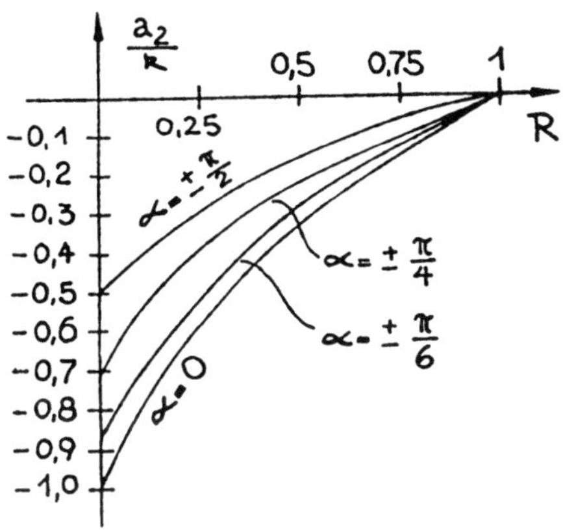

Figure 2: Relation between the coefficient a and the reflection coefficient R, assuming zero phase shift γ.

c) It will be assumed, that the incident waves arrive from infinity and the depth is constant outside the solution domain G. The velocity potential of the incident wave and, in case of a coastline, the potential of its reflected wave can be expressed as

$$\tilde{\Phi} = -\frac{ig\tilde{H}}{2\omega} e^{ik(x\cos\beta + y\sin\beta)}$$

$$\tilde{\Phi}_R = -\frac{ig\tilde{H}}{2\omega} R\, e^{ik(-x\cos\beta + y\sin\beta)} \tag{4}$$

The radiation condition, which implies that the scattered wave Φ_s must behave as an outgoing wave at infinity, can be described mathematically as follows (WYHL 1976 and SOMMERFELD 1949)

$$\lim_{r\to\infty} \sqrt{r}\left(\frac{\partial\Phi_s}{\partial r} - ik\Phi_s\right) = 0 \tag{5}$$

Supposing that there is a coastline, the wave potential along the open boundary is

$$\Phi = \Phi_s + \tilde{\Phi} + \tilde{\Phi}_R$$

with $\Phi_s = \sum_{n=0}^{\infty} b_n H_n(kr) \cos(n\Theta)$

b_n = unknown constant

H_n = Hankelfunction of the first order

Θ = arctan(x/y).

The basic equation (1) together with the boundary conditions along Γ_1, Γ_2 and Γ_3 constitute a boundary-value-problem which is illustrated in figure 3.

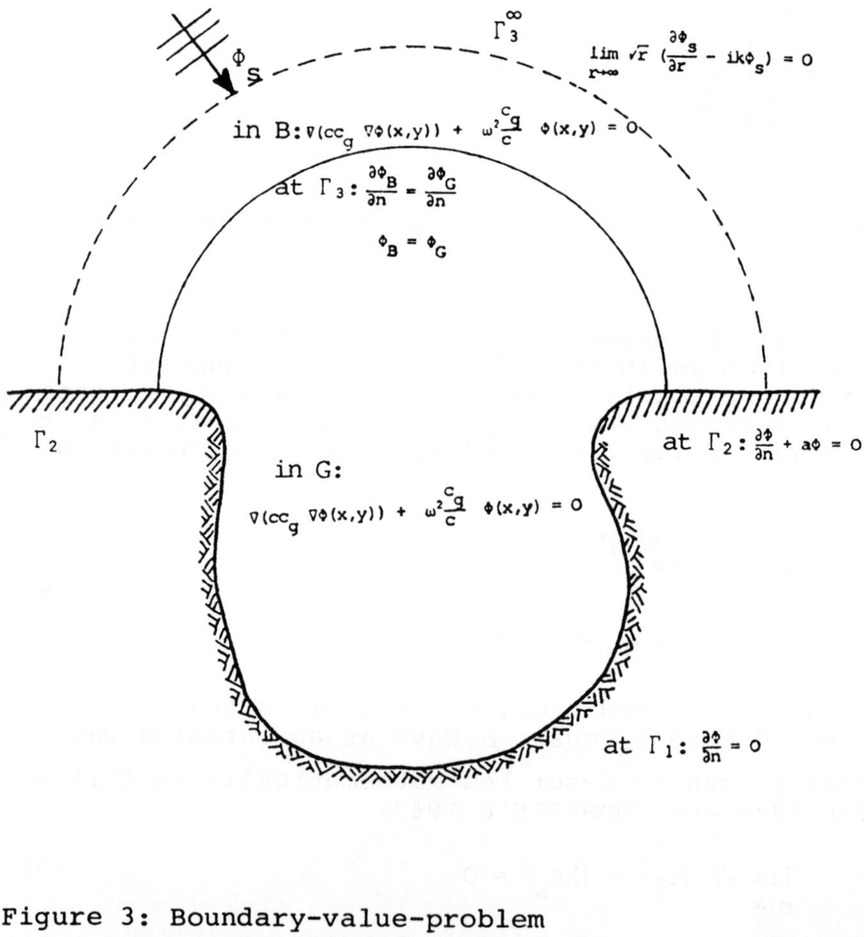

Figure 3: Boundary-value-problem

The equivalent variational formulation has the functional ($\Phi^I = \tilde{\Phi} + \tilde{\Phi}_R$)

$$J(\Phi) = \iint_G \left(cc_g (\nabla\Phi)^2 - \omega^2 \frac{c_g}{c} \Phi^2 \right) dG$$

$$+ \int_{\Gamma_2} cc_g\, a\, \Phi^2\, d\Gamma - \int_{\Gamma_3^\infty} ik\, cc_g\, (\Phi - \Phi^I)^2\, d\Gamma - \int_{\Gamma_3^\infty} cc_g \frac{\partial \Phi^I}{\partial n} \Phi\, d\Gamma$$

which can be rewritten as (MEI,1974)

$$J(\Phi) = \iint_G \left(cc_g (\nabla\Phi)^2 - \omega^2 \frac{c_g}{c} \Phi^2 \right) dG$$

$$+ \int_{\Gamma_2} cc_g\, a\, \Phi^2\, d\Gamma$$

$$+ \int_{\Gamma_3} cc_g\, (\Phi_B - \Phi^I) \frac{\partial(\Phi_B - \Phi^I)}{\partial n}\, d\Gamma - \int_{\Gamma_3} 2cc_g\, \Phi \frac{\partial(\Phi_B - \Phi^I)}{\partial n}\, d\Gamma$$

$$- \int_{\Gamma_3} 2cc_g\, \Phi \frac{\partial \Phi^I}{\partial n}\, d\Gamma + \int_{\Gamma_3} 2cc_g\, \Phi^I \frac{\partial(\Phi_B - \Phi^I)}{\partial n}\, d\Gamma$$

$$+ \int_{\Gamma_3} 2cc_g\, \Phi^I \frac{\partial \Phi^I}{\partial n}\, d\Gamma$$

The solving of the diffraction-refraction-equation with given boundary conditions corresponds to finding a function Φ for which the functional J has an extremum.

NUMERICAL TREATMENT

Because the variational formulation leads to an integral description of the problem, the finite-element-method is appropriate. It is applied to the functional J in its simplest form with triangular elements and linear approximation functions. The subsequent minimizing of the functional $J(\Phi)$ creates a system of complex linear equations:

$$\begin{aligned} A\vec{\Phi} + B\vec{\mu} &= R_1 \\ C\vec{\Phi} + D\vec{\mu} &= R_2 \end{aligned} \qquad (6)$$

or in explicit formulation

$$\vec{\Phi} = A^{-1}(R_1 - B(D - CA^{-1}B)^{-1}(R_2 - CA^{-1}R_1))$$

with $\vec{\Phi} = (\Phi_1, \Phi_2, \Phi_3, \ldots, \Phi_N)^T$.

The complex matrix A has a band structure, but its inversion still has high demands on computer storage. Therefore the calculation for large natural areas requires a special off-core version of the program.

TESTING WITH IDEALISED EXAMPLES

Open boundary condition

An open area is chosen with constant water depth and without any obstacle. Figure 4a displays the configuration of the elements, and figure 4b the result of a calculation with a water depth $h(x,y) = 3.0$ m and a period $T = 5.0$ s. The wave height proves to be constant throughout the whole area, with a maximum error of 4%.

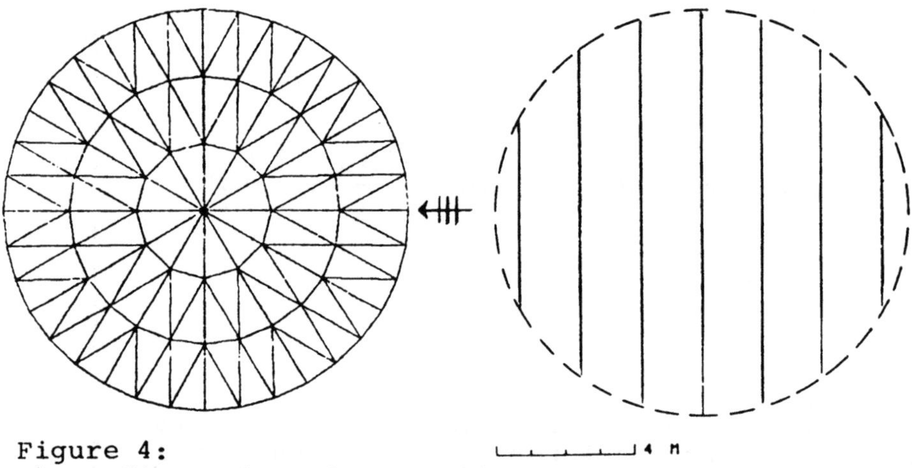

Figure 4:
a) configuration of elements
b) contours of phase angle

Reflecting boundary condition

In figure 5 to 9 some tests are illustrated, with a totally and a partially reflecting boundary and two different directions of the incoming wave. The comparsion between the numerical solution along the axis with the one-dimensional analytical result shows that the different kinds of reflection will be reproduced by the mathematical model.

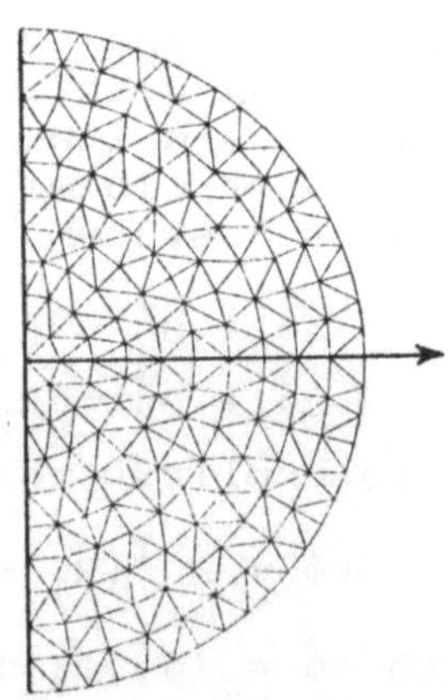

Figure 5:
configuration of elements
$h(x,y) = 10.0$ m, $T = 5.0$ s

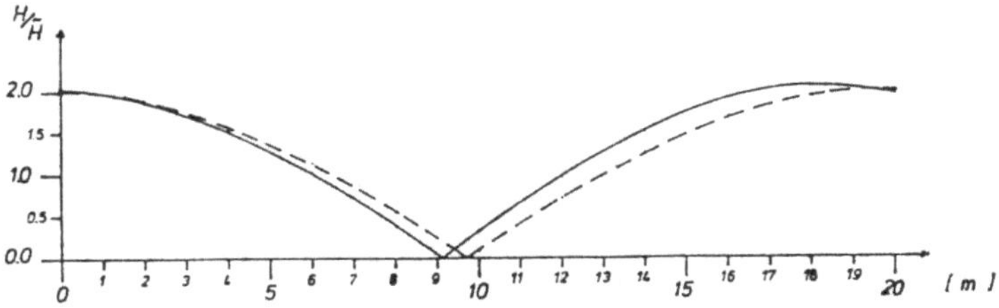

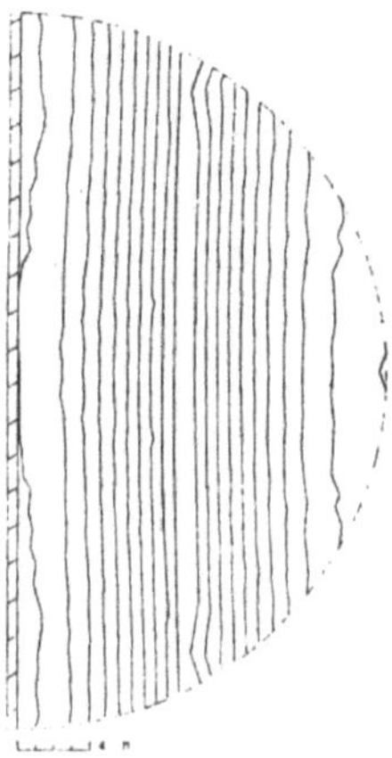

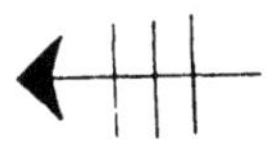

Figure 6: Total reflection

a) ——— finite-element-solution
   ----- analytical solution

b) contours of relative wave height

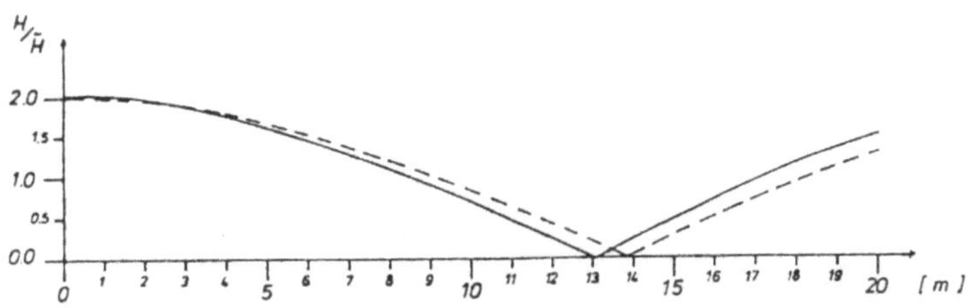

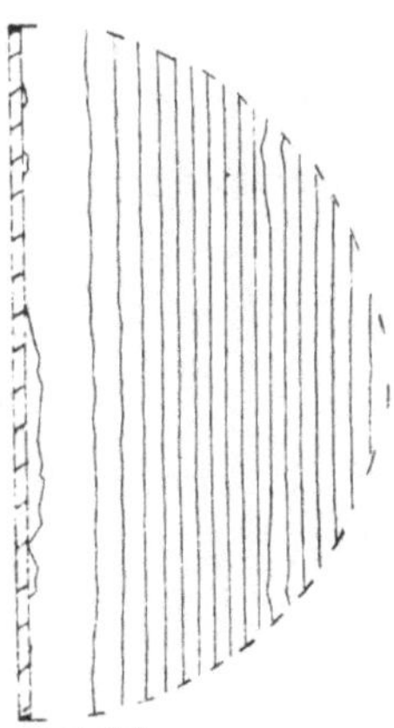

Figure 7: Total reflection

a) ——— finite-element-solution
   ----- analytical solution

b) contours of relative wave height

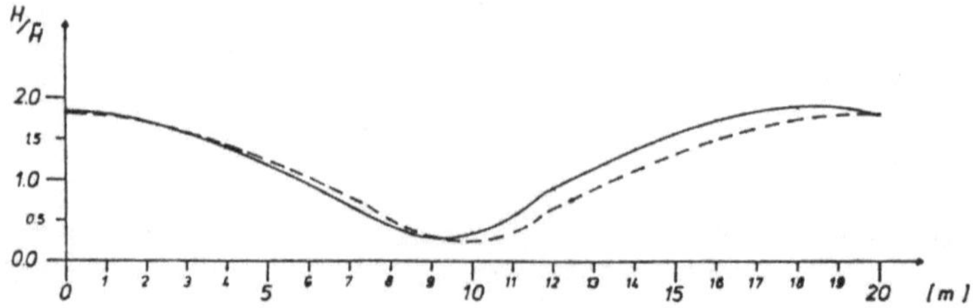

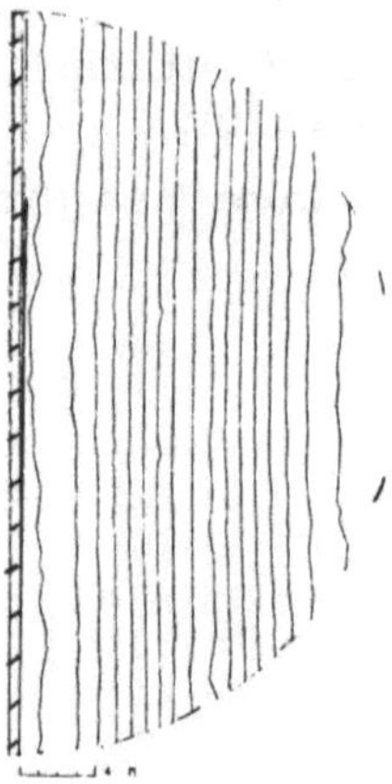

Figure 8: Partial reflection (R=0.8)

a) ——— finite-element-solution
----- analytical solution

b) contours of relative wave height

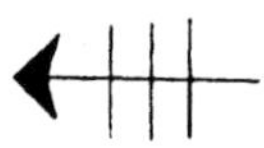

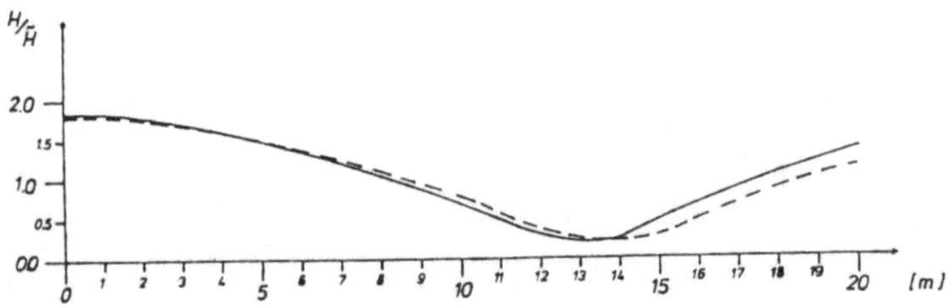

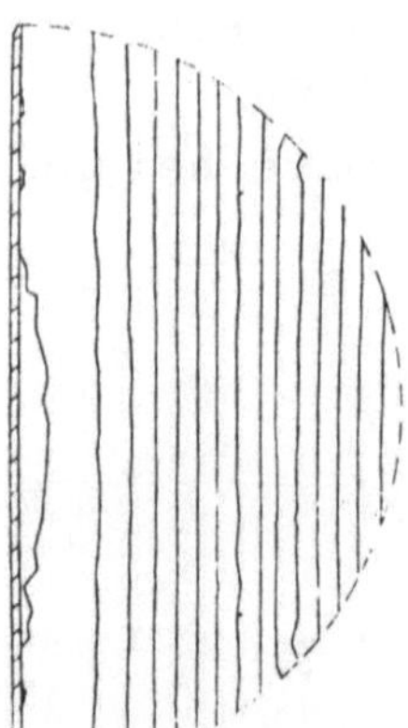

Figure 9: Partial reflection (R=0.8)

a) ——— finite-element-solution
----- analytical solution

b) contours of relative waveheight

Rectangular harbour

Considering only long waves MEI and ÜNLÜATA (1973) formulated an analytical solution for a rectangular harbour taking into account the diffraction effects at the entrance. Their solution which is valid only at a sufficient distance from the entrance, has been expanded here to cover intermediate and short waves, and is used to verify the finite-element-solution. The resulting equation for the wave distribution along the axis of the harbour is as follows:

$$\frac{H}{\tilde{H}} = \frac{\sqrt{gh}}{\sqrt{gL\,\tanh(kh)/(2\pi)}} \cdot \frac{\cos(k(x+l))}{\cos(kl) + \frac{kb}{\pi}\sin(kl)\,\ln\frac{\delta kb}{\pi e} + i\,\frac{kb}{2}\sin(kl)} \tag{7}$$

with e = 2.71828
and $\ln\delta$ = 0.577215 (Euler-Macheroni-constant).

A finite-element-grid for such harbours is displayed in figure 10. It is based on the dimensions of a small physical model by Ippen and Goda, presented by Mei et al. Calculated contours of wave height distribution are given in figure 11, while figure 12 shows the corresponding longitudinal section along the harbour axis. The three curves in figure 12 demonstrate the agreement between the finite-element-solution and the appropriate analytical solution,at a sufficient distance from the harbour entrance (where equation (7) is valid).They also emphasize the difference between the two analytical solutions.

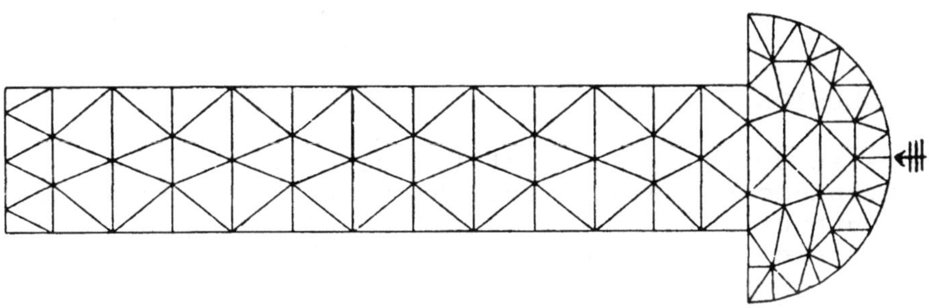

Figure 10: Definition sketch of the rectangular harbour
length l=0.31m, width b=0.06m,
water depth h=0.25725m, length of the incident wave L=1.44m

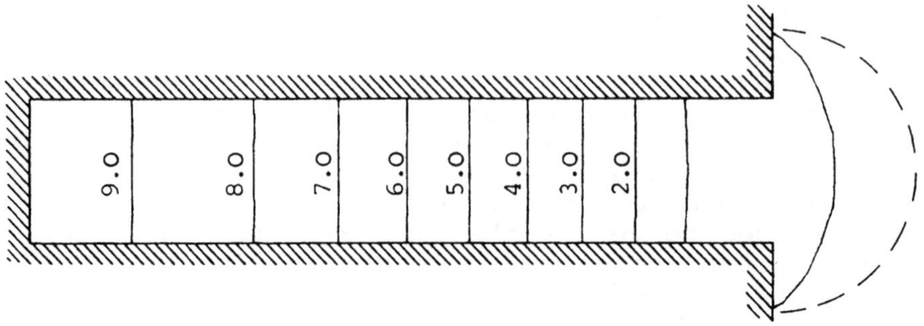

Figure 11: Contours of relative wave height, finite-element-solution

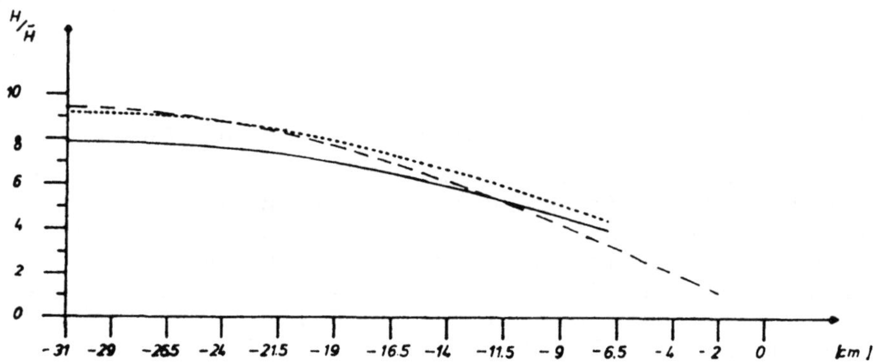

Figure 12: Longitudinal section along the harbour axis

——— analytical solution for long waves (Mei and Ünlüata)
..... analytical solution for intermediate waves (equation (7))
----- finite-element-solution

CONCLUSION

This study indicates that the developed finite-element-model permits the calculation of linear waves in harbours. In addition to irregular shapes of boundaries and variable water depth, totally and partially boundaries are included. This paper gives a review of the theoretical investigation and shows - on the basis of some simple tests and comparsions with analytical solutions - that the relevant phenomenon are correctly reproduced. That is a necessary prerequisite for a larger and more complicated application and a comparsion with experimental data which are currently being worked on.

REFERENCES

Berkhoff,J.C.W. (1976) Mathematical Models for simple Harmonic Linear Water Waves - Wave Diffraction and Refraction -. Delft Hydraulic Laboratory, Publication No. 163.

Chen,H.S. and Mei,C.C. (1974) Oscillation and Wave Forces in an Offshore Harbor. Ralph M. Parsons Laboratory for Water Resources and Hydrodynamics, Report No. 190, MIT.

Rottmann,W. and Zielke,W. (1983) FEM-Analysis of Combined Diffraction and Refraction in Harbours, Including Partially-Reflecting Boundaries, International Conference on Coastal and Port Engineering in Developing Countries, Colombo,Sri Lanka, Proc. Vol.II, pp.1319 ff.

Mei,C.C. and Ünlüata,Ü. (1973) Long wave Excitation in Harbours - An analytical study. Ralph M. Parsons Laboratory for Water Resources and Hydrodynamics, Report No.171, MIT.

Sommerfeld,A. (1949) Partial Differential Equations in Physics, Academic Press, New York.

Wyld,H.W. (1976) Mathematical Methods for Physics. W.A. Benjamin, Inc. Advancd Book Program, Reading, Massachusetts.

ACKNOWLEDGEMENT

The authors gratefully acknowledge the support of the "German Research Society" (Deutsche Forschungsgemeinschaft), which sponsors the work as a project in the Inter-Institutional Research Group on Coastal Engineering (Sonderforschungsbereich "Küsteningenieurwesen).

FINITE ELEMENT MODELS FOR THE STUDY OF ESTUARIES, LAGOONS, AND COASTAL SEAS

José Rodrigues Vieira

Laboratório Nacional de Engenharia Civil (LNEC)

ABSTRACT

The use of estuaries, lagoons and coastal regions plays an important role in human activity.

In these systems, where complex interactions between physical, chemical and biological processes are found, water circulation is one of the most important factors controlling those processes. Thus, the study of the environmental effects resulting from activities such as dredging works, port construction and pollutant discharges, imposes the previous knowledge and interpretation of the actual hydrodynamic regimen and the forecast of modifications induced in it.

In this regard and within the scope of activity of the Estuaries Division of the LNEC Hydraulics Department, the author has recently developed a computational system using finite elements, which basicaly consists of two coupled mathematical models, the first an hydrodynamic model, MHD2, the second a transport-dispersion model, MTD2.

In this paper the hydrodynamic two-dimensional model is presented, both in what concerns its numerical and theoretical basis. The application of the model to the study of the flow conditions of a lagoon is also included.

BRIEF DESCRIPTION OF THE HYDRODYNAMIC MODEL

Introdution

The hydrodynamic model allows one to solve the equations of motion and continuity integrated along the vertical, currently known as 'shallow water equations'.

It is possible to take into account varied flow and level conditions at points at the open boundaries, and, at the water surface, we can intoduce the wind effects and the atmospheric pressure variations.

In the solid boundaries, flow is taken as parallel to the boundary. The evaluation of the normal direction at each point is made using a sliding quadratic element. With this model it is also possible to take into account bottom friction, the Coriolis effect, and the occurence of density horizontal gradients.

<u>Hydrodynamic equations for swallow water</u>

The Reynolds three-dimensional equations and the continuity equation established in terms of the mean time values of the velocity components u, v and w, and of the pressure p, are the starting point for the study of incompressible turbulent flows, and can be written in the following form :

$$\rho\left(\frac{\partial u_i}{\partial t} + u_j \frac{\partial u_i}{\partial x_j}\right) = \rho F_i - \frac{\partial p}{\partial x_i} + \mu \nabla^2 u_i - \frac{\partial}{\partial x_j}\left(\overline{u_i' \cdot u_j'}\right) . \tag{1}$$

As regards estuaries and lagoons, in wich small relative depths are found, the vertical components of the mean local velocity are negligible, and therefore gravity and pressure forces are predominant over the vertical inertia forces, which amounts to assume that the pressure distribution follows a hydrostatic law, which is given as

$$\frac{1}{\rho}\frac{\partial p}{\partial z} + g = 0 . \tag{2}$$

Bearing in mind this condition, for well-mixed estuaries the following relation applies:

$$\frac{\partial p}{\partial x_i} = \frac{\partial p_a}{\partial x_i} + g\rho\frac{\partial \xi}{\partial x_i} + g\int_z^{\xi}\frac{\partial \rho}{\partial x_i}\,dz . \tag{3}$$

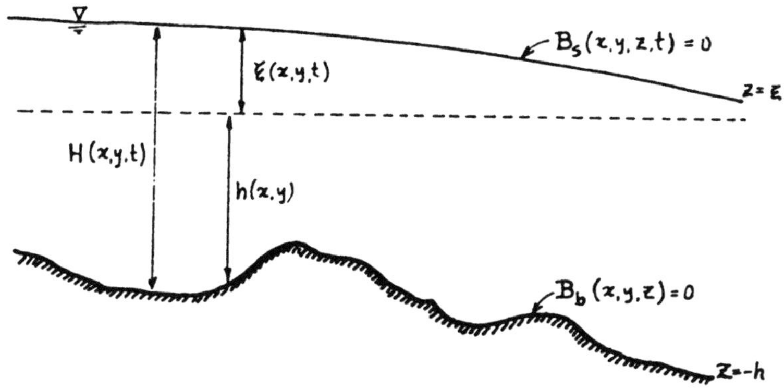

Fig.1 Schematic representation of an estuary section and notation employed.

This result shows the influence of the density horizontal gradients on the pressure forces and explains the possible ocurrence of velocity inversions along the vertical in the ebb, and the moving down of the point of maximum velocities at the flood, as shown in Fig.2.

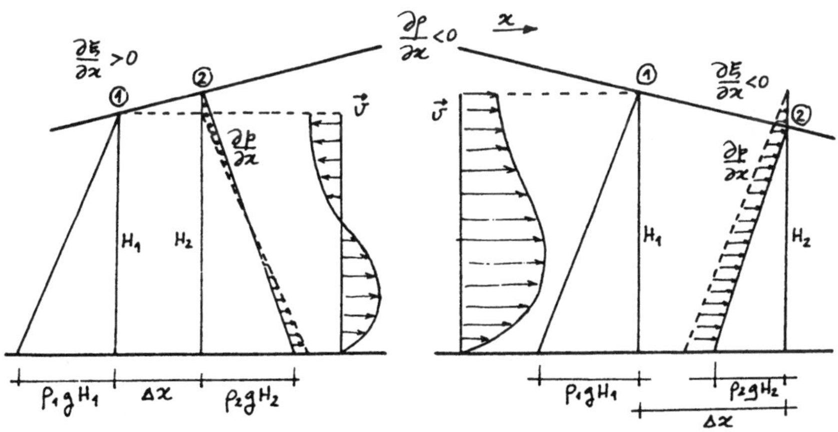

Fig.2 Schematic example of the vertical distribution of pressure and velocity in well-mixed estuaries

By substituting Eq.(3) in the horizontal momentum equations, by integrating the resulting equations in the vertical, and considering also that the lateral transport of momentum at the surface and at the bottom is negligible, we obtain in the x direction, for instance

$$\frac{\partial (HU)}{\partial t} + \frac{\partial}{\partial x}(\alpha_{uu} HUU) + \frac{\partial}{\partial y}(\alpha_{uv} HUV) = f HV - gH\frac{\partial \xi}{\partial x} - \frac{H}{\rho}\frac{\partial p_a}{\partial x} -$$
$$- \frac{g}{\rho}\frac{\partial \rho}{\partial x}\frac{H^2}{2} + \frac{1}{\rho}\left(\frac{\partial}{\partial x}\int_{-h}^{\xi}\tau_{xx}\,dz + \frac{\partial}{\partial y}\int_{-h}^{\xi}\tau_{yx}\,dz + \tau_{zx}\big|_{\xi} - \tau_{zx}\big|_{-h}\right). \qquad (4)$$

The α coefficients express the correlations in the vertical distributions of velocity, and the terms $\tau_{zx}|_{-h}$ and $\tau_{zx}|_{\xi}$ are related to the resistance to flow due to the bottom friction, and to the drag on the free surface due to the wind, and can be expressed by the empirical relations

$$\tau_{zx}\big|_{-h} = \rho g \frac{U\sqrt{U^2+V^2}}{C^2} , \qquad (5.1)$$

$$\tau_{zx}\big|_{\xi} = K W^2 \cos\theta , \qquad (5.2)$$

where C is the Chézy coefficient (m2/s), W the wind velocity, K a dimensionless coefficient and θ the angle between the vector of wind velocity and the x direction.

The integral terms give the lateral transport of momentum along the vertical, wich is associated with the Reynolds stresses. It is assumed that these stresses are proportional to the time rate of angular deformation (Boussinesq approximation), and thus

$$\tau_{ij} = -\overline{u_i' u_j'} = \nu_{ij}\left(\frac{\partial \hat{u}_i}{\partial x_j} + \frac{\partial \hat{u}_j}{\partial x_i}\right), \qquad (6)$$

where ν are the commonly designated turbulent viscosity coefficients and symbol ^ denotes mean temporal values. In practice, a mean coefficient along the vertical is used, ε, so that

$$\int_{-h}^{\xi}\tau_{xx}\,dz = \int_{-h}^{\xi} 2\nu_H \frac{\partial \hat{u}}{\partial x}\,dz = \varepsilon H \frac{\partial U}{\partial x} , \qquad (7.1)$$

$$\int_{-h}^{\xi}\tau_{yx}\,dz = \int_{-h}^{\xi} \nu_H \left(\frac{\partial \hat{u}}{\partial y} + \frac{\partial \hat{v}}{\partial x}\right) dz = \varepsilon H \frac{\partial U}{\partial y} . \qquad (7.2)$$

Although the influence of these terms is, as a rule, very small, their inclusion has usually the advantage of improving computational stability.

In the present version of the numerical model MHD2,

the influence of the vertical distribution of velocity in the inertia forces is not considered, that is to say, α was taken as equal to the unity. This approximation is justified by the fact that the terms of convective acceleration affected by α are, on average, of a lower order of magnitude than those of local acceleration, and, on the other hand, the value of α lies between 0.5 and 1.5 for extreme conditions, so its influence can be neglected for well-mixed flows.

Thus, the system of equations that constitute the hydrodynamic mathematical model for the study of well-mixed flows can be finally written in the following way:

$$\frac{\partial U}{\partial t} + U\frac{\partial U}{\partial x} + V\frac{\partial U}{\partial y} = fV - g\frac{\partial \xi}{\partial x} - \frac{1}{\rho}\frac{\partial p_a}{\partial x} - \frac{g}{\rho}\frac{\partial \rho}{\partial x}\frac{(h+\xi)}{2} + \frac{KW^2\cos\theta}{h+\xi} - \frac{gU\sqrt{U^2+V^2}}{(h+\xi)C^2} + \frac{\varepsilon}{\rho}\left(\frac{\partial^2 U}{\partial x^2} + \frac{\partial^2 U}{\partial y^2}\right), \tag{8.1}$$

$$\frac{\partial V}{\partial t} + U\frac{\partial V}{\partial x} + V\frac{\partial V}{\partial y} = -fU - g\frac{\partial \xi}{\partial y} - \frac{1}{\rho}\frac{\partial p_a}{\partial y} - \frac{g}{\rho}\frac{\partial \rho}{\partial y}\frac{(h+\xi)}{2} + \frac{KW^2\sin\theta}{h+\xi} - \frac{gV\sqrt{U^2+V^2}}{(h+\xi)C^2} + \frac{\varepsilon}{\rho}\left(\frac{\partial^2 V}{\partial x^2} + \frac{\partial^2 V}{\partial y^2}\right), \tag{8.2}$$

$$\frac{\partial \xi}{\partial t} + \frac{\partial[(h+\xi)U]}{\partial x} + \frac{\partial[(h+\xi)V]}{\partial y} = 0. \tag{8.3}$$

The numerical model

The analytical solution of these general equations is only possible by using strict simplifying hypotheses and considering very schematic geometries. To solve the equations in their general form and to take into account any kind of boundary conditions and geometric contours, it is then necessary to apply a numerical model.

The MHD2 numerical model is based on the finite element method for the discretization of the differential equations, using a formulation by weighted residuals for the approximation in each domain or element.

Owing to the existence of non-linear terms (convection and bottom friction), an iterative process has to be used to attain convergence for the true solution. However, these techniques require a long calculation time. It was therefore preferred to linearize the equations previously, which consists in adequately replacing some variables in the non-linear terms by the values calculated for these variables at the immediately preceding instant.

The approximation errors introduced by this method are unimportant because, as we have said, convection terms have little significance in the overall computation of the inertia forces and, besides, the system is slowly variable.

In what concerns the discretization in space and time, these domains are considered separately, which allows to reduce the number of simultaneous dependent variables and to employ the most adequate interpolation functions for the treatement of each one of them. Two-dimensional linear or quadratic elements are then used for space discretization, which leads to a system of first order differential equations in time. For the discretization of these equations, linear one-dimensional elements are used. The Galerkin method is employed in each case, which, for integration in time, leads to imconditionally stable implicit schemes.

The computation of elementary matrices is made by resorting to the Gauss-Legendre method. The storage of the coefficients in the overall matrices is made according to a profile defined by the effective height of columns, and the solving of the final system of linear equations by a Gauss direct elimination method.

The finite element method seems well suited for solving this type of problems, and in the last few years, it has to a large extent replaced the use of the finite difference method. In fact, the finite element method affords very efficient discretization of the domain, since it allows a great versality and simplicity in the construction of the meshes and in the choice of the shape and size of the elements. This not only gives better adaptation to the different gradients of the physical magnitudes, and consequently greater accuracy, but also more ease to guarantee overall continuity in the domain.

APPLICATION OF THE MODEL

The model is being used to study the hydrodynamics of several estuaries(Rio de la Plata, Aveiro, Sado), and it was first applied last year in a study of the Nador Lagoon, to forecast the behavior of a new artificial inlet designed to control the mean tidal prism, in order to guarantee the ideal conditions for the growth of several species of shellfish.

The lagoon is situated in the Mediterranean coast of Marocco, and its form is very regular and approximately oval, with a maximum lenght of a 25 Km along the coast and a maximum width of about 7 Km. The bottom is also very regular with depths of about 7 meters in the middle of tthe lagoon, and with an area attaining 115 Km2. The tidal amplitudes are very small, and vary between 0.1 and 0.2 meters for neap and spring tides respectively. Its behavior is typical of a small size lagoon with a horizontal water surface response, that is to say, the effect of propagation of the lagoon tide is negligible due to the small size of the basin relative to the tide wave lenght.

In this kind of lagoons the phase lags between high and low waters and the tide range ratio are controlled by the inlet characteristics, where the flow is a function of the level gradients along its channel.

Due to the variations of the area and hydraulic radius of the inlet along the tide period, and the integral volume of flow being equal in the ebb and flood, the following characteristics have also been found in the behavior of the lagoon:

-The lagoon tide is asymetric, with a mean level higher than the exterior mean level.

-The phase lags between the lagoon and the ocean in high water are larger than in low water, and in consequence the ebb is longer than the flood.

-The above phase lags grows with the damping of the lagoon tide, and their significance increases as the length and roughness of the inlet channel increase and its section decreases.

-The inversion of the currents occurs in the instants of high and low water in the lagoon.

Fisure 3 reProduces the assembly of the significant tide curves siven by the mathematical model when aPPlied to the lasoon with its existins natural inlet, and in Fis. 4 is shown the finite element mesh used.

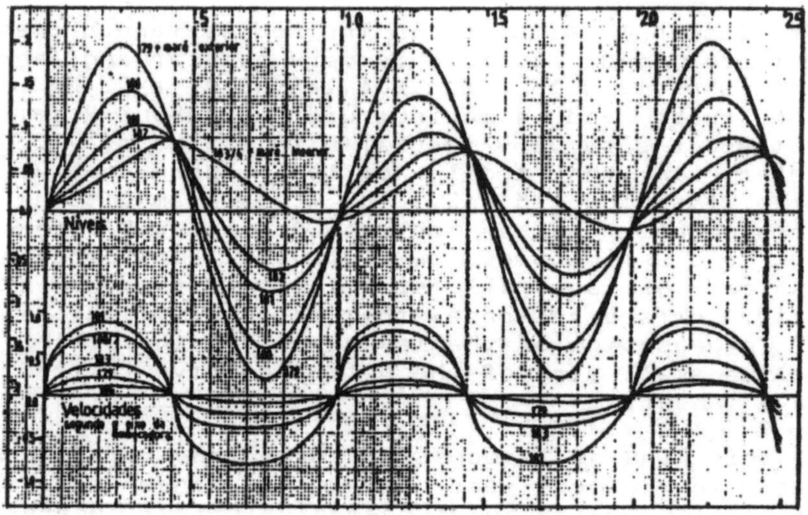

Fis.3 Tidal curves in the Nador Lasoon

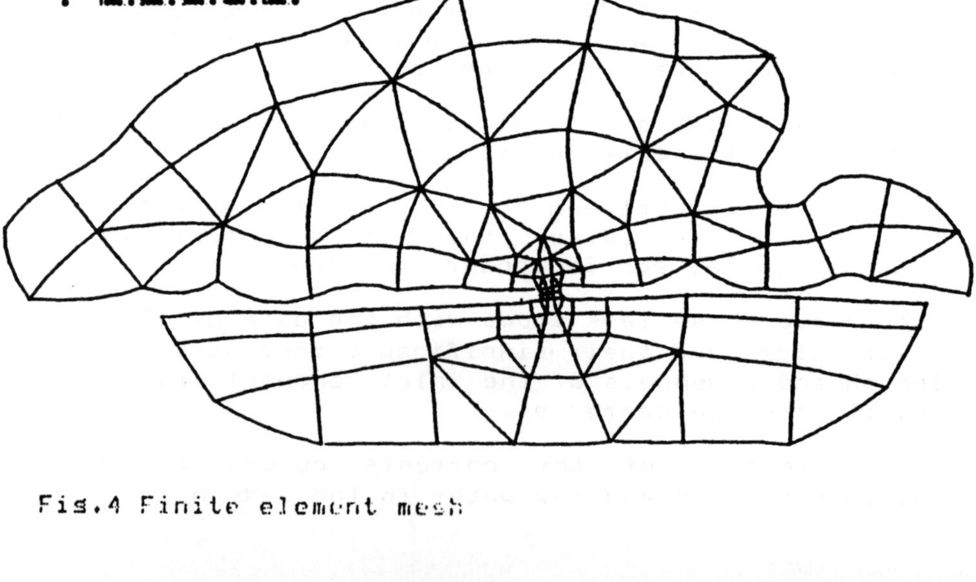

Fis.4 Finite element mesh

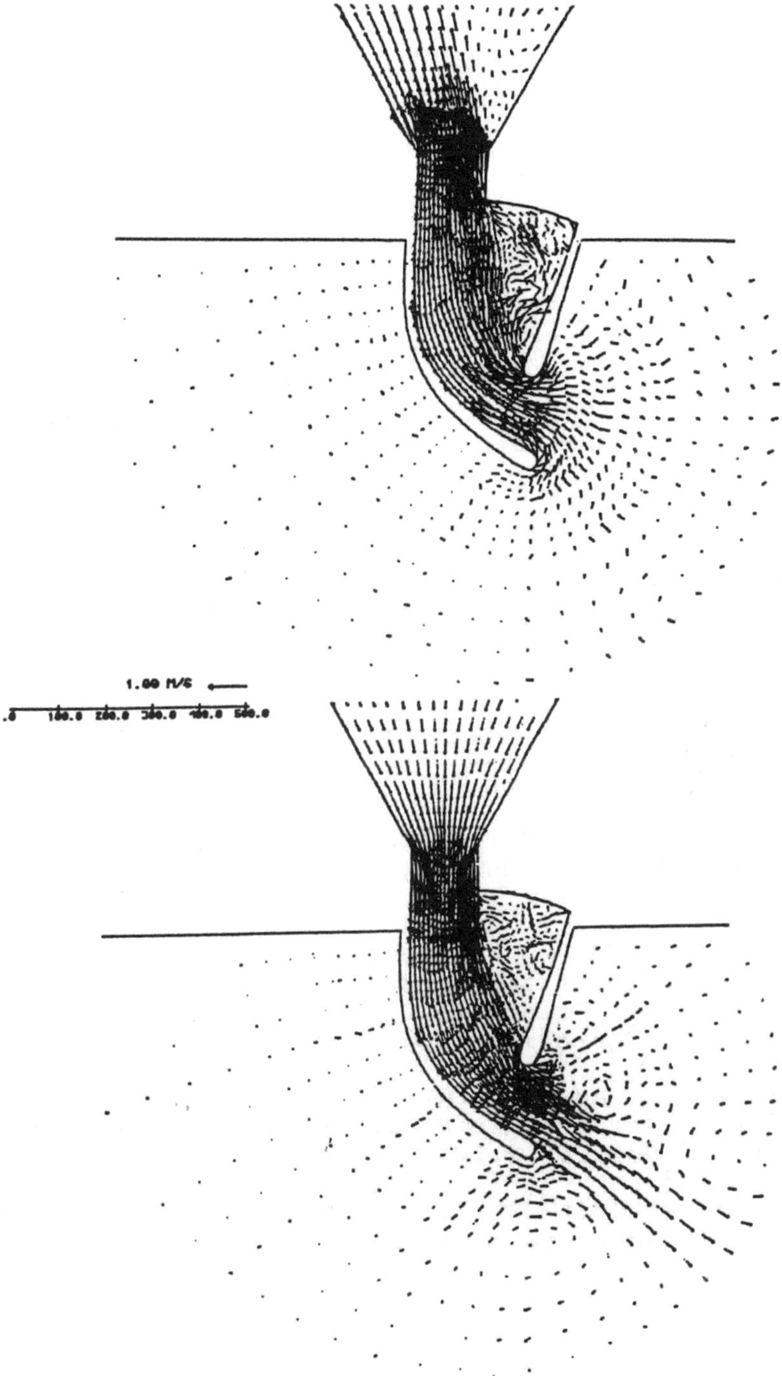

Fig.5 Computed fields of currents in the artificial inlet

REFERENCES

Abecasis,F.M.(1969) Hydraulic and Sedimentary Problems in Tidal Estuaries. Memória N.343, LNEC, Lisbon.

Bathe,K.J. and Wilson,E.L.(1976) Numerical Methods in Finite Element Analysis. Prentice-Hall, New Jersey.

Connor,J.J. and Brebbia,C.A.(1976) Finite Element Techniques for Fluid Flow. Newnes-Butterworths, London.

Falconer,R.A.(1980) Numerical Modelling of Tidal Circulation in Harbors. ASCE, vol. 106, N.WW1.

Holz,K.P. and Withum,D.(1976) Finite Element Applications for Transient Processes in Estuaries. U.S.- Germ. Symp. on F.E.M.

Oliveira,M.(1973) Embocaduras Lagunares. Contribuição para o Estudo das suas Condições de Equilíbrio. LNEC, Lisbon.

Pinder,G.F. and Gray,W.G.(1977) Finite Element Simulation in Surface and Subsurface Hydrology. Academic Press, London.

Vieira,J.R.(1983) Marchica à Nador. Etude sur Modèle Mathématique du Régime Hydrodynamique. LNEC, Lisbon.

Vieira,J.R.(1984) Mathematical Modelling of Estuaries, Lagoons and Coastal Regions. Theoretical and Numerical Foundations of a Finite Element Hydrodynamic Two-dimensional Model. LNEC, Lisbon.

Walters,R.A. and Cheng,R.T.(1980) Accuracy of an Estuarine Hydrodynamic Model Using Smooth Elements. Water Res., Vol.16, N.1.

Zienkiewicz,O.C.(1977) The Finite Element Method. McGraw-Hill Book Company, London.

APPLICATION OF CONSTRAINTS IN MODELLING TIDAL POWER GENERATION

Annette M. Harte, Department of Civil Engineering,
University College Galway, Ireland.

John F. McNamara, Department of Mechanical Engineering,
University College Galway, Ireland.

INTRODUCTION

Tidal energy has many advantages in that the tides are free and the supply is both predictable and renewable, nevertheless the high capital cost involved in the installation of tidal power plants has meant that only a small number of them have actually been constructed. These plants [1] are in France, the USSR and China, and all are operating satisfactorily at the present time. Experience gained from the construction of these plants and recent technological changes mean that many hitherto uneconomic schemes may now become viable.

Predicting the changes in the tidal dynamics resulting from the installation of a tidal barrage is a major aspect of the assessment of any tidal power project. This information is required in order to predict the energy potential of the project and to assess the environmental impact of such a scheme. This paper presents a finite element model which allows a continuous simulation of the fluid flow and power take-off during all phases of the tide.

The general flow in a bay or estuary is described using the shallow water equations as outlined in [2]. The flow through the sluices and turbines in the barrage is based on energy conservation requirements and the mechanical properties of these devices. All flow regimes are incorporated into a single model which represents the flow at all points both remote from and at the actual barrage itself. Flow through the barrage is simulated by special constraint equations between superimposed nodes at that location. This constraint technique is designed to accurately reflect the hydraulic characteristics of the barrage while calculating the barrage flows in a form compatible with the solution technique elsewhere in the bay or estuary.

Constraint applications have been used quite extensively in structural analysis as a means of introducing certain relationships or freedoms between displacements at adjacent points [3]. A special feature of the work presented here is the application of such constraint techniques to a problem in fluid dynamics. The application of the model to a proposed tidal power scheme in the Shannon Estuary, Ireland, is discussed and sample numerical values are obtained for the potential power available from the tidal flows.

MODELLING BARRAGE FLOW CONDITIONS AS CONSTRAINTS

In order to represent barrage flows in the model the finite element mesh contains two sets of parallel nodes at the barrage location, one set on the upstream face and one on the downstream face. These nodes are referred to as barrage nodes. Certain of these nodes correspond to the location of the sluices and turbines in the barrage and are referred to as sluice nodes and turbine nodes, respectively.

Three different sets of flow conditions occur at the barrage during the generation cycle. For an ebb generation scheme only the sluices are open during the flooding tide, and this is modelled by applying a constraint equation governing flow through the matching sluice nodes and considering all other barrage nodes as fixed impervious boundary nodes. The next stage of the cycle starts when the sluices are closed at full tide. Now there is no flow between the two sets of barrage nodes and all are treated as fixed boundary nodes. The third phase, or generation phase, occurs during the ebbing tide when the turbines are operating and power is being taken off. A constraint equation of turbine flow is applied to the matching turbine nodes and all other barrage nodes are treated as fixed boundary nodes. The change-over from one phase to the next is catered for in the model using a restart procedure so that there is a continuous simulation of flow over the entire tidal cycle.

The sluice and turbine constraint equations are developed from energy conservation principles. Flow through both these devices involves a loss of energy from the fluid body. In the case of sluices, this loss is made up from the sum of the minor losses such as entry, exit and contraction associated with the particular sluice. The loss incurred in flow through the turbines is made up from the energy loss due to the power generated by the turbomachine along with the minor losses.

Constraint Equations

A cross-section through the barrage at a sluice or turbine position is shown in Figure 1. The relationship describing the conservation of energy at sections 1 and 2 on either side of the barrage is expressed as follows:

$$H + h_1 + \frac{V_1^2}{2g} = H + h_2 + \frac{V_2^2}{2g} + \text{losses} \tag{1}$$

The velocity components along the longitudinal axis of the sluice or turbine are assumed to be zero at sections 1 and 2.

The following non-dimensional number π can be written for any turbomachine

$$\pi = {}^{P}/_{(Q\Delta H)} \tag{2}$$

where P is the power generated,
Q is the discharge through the device,
and ΔH is the head difference across the turbine.

For the purposes of this model, it is assumed that the flow rate through the turbines is constant during the generation phase. In this situation the power generated and consequently the power losses are directly proportional to the head difference across the machine as indicated by Equation (2).

Minor machine losses are generally written as a proportion of the velocity head ${}^{V^2}/_{2g}$ and it is further assumed that the upstream velocity head is small enough to be set to zero. Substituting for the power and minor losses and invoking the power take-off relation of Equation (2), Equation (1) is now written as

$$h_1 = h_2 + \frac{V_2^2}{2g} + k_1(h_1 - h_2) + k_2 \frac{V_2^2}{2g} \tag{3}$$

where k_1 and k_2 are power and minor loss coefficients, respectively. In the case of flow through sluices the power loss coefficient k_1 is zero. Rearranging terms in Equation (3) gives

$$h_1 = h_2 + kV_2^2 \tag{4}$$

where $k = (1 + k_2)/\{2g(1 - k_1)\}$

The above equation is a general relationship covering flow through sluices and turbines where the factor k is varied accordingly. The coefficient k must be determined for each individual machine or sluice.

Linearization of Constraint Equations

As noted in [2], quadratic velocity terms in the shallow water equations are eliminated by replacing the squared term as a product of the current value and the known value computed at the previous time step. A similar consistent procedure is adopted with respect to the general constraint relation of Equation (4) which is linearized as follows:

$$h_1 = h_2 + k\,\overline{V}_2\,V_2$$

or

$$h_1 = h_2 + \overline{k}\,V_2 \tag{5}$$

where $\overline{V}_2$ is the value of the variable V_2 from the previous time step and $\overline{k} = k\,\overline{V}_2$. The constraint equations are now in a form suitable for incorporation into the system flow equations following procedures developed mainly in the field of structural mechanics.

INCORPORATION OF FLOW CONSTRAINTS

Many methods [3] exist for incorporating a set of constraint equations into a general system equation including the elimination method, the method of Lagrange multipliers and penalty function methods. The method used here may be described as an extended elimination method and is adopted from [4]. It is a general technique which is applicable to many types of engineering problems.

Consider the general system equation

$$\underline{K}\,\underset{\sim}{q} = \underset{\sim}{f} \tag{6}$$

where $\underline{K}$ is the system matrix,
$\underset{\sim}{q}$ is the vector of nodal unknowns,
and $\underset{\sim}{f}$ is the corresponding nodal force vector.

In a constrained system a subset of the total number of system degrees of freedom, called slave degrees of freedom, are related to another subset, called master degrees of freedom, through the following equation

$$\underset{\sim}{q}^{\gamma} = \underline{A}^{\gamma\beta}\,\underset{\sim}{q}^{\beta} \tag{7}$$

where $\underset{\sim}{q}^{\gamma}$ is a vector of slave degrees of freedom,
$\underset{\sim}{q}^{\beta}$ is a vector of mutually independent master degrees of freedom,
and $\underline{A}^{\gamma\beta}$ is a matrix of multipliers.

In order to incorporate the constraint defined by Equation (7) into the system, Equation (6) is partitioned to give

$$\begin{bmatrix} \underline{K}^{\alpha\alpha} & \underline{K}^{\alpha\beta} & \underline{K}^{\alpha\gamma} \\ \underline{K}^{\beta\alpha} & \underline{K}^{\beta\beta} & \underline{K}^{\beta\gamma} \\ \underline{K}^{\gamma\alpha} & \underline{K}^{\gamma\beta} & \underline{K}^{\gamma\gamma} \end{bmatrix} \begin{Bmatrix} \underset{\sim}{q}^{\alpha} \\ \underset{\sim}{q}^{\beta} \\ \underset{\sim}{q}^{\gamma} \end{Bmatrix} = \begin{Bmatrix} \underset{\sim}{f}^{\alpha} \\ \underset{\sim}{f}^{\beta} \\ \underset{\sim}{f}^{\gamma} \end{Bmatrix} \tag{8}$$

where $\underset{\sim}{q}^{\alpha}$ refers to those degrees of freedom not involved in the constraint equations.

In the description of the matrix operations which follows the terms "row" and "column" of the partitioned system matrix refer to a row and column of submatrices. Equation (7) is first substituted into Equation (8). This is equivalent to post-multiplying the slave "column" by the constraint matrix $A^{\gamma\beta}$ and adding it to the master "column". The slave "row" is replaced by the constraint equation itself. The resulting system matrix is unsymmetric. In order to generate an equivalent symmetric system, the slave row of Equation (8) is premultiplied by $(A^{\gamma\beta})^T$ and added to the new master "row" and the new slave "row" is premultiplied by $-K^{\gamma\gamma}$. This procedure is detailed in [4] and yields the symmetric system

$$\begin{bmatrix} K^{\alpha\alpha} & K^{\alpha\beta}+K^{\alpha\gamma}A^{\gamma\beta} & 0 \\ K^{\beta\alpha}+(A^{\gamma\beta})^T K^{\gamma\alpha} & K^{\beta\beta}+K^{\beta\gamma}A^{\gamma\beta}+(A^{\gamma\beta})^T K^{\gamma\beta} & (A^{\gamma\beta})^T K^{\gamma\gamma} \\ 0 & K^{\gamma\gamma}A^{\gamma\beta} & -K^{\gamma\gamma} \end{bmatrix} \begin{Bmatrix} q^\alpha \\ q^\beta \\ q^\gamma \end{Bmatrix} = \begin{Bmatrix} f^\alpha \\ f^\beta+(A^{\gamma\beta})^T f^\gamma \\ 0 \end{Bmatrix} \qquad (9)$$

The row operations described above ensure that the static implications of the constraints are satisfied, while the column operations take care of the kinematic considerations. The solution of Equation (9) yields the values of all degrees of freedom for the constrained system.

A major consideration in adopting the above procedure was the fact that the overall dimensions of the master coefficient matrix do not vary when multipoint constraints are included. An active column profile equation solver was already in use and adjustments to the profile had to be designed such that constraints could be incorporated. A further complication exists in that a varying number of constraints are active at the barrage nodes for each of the flow phases covering sluicing, waiting and turbining. Although the general procedure is described in submatrix operations by Equation (9), the profile adjustments must be carried out on a term by term basis and extensive programming is required in order to incorporate this algorithm.

APPLICATION OF MODEL TO THE SHANNON ESTUARY

The model developed above is applied to a single effect ebb-generation tidal power scheme in the Shannon Estuary on the West Coast of Ireland; the selected study area is indicated on Figure 2. This area is discretized using 488 linear triangular elements and a total of 333 nodes. The two sets of barrage nodes are node numbers 241-246 on the downstream side and 247-252 on the upstream side. Flow through the barrage is described by imposing constraint relationships on those barrage nodes corresponding to the sluice and turbine positions as described above. Node numbers 1-8 represent the ocean boundary and the wave height forcing function at this boundary is specified as having an amplitude of 1.9 m. The inflow from the river Shannon is fixed at 282 m^3/s and this is applied as a uniform flux across the boundary between nodes 331 and 333.

The constraint equations of flow which are applied to the sluice and turbine nodes include a parameter k which is dependent on the hydraulic properties of the particular sluices and turbines being used in the scheme. In this case, twenty sluices and thirty turbines are used. This particular combination is one which has been recommended by Wilson [5] for this site. The sluices are 12.2 m square and are of the submerged Venturi type with an assumed coefficient of discharge of 1.5. The turbines are of the variable distributor, fixed blade type as supplied by Escher Wyss, A.G. of Zurich. The turbine characteristics are as follows:

Runner diameter	7.6 m	Rotational speed	44.78 rpm
Rated head	3.5 m	Blade angle	14°
Minimum head	1.01 m	Rated power capacity	10.60 MW
		Generator efficiency	95%

The head-flow curve for this machine is shown in Figure 4 [5]. From these characteristics the precise form of the constraint equation is determined for each phase.

The results of applying the model to this problem are given in Figure 5 where the variation in sea and basin levels during the entire tidal cycle is shown. The most dramatic change occurs upstream of the barrage in the basin area where the variation in levels over a tidal cycle is computed as 1.6 m to 0.6 m as compared to approximately 2.36 m to -2.3 in the existing situation. Downstream of the barrage the tidal range is reduced by approximately 0.35 m to 4.3 m, and there is a phase difference of approximately 4 minutes caused by the presence of the barrage. Velocity plots for the phase of operation when the the turbines are active is given in Figure 6. These velocities are average values for each element and, therefore, velocities in the barrage region do not represent the exit velocities from the sluices or turbines. Comparison of the overall flow patterns predicted by the model with those computed for the existing situation show a very slight reduction in velocities between the mouth of the estuary and Kilrush but quite a dramatic reduction in the area between Kilrush and the barrage site and in the entire upstream basin areas.

The power output curve for the generation phase is shown in Figure 7 and the maximum power output during the tidal cycle is approximately 290 MW. These results demonstrate the effects of an operational barrage under one particular set of tidal flow conditions in the estuary. For a full assessment of the scheme, the model may be applied over the full range of known tidal conditions that occur in the estuary and power output over the useful life of such a scheme may be estimated.

CONCLUSION

The utility of using constraint equations with a finite element flow model for simulating a variety of physical conditions such as turbine or sluice action, solid barrage and computation of power output is demonstrated. A major feature of this approach is the ability to continuously model the flow regimes over the changing phases of a tidal cycle and corresponding physical adjustments at the barrage site. The model is also capable of producing more accurate results with respect to flow and water levels for considering the environmental impact of a tidal power barrage on an estuary.

REFERENCES

1. Charlier, R.H. (1982) Tidal Energy, Van Nostrand Reinhold.

2. Brebbia, C.A. and Partridge, P. (1976) Finite Element Models for Circulation Studies. Mathematical Models for Environmental Problems. Pentech Press. 141-159.

3. Barlow, J. (1982) Constraint Relationships in Linear and Nonlinear Finite Element Analysis. Int. J. Num. Meth. Eng, 18; 521-533.

4. Abel, J.F. and Shephard, M.S., (1969) An Algorithm for Multipoint Constraints in Finite Element Analysis. Int. J. Num. Meth. Eng., 14, 464-467.

5. Wilson, E.M. et al (1981). Report on the Tidal Energy Potential of the Shannon Estuary. Report to Electricity Supply Board, Dublin.

ACKNOWLEDGEMENTS

The authors wish to express their gratitude to Mr. R.M. O'Flaherty, Divisional Manager, Civil Works Department of the Electricity Supply Board for his interest in this project. The views expressed in this paper, however, are solely those of the authors and do not represent those of the ESB. The assistance of Ann Cullinane in typing and Colm Walsh for assistance with the figures is gratefully acknowledged.

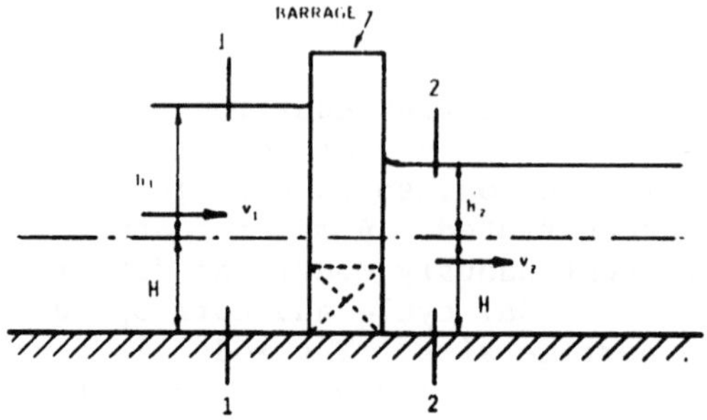

FIGURE 1 CROSS-SECTION THROUGH BARRAGE

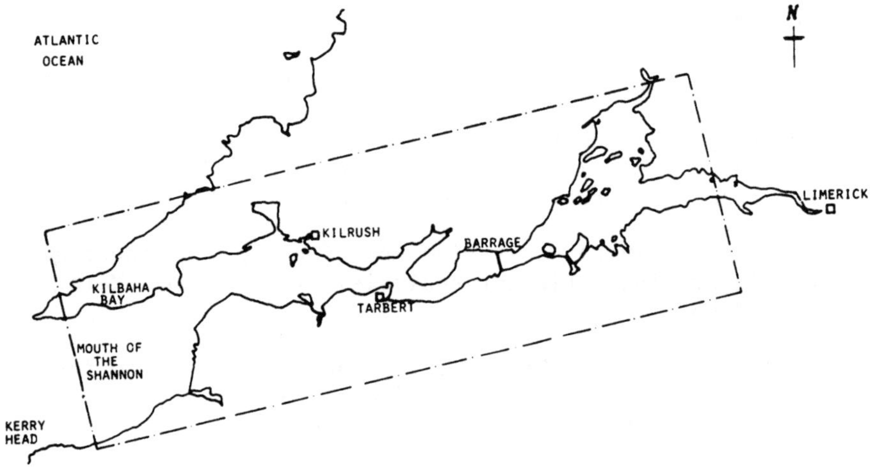

FIGURE 2 SHANNON ESTUARY - STUDY AREA

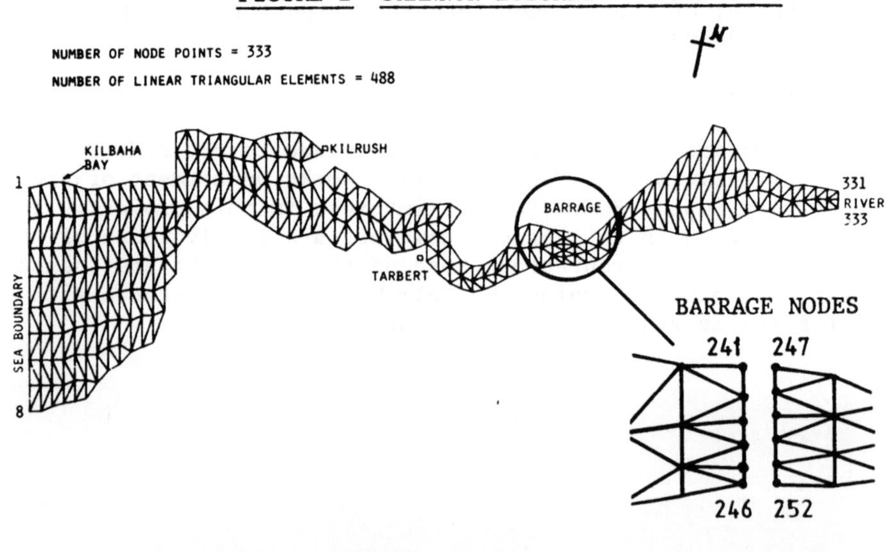

FIGURE 3 SHANNON ESTUARY - FINITE ELEMENT MESH

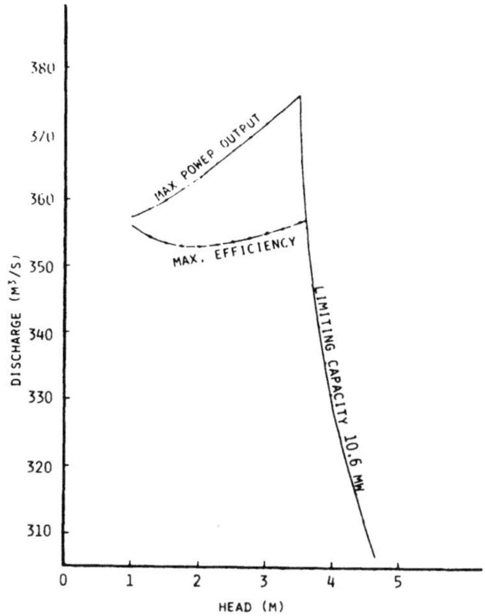

FIGURE 4 TURBINE CHARACTERISTIC CURVE

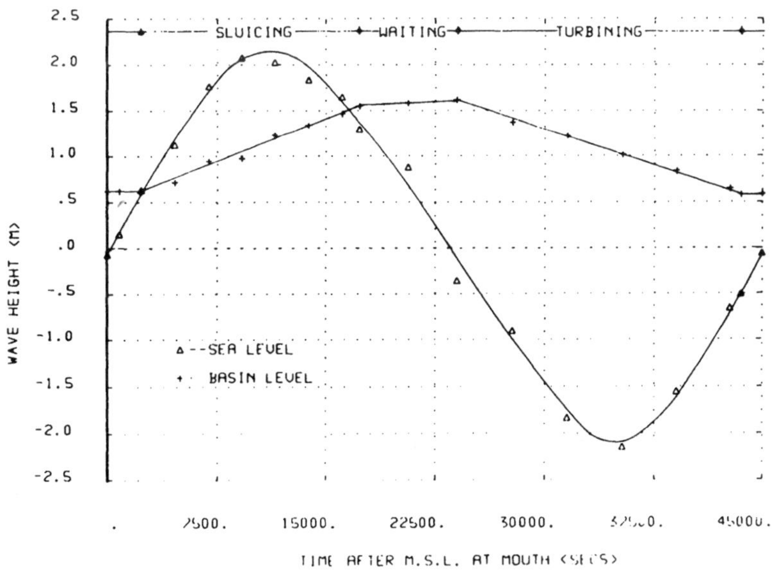

FIGURE 5 WAVE HEIGHT ABOVE AND BELOW BARRAGE

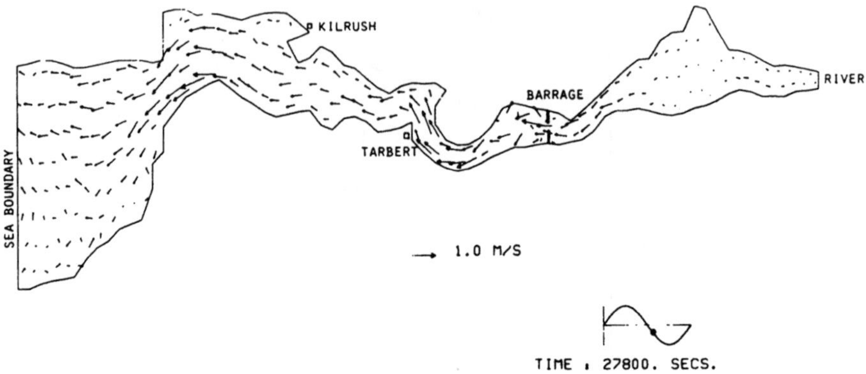

FIGURE 6 FLUID VELOCITIES DURING TURBINING PHASE

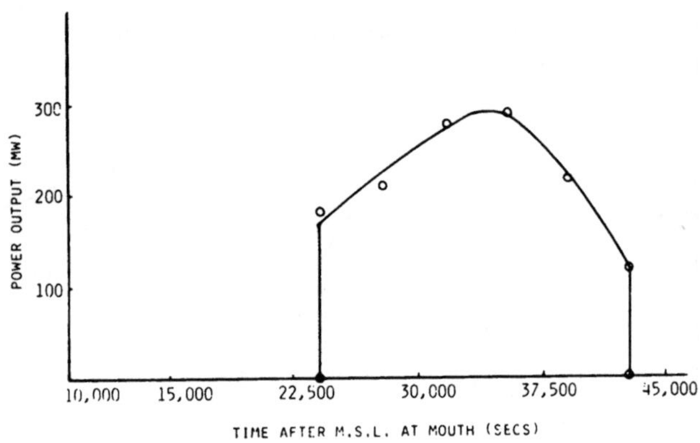

FIGURE 7 POWER OUTPUT CURVE

SHALLOW-WATER WAVES AT URSELL NUMBERS OF ORDER UNITY

N. D. Katopodes and C.-T. Wu

Department of Civil Engineering, The University of Michigan

INTRODUCTION

In a variety of engineering problems involving free-surface flow, certain combinations of initial and boundary conditions give rise to long waves of moderate amplitude, which are not accurately described by the zero-order shallow-water equations. Although these are predominantly nonlinear waves, the pressure distribution in the vertical is not hydrostatic and significant wave dispersion is manifested by oscillations of the free surface. Since such waves are often observed in the transition from a constant state to that of nonlinear nondispersive waves, as in the case of the sudden opening of a sluice gate in a channel, the proper formulation and efficient solution of the associated flow conditions is of great interest.

A more formal identification of the above flow condition is made by means of the Ursell number U (Ursell, 1953). If the wave motion is described by a characteristic horizontal length L and a characteristic wave height H, the relative importance of nonlinearity and dispersion is measured in terms of the following dimensionless parameters

$$\alpha = \frac{H}{D} \tag{1}$$

$$\beta = \frac{D^2}{L^2} \tag{2}$$

in which D is the undisturbed water depth. In most cases of practical interest the wave length is large compared to the depth and, in addition, the wave height is small compared to the depth. It is obvious that α is a measure of the nonlinear effects, while β represents the magnitude of the dispersive effects in the wave.

Although α and β are completely independent, pertinent physical relations can only be formulated if their relative magnitude is of order unity. Otherwise the order of approximation of the velocity variation in the vertical is not consistent. This accomplished by means of the Ursell number

$$U = \frac{\alpha}{\beta} = \frac{HL^2}{D^3} \tag{3}$$

For large values of U, the nonlinear effects completely dominate the flow and the zero-order shallow-water equations describe the wave motion in its entirety. For small values of U, the theory of linear, dispersive waves is applicable. For Ursell numbers of order unity both nonlinear and dispersive effects are important. The equations governing such flows were derived by Boussinesq (Whitham, 1967)

$$\frac{\partial \eta}{\partial t} + \frac{\partial}{\partial x}\{(1+\alpha\eta)u\} - \frac{1}{6}\beta\frac{\partial^3 u}{\partial x^3} = 0 \tag{4}$$

$$\frac{\partial u}{\partial t} + \alpha u\frac{\partial u}{\partial x} + \frac{\partial \eta}{\partial x} - \frac{1}{2}\beta\frac{\partial^3 u}{\partial x^2 \partial t} = 0 \tag{5}$$

in which η is the dimensionless wave height, u is the dimensionless velocity at the channel invert and x, t represent dimensionless distance and time, respectively. In terms of the depth averaged velocity Eqs. 4 and 5 can be written

$$\frac{\partial \eta}{\partial t} + \frac{\partial}{\partial x}\{(1+\alpha\eta)\,\bar{u}\} = 0 \tag{6}$$

$$\frac{\partial \bar{u}}{\partial t} + \alpha\bar{u}\frac{\partial \bar{u}}{\partial x} + \frac{\partial \eta}{\partial x} - \frac{1}{3}\beta\frac{\partial^3 \bar{u}}{\partial x^2 \partial t} = 0 \tag{7}$$

Equations 6 and 7 are first-order shallow-water equations, which allow for a linear variation of velocity in the vertical. The higher order derivative in Eq. 7 represents a deviation from hydrostatic pressure distribution, since for $\beta \ll 1$, this equation reduces to the zero-order equation of motion.

CONSERVATION LAWS FOR NONLINEAR DISPERSIVE WAVES

Equations 6 and 7 may be written in dimensional conservation form by introducing the total flow depth $h = D + \eta$ and the volumetric flow rate per unit width $q = uh$. Then

$$\frac{\partial h}{\partial t} + \frac{\partial q}{\partial x} = 0 \tag{8}$$

$$\frac{\partial q}{\partial t} + \frac{\partial}{\partial x}\left(\frac{q^2}{h} + g\,\frac{h^2}{2}\right) - \frac{h^3}{3}\,\frac{\partial^3(q/h)}{\partial x^2 \partial t} = 0 \tag{9}$$

Equations 8 and 9 are easily recognized as the conservation laws for mass and momentum in a horizontal frictionless channel, accounting for non-hydrostatic pressure distribution in the vertical. It is convenient to rewrite the system of Eqs. 8 and 9 in matrix form as follows

$$\frac{\partial U}{\partial t} + A\,\frac{\partial U}{\partial x} - B^T Q = 0 \tag{10}$$

where

$$A = \begin{pmatrix} 0 & 1 \\ c^2-u^2 & 2u \end{pmatrix} \tag{11}$$

$$U = (h \quad q)^T \tag{12}$$

$$B = (0 \quad h^3/3)^T \tag{13}$$

$$Q = (0 \quad q/h)^T \tag{14}$$

in which $c = (gh)^{1/2}$, is the maximum celerity of elementary gravity waves.

The system represented by Eq. 10 is known to be of the hyperbolic-parabolic type and possesses four families of real characteristics given by (Abbott and Rodenhuis, 1975)

$$\frac{dx}{dt} = \bar{u} \pm c, \qquad \frac{dx}{dt} = \bar{u} \qquad \text{and} \qquad \frac{dx}{dt} \to \infty \tag{15}$$

The numerical solution of Eq. 10 is not an easy task, because the well-known difficulties associated with the zero-order equations are further amplified by the presence of the higher derivatives in the momentum equation. In fact, it is not difficult to realize that the truncation error of any conventional method used for the lower order equations will contaminate the true significance of the dispersion term.

The characteristic directions given by Eq. 15 on the other hand, are strongly reminiscent of those corresponding to the Saint-Venant and Burger's equations. Both of these equations are approximated with remarkable accuracy by certain Petrov-Galerkin scemes(Baker, 1979; Hughes et al, 1982; Katopodes,1984)

The solution vector is first approximated by means of the classical chapeau function

$$U = \langle N \rangle \{U\} \tag{16}$$

where $\{U\}$ represents the nodal values of U. The C^{-1} Petrov-Galerkin approximation to Eq. 10 reads

$$\int_L \left[\langle N \rangle + \varepsilon A^T \frac{\partial \langle N \rangle}{\partial x}\right]^T \left[\frac{\partial U}{\partial t} + A \frac{\partial U}{\partial x} - B^T Q\right] dx = 0 \tag{17}$$

where

$$\varepsilon = \frac{\Delta x}{\rho(A)\ 15} \tag{18}$$

and ρ (A) represents the spectral radius of A. This formulation is known to lead to a fourth-order accurate approximation of the zero-order equations accompanied by highly selective dissipation of wave amplitude. By representing the continuity equation in Eq. 8 by E_1 and the first two terms in Eq. 9 by E_2, Eq. 17 may be written in scalar form as follows

$$\int_L (\langle N \rangle^T E_1 + \varepsilon \langle N \rangle^T E_2)\, dx = 0 \tag{19}$$

$$\int_L \{(\langle N \rangle + 2\varepsilon u \frac{\partial \langle N \rangle}{\partial x})^T E_2 + \varepsilon\, (c^2 - u^2) \frac{\partial \langle N \rangle^T}{\partial x} E_1 + [\langle N \rangle^T h^2 + \frac{\partial \langle N \rangle^T}{\partial x} (\frac{h^3}{3})] \frac{\partial^2}{\partial x \partial t} (\frac{q}{h})\}\, dx - \int_\Gamma \langle N \rangle^T \{(\frac{h^3}{3}) \frac{\partial^2}{\partial x \partial t} (\frac{q}{h})\} d\Gamma = 0 \tag{20}$$

The last term in Eq. 20 vanishes on all boundaries where conditions of practical interest are specified.

The remaining crucial step in the numerical solution of the Boussinesq equation focuses on the approximation of the time derivatives involved. A simple, central-difference scheme is adopted, which leads to the following discrete equations for continuity and momentum

$$\int_L \{\langle N\rangle^T(ah-c_1+\frac{\partial q}{\partial x})+\varepsilon\langle N\rangle^T[aq-c_2+\frac{\partial}{\partial x}(\frac{q^2}{h}+g\frac{h^2}{2})]\}dx = 0 \qquad (21)$$

$$\int_L \{((\langle N\rangle+2\varepsilon u\frac{\partial\langle N\rangle^T}{\partial x})[aq-c_2+\frac{\partial}{\partial x}(\frac{q^2}{h}+g\frac{h^2}{2})]+\varepsilon(c^2-u^2)\frac{\partial\langle N\rangle^T}{\partial x}$$

$$(ah-c_1+\frac{\partial q}{\partial x})+[\langle N\rangle^T h^2+\frac{\partial\langle N\rangle^T}{\partial x}(\frac{h^3}{3})](a\frac{q}{h}-c_3\frac{\partial\langle N\rangle}{\partial x}\}dx=0 \qquad (22)$$

where $a = \frac{2}{\Delta t}$ and c_1, c_2, c_3 are constants evaluated at $t = t_0-\Delta t$. It is apparent that Eqs. 21 and 22 contain no derivatives of order higher than one and therefore the C^0 basis functions suggested earlier for the numerical approximation of the equations should be adequate.

THE TRANSITION TO NONLINEAR NON-DISPERSIVE WAVES

The scheme suggested by Eqs. 21 and 22 is considered satisfactory for most situations characterized by long waves ($\beta<1$) of moderate amplitude ($\alpha<0.3$) even when the dissipation interface induced by $\varepsilon>0$ is removed. On the other hand, as the wave height becomes larger than about 30% of the undistributed water depth, the nonlinear effects begin to play a more significant role in the flow. Laboratory observations (Binnie and Orkney, 1955) indicate that for $0.3 < \alpha < 0.7$ a gradual steepening of the wave front takes place. In this transition zone both nonlinear and dispersive effects are important, therefore there is no reason to suspect the validity of the Boussinesq system. The latter does become invalid for $\alpha > 0.7$, when all dispersive effects are suppressed and the wave motion is characterized by a steep front in the form of a hydraulic base.

Previous attempts to model the transition from dispersion dominated nonlinear waves to nonlinear non-dispersive wave motion have shown that the governing equations are extremely sensitive to both phase and amplitude dissipation. Although the Boussinesq system is expected to simulate the flow for values of α up to unity, most available models become unstable for $\alpha > 0.3$, which is the value at which characteristics of the family $dx/dt = u$ intersect. It is now understood that unless a very accurate numerical scheme is employed, the wave dispersion is contaminated by the truncation error and any form of artificially induced dissipation will not be effective for the Boussinesq system. On the other hand, as shown by Eq. 20, the dissipative interface introduced in the present scheme is two orders of magnitude higher than the dispersion term written in weak form. Since the C^{-1} weighting does not affect the dispersion term itself at all, both the order zero and order one systems are solved to optimum accuracy, as far as linear basis functions are concerned.

SOLITARY WAVE TESTS

The accuracy of the shallow-water equations to order one approximation is tested by simulation of a solitary wave. For such a test the dissipative interface is not necessary, as the nonlinear effects are exactly balanced by the physical dispersion. In the test problem the undisturbed depth D is taken equal to 1 m. The initial condition consists of a Boussinesq solitary wave (Daily and Stephan, 1952) in which

$$\eta(x,t) = H \operatorname{sech}^2 \sqrt{\frac{3}{4}\frac{H}{D^3}}\,(x-ct) \tag{23}$$

$$u(x,t) = \sqrt{\frac{g}{D}}\left(\eta - \frac{1}{4D}\eta^2 + \frac{1}{6}D^2\frac{\partial^2\eta}{\partial x^2}\right) \tag{24}$$

where H is the wave amplitude. The computational test involves 40 uniform elements with q fixed at zero at both ends of the channel. Figures 1 through 3 show a comparison of the analytical and numerical solutions. Although the solitary wave is not an exact solution of the Boussinesq system the agreement for the cases shown is very satisfactory. There is no evidence of dissipation or phase discrepancies and the overall characteristics of the model with respect to convergence and computation time are very good.

SUDDEN WATER RELEASES

The purpose of the present model is to provide a means for continuous computation as α approaches unity and the nonlinear effects become dominant. The following water release tests were simulated. In a 10 m long channel the water is initially at rest at a 10 cm depth. Then, a step increase in depth takes place at the upstream end sending a water wave into the channel. The zero-order shallow-water theory predicts instantaneous formation of a hydraulic bore regardless of the value of the upstream pulse. On the other hand the first-order shallow-water theory shows a gradual approach to the bore condition. The results shown in Fig. 4 correspond to three different values

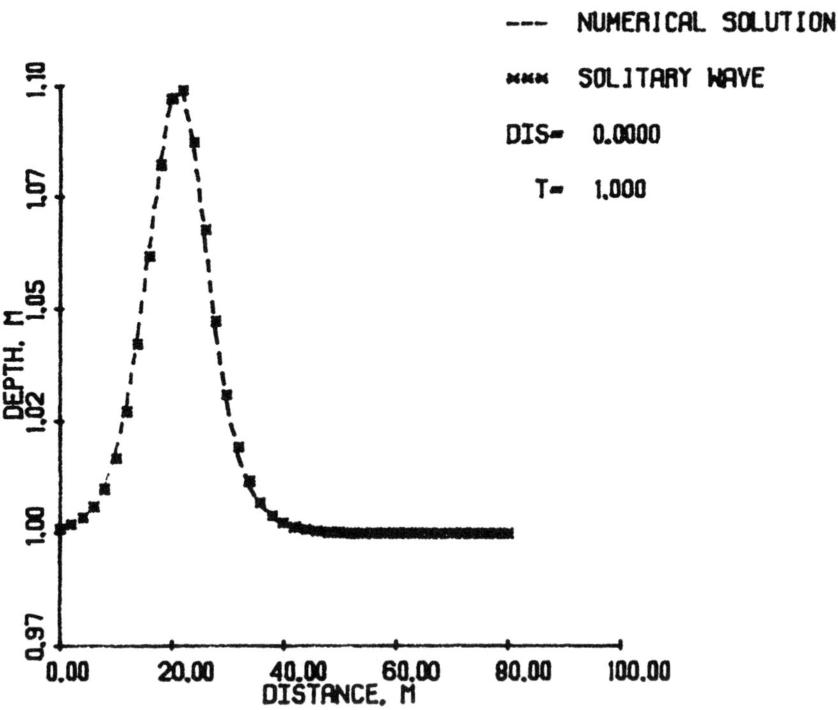

Figure 1. Solitary wave propagation t = 1 sec.

of upstream depth corresponding to 20, 50 and 80 percent increase in depth and therefore corresponding to the observed waves of Binnie and Orkney (1955). The results correspond to computed hydrographs 10 m downstream of the pont of the wave release. The qualitative behavior of the model is excellent, indicating a pure undular wave for the first case and a hydraulic bore for the last. The intermediate test is also in agreement with experimental findings showing both a broken wave front and a series of undulations. Although no quantitative tests were made for the water release problems in here, the model behaves qualitative as anticipated, that is, it successfully simulates water-wave motion for $0 < \alpha < 1$. A quantitative analysis and certain interesting laboratory findings for the same case studies will be presented in another publication, but it suffices to say that in its present state the model gives results with satisfactory accuracy for very small and very large values of α, while at intermediate values tends to overemphasize the nonlinear wave behavior.

The fact that the model simulates successfully the transition between dispersion balanced nonlinear wave motion and purely nonlinear wave propagation, can only be viewed with optimism and should lead to a unified numerical theory for shallow-water flow for both zero and first order effects.

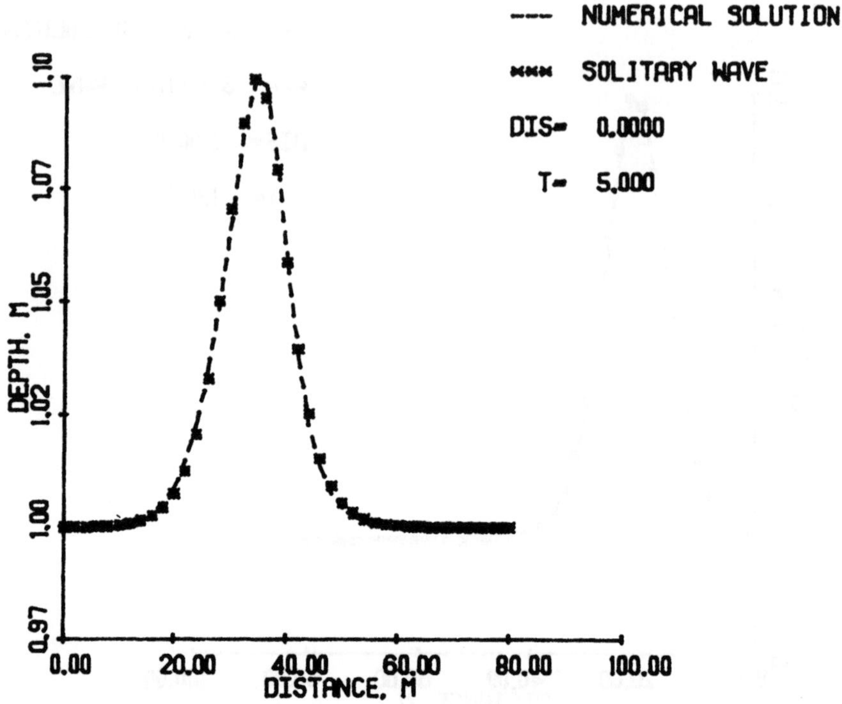

Figure 2. Solitary wave propagation t = 5 sec.

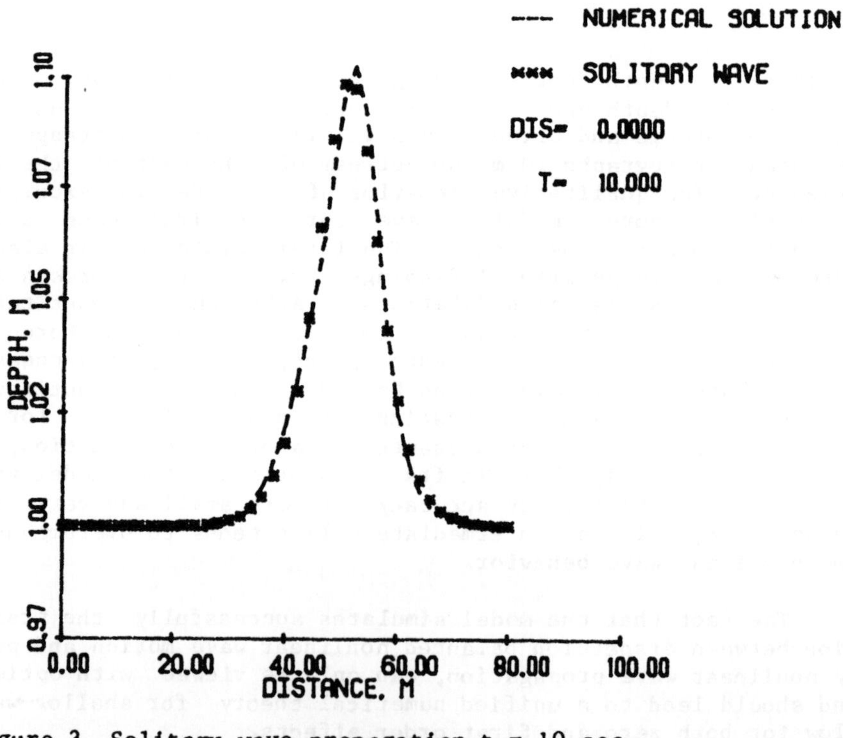

Figure 3. Solitary wave propagation t = 10 sec.

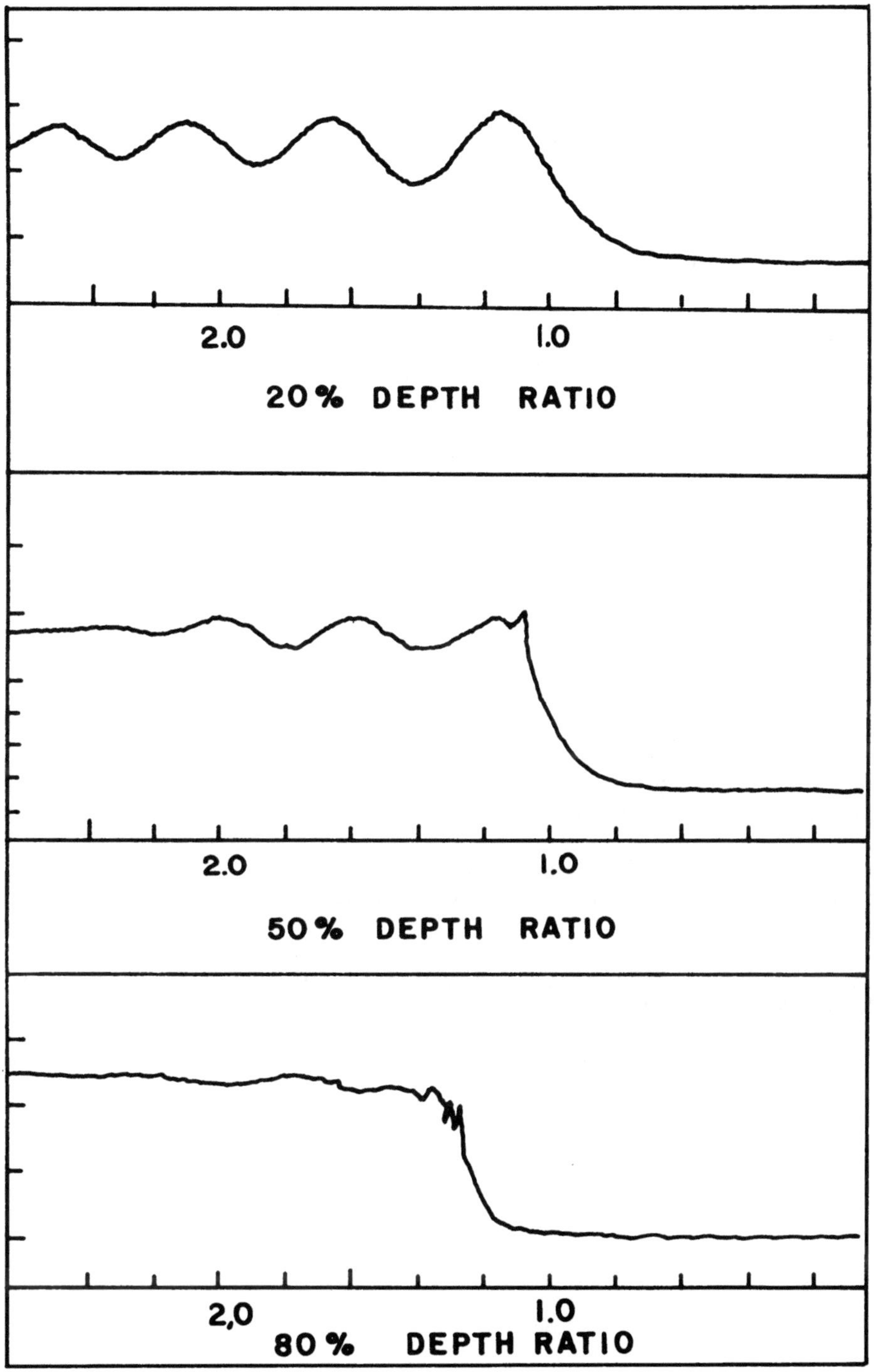

Figure 4. Sudden water releases. Depth hydrographs at x = 10 m.

REFERENCES

Abbott, M.B., Petersen,H.M. and Skovgard, O. (1978), "On the Numerical Modeling of Short Waves in Shallow Waters," Journal of Hydraulic Research, No.3, pp.173-203

Abbott, M.B. and Rodenhuis, G.S., (1971), "A Numerical Simulation of the Undular Hydraulic Jump," Journal of Hydraulic Research, No.2, pp.239-257

Baker, A.J. and Soliman, M.D., (1981) "On the Accuracy and Efficiency of a Finite Element Tensor Product Algorithm for Fluid Dynamics Application," Computer Methods in Applied Mechanics and Engineering, Vol. 27, pp.215-237

Binnie, A.M. and Orkney, J.C., (1955),"Experiments on the Flow of Water from a Reservoir Through an Open Horizontal Channel," Proc. Roy. Soc. A., pp.237-246

Daily, J.W. and Stephan, S.C, (1952), "The Solitary Wave" Technical report no. 8, Hydrodynamics Laboratory, MIT

Hughes, T.J.R. and Brooks, A.N., (1982), "A Theoretical Framework for Petrov-Galerkin Methods with Discontinuous Weighting Functions," Finite Elements in Fluids, Gallagher, R.H., ed. vol. 4, Wiley, London

Katopodes, N.D., (1984),"A Dissipative Galerkin Scheme for Open-Channel Flow," Journal of Hydraulic Engineering, ASCE, Vol.110

Ursell, F.(1953), "The Long Wave Paradox in the Theory of Gravity Waves," Proceedings of the Cambridge Philosophical Society, vol.49

Whitham, G.B., (1967), "Nonlinear Dispersive Waves," Proc. Roy. Soc., Series A, Vol. 283.

A TWO DIMENSIONAL ANALYSIS OF THE WAVE EQUATION MODEL FOR FINITE ELEMENT TIDAL COMPUTATIONS

M.G.G. Foreman

Institute of Ocean Sciences
Sidney, B.C., Canada

INTRODUCTION

Recently Gray and Lynch (1977, 1978) introduced a promising finite element method for solving the shallow water equations. Rather than working with the governing equations in conservation form, their 'wave equation' method (WEM) involves transforming the continuity equation to a second order partial differential equation. The revised system of equations is then solved with a Galerkin FEM, piecewise linear basis functions, and centred time stepping.

Dispersion analyses of the one dimensional WEM and its lumped counterpart, the LWEM, were presented in Foreman (1983). This study extends many of these results to two configurations of triangular elements. An asymptotic analysis for small wavenumbers is used to determine the time stepping method which most accurately represents phase and group velocity. This is done for both the WEM and LWEM. A linear stability analysis assuming zero Coriolis force is also given. Accuracy is shown to be independent of wave direction for a mesh of equilateral triangles. Furthermore, with equilateral triangles and an appropriate time stepping method, it is shown that lumping need not cause an accuracy loss in phase and group velocity.

Many of the algebraic details omitted in this presentation are found in Foreman (1984).

MATHEMATICAL BACKGROUND

The two dimensional linearized wave equations are

$$\frac{\partial^2 z}{\partial t^2} + \tau \frac{\partial z}{\partial t} - gh\left(\frac{\partial^2 z}{\partial x^2} + \frac{\partial^2 z}{\partial y^2}\right) + hf\left(\frac{\partial v}{\partial x} - \frac{\partial u}{\partial y}\right) = 0 \quad (1a)$$

$$\frac{\partial u}{\partial t} + g\frac{\partial z}{\partial x} - fv + \tau u = 0 \quad (1b)$$

$$\frac{\partial v}{\partial t} + g\frac{\partial z}{\partial y} + fu + \tau v = 0 \quad (1c)$$

where $z(x,y,t)$ = elevation above mean sea level,
$u(x,y,t)$ = x component of the velocity,
$v(x,y,t)$ = y component of the velocity,
$h(x,y)$ = mean sea depth,
g = gravity,
$f(x,y)$ = Coriolis coefficient,
τ = linear bottom friction coefficient.
In this analysis, the effects of both boundary conditions and initial conditions are ignored.

Assuming constant values for the depth and Coriolis coefficient, plane wave solutions of the form

$$\begin{pmatrix} z \\ u \\ v \end{pmatrix} = \mathrm{Re}\left\{\begin{pmatrix} z_o \\ u_o \\ v_o \end{pmatrix} e^{i\left[k_1 x + k_2 y - \omega t\right]}\right\} \quad (2)$$

can be found for Equation 1. ω is frequency and

$$\underline{k} = (k_1, k_2) \quad (3a)$$

are the (x,y) components of wavenumber. The distance between successive wave crests is the wavelength

$$L = 2\pi/|\underline{k}| \; . \quad (3b)$$

For nontrivial plane wave solutions, the following quartic

$$(-i\omega+\tau)\left(\omega^3+2i\tau\omega^2-\omega\left[\tau^2+f^2+gh(k_1{}^2+k_2{}^2)\right]-igh\tau(k_1{}^2+k_2{}^2)\right) = 0 \quad (4)$$

must be satisfied. When, for some functions W_j, each of the four roots is expressed in the form

$$\omega_j = W_j(k_1,k_2,f,\tau,g,h) \quad \text{for} \quad j = 1, 4 \quad (5)$$

they are said to define a dispersion relationship. Such relationships describe the behaviour of the wave solutions. Of the four solutions to Equation 4, only two are associated with propagating solutions. These are the gravity modes. A third solution would describe a planetary wave if either f or h were nonconstant. However, when both are constant, as is assumed here, the solution no longer propagates. It simply decays (or grows) in time. The fourth solution is a consequence of transforming the continuity equation to a second order partial differential equation. This mode is also stationary and decays in time when $\tau>0$. In the subsequent analysis, only numerical gravity waves are studied.

Phase and group velocity are defined (LeBlond and Mysak (1978)) as

$$\underline{C} = \mathrm{Re}(\omega)\underline{k}/|\underline{k}|^2 \tag{6a}$$

$$\underline{C} = \mathrm{Re}\left(\frac{\partial\omega}{\partial k_1},\frac{\partial\omega}{\partial k_2}\right). \tag{6b}$$

Whereas phase velocity describes the speed and direction of individual waves, group velocity describes the speed and direction of energy propagation. The importance of group velocity in numerical methods is surveyed by Trefethen (1982). Although shallow water waves have virtually the same phase and group velocity, their numerical model representations may not.

As might be expected, the particular shape and configuration of the spatial discretization affects the accuracy of a FEM implementation. Platzman (1981) examines two triangular meshes in his study of FEM tidal models, while Mullen and Belytschko (1982) study the effects of four meshes on spatial discretizations of the wave equation. Both investigations assume linear basis functions.

The two configurations studied here are meant to be illustrative rather than comprehensive. They are shown in Fig. 1. Both contain six equal triangles with the three variables z, u, and v defined at each vertex or node. The first mesh contains right isoceles triangles. A mirror image of this case is examined by Mullen and Belytschko (1982).

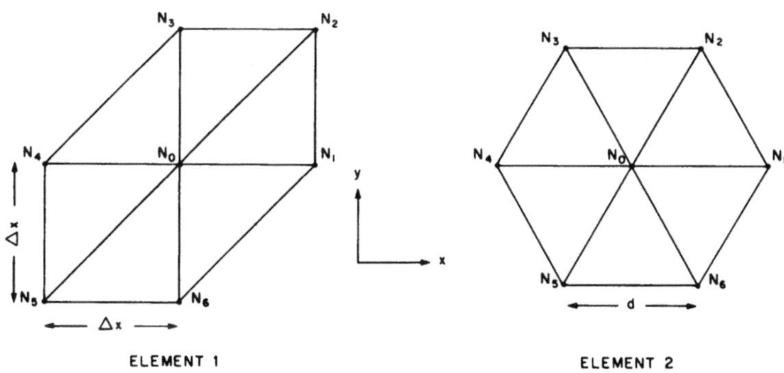

Figure 1. Triangular element configurations for the analysis (adapted from Foreman (1984)).

The second mesh contains equilateral triangles and is also considered by Mullen and Belytschko (1982). Because of their symmetry, one would intuitively expect equilateral triangles to be more accurate. Indeed, Mullen and Belytschko conclude that for their problem, this arrangement almost removes the directional dependence of phase velocity. Numerical experiments by Hinsman et al. (1982) have also demonstrated that equilateral triangles are more accurate than right triangles.

THE ANALYSIS

Lynch and Gray (1978) use the following approximations for the partial time derivatives in Equation 1

$$\frac{\partial}{\partial t} z_j(n\Delta t) \simeq \frac{z_j^{n+1} - z_j^{n-1}}{2\Delta t} \tag{7a}$$

$$\frac{\partial^2}{\partial t^2} z_j(n\Delta t) \simeq \frac{z_j^{n+1} - z_j^{n} + z_j^{n-1}}{\Delta t^2}. \tag{7b}$$

Terms which do not involve time derivatives are treated in one of three ways. The gravity terms in all equations are time averaged as

$$z_j(n\Delta t) \simeq \tfrac{1}{2}\theta\left(z_j^{n+1} + z_j^{n-1}\right) + (1-\theta)z_j^{n}. \tag{7c}$$

For all values of θ, this averaging is centred over time level n.

The friction term in the momentum equations is treated similarly but with the weighting parameter α. Separate averaging of the gravity and momentum terms permits a relaxation of the stability constraints.

All Coriolis terms are evaluated directly at time level n. Consequently, the fully discretized wave equation does not involve velocities at time level n+1, although the momentum equations do require elevations at that level. This permits a computational time saving since the continuity and momentum equations may now be solved sequentially rather than simultaneously.

Dispersion relationships, phase velocities, and group velocities can be calculated for both the spatially and fully discretized numerical solutions. An accurate spatially discretized solution does not always produce an accurate fully discretized solution. The time stepping method may either augment, or partially cancel, errors arising from the spatial discretization.

The fully discretized equations for both elements 1 and 2 have a characteristic equation which is a sixth order polynomial in the eigenvalue

$$\lambda = e^{-i\omega\Delta t} \ . \qquad (8)$$

ω, $\underline{C}$ and $\underline{G}$ are calculated from the roots of this polynomial. Two roots are associated with gravity waves. They are called principal roots. The remaining four roots are either artifacts of the numerical solution or approximations to the analytic stationary modes. We shall refer to all these roots as spurious. They are important only insofar as they can contaminate the principal numerical solution.

A linear stability analysis of the WEM is difficult when f is nonzero. However, necessary stability conditions with realistic nonzero values of f should only be perturbations of the conditions derived by assuming f = 0. Therefore, the restrictions obtained by assuming f = 0 should be close to those required with nonzero Coriolis. Numerical computations confirm this.

When f = 0, the characteristic polynomials for both elements in Fig. 1 have a form that is similar to the one dimensional characteristic equation. Consequently, the stability analysis is also similar.

In one dimension, spurious roots are stable when (Lynch and Gray (1978))

$$\alpha \geq \tfrac{1}{2} \ . \qquad (9a)$$

When f = 0, the two dimensional spurious roots are identical to the one dimensional roots. However, each root now has multiplicity two. This means that Equation 9a should be replaced by

$$\alpha > \tfrac{1}{2} \ . \qquad (9b)$$

We also wish to choose a value of α so that the spurious roots have magnitudes that are less than the principal roots (see Foreman (1983)). This is not possible for all $\tau\Delta t$, but $\alpha = 1$ seems to be a good compromise.

Define the following Courant numbers

$$f_2 = (gh)^{\frac{1}{2}}\Delta t/\Delta x \qquad (10a)$$

$$f_2' = (gh)^{\frac{1}{2}}\Delta t/d \qquad (10b)$$

for element 1 and 2. The principal roots for element 1 are stable when

$$\theta \geq \tfrac{1}{2}\left(1 - \frac{1}{6.4641\ f_2{}^2}\right) \qquad (11a)$$

while for element 2, they are stable when

$$\theta \geq \tfrac{1}{2}\left(1 - \frac{1}{5.5783(f_2')^2}\right). \tag{11b}$$

In one dimension, it was possible to choose a value of the time stepping parameter θ which provided optimal accuracy of the WEM dispersion relationship. The same can be done in two dimensions.

Assume $f = \tau = 0$. The most accurate value of θ is found by matching terms in expansions of the analytic and numerical dispersion relationships. Since numerical models are usually designed so that wavelengths are at least $20\Delta x$ (i.e., $k\Delta x<.314$), we expand for small $k\Delta x$. This means that the optimal θ may only be valid for long waves. However, this is acceptable. Principal waves which are shorter than $20\Delta x$ in some model regions suggest the need for mesh refinement. Short waves (e.g. $2\Delta x$ waves) that have been generated by boundary conditions, interfaces, and round-off errors may exist in a model and may be important insofar as they can contaminate the desired waves. However, it is not important that they be modelled accurately, only that their growth be controlled.

With $k_2 = sk_1$, matching terms (to order 4) in the analytic and element 1 expansions of $\cos(\omega\Delta t)$ reveals that

$$\theta = 1/6 + \frac{1}{3f_2^2} + \frac{1}{6f_2^2}\left[\frac{2s}{1+s^2} - \frac{1+s^4}{(1+s^2)^2}\right] \tag{12}$$

is the most accurate time stepping parameter. Notice that this value is dependent on wave direction. With $s = 0$ or $s = \infty$, Equation 12 reduces to the one dimensional result

$$\theta = 1/6\left(1 + \frac{1}{f_2^2}\right). \tag{13}$$

This value produces the most accurate representation of wave propagation along either axis of element 1 in Fig. 1. However for waves propagating along the diagonal (i.e., $s = 1$), the most accurate value is

$$\theta = 1/6 + \frac{5}{12f_2^2}\ . \tag{14}$$

With $f_2 = (2)^{-\frac{1}{2}}$, these two optimal values are appreciably different, namely $\frac{1}{2}$ and 1.

Accuracy which varies with wave direction is clearly undesirable. It implies that grid orientation affects model accuracy, and that by simply changing direction, a wave may be less accurately represented. Fortunately, with the preceding simplifying assumptions this directional dependence can be avoided with equilateral triangles.

Matching terms in the element 2 and analytic expansions of $\cos(\omega\Delta t)$ yields

$$\theta = 1/6 + \frac{1}{8(f_2')^2} \tag{15}$$

as the time stepping parameter value which most accurately approximates analytic wave propagation. Unlike Equation 12, it is not directionally dependent. Furthermore, with the substitution $\Delta x = ((3)^{\frac{1}{2}}/2)d$, Equation 15 reduces to the one dimensional result given by Equation 13. This substitution is equivalent to projecting element 2 onto the y axis.

Lynch and Gray (1978) also discuss an appropriate quadrature rule which has the effect of lumping their equations. Necessary stability restrictions for the LWEM can also be found when $f = 0$. The spurious eigenvalues are identical to those for the WEM and are thus governed by the same stability conditions. The principal eigenvalues are stable when

$$\theta \geq \tfrac{1}{2}\left(1 - \frac{1}{2f_2^2}\right) \tag{16a}$$

for element 1, and

$$\theta \geq \tfrac{1}{2}\left(1 - \frac{1}{1.47218(f_2')^2}\right) \tag{16b}$$

for element 2.

The one dimensional analysis in Foreman (1983) found an optimal value of θ for the LWEM. Specifically $\theta = 0$ was seen not only to avoid an accuracy loss through lumping, but also to most accurately represent both wave propagation and wave amplitudes. A similar analysis can be applied in two dimensions.

Assume $f = \tau = 0$. With element 1 and $k_2 = sk_1$, matching the analytic and element 1 expansions of $\cos(\omega\Delta t)$ for small wavenumber yields

$$\theta = 1/6\left(1 - \frac{1 + s^4}{f_2^2(1+s^2)^2}\right). \tag{17}$$

As with the WEM, $s = 0$ or $s = \infty$ produces the one dimensional result (Foreman (1983)),

$$\theta = 1/6\left(1 - \frac{1}{f_2^2}\right). \tag{18}$$

For element 2, similar matching yields

$$\theta = 1/6 - \frac{1}{8(f_2')^2}\ . \tag{19}$$

Again, this optimal value is not directionally dependent. And with the substitution $\Delta x = ((3)^{\frac{1}{2}}/2)d$, it reduces to the one dimensional result given by Equation 18.

Fig. 2 illustrates the stability regions and the most accurate values of θ for the WEM and LWEM over a configuration of equilateral triangles. Values for f_2' and θ should be chosen so that the resultant numerical method is stable. The particular choice may be a compromise between accuracy and time step size. Large values of Δt (or f_2') result in less computation cost but may be less accurate. Computationally, the explicit LWEM ($\theta = 0$) should be most economical. Unfortunately, the associated f_2' value which most accurately represents phase and group velocity ($f_2' = .866025$) is outside the stability region. $f_2' = .824175$ is the most accurate and stable choice.

Figure 3 shows the phase and group velocity accuracy of the explicit ($\theta = 0$) LWEMs for element 1 and element 2 when $f = \tau = 0$. Symmetries in the principal dispersion relationships mean that only positive k_2 values need be shown. f_2 and f_2' are the maximum stable values dictated by Equation 16, namely $f_2 = .7071$ and $f_2' = .824175$. The element 1 value is most accurate for wave propagation along the directions $k_2=\pm k_1$, while the element 2 value is almost directionally independent for small wavenumbers.

All phase velocity directions are exact so only phase speed errors are shown. Negative errors denote waves travelling too slowly. For example, -.01 denotes a numerical wave speed which is 1% too slow. With element 1, notice the deterioration of phase speed accuracy when waves are not propagating along the favoured directions. With element 2, error contours for small wavenumbers are virtually circular. This implies no favoured direction. Element 1 phase speeds are more accurate than those of element 2 in some wavenumber regions. However, because of its generally high accuracy in all directions, element 2 is preferable.

Group velocities are not shown for the case $(k_1,k_2) = (0,0)$ which represents infinite sampling per wavelength. However consistency of the numerical solution indicates that for this limit, the LWEM values approach the analytic. Notice that numerical group velocity directions can vary considerably from the analytic directions (k_1,k_2). Consequently, phase and group velocity are not co-directional as they should be for shallow water waves. This means that energy travels in a different direction than the wave crests.

In one dimension, $\theta = 0$ was seen to produce the most accurate approximation of wave amplitudes. The same is true in two dimensions when $f = 0$ and $\tau > 0$. In fact, it is true for both element 1 and element 2, and the WEM and LWEM. For the special case $f = \tau = 0$, all values of θ produce exact amplitudes.

Denote the optimal parameter values of Equations 15 and 19 by θ^*, and set $x = k_1 d$, $y = k_2 d$. Dispersion relationships for both the WEM and LWEM, when expanded to terms of order 6, are

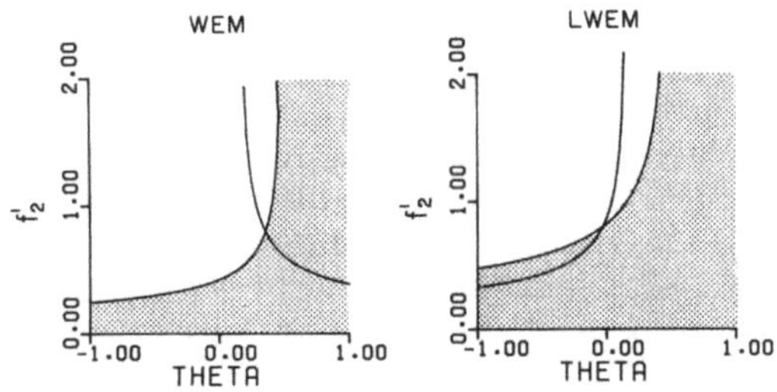

Figure 2. Stability and accuracy for the WEM and LWEM over equilateral triangles. Shaded regions denote stability. Solid lines designate the most accurate values of θ and f_2' for wave propagation (from Foreman (1984)).

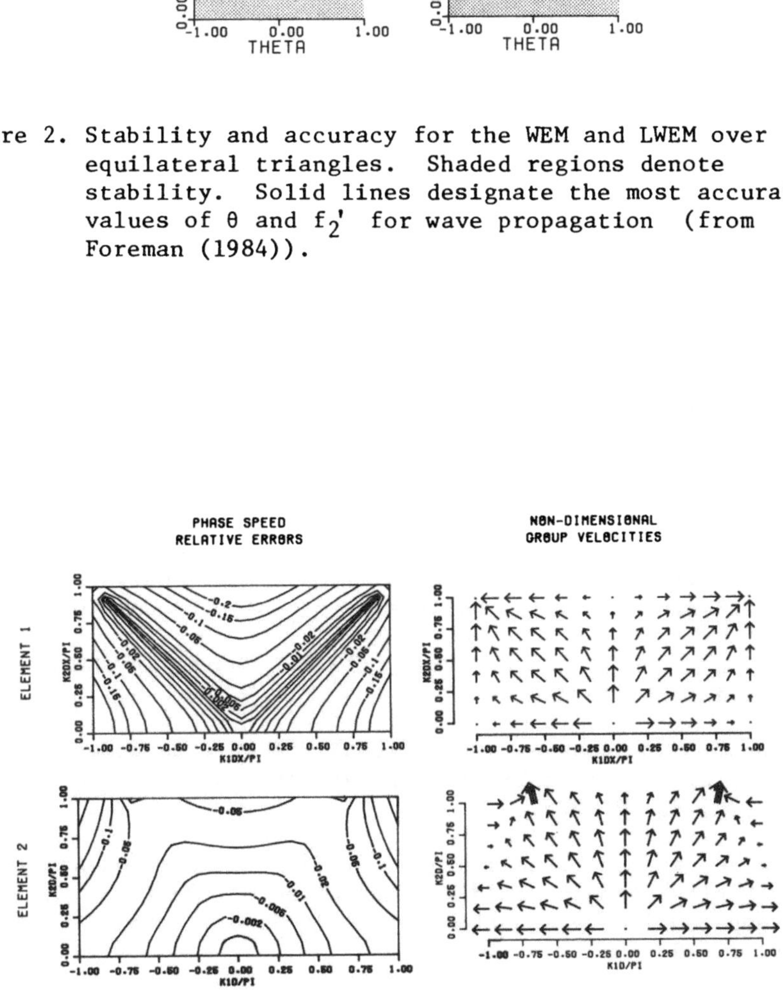

Figure 3. Phase speed errors and group velocities for the LWEM with τ=f=0. Parameter values are θ=0, α=1, f_2=.7071 for element 1, and f_2'=.824175 for element 2. Each full shaft of multi-shafted group velocity vectors has the magnitude $(gh)^{\frac{1}{2}}$. The smallest phase speed error contour is -.0005 (adapted from Foreman (1984)).

$$\cos(\omega\Delta t) \simeq 1 - \tfrac{1}{2}(f_2')^2(x^2+y^2) + \tfrac{1}{4}(f_2')^4(x^2+y^2)^2\left[1/6 + \theta - \theta^*\right]$$
$$- 1/8\ (f_2')^2(x^2+y^2)^3\left[(f_2')^4(\theta-\theta^*+1/6)^2 - 1/64\right]$$
$$- 1/256\ (f_2')^2\left[11/45\ x^6 + 1/5\ y^6 + x^2y^4 + 1/3\ x^4y^2\right]. \quad (20)$$

This implies that around their respective θ^* values, both the WEM and LWEM have the same accuracy deterioration for $\cos(\omega\Delta t)$. As was found in one dimension, the best time stepping method for the lumped scheme produces the same propagation accuracy (to order 8 in $(k_1 d, k_2 d)$) as the best time stepping method for the unlumped method.

REFERENCES

1. Foreman, M.G.G. (1983) 'An Analysis of the "Wave Equation" Model for Finite Element Tidal Computations', J. Comput. Phys., 52, 290-312.

2. Foreman, M.G.G. (1984) 'A Two Dimensional Dispersion Analysis of Selected Methods for Solving the Linearized Shallow Water Equations', J. Comput. Phys., to appear.

3. Gray, W.G. and D.R. Lynch (1977) 'Time-Stepping Schemes for Finite Element Tidal Model Computation', Advances in Water Resources, 1, 83-95.

4. Hinsman, D.E., R.T. Williams, and E. Woodward (1982) 'Recent Advances in the Galerkin Finite Element Method as Applied to the Meteorological Equations on Variable Resolution Grids', in "Finite Element Flow Analysis" (ed. Tadahiko Kawai), University of Tokyo Press, Tokyo.

5. LeBlond, P.H. and L.A. Mysak (1978) "Waves in the Ocean", Elsevier, Amsterdam.

6. Lynch, D.R. and W.G. Gray (1978) 'A Wave Equation Model for Finite Element Tidal Computations', Computers and Fluids, 7, 207-228.

7. Mullen, R. and T. Belytschko (1982) 'Dispersion Analysis of Finite Element Semidiscretizations of the Two-Dimensional Wave Equation', Int. J. Numer. Methods Engrg., 18, 11-29.

8. Platzman, G.W. (1981) 'Some Response Characteristics of Finite-Element Tidal Models', J. Comput. Phys., 40, 36-63.

9. Trefethen, L.N. (1982) 'Group Velocity in Finite Difference Schemes', SIAM Review, 24, 113-136.

FINITE ELEMENT COMPUTATION OF TWO-DIMENSIONAL UNSTEADY FLOW FOR RIVER PROBLEMS

Zia Hosseinipour
Shaw University, Raleigh, North Carolina

Michael Amein
North Carolina State University, Raleigh, North Carolina

SUMMARY

This paper presents the results of studies on the development and application of a generalized finite element model for the computation of two-dimensional unsteady free surface flows with emphasis on river problems. The model formulation is based on the complete depth-integrated shallow water equations. The finite element procedure employs linear triangular elements and linear shape functions with Galerkin's method of weighted residuals. The time integration is carried out implicitly and the scheme is unconditionally stable. The computed results are in good agreement with published data and analytical solutions. The model has been tested on different problems and has been found to be very efficient.

INTRODUCTION

Many problems of hydraulic engineering require information concerning water heights and currents in the two-dimensional horizontal domain. Typical cases involve bays, estuaries, harbors and wide rivers. Recent advances in computation technology allow us to use sophisticated numerical methods for the analysis of such problems. Because of the relative ease and economy of computations on one hand and the ever growing demand for reliable information on the other hand, mathematical modeling has become a useful tool in the field of hydraulic engineering. In the beginning, the modeling was carried out by finite difference methods. But the complex topography and irregular boundaries of two-dimensional cases require a better and more reliable scheme if detailed information is desired. The finite element technique meets the above requirement. In recent years, the finite element method has been used to develop two-dimensional models for the simulation of bays, estuaries, harbors and lakes. Few models have been constructed for rivers. The model presented in this work emphasizes the application of FEM to unsteady river problems.

THE GOVERNING EQUATIONS

The depth-averaged hydrodynamic equations used in the formulation of the finite element model are based on the assumptions that a) the medium is homogeneous and incompressible, b) all shear stresses except vertical shear are negligible, c) pressure distribution is hydrostatic, d) friction losses in unsteady flow are equivalent to those of steady uniform flow and e) the channel bed is fixed. The equations are [Reid & Bodine (1968), Leendertse (1970)]:

$$\frac{\partial u}{\partial t} + u\frac{\partial u}{\partial x} + v\frac{\partial u}{\partial y} + g\frac{\partial z}{\partial x} + g.u.\frac{|V|.n^2}{C_r h^{4/3}} - \frac{C_d \rho_a V_w^m \cos\beta}{\rho h} = 0 \qquad (1)$$

$$\frac{\partial v}{\partial t} + u\frac{\partial v}{\partial x} + v\frac{\partial v}{\partial y} + g\frac{\partial z}{\partial y} + g.v.\frac{|V|.n^2}{C_r h^{4/3}} - \frac{C_d \rho_a V_w^m \sin\beta}{\rho h} = 0 \qquad (2)$$

$$\frac{\partial z}{\partial t} + \frac{\partial (uh)}{\partial x} + \frac{\partial (vh)}{\partial y} = q_I$$

where

$$|V| = \sqrt{u^2 + v^2}$$

x,y = coordinate direction
u,v = x and y components of the vertically integrated velocities
g = the acceleration due to gravity
z = water surface elevation from a chosen datum
n = Manning's friction coefficient
C_r = a constant equal to 2.20 in the English system and 1 in metric system
h = water height from the bed to surface
ρa, ρ = densities of air and water, respectively
V_w = wind velocity
m = an empirical constant
β = the angle of wind velocity with the x - direction
q_I = rate of distributed inflow or outflow per unit area
t = time

The Coriolis force was omitted because of its relative insignificance in most engineering applications.

FINITE ELEMENT FORMULATION

The formulation of the finite element procedure presented in this paper is based on Galerkin's method of weighted residuals.

In this method, the governing equations are multiplied by weighting functions and integrated by parts using Green's Theorem. The general behavior of the field variables is represented by some functional approximately to satisfy the differential equations and boundary conditions. Substitution of this functional in the original equations results in some error called a residual; the residual is required to vanish in some average sense over the solution domain. When identical polynomials are used for both the weighting functions and the unknown variables, the scheme is called Galerkin's method. Details of the method are given in many standard textbooks, e.g., Huebner (1975). Applications of the FEM to hydrodynamic problems are given by Thienpont and Berlamont (1980), Gray and Lynch (1977), Koutitas (1978), Katapodes (1980) and others.

The work reported in this paper concentrates on the treatment of the nonlinear friction and convective terms. For the friction terms, two different alternatives have been adopted, namely inclusion of these terms in the left hand side or right hand side matrices. Depending on the significance of the terms and the problem being analyzed, one or the other mode may be selected. In the first alternative, each of the friction terms, which involves three basic variables, is treated as a single variable. Thus, $SF_x = \frac{u|V|}{h^{4/3}}$ and $SF_y = \frac{v|V|}{h^{4/3}}$ are treated as single functionals. The integration is carried out to evaluate the functionals on the right hand side of the system matrix, which in effect is a linearization of the shear terms.

The second alternative involves treatment of the friction term as the product of a velocity component and the function $SF = \frac{|V|}{h^{4/3}}$, which translates into a nonlinear integration and inclusion of the shear terms in the left-hand side matrix. Using this approach, faster convergence is achieved in problems in which friction is dominant.

For the convective terms, several alternatives are used. In the first alternative, a convective term, such as Uh is treated as a single variable. The convective terms then will appear as columns on the right-hand side.

In the second alternative, one of the variables in the convective term is treated as a constant. Thus, $\frac{\partial}{\partial x}(uh)$ is represented by $h \frac{\partial u}{\partial x}$, the water depth h being taken from the previous time step.

The third alternative would apply the interpolation function to each variable in the convective term. This approach follows the behavior of the original differential equation most closely, however, the finite element formulation becomes more involved compared to the other alternatives. The model has provisions for all of the above cases and by invoking different subroutines one or the other mode may be used. Details of the methods are given by Hosseinipour (1983).

SOLUTION PROCEDURE

Global conveyance and flow matrices (stiffness and force in structural engineering) are obtained using the method of assembly by elements. Assembling the elemental equations for the entire region, the system of finite element equations is written as:

$$[C]\,\{\dot{\phi}_t\} + [A(\phi)]\,\{\phi\} = \{R\} \tag{4}$$

where ϕ is the vector of unknowns (velocities and height) at each time step in the flow domain, ϕ_t is the time derivative of ϕ, [C] is a matrix of constants, $[A(\phi)]$ is a matrix containing the non-linear convective and shear terms, and {R} is the right hand side column vector. The time integration of the system of equations is done by am implicit finite difference procedure as follows:

$$[C]\left(\frac{\phi^{t+1}-\phi^{t}}{\Delta t}\right) + (1-\Theta)\left[(1-\sigma)A^{t+1}_{(\phi)_{m+1}} + \sigma A(\phi)^{t+1}_{m}\right]\{\phi^{t+1}\} +$$

$$\Theta\left[(1-\sigma)A^{t+1}_{(\phi)_{m+1}} + \sigma A^{t+1}_{(\phi)_m}\right]\{\phi\}^{t} = \{R\} \tag{5}$$

where Θ is a time level weighting coefficient, subscript m is the iteration counter, superscript t refers to time levels and σ is the weighting coefficient for the evaluation of the non-linear terms in the iterative procedure. Upon simplification, the above equations reduce to

$$\left[[C] + \Delta t.(1-\Theta).\left[(1-\sigma)A^{t+1}_{(\phi)_{m+1}} + \sigma.A^{t+1}_{(\phi)_m}\right]\right]\{\phi^{t+1}\} =$$

$$\left[[C] - \Delta t.\Theta.\left[(1-\sigma)A^{t+1}_{(\phi)_{m+1}} + \sigma.A^{t+1}_{(\phi)_m}\right]\right]\{\phi^{t}\} + \Delta t.\{R\} \tag{6}$$

Since $\{\phi^t\}$ is a known vector at each time step and the coefficient matrix associated with it is also known at every iteration, therefore, the whole right hand side becomes a column vector. For the sake of simplicity, we will give the matrix coefficient of $\{\phi^{t+1}\}$ a new name, $[A_1]$ and we will express the right hand side as a known vector $[C_1]$, then

$$[A_1]\{\phi^{t+1}\} = [C_1] \tag{7}$$

In this form, $[A_1]$ is called the conveyance matrix and $[C_1]$, the flow matrix. To deal with the problem of non-linearity in the solution process, an iterative procedure that changes the conveyance and flow matrices at each time step and each iteration is used. To start the solution cycle, both ϕ components of the matrices will have the same values at time step t, i.e., ϕ_0^{t+1} and ϕ_1^{t+1} both are equal to ϕ^t. Then as the solution cycle progresses, the calculated values of ϕ at time step t+1 and iteration steps m and m+1 are used to find the solution at t+1 and m+2. The iteration cycle is continued until a tolerance vector $\{\varepsilon\}$ is satisfied. This scheme of evaluating the conveyance and flow matrices based on ϕ values obtained in the last iteration cycle rather than the ϕ values of the previous time step helps to expedite the convergence of the iterative scheme to a great extent, and hence, it is an alternative to under-relaxation or over-relaxation procedures which some researchers have used in the past {King et al (1975}. The solution scheme proved to be unconditionally stable for all values of $o \leq \theta \leq 1$ and $o \leq \sigma \leq 1$ and is most efficient at t $\theta = ½$ and $\sigma = ½$ (Crank-Nicholson scheme). The model is tested against the experimental published data of harbors, reservoirs and channels as well as an analytical solution of flow in a canal. In each case, the computed flow profiles agree very well with the experimental data and are very close to the analytical solutions. Some of the experiments are discussed below. Experiments conducted generally concern flow patterns rather than the exact numerical values, when the exact test data are not available.

EXPERIMENTS IN A RECTANGULAR PRISMATIC CHANNEL

This test problem is taken from Wang and Connor (1975). The model was tested over an extended period of time (several wave cycles) in a rectangular channel with a constant water depth under the action of a wave imposed at the open end. The input wave equation is given by

$$\eta = a \sin wt$$

where a is the amplitude, w is the wave frequency, t is the time, and η is the wave height. The analytical solutions of the above problem using the method of separation of variables are:

$$u = \frac{-a\sqrt{gh}}{h\cos w\frac{L}{\sqrt{gh}}} \sin \{w \frac{L}{\sqrt{gh}} (\frac{x}{L} - 1)\} \cos wt \tag{9}$$

and

$$\eta = \frac{a}{\operatorname{Cos} w \frac{L}{\sqrt{gh}}} \cdot \cos \left\{ \frac{wL}{\sqrt{gh}} \left(\frac{x}{L} - 1 \right) \right\} \sin wt \qquad (10)$$

where, L = total channel length, h = water height at rest, g = gravity constant and a = wave amplitude. The finite element grid network of the test channel, the analytical solutions and the model results are shown on figures 1-3. For the numerical solution, a small relative amplitude ($\frac{a}{h}$) was chosen, since the analytical solution was obtained using the linearized equation. The numerical solutions were obtained with L = 200m, w = .03142 rad/sec, channel width = 100 m, g = 9.81 m/s^2, h = 4 m and a = 0.1 m. The computations were carried out with time step sizes of 5 and 10 seconds. The computed wave heights at the closed end (node #15) are plotted along with the analytical solutions in Figure 2. The longitudinal velocity at the entrance (node #13) is shown in Figure 3. For all practical purposes, the model results agree very well with the analytical solutions.

TIDAL PROPAGATION IN AN S-SHAPED CANAL

This test problem is taken from Benque' et al (1982). The model was tested on a canal with an S-shaped depression one meter deeper than the rest of the canal. The canal was closed on three sides and was subjected to a tidal wave. The calculations started with at rest conditions and water heights of 6 and 7 meters, in the shallow and deep regions, respectively. As expected, higher currents were observed in the deeper region; the currents followed the S-shaped channel. This observation paralleled those of Benque' et al (1982). In their work, an ADI (Alternate Direction Implicit) finite difference with high and low Courant numbers was used. In another experiment of this research, the canal problem was simulated for a steady state flow. Upstream and downstream boundary conditions were assigned by a slanted (flat) parabolic velocity distribution. These conditions were maintained during the simulation time. Steady state conditions were reached with higher velocity gradients in the deeper regions. The flow patterns were very close to uniform conditions which is expected in a flume shaped canal. The velocity patterns are displayed on Fig. 4.

EXPERIMENTS WITH A SQUARE HARBOR

The model was tested in a square harbor under the influence of tidal action to check the applicability to similar problems. The simulation starts with a flat water surface, a uniform depth and at rest conditions. Boundary conditions are specified as a) zero velocities at the walls; and b) a sinusoidal wave at the harbor mouth. The circulation and water height patterns are shown in Fig. 5. The model results compare very

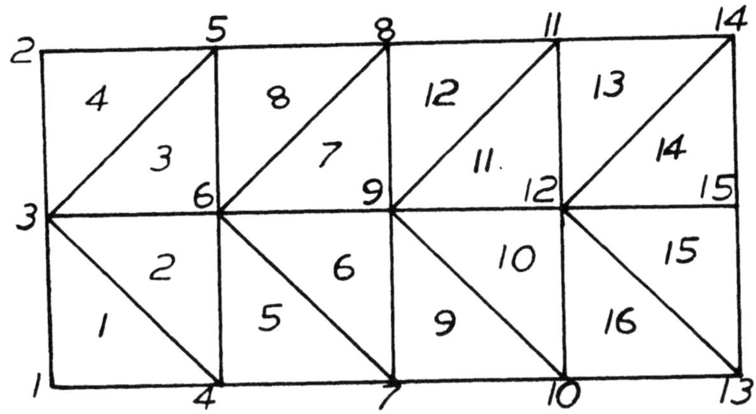

Figure 1 Channel grid layout

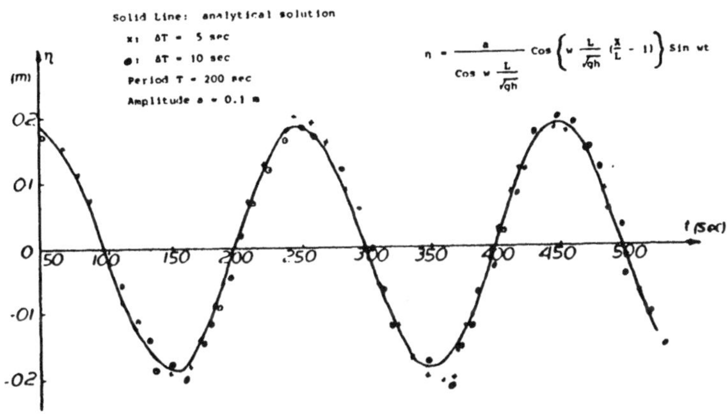

Figure 2 Wave height at x = L (node #15)

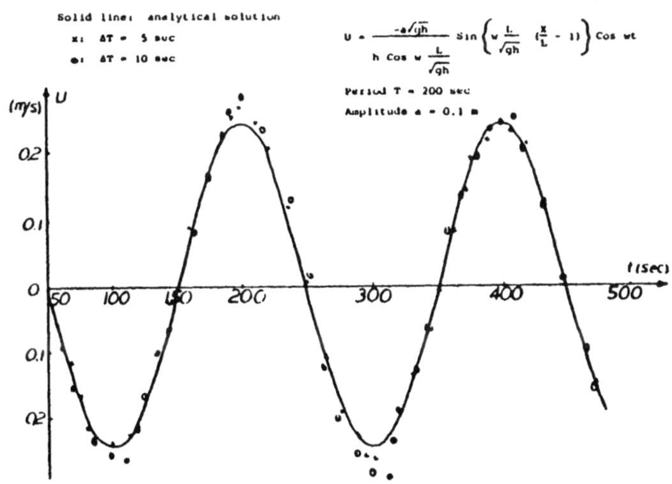

Figure 3 Water velocity at entrance x = 0 (node #3)

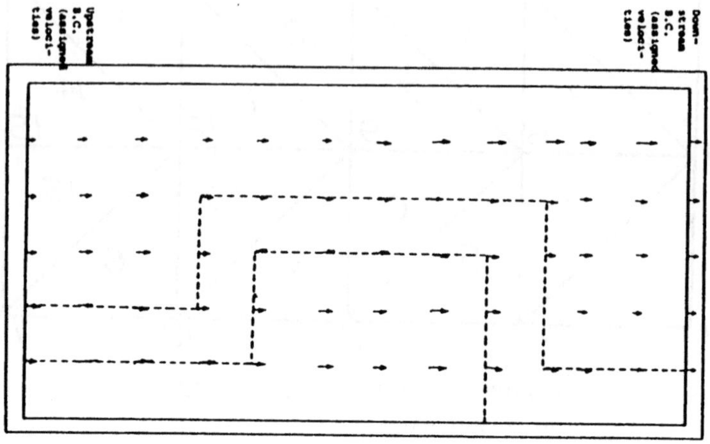

Figure 4   Steady state test in an S-shaped channel

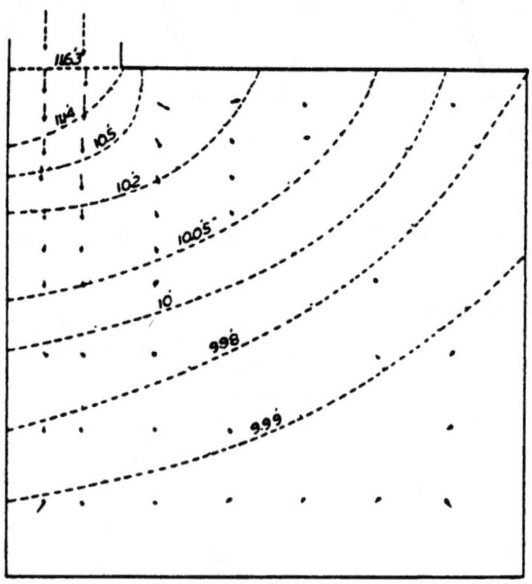

Figure 5   Velocity patterns and water heights in tidal circulation in a square harbor

well with the experimental results of laboratory model, and an ADI numerical model [Falconer (1980)]. In other tests, the same problem was attacked under a different time stepping mechanism to check for stability. The calculations were started with a time step size of $\Delta t = 30$ seconds, and then Δt was doubled in consecutive solutions up to 16 minutes. The scheme proved to be unconditionally stable. With increased size of time step, the number of iterations required to arrive at a solution stage increased accordingly. For this problem, a time step size of 4 minutes proved to be the most efficient. This test revealed that there is a relationship between the size of time step and the number of iterations for the convergence of the solution. Therefore, although the scheme is unconditionally stable, for a particular problem an optimum time step size can be obtained which is computationally most efficient. In other tests, with the square harbor with a dock in the middle and on a circular reservoir, the computed results compared very well with those of the laboratory experiments and ADI finite difference models.

CONCLUSIONS

A finite element model was developed for the study of two-dimensional problems. The model was tested on different problems with different boundary conditions. In all tests, the results were very satisfactory. It is believed that an efficient finite element algorithm has been developed which is accurate, fully stable and applicable to a variety of shallow water problems. Further studies are desirable in the testing of the model against field data over an extended period of time, and the extension of the model to include transport equations for the simulation of pollution and sediment transport.

REFERENCES

1. Benque', J. P., Cunge, J. A., Feuillet, J., Hauguel, A., and Holley, F. M. (1982) "New Methods for Tidal Current Computations," Journal of Waterway, Port, Coastal and Ocean Division, ASCE 108 (WW3): 396-467.

2. Falconer, R. A., (1980) "Numerical Modeling of Tidal Circulations in Harbors," Journal of the Waterway, Port, Coastal and Ocean Division, ASCE 106 (WW1): 31-48.

3. Gallagher, R. H., Zienkiewicz, J. T., Oden, M. M., Cecchi and Taylor, C., eds. (1974) Finite Elements in Fluids, Vols. 1, 2 and 3, John Wiley and Sons, New York.

4. Gray, G. W. and Lynch, D. R. (1977) "Time Stepping Schemes for Finite Element Tidal Model Computation," Advances in Water Resources, 1(2), pp. 83-95.

5. Hosseinipour, Z. (1983) "A Comparative Two-Dimensional Mathematical Modeling of Free Surface Flow with Special Application to Wide Rivers." Ph.D. Dissertation, Department of Civil Engineering, North Carolina State University at Raleigh, North Carolina.

6. Huebner, K. H. (1975) The Finite Element Method for Engineers, John Wiley and Sons, New York.

7. Katapodes, N. D. (1980) "Finite Element Model for Open Channel Flow Near Critical Conditions," Finite Elements in Water Resources, III, Yang, S. Y. and Brebbia, C. A. Editors, The University of Mississippi, University, Mississippi, pp. 5.37-5.46.

8. King, I. P., Norton, W. R., and Orlob, G. T. (1975) "A Finite Element Solution for Two-Dimensional Density Stratified Flow," U.S. Department of Interior, Office of Water Resources Research, Water Resources Engineers.

9. Koutitas, C. G. (1978) "Numerical Solution of the Complete Equations of Nearly Horizontal Flows," Advances in Water Resources, 1(4), pp. 213-217.

10. Leendertse, J. J. (1970) "A Water Quality Simulation Model for Well Mixed Estuaries and Coastal Water," Vol. 1, Principles of Computation, Report RM-6230-RC, The Rand Corp., Santa Monica, California.

11. Reid, R. O. and Bodine, B. R. (1968) "Numerical Model for Storm Surges in Galveston Bay," J. of Waterways, Harbors and Coastal Engineering Division, Proc. ASCE, Vol. 94, No. WW1, pp. 33-37.

12. Thienpont, M. and Berlamont, J. (1980) "A Finite Element Solution for Depth Averaged 2D Navier-Stokes Equations for Flow in Rivers and Canals." Finite Elements in Water Resources, III, Yang, S. Y. and Brebbia, C. A., Editors, The University of Mississippi, University, Mississippi, pp. 5.28-5.36.

13. Wang, J. D. and Connor, J. J. (1975) "Mathematical Modeling of Near Coastal Circulation," Report No. 200, Massachusetts Institute of Technology, Cambridge, Massachusetts, 272 pp.

FOURIER ANALYSIS FOR TESTING A FINITE ELEMENTS METHOD IN SHALLOW WATER PROBLEMS

M. Vansnick and Y. Zech

Catholic University of Louvain, Department of Civil Engineering, Louvain-la-Neuve, Belgium.

SUMMARY

A finite element model for transient open channel flow is presented. The time integration scheme, a semi-implicit method without iteration, gives the possibility of simulating any problem, even without energy dissipation. Application of Fourier analysis may improve the choice of the finite element grid density. If this analysis is applied to in situ measurements or to the theoretical solution, it is possible to predict the accuracy of the model with regards to the finite element density. If it is applied to the response of the model, it helps to define the causes of the lack of accuracy (the amplification or damping of integration process). Application to a theoretical example is presented.

INTRODUCTION

Since many years, people try to apply the finite element to the hydraulic modelling of transient flow. The most presented models are very expensive in computer time. To make it realizable, it appears necessary to develop a model without iteration and to find a technique to guide the choice of the finite element grid. Use of Fourier series gives us the possibility to compare various algorithms and to improve the choice of our finite element grid.
In the most practical problems of hydraulic modelling of tidal flow, the engineer disposes of a lot of measurements to compare the model with the reality. Data are used in a first stage to choose the model, and in a second stage, after building this, for calibration.
This preliminary investigation may ameliorate the choice of a finite element grid or a finite difference discretization.
This method gives the possibility to analyse the noise due to time integration process, to detect its causes and to foresee the theoretical accuracy of the results.

Errors due to the discretization process may be distinguished from noise due to the time integration scheme. Such an analysis may be used to test a new algorithm in classical cases with well known solutions.

Description of the mathematical model

When the vertical velocity components can be neglected, transient shallow water flow in an open channel can be mathematically described by the Saint-Venant equations. The two first equations are the well known momentum equations :

$$\frac{1}{g}\left(\frac{\partial u}{\partial t} + u\frac{\partial u}{\partial x} + v\frac{\partial u}{\partial y}\right) + \frac{\partial h}{\partial x} - i_x - \frac{\omega v}{g} + \frac{u\sqrt{u^2+v^2}}{c^2\ R} + \frac{W_x}{gh} = 0$$

$$\frac{1}{g}\left(\frac{\partial v}{\partial t} + u\frac{\partial v}{\partial x} + v\frac{\partial v}{\partial y}\right) + \frac{\partial h}{\partial y} - i_y + \frac{\omega u}{g} + \frac{v\sqrt{u^2+v^2}}{c^2\ R} + \frac{W_y}{gh} = 0$$

where u and v are the mean velocity components along the ox and oy axes

h	the water depth
i_x and i_y	the bottom slope parallel to the ox and oy axes
ω	the constant of Coriolis
W_x and W_y	the wind friction coefficients parallel to ox and oy axes.

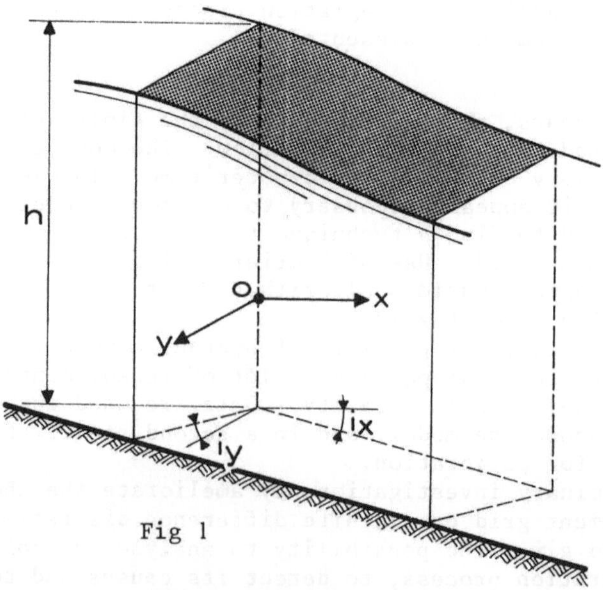

Fig 1

The third equation is the continuity equation :

$$\frac{\partial h}{\partial t} + \frac{\partial (uh)}{\partial x} + \frac{\partial (vh)}{\partial y} = o$$

Finite element formulation

To solve this differential equations, it's possible to obtain a finite element scheme by two different ways. In the first one, the spatial and time variables are treated by the finite element method. The use of this time-space elements is very expensive in computer time. The second way associates a spatial finite element grid with a separate time integration method.
This scheme allows to test various finite difference methods in time and gives the possibility to separate the time variable from the spatial x and y variables.
Two types of finite elements were used : rectangular and triangular element.
To obtain the finite element scheme, it is necessary to develop the different unknows by a set of trial functions.

$$h = \sum_{i=1}^{N} H_i \ (t) \ \phi_i \ (x,y)$$

$$u = \sum_{i=1}^{N} U_i \ (t) \ \phi_i \ (x,y)$$

$$v = \sum_{i=1}^{N} V_i \ (t) \ \phi_i \ (x,y)$$

where H_i, U_i, V_i are the value of water depth and velocity components at the nodes : these values depend on the time. This scheme separates completely the time variable from the other ones.
Assuming this approximate values of u,v and h we can use a Galerkin procedure to obtain three variational equations for each finite element :

$$\iint \frac{1}{g} \left(\frac{\partial u}{\partial t} + u \frac{\partial u}{\partial x} + v \frac{\partial u}{\partial y}\right) + \frac{\partial h}{\partial x} - i_x - \frac{\omega v}{g} + \frac{u\sqrt{u^2+v^2}}{C^2 \ R} + \frac{W_x}{gh}) \ \delta u \ d\Omega = o$$

$$\iint \frac{1}{g} \left(\frac{\partial v}{\partial t} + u \frac{\partial v}{\partial x} + v \frac{\partial v}{\partial y}\right) + \frac{\partial h}{\partial y} - i_y + \frac{\omega u}{g} + \frac{v\sqrt{u^2+v^2}}{C^2 \ R} + \frac{W_y}{gh}) \ \delta v \ d\Omega = o$$

$$\iint \left(\frac{\partial h}{\partial t} + \frac{\partial (uh)}{\partial x} + \frac{\partial (vh)}{\partial y} \right) \ \delta h \ d\Omega = o$$

where δu, δv and δh are the variations of u, v and h. We may develop these in function of the variation of the node values, for example

$$\delta u = \sum_{i=1}^{N} \delta U_i \ \phi_i \ (x,y)$$

When developped, a set of three linear equations systems may be obtained

$$[R] \ [\frac{\partial U}{\partial t}] = [F_x]$$

$$[R] \ [\frac{\partial V}{\partial t}] = [F_y]$$

$$[R] \ [\frac{\partial H}{\partial t}] = [F_h]$$

where $R_{i\,j} = \iint \phi_i \ \phi_j \ d\Omega$

$$F_{x_i} = \sum_{k=1}^{N} \iint \{ \sum_{j=1}^{N} \frac{1}{g} (U_k U_j \ \phi_k \frac{\partial \phi_j}{\partial x} + V_k U_j \ \phi_k \frac{\partial \phi_j}{\partial y}) +$$

$$H_k \ \frac{\partial \phi_k}{\partial x} \} \phi_i d\Omega \ \ \iint (i_x - \frac{uV}{c^2 h}) \ \phi_i d\Omega$$

$$F_{y_i} = \sum_{h=1}^{N} \iint \{ \sum_{j=1}^{N} \frac{1}{g} \ (V_k V_j \ \phi_k \frac{\partial \phi_j}{\partial y} + U_k V_j \ \phi_j \frac{\partial \phi_k}{\partial y}) +$$

$$H_i \ \frac{\partial \phi_k}{\partial y} \} \phi_i \ d \ \Omega - \iint \ (i_y - \frac{vV}{C^2 h}) \ \phi_i \ d \ \Omega$$

$$F_{H_i} = \iint (\sum_{k=1}^{N} \ \sum_{j=1}^{N} \ U_k \ \ H_j \ \frac{\partial (\phi_k \phi_j)}{\partial_x} + V_k H_j \ \frac{\partial (\phi_k \phi_j)}{\partial_y}) \phi_i d \ \Omega$$

This result is very favourable because the elementary characteristic matrix R is independent of the time. We can assemble this matrix to obtain characteristic matrix for all the system, independant of the time. That allows to compute and triangularize it before time integration.
Introduction of boundary conditions may influence the aspect of the three matrices. If there is no boundary node, where the velocity is defined non-parallel to the x or y axis, the problem can be directly solved without precaution. If there exist nodes with a defined normal velocity non-parallel to

one of the x or y axis, it remains possible, as demonstrated by Gray and Lynch, to use a rotation of axes to maintain three separate systems.

Time integration

Any finite difference scheme may be used. An iterative algorithm is too expensive in time computation. However the simplest explicit scheme is not stable.
Due to the separation of the three equation systems, it is possible to develop a semi-implicit scheme without iteration :

1ste step : computation of U_i $(t + \Delta t)$ and V_i $(t + \Delta t)$ at a time $t + \Delta t$

$$\text{by } [R] \left[\frac{U_i(t+\Delta t) - U_i(t)}{\Delta t}\right] = F_{X_i}^{(t)} \; (H_i^t, U_i^t, V_i^t)$$

$$[R] \left[\frac{V_i(t+\Delta t) - V_i(t)}{\Delta t}\right] = F_{y_i} \; (H_i^t, U_i^t, V_i^t)$$

2sd step : computation of $H_i^{(t+\Delta t)}$

$$[R] \left[\frac{H_i^{(t+\Delta t)} - H_i^{(t)}}{\Delta t}\right] = F_{h_i} (H_i^t, U_i^{t+\Delta t}, V_i^{t+\Delta t})$$

This algorithm is stable as long as the Levy-Friedrich - Courant condition is respected, depending on the choice of the time step. The theoretical flow even without energy dissipation may be simulated.

Fourier Analysis to define the accuracy of mathematical model

All periodic functions can be approximated by a serie of sine and cosine functions. When the function is defined by a set of points with a constant interval, it can be written :

$$F\,((i-1/2)\,t) = B_o + \sum_{j=i}^{L} A_j \,\text{Sin}\, \frac{\pi(2i-1)j}{N} + \sum_{j=i}^{L} B_j \text{Cos}\, \frac{\pi(2i-1)j}{N}$$

where F is a periodic function that must be defined by a set of N values with a constant time interval Δt (Period T of the first function is N Δt, the first point is supposed at $\Delta t/2$)

B_o is the mean value of the function

A_j, B_j are the computed coefficients of sine and cosine function.

If $N.\Delta t$ is chosen equal to the period of the function, A_1 Sin $\frac{\pi(2i-i)}{N}$ is the fundamental wave. To simplify the analysis, it is interesting to choose the initial time to have B_1 equal to zero. The cosine function defines the phase of the function, in this case the fundamental function starts without dephasing.

This development may be used to observe two different aspects of mathematical modelling : the choice of the finite element grid density and the fitting control of theoretical characteristics of flow such as wave height and celerity.

Choice of finite element grid density

For most problems of hydraulic modelling, engineers dispose of measurements of water height or velocity to calibrate their model. In this case, it is possible to analyse this measurements by a Fourier analysis, to determine which is the last harmonic wave with a significant influence.

The wave length of this last harmonic determines in fact the density of nodes needed by the model. This method gives the magnitude of the theoretical approximations associated with the chosen grid.

Control of the response of an hydraulic model

Before using a mathematical model of flow, it is necessary to test it in theoretical examples to control its ability of simulating a type problem and to compare the results of this computation with the theoretical solution. In this case, it is important to know what is the limit of the model. To illustrate the Fourier analysis in such a test case, we have chosen an example of a wave reflection in a rectangular basin where water is initially motionless (flat condition). At the initial time water level at the begin of the basin increases and creates a progressive wave that must be reflected at the end of the basin. If the energy dissipation is neglected, the theoretical solution of water height at the end of the basin is shown in figure 2.

Development of this solution by a Fourier serie show a sum of sine functions without dephasing (amplitude of even harmonic functions are zero).

TABLE 1

Order of harmonic wave	Amplitude	Order of harmonic wave	Amplitude
1 (fundamental)	0.12739	15	0.50958
		17	0.00876
3	0.04264	19	0.00816
5	0.02580	21	0.00773
7	0.01867	23	0.00743
9	0.01477	25	0.00725
11	0.01234	27	0.00715
13	0.01234		

This analysis shows how difficult it is to simulate this problem : the 27th harmonic is not neglectible in respect to the 3^{d} of 5th. This explains why it is practically impossible to obtain a very good accuracy. However, it may be very useful to put forward the possibilities of different finite element grids.

For each function, it is possible to define the wave length

$$L_j = c \frac{N\Delta t}{j}$$

where c is the hydraulic celerity.

A first analysis shows that it is necessary to have at least three points on each half wave length to obtain a grid able to simulate an harmonic wave. The accuracy increases with the number of points on each half wave lenght : by a linear interpolation, as in a triangular finite element, the theoretical accuracy of the integral of a sine function, computed by a trapezoïdal rule, is shown in the following table

TABLE 2

Number of points at each half wave length	Minimum relative accuracy of sine integral
3	0.785
4	0.907
5	0.948
6	0.967
7	0.977
9	0.987
25	0.998

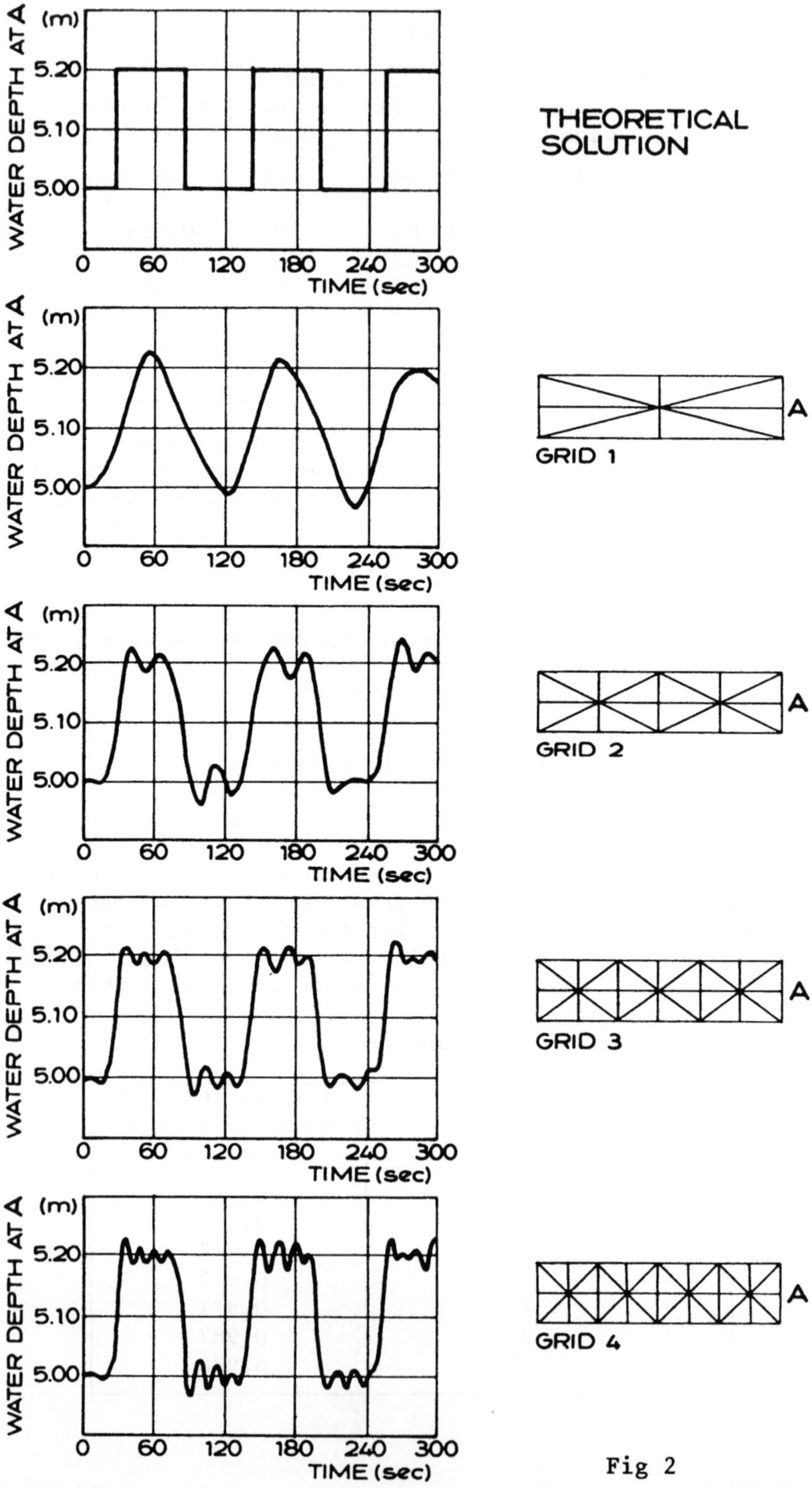
WATER DEPTH AT A (m)
5.20
5.10
5.00
0 60 120 180 240 300
TIME (sec)
THEORETICAL SOLUTION
A
GRID 1
GRID 2
GRID 3
GRID 4

Fig 2

TABLE 3

Coefficients of harmonics	Theoretical solution	Response of 1st grid	Response of 2nd grid	Response of 3th grid	Response of 4th grid
mean value	5.10	5.099	5.100	5.101	5.101
1st harmonic	0.12739	0.11494	0.12481	0.12726	0.12766
2nd harmonic	0	0.02070	- 0.00275	0.00027	0.00029
3th harmonic	0.04264	- 0.0098	0.03311	0.03965	0.04062
4th harmonic	0	0.0001	- 0.01016	0.00130	0.00129
5th harmonic	0.02580	0.00012	0.00809	0.01210	0.02247
6th harmonic	0	0.00013	0.00125	- 0.0043	0.00199
7th harmonic	0.1867	0.00013	0.00038	0.00729	0.00181

Choice of the last harmonic that must be simulated and expected accuracy on its amplitude define the maximum interval between two nodes.
In our example, to be able to simulate the ith harmonic wave, in a rectangular basin with water depth H, it is necessary not to exceed an interval lenght

$$\Delta L = \sqrt{gH}\,\frac{T_i}{6_i},$$

where T_i is the period. In this hypothesis, the relative accuracy on the amplitude is 78.5 %. If this accuracy is not sufficient and must be increased to 96 %, the maximum node interval is reduced to $\Delta L = \sqrt{gH}\,\frac{T}{12i}$.

Fig 2 describes the evolution of the water depth at the end of the basin. Following the theoretical solution, the results of four computations are shown with gradually increasing grid density.
Table 3 gives the results of a Fourier analysis. The values of the significant amplitudes, i.e., the harmonic waves which are simulated with a maximum error of 0.005 m are underlined. It appears a good correlation with table 2; the harmonic waves, with a non-zero amplitude, are simulated as long as there remain three points on the half wave lenght.
This analysis confirms the theoretical basis of the model. Altough the computation of time integration is not made explicitely, there appears no amplification. Observation of the results might be misinterpreted as a wave amplification.
The error in wave magnitude is not an amplication but is due to the lack of some harmonic waves.
The Fourier analysis shows, as theoretically predicted, an approximation of the functions with too small values of amplitude. The value of each harmonic wave amplitude approaches the theoretical value if the finite element density increases.

CONCLUSION

The simple time integration scheme used in this finite element model is able to simulate wave propagation without damping or amplification. Control of the algorithm permits to show the good fit of the model and to define a theoretical accuracy, depending on the finite element density.
If engineers dispose of preliminary measurements it is possible by Fourier series to analyse the tidal flow to specify the most important harmonic wave lengths. This preparatory computation allows to ameliorate the choice of the finite element grid.
In an other way, it is possible to control noise on a new finite element algorithm by the same analysis. Computation of theoretical examples allows to detect amplification or damping due to the integration process. This phenomenon cannot be reduced with a greater node density.

A FINITE ELEMENT METHOD FOR STORM SURGE AND TIDAL COMPUTATION

P. Esposito, A. Hauguel and B. Latteux

Laboratoire National d'Hydraulique
Electricité de France

INTRODUCTION

Long waves in coastal areas, when water depth is small compared to the wave length, can be numerically simulated by the resolution of the shallow water equations. The use of a finite element method is then specially well fitted when the shape of the domain to deal with is rather complex, as it often occurs in the field of maritime hydraulics, or when some areas of particular interest require a finer description (entrance of a harbour, for example).

Such a model has been developped, using a special procedure to reduce the storage requirement by deriving from the basic equations separate problems on each of the scalar variables. The implementation of a semi-implicit scheme makes it possible to avoid wave damping when simulating wave propagation on broad areas with a limited number of nodes in a wave length. Finally the possible use of an incident wave condition, as well as the consideration of the atmospheric forcing terms, qualify the model to examine the generation and propagation of storm surges.

EQUATIONS AND GENERAL DESCRIPTION OF THE METHOD

The shallow water equations are the following :

$$\frac{\partial z}{\partial t} + \operatorname{div} \vec{Q} = 0 \tag{1}$$

$$\frac{\partial Q_i}{\partial t} + \operatorname{div} (Q_i \vec{u}) + gh \frac{\partial z}{\partial x_i} - K \Delta Q_i = F_i \quad (i = 1,2) \tag{2}$$

where
- g = acceleration due to gravity
- h = depth of water
- K = dispersion coefficient
- $\vec{Q} \begin{cases} Q_1 \\ Q_2 \end{cases}$ = flow rate per unit length
- $\vec{u} \begin{cases} u_1 \\ u_2 \end{cases}$ = velocity field
- z = free surface elevation
- $\vec{F} \begin{cases} F_1 \\ F_2 \end{cases}$ = forces (Coriolis, bottom friction, wind, atmospheric pressure gradient).

Fractional step algorithm for the convection terms.
Taking a simple first order time discretisation of equations (1) and (2) between two time steps t^n and t^{n+1} the discretised unknown fields $\vec{Q}^n$, z^n, $\vec{u}^n$ and $\vec{Q}^{n+1}$, z^{n+1}, $\vec{u}^{n+1}$ are introduced.

Then, using an auxiliary unknown field $\vec{Q}^*$, equations (1) and (2) can be decomposed into a pure convection system and the remainder.

* convection step : $\frac{\partial Q_i}{\partial t} + \operatorname{div} (Q_i \vec{u}^n) = 0 \ (i=1,2)$ (3)

this step gives $\vec{Q}^*$

* diffusion and propagation step :

$$\begin{cases} \frac{\partial z}{\partial t} + \operatorname{div} \vec{Q} = 0 \\ \frac{\partial Q_i}{\partial t} - K \Delta Q_i + gh \frac{\partial z}{\partial x_i} = F_i \ (i=1,2) \end{cases} \tag{4}$$

with initial conditions z^n, $\vec{Q}^*$ at t^n.

The convection step is solved using characteristics method.

The main advantages of this method are :

- unconditional stability
- no non-symmetric system to assemble and solve at each time step
- good accuracy (reduced numerical diffusion) when using quadratic shape functions (at least).

The main drawback of classical characteristic method is probably the fact that conservativity is not ensured by the scheme. But a new approach using a weak formulation of equations (3) can be used in order to achieve conservativity (BENQUE, LABADIE and RONAT - 1983). Moreover convection is not the main phenomenon in problems such as storm surges or tidal computations in the English Channel or the North Sea and the results presented in this paper are not so bad as for as conservativity is concerned.

Diffusion-Propagation step

A semi-implicit discretisation has been chosen for system (4) because it has been shown that pure implicit schemes have poor propagation characteristics when used on a coarse mesh (figure 1 and ESPOSITO 1983).

The non linear term g h $\vec{grad}$ z is first splitted up into two terms :

$$g\, h_m\, \vec{grad}(z) + g(h - h_m)\, \vec{grad}(z)$$

h_m being a reference depth of water dependant on the location of the point but independant of the time.

Then introducing
$$\begin{cases} C = gh_m \text{ : celerity of waves} \\ \alpha = \dfrac{1}{\Delta t} \\ \theta,\ \beta \quad]0.5,\ 1] \text{ two coefficients} \end{cases}$$

the following discretisation :

$$\begin{cases} \dfrac{\partial z}{\partial x_i} = \beta \dfrac{\partial z^{n+1}}{\partial x_i} + (1-\beta) \dfrac{\partial z^n}{\partial x_i} & (5) \\ \dfrac{\partial Q_j}{\partial x_i} = \theta \dfrac{\partial Q_j^{n+1}}{\partial x_i} + (1-\theta) \dfrac{\partial Q_j^n}{\partial x_i} & (6) \end{cases}$$

leads to the system to be solved :

$$\begin{cases} \alpha^2\, z^{n+1} - K\theta\alpha\Delta z^{n+1} - \theta \operatorname{div}(C^2\, \beta \vec{grad}\, z^{n+1}) = W & (7) \\ \alpha\, \vec{Q}^{n+1} - K\theta\Delta\, \vec{Q}^{n+1} + C^2\, \beta \vec{grad}\, z^{n+1} = \vec{T} & (8) \end{cases}$$

where equations (7) comes from α (1) + θ div (4) and $(W,\vec{T})$ are known at each time step.

It must be noticed that equation (7) can be solved easily if z^{n+1} is given on the boundary. Then, $\vec{Q}$ can be computed from equation (8) where the term C^2 ß $\vec{grad}\ z^{n+1}$ is now a known function and can be rejected into the right hand side.

The problem then amounts to the determination of the trace of z on the boundary. The procedure used for this purpose is an extension of the technique proposed by Glowinski-Pironneau (1978) for the Stokes problem. It is described in detail in BENQUE, LABADIE and LATTEUX (1981) and LABADIE and LATTEUX (1982).

Finite Element Discretisation

All the numerical results presented in this paper have been carried out using a quadratic triangular mesh for the flow rate Q and for the surface elevation z. The elliptic problems of equations (7) and (8) lead to the resolution of 3 symmetric definite matrix problems (one for z and one for each component of the flow rate). At each time step the boundary conditions for z is computed through the resolution of 3 problems (similar to the 3 previous one) and the resolution of a boundary problem (the size of its definite matrix beeing the number of boundary points).

This technique is advantageous on the point of view of computational efficiency. Two "domain" matrices have to be stored, one for the variable z, another for each component of $\vec{Q}$ (it is the same matrix for the two components). They are symmetric and sparse and small enough to be processed by an in-core solver such as an incomplete Choleski preconditioned conjugate gradient. In addition, the matrix of the boundary operator has to be factorized and stored.

Boundary conditions

On the coast a simple prescribed flux condition can be prescribed : $\vec{Q} = \vec{0}$.

On the open boundaries, a specified flux condition (both components) could be too reflexive. In order to specify a wave rather than a flow rate and above all to let waves generated inside leave the domain easily, an incident wave boundary condition may be used. Such a condition, described in details in LABADIE and LATTEUX (1982), is written in the following way :

$$\left\{\begin{array}{ll} - C^2\, z + K\, \dfrac{\partial \vec{Q}}{\partial n} \cdot \vec{n} + c\, \vec{Q} \cdot \vec{n} = (\vec{n} \cdot \vec{u} - 1)\, c^2\, \psi & (9) \\[2ex] K\, \dfrac{\partial \vec{Q}}{\partial n} \cdot \vec{\tau} + c\, \vec{Q} \cdot \vec{\tau} = (\vec{\tau} \cdot \vec{u})\, c^2\, \psi & (10) \end{array}\right.$$

Where $\vec{n}$ and $\vec{\tau}$ denote boundary exterior normal and tangent vectors, $\vec{u}$ is a unit vector indicating the direction of the wave to specify and ψ is a scalar function of space and time describing the incident wave entering the domain through the open boundary (z and $\vec{Q}$ must be discretised by equations (5) and (6)). The conditions (9) and (10) allow any wave normal to the boundary to leave the domain without any reflexion. In the simple study of storm surges described further, ψ can be chosen as follows :

$$\psi = \psi_o = \frac{P - P_o}{\rho g}$$

with P : atmospheric pressure

P_o : reference atmospheric pressure : 1013 mb

This expression denotes simple static equilibrium at the edge of the continental shelf. Other terms may be added in order to take into account tidal forcing.

NUMERICAL RESULTS

Dunkirk outer harbour

The first realistic application concerns the computation of the tidal flow pattern in the vicinity of the new outer harbour of Dunkirk, located on the french coast of the North Sea.

This example has been chosen for two main reasons :

- the shape of the domain and the complexity of the laying out of the limits give the typical case where a finite element method turns out to be specially suitable (good boundary representation and local refinement facilities) and provide a rather severe test of the model ;

- this case has been previously studied by means of a scale model and of a finite difference model : so there exist references to which compare the results of the finite element model.

The finite element grid (fig. 2) has been generated automatically by a program developed in our laboratory. As it can be noticed from this illustration, a special mesh refinement has been introduced near the limits, in order to obtain a better consideration of the boundary conditions (which is particularly useful for the land boundaries).

In this experiment a piecewise quadratic approximation has been used for each variable ; there are about nine hundred triangular elements and two thousand calculation nodes.

The characteristics of the computation are as follows :

- tide range is about five meters ;
- maximum velocity is about 1,5 m/s ;
- tide and current are in phase (that is to say that the tide wave is here quite purely progressive) ;
- eddy viscosity coefficient is equal to 5 m^2/s ;
- time step is sixty seconds and Courant number for propagation varies locally from 1 to 5.

Finally both components of the flow discharge are prescribed on the boundaries ; these components are deduced from the results of a previous computation carried out with a finite difference method on a larger domain.

The computation has been worked out during more than one tide, and the results have been compared to the data obtained by photographs of surface floats on the scale model, built about twelve years ago for the study of the new outer harbour.

Fig. 3 displays this comparison at the beginning of the flood : an eddy appears in a similar way on both models, just behind the end of the west jetty, while the harbour fills in ; half an hour later, both eddies have shifted and spread out likewise.

Concerning the mass conservation, the global error on the free surface elevation over one tide was found to be about 3 % of the tide range ; this discrepancy, which seems to be due to the steepness of the access channel (about 1:10), is in fact rather small for such a narrow domain.

<u>Storm surge of 19 and 20 november 1973</u>

A series of storm surge computation models of the North-West European Continental Shelf have been developped since 1969 at the I.O.S. (Institute of Oceanographic Sciences, BIDSTON, UK) using finite difference discretisations and a review of this work is given by HEAPS (1983). As a test for an extended prediction covering a period of 44 days, FLATHER (1976) computed the surges from 4 november to 18 december 1973 using an improved model including tidal influence. Using our model on the spherical coordinate mesh of figure 4 and taking meteorological data from a tape furnished by JONSMOD (JOINT NORTH SEA MODELLING GROUP) we computed the surge of 19 and 20 november.

Figure 5 presents an exemple of surface elevation and currents just before the very high surge of 19 november 13 h TU. It is an instant when the currents reach their maxima. Although the meteorological data have different origins, the results of this computation and previous results from FLATHER (1976) are very similar. The comparison between measurements, FLATHER's results and this finite element computation is done by means of the plots of figure 6a and 6b. In order to simulate the surge of 19-20 november, the computation was started 3 days before (16/11/73 at 3 h). It can be shown that a few hours later, the behaviour of the model is specially good at CUXHAVEN where the peak of 17 november is very well reproduced. Among all the plots, many features of the surge are correctly predicted and only one result seems to be very bad, at SOUTHEND. The general tendancy to oscillations of the computation is probably related to semi-implicitation. This particular aspect is at the moment under examination. This computation is for us a first trial in the field of storm surge prediction with a finite element model. The good agreement of the results obtained is very encouraging since we can improve our model using tidal influence, better meteorological data, optimized mesh...

Tidal simulation in the English Channel

The tidal phenomenon in the English Channel can be considered as an interesting case for the application of the model on a large domain : tide and currents are rather well known, and the high tide ranges, small depths and strong currents provide a good test for the treatment of the non-linearities of the equations.

The domain extends westwards till to the edge of the continental shelf, and covers on the north-east a small part of the North Sea.

Computation has just begun, under simple conditions : flow discharge prescribed on the boundaries, corresponding to a mean spring tide, and totally implicit numerical scheme. Besides, the advection step is not considered here.

The mesh is rather coarse for this first trial on the English Channel (fig. 7) : there are 220 triangular elements, and 521 nodes of discretization for the quadratic approximation. This rough discretization allows a very fast computation : the whole tide is simulated within less than 10 minutes C.P.U. on a CRAY 1 computer.

Fig. 8 displays the tidal current pattern at High Water (Brest). Fig. 9 exhibits the comparison between computed and measured tide range over the English Channel. The main features of the complex tidal pattern in this region, as the significant difference between english and french coast, due to the rotation of the Earth, and the wave amplification in the Gulf of Saint-Malo, are qualitatively well reproduced. However a damping of the computed tidal wave is obvious, coming from the use of an implicit scheme : with only 40 nodes in a wave length, the tide range has generally been weakened by 10 to 20 %.

Further computations performed with a semi-implicit scheme, and possibly with a finer mesh, will probably strongly improve these results.

CONCLUSION

A finite element method has been developped to solve the shallow water equations, and has proved to be particularly well fitted for the simulation of hydrodynamics in complex shaped domains, whatever their size, with reasonable in core memory requirements.

The use of a semi-implicit scheme enables the model to avoid wave damping even with a coarse discretization of a wave length ; the implementation of an incident wave condition and the consideration of the atmospheric forcing terms make it quite convenient for the study of the generation and propagation of storm surges on broad areas, just as the north-west european continental shelf : first results concerning the simulation of a measured storm surge has turned to be very satisfactory.

Next steps will concern the tuning of the model on other observed storm surges and the examination of tide-storm surge interaction.

REFERENCES

BENQUE J.P., LABADIE G., LATTEUX B. (1981)
A finite element method for the shallow water equations. 2nd International Conference on numerical methods for laminar and turbulent flow, VENICE, Italy, July 13-16 1981.

BENQUE J.P., LABADIE G., RONAT J. (1983)
A finite element method for the Navier Stokes equations coupled with temperature equation, submitted to the International Journal for Numerical Methods in Fluids.

BRISTEAU M.O., GLOWINSKI R., PERIAUX J., PERRIER P., PIRONNEAU O., POIRIER G. (1978)
Application of optimal control and finite element methods to the calculation of transonic flows and incompressible viscous flows. Rapport de recherche 78-294. INRIA LABORIA, 78150 LE CHESNAY, FRANCE.

ESPOSITO P. (1983)
Utilisation des éléments finis pour le calcul des surcotes en Manche. Rapport E.D.F. HE/41/82.55.

FLATHER R.A. (1976)
Results from a storm surge prediction model of the North-West European Continental Shelf for April, November and December 1973. Report n° 24. Institute of Oceanographic Sciences, BIDSTON, U.K.

HEAPS N.S. (1983)
Storm surges 1967-1982 Geophys J.R. astr. Soc. (1983) 74, 331-376.

LABADIE G., LATTEUX B. (1982)
Résolution des équations de St Venant par une méthode d'éléments finis. Rapport E.D.F. HE/41/82.15 and HE/42/82.34.

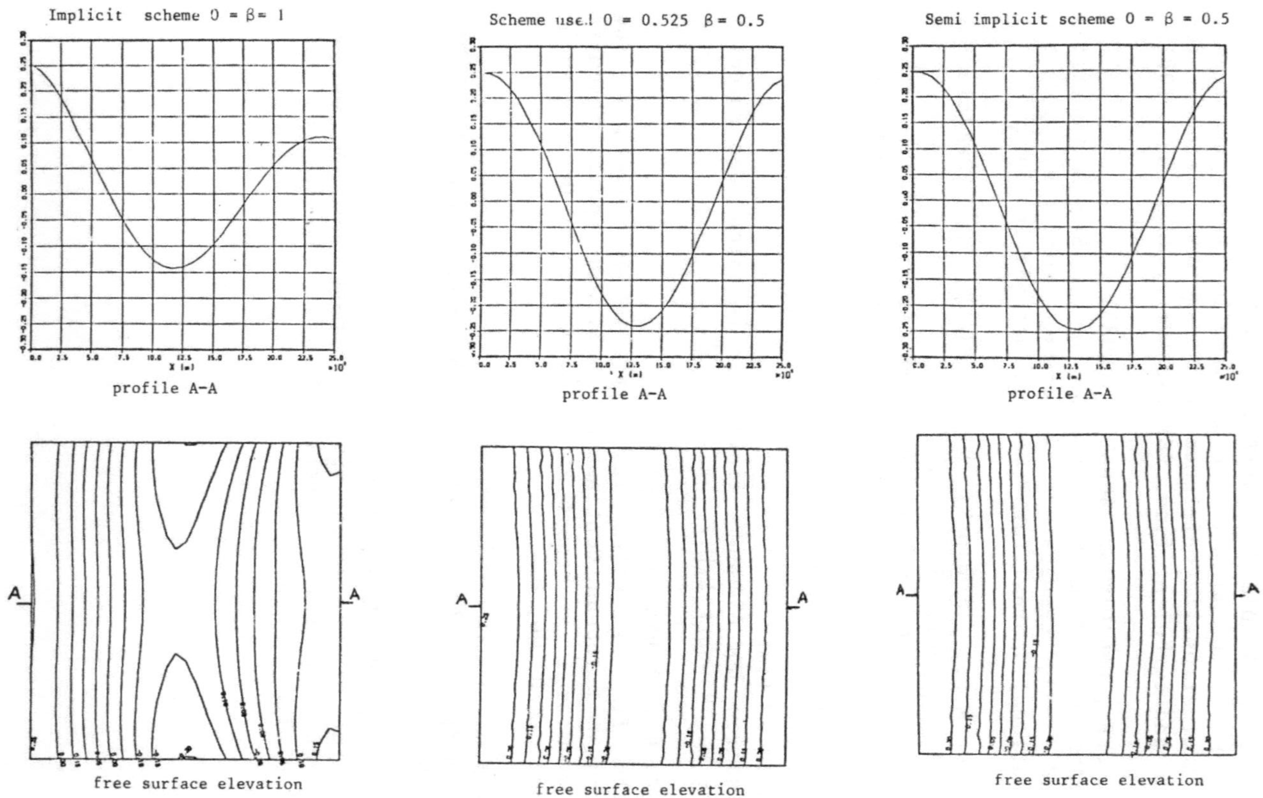

FIGURE 1 · Influence of semi-implicitation. A plane wave moves from left to right in a square vessel with 17 nodes per wave length.

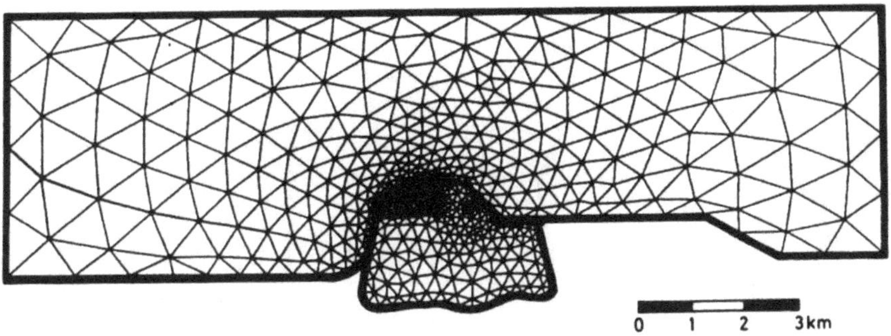

a. Mesh of the whole area

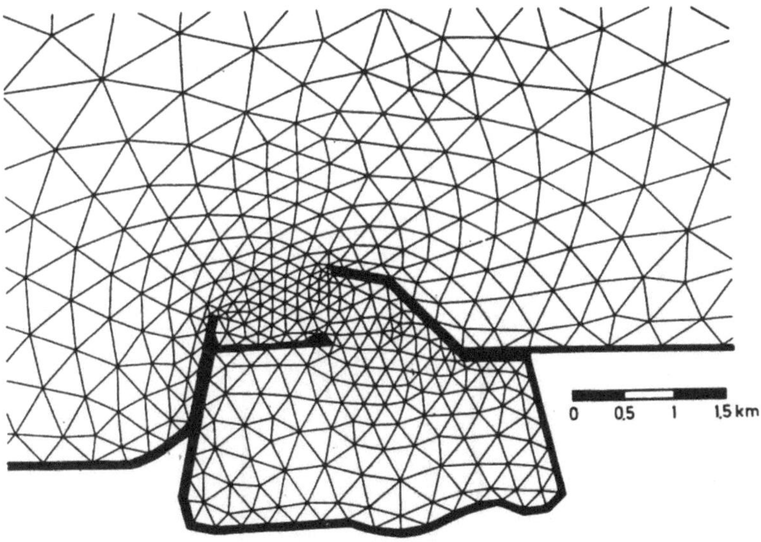

b. Detail of the harbour

Fig.2 _FINITE ELEMENTS MODEL OF DUNKIRK OUTER HARBOUR

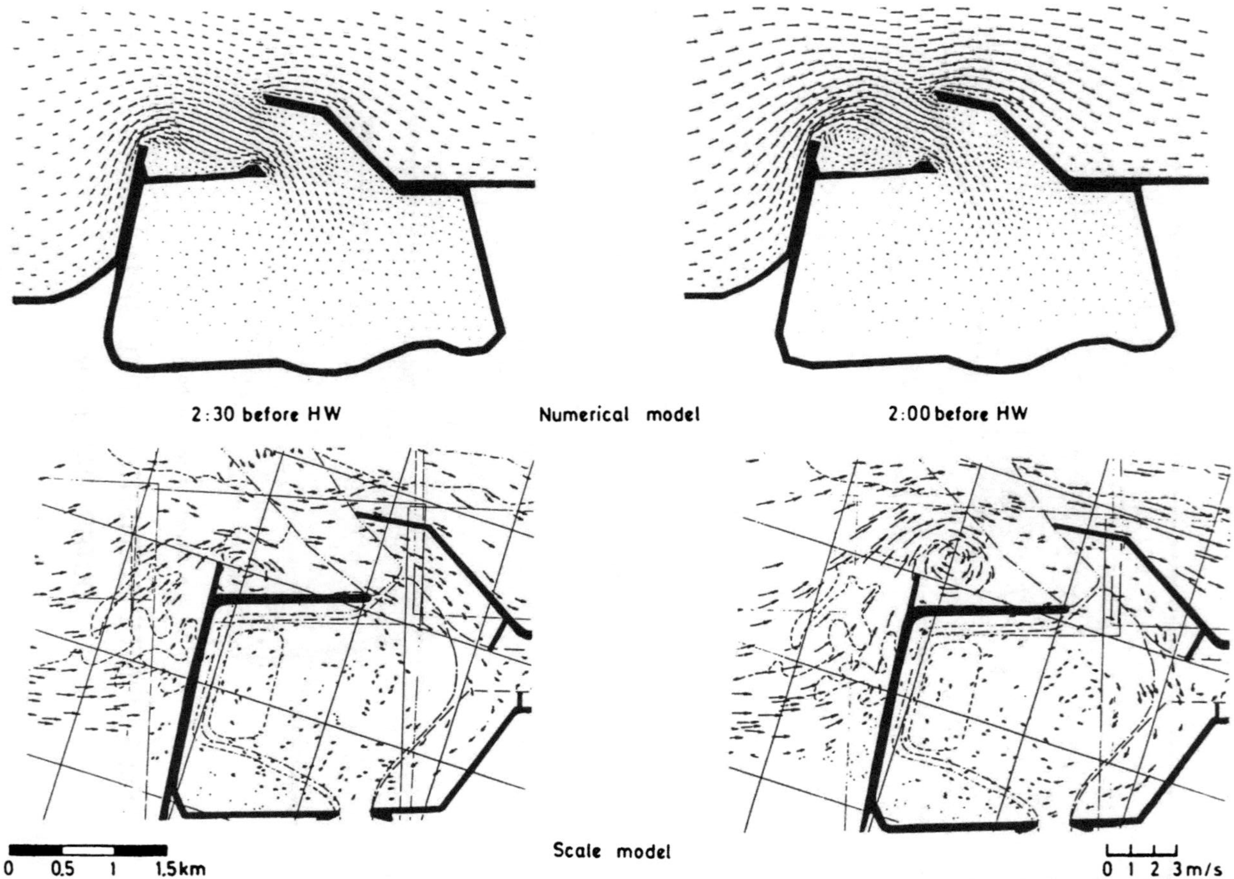

Fig.3 _ COMPARISON OF THE RESULTS OF NUMERICAL AND SCALE MODELS

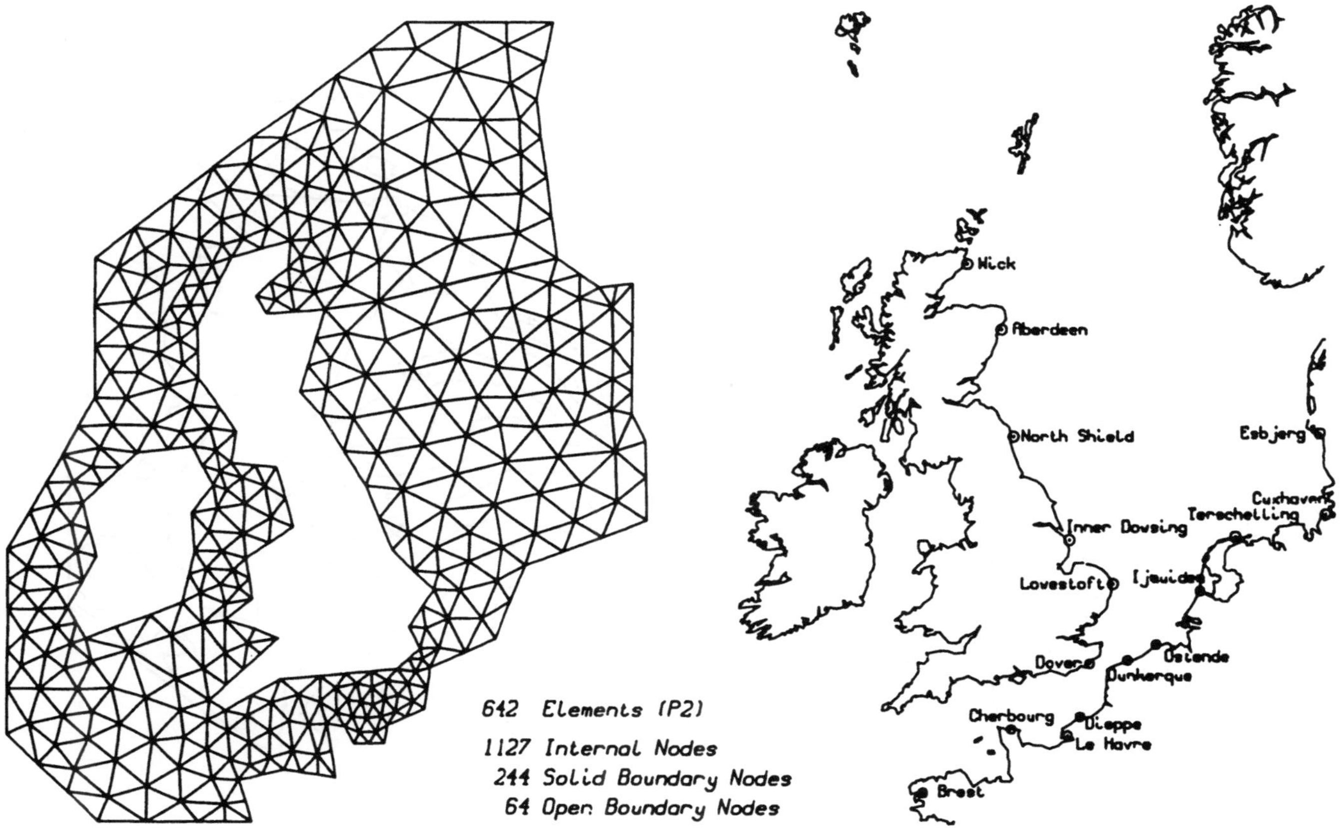

Figure 4 : Finite Element Mesh of the North-West European continental shelf and position of ports

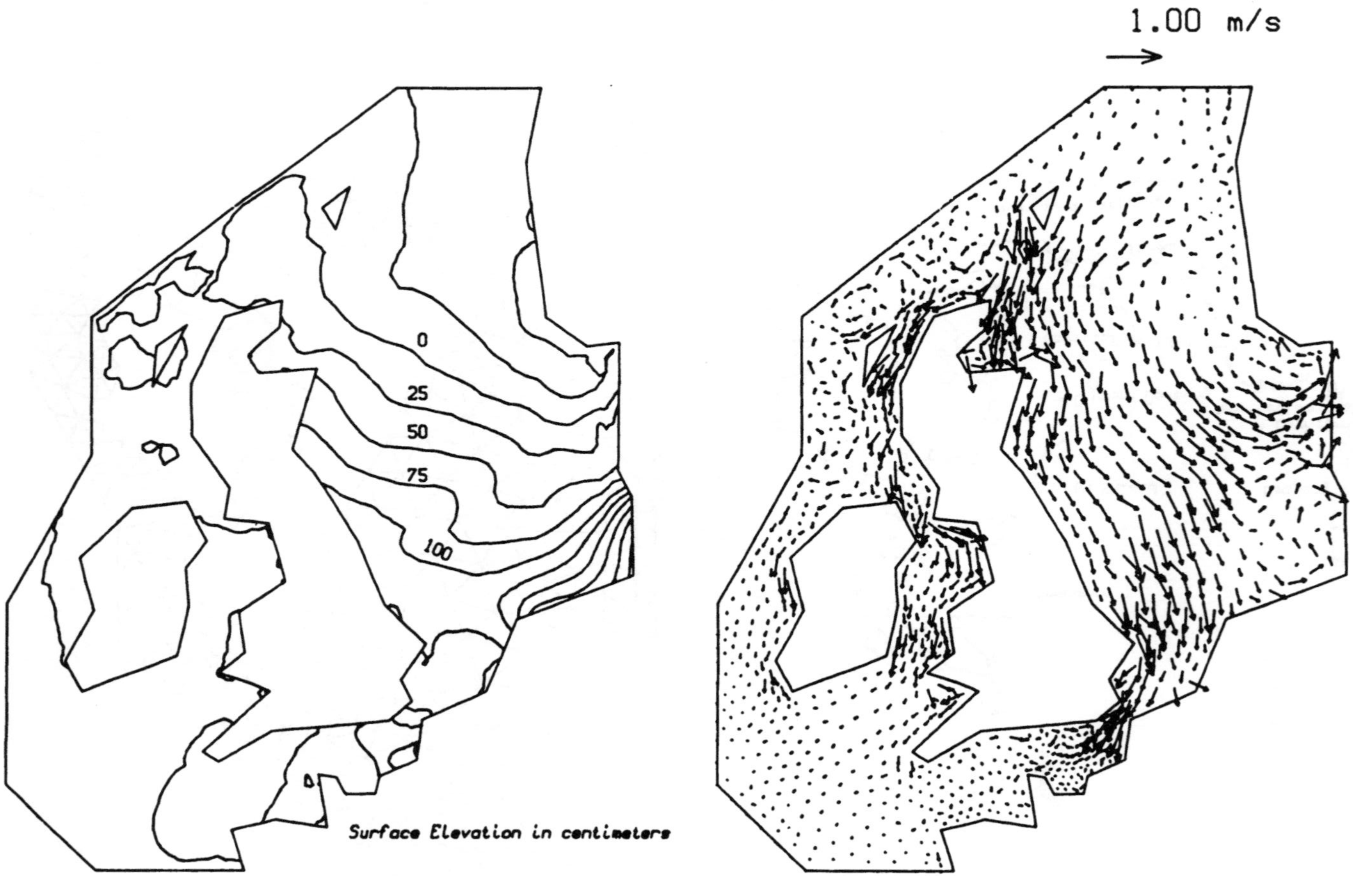

Figure 5 : Surface Elevation and Currents - 19/11/73 at 13H

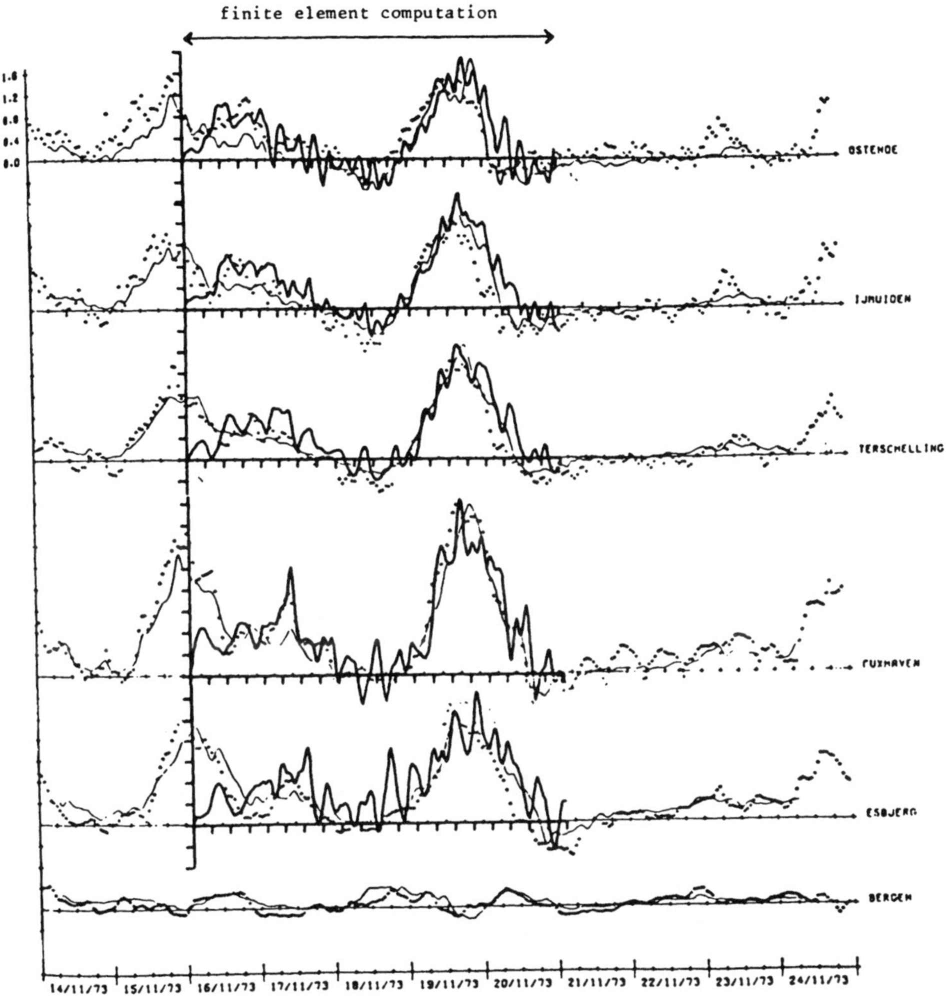

FIGURE 6a : STORM SURGES AT CONTINENTAL PORTS

dot line = observed surges
fine line = results from FLATHER (I.O.S.)
thick line = finite element computation

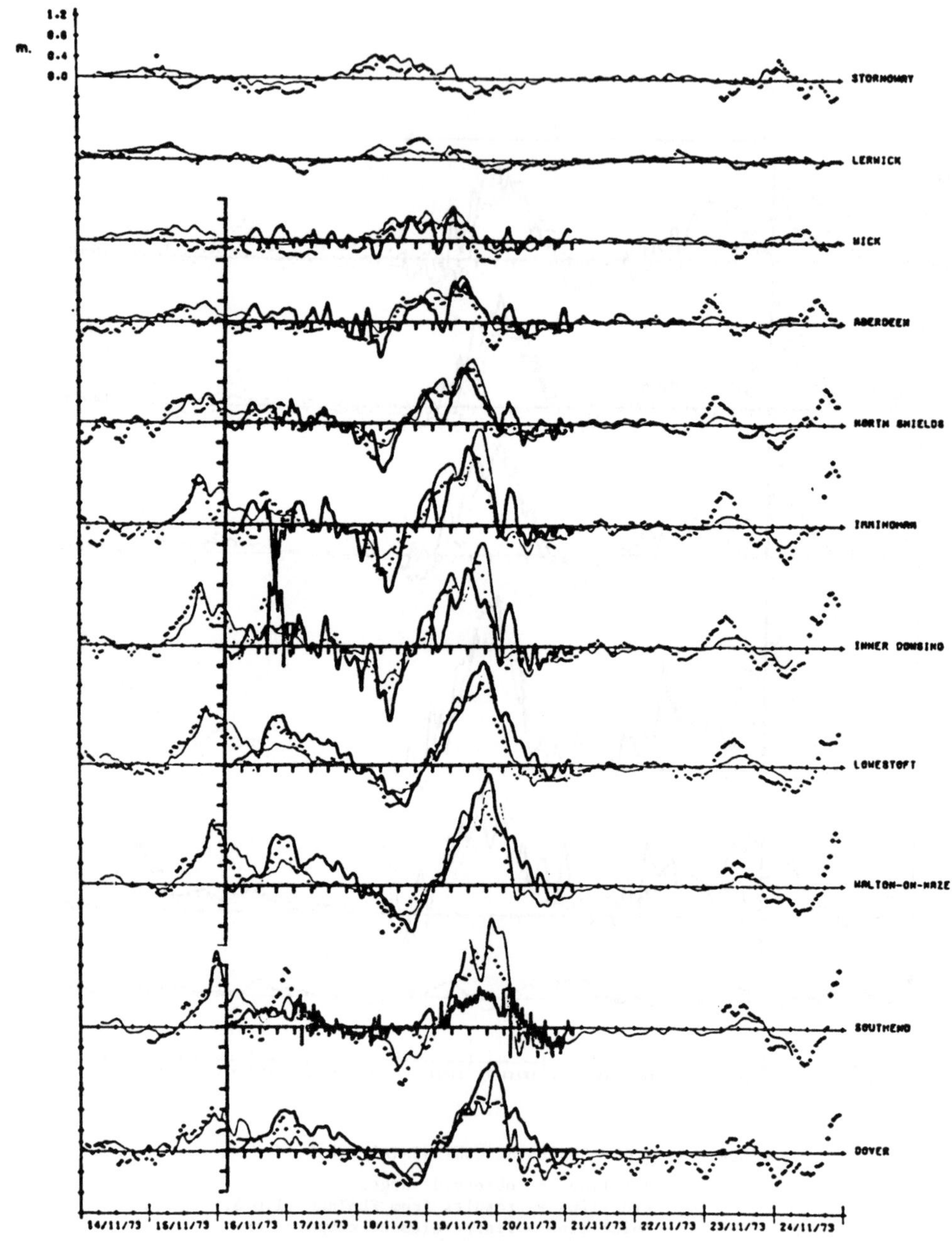

FIGURE 6b : STORM SURGE AT BRITISH PORTS

dot line = observed surges
fine line = results from FLATHER (I.O.S.)
thick line = finite element computation

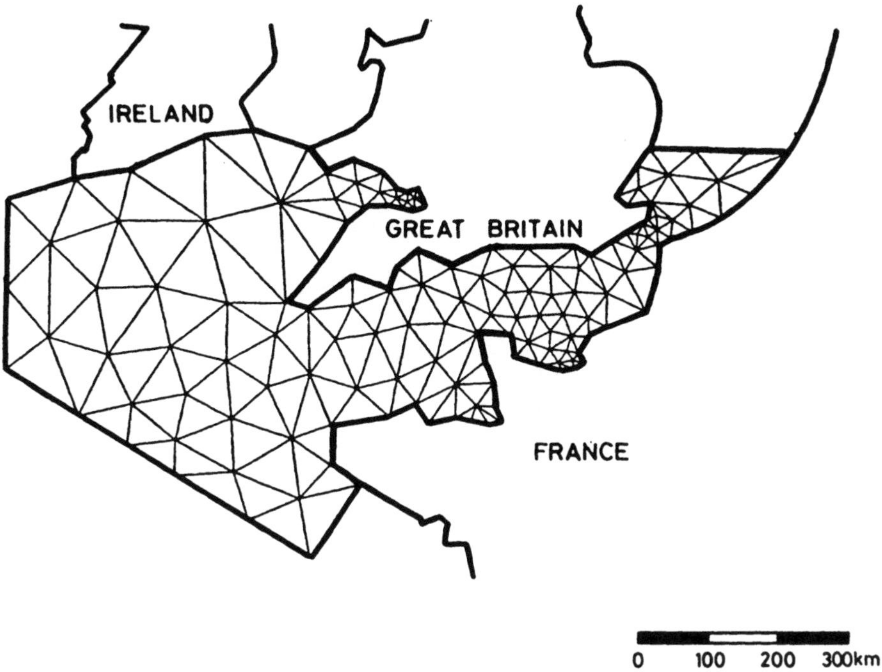

Fig.7 _ Finite element grid of the English Channel

Fig.8 _ Computed tidal currents at H.W. at BREST

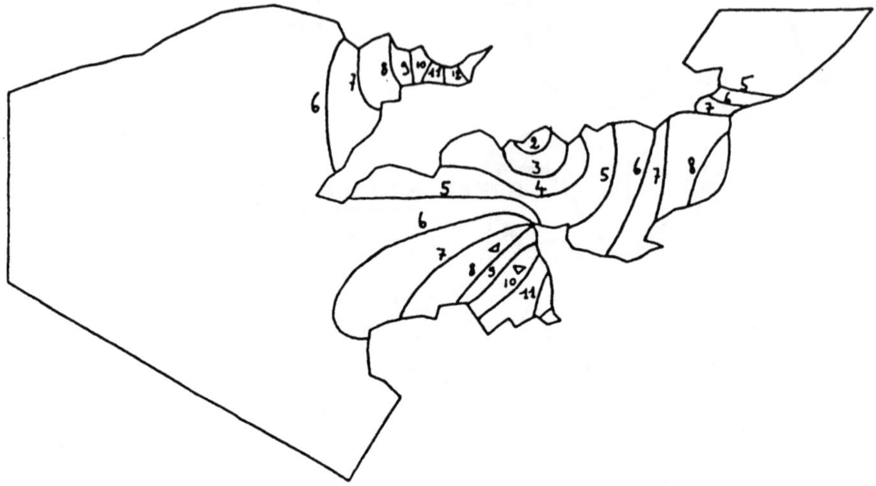

b) Measured co - tidal lines

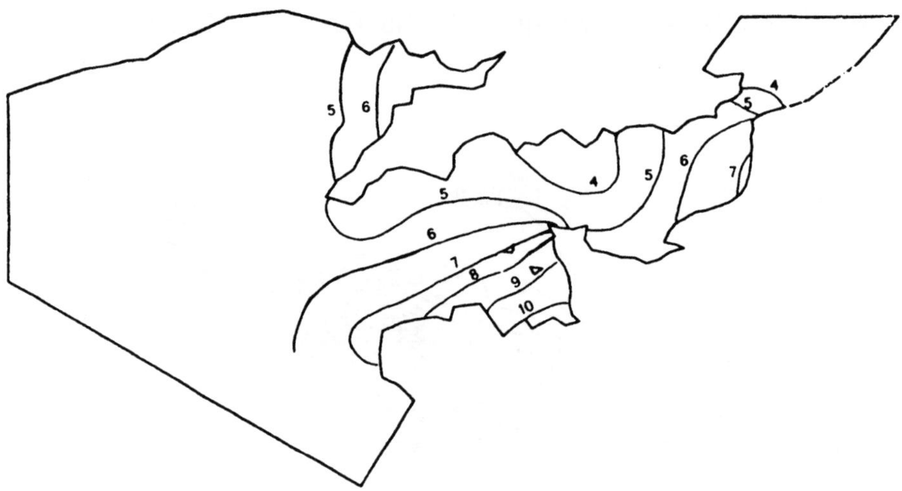

a) Computed co- tidal lines

Fig. 9 _ Comparison between computed and measured tide ranges in the English Channel (Mean Spring Tide)

A Time Variant Computational Mesh Technique To Simulate A Large Scale Ponding Test

G. Schmitz

The Technical University of Munich
F. R. of Germany

M. Vauclin

Université Scientifique et Medicale de Grenoble
France

G. J. Seus

The Technical University of Munich
F. R. of Germany

SUMMARY

A ponding test in a 1600 m long reach of a trapezoidal irrigation canal, cross sectional area of max. 244 m², has been performed to determine the hydraulic properties of the lining as well as the water losses due to infiltration and evaporation.

The experimental results were then used to validate a numerical model based on the 2D-Richards equation. The model takes into account the governing equation nonlinearity through the use of a time variant computational grid. This "breathing" grid automatically reduces its mesh size in regions of high gradients in hydraulic head, while in regions of low gradients the grid expands.

INTRODUCTION

Simulating transient unsaturated-saturated subsurface flow has occupied a great number of researchers. Among others, the models of RUBIN (1968), TAYLOR and LUTHIN (1969), HORNBERGER et al. (1969), VERMA and BRUTSEART (1970), COOLEY (1971), FREEZE (1969, 1971), and VAUCLIN et al. (1975) were based upon the Richards equation using for its solution a fixed computational grid. PIKUL et al. (1974) consider this single equation approach to be the most general and rigorous one, which has, however, still the great difficulty of adapting the highly nonlinear governing equation to an efficient numerical technique.

In this study a numerical model for solving the 2D-Richards equation is presented which takes into account the governing equation nonlinearity through the use of a time variant computational grid. According to locally strong or weak changes in hydraulic head, the irregular grid is automatically refined or extended.

The model is applied to simulate subsurface flow from an earth lined canal with time dependent wetted perimeter.

FORMULATION OF THE PROBLEM

Transient unsaturated-saturated subsurface flow can be described by the so-called Richards equation. For the case of infiltration and aquifer recharge from a long canal, subsurface flow in the axis of the canal can be regarded as negligible. Thus, the Richards equation is written in two dimensions (VAUCLIN, 1975) as

$$C(\psi)\frac{\partial H}{\partial t} = \frac{\partial}{\partial x}\left[K(\psi)\cdot\frac{\partial H}{\partial x}\right] + \frac{\partial}{\partial z}\left[K(\psi)\cdot\frac{\partial H}{\partial z}\right], \qquad (1)$$

where
- $H(x,z,t)$ .... hydraulic head;
- $\psi(x,z,t)$ .... capillary pressure head H-z;
- $K(\psi)$ .... hydraulic conductivity;
- Θ .... volumetric soil moisture content;
- $C(\psi) = \frac{\partial\Theta}{\partial\psi}$ ...... specific moisture capacity;
- z .... vertical space coordinate;
- x .... horizontal space coordinate;
- t .... time.

(For the orientation of x and z see Fig. 1)

The following assumptions are made:

(1) Two dimensional subsurface flow;
(2) air phase moves freely at atmospheric pressure;
(3) rigid soil matrix;
(4) total soil water potential is the sum of gravitational and soil water pressure potential;
(5) nonhysteretic soil properties;
(6) soil properties hold for dynamic flow;
(7) no chemical reactions between soil and water;
(8) isothermal flow.

Unsaturated Soil Properties

The values of $C(\psi)$ and $K(\psi)$ must be known in order to solve eq. (1). For this study a soil model adapted from VERMA and BRUTSEART (1970) is used. Therewith the soil properties are represented by

$$K(\psi) = K_{sat} \cdot \left[\frac{A}{A+|\psi|^B}\right]^N ; \tag{2}$$

$$\Theta(\psi) = (\Theta_{sat} - \Theta_{res}) \cdot \frac{A}{A+|\psi|^B} + \Theta_{res} ; \tag{3}$$

$$C(\psi) = \frac{\partial\Theta}{\partial\psi} = \frac{(\Theta_{sat}-\Theta_{res}) \cdot AB|\psi|^{B-1}}{(A+|\psi|^B)^2} , \tag{4}$$

where K_{sat} and Θ_{sat} are the conductivity resp. the soil moisture content at saturation.
A, B, N are parameters depending upon the soil type and Θ_{res} is the "residual" soil moisture content when ψ has a very large negative value.

NUMERICAL MODEL

In order to generate an irregular computational grid, eq.(1) is approximated by

$$C_{i,j}^{k+1/2,2m} \cdot \frac{H_{i,j}^{k+1,2m+1} - H_{i,j}^{k}}{\Delta t} + R^m \cdot \bar{K}_k \cdot (H_{i,j}^{k+1,2m+1} - H_{i,j}^{k+1,2m}) =$$

$$= \frac{K_{i,j+1/2}^{k} \cdot (H_{i,j+1}^{k+1,2m} - H_{i,j}^{k+1,2m})}{1/2(\Delta x_{j-1} + \Delta x_j) \cdot \Delta x_j} - \frac{K_{i,j-1/2}^{k} \cdot (H_{i,j}^{k+1,2m} - H_{i,j-1}^{k+1,2m})}{1/2(\Delta x_{j-1} + \Delta x_j) \cdot \Delta x_{j-1}} +$$

$$+ \frac{K_{i+1/2,j}^{k} \cdot (H_{i+1,j}^{k+1,2m+1} - H_{i,j}^{k+1,2m+1})}{1/2(\Delta z_{i-1} + \Delta z_i) \cdot \Delta z_i} -$$

$$- \frac{K_{i-1/2,j}^{k} \cdot (H_{i,j}^{k+1,2m+1} - H_{i-1,j}^{k+1,2m+1})}{1/2(\Delta z_{i-1} + \Delta z_i) \cdot \Delta z_{i-1}} , \tag{5}$$

where k = time index;
m = index of iteration;
i = horizontal space index;
j = vertical space index,

and

$$C_{i,j}^{k+1/2,2m} = C[0{,}5\cdot(H_{i,j}^{k+1,2m} + H_{i,j}^{k,2m})] \; . \tag{6}$$

The internodal conductivities which appear in eq.(5) were calculated as geometric averages, since VAUCLIN et al. (1979) showed that this method of weighting gives the most accurate numerical solution (better than harmonic or arithmetic mean values).

$$K_{i,j\pm1/2} = \sqrt{K_{i,j\pm1} \cdot K_{i,j}} \; ; \tag{7}$$

$$K_{i\pm1/2,j} = \sqrt{K_{i\pm1,j} \cdot K_{i,j}} \; ; \tag{8}$$

$$\bar{K}_k = \sqrt[4]{K_{i+1/2,j}^{k} \cdot K_{i-1/2,j}^{k} \cdot K_{i,j+1/2}^{k} \cdot K_{i,j-1/2}^{k}}$$

Most terms in eq.(5) are strongly dependent on space discretization. Important variations of the discretization steps therefore can lead to an ill conditioned matrix endangering stability. In order to avoid this difficulty for the case of an applied ADI iterative procedure, eq.(5) is, for the vertical sweeping, multiplied by

$$\Delta z_i \cdot \Delta z_{i-1}$$

and for the horizontal sweeping by

$$\Delta x_i \cdot \Delta x_{i-1}$$

The expression

$$R^{m} \cdot \bar{K}_k \cdot (H_{i,j}^{k+1,2m+1} - H_{i,j}^{k+1,2m})$$

is added to accelerate the convergence. Because of

$$R^{m} \leqq 0{,}2$$

and

$$(H_{i,j}^{k+1,2m+1} - H_{i,j}^{k+1,2m}) \rightarrow 0 \qquad \text{for increasing } m,$$

that term becomes negligible after only a few iterations. The algebraic system (5) was proved to be consistent with the differential equation (1).

AUTOMATIC MESH VARIATION

High values of the gradients $\partial H/\partial z$ and $\partial H/\partial x$, respectively, usually occur in just relatively small subdomains of the investigated area. These special properties of the unsaturated-saturated flow could be taken into account by implementing an irregular computational grid; however, when treating a transient problem, the wetting front does not stay at a fixed place but moves with time.

To overcome this difficulty, a "breathing" computational net is developed which for each time step automatically concentrates the computational points in areas of strong changes of hydraulic head, while in subdomains with low gradients in head, the mesh is extended.

Thus, most of the grid points in the computation are not permanent. However, contrary to these transient computational points, the meshpoints of a basic coarse grid always stay at the same place during the computation. This fixed basic grid has to be defined before applying the model; its mesh size represents the maximum possible discretization steps max Δx and max Δz. The solution of the model always refers to the basic grid, thus providing the possibility of comparative studies, sensitivity analysis, etc.

Before starting the computation with the first time step, the model itself rearranges the grid according to the initial conditions if necessary. Then, during the computation, it is checked every k time steps, if a grid reorganization is needed. The first cycle of iterations at k+1, still based on the mesh configuration at k, already provides a good approximation of the variables at k+1. Using these values of hydraulic head in eq.(9) together with criterion (10) the model decides whether new computational lines or columns need to be added or removed from the grid, or nothing needs to be changed.

In principle, the net can be adapted to the physical process for every time step. Nevertheless it is sometimes suitable to choose, according to the soil characteristics (e.g. low hydraulic conductivity), a longer period for the variation of the grid.

Let us consider for example the mechanism for removing or adding computational lines. The values of hydraulic head of the successive lines (i) and (i+1) are compared in order to determine the maximum of

$$\Delta H = \max_{\forall j}\{|H_{i-1,j} - H_{i,j}|\} \quad . \tag{9}$$

Applying eq.(9) to all lines i = 1,2, ...,n-1, each Δz_i of the old mesh configuration is associated with a ΔH_i. A possible reorganisation of the grid is executed conforming the following scheme.

$$\Delta H_i \cdot \Delta z_i \leq \delta_1 \ldots$$ line (i) is removed, if it does not belong to the basic grid:

$$\delta_1 < \Delta H_i \cdot \Delta z_i \leq \delta_2 \ldots$$ no change;

$$\delta_2 < \Delta H_i \cdot \Delta z_i$$ l lines between (i) and (i+1) are added with (10)

$$l = \text{INT}\left(\frac{\Delta H_i \cdot \Delta z_i}{\text{const}}\right) .$$

Now the treatment of the columns has to be done in an analogous way, thus generating the new grid.

For the successive computation initial values of the variables have to be defined at every mesh point. The values of hydraulic head calculated from the old grid configuration at time k are refered to the newly introduced lines resp. columns by linear interpolation, thus providing the complete initial conditions for the next cycle of iterations.

Then the same procedure starts again, until the number and location of computational lines and columns finally remains unchanged. Except for the initial state, however, convergence mostly occurs after one grid reorganisation. When the iterative procedure yields convergence, the model moves to the next time step and the computation in the irregular grid continues "normally" until, after k time steps, the net is checked again.

APPLICATION

The model was applied to simulate a ponding test which has been executed in a trapezoidal irrigation canal. Fig. 1 shows the cross section of the canal as well as the unsaturated-saturated flow system.

Time Dependent Wetted Perimeter

Investigations of FREYBERG (1983) demonstrate the significant impact variations the wetted perimeter can have on cross sectional infiltration rates. Therefore considering the relatively small slope of the canal banks together with the varying water level, it was necessary to treat the domain of computation notas rectangular, but instead to calculate infiltration from a time dependent wetted perimeter.

In the model a subroutine GEO adjusts for the type of cross section (e.g. circular, trapezoidal, etc.) by defining the suitable basic grid; it also controls the moving boundary conditions, caused by the computation.

Soil Characteristics

Drilling has revealed that the soil consists of sandy and clayey silt containing some gypsum, involving rather high densities (mean value 1.65 gcm^{-3}, average minimum 1.50 gcm^{-3}) and relatively high permeability at saturation (see Fig. 2).

The values of the parameters in eq.(2) and eq.(4), respectively, were determined from field data as

$K_{sat} = 0{,}36$ cm/h $A = 110$

$\Theta_{sat} = 0{,}41$ cm³/cm³ $B = 1{,}15$

$\Theta_{res} = 0{,}288$ cm³/cm³ $N = 2{,}7$

In the model the soil is approximated as homogenous with the only exception of the canal lining which has about 10 % of the permeability at saturation shown in Fig. 2.

Initial and Boundary Conditions

Assuming equilibrium in the region under investigation at the beginning of the filling period (canal still dry), the initial conditions are determined by the ground water level measured at t = 0 (see Fig. 1) at approximately 19 m under the ground. Thus the values of hydraulic head were initialized in all grid points to

$$H_{i,j} = -1900 \text{ cm}, \quad \forall_{i,j} \tag{11}$$

The boundary conditions in the canal are determined by the water level h(t):

$$\begin{aligned} 0 \text{ cm} &\leqq x < 2050 \text{ cm} ; \quad H(x) = h(t); \\ 2050 \text{ cm} &\leqq x < 2770 \text{ cm} ; \quad H(x) = h(t) - (x-2050)\cdot \mathrm{tg}\alpha . \end{aligned} \tag{12}$$

The rise of the ground water table at x = L is assumed to be negligible; this also yields a condition of the Dirichlet type (given head):

$$1900 \text{ cm} \leqq z \leqq 2700 \text{ cm} ; \quad H(z) = -1900 \text{ cm} . \tag{13}$$

All other boundaries are characterized by a Neumann condition. The dry surface of the soil as well as the impermeable stratum at z = 19 m are described by

$$\frac{\partial H}{\partial z} = 0 . \tag{14}$$

The symmetry boundary in the middle of the 1600 m long canal reach is regarded as an equivalent impermeable (no flow) boundary. On this line in the unsaturated zone at x = L the boundary condition is

$$\frac{\partial H}{\partial x} = 0 . \tag{15}$$

The basic grid, defined for the subsequent computation, was chosen to have maximal mesh widths of $\Delta x = 4{,}10$ m and $\Delta z = 3{,}20$ m.

PONDING TEST

The field experiment has been carried out from the beginning of April to mid July 1981 at a location in the Middle East.

Filling Phase

Flooding of the initially dry canal bed started on April 1st and the filling of the entire reach ended on May 25th, when the water level in the canal was 2,40 m above bottom. The filling operation obviously caused a significant change in the subsurface flow pattern. It was therefore necessary to compute realistic initial conditions for simulating the recession phase.

This Computation was executed, using in eq.(12) the recorded water levels from April 1st to May 25th (see Fig. 3). The initial conditions (11), namely the uniform distribution of hydraulic head, allowed a reduction in computing time by extending the computational grid several nodes beyond the furtherst node whose change of pressure potenial in one time step exceeds some small critical value.
This must not be confused with the variable mesh technique; it is merely an extension of the computational domain with the propagation of the infiltration front, as already applied for example by REEDER et al. (1980) for the 1D-case.

The results of the calculation after 1320 hours of filling satisfied the law of mass conservation to 0,15 %. The mass balance error is defined as the change in storage over the whole computational area from the beginning less the accumulated net inflow across the model boundaries, all divided by the change in storage.
The calculated subsurface flow pattern which represents the initialconditions for the simulation of the recession phase is shown in Fig. 4.

Calibration Phase

On May 25th the water level in the canal reached 2,40 m which terminated the filling operation. The total amount of water impounded was about 180 000 m³; this volume was then affected only by evaporation and infiltration losses.

The recorded water level in the canal was no longer used as the boundary condition in the computation; instead the model itself determined the transient boundary condition $h(t_k)$.

The simulation started with the computed initial conditions (see Fig. 4) and the water level of 2,40 m in the canal. Thus infiltration losses during the first time step $k = 1$ were calculated and, together with the information about the canal geometry and recorded evaporation rates, used to determine the boundary condition $h(t_{k=2})$ for the next time step.

Employing time steps of one hour, the subsequent computation from May 25th to the arbitrarily chosen date of June 10th (360 hours) provided a hydrograph of the receding water level.

The measured values of h(t) were then compared to the computed hydrograph recession line. In order to minimize the differences between them for the calibration period May 25th to June 10th, the hydraulic conductivity of the lining, K_{lin}, was adjusted by trial and error.
Obviously it was necessary to recalculate the initial conditions for every new value of K_{lin} resulting from the calibration. This procedure consumes considerable CPU-time and thus gave reason to apply an interrupted, guided trial and error strategy.

The calibration showed the lining hydraulic conductivity at saturation, K_{lin}, to be a function of time, possibly due to the gypsum content in the soil. Under pressure, the gypsum might have been partially washed out, thus affecting soil characteristics of the lining. From the beginning of the filling phase up to 480 hours, the temporal variation of the conductivity K_{lin} was still assumed to be negligible. Subsequently, because of pressure increase in most of the canal lining, the lining hydraulic conductivity at saturation was considered to be a function of time. The calibration, carried out in about five trials provided the following results:

For $T < 480$ h , $K_{lin} = 0{,}025$ [cm/h]

For $T > 480$ h , $K_{lin} = 0{,}025\,(T \cdot a + b)$ [cm/h]

with $a = 1/9oo$ and $b = 1 - 480/9oo$;

T = time in hours.

Verification phase

The calibrated model was now applied to calculate the graph h(t) of the falling water table in the canal, until the canal fell dry. A comparison between both recorded and calculated hydrographs during the verification phase (see fig. 5, June 10 - July 10) shows relatively good correspondance. The calculated time for the channel to become dry exceeded the recorded time by approximately 4 days, a 8,5 % error for the emptying cycle.

CONCLUSIONS

The simulation brought up some interesting results:

- With the "breathing" grid, in areas of low gradients in hydraulic head the model used maximal discretisation steps of $\Delta x = 4.10$ m and $\Delta z = 3.20$ m. The minimum mesh widths around the wetting front became $\Delta x = 51$ cm and $\Delta z = 22$ cm.
- The value of K_{lin} was determined to be a function of time, increasing during the simulation from 0,025 cm/h to about 0,078 cm/h; possibly due to the washing out of gypsum.

- More iterations were needed after a significant rise of the groundwater table together with a water level in the canal in excess of almost 2 m.
- The total CPU-time used for one time step was approximately 1,5 seconds on a CDC 175.

The difficulties of basing hydraulic conductivity values on laboratory tests of point samples of soil are well known. The model presented provides an additional tool to obtain reliable aerial values of hydraulic conductivity to use for earth lined canal system calculations.

ACKNOWLEDGEMENTS

The authors wish to express sincere thanks to the German Research Association (Deutsche Forschungsgemeinschaft / DFG) for providing financial support. This paper is based on DFG Grants Se 316/7-1.
All computations were carried out free of charge on the CDC CYBERS 175 DP of the LEIBNIZ computing center of the Bavarian Academy of Sciences. This valuable assistance is gratefully acknowledged.

REFERENCES

Bear, J., (1972), "Dynamics of Fluid in Porous Media", American Elsevier, New York.

Bear, J., (1979), "Hydraulics of Groundwater", Mc Graw-Hill.

Cooley, R.L., (1971), "A finite difference method for unsteady flow in variably saturated porous media, Application to a single pumping well, Water Resour.Res. 7(6), 1607-1625.

Freeze, R.A., (1969), "The mechanism of natural groundwater recharge and discharge, 1, One-dimensional vertical unsteady unsaturated flow above a recharging or discharging groundwater flow system, Water Resour.Res. 5(1), 153-171.

Freeze, R.A., (1972), "Role of subsurface flow in generating surface runoff, 1, Base flow contributions to channel flow. Water Resour.Res. 8(3), 609-623.

Freyberg, D.L., (1983), "Modeling the Effects of a Time-Dependent Wetted Perimeter on Infiltration from Ephemeral Channels. Water Resources Research, Vol. 19, No. 2, 559-566.

Hornberger, G.M., I.Remson, and A.A. Fungaroli, (1969), Numerical studies of a composite soil-moisture groundwater system. Water Resour.Res., 5(4), 797-802.

Pikul, M.F., R.L. Street, and I.Remson, (1974), "A numerical model based on coupled one-dimensional Richards and Boussinesq equations", Water Resour.Res., Vol.10, No. 2, 295-302.

Reeder, J., Freyberg, D., Franzini, J., Remson,I., (1980) "Infiltration under Rapidly Varying Surface Water Depths", Water Resources Research, Vol. 16, No. 1, pp. 97-104.

Richards, L.A. (1931), "Capillary conduction of liquids through porous mediums, Physics.1(5), 318-333.

Rubin, J., (1968), "Theoretical analysis of two-dimensional, transient flow of water in unsaturated and partly unsaturated soils, Soils Sci.Soc. Amer.Proc. 32, 607-615.

Taylor, G.S. and J.N. Luthin, (1969), "Computer methods for transient analysis of water table aquifers, Water Resour. Res. 5(1), 144-152.

Vauclin, M., R., Haverkamp, and G. Vachaud, (1979), "Etudes de la Résolution Numérique de l'Equation d'Eau en Milieu Nonsaturé", 165 pp., Presses Universitaires de Grenoble.

Vauclin, M., (1975), "Etude expérimentale et numérique du drainage de nappes à surface libre. Influence de la zone non saturée", Ph. D. thesis, 196 pp., Univ.Sci. et Méd. de Grenoble, Grenoble, France.

Vauclin, M., Khanji, D. and Vachaud, G., (1979), "Experimental and Numerical Study of a Transient Two-Dimensional Unsaturated-Saturated Water Table Recharge Problem ", Water Resources Research, Vol. 15, No. 5, Oct., pp. 1089-1101.

Verma, R.D., and W. Brutsaert, (1970), "Unconfined aquifer seepage by capillary flow theory, J. Hydraul.Div.Amer.Soc. Civil Eng., 96(HY6), 1331-1334.

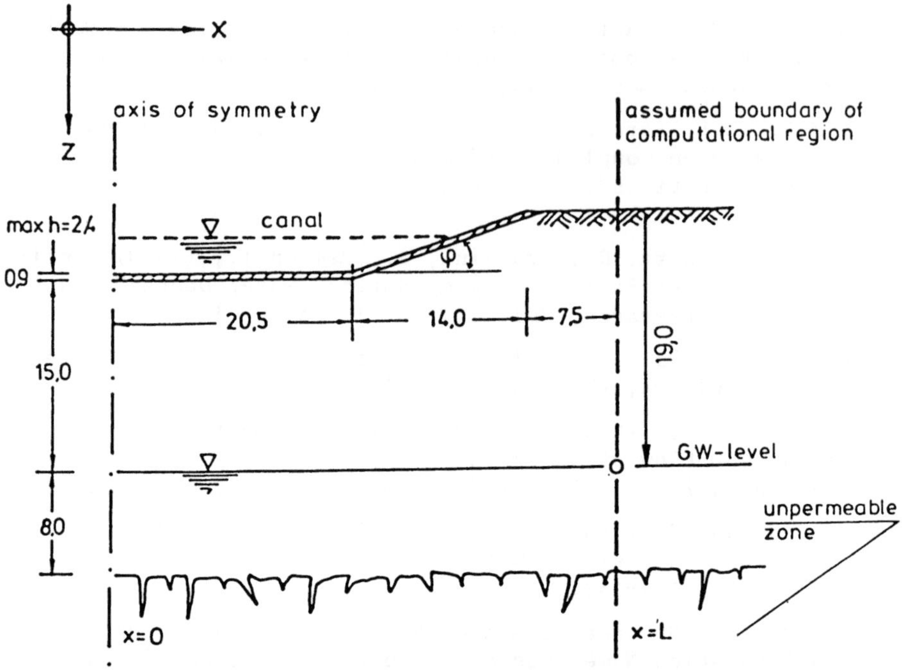

Fig. 1: Cross section of the irrigation canal at t=0 (beginning of the filling period) with subsurface flow system.

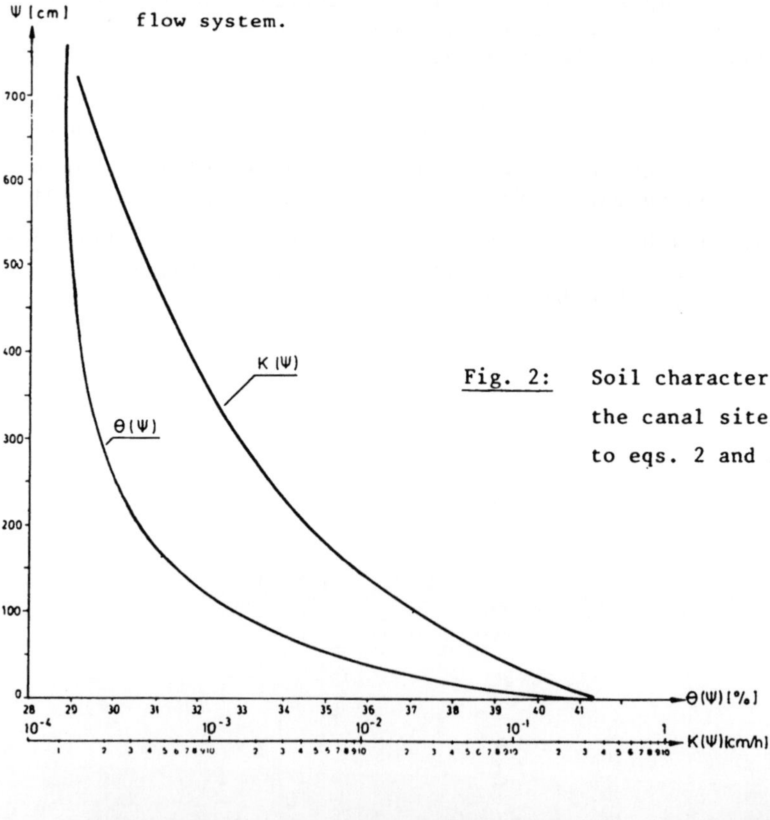

Fig. 2: Soil characteristics of the canal site (according to eqs. 2 and 3).

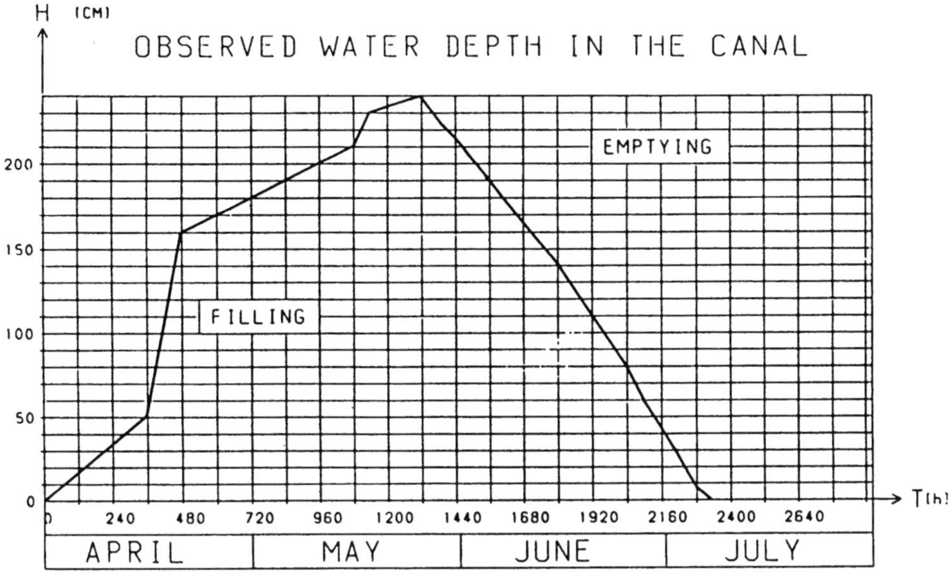

Fig. 3: Recorded water levels in the canal during the ponding test.

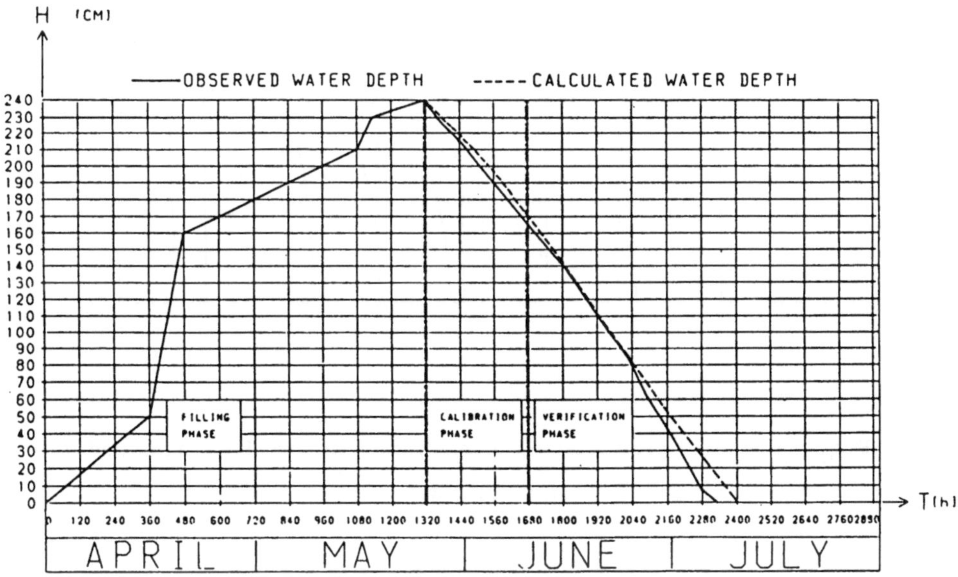

Fig. 5: Comparison of recorded and calculated stage hydrograph.

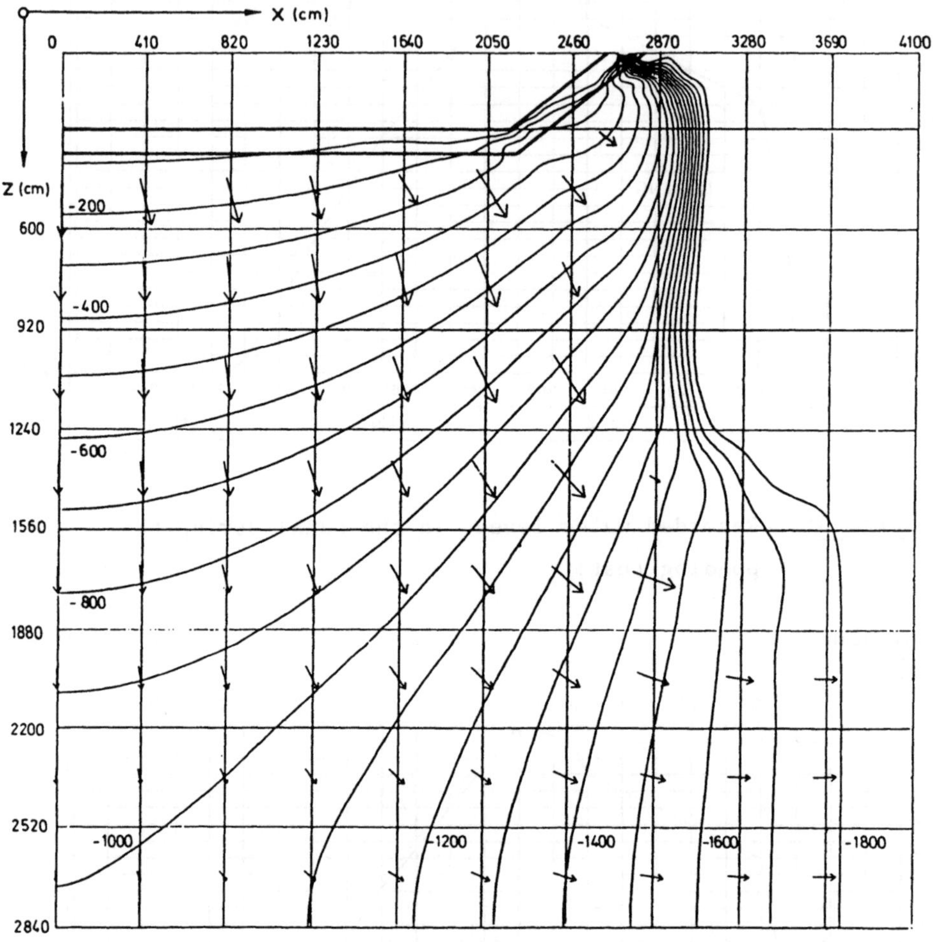

Fig. 4: Distribution of the hydraulic head at the end of the filling period.

PREDICTION OF MAIN CHANNEL/FLOOD PLAIN FLOW INTERACTION WITH FEM

P. Prinos, R.D. Townsend

Department of Civil Engineering, University of Ottawa, Ottawa, Canada

1. INTRODUCTION

Momentum transfer phenomena associated with flows in compound open channels have been studied experimentally by Myers 1972, Rajaratnam 1979, Wormleaton et al 1981 and Prinos and Townsend 1983 among others.

These phenomena are generated, for example, when flow in a deep central channel just inundates flood plain areas on either side of the former, thereby producing intense shearing zones in the junction regions between the deep and shallow sections.

The transfer mechanisms (vorticies) on one hand are found to produce local increases in flow velocity and boundary shear stress on the flood plains, while at the same time reducing local velocities and boundary shear stress in the main channel. The principal finding, however in all studies was the mechanism's overall negative effect on system conveyance. In general when the stage-discharge curve for a compound channel was compared with that obtained for the equivalent "separate channels" system, the former indicated reduced system efficiency, at channel depths just greater than bankfull depth.

Although these transfer mechanisms can distort significantly the flow patterns (and hence shear stress distributions) in "narrow" channel systems, their influence is largely limited to the relatively narrow mixing regions in the case of river systems having "wide" main channels. Accordingly velocity distribution laws developed for two-dimensional flow fields can be applied in those portions of the cross section remote from the mixing regions. Hence average values of velocity and boundary shear stress, for both main channel and flood plain zones, can be calculated using conventional formulas.

The aforementioned complex nature of compound channel flows, especially for "narrow" channel systems, has limited to some extent the development of predictive numerical models for discharge computation. Radojkovic and Ivetic's model (1982) applies to fully-developed flow in compound channels. The governing equations are simplified by combining convection and shear stress in an "effective shear stress" parameter which is calculated from a length-scale model. The resulting differential equation is then solved by a finite element technique.

Vreugdenhil and Wijbenga (1982) use a finite difference method of the ADI type to solve their depth-averaged equations and eddy viscosity is calculated as part of the calibration process.

Demuren (1983) uses a three-dimensional numerical procedure for computing "developing" flow in channels with flood plains. The partial differential equations governing the steady incompressible flow are solved in conjunction with the k-ε model of turbulence. This model was applied to Rajaratnam's data (1979) with satisfactory results but it was found to be computationally very expensive.

Prinos (1984) uses the depth-averaged version of the k-ε model along with the Navier-Stokes equations to predict depth-averaged velocities in his laboratory model. The physical model consisted of a trapezoidal-shaped central channel with "wide" flood plain sections on either side of the former. The base width of the central channel could be adjusted to produce both "wide" and "narrow" channel conditions. The numerical model gave satisfactory results for "wide" conditions in the experimental channel however, for "narrow" conditions, an additional boundary condition had to be introduced at the main channel/flood plain junctions. The results were then satisfactory for the conditions examined.

This paper describes a finite element model (FEM) for predicting velocity distributions in fully-developed compound channel flows. A FEM based on variational principles, similar to those employed by Delleur and Sooky 1961, Zinkiewicz and Cheung 1965 and Gerard 1974, was used in the present study. In essence our model employs the Ritz method in conjunction with an algebraic expression used in determining the eddy-viscosity distribution. Predicted velocity distributions are compared with those determined in the experimental program.

2. THE PHYSICAL MODEL

A compound channel having a central trapezoidal-shaped deep section (2V:1H side slopes) with two adjoining shallow berms (flood plains) was used in the study. The experimental program was part of a larger study dealing with momentum transfer phenomena in compound channels (Prinos, 1984). While differential boundary roughness was a major consideration in the larger study, this paper will address the "smooth boundary" case only.

The channel's trapezoidal main section was 10.2 cm deep with base width settings of 20.3, 30.5, 40.6 and 50.8 cm. Flood plain sections, each 38.1 cm wide, were located on either side of the main section. These particular dimensions for the system represented the "narrow channels" category of problem examined. A 7.2 cm high "false floor" unit was subsequently placed in the central channel section to produce "wide" conditions in the latter for the second series of experiments dealing with two-dimensional flows.

Once reasonably uniform flow was established in the channel, local velocities were measured using a miniature current flowmeter system and a commercial Pitot tube (4mm OD) connected to a differential pressure transducer (Validyne DP-103) with a digital indicator (Validyne CD-23). With the system in place very low velocities, such as those present in the flood plain zones could be measured accurately. The Pitot tube was used to measure velocities close to both the channel boundaries and free water surface and also on the flood plain zones at low flood plain depths. Because of its size the propeller current meter was largely restricted to measurements in the main body of the channel flow. Periodic calibration of both probes was made against a known pressure source. Further details of the experimental setup and program of experiments are available elsewhere (Prinos, 1984).

3. MEAN FLOW EQUATIONS

For steady, incompressible open channel flow the distribution of velocity is governed by continuity and momentum (Navier-Stokes) equations. The momentum equation in the streamwise direction (x-direction) is written as:

$$U\frac{\partial U}{\partial x} + V\frac{\partial U}{\partial y} + W\frac{\partial U}{\partial z} = \frac{\partial}{\partial y}\left(\rho\nu_t\frac{\partial U}{\partial y}\right) + \frac{\partial}{\partial z}\left(\rho\nu_t\frac{\partial U}{\partial z}\right) + gS \qquad (1)$$

where U, V, W = mean velocity components in the x, y, z, directions respectively;

ν_t = eddy-viscosity (turbulent diffusivity of momentum);

ρ = density of water;

g = acceleration due to gravity;

S = slope of the channel.

In the above equation the concept of eddy-viscosity and the Boussinesq approximation are used, therefore the Reynolds stresses can be considered proportional to the mean velocity gradients. The eddy viscosity however depends strongly on the turbulence level and its distribution has to be determined.

If we consider uniform flow ($\frac{\partial U}{\partial x} = 0$) and V=W=0, equ. (1) is simplified as follows:

$$\frac{\partial}{\partial y}\left(\rho\nu_t\frac{\partial U}{\partial y}\right) + \frac{\partial}{\partial z}\left(\rho\nu_t\frac{\partial U}{\partial z}\right) + gS = 0 \tag{2}$$

Using variational principles, a governing function for equ. (2) has to be found and then minimized with respect to U. The distribution of U which minimizes the governing function will also satisfy equ. (2).

The governing function for equ. (2) has been provided by Helmoltz, as reported by Lamb (1932), and Euler's theorem from the calculus of variation can be used to show that the velocity U which minimizes the functional also satisfies equ. (2). The governing function for equ. (2) can be written as follows:

$$\Phi = \int_A \left\{ \frac{1}{2}\rho\nu_t\left[\left(\frac{\partial U}{\partial y}\right)^2 + \left(\frac{\partial U}{\partial z}\right)^2\right] - gSU\right\} dA \tag{3}$$

where A = total cross sectional area.

When equ. (3) is minimized with respect to U, the distribution of U in the vertical and lateral directions (y and z respectively) can be determined.

The above theory can be applied to "wide" channels where the V and W components of velocity are approximately zero and therefore secondary currents do not significantly influence the primary flow field.

When the secondary currents are strong, i.e. the lateral and vertical components of velocity alter significantly the flow field ("narrow" channels, "interaction region" in compound channels), then V and W components cannot be ignored.

In this case a new functional applies which, when minimized will give a new function that satisfies equation 1 (with $U\frac{\partial U}{\partial x} = 0$). Such a function is unknown and therefore the method cannot be applied.

Gerard (1974) proposed the Galerkin method of weighted residuals or the Ritz approach to overcome this problem. In the latter approach the convective terms of equ. (1) were combined with the known pressure term (i.e. gS) and were prescribed as additional pressure terms on an iterative basis.

4. FINITE ELEMENT APPROACH

The application of FEM implies subdivision of the flow field into representative elements. Within the domain of each element the distribution of velocity can be expressed as:

$$U = a_1 + a_2 y + a_3 z + a_4 y^2 + a_5 yz + a_6 z^2 \tag{4}$$

where a_1, a_2 ..., a_6 = undetermined parameters.

With the velocity function expressed by equ. (4) a 6-node triangular element is applied and equ. (4) can now be written in matrix notation as:

$$\{U\} = [N_1]\ \{a_i\} \quad , \quad i = 1,6 \tag{5}$$

Also the derivatives of U with respect to y and z are formulated as:

$$\begin{Bmatrix} \frac{\partial U}{\partial y} \\ \frac{\partial U}{\partial z} \end{Bmatrix} = [N_2]\ \{a_i\} \tag{6}$$

Combining equs. (5),(6) and (3) based on integrals evaluated over the elemental area, the elemental function can be written as:

$$\Phi_e = \{U_e\}^T[K_e]\{U_e\} - 2\{F_e\}^T\{U_e\} \tag{7}$$

where $\{U_e\}$ = vector of unknown nodal velocities;

$\{K_e\}$ = known function of nodal coordinates and eddy viscosity;

$\{F_e\}$ = known function of the constant pressure gradient and nodal coordinates.

The functional expressed by equ. (7) is now minimized with respect to the undetermined parameters, or nodal velocities. Therefore for each element we have:

$$\frac{\partial \Phi_e}{\partial U_e} = [K_e]\{U_e\} + \{F_e\} \tag{8}$$

Using the fact that the functional for the whole domain is equal to the summation of the functionals for the N elements used to subdivide the domain, equ. (8) becomes:

$$\{\frac{\partial \Phi}{\partial U}\} = [K]\{U\} + \{F\} \tag{9}$$

For the functional to be a minimum, equ. (9) should be made equal to zero

$$[K]\{U\} + \{F\} = 0 \tag{10}$$

and hence the above system of equations can be solved when the appropriate boundary conditions are introduced.

5. BOUNDARY CONDITIONS

The necessary boundary conditions for velocity U and eddy viscosity have to be introduced into equ. (10) for the system bed and centre line, and free water surface.

Both the centre line and free water surface can be considered as planes of symmetry and therefore the velocity function which minimizes the functional, will satisfy these boundary conditions automatically. For the channel walls, known boundary conditions have to be imposed. Instead of using the "zero velocity" boundary condition at the wall, (which requires very small elements near the wall region), boundary conditions derived from "wall region" laws are applied.

Near the wall the flow can be considered two-dimensional and the well-known logarithmic velocity distribution can be used, namely:

$$\frac{U}{U_*} = 5.75 \log \frac{yU_*}{\nu} + 5.5 \tag{11}$$

where U_* = shear velocity;

y = distance from the wall;

ν = kinematic viscosity.

Equation (11) applies to smooth wall conditions and may be used in the "inner" region of flow $(5<[(yU_*)/\nu]<30)$ to match the "wall" conditions with the "outer" region of flow which is solved by the finite element method.

The shear velocity, U_* is defined as:

$$U_* = \sqrt{gRS} \tag{12}$$

where R = hydraulic radius of the channel.

This can be used as a first approximation to calculate the velocity from equ. (11). If the velocity is known at a node of the "wall region", the shear velocity becomes:

$$U_*^2 = c_f U^2 \tag{13}$$

where c_f = friction coefficient.

Equations (11), (12) and (13) have been used as boundary conditions in the iteration process of the solution procedure.

The distribution of the eddy-viscosity can be determined by various formulae available for this purpose. Van Driest (1956) derived an expression for eddy-viscosity based on Prandtl's mixing length hypothesis for smooth and rough walls. The expression adopted in this study is the following:

$$\nu_t = \nu \; \frac{-1+\sqrt{1+4k^2 {y_*}^2 [1-\exp(-y_*/A_*)]}}{2} \tag{14}$$

where $y_* = \frac{yU_*}{\nu}$

k = von Kármán's universal constant;

A_* = constant characterizing the thickness of the "inner" region of flow.

Equation (14) can be used in determining the eddy-viscosity distribution within the "inner-region" of the wall. In the "outer-region", the distribution can be assumed constant and equal to the one determined by equ. (14) for $y_*= 30$.

The above assumption and the use of equ. (14) are discussed in the last section where numerical and experimental results are compared.

6. SOLUTION PROCEDURE

Velocity, boundary shear stress and eddy-viscosity are interrelated and therefore an iteration process is required. Based on an initial guess for one of the parameters, an iteration solution can be used to calculate the final distributions of the parameters. The boundary shear velocity was chosen as one of the parameters requiring an initial guess.

For all boundary node points the boundary shear velocity is calculated from equ. (12) for the first iteration and hence the velocity and eddy-viscosity can be calculated at the same nodes using equs. (11) and (14) respectively. Point velocities can be also obtained from flow calculations by the finite element method. If the two velocities are different, a new value for local boundary shear stress is calculated (equ. (13)) and used to give the average boundary shear stress (equ. (12)) after integration over the total wetted perimeter.

The new boundary shear velocities are then employed in the next iteration. This procedure continues until the values obtained in last two iterations differ only slightly.

7. APPLICATION OF THE MODEL

The model was applied to four representative experimental tests. The first two tests were performed under "wide channel" conditions with relative depth of flow $Y_r = (y_f/y_t) = 0.33$ and 0.5 respectively, (where y_f is the flood plain depth and y_t is the total flow depth). Also the relative width W_r (the ratio of the flood plain width over the main channel top width) and the relative roughness n_r (the ratio of flood plain roughness over main channel roughness) were kept constant at 1.25 and 1.0 respectively.

For these particular boundary conditions the transfer mechanism was found to reduce, only slightly, local velocities in the mixing region and not at all in the central region of the channel. Moreover maximum velocities were measured at (or near) the water surface, even in the narrow mixing region. For these conditions the model appeared to work reasonably well, although overestimating slightly velocities in the main channel while underestimating those in the flood plain (Figure 1).

In general, calculated isovel patterns were found to agree

favorably with those observed. Average velocities for the main channel and flood plain zones were also compared (for both measured and calculated local velocities) and were found to be in close agreement. Both main channel average velocities (measured and calculated by FEM) were approximately 6% less than that calculated by Manning's equation indicating the relatively small effect of the transfer mechanism on average velocity. For the two flow conditions considered, measured velocities in the flood plain were 21% and 13% greater respectively, than those calculated using Manning's equation (the corresponding FEM values were 17% and 10% greater respectively).

The last two experiments were performed under "narrow main channel" conditions with relative depths of flow equal to 0.16 and 0.28. Relative width and roughness were set at 2.5 and 1.0 respectively. As expected, because of the narrow main channel conditions here, the maximum velocity filament was located below the free surface. The degree of submergence, however, was greater than normal— an effect produced by the presence of the strong interaction mechanism at the water surface (mixing region). With regard to velocity distribution in the system the interaction mechanism had a twofold effect: it reduced the channel mean velocity (by 19% and 13% for relative depth values of 0.16 and 0.28 respectively) while producing related increases in flood plain mean velocity (34% and 19% respectively).

For "narrow channel" conditions the model does not predict local and average velocities satisfactorily (Figure 2). The model, since it does not properly account for secondary motion, is unable to simulate the depressed position of the maximum velocity filament below the free surface. The strong secondary motion in the mixing region is directed away from the channel walls so that it transports fluid with relatively low longitudinal momentum towards the central region of the channel, thereby producing lower velocities at the free surface. As a result of this limitation the model overestimates average velocity in main channel and underestimates flood plain average velocity. The average velocities for both subsections were in close agreement with those calculated using Manning's equation (Error ± 3%).

This modelling limitation arises from assumptions made in the formulation of the problem, namely that velocity components in the y-z plane are zero and the eddy-viscosity distribution is fully described by equ. (14). An isotropic eddy-viscosity applied to the "outer region" of flow does not reproduce any secondary motion. Also the distribution of the Reynolds stresses is not influenced by mean velocity gradients alone (Boussinesq's approximation) in such instances but by other physical processes too (e.g. convection, diffusive transport, turbulence production, viscous dissipation, etc.).

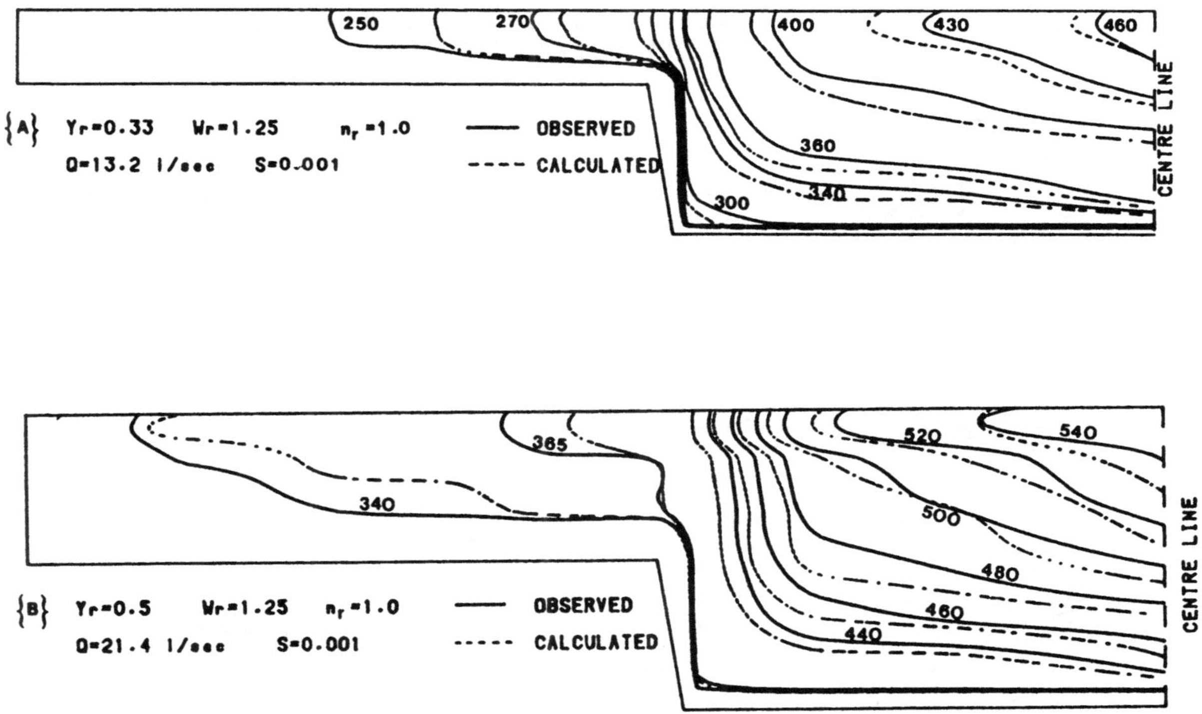

FIG. 1: COMPARISON OF CALCULATED AND OBSERVED ISOVEL PATTERNS FOR "WIDE" CHANNEL CONDITIONS (isovel units in mm/sec).

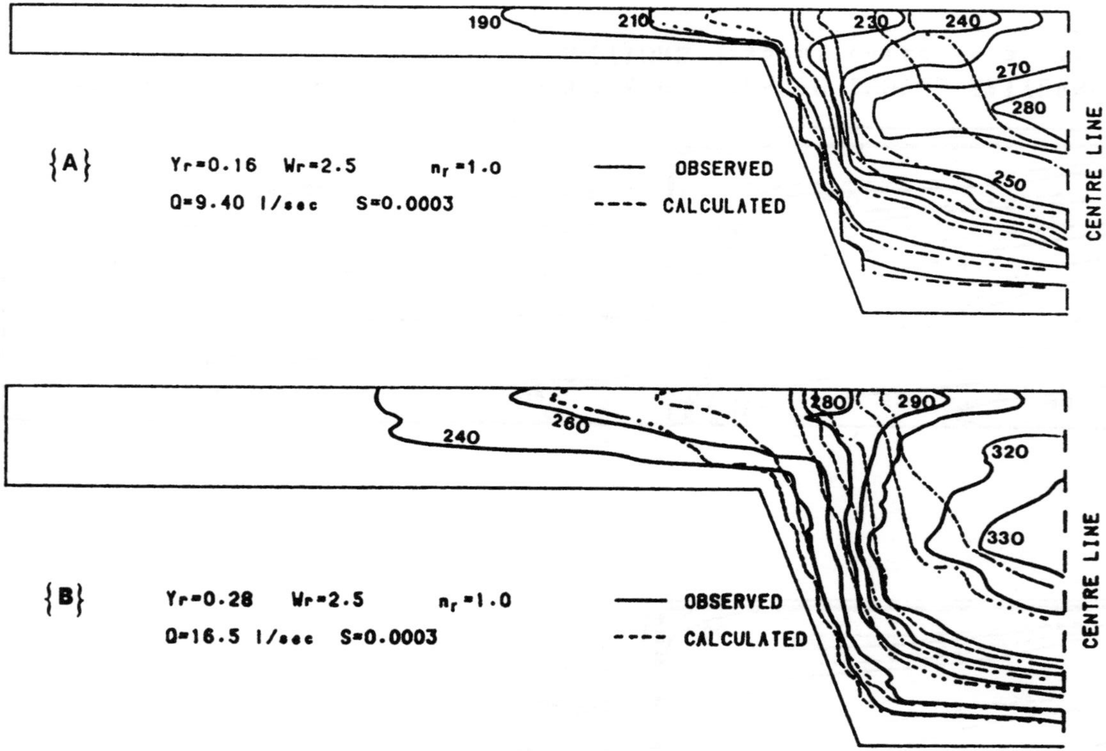

FIG. 2: COMPARISON OF CALCULATED AND OBSERVED ISOVEL PATTERNS FOR "NARROW" CHANNEL CONDITIONS (isovel units in mm/sec).

8. CONCLUSIONS

1. The numerical model appears to work reasonably well for wide conditions in the main channel of a compound channel system. While certain differences arise between the predicted and observed velocity distributions, these differences are considered relatively unimportant from a practical viewpoint.

2. When narrow channel conditions are examined the inherent errors in simulation are much more apparent. Firstly the numerical model cannot reproduce satisfactorily the observed isovel patterns which, in general, indicate depressed positions for the maximum velocity filaments. Secondly the model tends to overestimate mean velocity in the main channel while underestimating mean velocity on the flood plains. Both effects are largely due to presence of the strong momentum transfer mechanism in the mixing region which is only partly accounted for in the numerical model.

3. Better simulation of flow in narrow compound channels will require the development of more sophisticated numerical models that properly account for the complex 3-dimensional flow field within the mixing regions. In this endeavour special attention should be given to modelling correctly those stress parameters causing intense secondary motion in the mixing regions.

References

Delleur, J.W. and Sooky, A.A. (1961) "Variational methods in Fluid Dynamics". Journal of the Engineering Mechanics Division, ASCE, Vol. 87, No. EM6, pp. 1-23.

Demuren, A.O. (1983) "Prediction of flow in channels with flood plains". Proceedings of Engineering Mechanics Conference, ASCE, Lafayette, Indiana, Vol. 2, p. 1212-1215.

Gerard, R. (1974) "Finite Element solution for flow in non-circular conduits". Journal of Hydraulics Division, ASCE, Vol. 100, No. HY3, pp. 425-441.

Lamb, H. (1932) "Hydrodynamics". 6th edition, Cambridge University Press, Cambridge, England.

Myers, W.R.C. (1978) "Momentum Transfer in a Compound Channel". Journal of Hydraulic Research International Association for Hydraulic Research, Vol. 16, No. 2, pp. 139-150.

Prinos, P. and Townsend, D.R. (1983) "Estimating discharge in compound open channels". Proceedings of 6th Canadian Hydrotechnical Conference, Vol. 1, pp. 129-146.

Prinos, P. (1984) "A study of momentum transfer phenomena in compound sections having differential roughness". To be submitted to the School of Graduate Studies for the degree of Doctor of Philosophy in Civil Engineering, University of Ottawa, Ottawa, Canada.

Radojkovic, M. and Ivetic, M. (1982) "Finite element analysis of momentum transfer in rivers of complex cross section". Proceedings of 4th Int. Conference on Finite Element in Water Resources, Hannover.

Rajaratnam, N. and Ahmadi, R.M. (1979) "Interaction between Main Channel and Flood Plain Flows". Journal of the Hydraulics Division, ASCE, Vol. 105, No. HY5, pp. 573-588.

Van Driest, E.R. (1956) "On turbulent flow near a wall". Journal of Aeronautical Science, Vol. 23, pp. 1007-1011.

Vreugdenhil, C.B.and Wijbenga, J.H.A. (Computation of Flow Patterns in Rivers". Journal of Hydraulics Division, ASCE, No. HY11, pp. 1296-1309.

Wormleaton, P.R.,Allen, J. and Hadjipanos, P. (1982) "Discharge Assessment in Compound Channel Flow". Journal of the Hydraulics Division, ASCE, Vol. 108, No. HY9, pp. 975-993.

Zienkiewicz, O.C. and Cheung, Y.K. (1965) "Finite Elements in the Solutions of Field Problems". The Engineer, pp. 507-510.

ON THE APPLICATION OF DIGITAL SURFACE REPRESENTATIONS IN HYDRODYNAMIC MODELLING

I.J. Westwood, K.-P. Holz and R. Ratke

Hannover University

INTRODUCTION

For the purposes of modelling flow and transport processes in estuaries and the open sea, it is normal practice to evaluate variable quantities such as surface water levels, flow velocities and concentrations at nodes in a computational grid. These may either be of the rectilinear type, in which depths are defined at nodes in a square or rectangular grid, or irregular, in which case the bathymetry is usually represented by a triangular mesh. Although regular grids have the advantage that the storage of X, Y cartesian coordinates may be avoided and that the formulation of first and higher derivatives is relatively straightforward, irregular grids require fewer nodes for the same resolution of bathymetry. Since computing costs for explicit schemes rise in proportion to the number of grid nodes, considerable economies may therefore be gained by adopting the method of finite elements with an irregular mesh.

In the past, it has been common practice to develop finite element meshes manually, based upon various forms of hydrographic data. In addition to the fact that manual methods are cumbersome and time-consuming, the intuitive selection and triangulation of nodes often leads to erroneous representation of the bathymetry. With the objective of overcoming many of the problems associated with manual methods, the authors present an automatic mesh generation technique involving the application of a digital terrain model (DTM).

DIGITAL TERRAIN MODEL

As a basis for automatic mesh generation, a DTM (Program System TASH) developed by the Department of Topography and Cartography at Hannover University (Kruse, 1979) was employed. Input to the DTM consists in the three-dimensional coordinates of soundings from hydrographic charts, which are automatically registered by digitisation of either stochastically distributed values or depth contours. By selecting the most suitable form of surface to describe the local curvature between individual nodes and their nearest neighbours ("gliding surface method"), depths are interpolated at nodes in a square grid with a preselected mesh spacing. By adopting various reference numbers for input nodes defining special bathymetric features, such as ridge and structure lines, a suitably modified form of the normal interpolation procedure is applied in order to avoid undesirable smoothing of such features.

Interpolated depths in the DTM grid are stored row by row on a random access file, with dummy values at nodes outside the prescribed boundary or within bounded internal regions representing such features as islands or artificial structures. With the aid of additional programs, the stored grid may either serve as input for the calculation of depth contours or as a basis for the evaluation and graphical display of depth differences between corresponding nodes in two or more digital representations of a given bathymetry.

GENERATION OF FINITE ELEMENT MESH

General Considerations

In recent years, much attention has been given to the automatic generation of irregular computational grids. A fairly recent review of the methods which have been applied is given by Thacker (1980) and a symposium specifically relating to this subject was recently held in Houston (1983). Although the application of coordinate transformation techniques has proved most successful as a means of generating meshes for purely two-dimensional flow problems, the inclusion of a third spatial dimension introduces considerable difficulties, owing to the often conflicting requirements of hydrodynamic and topographical criteria. This problem is particularly relevant to explicit schemes, in which the minimum permissible segment length determined by the Courant stability criterion may be too large to adequately represent local bathymetric variations. Since, however, a priori knowledge of flow characteristics is seldom available, such considerations must often of necessity be dealt with in a post-processing exercise, based upon the initial results of a hydrodynamic model.

In the case of the present study, the authors are primarily concerned with flow systems, such as estuaries, in which bathymetric variations generally play a predominant role in determining overall flow patterns. Confining attention to bathymetric considerations, the following two basic requirements for automatic mesh generation are identified:

- optimum selection of reference nodes for the desired degree of resolution of a given bathymetry.
- optimum triangulation of reference nodes so as to faithfully

reproduce bathymetric features, whilst avoiding highly distorted elements, which can lead to numerical instability and inaccuracies in hydrodynamic models.

In the majority of mesh generation studies which include bathymetric variations, the primary objective concerns the optimum triangulation of a given set of nodes, based upon various geometric criteria such as the requirement of minimum connection lengths or the nearest approximation to equilateral triangles. Reduction of the node population from a stochastically distributed set with high redundancy, however, is far more complex owing to the three-dimensional nature of the necessary filtering algorithms, which make high demands on computer runtime and storage. By means of a DTM, however, the authors were able to apply a rapid filtering algorithm requiring low store core memory. A description of the filtering technique is given in the following.

Filtering of Nodes

Filtering of nodes from the DTM grid was carried out by approximating the bed profile along each row and column by a series of straight line segments, according to the smoothing criteria shown in Figure 1. In the first of these (Figure 1a), a length of bed profile is replaced by an approximation segment AB, provided that the ratio of the truncated profile area Ai to the trapezoidal area below a given reference level Ao does not exceed a prescribed critical error value E. The reference level Zo must be large enough to ensure that similar smoothing is attained in both shallow and deep regions of the bathymetry. For a closely-spaced grid, in which successive profiles along a given axis vary only gradually in form in the orthogonal direction, the application of areal error criteria provides a convenient means of approximately representing local volumetric errors. Consequently, the complexities and high computational demands of a truly three-dimensional smoothing procedure are avoided.

For a gradually undulating bed profile, in which the above condition seldom applies, an additional smoothing criterion was necessary. In this case (Figure 1b), an approximation segment AB is accepted when a close volumetric (areal) balance exists between the truncated and augmented portions of the replaced profile. In applying the above criteria, repetitive area computations were avoided by expressing the area of truncated polygons as the cumulative sum of the areas of successive triangles, with a common node at the left-hand origin. For this purpose, the Stoke's boundary integral method was employed.

With the objective of maintaining low store core memory requirements and minimum disc transfer time, filtering of the DTM grid was carried out simultaneously in the X and Y directions, rather than sequentially. By this means, it was only necessary to hold three rows of data in temporary storage at any given time, which were successively shifted forward by one row following each iteration. In order to ensure that the DTM grid remained intact for further possible applications with different smoothing criteria, the results of the filtering program were stored separately.

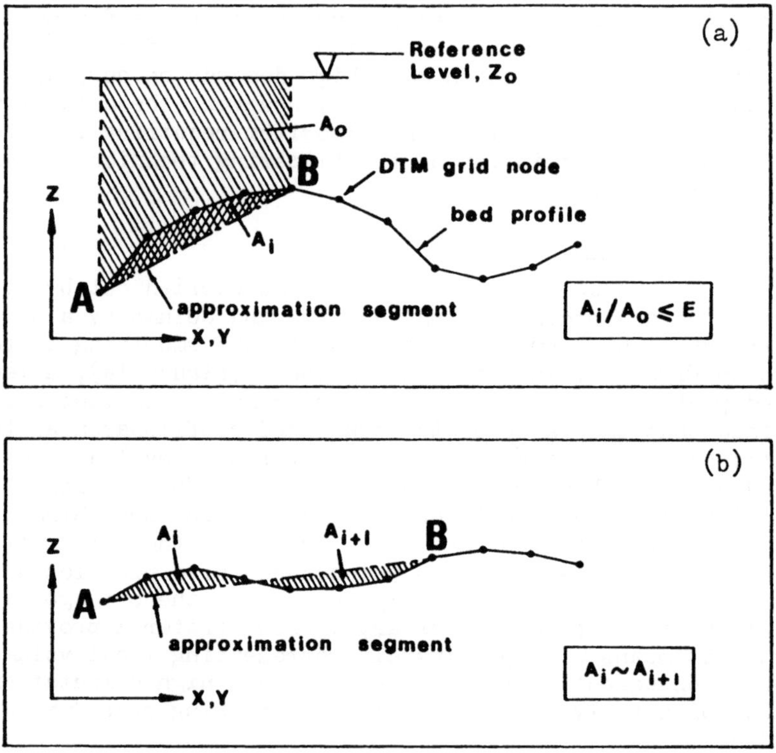

Figure 1. Filtering of nodes from the DTM grid , (a) truncated profile criterion, (b) smoothing of gentle undulation.

As a consequence of the filtering technique used, isolated clusters of retained nodes at neighbouring grid points often resulted. Upon further examination, it was found that further thinning-out of such clusters was possible without any significant loss of accuracy. With regards to boundary nodes, which were all initially retained, manual smoothing was subsequently undertaken, based upon the local curvature of the boundary and the characteristic node density adjacent to it. In order to test how well the natural bathymetry is reproduced for a particular number of retained nodes, a new DTM grid may be generated with the latter as input and compared with the original. Provided interpolated depths in the original grid closely represent natural bathymetry (sufficient input data), such a comparison provides a valuable means of quantifying and correcting errors.

Triangulation Technique

The triangulation procedure comprises four basic stages, as illustrated by way of example in Figure 2 for an imaginary region with 22 boundary nodes and 21 internal nodes. In stage I, the region is subdivided into macro-elements by connecting boundary nodes only. Commencing with the last boundary node as origin, successive elements are constructed with sequential node pairs in the boundary set. Once a new diagonal intersects the previous boundary line segment, the last but one boundary node becomes the origin. By ignoring all boundary nodes already cut-off by triangles, the above iteration continues until no more boundary nodes are available.

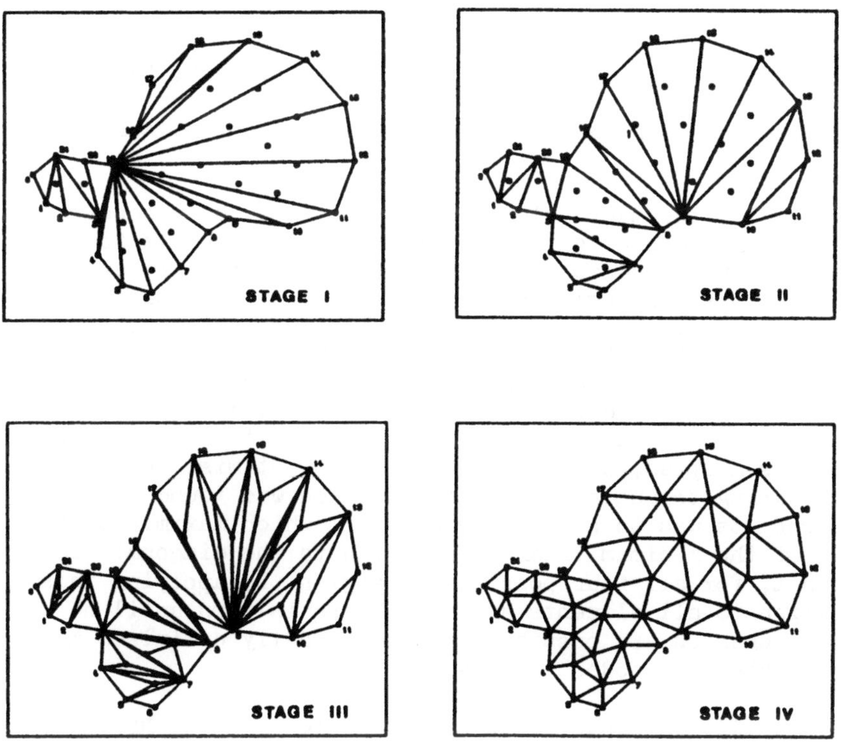

Figure 2. Basic stages of triangulation for an imaginary region.

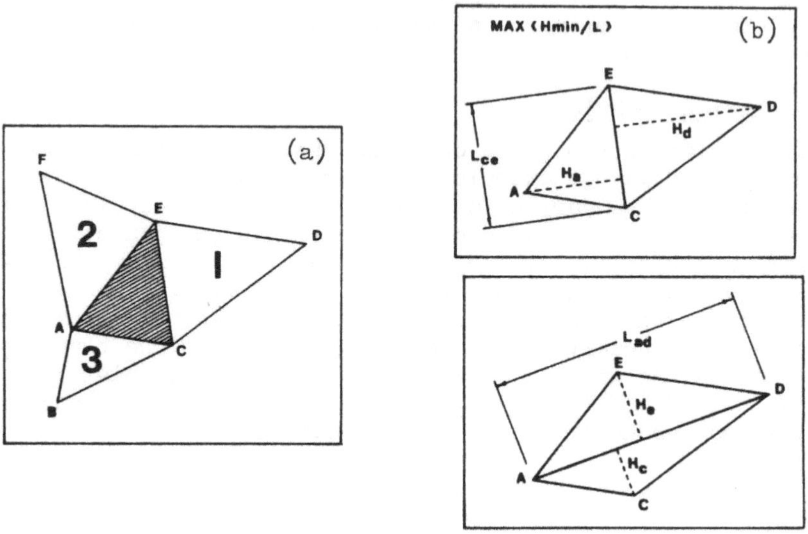

Figure 3. Optimisation of finite element mesh, (a) base triangle with 3 possible neighbours, (b) criterion for selecting diagonals within quadrilaterals (eg. ACDE).

In stage II, the macro-elements are optimised by reconstructing diagonals between neighbouring triangles. As a basis for the optimisation procedure (Figure 3a), the geometric relationships between a base triangle (shaded) and its three possible neighbours (1,2,3) are examined. Considering, for example, the quadrilateral ACDE formed between the base triangle and neighbour 1 (Figure 3b), the existing diagonal CE may either be retained or replaced by the new diagonal AD. In order to select the most suitable diagonal, a minimum connection length criterion was adopted, with modifications to optimise the shape of neighbouring triangles. A condition which was found to give good results was to select the diagonal which maximises the ratio of the minimum triangle height H to the diagonal (base) length L, for all four possible triangles. By applying this criterion to all elements and neighbours in turn until no further changes are possible, a geometrically optimised mesh for the boundary nodes is obtained.

In stage III, internal nodes are added to the mesh by simple subdivision of the macro-elements in which they occur. Finally in stage IV, the optimisation technique is applied to the stage III mesh in order to obtain the optimised network for the complete set of nodes. By extending the external boundary via a dummy segment into the interior region, the same procedure may be applied for triangulating regions with islands or artificial structures.

APPLICATION IN THE ELBE ESTUARY

General Considerations

In order to test the application of the mesh generation technique to a hydrodynamic model, a 500 km^2 study area comprising the lower 50 km of the Elbe Estuary was chosen. Input to the DTM consisted in 12000 digitised depth soundings from hydrographic charts, which formed the basis of the contour plot shown in Figure 4a. As indicated in the Figure, the reach is characterised by a wide expanse of mudflats to the north with a deep-water channel adjacent to the southern boundary, which deviates northwards towards the estuary mouth beyond Cuxhaven. Additional features include a 10 km long training wall extending northwards from Cuxhaven, and a complex channel system to the south of the latter, which was idealised in the model by a closed rectangular basin. Open boundaries are located at Grosser Vogelsand and Zehnerloch at the seaward end and at Brunsbüttel at the landward end.

Finite Element Mesh

Based upon the 12000 digitised depth soundings, a DTM grid with a mesh spacing of 200 m and containing approximately the same number of nodes was generated. Since it was proposed to carry out a pilot study of flow characteristics in the estuary, cost considerations were dominant in deciding the maximum permissible number of nodes which could be handled in a hydrodynamic model. By suitably adjusting the error criterion in the filtering routine, it was possible to reduce the number of nodes to between 400 and 500, which enabled an existing hydrodynamic model to be used on a small computer. By generating a second DTM grid with the retained nodes as input and comparing this to the original (Figure 4b), local repositioning and addition of nodes was carried out in regions showing large deviations. In the case of the training wall, which represents a discontinuity in the natural bathymetry, manual insertion of nodes was necessary to ensure proper definition.

The final finite element mesh, containing 457 nodes, is shown in Figure 4c. Assuming linear variations in depth between adjacent nodes in the mesh (in compliance with the adoption of linear shape functions in the hydrodynamic model), a finite element approximation of the bathymetry, as represented by the depth contours of Figure 4d, was obtained. In comparing the latter contours with those of Figure 4a, it may be seen that apart from small local deviations, the essential features of the bathymetry are faithfully reproduced.

Hydrodynamic Modelling

For the purpose of computing time-varying velocities, a depth-averaged flow model developed by Holz and Nitsche (1982) for

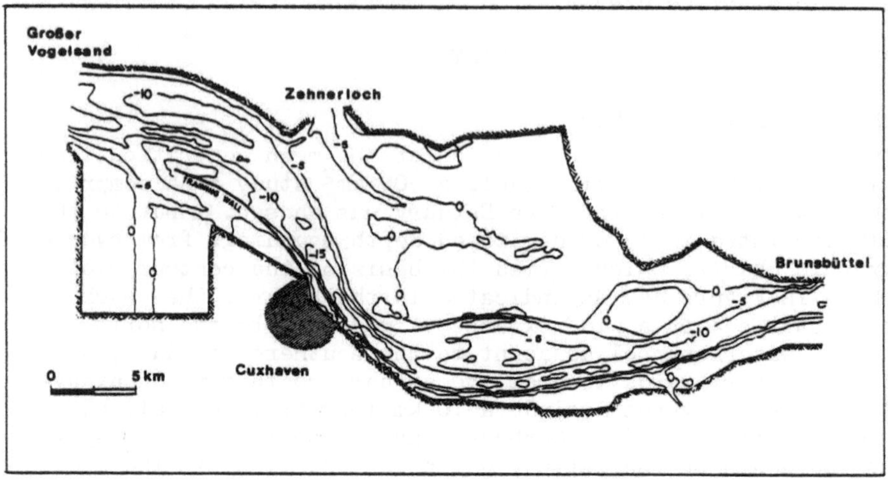

(a) Natural bathymetry computed from 12000 depth nodes.

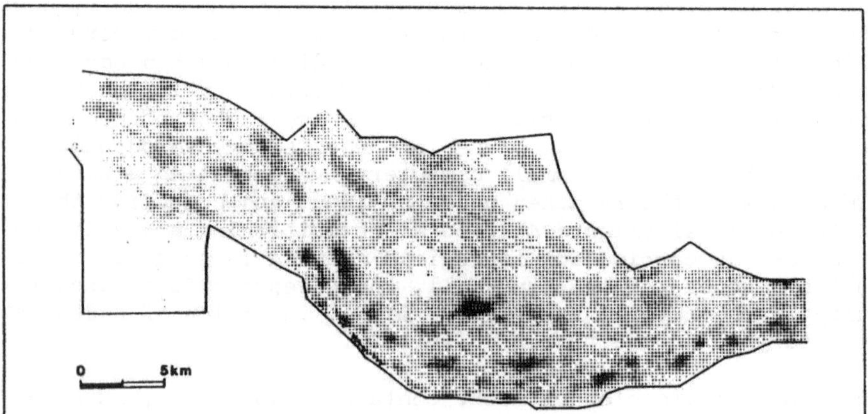

(b) Bathymetric approximation errors.

Figure 4. Generation of finite element mesh for the Lower Elbe Estuary.

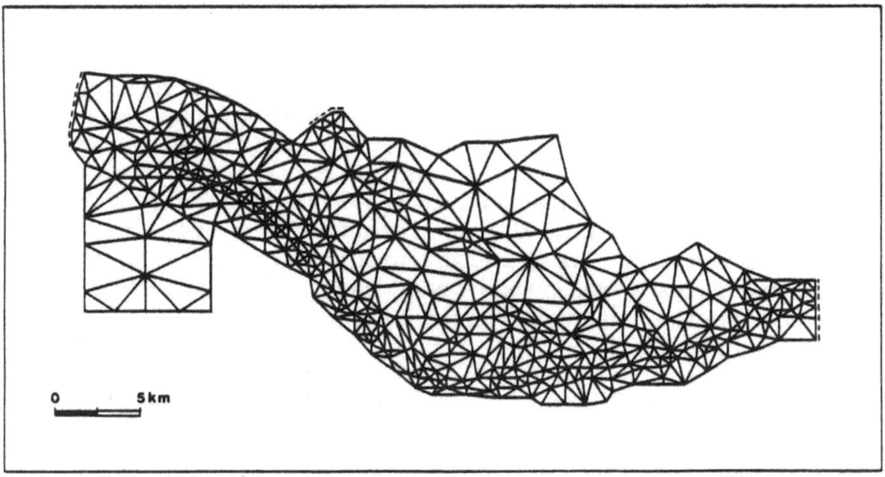

(c) Finite element mesh with 457 nodes.

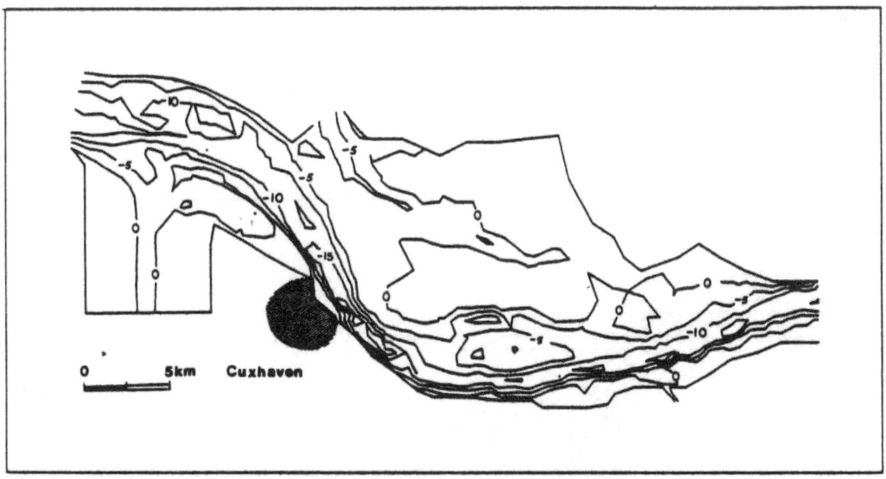

(d) Approximated bathymetry with linear shape functions.

Figure 4./cont'd/ Generation of finite element mesh for the Lower Elbe Estuary.

estuaries with intertidal flats was used. The simulation was for a mean tide, with zero normal discharge at the closed boundaries and prototype tidal variations at the three open boundaries. As the flow model employs an explicit scheme, the time step was limited to 20 sec to ensure stability for the particular mesh generated. The model was calibrated by adjusting the bed friction coefficient until satisfactory agreement was obtained between computed and observed tidal elevations at 6 longitudinal stations. With uniform friction over the reach, the maximum error in tidal height predictions was less than 5 per cent of the tidal range (mean range 3 m), which is comparable to the expected degree of accuracy for manually generated meshes.

Differences between computed and measured velocities at the same stations, however, were larger, with maximum errors of between 20 and 30 per cent of the velocity amplitude (1.5 m/s). Considering that the average node density in the simulation was only of the order of $1/km^2$, the latter deviations are most probably due to the coarseness of the mesh. In qualitative terms, however, computed flow patterns, as shown by way of example in Figure 5, agreed well with observed tendencies and clearly indicate the influence of bathymetry.

As part of an ongoing programme of research, it is hoped to improve the resolution of the velocity field by repeating the simulation in a more intensive investigation with a significantly larger number of nodes. In addition, further development of the mesh generation technique is proposed, with particular emphasis upon the application of hydrodynamic as well as bathymetric criteria.

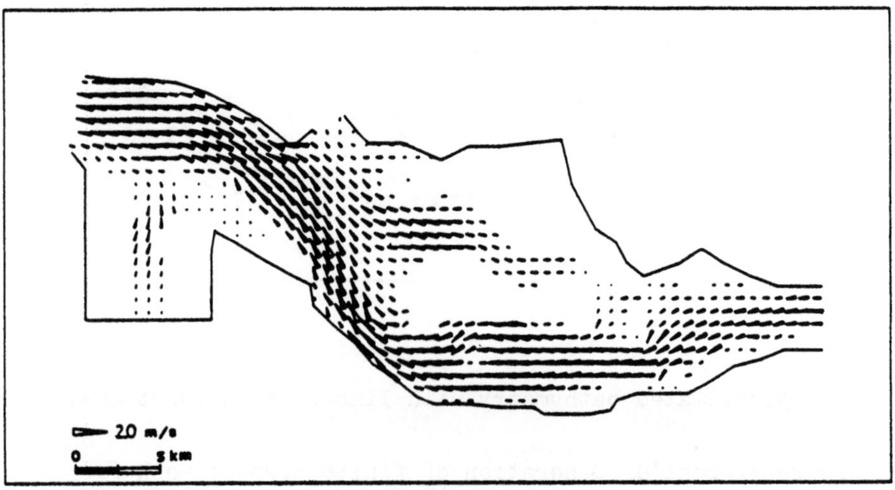

Figure 5. Typical velocity distribution in the Elbe, as computed with an automatically generated mesh.

CONCLUSIONS

The successful application of a digital terrain model for automatically generating irregular computational grids for hydrodynamic models has been demonstrated. The technique is of particular relevance to a highly variable bathymetry, in which local depth variations significantly influence the flow field. The technique has the distinct advantage over manual methods in that the original DTM grid, once generated, remains unaffected by the filtering and triangulation algorithms, and is thus readily available for repeated triangulations with differing node densities. The use of a DTM for evaluating depth differences at corresponding nodes in two or more digital representations of a bathymetry has also been shown to assist greatly in quantifying errors between a natural bathymetry and its finite element approximation. The authors anticipate that this application will also prove to be most useful in the future analysis of the results from movable-bed sediment transport models.

ACKNOWLEDGEMENTS

The authors gratefully acknowledge the financial support of the German Research Council (DFG) throughout the investigation. They would also like to thank members of the Department of Topography and Cartography, Hannover University, for their advice and assistance regarding the digital terrain model, and staff of the Harbour Authorities (WSA), Cuxhaven, for their interest and assistance with field data.

REFERENCES

Crotogino, A., Holz, K.-P. (1984) Numerical Movable-Bed Models for Practical Engineering, Appl. Math. Modelling, Vol.8,45-49.

Holz, K.-P., Nitsche, G. (1982) Tidal Wave Analysis for Estuaries with Intertidal Flats, Adv. Water Res., Vol. 5, 142-148.

Kruse, I. (1979) TASH- Ein System zur EDV-unterstützten Herstellung topographischer Grundkarten, Nachr. Karten- u. Vermessungswesen, Reihe 1, Heft 79, Frank. a. M., 95-107.

Proc. Mini-Symposium on Advances in Grid Generation, ASME Fluids Eng. Conference, Houston, Texas, June 1983.

Thacker, W. C. (1980) A Brief Review of Techniques for Generating Irregular Computational Grids, Int. Journ. for Num. Math. in Eng., Vol. 15, 1335-1341.

AN IMPLICIT WAVE EQUATION MODEL FOR THE SHALLOW WATER EQUATIONS

Ingemar P.E. Kinnmark and William G. Gray

Department of Civil Engineering, University of Notre Dame, Notre Dame, Indiana 46556

INTRODUCTION

In this paper an implicit numerical solution procedure for the shallow water equations is presented. It achieves efficiency through replacement of decompositions of a time-varying matrix by back substitutions for the solution of the equation system.

Explicit time marching procedures offer efficient computation of each time step but usually stability constraints, rather than accuracy considerations, limit the size of the time step and therefore lead to costly simulations. An implicit time marching scheme eliminates or strongly reduces the restriction on the time step imposed by the stability requirement. The computational effort per time step is however in general increased due to the following factors

-- A matrix equation is generated which requires the simultaneous solution of elevation and both horizontal velocity components.
-- The matrix of the equation system has to be reformulated at every time step.
-- The matrix has to be decomposed into a lower and upper triangular matrix at every time step.
-- The two resulting triangular equation systems must be solved using forward solution and back substitution at every time step.

The present study concentrates on presenting methods to eliminate or strongly reduce the computational effort involved in performing the first three of these steps. Irrespective of whether an explicit or implicit method is used, the numerical scheme must be capable of suppressing node-to-node oscillations in both space ($2\Delta x$ oscillations) and time ($2\Delta t$ oscillations).

Lynch and Gray (1979) have shown that the use of a wave continuity equation instead of the standard primitive continuity equation strongly suppresses $2\Delta x$ oscillations. Kinnmark and Gray (1982) showed that a proper time-weighting of the momentum equations effectively suppresses $2\Delta t$ oscillations. The need to simultaneously solve for surface elevation and velocity in the wave continuity and momentum equations, can be eliminated by treating the Coriolis terms explicitly in an implicit wave continuity equation. Kinnmark and Gray (1984) showed with linear analysis that this method produces an unconditionally stable scheme in which the surface elevation at the new time level is the only unknown solved for using the wave equation. Subsequently the momentum equations may be solved for velocity from a diagonal matrix.

EQUATION FORMULATION

Horizontal circulation in a vertically-homogeneous, one-layer, water body with constant density can be described by the shallow water equations for long wavelengths. The primitive forms of these are

Continuity Equation

$$L(\zeta,\underset{\sim}{v};h) \equiv \frac{\partial \zeta}{\mu t} + \nabla\cdot(H\underset{\sim}{v}) = 0 \qquad (1)$$

Non-Conservative Momentum Equations

$$\underset{\sim}{M}(\zeta,\underset{\sim}{v};h,\tau,f,g,\underset{\sim}{A}) \equiv \frac{\partial \underset{\sim}{v}}{\partial t} + \underset{\sim}{v}\cdot\underset{\sim}{\nabla}\underset{\sim}{v} + \tau\underset{\sim}{v} + f\underset{\sim}{k}\times\underset{\sim}{v} + g\underset{\sim}{\nabla}\zeta - \frac{\underset{\sim}{A}}{H} = 0 \qquad (2)$$

where
$\zeta(x,y,t)$ is elevation above a datum (L),
$h(x,y)$ is depth to bottom from a datum (L),
$H(x,y,t)=h(x,y)+\zeta(x,y,t)$ is total fluid depth (L),
$\underset{\sim}{v}(x,y,t)$ is vertically averaged velocity in eastward (x) and northward (y) directions (L/T),
$\tau(x,y,t)$ is the nonlinear bottom friction (1/T),
f is the Coriolis parameter (1/T),
g is acceleration due to gravity (L/T^2),
$\underset{\sim}{A}(x,y,t)$ is wind forcing in eastward (x) and northward (y) directions (L^2/T^2),
x is positive eastward (L),
y is positive northward (L), and
t is time (T).

Lynch and Gray (1979) found that a leapfrog approximation of the primitive continuity and momentum equations (1) and (2), developed spurious node-to-node oscillations. Therefore they replaced the primitive continuity equation (1) with a wave continuity equation which suppressed the node-to-node oscillations. The wave equation may be obtained in operator form from equations (1) and (2).

Operator Form of Wave Continuity Equation

$$W \equiv \frac{\partial L}{\partial t} - \underset{\sim}{\nabla} \cdot (L\underset{\sim}{v}) - \underset{\sim}{\nabla} \cdot (H\underset{\sim}{M}) + \tau L = 0 \tag{3a}$$

Substitution of (1) and (2) into this equation and rearrangement yields

$$\frac{\partial^2 \zeta}{\partial t^2} + \tau \frac{\partial \zeta}{\partial t} - \underset{\sim}{\nabla} \cdot [gh\underset{\sim}{\nabla}\zeta] = \underset{\sim}{\nabla} \cdot [\underset{\sim}{\nabla} \cdot (H\underset{\sim}{v}\underset{\sim}{v}) + f\underset{\sim}{k} \times H\underset{\sim}{v} + g\zeta\underset{\sim}{\nabla}\zeta - \underset{\sim}{A}]$$
$$+ H\underset{\sim}{v} \cdot \underset{\sim}{\nabla}\tau \tag{3b}$$

All the terms on the right side are evaluated at the known time level while centered approximations using levels $t - \Delta t$, t and $t + \Delta t$ are used for the terms on the left side.

REPEATED BACK SUBSTITUTIONS REPLACING DECOMPOSITIONS

A band solver decomposes a band matrix in approximately $np(p+1)$ operations (see Dahlquist et. al. (1974)) where n is the number of unknowns and p is the halfband width. Back substitutions require approximately $n(2p+1)$ operations. Therefore the computational effort involved in performing the decomposition will dominate that of the back substitution for large size applications. Also for sparse matrix solvers, such as that of Eisenstat et. al. (1977) used herein, the computational effort associated with the decomposition dominates the back substitution for large size applications. Subsequently a method is presented which replaces decompositions with back substitutions.

The only reason for the time-dependence of the matrix, generated from equation (3b) is the term containing the friction coefficient τ. However the matrix problem may be formulated such that time-dependent terms only appear on the diagonal of the matrix. Therefore the method of replacing decompositions by back substitutions will be especially simple.

Consider a linear equation system

$$\underset{\approx}{A}\underset{\sim}{w} = \underset{\sim}{b} \tag{4}$$

where $\underset{\approx}{A}$ depends on time only through its diagonal elements. These time-dependent diagonal elements are denoted by $\underset{\sim}{d}_t$. It is possible to express $\underset{\approx}{A}^{-1}$ in terms of $\underset{\approx}{A}_o^{-1}$ where $\underset{\approx}{A}_o$ is the matrix $\underset{\approx}{A}$ evaluated at $t = t_o$. A multivariable Taylor series expansion of $\underset{\approx}{A}^{-1}$ around $\underset{\approx}{A}_o^{-1}$ with respect to each one of the components in $\underset{\sim}{d}_t$ yields

$$\underset{\approx}{A}^{-1} = \underset{\approx}{A}_o^{-1} + \sum_{k=1}^{n} \frac{\partial \underset{\approx}{A}_o^{-1}}{\partial d_{ok}} \cdot (d_{tk} - d_{ok}) + \ldots \tag{5}$$

To use this formula effectively, the derivatives of the inverse matrix $\underset{\approx}{A}_o^{-1}$ must be expressed in terms of derivatives of the matrix $\underset{\approx}{A}_o$. From the definition of an inverse

$$\underset{\approx}{A}_o \cdot \underset{\approx}{A}_o^{-1} = \underset{\approx}{I} \tag{6}$$

calculation of the derivative of both sides of equation (6) with respect to d_{ok} and premultiplication by $\underset{\approx}{A}_o^{-1}$ yields after rearrangement

$$\frac{\partial \underset{\approx}{A}_o^{-1}}{\partial d_{ok}} = - \underset{\approx}{A}_o^{-1} \frac{\partial \underset{\approx}{A}_o}{\partial d_{ok}} \cdot \underset{\approx}{A}_o^{-1} \tag{7}$$

The derivative term $\frac{\partial \underset{\approx}{A}_o}{\partial \underset{\sim}{d}_o}$ is the three-dimensional tensor with 1 on the diagonal and zeroes elsewhere. Similar expressions may be derived for higher order derivatives. Using formulas (5) and (7), one finally arrives at the solution to equation system (4).

$$\underset{\sim}{w} = \underset{\approx}{A}^{-1} \cdot \underset{\sim}{b} = \sum_{j=o}^{\infty} [\underset{\approx}{A}_o^{-1} \cdot \underset{\approx}{D}]^j \cdot \underset{\approx}{A}_o^{-1} \cdot \underset{\sim}{b} \tag{8a}$$

or

$$\underset{\sim}{w} = \sum_{j=o}^{\infty} \underset{\sim}{c}_j \tag{8b}$$

where $\underset{\approx}{D}$ is a time dependent diagonal matrix with elements $D_{ii} = d_{oi} - d_{ti}$. An important computational feature of the terms in equation (8) which greatly reduces the effort required to evaluate the series is the recursion relation

$$\underset{\approx}{A}_o \cdot \underset{\sim}{c}_o = \underset{\sim}{b} \tag{9a}$$

$$\underset{\approx}{A}_o \cdot \underset{\sim}{c}_{j+1} = \underset{\approx}{D} \cdot \underset{\sim}{c}_j \qquad (j \geq 0) \tag{9b}$$

Thus the actual solution process involves solution of equations (9) with the time invariant coefficient matrix $\underset{\approx}{A}_o$ for the unknown vectors $\underset{\sim}{c}_j$. This is readily accomplished by back substitution after an initial decomposition of $\underset{\approx}{A}_o$. The accuracy of the solution for $\underset{\sim}{x}$ is realted to the number of terms in the series (8b) which are retained. If the series (8) is truncated after T terms and n is the number of unknowns with p the halfband width of $\underset{\approx}{A}$, the method requires

$$2Tn(p + 1) \tag{10}$$

operations per time step during which $\underset{\approx}{A}$ is not updated. The standard decomposition and back substitution procedure, however requires n[p(p + 3) + 1] operations. Therefore, used in conjunction with a bandsolver, the new method requires less computational effort if

$$T < \frac{p(p + 3)}{2(p + 1)} \tag{11}$$

Effective use of this method requires careful selection of T, the truncation limit of the series, and t_o, the time at which the matrix is decomposed.

T is related to the accuracy of the solution. The highest meaningful upper limit for T is dictated by the number of significant digits used to represent a floating point number in the computations. A limit substantially lower than that will in many cases provide good accuracy in the final results of the computation. The choice of t_o clearly affects the magnitude of $d_{oi} - d_{ti}$ which controls the magnitude of the higher order terms in the series in equation (8). To keep $d_{oi} - d_{ti}$ small, t_o may be updated at regular intervals necessitating a new decomposition of $\underset{\approx}{A}$ at these intervals.

Also of concern is the fact that this algorithm will be useful only if the series (8) converges. Convergence is assured if

$$||\underset{\approx}{A}_o^{-1} \cdot \underset{\approx}{D}|| < 1 \tag{12}$$

Equation (12) is satisfied if

$$||\underset{\approx}{A}_o^{-1}|| \cdot ||\underset{\approx}{D}|| < 1 \tag{13}$$

If we use $\kappa(\underset{\approx}{A}_o) = ||\underset{\approx}{A}_o|| \cdot ||\underset{\approx}{A}_o^{-1}||$, the condition number of $\underset{\approx}{A}_o$, this is equivalent to

$$||\underset{\approx}{D}||/||\underset{\approx}{A}_o|| < 1/\kappa(\underset{\approx}{A}_o) \tag{14}$$

If the matrix norm is chosen to be the maximum norm, equation (14) is satisfied if

$$\mathrm{Max}\{|\tau_{oi} - \tau_i|\Delta t/2\} < 1/\kappa(\underset{\approx}{A}_o) \tag{15}$$

where τ_i is the nonlinear friction coefficient causing the temporal dependence of the diagonal element in row i of $\underset{\approx}{A}$, and

τ_{oi} is the value of τ_i at $t = t_o$.

APPLICATION TO THE SOUTHERN PART OF THE NORTH SEA

The performance characteristics of the model were determined in an application to the Southern Part of the North Sea. The data used here is based upon information provided in Verboom (1982a,b) and Praagman (1980). The coastlines of the region and the imposed mesh of isoparametric quadratic Lagrangian finite elements are shown in figure 1. The mesh consists of 265 nodes and 58 elements, a considerably coarser approximation than the one used in Verboom (1982a). Bathymetry values have been assigned to each node through linear interpolation of the original data provided. A constant Chèzy-coefficient of 60 $m^{1/2}/s$ throughout the region has been used to account for bottom friction. Open boundary conditions have been chosen as the superposed M_1-, M_2-, M_4- and M_6-tides.

The first version of the implicit model to be considered is the direct method requiring updating and decomposition of the matrix at every time step. Computed results will be compared with those in Verboom (1982a). Figures 2 through 4 demonstrate good agreement for the tidal elevations.

Next results for the Taylor expansion method will be compared to the direct method. The cheapest way to proceed would use the Taylor series in equation (8) truncated after the first term along with a minimal number of decompositions of the matrix. Computer tests on the North Sea data revealed an inexpensive bur unfortunately very low accuracy method which produces unphysical negative depths if the matrix is decomposed fewer than every second time step. However results obtained retaining 2 terms (linear correction) or 3 terms (quadratic correction) in the series provided excellent results essentially indistinguishable from the direct results at approximately 20% lower cost. Of course savings will be larger for larger size applications. When using 3 terms (i.e. 3 back substitutions), the results were accurate with only one matrix decomposition, and therefore there is no need to consider higher order corrections. The CPU-time, using a FORTRAN 4 compiler on an IBM 3081, for cases with different frequency of decompositions and number of terms retained in the Taylor series is summarized in table 1. In comparison with an explicit code, the most favorable scheme in table 1 required only 30% more execution time per time step while avoiding Courant stability constraints.

NBS/Δt	1	1	1	2	2	3
D	0	1	2	4	7	9
NET	1.0	0.84	X	0.79	0.78	0.84

Table 1. Characterization by execution time of the Taylor expansion method. NBS/Δt is the number of back substitutions per time step. D is the number of time steps between decompositions. NET is the normalized execution time.

Another important parameter is the time step. Although an implicit formulation allows larger time steps and still remains stable, the truncation error will grow with increasing time step. This is manifested in the present application through an increase in the variation between solutions on consecutive tidal cycles. For a time step of 10 minutes the variation is typically less than 0.1% while for a time step of 12.5 minutes the variation is typically 2%. Use of too large a time step makes determination of achievement of a dynamic steady state very difficult.

Another interesting fact is the sensitivity to boundary location. Originally the southern boundary was located between Dover and Cap Gris Nez. However this location resulted in instabilities in the southern North Sea. Repositioning of the boundary, with appropriate modification of the forcing function, to the line Ramsgate-Dunkerque produced stable simulations. One important difference between these locations is that the amplitude of the tidal forcing along Ramsgate-Dunkerque is lower than that of the Dover-Cap Gris Nez line. It is interesting to notice that a common practice (Grotkop, 1973; Brebbia and Partridge, 1976; Verboom, 1982a) seems to be to locate this southern boundary exactly in a region where the tides are high. In other modeling fields, however, it is a practice to try to locate a boundary as far away as possible from a region where a substantial variation in the dependent variables occurs. It is possible that Verboom (1982a) also may have encountered a difficulty with the location of the southern boundary which resulted in a repositioning from the line Dover-Calais to Dover-Cap Gris Nez.

CONCLUSIONS

An implicit method for solving the shallow water equations was shown to be an efficient alternative to an explicit method in application to the Southern part of the North Sea. Further savings were realized, in this application amounting to 20%, by the introduction of a Taylor expansion method which replaces decompositions by back substitutions. The savings will increase for larger problems.

ACKNOWLEDGEMENT

The authors are grateful to N. Praagman and G.K. Verboom for their generosity in supplying the data set. This work has been supported in part by grant number CME-7921076-03 obtained from the United States National Science Foundation.

REFERENCES

Brebbia, C.A. and P.W. Partridge (1976) Finite element simulation of water circulation in the North Sea, Appl. Math. Modelling, 1. 2:101-107.

Dahlquist, G., Å. Björek and N. Anderson (1974) Numerical Methods, Prentice Hall, Englewood Cliffs, NJ.

Eisenstat, S.C, M.C. Garsky, M.H. Schultz and A.H. Sherman (1977) Yale Sparse Matrix Pakcage: I. The Symmetric Codes, Yale Computer Science Report, 112.

Grotkop, G. (1973) Finite element analysis of long-period water waves, Comput. Methods Appl. Mech. Eng., 2, 2:147-157.

Kinnmark, I.P.E. and W.G. Gray (1982) Time-Weighting of the Momentum Equations in Explicit Wave Equation Models of Surface Water Flow, Proc. 4th Intl. Conf. on Finite Elements in Water Resources, Hannover, Springer-Verlag, Berlin, 5.67-5.77.

Kinnmark, I.P.E. and W.G. Gray (1984) A Two-dimensional Analysis of the Wave Equation Model for Finite Element Tidal Computations, Intl. J. Numer. Methods. Eng. (in press).

Lynch, D.R. and W.G. Gray (1979) A Wave Equation Model for Finite Element Tidal Computations, Computers and Fluids, 7, 3:207-228.

Praagman, N. (1980) A comparison of discretization methods for the shallow water equations, Dept. of Math., Univ. of Technology, Delft, Report, NA-33.

Verboom, G.K. (1982a) een getijberekning voor de zuidelijke Nordzee met verschillende numericke modellen, Waterloopkundig Laboratorium, Delft Hydraulics Laboratory, Report, R1718.

Verboom, G.K. (1982b) Private communication.

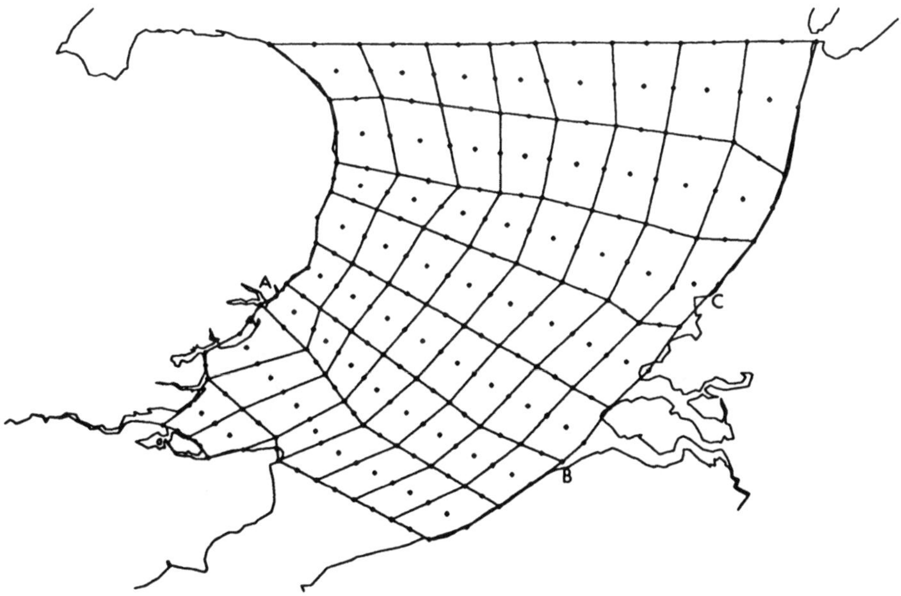

Figure 1. Quadratic Lagrangian isoparametric discretization of the southern North Sea. Tidal forcing is specified at open boundaries, no flow otherwise. Solutions obtained at points A, B, and C are in subsequent figures.

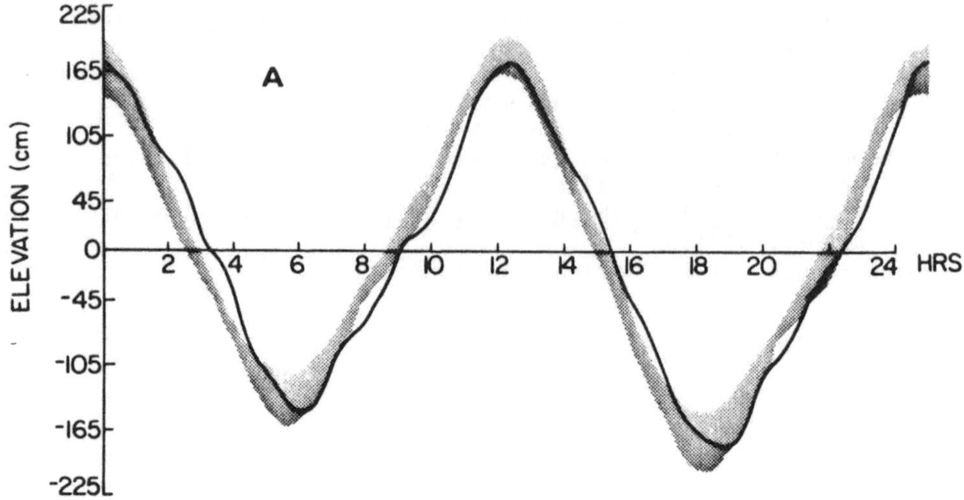

Figure 2. Comparison of computed elevations (——) at Harwich (A) with range of values obtained by Verboom (1982a) (shaded region). Both studies use uniform $C = 60\ m^{\frac{1}{2}}/sec$.

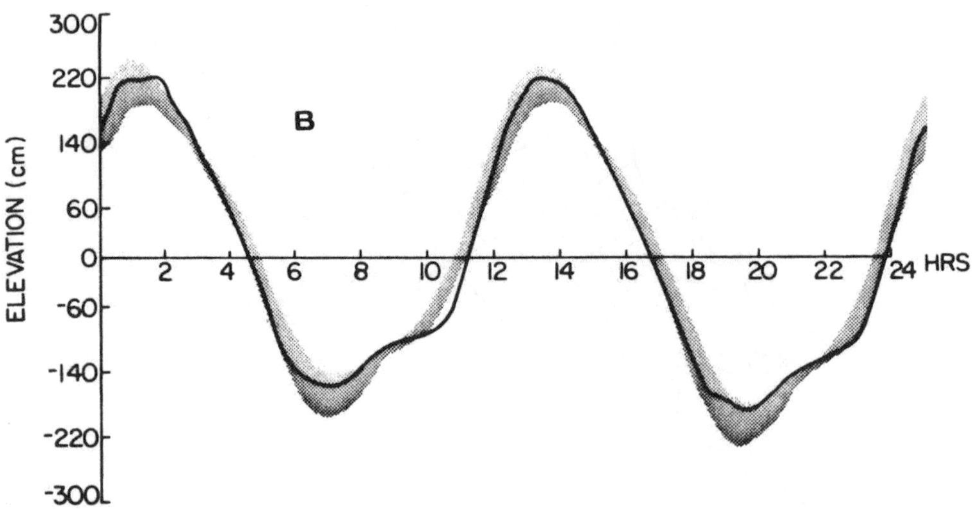

Figure 3. Comparison of computed elevations (——) at Zeebrugge (B) with range of values obtained by Verboom (1982a) (shaded region). Both studies use uniform $C = 60\ m^{\frac{1}{2}}/sec$.

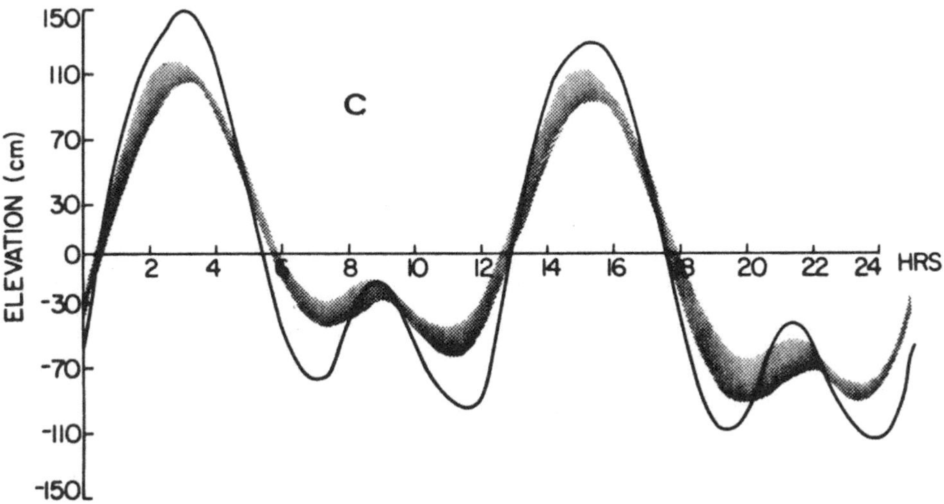

Figure 4. Comparison of computed elevation (——) at Hoek van Holland (C) with range of values obtained by Verboom (1982a) (shaded region). Both studies use uniform $C = 60\ m^{\frac{1}{2}}/sec$.

A REVIEW OF STRATEGIES FOR FINITE ELEMENT MODELING OF THREE DIMENSIONAL HYDRODYNAMIC SYSTEMS

Ian P. King

University of California, Davis
Resource Management Associates, Lafayette

INTRODUCTION

The methodology for three dimensional finite element analysis of hydrodynamic systems has been readily available since the basic development of two dimensional models. However, the simulation of a system composed entirely of three dimensional elements is computationally very expensive and little practical analysis has been undertaken. In an effort to reduce cost the author in a previous paper presented a simplified three dimensional model that used the hydrostatic pressure approximation to eliminate the vertical momentum equation and a transformation of the geometric system to a unit vertical dimension to directly solve for the free surface profile (King, 1982).

This paper will examine two types of approaches to the reduction of simulation costs within the framework of the original approach.

a) Restructuring of the model to accommodate one, two and three dimensional elements that are appropriately coupled at their interfaces. This approach has the advantage of fewer equations and simpler elements, and a consequent computational economy in areas where the various approximations are appropriate, it allows the user to limit the three dimensional analysis to the area where it is really needed.

b) The use of the simplification and transformation referred to above, but revised to utilize an alternative form of the transformation. The methods will be compared in order to evaluate their effectiveness, economy and range of validity. In this context, particular attention will be paid to the influence of the bottom profile on the solution. In order to establish a baseline, results from a full analysis that does not include the hydrostatic pressure assumption will also be presented.

COUPLING OF MODEL APPROXIMATIONS

The most comprehensive approach to three dimensional analysis of estuarine systems would utilize some version of the full three dimensional momentum equations coupled with the continuity equation, thus creating equations in four dependent variable (three velocity components and pressure). Such an analysis scheme, whilst theoretically possible would lead to extremely expensive computer models, particularly when the free surface geometric variation is incorporated. The most common simplification in hydraulic systems is to neglect the influence of vertical momentum effects and thus effectively eliminate the vertical velocity as a coupled dependent variable. This method - the hydrostatic pressure assumption - replaces the vertical pressure distribution with a constant head on any vertical line. Indeed, solutions that utilize this approach give acceptable verification of prototype systems (see for example Leschziner and Rodi, 1979). However, even with this simplification a three dimensional geometric description is necessary and a large number of total degrees of freedom are created and the description of the free surface poses a complex problem. Additionally, in the case of finite element method, it is necessary to use computationally costly integration over three dimensional elements.

In order to reduce computer time additional simplification or model restructuring is desirable. It is the purpose of this section to describe one approach to this difficulty.

In most practical applications it is apparent that a full three dimensional geometry is not entirely necessary. As Figure 1 shows a typical geometry may have rivers that feed into the main bay and shallow areas, in these areas different approximations are appropriate. In fact, depending upon the exact area of interest four different approximations may be appropriate.

1. One dimension - for upstream sections of rivers where effects of flow variation across the river or with depth can be neglected but the tidal storage prism of the river requires its incorporation.

2. Two dimensions (vertical elevation - laterally averaged) close to the mouth of the river, near to the area of interest.

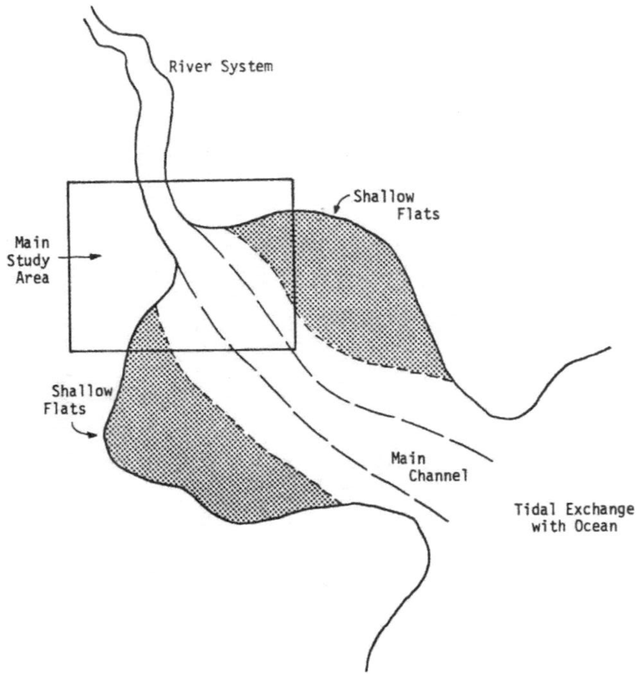

Figure 1. Typical Estuarial System

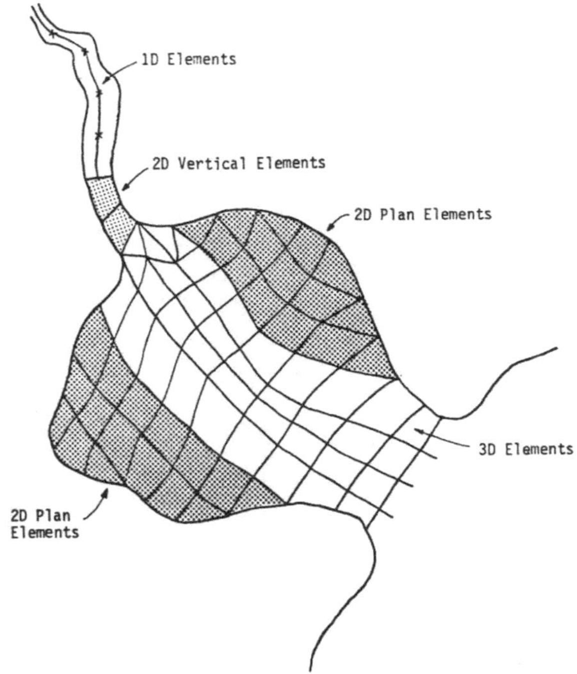

Figure 2. Mixed Element Representation

3. Two dimensions (horizontal plan - vertically averaged) - shallow areas of the bay where variation with depth can be ignored.

4. Three dimensions - main study area where bottom profiles change or sharp boundaries induce secondary currents.

A model must therefore be capable of simplification to the three simplest forms in the various areas with consistency between the interfaces. It is important in finite element applications as complex as this not to default to the one dimensional mode and still integrate over a three dimensional space because of the unnecessary computing effort associated with three dimensional numerical integration. Furthermore, in order to assure compatibility with depth averaging, a three-dimensional approach that uses a constant geometry is desirable. Constant unit vertical dimension is in fact implied in most depth averaged models. The scheme described previously (King, 1982) and discussed later satisfies these requirements, the transformation to unit dimensions assures the constant geometric space and is exactly consistent with the usual depth averaged formulation. The transformation to lateral averaging is straightforward and requires only a consistent geometric layout that allows a smooth transition. The transitions between approximation orders for three dimensions and depth averaging provides an opportunity for the user to specify other than constant velocity distributions over depth. In fact, the depth averaged formulation is easily adjusted to assure a functional distribution with depth. Such a capability greatly expands the area where two dimensional models can be used (see Figure 2) and allows smooth interfacing.

A second aspect of an economic computer model for three dimensional applications is the capacity to use the model in a way that excludes the vertical dimension, in a preparatory two dimensional analysis. This capability provides a preliminary verification without the computer cost of using all three dimensions. In the model described here this is achieved by automatically generating the three dimensional elements from an input plan view of one and two dimensional elements. This allows the user to start with the two dimensional analysis and then expand by specifying the number of vertical elements at each plan node point. The user may specify uniformly spaced points or define any kind of irregular configuration, the model automatically generates the three dimensional elements and the interface locations, as appropriate.

BASIC EQUATION FOR THREE DIMENSIONAL MODEL

The basic formulation for this model uses three dimensional momentum equations incorporating eddy viscosity both to account for turbulent energy losses and assure smooth numerical behavior of the finite element model. The writer recognizes that this form of the equations has theoretical shortcomings, but the field performance in actual application to estuaries and streams of two dimensional models using the same basic form serves as a justification for their incorporation in the three dimensional model.

Recognizing that vertical momentum will in general be much smaller than its horizontal counterpart the hydrostatic assumption is made, effectively eliminating vertical momentum transport.

The final form of the equations is thus:*

$$\frac{\partial u}{\partial t}+u\frac{\partial u}{\partial x}+v\frac{\partial u}{\partial y}+w\frac{\partial u}{\partial z}-g\frac{\partial(h+a)}{\partial x}-\{\frac{\partial}{\partial x}(\varepsilon_{xx}\frac{\partial u}{\partial x})+\frac{\partial}{\partial y}(\varepsilon_{xy}\frac{\partial u}{\partial y})+\frac{\partial}{\partial z}(\varepsilon_{xz}\frac{\partial u}{\partial z})\} = 0 \quad (1)$$

$$\frac{\partial v}{\partial t}+u\frac{\partial v}{\partial x}+v\frac{\partial v}{\partial y}+w\frac{\partial v}{\partial z}-g\frac{\partial(h+a)}{\partial y}-\{\frac{\partial}{\partial x}(\varepsilon_{yx}\frac{\partial v}{\partial y})+\frac{\partial}{\partial y}(\varepsilon_{yy}\frac{\partial v}{\partial y})+\frac{\partial}{\partial z}(\varepsilon_{yz}\frac{\partial v}{\partial z})\} = 0 \quad (2)$$

plus an integrated continuity equation:

$$\int_a^{a+h} (\frac{\partial u}{\partial x} + \frac{\partial v}{\partial y} + \frac{\partial w}{\partial z})dz = 0 \quad (3)$$

where the usual definitions apply, note that a = bottom elevation from some datum and h = depth of water.

In a previous paper the writer transformed these equations from the variable vertical dimensional 'a' to 'a+h' to the constant vertical dimension "0" to "1", a transformation also used in finite difference models. By this device the variable geometric problem is eliminated and water surface elevation (a+h) can be directly developed during the solution. This of course means that the vertical coordinate locations whilst constant from a finite element point of view are not values at a constant spatial location. The network is subject to vertical stretching. In order to describe the vertical velocity component, weak coupling was assumed and a two stage procedure used that first used equations (1), (2) and (3) to solve for u, v and h, then later using these known values the basic continity equation was written in a boundary value form as:

---

* External tractions such as bottom friction and wind shear, and the coriolis term have been omitted. They are actually incorporated in the model, but for simplicity of presentation they are omitted in these equations.

$$\frac{\partial^2 w}{\partial z^2} = - \frac{\partial}{\partial z}\left(\frac{\partial u}{\partial x} + \frac{\partial v}{\partial y}\right) \tag{4}$$

The solution for equation (4) then uses specified values for w at the bed and water surface derived from the horizontal flow components combined with the bed and water surface slopes at these locations. Note that because of inconsistencies in bottom slope and surface slopes implied by the linear approximations used for head and bottom elevation, average values are used for particular node.

Equations (1) through (4) are solved using a Galerkin finite element scheme with mixed interpolation functions (quadratic velocities, and linear heads), operating on several types of isoparametric elements. A Newton-Raphson iterative scheme is incorporated to resolve the nonlinearity and an implicit finite difference method is used to progress through time.

By the nature of the vertical integration of continuity for equation (3) it became necessary to enforce the requirement that nodes lie on vertical lines projected from the nodes of a single horizontal plan view. This restriction in turn makes possible a relatively simple specification of the geometry of the system and permits automatic generation of 3-D elements from a basic 2-D system.

Results from test analyses using this model showed generally good agreement with available experimental data both in homogeneous flow and in a companion stratified flow model.

Some questions however arise as to the continuity of velocity vectors particularly at the bottom in the vicinity of changes in bottom slope. By the nature of the transformed system, velocity vectors at the bottom stay parallel to the bottom and discontinuities of direction occur at points of changed bottom slope. This forced redirection of momentum appeared to cause the flow to hug the bottom and not break away to follow its momentum direction at these times. The extent and significance of this effect does, however, depend on the magnitude and irregularity of the bottom slopes and in many cases would not have a significant impact on the solution.

In order to evaluate this effect a revised transformation scheme was developed that preserved the original bottom profile but enforced a horizontal upper dimension for the system. Thus the transformation space now covers the vertical dimension from 'a' to 'b'.

The original transformation took the form:

$$z' = (z-a)/h \tag{5}$$

and the revised transformation is written:

$$z' = a + \frac{(z-a)}{h}(b-a) \tag{6}$$

where b is a fixed vertical location of the geometric system. For practical purposes it is selected so that b is approximately at the water surface, i.e., b approximately equals a+h.

In the unit transformation the problem of parallel flow boundary conditions the water surface at the bottom neatly - too neatly perhaps - eliminate themselves. However, in the revised approach provisions must be made to ensure overall flow continuity. Thus at points on the bottom with discontinuities of slope the scheme described by Engleman et al. (1982) was incorporated to assure that flow lost through one element face exactly matched that gained on an adjacent element. As will be seen in the example this does cause some differences in total flow across internal faces but in an oveall sense continuity is still maintained.

EXAMPLE PROBLEM

In order to evaluate the impact of the alternate transformations and to pin down where bottom slopes may be important a previously analyzed (King et al., 1975) problem of two dimensional flow over a submerged weir was computed using (a) the basic transformation, (b) the modified transformation, and (c) the original 2-D vertical model that does not incorporate the hydrostatic pressure assumption but uses a rigid lid to treat the free surface. The same basic network was used for all three cases, see Figure 3. With a bottom slope of 1:1.5 the vertical velocities can be expected to be significant and the validity of the hydrostatic pressure assumption for (a) and (b) is open to question. However, as the schematic of horizontal velocity component at two locations downstream of the weir shows, see Figure 4, the influence of the flat bottom transformation is so strong that the downstream eddy is completely eliminated. Whereas the other model results show it strongly apparent. The case (b) solution also compares reasonably well with the extent of the eddy measured in the prototype and computed using the nonhydrostatic pressure algorithm developed earlier. It was of course not anticipated that the three dimensional model would be applied in situations with such steep bottom slopes, therefore the analysis was repeated with the entire system stretched by a factor of 5 horizontally, i.e., a weir with slope 1:7.5. As might be expected the extent of the eddy (see Figure 5) is much reduced, but the velocity distributions for the two transformations are still significantly different at this point. As a final test the weir slope was reduced by a factor of 50 from the original,

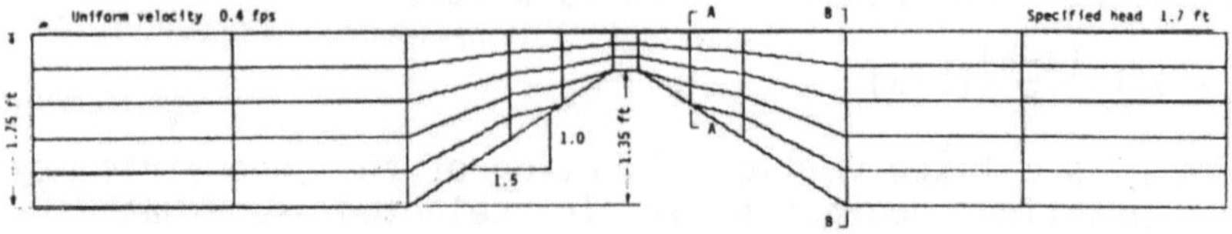

Figure 3. Finite Element Network for Submerged Weir (not to scale) (weir crest length 0.25 feet)

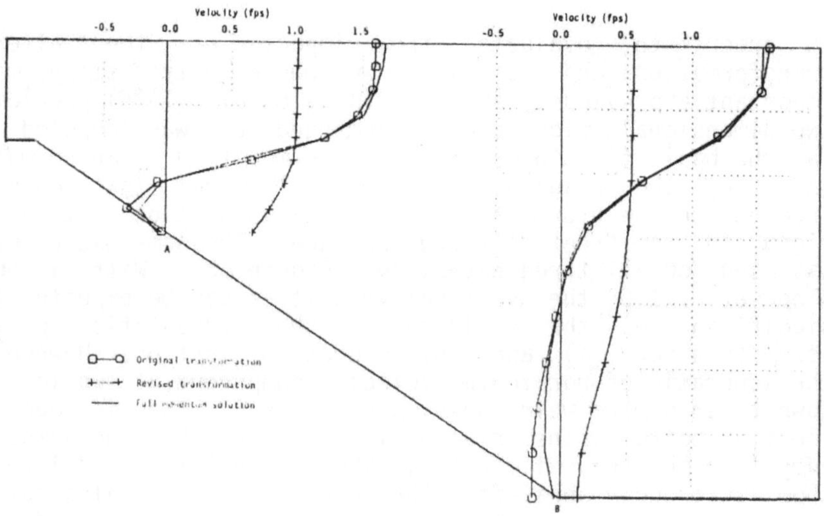

Figure 4. Horizontal Velocities at Sections A and B Weir Slope 1:1.5 (weir slopes not to scale)

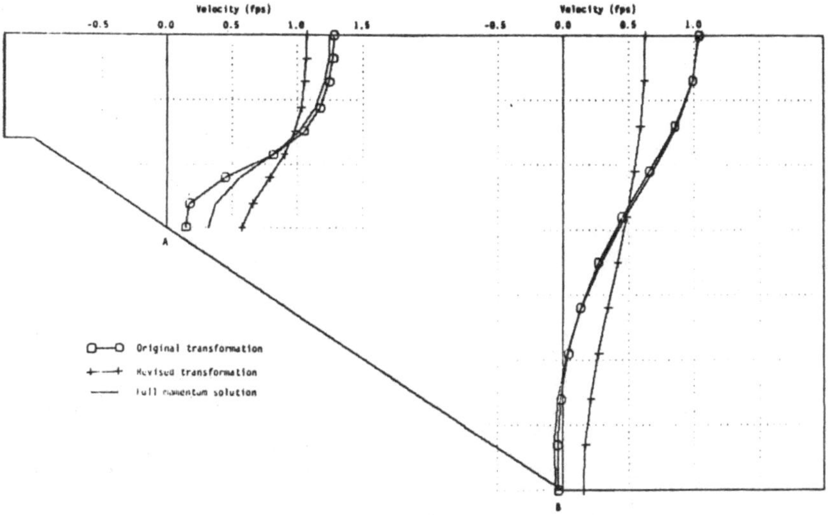

Figure 5. Horizontal Velocities at Sections A and B Weir Slope 1:7.5 (weir slopes not to scale)

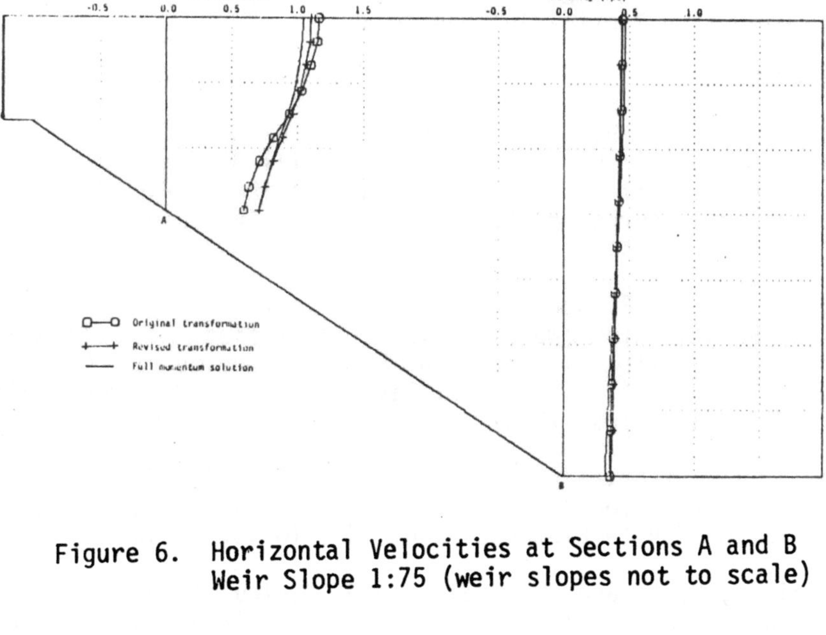

Figure 6. Horizontal Velocities at Sections A and B Weir Slope 1:75 (weir slopes not to scale)

i.e., 1:75 and the cases repeated. As Figure 6 shows, the velocities for this case are very similar.

From this analysis two preliminary observations can be made, 1) the use of the unit dimension transformation can lead to erroneous results in areas where the bottom slope is in the range of less than 10:1, and 2) despite the approximations introduced by the hydrostatic pressure assumption, head pressures (vary by approximately 5/100 feet of water in the nonhydrostatic model) a hydrostatic approximation can model the main features of the flow regime.

CONCLUDING REMARKS

The first observation suggests that the modified transformation, whilst it is less elegant and does not lead to a convenient rectangular system of elements is desirable if there is a strong influence of bottom profile on the system. It is entirely possible that the effect is created by the linear approximation for the bottom profile and would be improved if smooth transition between elements was created. Such a task is not straightforward in the general three dimensional case and introduces further difficulties with respect to the location of midside nodes and consistency between depth and bottom profile.

The latter observation indicates that the range of application of methods that use the hydrostatic pressure assumption is large enough that these models can usefully be applied to many three dimensional estuarial problems.

REFERENCES

Engleman, M. S., R. L. Sani and P. M. Gresho, 1982, "The Implementation of Normal and/or Tangential Boundary Conditions in Finite Element Codes for Incompressible Fluid flow," International Journal of Numerical Methods in Fluids, Volume 2, No. 3.

King, I. P., W. R. Norton and K. R. Iceman, 1975, "A Finite Element Solution for Two-Dimensional Stratified Flow Poblem," Chapter 7, Finite Elements in Fluids, Volume 1, edited by R. H. Gallagher, J. T. Oden, C. Taylor and O. C. Zienkiewicz, John Wiley and Sons, 1975

King, I. P., 1982, "A Three Dimensional Finite Element Model for Stratified Flow," presented at the 4th International Sumposium on Finite Elements in Fluids, Tokyo, Japan.

Leschziner, M. A. and W. Rodi, 1979, "Calculation of Strongly Curved Open Channel Flow," Journal of the Hydraulics Division, ASCE, Volume 105, No. HY10.

Simulating Flood Flows By Coupled 1-D and 2-D Models

G. Schmitz and G.J. Seus
The Technical University of Munich
F. R. Germany

SUMMARY

The paper presents a model system developed especially as a decision aid for flood control measures and/or flood plain management. Simulating flood events in regions with complex boundaries necessitates a mathematical model of high accuracy and flexibility as well as economic efficiency. The approach couples 1-D and 2-D submodels.

Both parts of the coupled model, the 1-D as well as the 2-D, are based on the theory of characteristics.
In the 1-D case the ordinary differential equations are integrated along the characteristic lines. Accordingly, for the 2-D approach, this integration is replaced by one executed over the entire surface of the characteristic cone. Thus our 2-D method of characteristics differs significantly from the so-called quasicharacteristic methods, which carry out the integration of the ordinary differential equations along only four arbitrarily chosen lines on the surface of the cone.

Starting from the exact solution of the wave equation (Poisson-Parseval solution), a semi-analytical solution of the depth-averaged equations in the remaining two spatial dimensions, including nonlinear terms, is developed.

Both of our characteristic methods are independent of the CFL criterion.
The model presented is able to calculate changes from subcritical to supercritical flow and vice versa. In general it performs well for Froude numbers in the vicinity of one.

Modelling complex flow domains, such as flood flows in urban areas, requires an irregular space discretisation to keep truncation errors small and to represent the geometric boundaries properly.

Simulating flows as in narrow, nearly straight roads, however, leads to an enormous quantity of grid points and thus to high computer costs. To avoid these disadvantages and to gain computational efficiency the 1-D model is employed for those regions having 1-D flow regime. Thus internal boundaries are defined wherever the general 2-D description of the flow is replaced by a 1-D approach. In the numerical model a set of boundary and transition conditions perform the coupling of the 1-D and the 2-D model parts.

Another important feature of the approach is that time varying geometric boundaries can be treated. They are important when initially dry regions are subject to rising and receding flood flows.

These advantages make the model a suitable tool for handling flood control problems under steady and unsteady flow conditions. The model's advantages will be demonstrated by calculating the complex flow patterns for a partially flooded Bavarian community.

INTRODUCTION

The flow regime in flooded areas and communities can be improved by some engineering measures. These are usually structural changes consisting of alterations in the river bed geometry in lateral and longitudinal sections. The effects of such efforts and their impacts have to be predicted for the future and compared to the flow regime which existed before any measures were undertaken.

To carry out such comparisons and predictions physical models have principally been used until now. But as in many other fields of engineering, mathematical models are becoming more and more important as predictive tools.

Encouraged by the success other authors gained in modelling coastal areas, estuaries etc., we started to develop a mathematical model for regions with highly complex geometrical boundaries.

In contrast to the conditions governing numerical modelling of flows in large areas like coastal regions, bays, and estuaries, the flow pattern of a river flooding urban areas principally governed and controlled by the irregularly structured geometrical boundaries.

This possibility of controlling the flow pattern is actually very welcome, because a certain change of boundaries can be used to improve the flow regime. This aspect then points to the most important difference between modelling coastal and urban areas: the "marines" try to predict the effects of different hydrological loads of a region under constant geometrical conditions; the "landlubbers" want to change the geometrical conditions in an optimal direction in order to improve the flow conditions of a region under invariant given hydrological load.

Developing mathematical models for flood flows in urban areas is thus a special task, because such models need a specific adaptation to mathematical and numerical problems caused by the complex geometrical boundaries not encountered when computing flow patterns of bays or estuaries.

BASIC EQUATIONS AND THEIR NUMERICAL TREATMENT

The complete theory of both the 1-D and 2-D model parts has already been treated in previous publications; we therefore present only the essential results here. For additional information, derivations, and presentation of a consistent theory we refer the reader to the references listed.

The 1-D Approach

The basis of the 1-D approach employed in the coupled model consists of the well known Saint-Venant equations, using the wetted cross sectional area $A(x,t)$ and the mean velocity $u(x,t)$ as the two dependent variables.

$$\left.\begin{aligned} \frac{\partial A}{\partial t} + u\frac{\partial A}{\partial x} + A\frac{\partial u}{\partial x} &= q \; ; \\ \frac{\partial u}{\partial t} + u\frac{\partial u}{\partial x} + g\frac{\partial h}{\partial x} &= g\,(S_o - S_f) - \frac{q}{A}(u - u_q \cos\alpha) \; . \end{aligned}\right\} \quad (1)$$

(For the meaning of the variables, please, see the NOMENCLATURE.)
Recently we showed (SCHMITZ, EDENHOFER, CIRWITZKY; 1982) equations (1) to be analytically integrable to a form where no more derivatives of the dependent variables occur.

With the integral function $H(x,A)$ to be determined in advance from the river geometry (for example by polygonial courses) eqs. (1) yield

$$\left.\begin{aligned} &[(-1)^{i+1}\cdot H(x,A) + u(x,t)]_{\substack{x=x_i(t)\\ t=t}} = (-1)^{i+1}\cdot H(x,A) + \\ &+ u(x,t)]_{\substack{x=x_i(t_i)\\ t=t_i}} + \int_{t_i}^{t} [(-1)^{i+1}\sqrt{\frac{\partial h^*}{\partial A}\frac{g}{A}}\cdot q + g(S_o - S_f) - \\ &- \frac{q}{A}(u - u_q\cdot\cos\alpha) - g\frac{\partial h^*}{\partial x} + (-1)^{i+1}\cdot \dot{x}_i\frac{\partial}{\partial x}H(x,A)]_{x=x_i(\tau)}d\tau \; ; \end{aligned}\right\} \quad (2)$$

with $i = 1, 2$;

$$\dot{x}_i = u \pm \sqrt{gA \cdot \frac{\partial h^*}{\partial A}} \quad ;$$

$$\frac{\partial h}{\partial x} = \frac{\partial h^*}{\partial A} + \frac{\partial h^*}{\partial A} \cdot \frac{\partial A}{\partial x} \quad ;$$

h^* being only a function of the river geometry.

Subjecting eqs. (2) to the principles of the method of characteristics displayed in Fig.1 results in the following algebraic working equations

$$\left.\begin{aligned}
V_1(P_{jk}) &= V_1(P_{1jk}) + \frac{x_j - x_{j-1}}{2}\left[\frac{F_1(P_{1jk})}{\dot{x}_1(P_{1jk})} + \frac{F_1(P_{jk})}{\dot{x}_1(P_{jk})}\right]; \\
V_2(P_{j-1,k}) &= V_2(P_{2,j-1,k}) - \frac{(x_j - x_{j-1})}{2}\left[\frac{F_2(P_{2,j-1,k})}{\dot{x}_2(P_{2,j-1,k})} + \right. \\
&\left. + \frac{F_2(P_{j-1,k})}{\dot{x}_2(P_{j-1,k})}\right] \; ; \quad j = 2,\ldots,n
\end{aligned}\right\} \quad (3)$$

with $V_i(x,t) = (-1)^{i+1} \cdot H[x,A(x,t)] + u(x,t).$

Point P_{ijk} usually is determined by an iteration procedure approximating the characteristics by straight lines.

This new method (IMOC) to solve the Saint-Venant equations based on the theory of characteristics has been proven to be unconditionally stable (EDENHOFER, SCHMITZ; 1981). To compute the unknown variables an implicit scheme with a rectangular grid is employed, thus providing a remarkably simple and fast algorithm without the cumbersome need to satisfy the CFL-criterion.
This specific quality turns out to be of great advantage when coupling the 1-D and 2-D models.

The 2-D Approach

Two dimensional (2-D) depth averaged flow problems can be described by the following set of partial differential equations:

$$\left.\begin{aligned}
u_t + uu_x + vu_y + gh_x &= -g\cdot(s_x + u\sqrt{u^2+v^2}/C^2h) \; ; \\
v_t + uv_x + vv_y + gh_y &= -g\cdot(S_y + v\sqrt{u^2+v^2}/C^2h) \; ; \\
h_t + uh_x + vh_y + hu_x + hv_y &= 0 \; .
\end{aligned}\right. \quad (4)$$

These eqs. (4) consider the influence of the turbulent shear stress only by means of the modified Chézy formula.

Applying a certain transformation of the variables (SCHMITZ, SEUS, CIRWITZKY; 1983) the authors converted eqs. (4) to the wave equation. A well known solution to this problem is the Poisson-Parseval solution (COURANT, HILBERT; 1968) given by

$$\bar{h}(\bar{x},\bar{y},\bar{t}) = \frac{\partial}{\partial t}\{I_o(\bar{h})\} + I_o(\bar{h}_{\bar{t}}) + \int_o^{\bar{t}} I(\psi)\cdot d\tau \quad , \tag{5}$$

where

$$I_\tau(f) := \frac{1}{2\pi\cdot c} \iint_{\alpha^2+\beta^2\leq c^2(\bar{t}-\tau)^2} \frac{f(\bar{x}+\ ,y+\beta,\tau)}{\sqrt{c^2(t-\tau)^2-\alpha^2-\beta^2}}\cdot d\alpha\cdot d\beta \ . \tag{6}$$

Substituting $\alpha = c\cdot(\bar{t}-\tau)\cdot\sin\psi\cdot\cos\phi \ ; \quad 0 \leq \psi \leq \frac{\pi}{2} \quad ;$

$\beta = c\cdot(\bar{t}-\tau)\cdot\sin\psi\cdot\cos\phi \ ; \quad 0 \leq \psi \leq 2 \quad ;$

in eq. (6) yields

$$I_\tau(f) := \frac{(\bar{t}-\tau)}{2\pi}\int_o^{2\pi}\int_o^{\pi/2} f[\bar{x}+c(\bar{t}-\tau)\cdot\sin\psi\cdot\cos\phi,\ \bar{y}+c(\bar{t}-\tau)\cdot\sin\psi\cdot\cos\phi,\tau] \ldots]\cdot\sin\psi\cdot d\psi\cdot d\phi \ . \tag{7}$$

Eq. (5) represents an exact solution for the transformed variable $\bar{h}$ of the two dimensional problem of eqs. (4). It is easy to retransform to $h(x,y,t)$ the water depth at point (x,y) at time (t).

Equation (5) consists of three integrals; the first and the second describe an integration over the domain of dependence (base circle of the characteristic cone); the third integral ranges over the volume of the characteristic cone (see fig.2). Thus our (exact) approach is quite different from so-called quasi-characteristic methods which only choose some favoured lines on that cone to integrate along.

In order to solve eq. (5) numerically we employ a rectangular grid with meshpoints $P_{i,j,k}$. Calculating the unknown variable $h(x_i,y_j,t_{k+1})$ at point $P_{i,j,k+1}$ the integrals of eq. (6) are evaluated using values of the dependent variables of 13 grid-points at a time level $t_k \leq t_o < t_{k+1}$ (see Fig. 3).

Now we express the variable $h(p_{i,j,k+1})$ by a weighted sum

$$\left.\begin{aligned} h(P_{i,j,k+1}) = \sum_{l^2+m^2\leq 4} \{ & a_{l,m}\cdot h(x_{i+l},Y_{j+m},t_o) + \\ & + b_{l,m}\cdot h_t(x_{i+l},Y_{j+m},t_o)\} \ . \end{aligned}\right\} \tag{8}$$

Using Taylor's formula we expand eqs. (5) and (8) around point M see Fig. 2).

Comparing the coefficients associated with the partial derivatives of the "Taylored" eqs. (5) resp. (8) enables us to determine the values of $a_{l,m}$ and $b_{l,m}$ of eq. (8).

Intergrating the momentum equations of eq. (4) and substituting the corresponding derivatives of $\bar{h}$ (see eq. 5) results in an analytical expression for $\bar{u}$ and $\bar{v}$. The numerical solution of the entire system (4) is obtained by following the same approach as with h, initially starting from a time level $t_k \leq t_o < t_{k+1}$.

Finally the values of the variables u,v,h at t_o are expressed by interpolation between the known values at time level t_k and the unknown at t_{k+1} (see Fig. 2). Thus an implicit numerical scheme is established not bound to the CFL-criterion; this means there is no limitation to the size of the time step (similar to the 1-D approach).

COUPLING OF 1-D AND 2-D MODEL PARTS

Our approach of coupling 1-D and 2-D models refers to the 1-D treatment of special subsystems like roads e.g. lying within the entire system to be modeled in two dimensions (see Fig. 4). Principally, the boundary condition of the 2-D model and the boundary condition of the 1-D model at one subsystem/system intersection have to be of different type, thus allowing the necessary feed-back between both modules.

Let us first consider the (internal) outflow boundary of the 2-D domain, which also represents the (external) inflow boundary of the 1-D subsystem (see Fig. 3). When starting the simulation, the boundary condition for the 2-D model, namely the water level h_i at every gridpoint (i) along the subsystem/system intersection is given by the initial conditions. After the first iteration the 2-D computation then provides the flow velocities u,v along the intersection line. The whole discharge through the cross section at this intersection is then determined by

$$Q(r = 0,t) = \int h(s)\cdot\bar{u}\, ds \qquad (9)$$

with $\bar{u}$ = value of flow velocity vector in direction of the axis of 1-dimensional flow.

This computed discharge Q is transferred to the 1-D model as its inflow boundary condition for the subsequent iteration, which yields as one of its results the water level at the upstream boundary of the 1-D reach. This water level becomes the boundary condition for the next 2-D iteration, when assuming a horizontal water table at the intersection line.

At the 2-D inflow boundary (downstream boundary of the 1-D subsystem), the initially given specific discharges q,p are used as (internal) boundary conditions. The 2-D calculation then provides the water level h as the outflow boundary condition for the subsequent 1-D computation. To close the circle, this 1-D computation yields, at its downstream end, the global discharge Q, which then is transformed to the inflow boundary condition, namely the specific discharges p,q, for the 2-D model by assuming an uniform velocity distribution over the cross section.

The different numerical properties of both the 1-D and 2-D model parts require a weighting of the variables to be transferred at the subsystem/system intersections from 1-D to 2-D or vice versa

$$\bar{a}^{m+1} = a^{m+1} \cdot x + a^{m} (1-x) \tag{10}$$

with $\bar{a}$ = generalized boundary condition value

x = relaxation parameter; in the computation a value of $x = 0.4$ was used.

Convergence of the iterative procedure is achieved if

$$\max |h_{i,j}^{m+1} - h_{i,j}^{m}| < \varepsilon \quad \forall_{i,j} \quad . \tag{11}$$

WATER LEVEL - GEOMETRIC BOUNDARY FLUCTUATION

When modelling unsteady flow in natural systems, the numerical treatment of time varying geometric boundaries gives results, when initially dry regions are subject to rising and receding flood flows.

In our model, the decision if a point P* in the computational domain becomes dry is made in reference to the characteristic cone in P*. When water depth together with flow velocity at P* become very small, the domain of dependence diminishes and the hull of the characteristic cone recedes significantly from the surrounding gridpoints. Thus, the influence of these surrounding points with respect to the solution at P* is negligible and P* is defined to be dry. Expressing this criterion in mathematical terms provides

$$\max \{|u|,|v|\} + \sqrt{gh} \leq \frac{1}{\alpha}\frac{\Delta x}{\Delta t} \quad ; \tag{12}$$

with the weighting parameter α .

Using a value of $\alpha = 2$ in our investigations led to a critical water depth h* of about 3 cm

In the model a point P* gets flooded again when the water level at one of the surrounding points rises to more then h* + 10 % over the bottom height of the "dry" point.

APPLICATION

Modelling complex geometric boundaries requires a higher density of meshpoints in the numerical grid than the discretisation of relatively undisturbed flow domains. Our model is flexible enough to deal with nearly arbitrary mesh refinements simultaneously. However, simulating some flows, as in narrow roods, leads to an enormous quantity of gridpoints. In order to gain computational efficiency, the general 2-D description is replaced in those areas having roughly 1-D flow regime by a 1-D description of the flow.

The model was applied to simulate the flow pattern of the Iller river in the town of Kempten/FRG. The area investigated is about 5oo m long, the maximum width is about 2oo m. The flow domain is partially covered by houses and roads. One long, narrow road, the Brennergasse, was modelled one - dimensionally. A large side weir extending obliquely from near the left bank across the river is another problem for the numerical simulation.

As a possible initial condition for the computations we chose a horizontal water level $H(x,y,t_o) = H_o$ over the whole flow domain; velocities $u(x,y,t_o) = u_o(x,y)$ resp. $v(x,y,t_o) = v_o(x,y)$ were zero.

During the computation, the flow velocities at the inflow boundary were increased up to the corresponding design flood of 750 m³/s. Simultaneously at the right boundary, starting from H_o, we lowered the height of the water level step by step. When finally the given water level H_r was attained, a number of gridpoints had fallen dry and we came up with steady state conditions after 20 time steps, $\Delta t = 20$ sec.

The river domain investigated and the calculated flow pattern for the design flood of 750 m^3/s^{-1} are displayed by a computer graph in fig. 4 (showing amount and direction of flow velocities).

RESULTS

In order to investigate the efficiency of the coupled 1-D/2-D approach, the same natural system has been simulated by using only the 2-D model.
The rather coarse 2-D discretisation of only the "Brennergasse" ($\Delta x = \Delta y = 2.50$ m), which has an average width of 10 m, required an additional 300 gridpoints, that is about 17 % in excess of the total 1725 gridpoints used in the coupled simulation.

A comparison of both the coupled and single 2-D simulation provided some interesting results:

- The 2-D modelling of the narrow Brennergasse, located obliquely in the 2-D network, resulted in "numerical edges" (see fig. 5), which together with truncation errors caused in this domain a mass balance error of about 5 %, whereas the 1-D computation satisfied mass conservation completely.
- The 2-D flow pattern in both simulations, including the areas about the subsystem/system intersection is principally the same (see fig. 4 and 5).

- The coupled approach led to a saving of computer time of more than 15 %.

CONCLUSIONS

Applied to a complex flow situation the coupled 1-D/2-D model turns out to be a fairly comprehensive tool for engineering studies for such problems. This is mainly due to the following features.

- Areas with dominant 1-D flow regime inside a complex 2-D flow domain can be modeled by a 1-D approach, without any loss of accuracy.
- The coupled approach leads to considerable saving in computer time and storage requirements.
- The numerical treatment of geometric boundaries varying with time is incorporated in the model. This situation occurs when initially dry regions gradually get flooded and fall dry again.
- Steady and unsteady flow conditions can be simulated disregarding Froude numbers greater than one.
- The model allows refinements of the grid wherever it is required by the consistency of information.
- The model is capable of using large time steps independent of the size of the Courant step.

ACKNOWLEDGEMENTS

The authors wish to express sincere thanks to the German Research Association (Deutsche Forschungsgemeinschaft / DFG) for providing financial support. This paper is based on DFG Grants Se 326/6-1 and Se 316/6-2.
All computations were carried out free of charge on the CDC CYBERS 175 DP of the LEIBNIZ computing center of the Bavarian Academy of Sciences. This valuable assistance is gratefully acknowledged.

REFERENCES

Courant, R. and Hilbert, D.: "Methoden der mathematischen Physik II". Berlin-Heidelberg-New York, Springer, 1968.

Edenhofer, J. and Schmitz, G.: "Ein implizites Charakteristikenverfahren zur Lösung von Anfangsrandwertproblemen bei hyperbolischen Systemen". Computing, 26, 1981, pp. 257-264.

Schmitz, G.; Seus, G.; Czirwitzky, H.J.: "Simulating Two Dimensional Flood Flow". Proc. of the Intern. Conference on Hydraulic Aspects of Floods & Flood Control, London, Sept. 1983, pp. 195-206.

Schmitz, G.; Edenhofer, J.; Czirwitzky, H.J.: "An analytical and numerical solution of Saint Venant equations". In: La modélisation fine des écoulements C.R. du Symp., Int., Paris, Sept. 1982, Vol. 2, pp. 475-490.

NOMENCLATURE

t	time
x,y	coordinates in the plane
u,v	velocities in the x- resp. y-direction
h	water depth
A	cross sectional area
S_o	bottom slope
S_f	friction slope
H	height of water level above datum
S_x, S_y	bottom slope in the x- resp. y-direction
C	Chézy coefficient
g	gravitational constant (= 9,81 ms^{-2})
P_{ijk}	mesh point in the xyt-grid
$I_\tau(f)$	integral operator at time τ with argument function f
c	celerity of wave propagation
i,j	space indices ref. to x,y
k	time index
$a_{l,m}$, $b_{l,m}$	weighting coefficients in eq.(8)
Q	discharge
r,s	coordinates of an inner 1-D subsystem

Transformed variables are marked by a bar.

Indices x,y,t indicate partial derivation of the variables in the x,y,t - direction.

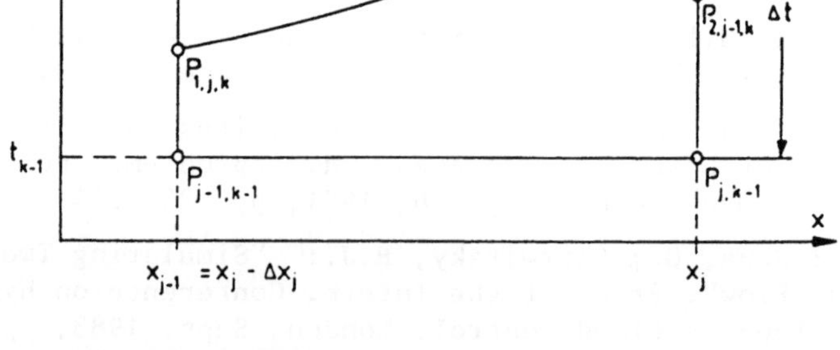

Fig. 1: Implicit Scheme, with $P_{jk}(x_j, t_k)$ as discretisation and P_{ijk} as intersection points of the characteristics with the grid

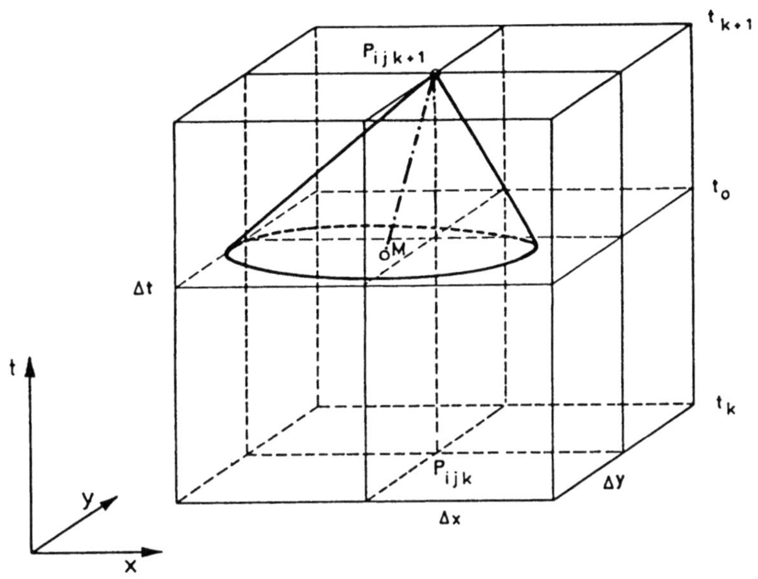

Fig. 2: Characteristic cone in the xyt-grid

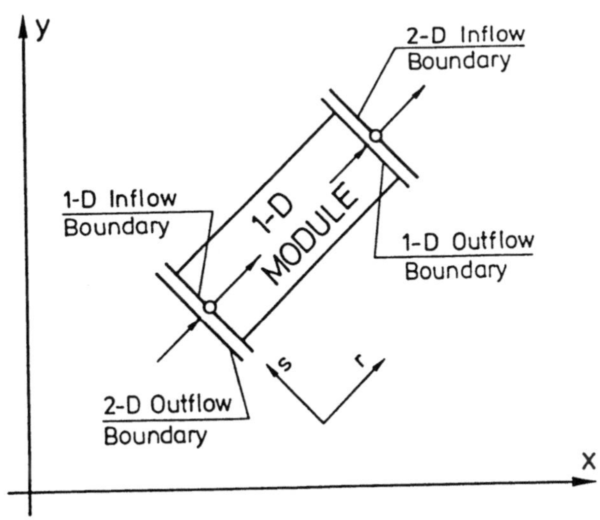

Fig. 3: Subsystem/System Boundaries

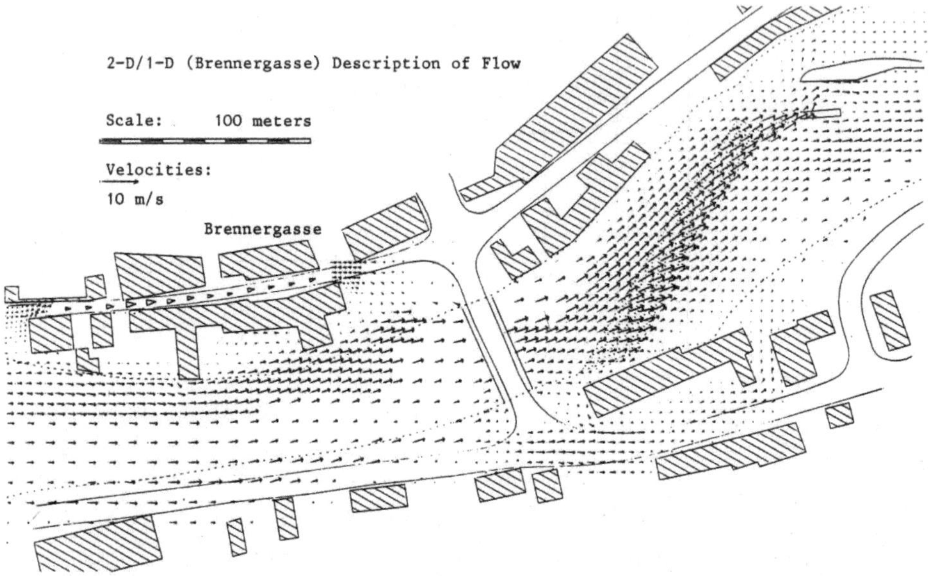

Fig. 4: Flow pattern of the design flood of the Iller river at Kempten/FRG, 1-D/2-D description of flow

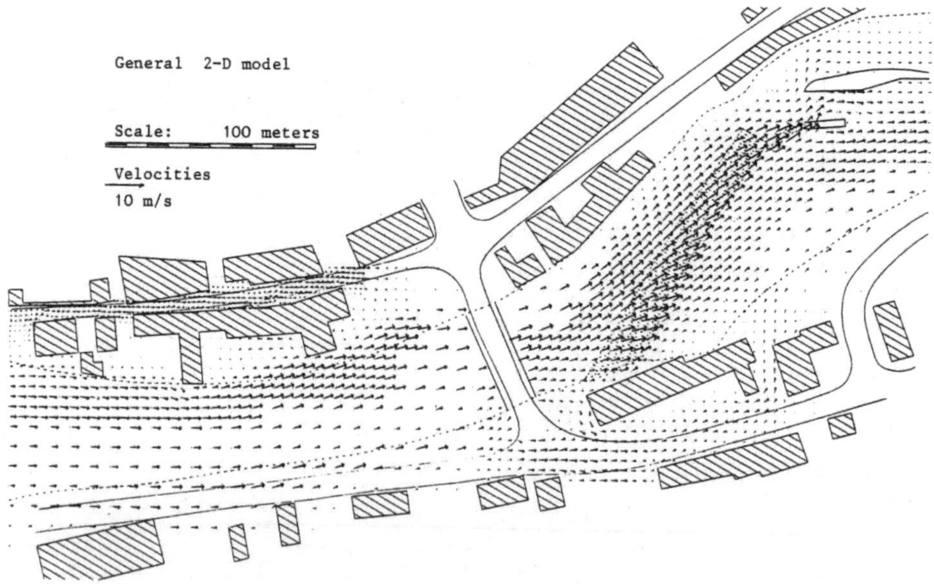

Fig. 5: Flow pattern of the design flood of the Iller river at Kempten/FRG, general 2-D description of flow

AN APPLICATION OF FINITE ELEMENT METHODS FOR MODELLING WIND DRIVEN CIRCULATIONS IN A STRATIFIED OCEAN

C. LE PROVOST

Institute of Mechanics, Grenoble, France

INTRODUCTION.

Since the middle of this century, considerable emphasis has been placed upon the understanding of the role of the wind as one of the major driving forces of the large scale ocean circulation. The pioneering investigations were carried analytically by Sverdrup (1947), Stommel (1948) and Munk (1950) who considered an ideal square box ocean of constant depth, filled with a fluid of constant density moving on a beta plane. These solutions provided valuable insight. However the difficulties of analytical treatment of the inertial and unsteady characteristics of the problem limited the extent of these investigations and motivated the development of numerical models in the beginning of the sixties. The first attempts, by Bryan (1963), Veronis (1966), and Blandford (1971), were direct extensions of the early analytical wind driven circulation models to the general time dependant non linear case. Their results showed the power and the usefulness of even simple numerical models and brought important contributions to understanding the dynamics of the ocean, especially the evidence of the inherent time dependency of the flows owing to barotropic instabilities. As real ocean flows are known to be significantly baroclinic (statistically, the ratio of barotropic to baroclinic kinetic energy has been estimated at approximately 0.5, cf. Mc Williams (1976)), it was necessary to take into account the stratification of the ocean. The pioneering works in development of prognostic wind -and thermohaline- driven circulation models has been carried out by Sarkisyan (1962), and Bryan and Cox (1967). As these models seek to reproduce the three dimensional density distribution, as well as the three dimensional flow field within the basins, they were very computer time consuming, and consequently difficult to use for process studies; ;they were consequently restricted to coarse space resolutions with a simple turbulent viscosity parametrisation of the sub grid scales effects, which, now, is known to be non correct. Indeed the baroclinic instability processes, which appear only quite recently to be an important mechanism of the ocean dynamics, need the use of horizontal grid sizes smaller than the internal Rossby radius of deformation, of the order of 40 km ! This brings the modellers to develop what is now called EGCM, i.e. Eddy resolving General Circulation Models, applied to ideal box-shape oceans, with high resolution and two layers on the vertical for Holland and Lin (1975), Holland (1978) and Schmitz and Holland (1982), or several levels in the vertical for Robinson and al (1977), and Semtner and Mintz (1977).

All these models have been developped on the basis of classical finite difference technics. However, Fix (1975) investigated the properties of the finite element technics for modelling ocean dynamic problem in the quasi geostrophic approximation and showed their interests on several points : precision, conservation of energy and enstrophy, natural treatment of boundary conditions, flexibility of triangulation for the resolution of complex areas. He proved the stability and convergence of the semi discrete finite element formulation of the problem, and established the conservative properties of its numerical approximation for energy, vorticity and enstrophy, independant of the irregularity of the grid used for spatial integration. Given these conclusions, some attempts have been made recently to introduce F.E.M. in the field of ocean modelling. As before with the development of finite difference technics, the barotropic problem has been first investigated. Haidvogel, Robinson and Schulman (1980) compared the precision of a finite difference model, a finite element model, and a spectral model, for applications to open ocean problems, and showed the interest of F.E.M.. Dumas, Le Provost and Poncet (1982) investigated the performances of a F.E. model for solving the wind driven oceanic problem in a closed basin, and compared favourably the precision of their results to the analytical solutions of Stommel and Munk for linear cases, and to the previous results obtained by finite difference technics for the non linear cases. As these solutions typically include a western boundary layer with intense velocity gradients, the interest of refinement of the triangulation along the western wall of the basin and the superiority of quadratic elements were clearly shown. Moreover, tests on computer requirements proved that such a model is ecomically competitive, if compared to classical finite difference relaxation technics.

But these studies were limited to barotropic flows. As noticed before, the ocean is stratified in density, and the real oceanic flows are significantly baroclinic. Thus, the next step has been to introduce in the previous models vertically variable density and velocity distributions. Miller, Robinson and Haidvogel (1983) have developped two methods of treating the depth dependance of the flow (a finite difference method, and a collocation method) and demonstrated their feasability and efficiency for modelling realistic mid-oceanic mesoscale eddy flow regimes.

The aim of this paper is to present an extension of the earlier barotropic F.E. model of Dumas, Le Provost and Poncet to deal with mixed barotropic and baroclinic dynamics in relation with the general wind driven oceanic circulations in closed basins. The presentation will be limited to a two layer quasigeostrophic version of this new baroclinic model, but it can be generalized without any difficulty to a multi layer system.

MODELING EQUATIONS AND METHODS.

<u>The quasi-geostrophic model</u>.

The ocean model is taken as a square basin of constant depth H on a beta plane which is a projection of the sphere on a tangent plane where the Coriolis parameter is taken to be $f = f_o + \beta y$, with f_o and β constant, and y the north south coordinate. The vertical structure is modelled by dividing the ocean into two homogeneous layers with density contrast $\Delta\rho$ between them.

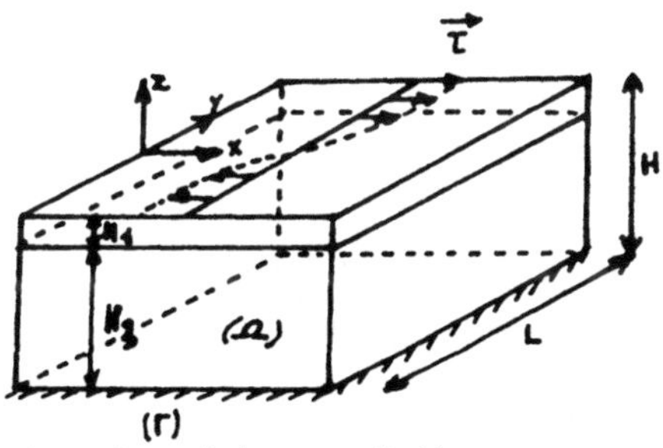

Fig. 1 : Schema of the ocean model over the area (Ω), bounded by the solid walls (Γ).

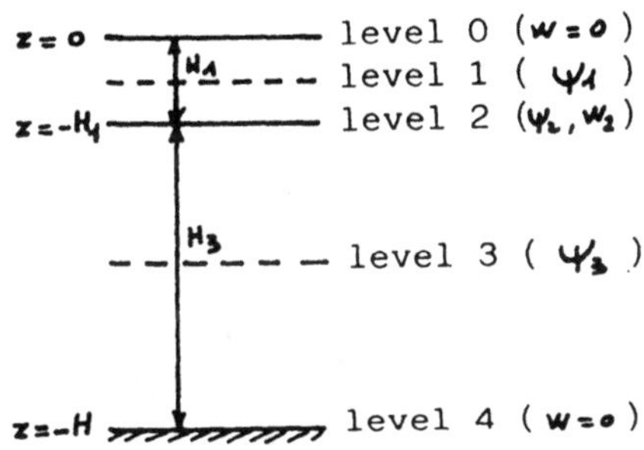

Fig. 2 : Diagram of the vertical structure of the two-layer quasi-geostrophic model.

The upper layer is driven by a zonal wind stress which is steady in time and varies sinusoïdally with latitude, as a first approximation of the westerlies and trades : $\tau(y) = -\tau_0 \cos \pi y/L$
Some dissipation process is necessary : although we know that horizontal "eddy" viscosity is not adequat, we use a laplacian friction, and consider this viscosity as a parameterization of subgrid-scale physical processes. We consider also a linear bottom friction. No interfacial stresses are included.

The equations to be solved are the quasi geostrophic vorticity equations for these two layers of constant thickness and a thermodynamic equation applied at the interface, cf. Phillips (1956) and Pedlosky (1964) :

$$(1) \quad \frac{\partial}{\partial t} \nabla^2 \psi_1 = J(\nabla^2 \psi_1, \psi_1) - \beta \frac{\partial \psi_1}{\partial x} - \frac{f_o}{H_1} w_2 + F_1 + \frac{1}{H_1} \mathrm{curl}_z \tau$$

$$(2)\quad \frac{\partial}{\partial t}\nabla^2\psi_3 = J(\nabla^2\psi_3,\psi_3) - \beta\frac{\partial\psi_3}{\partial x} + \frac{f_0}{H_3}w_2 + F_3 + B_3$$

$$(3)\quad \frac{\partial}{\partial t}(\psi_1-\psi_3) = J(\psi_1-\psi_3,\psi_2) - \frac{g'}{f_0}w_2$$

with $F_i = A_m\nabla^4\psi_i$, $B_3 = -\varepsilon\nabla^2\psi_3$ and $g' = g\Delta\rho/\rho_0$. ψ_i are the quasi geostrophic streamfunctions at the various levels, and w_2 the vertical velocity at the interface. Following Phillips (1956), the interfacial stream function ψ_2 is evaluated as the weighted average of ψ_1 and ψ_3 , $\psi_2 = (H_1\psi_3 + H_3\psi_1)/H$. The velocity components in the two layers are $u_i = -\partial\psi_i/\partial y$ and $v_i = \partial\psi_i/\partial x$. Eqs (1) and (2) express the fact that the local rate of change of relative vorticity in each layer is produced by horizontal advection, stretching effects of vortex lines, frictional effects, and vorticity input by the wind stress curl. Eq (3) can be considered as a thermodynamic equation, or equivalently as a continuity equation predicting the deviation of the interface from its equilibrium position $(h_2 = f_0(\psi_1-\psi_3)/g')$.

The method of resolution.

Rather than ψ_1 and ψ_3, following Holland (1978), we solve the baroclinic and barotropic modes.

* Taking $\psi = \partial/\partial t\,(\psi_1-\psi_3)$, and substracting (2) from (1), we obtain with (3), for the baroclinic mode :

$$(4)\quad (\nabla^2-\lambda^2)\,\psi = \gamma_a$$

with
$$\gamma_a = J(\nabla^2\psi_1,\psi_1) - \beta\frac{\partial\psi_1}{\partial x} - J(\nabla^2\psi_3,\psi_3) + \beta\frac{\partial\psi_3}{\partial x} - \lambda^2 J(\psi_1-\psi_3,\psi_2) + \frac{1}{H_1}\,\mathrm{curl}_z\,\tau + F_1 - F_3 - B_3$$

and $\lambda^2 = H f_0^2/(g' H_1 H_3)$. λ^{-1} is the internal radius of deformation. Along the side walls (Γ), $\psi_1-\psi_3$ is constant, but must be a function of time to satisfy continuity equation (3). As $\iint w\,dx\,dy = 0$, $\iint \psi\,dx\,dy = 0$. To solve (4), taking into account that condition, let us consider :

(5) $\psi = \psi_a + c(t)\,\psi_b$ where ψ_a is solution of equation :

(A) $(\nabla^2-\lambda^2)\,\psi_a = \gamma_a$ with $\psi_a/_\Gamma = 0$, and ψ_b of equation :

(A_1) $(\nabla^2-\lambda^2)\,\psi_b = 0$ with $\psi_b/_\Gamma = 1$.

Thus c(t) can be computed at every time by

$$(6)\quad c(t) = -\iint\psi_a\,dx\,dy \Big/ \iint\psi_b\,dx\,dy$$

Note that the time independant field ψ_b needs to be determined only once.

* Taking $\phi = (\partial/\partial t)(H_1\psi_1 + H_3\psi_3)/H$, and multiplying (1) by H_1/H, (2) by H_3/H, and adding, we obtain for the barotropic mode :

(B) $\nabla^2\phi = \gamma_b$ with $\phi/_\Gamma = 0$

and
$$\gamma_b = \frac{1}{H}\left[H_1 J(\nabla^2\psi_1,\psi_1) - \beta H_1\frac{\partial\psi_1}{\partial x} + H_3 J(\nabla^2\psi_3,\psi_3) - \beta H_3\frac{\partial\psi_3}{\partial x}\right] + \frac{1}{H}\left[\mathrm{curl}_z\,\tau + H_1F_1 + H_3F_3 + H_3B_3\right]$$

We have thus to solve the equations A and B, but they contain fourth order term ($J(\psi_i, \nabla^2\psi_i)$ and $\nabla^4\psi_i$). Consequently, we introduce the relative vorticity : $\zeta_i = \nabla^2\psi_i$, so that (A) and (B) are reduced to second order equations in ψ_i and ζ_i, and we consider complementary equations deduced from (1) and (2) which allow to step forward in time the new variables :

(C) $$\frac{\partial}{\partial t}\zeta_1 = J(\zeta_1,\psi_1) - \beta\frac{\partial\psi_1}{\partial x} - \frac{f_0}{H_1} w_2 + \frac{1}{H_1}\,\mathrm{curl}_z\,\tau + A_m\nabla^2\zeta_1$$

(D) $$\frac{\partial}{\partial t}\zeta_3 = J(\zeta_3,\psi_3) - \beta\frac{\partial\psi_3}{\partial x} + \frac{f_0}{H_3} w_2 - \varepsilon\zeta_3 + A_m\nabla^2\zeta_3$$

with (7) : $$w_2 = \frac{f_0}{g'}\left[J(\psi_1-\psi_3,\psi_2) - \psi\right]$$

In addition to the boundary conditions given above (no flow through the boundary) for the barotropic and baroclinic modes, other conditions are required for ζ_i : we take slip boundary conditions $\zeta_i/_\Gamma = 0$.

NUMERICAL METHOD.

The time discretization.

We use a Leap-Frog scheme. Time is divided into increments dt, and for t = ndt we note f(t) as f^n. Thus, a discrete formulation of the vorticity equations (C) and (D) is :

(F) $$\frac{\zeta_1^{n+1}-\zeta_1^{n-1}}{2\,dt} = J(\zeta_1^n,\psi_1^n) - \beta\frac{\partial\psi_1^n}{\partial x} - \frac{f_0}{H_1}w_2^n + \frac{1}{H_1}\,\mathrm{curl}_z\,\tau + A_m\left(\gamma\nabla^2\zeta_1^{n+1} + (1-\gamma)\nabla^2\zeta_1^{n-1}\right)$$

(G) $$\frac{\zeta_3^{n+1}-\zeta_3^{n-1}}{2\,dt} = J(\zeta_3^n,\psi_3^n) - \beta\frac{\partial\psi_3^n}{\partial x} + \frac{f_0}{H_3}w_2^n - \varepsilon\frac{\zeta_3^{n+1}+\zeta_3^{n-1}}{2} + A_m\left[\gamma\nabla^2\zeta_3^{n+1} + (1-\gamma)\nabla^2\zeta_3^{n-1}\right]$$

with $\gamma \in [0,1]$.

This scheme is consistant and of order 1, or even order 2 if $\gamma = 1/2$; it is unconditionally stable for $\gamma \geq 1/2$.

But, to solve these equations, it is necessary to know w_2^n, which can be obtained through (7), if ψ^n is determined. Consequently, it is necessary to solve previously eq (A) at the n time step. If the solution Φ^n of eq (B) is obtained by the same time, it is then possible to step forward the stream functions ψ_1 and ψ_3 through the system :

(H) $$\begin{cases} \psi_1^{n+1} - \psi_3^{n+1} = 2\,dt\;\psi^n + (\psi_1^{n-1} - \psi_3^{n-1}) \\ H_1\psi_1^{n+1} + H_3\psi_3^{n+1} = 2\,H\,dt\;\Phi^n + (H_1\psi_1^{n-1} + H_3\psi_3^{n-1}) \end{cases}$$

The variational formulation.

For each time step, we have to solve at least equations (A) and (B), two second order elliptic equations, if $\gamma = 1$, i.e. when (F) and (G) are taken explicitly ; when $\gamma \neq 1$, four second order elliptic equations must be solved (A), (B), (F), (G). We look for solutions in a weak sense, i.e. in the space $H_0^1(\Omega)$. Integrating (A) with a testing function $\psi_a' \in H_0^1(\Omega)$ over the domain Ω, we have :

$\{(\nabla^2-\lambda^2)\psi_a^n , \psi_a'\} = \{\gamma_a^n(\psi_i^n,\zeta_i^n) , \psi_a'\}$ with the notations :

and $\{a , a'\} = \iint_\Omega a \cdot a' d\Omega$, $\gamma_a = f_a^n + A_m(\zeta_1^n - \zeta_3^n)$

$$f_a^n = J(\zeta_1^n,\psi_1^n) - \beta\frac{\partial\psi_1^n}{\partial x} - J(\zeta_3^n,\psi_3^n) + \beta\frac{\partial\psi_3^n}{\partial x} + \frac{1}{H_1}\mathrm{curl}_z\tau + \varepsilon\zeta_3^n - \lambda^2 J[\psi_1^n-\psi_3^n, (H_1\psi_3^n + H_3\psi_1^n)/H]$$

Using the Green formula :

$$\{\nabla^2 a, a'\} = -\{\nabla a, \nabla a'\} + \oint_\Gamma \frac{\partial a}{\partial n} a' d\Gamma ,$$

and noting that the curvilinear integral is null, because $\zeta'/_\Gamma = 0$, we deduce the variational formulation of (A) :

$$(\bar{A}) \quad \{\nabla\psi_a^n , \nabla\psi_a'\} + \lambda^2\{\psi_a^n,\psi_a'\} = -\{f_a^n,\psi_a'\} + A_m\{\nabla(\zeta_1^n-\zeta_3^n),\nabla\psi_a'\}$$

The same kind of formulation can be derived for eq (B), (F) and (G).

The finite element approximation.

The domain Ω is divided into triangles, and we consider a finite element space S^Δ on that grid, where any function can be written :

$$\alpha^\Delta(x,y,t) = \sum_{j=1}^{N} \alpha_j^\Delta(z_j,t)\ \varphi_j(x,y)$$

where z_j, ($j = 1, N$) are the nodes of the elements, and φ_j the basic functions ; the problem is to determine the weights $\alpha_j^\Delta(t)$ for each time step by using the variational formulation corresponding to that function α . Using the matrix notation :

$M = \{\varphi_i,\varphi_j\}$, the mass matrix, and

$K = \{\nabla\varphi_i,\nabla\varphi_j\}$ the stiffness matrix,

$(\bar{A})$: $(K+\lambda^2 M)\ \psi_a^{n,\Delta} = -F^{n,\Delta} + A_m K(\zeta_1^{n\Delta}-\zeta_3^{n\Delta})$ with $F^{n\Delta} = \{f_a^n(\psi_i^{n\Delta},\zeta_i^{n\Delta}),\psi_a'\}$

The matrices M and K are built at the beginning of the iterations, and do not vary while iterations occur. Thus the matrix $K + \lambda^2 M$ is inverted only once, at the beginning of the computation, and the problem reduces at each time step to compute the second member of (9) and solve the system by direct substitution. The same remarks are valid for the functions : ϕ, ζ_1 and ζ_3 , leading to similar variational formulations ($\bar{B}$), ($\bar{F}$) and ($\bar{G}$).

The practical resolution process.

Knowing the solutions ψ_i and ζ_i and their first derivatives at time steps n - 1 and n, we compute the new values at time step n + 1, as follows :

1 - ψ_a^n is obtained from ($\bar{A}$), $c^n(t)$ from (6) and thus ψ^n from (5).
2 - ϕ^n is obtained from ($\bar{B}$).
this gives ψ_1^{n+1} and ψ_3^{n+1} through (H).
3 - w_2^n is deduced from (7).
4 - ζ_1^{n+1} and ζ_3^{n+1} are obtained from ($\bar{F}$) and ($\bar{G}$).

Initially, we start the integration from rest ($\psi_i = \zeta_i = 0$), so that a two level scheme is necessary to compute the first step solutions ψ_i^1 and ζ_i^1 . An automatic code is used to built a triangulation of the domain Ω, and to solve the variational problems $(\bar{A})$, $(\bar{B})$, $(\bar{C})$ and $(\bar{D})$, cf. Poncet (1979).

NUMERICAL TEST.

To test this model, we have reproduced an experiment already done by Holland (1978) with a finite difference model. The corresponding numerical values of the parameters are the following : we consider a two layer square basin L = 1000 km large, of constant depth H = 5000 m, with an upper layer H_1 = 1000 m deep, and thus a lower layer H_3 = 4000 m thick. The reduced gravity g' = 0.02 ms^{-1} caracterizes the density stratification. The rotation parameters are taken as $f_o = 8.3\ 10^{-5}\ s^{-1}$ and $\beta = 2 \times 10^{-11}\ m^{-1}\ s^{-1}$. The wind stress amplitude τ_o is 10^{-4} $m^2\ s^{-2}$. Lateral eddy viscosity A_m is 330 $m^2\ s^{-1}$; bottom friction is ignored : $\varepsilon = 0$.

The triangulation.

Fig. 3 : The triangulation used : 493 triangles
typical size of the mesh :
east and south : 200 km
west and north : 50 km

We know from the physics of the problem that a narrow boundary current will develop along the western and northern walls (similar to the Gulf Stream), and that an unstable westward recirculation pattern will take place in the northern half of the basin. We know also that baroclinic instabilities will occur, at the scale of the internal radius of deformation, which is here Rd = 48 km. We have thus generated the triangulation presented on figure 3. When using piecewise quadratic elements (P_2), the spatial resolution will be less than 25 km, and thus fine enough (Holland (1978)).

The energetic control parameters.

Given the volume of information obtained from this test simulation, we have choosen to check our solution on energetic integral quantities, typical of the dynamic of this experiment. These are, with the notation $\psi_i^* = \psi_i - \psi_{i/r}$ and $d\Omega = dx\ dy$:

$K_i = 1/2\ H_i \iint (\nabla\psi_i)^2\ d\Omega$: kinetic energy of layer i

$$P = (f_0^2/2g') \iint (\psi_1^* - \psi_3^*)^2 \, d\Omega$$: potential energy

$$(\tau \to K_1) = -\iint \psi_1^* \, \mathrm{curl}_z \tau \, d\Omega$$: energy transfer rate from the wind to the upper layer

$$(K_i \to P) = -f_0/H \iint w_2 H_i (\psi_1^* - \psi_3^*) \, d\Omega$$: work done by buoyancy forces in layer i

$$(K_1 \to K_3) = -f_0 \iint w_2 \psi_2^* \, d\Omega$$: downward transfert of energy through the work done by pressure at the interface

$$(K_i \to F_i) = H_i \iint \psi_i^* F_i \, d\Omega$$: energy dissipation by lateral friction in layer i.

Preliminary results.

The model has been run for only a hundred of days, with a time step of 12 h. Time history of the spatial average energies are presented on figure 4, and of energy conversion rates in the upper layer on figure 5. Instantaneous streamfunction map of the upper and lower levels at t = 100 days are also given as an illustration of the circulation at that time on figure 6.

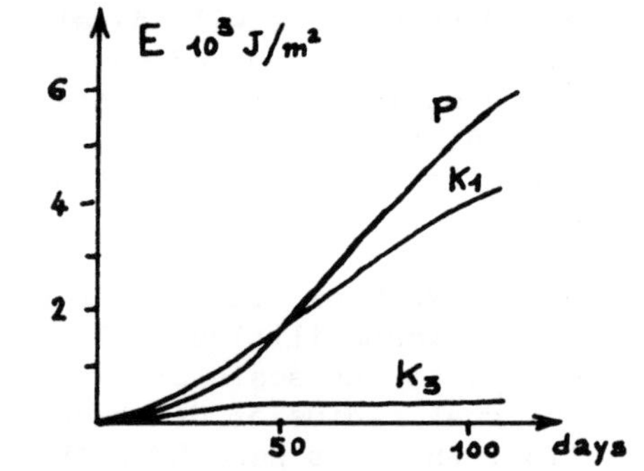

Fig. 4 : Time history of spatially average energy per unit of area

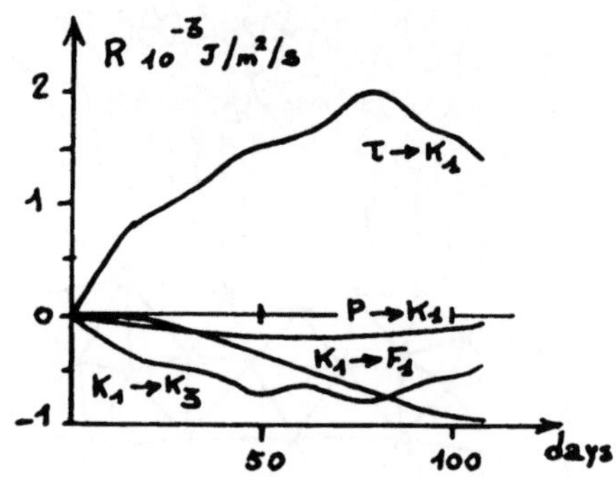

Fig. 5 : Time history of energy rates for the upper layer

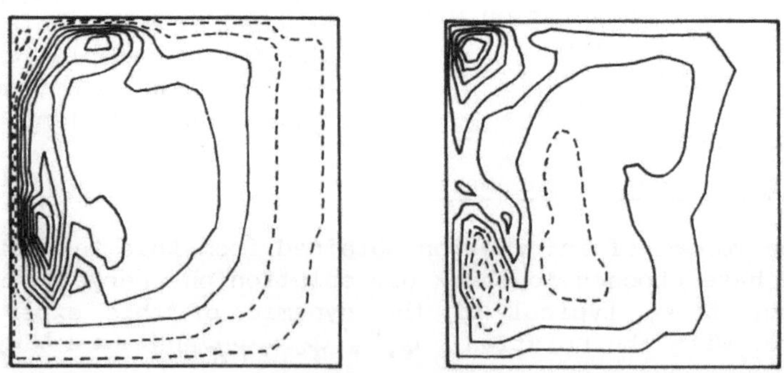

Fig. 6 : The quasi geostrophic stream functions ψ_1 (left) and ψ_3 (right) with intervals of 2700 m^2/s for ψ_1 and 775 m^2/s for ψ_3.

Of course, this is only the early stages of that experiment, but by reference to the results of Holland (1978) experiment, we can see that the different physical processes are going on correctly. We observe an increase in kinetic energy in the upper layer, driven by the wind, and in potential energy (the interface deforms), while the lower layer remains nearly motionless. This occurs because all the downward transfer of energy through the work done by pressure at the interface ($K_1 \rightarrow K_3$) is immediatly converted in available potential energy. At t = 100 days, we see however that the energy transfer by the wind to the upper layer ($\tau \rightarrow K_1$) seems to stabilize, just like in Holland and Lin (1975) and Holland (1978). The streamfunction maps of figure 6 show for the upper layer the existence of the western boundary layer, and the establishment of the northern B.L., with its southward inertial recirculation pattern ; we can observe also two large eddy structures in the southern part of the basin. In the lower layer, motions are considerably weaker : superposed to a large basin scale circulation, we can notice in the northwest corner the initiation of a gyre which is extending eastward and in the southern part of the basin, qualitatively, the same eddy pattern as in the upper layer.

We know from Holland's studies that it will take about 500 days before the vertical shears built up sufficiently for the flow to become baroclinically unstable, and generate spontaneously vigourous mesoscale eddies. This finite element experiment needs to be carried on for a longer period, but these preliminary results prove that the model works correctly.

REFERENCES.

BLANDFORD, R.R. (1971). Boundary conditions in homogeneous ocean models. Deep Sea Research 18, 739-751.

BRYAN, K. (1963). A numerical investigation of a non linear model of wind driven ocean. J. of Atmosph. Sciences, 20 : 594-606.

BRYAN, K., and COX, M.D. (1967). A numerical investigation of the oceanic general circulation. Tellus, 19 : 54-80.

DUMAS, E., LE PROVOST, C. and PONCET, A. (1982). Feasability of finite element methods for oceanic general circulation modelling. 4th Int. Conf. on finite element in Wat. Res. Hannover, Springer Verlag, 5-43-55.

FIX, G.J. (1975). Finite element models for ocean circulation problems. SIAM J. Appl. Math. 29-3 : 371-387.

HAIDVOGEL, D.B., ROBINSON, A.R. and SCHULMAN, E.E. (1980). The accuracy efficiency and stability of three numerical models with application to open ocean problems. Journal of computational Physics, 34 : 1-53.

HOLLAND, W.R. and LIN, L.B. (1975). On the generation of mesoscale eddies and their contribution to the oceanic general circulation. J. of Physical Ocean. 5 : 642-669.

HOLLAND, W.R. (1978). The role of mesoscale eddies in the general circulation of the ocean. J. of Phys. Ocean. 8-3 : 363-392.

MC WILLIAMS, J.C. (1976). Maps from the mid-ocean dynamics experiment. J. of Physical Ocean. 6-810-827.

MUNK, W.H. (1950). On the wind driven ocean circulation. J. of Meteorol. 7 : 79-93.

PEDLOSKY, J. (1964). The stability of currents in the atmosphere and the ocean. Part 1 : J. Atmosph. Sci. 21 ; 201-219.

PHILLIPS,N.A.. (1956). The general circulation of the atmosphere : a numerical experiment. Quart. J. Roy. Meteor. Soc. 82 : 123-164.

PONCET, A. (1979). Autour de l'écriture d'un code d'éléments finis. Thèse. I.M.A.G. Grenoble, 474 p.

ROBINSON, A.R., HARRISON, D.E., MINTZ, Y. and SEMTNER, A.J. (1977). Eddies and general circulation of an idealized oceanic gyre. Journal of Phys. Ocean. 7 : 182-207.

SARKISYAN, A.S. (1962). On the dynamics of the origin of wind driven currents in the baroclinic ocean. Okeanologie, 11 : 393-409.

SCHMITZ, W.J. and HOLLAND, W.R. (1982). A preliminary comparison of selected numerical eddy-resolving general circulation experiments with observations. J. of Marine Research, 40-1 : 75-117.

SEMTNER, A. and MINTZ, Y. (1977). Numerical simulation of the Gulf Stream and mid-ocean eddies. J. Phys. Ocean. 7 : 208-230.

STOMMEL, H. (1948). The westward intensification of wind driven ocean currents. Trans. of the American Geoph. Union, 29 : 202-206.

SVERDRUP, H.U. (1947). Wind-driven currents in a barotropic ocean. Proc. U.S. Nat. Acad. Sc. 33 : 318-326.

VERONIS, G. (1966). Wind driven ocean circulation. Deep Sea Research, 13 : 17-29, 31-55.

ACKNOWLEDGMENTS

The author is indebted to P. Chardaire and C. Rivière for their contribution in programming the model. This work has been supported by the Institute of Applied Mathematics of Grenoble and the Center of National Research (C.N.R.S.).

ON THE AUTOMATIC CALIBRATION OF TIDAL AND TRANSPORT MODELS

U. Meissner, M. Narten
University Hannover, Germany

INTRODUCTION

When a numerical model is used for the investigation of natural phenomena usually a calibration becomes necessary in order to adjust the response from the model to a known state of the natural system. Depending on the size of the problem and the amount of relevant field data the task may be quite expensive with respect to personal and computer costs as adequate strategies presume sufficient experience from comparable investigations. In finite element models for tidal flow and transport problems a great amount of discrete parameters associated to each element may be varied within the range of validity, e.g. bottom friction and wind shear coefficients, exchange parameters due to turbulent dispersion or due to energy flux through free surfaces. Even when all boundary conditions are well-known in most cases the calibration problem has no unique solution.
The calibration is a typical optimization problem for which the errors between the calculated data and the measured field data ought to be minimized by a suitable mathematical algorithm which performs the variation of parameters in an efficient manner. In order to hold down computer costs the authors developed a method which approaches the problem in a straight forward, deterministic manner. From the first applications to a tidal river presented in /4/ good experiences were made by use of this procedure. Several applications proved that the computer is self-acting able to solve the calibration for this class of problems reliably, with high accuracy and more economically than it can be done by trial and error methods. The improved basic theory is summarized briefly in the following and a typical application presented as an example.

CALIBRATION ALGORITHM

The algorithm chosen to minimize the errors between calculated and measured data is based on the following least square formulation

$$E(p_\alpha) = \frac{1}{T}\int_{t_1}^{t_0} [M_i - C_i\ (p_\alpha)]\ [M_i - C_i\ (p_\alpha)]\ dt\ , \qquad (1)$$

where M_i are measured reactions at locations i and C_i are calculated reactions from the numerical model, depending on a set p_α of N unknown parameters, which initial values are usually estimated from the start of any calculation. By variation of the parameters p_α the error function $E(p_\alpha)$ is to be minimized by the algorithm taking into account some relevant time interval T between moments t_0 and t_1 of the tidal process (Figure 1).

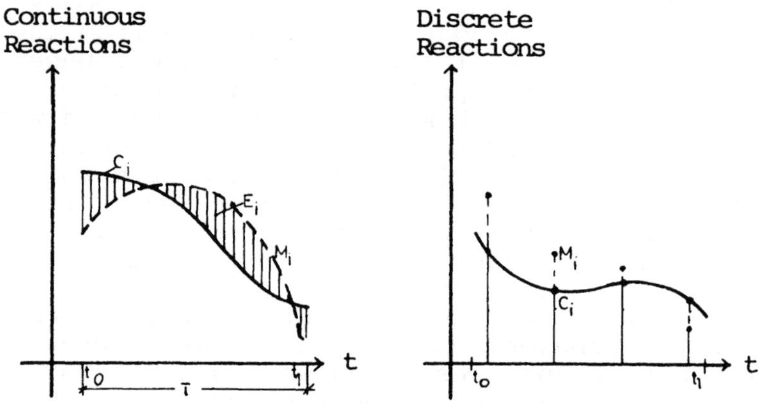

Figure 1: Measured (M_i) and Calculated (C_i) Reactions at Location i

From equ. (1) it is obvious that no special differential equations about the tidal process nor any functional relations between the state variables and the parameters are considered for the calibration problem.

In order to minimize the error function the necessary condition

$$\delta E = \frac{\partial E}{\partial p_\alpha}\delta p_\alpha = E_{,\alpha}\ \ \delta p_\alpha = 0 \qquad (2)$$

has to be satisfied. From equ. (1) this condition results in the basic equation

$$\delta E = -\frac{2}{T}\int_{t_0}^{t_1} (M_i - C_i)\, C_{i,\alpha}\, dt\, \delta p_\alpha = 0 . \tag{3}$$

From the second variation

$$\delta^2 E = \frac{2}{T}\int_{t_0}^{t_1} \left[C_{i,\alpha} C_{i,\beta} - (M_i - C_i)\, C_{i,\alpha\beta}\right] dt\, \delta p_\alpha\, \delta p_\beta > 0 \tag{4}$$

the sufficient condition results, which has to be met in order to ensure that a minimum will be found.

As the transient processes considered imply in general a non-linear relationship between model reactions and parameters a solution of equ. (3) can only be achieved by means of a numerical method. Therefore two important assumptions have to be introduced at this stage of the formulation: a) the solution can be found by solving a consecutive sequence of linearized equations, b) the derivatives of C_i with respect to p_α can be approximated numerically with sufficient accuracy.

When the numerical iteration procedure is started with an initial choice p_α^0 of the parameters, the initial error E^0 results from the difference of measured (M_i) and calculated (C_i^0) reactions specified by equ. (1). An improvement of the parameter set

$$p_\alpha = p_\alpha^0 + \Delta p_\alpha \tag{5}$$

will be gained by use of the following expansion of equ. (3):

$$\delta E = \delta E^0 + \frac{\partial}{\partial p_\beta}\delta E^0 \Delta p_\beta + \ldots = 0 . \tag{6}$$

Taking into account only the first two terms of the series, this leads to the basic linearized equation

$$\delta p_\alpha \quad \{\frac{2}{T}\int_{t_0}^{t_1} \left[C_{i,\alpha}^0\, C_{i,\beta}^0 - (M_i - C_i^0)\, C_{i,\alpha\beta}^0\right] dt \cdot \Delta p_\beta \tag{7}$$

$$- \frac{2}{T} \int_{t_0}^{t_1} (M_i - C_i^0)\, C_{i,\alpha}^0 \; dt\} = 0.$$

For the algorithm proposed herein the derivatives of C_i^0 are calculated numerically by N separate test computations

$$C_i^{(\alpha)} = C_i \; (p_\alpha^0 + \delta p_\alpha) \tag{8a}$$

for each parameter and by an additional control computation

$$C_i = C_i \; (p_1^0 + \delta p_1, \; \dots \; , \; p_N^0 + \delta p_N) \tag{8b}$$

of the non-linear system. These test computations produce specific system reactions at the locations i, from which the numerical values of all first order derivatives can be estimated.

$$C_{i,(\alpha)}^0 \, \delta p_{(\alpha)} \simeq C_i^{(\alpha)} - C_i^0 \tag{9}$$

From the comparison between the non-linear result (8b) and the linear superposition of all test computations (8a)

$$R = C_i - C_i^0 - \sum_{\alpha=1}^{N} (C_i^{(\alpha)} - C_i^0) \tag{10a}$$

it becomes obvious that the residuum R depends mainly on the second order derivatives

$$R = \frac{1}{2} \, C_{i,\alpha\beta}^0 \, \delta p_\alpha \, \delta p_\beta \; + O_2 \; . \tag{10b}$$

As it is very costly to compute these from additional cross computations the hypothetical assumption is made that all second order derivatives cause an equal contribution to the residuum R. By further Taylor expansions for $C_i^{(\alpha)}$ and $C_i^{(\alpha\beta)}$ it can be shown that this assumption leads to the following approximations

$$C_{i,(\alpha\beta)} \, \delta p_{(\alpha)} \, \delta p_{(\beta)} \simeq C_i^{(\alpha\beta)} - C_i^{(\alpha)} - C_i^{(\beta)} + C_i^0 \simeq 2R/(N \cdot N)$$

$$\frac{1}{2} \, C_{i,(\alpha\alpha)} \, \delta p_{(\alpha)}^2 \simeq C_i^{(\alpha)} - C_i^0 - C_{i,(\alpha)} \, \delta p_{(\alpha)} \simeq R/(N \cdot N) \tag{11}$$

from which the second order derivatives are estimated in a crude, but effective manner.

Finally the evaluation of equ's (9), (11) and the substitution into equ. (7) leads to the symmetric algebraic equation system.

$$[K_{i\alpha}\, K_{i\beta} - (M_i - C_i^0)\, K_{i\alpha\beta}]\, \Delta p_\beta = K_{i\alpha}\, (M_i - C_i^0) \quad (12)$$

which has to be solved for the improved vector Δp_β of the parameters. Due to the non-linear character of the problem it might be necessary to set up equ. (12) completely anew within the iterations of the algorithm. Obviously this is quite expensive as the test computations (8) are costly. But depending on the influence of each parameter on the total response of the system further modifications are possible. For the type of applications presented in /4/ and herein, however, the authors found that it was unnecessary to recalculate the coefficients of matrix (12) at all; even when the start vector p_α^0 was empty. Sufficient improvements could be achieved by a few iterations over the two terms of this equation which contain the error $(M_i - C_i^0)$, when only these terms were updated.

As in general the possible number of parameters (α) is larger than the number of locations (i) where measurements are available, difficulties may arise from solving the equation system (12). As discussed in /4/ the problem can than be transformed into a subspace by substituting the parameters Δp_β by a set of generalized parameters Δx_K

$$\Delta p_\beta = \psi_{\beta K}\, \Delta x_K \quad (13)$$

resulting in an associated Eigenvalue problem.

EXAMPLE OF APPLICATION

The practical use of the automatic calibration procedure was tested by applications to a finite element model of the German tidal river Weser illustrated in Figure 2. This river has a length of about 65 km from a weir (Weserwehr) to the river mouth (Bremerhaven) at the North Sea and has an average depth of 9 m at low water. The tidal boundary condition causes water level variations of 3,7 m at Bremerhaven, of 3,5 m at the nuclear power plant KK Unterweser and of 3,3 m at Brake. At the weir a constant dis-

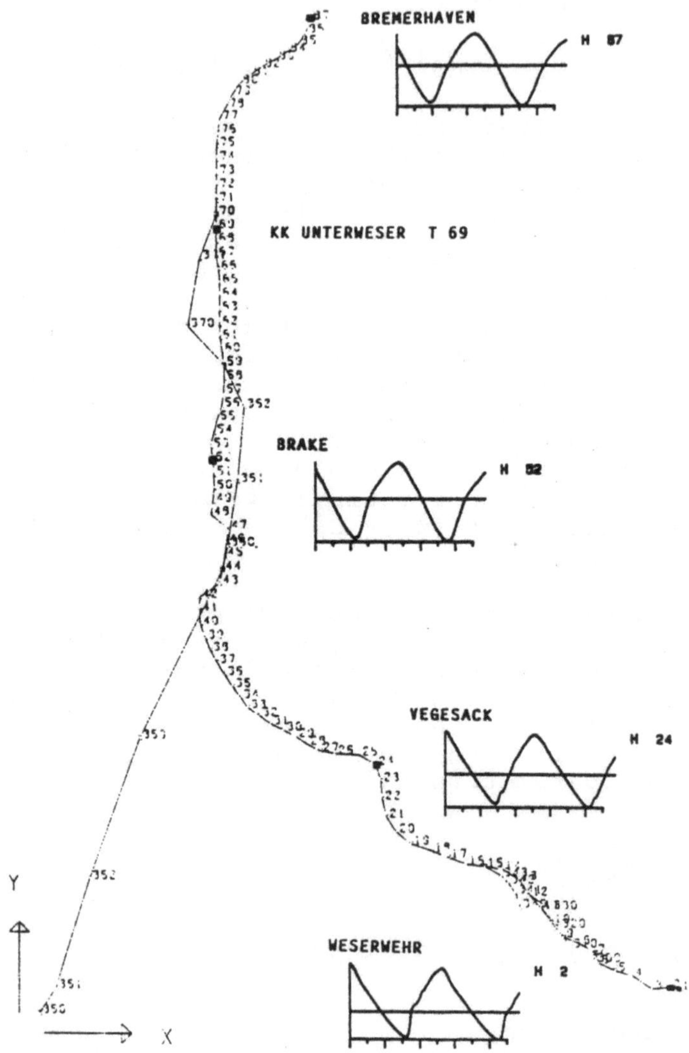

Figure 2: Finite Element Discretization of River Weser in Germany

charge of 257 m^3/s flows into the system. Details about the one-dimensional cross-section-averaged finite element models used for the analysis of the tidal flow and the heat transport are given in /3/,/5/,/6/. The publication /4/ covers results about the tidal flow analysis and the application of a previous calibration procedure. In this paper some test studies about the heat transport within the river are presented.

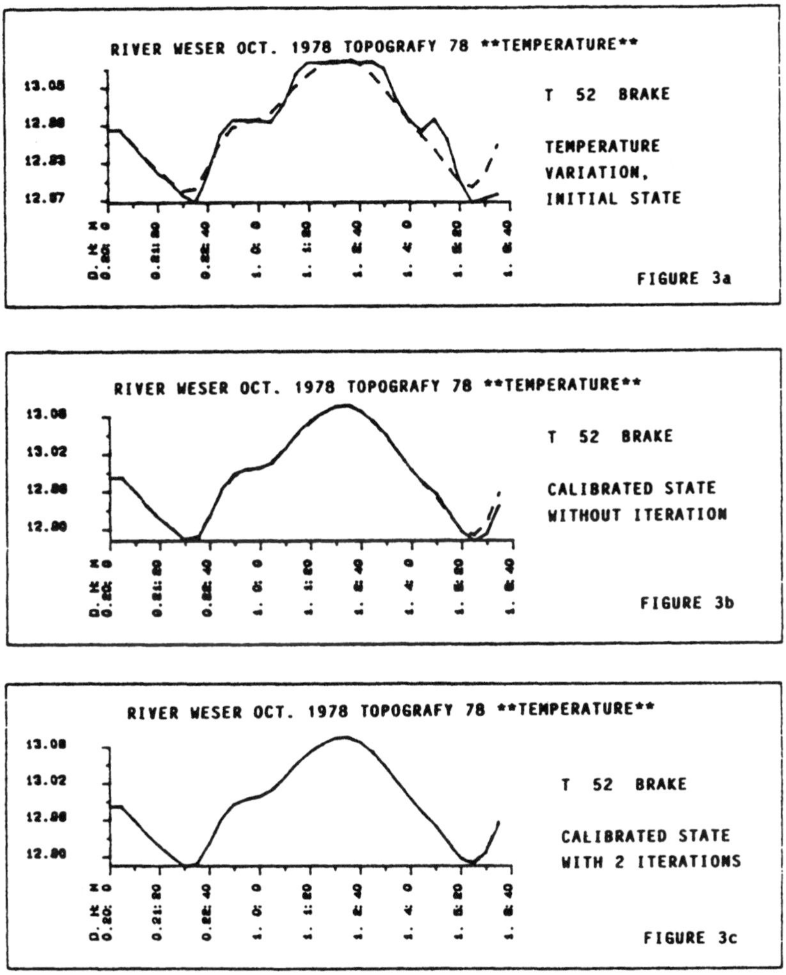

Figure 3: Temperature Distribution at Location Brake

Figure 3 illustrates a numerical test of the calibration algorithm under real tidal conditions. By choice of an artificial temperature distribution in the river and by the assumption of one turbulent dispersion coefficient and one surface exchange coefficient the transport model produced the dotted time response over one tidal cycle at the shown location Brake. For the test the numerical results at two different locations (nodes 44, 52) were taken as "measured" data for the subsequent calibration which was started off with the choice of zero-valued exchange parameters. Fig. 3a illustrates the initial error of the temperature distribution produced in this way. For the parameters included two test calculations over the tidal cycle were neccessary to set up equ. (12). The first parameter improvement led to the result of Fig. 3b, which is very close to the expected temperature distribution. The remaining errors vanished (Fig. 3c) after two iterations over the error term of equ. (12).

Further applications especially about transport phenomena are still under research. For this class of problems the calibration of a numerical model depends very on the quality and reliability of measured field data. This becomes obvious from examples like the one of Fig. 4.

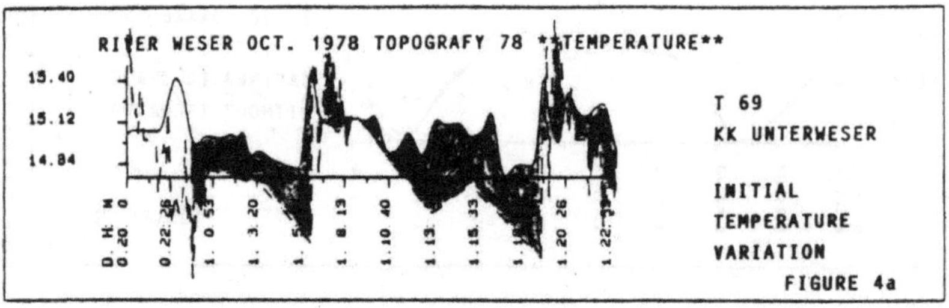

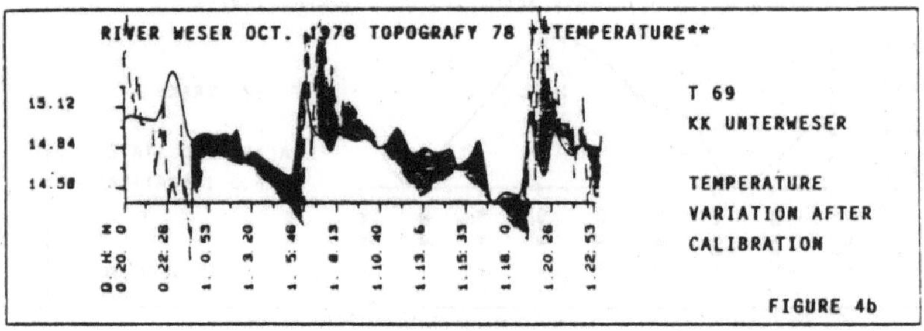

Figure 4: Measured and Calibrated Temperature Distribution at Location KK Unterweser

The dotted lines represent a natural daily temperature variation in the river at node 69 (KK Unterweser) shown in Fig. 2 due to a discharge of 61 m^3/s cooling water with a temperature gradient of 1o K near the condenser of the nuclear power plant. From these field measurements taken by GKSS /1/ only the one time variation, which is partly incomplete between 9 a.m. and 1 p.m., was available for a calibration of the numerical model. For this study only one surface exchange coefficient was included.

Fig. 4a shows the initial error between the measurement and the uncalibrated numerical model without any parameter influence. Inspite of the bad quality of data Fig. 4b demonstrates that the calibration procedure is able to minimize the error between measured and computed data although the numerical model in this case is not capable to reproduce the natural response sufficiently exact.

CONCLUSION

The basic idea of the presented calibration algorithm is to construct an optimization procedure into which as much physical informations as necessary about the systems real reaction behaviour enter in order to find an economical approach to the error functions minimum. Thus the algorithm differs from other optimization methods as described in /2/. By means of test computations for each parameter mutual influence coefficients enter into the equation system (12) and establish a method with deterministic character. As the influence coefficients can only be found by test computations the computer costs depend directly on the number of independent parameters.

REFERENCES

/1/ J. Häuser, D. Eppel, A. Müller, A. Nehlsen, F. Tanzer: "A Thermal Impact Assessment Model with Measured Field Data Applied to the Tidal River Weser", GKSS-Forschungszentrum Geesthacht GmbH, Third Waste Heat Management and Ultilization Conference, Miami Beach, Florida, USA, 1981.

/2/ U. Januszewski: "Automatische Eichung für ein- und zweidimensionale hydrodynamisch-numerische Flachwassermodelle", Dissertation Universität Hannover, Fortschr.-Ber. VDI-Zeitschriften, Reihe 4, Nr. 58, 198o.

/3/ U. Meissner: "An Explicit-Implicit Water-Level Model for Tidal Computations of Rivers", Comp. Mech. Appl. Mech. Engrg., Vol. 13, Nr. 2, 1978, pp. 221 - 232.

/4/ U. Meissner, M. Narten, R. Ratke: "Numerical Models and their Calibration for the Analysis of Flow and Heat Transport Problems in Rivers", Finite Element Flow Analysis, University of Tokyo

Press, 1982, pp. 595-6o1.

/5/ U. Meissner, R. Ratke: "Collocation Formulations for the Finite Element Analysis of Tidal and Transport Processes in Estuaries", Proc. 3rd. Int. Conf. FE in Water Resources, Univ. Mississippi, USA, 198o, pp. 2.167-2.176.

/6/ D. Withum, K.-P. Holz, U. Meissner: "Finite Element Formulations for Tidal Wave Analysis", Comp. Meth. Appl. Mech. Engrg., Vol 17/18, 1979, pp. 699-716.

FINITE ELEMENT SOLUTION METHODS FOR CIRCULATION IN ESTUARIES

Roy A. Walters

U.S. Geological Survey

INTRODUCTION

In this paper, the shallow water equations are used to approximate the depth-mean circulation in estuaries. The time-scales of the motions can be conveniently divided into three ranges: 1) low-frequency (residual) variations with periods of two days or longer, 2) tidal-frequency variations, and 3) high-frequency variations with periods of an hour or shorter. The emphasis here will be on the tidal-period variations that are characterized by line spectra and thus allow a harmonic decomposition of the governing equations.

The finite element method is used for the spatial discretization. For this problem, as with many applications, the true power of the finite element method is the ease with which complex spatial domains can be represented, including variations in both depth and boundary shapes. However, the application of traditional finite element methods to the solution of the shallow water equations in primitive form generally leads to unsatisfactory results due to the presence of numerical oscillations and high noise levels (Gray, 1982; Gray and Lynch, 1979). There is now a substantially better understanding of the discretized shallow water equations than existed a few years ago. Various studies of free oscillations (Walters and Carey, 1983) and forced oscillations (Gray and Lynch, 1979; Lynch and Gray, 1979; Platzman, 1981; Walters, 1983) have examined these problems from several different perspectives and have described methods with which to gain satisfactory solutions.

One of these methods involves filtering or smoothing the input and/or the output data (e.g., smoothing bottom bathymetry). Except where smoothing is attained by network refinement, this approach is not satisfactory because the numerical problem will no longer represent the physical problem. Two other methods involve the creation of elements that do not lead to a folded dispersion relation (staggered node schemes), and the creation of modified governing equations (wave equation formulations).

In the process of developing a circulation model, I have examined several of these methods using a variety of different elements. In the next few sections I describe the governing equations followed by the test problems and the numerical results. Finally, I use these results to choose a model to apply to a highly irregular network, that for South San Francisco Bay.

GOVERNING EQUATIONS

Because the amplitudes of the tides are of specific interest, it is advantageous to use tidal harmonics directly rather than employ discrete time-stepping methods. First, the dependent variables (sea level and velocity) are expanded in terms of a mean plus a sum of periodic components whose frequencies are known from tidal theory (see Snyder et al, 1979; Jamart and Winter, 1980).

$$\eta(\underset{\sim}{x},t) = \eta_o(\underset{\sim}{x}) + \tfrac{1}{2}\sum \eta_n(\underset{\sim}{x}) \exp(i\omega_n t) \tag{1}$$

$$\underset{\sim}{u}(\underset{\sim}{x},t) = \underset{\sim}{u}_o(\underset{\sim}{x}) + \tfrac{1}{2}\sum \underset{\sim}{u}_n(\underset{\sim}{x}) \exp(i\omega_n t) \tag{2}$$

where η is sea level measured from its mean, and $\underset{\sim}{u}$ is velocity in the x,y plane. These expansions are introduced into the primitive shallow water equations, and the various frequencies are separated by harmonic decomposition:

$$i\omega_n \eta_n + \nabla\cdot(H\underset{\sim}{u}_n) = W_n \tag{3}$$

$$i\omega_n \underset{\sim}{u}_n + \underset{\sim}{f}\times\underset{\sim}{u}_n + \tau\underset{\sim}{u}_n + g\nabla\eta_n = g\underset{\sim}{T}_n \tag{4}$$

where ω_n is the angular frequency of the nth tidal constituent, η_n and $\underset{\sim}{u}_n$ are the amplitudes of this constituent, $\underset{\sim}{f}$ is the Coriolis parameter, τ is the friction coefficient, g is gravitational acceleration, H is the mean-water depth, W_n involves correlations between η and $\underset{\sim}{u}$, and $\underset{\sim}{T}_n$ represents a tidal-generation term (see Snyder et al, 1979). The nonlinearities in this system are contained in τ, W_n, and $\underset{\sim}{T}_n$ and are responsible for generating tidal harmonics. For the test problems, τ = constant and W_n and $\underset{\sim}{T}_n$ are zero.

Eliminating $\underset{\sim}{u}_n$ from equations 3 and 4, one can derive an elliptic equation

$$i\omega_n \eta_n - \nabla \cdot \left\{ \left(\frac{gH}{q^2 + f^2} \right) \left[q(\nabla \eta_n - \underset{\sim}{T}_n) - \underset{\sim}{f} \times (\nabla \eta_n - \underset{\sim}{T}_n) \right] \right\} = W_n \quad (5)$$

where $q = i\omega_n + \tau$. Equations 5 and 4 are then used to solve for η and $\underset{\sim}{u}$. The resulting equations are uncoupled iteratively and solved sequentially.

A variational statement and finite-element approximation is constructed for equations 3 and 4, primitive equations, and for equations 4 and 5, wave equation (this method is denoted "wave equation" in keeping with terminology in the literature. In its time dependent form, equation 5 is a wave equation.) The elements used in the various tests are described in the next section.

There are two types of boundary conditions specified. One is a specification of the real and imaginary amplitudes for the nth harmonic at the open boundary. The other is the specification of zero flow normal to the solid boundaries. For the primitive equations, the latter condition is applied by removing one degree of freedom from the velocity such that only the tangential component appears in the solutions. For the wave equation, the zero-flow condition is a natural boundary condition to the variational statement.

The harmonic amplitudes are specified by removing equation 3 or 5 at the open boundary and replacing it with the boundary condition. Because the governing equation is not enforced at the boundary, the solution can be inaccurate, more so with lower order elements. However, there are methods to remove this problem. For the primitive equations, sea level can be specified as a boundary-forcing term in equation 4 (Walters and Cheng, 1980, p. 194). For the wave equation, Lynch (1984) describes a technique to enforce mass conserving boundary conditions.

TEST PROBLEMS

Each of the numerical models was examined following a three-part test sequence. First, solutions were found for a single constituent in a rectangular basin of constant depth in order to verify that the computer code was working properly. Next, solutions were found for the rectangular basin with quadratic bathymetry as described in Gray and Lynch (1979). Most of these results are reported in Walters (1983) and complement the results of Gray and Lynch (1979) and Lynch and Gray (1979). In general, I found that short-wavelength oscillations were present in most solutions when using elements with equal-order interpolation (quadratic) and the primitive shallow water equations. Mixed-interpolation (quadratic velocity, linear sea level) showed improved results for sea level and worse results for velocity. Wave equation formulations performed best, with some discrepancies in velocity, a point to be brought up in the results section.

The third test sequence is presented here. This network is the polar quadrant with quadratic bathymetry (Gray and Lynch, 1979) as shown in Figure 1. This network contains 63 nodes with a radial grid spacing dr of 15,240 m (5×10^4 ft), an inner radius of 4 dr and an outer radius of 10 dr. The friction coefficient τ was taken to be $10^{-4} s^{-1}$, and a sea-level amplitude of .03048 m (.1 ft) with a period of 12.4 hours was specified on the outer radius. The remaining boundaries have the normal component of velocity specified as zero.

Two methods were used to discretize the network into triangular elements. The first network (P1) has a uniform distribution of elements. The second (P2) has a configuration where alternately 4 and 8 triangles meet at the vertices. Five separate combinations of governing equations were tested. These are primitive equations with quadratic equal-order elements (PQ), with linear equal-order elements (PL), and with mixed-interpolation elements having quadratic velocity and linear sea level (PM); wave equation with quadratic elements (WQ), and with linear elements (WL). Two of the solutions above (PL/P2 and WL/P2) are reported in the literature (Lynch and Gray, 1979; Gray and Lynch, 1979). The results I present here complement these earlier results by including higher-order elements and mixed-interpolation elements, and by considering an alternate network configuration. The latter is prompted by observations that certain element layouts are very sensitive to short-wavelength noise.

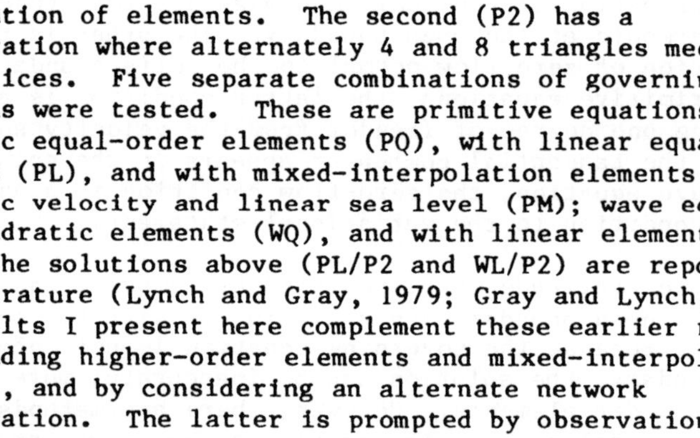

Figure 1. Polar networks: (a) P1 network with quadratic elements, (b) P2 network with linear elements.

TEST RESULTS

The analytical solution for the test problem may be found in Lynch and Gray (1978). Using this solution and the numerical solutions, I have computed the root-mean-square errors for the various solution methods (Table 1). The sea level values are in units of 10^{-3} m, and the velocity values are in units of 10^{-3} m/s. To normalize the errors, divide by the rms values of the dependent variables at the top of the table.

In general, the most accurate results were obtained with the

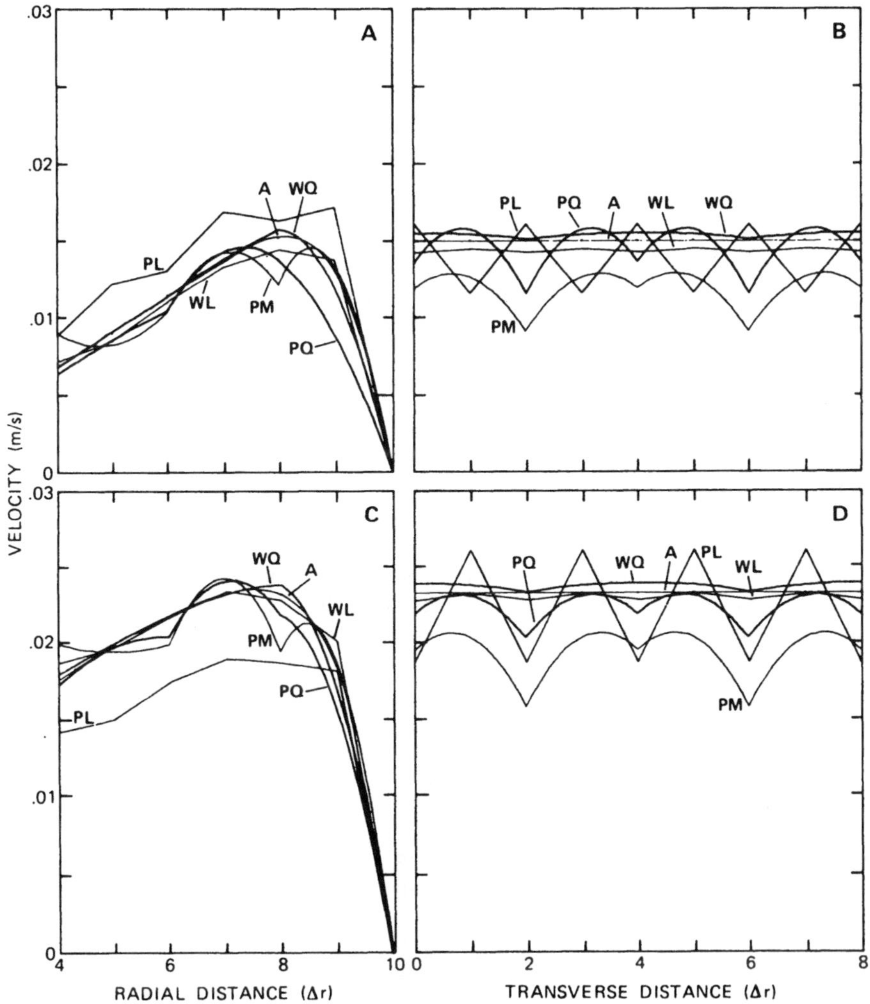

Figure 2. Results for the polar network: (A) real component of velocity along centerline of network, (B) real component along transverse 2 dr in from inner radius, (C) imaginary component of velocity along centerline, and (D) imaginary component along transverse.

wave equation methods (Figure 2). The primitive equation methods were generally noisy and the accuracy decreased as the order of the element (PQ, PM, PL). The results for the irregular network (P2) were poor for the linear element (PL) which had very large internode oscillations. In addition, the configuration P1 seems to excite the spurious mode for PQ.

The results can be interpreted in terms of the form of the dispersion relation for the discretized equations. The key idea is the existence of a folded dispersion relation for some of these equations (Platzman, 1981). Consider the plot of the dispersion relation (the relation between ω, the angular frequency, and k, the wavenumber) for various elements (Figure 3). In the continuum, $\omega = ck$ where c is the phase speed for shallow-water waves (this is the straight line marked A). For elements that use the same node for velocity and sea level, the phase speed decreases to zero at a wavelength equal to twice the grid spacing, an artifact caused by the finite discretization. Thus, with a particular forcing frequency, there are two wavelengths that are excited (the long "real" wave, and a short "spurious" wave). The partitioning of energy between these modes is found from their relative admittances. For irregular grids, there is substantial forcing of the short-wave component because of small-scale variations in the network.

All the primitive equation methods have folded dispersion relations such that there is a short-wavelength aliased solution. The polar quadrant with quadratic bathymetry has sufficient small-scale variations to excite the spurious modes directly, independently from any spurious mode-resonances as discussed by Platzman (1981). Note in particular PL/P2 and PQ/P1. In the former, there is an

TABLE 1

Root-mean-square errors (x 10^{-3})

Method/Network	real	imag	Ureal	Uimag
RMS	39.33	20.36	10.88	19.33
PL/P1	0.256	1.838	2.232	2.197
PM/P1	0.327	0.700	2.141	2.638
PQ/P1	2.932	1.406	0.990	1.483
WL/P1	0.168	0.542	0.502	0.490
WQ/P1	0.040	0.098	0.398	0.496
PL/P2	9.863	1.325	2.755	4.250
PM/P2	0.312	0.693	1.834	2.203
PQ/P2	0.256	1.345	1.464	1.127
WL/P2	0.137	0.528	0.522	0.568
WQ/P2	0.042	0.121	0.384	0.485

alternating 4 and 8 element support at the corner nodes, a configuration that is spatially the same as the spurious oscillation mode. Numerical experience has shown that in these cases, the spurious mode is excited, as may be seen in Figure 2 (PL) and Table 1 (column 1, PL/P2). In the case of PQ/P1, all the corner nodes have the same 6 element support, while the midside nodes have 2 element support. Again this network irregularity has the same spatial layout as the spurious mode and the mode is excited (Table 1, column 1, PQ/P1). Using mixed-interpolation, the sea level mode is removed, unfortunately, at the expense of the velocity solution (compare PM and PQ).

The wave equation methods do not contain aliased modes and, hence, display lower noise levels. The disperson relation is similar to LC in figure 3 where there is no folding. The errors are then due to the inadequate resolution of the main mode because of the finite-spatial discretization (Figure 2).

As yet, there are some unresolved questions pertaining to how velocity is calculated when using the wave equation formulation. In practice, sea level is calculated using equation 5. Following this, velocity is calculated from equation 4. The problem which arises is that velocity depends upon the gradient of sea level, a lower-degree interpolation. With the quadratic elements (WQ), there is a tendency for the velocity at the midside node to be a linear interpolation of the velocities at the adjacent corner nodes. This behavior is seen in the test results for the

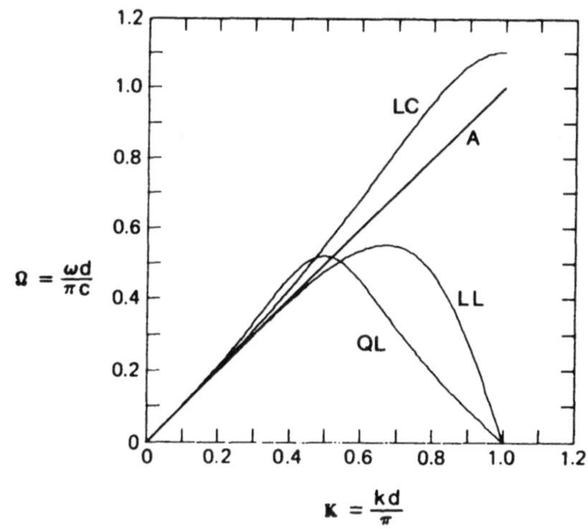

Figure 3. One-dimensional dispersion relation for various schemes: A continuum: LL, element with linear velocity and linear sea level; LC, element with linear velocity and piecewise constant sea level; QL, element with quadratic velocity (three nodes) and linear sea level (two end nodes).

rectangular basin (Walters, 1983; Figures 4 and 5) and for the polar quadrant (Figure 2 herein). For the linear elements, sea level gradient is interpolated as a piecewise constant on each element. One method is also to interpolate velocity as a piecewise constant function. However, for my purposes, both velocity and sea level need to be specified at the same nodes. Hence, I have chosen to solve the finite element system derived from equation 4. This method effectively interpolates velocity back to piecewise-linear functions at the nodes. The net result is that linear elements can better represent small-scale variations in the network (for an equivalent number of nodes). Unfortunately, the rate of convergence as a function of mesh size is lower for the linear elements than for the quadratic elements.

FURTHER DEVELOPMENTS

On the basis of the test sequence, WQ and WL were chosen to examine the tidal circulation in a field-scale network -- that for South San Francisco Bay (Figure 4).

A network describing South San Francisco Bay was generated using six-node triangular elements for WQ, and each element was subdivided into four linear elements for WL. This network contains 111 elements (WQ) and 270 nodes. Sea level is specified on the open boundary across the upper edge of the spatial domain. The solid boundaries are approximated by C^1 continuous piecewise polynomials (WQ) or C^0 piecewise-linear functions (WL). The normal velocity is zero as a natural boundary condition to the variational problem.

The difficulty of the South Bay network is due to a deep channel (depth = 15m) which lies in the western part of the bay (mean depth = 2m). The scheme WQ failed to represent

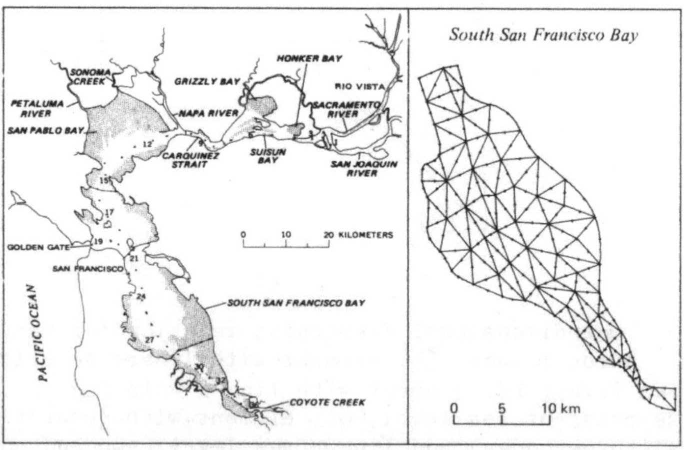

Figure 4. Index map for San Francisco Bay (left). Finite element network (right).

the high velocity associated with the channel. The reason for this is as follows: At this level of refinement there is one quadratic element across the channel. For reasons explained in the last section, the velocity is constrained to be "almost" linear on the element. Thus the high velocity at the midside node is reduced. However, the case is different for the scheme WL. Numerical results show that this scheme reproduces the larger channel velocities which are about 50% larger than those using WQ. Sample results (WL) are presented in Figure 5 for the real (inphase with sea level forcing) and imaginary (90° phase lag) amplitude of the M_2 (lunar semidiurnal) tidal constituent.

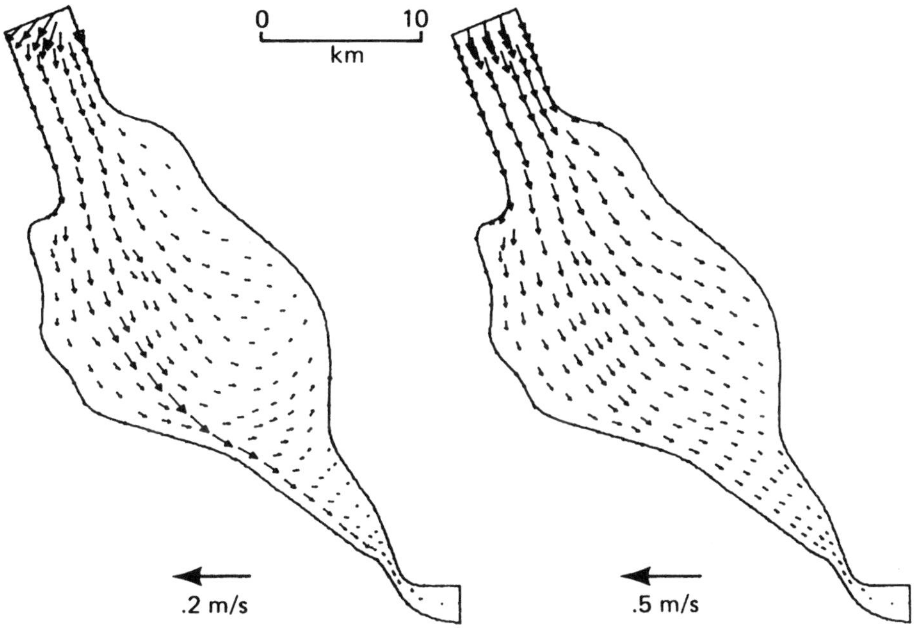

Figure 5. Results for South Bay using WL: (a) real (in phase with sea level at the open boundary) component of velocity, (b) imaginary (90° phase lag) component of velocity.

CONCLUSIONS

The harmonic decomposition of the governing equations is a very efficient way to solve for tidal-period variations, including harmonics. The wave equation schemes (WQ, WL) have much lower noise levels and, hence, greater accuracy than the primitive equation schemes (PQ, PM, PL). However, there is a problem with WQ related to the lower-order approximation for velocity. Although WL converges at a lower rate as a function of mesh size, it is better able to represent small-scale variations in the network.

REFERENCES

Gray, W.G. (1982) Some inadequacies of finite-element models as simulators of two-dimensional circulation. Advances in Water Resources, 5:171-177.

Gray, W.G. and D.R. Lynch (1979) On the control of noise in finite element tidal computations: a semi-implicit approach. Computers and Fluids, 7:47-67

Jamart, B.M. and D.F. Winter (1980) Finite element computations of the barotropic tides in Knight Inlet, British Colombia. Fjord Oceanography, Freeland et al., eds., Plenum Press, New York:283-289

Lynch, D.R. (1984) Mass balance in shallow water simulations. Proc. Fifth Int. Symp. on Finite Element in Flow Problems. January 23-26, 1984, University of Texas of Austin, Austin, Texas:1-5.

Lynch, D.R. and W.G. Gray (1978) Analytic solutions for computer flow model testing. Journal of the Hydraulics Division, ASCE, 104, HV 10:1409-1428.

Lynch, D.R. and W.G. Gray (1979) A wave equation model for finite element tidal computations. Computers and Fluids, 7:207-228.

Platzman, G.W. (1981) Some response characteristics of finite-element tidal models. J. Comput. Phys., 40:36-63.

Snyder, R.L., M. Sidjabat, and J.H. Filloux (1979) A study of tides, setup and bottom friction in a shallow semi-enclosed Basin. Part II: Tidal model and comparisons with data. J. Phys. Oceanogr., 9:170-188.

Walters, R.A. (1983) Numerically induced oscillations in finite element approximations to the shallow water equations. Int. J. Num. Meth. in Fluids, 3:591-604.

Walters, R.A. and G.F. Carey (1983) Analysis of spurious oscillation modes for the shallow water and Navier-Stokes equations. Computers and Fluids, 11:51-68.

Walters, R.A. and R.T. Cheng (1980) Accuracy of an estuarine hydrodynamic model using smooth elements. Water Resources Research, 16:187-195.

NESTED MODELS: APPLICATIONS TO PRACTICAL PROBLEMS

Verboom G.K., de Vriend H.J., Akkerman G.J., Thabet R.A.H., Winterwerp J.C.
Delft Hydraulics Laboratory, Delft, The Netherlands

SUMMARY

Nested models are used in hydraulics to solve practical problems upto very fine scales. If the detail model has several open boundaries it can be a difficult task to calibrate the model. In this paper several applications are discussed and some guidelines are formulated by which different kinds of problems can be solved along the same lines.

INTRODUCTION

To solve hydraulic engineering problems one can use hydraulic scale models and mathematical models. For many years the latter were mainly used for preliminary (flow) information, whereas details of the flow were taken from hydraulic scale models. In recent years, however, mathematical models are applied to smaller and smaller scale problems. Mathematical solution procedures and the formulation of certain physical phenomena have been improved sofar that even such difficult flow problems as recirculating flow can be computed with an accuracy as required by practical applications.
Small scale applications inevitably means a small grid increment in the numerical solution procedure: for this one can use a local grid refinement technique or a series of nested models. A model is called nested if it covers just a part of the problem area and if it gets its boundary data from a model that covers a larger area. At the Delft Hydraulics Laboratory we have chosen for nested models because one can use the existing codes with only minor adjustment to facilitate easier data transfer and because one can use full storage capacity for the detail model. For solving nearly horizontal vertically averaged flow DHL has the disposal of several program systems based on finite difference or finite element methods. For intermediate and large scale (tidal) flow computations we use the program system WAQUA, an implicit finite difference program; for full details the reader is referred to Stelling

1983. For smaller scale unsteady flow computations, aiming at providing boundary conditions for e.g. morphological or nautical studies, we use (again) WAQUA or ODYSSEE, Alfrink, et al 1983. For small scale steady flow computations use can be made in addition-of the finite element program FRIMO.
When nested models are used certain problems that exist in local mesh refined models vanish, but others remain and new ones are created. In detail models in- and outflow boundaries are only a short distance apart, dissipation by bottom friction is rather inactive, as a result of which disturbances included in the boundary conditions may stay for quite a long time in the area if no special measures are taken. In addition, it is not obvious what kind of boundary conditions to use and in what kind of combination if various open boundaries are present. Another problem concerns the topography: can it be taken from the coarse grid model by some kind of interpolation or must we use the original data; must we change the boundary conditions if the cross-sections in the two grids differ substantially and if so in what sense?
These questions are dealt with in this paper for flow and morphological problems. Results are given for four practical applications, two of which are discussed in more detail: a harbour approach channel and a cooling water in- and outlet system. No generally valid rules can be given, but we discuss some guidelines by which different kinds of problems can be solved succesfully along the same lines.

NESTED MODELS, DISCUSSION OF PROBLEMS

Nested models are used to get detailed information in areas where this is required without having to bother about the surroundings. Unfortunately, boundary conditions for a detail model can not be obtained from field measurements as shown by the following figures: for a progressive (tidal) wave in a typical coastal area the longitudinal surface gradient and phase difference are given by $\Delta\zeta = \omega\hat{u}\, L/g$ and $\Delta\phi = L/\sqrt{gH}$, where ω is the tidal frequency, $\hat{u}$ is the maximum velocity, g is the gravitational acceleration and H is the depth. With $\hat{u}$ = 1 m/s, L = 4 km, and H = 10 m one finds $\Delta\zeta$ = 0.06 m and $\Delta\phi$ = 400 s. For a reliable computation these quantities must be known with an inaccuracy that is significantly less: obviously, this can not be accomplished under general prototype conditions. Large scale computations even becomes inevitable if the overall water motion changes due to the human intervenence under study.
If the boundary conditions for the detail model are taken from the result of a coarse grid model we might expect the accuracy of the main flow in the detail model to be of the same order as in the coarse grid model. The main advantage of the nesting procedure is that the boundary data for the detail mnodel are internally consistent, i.e. surface elevation and discharges satisfy the continuity equations (for mass and constituents)

exactly. One might expect that only minor adjustments are necessary to calibrate the detail model. From literature, Boulot 1980, and from our own experience we know this not to be true in general, but adjustments can be kept to a minimum if certain guidelines are taken into account. Though the accuracy of the main flow of the detail model is not improved, more details can be represented with the finer grid: thus accuracy is increased but more in a general sense. To value these details, general information of the flow in prototype is often more useful than point measurements. Float tracks and discharge distributions are very useful in this respect though the latter is difficult to measure accurately.
Given the coarse grid model, where to apply the boundaries of the detail model, how to generate the topography and what kind of boundary conditions are best suited for the various open boundaries? These questions are discussed in the following sections.

Layout of the detail model

The dimensions of the detail model are quite often to be determined from a set of conflicting demands. The minimal overall dimensions are determined by the engineering problem under study and the physical phenomenon therein. The grid size must be tuned to the smallest relevant topographic details and should be kept below one half of the topographic length scale. Unfortunately, the resulting total number of grid points is not seldomly too large; not so much because of the available computer capacity, but because of budget constrains. In such a case, it is preferred to apply an intermediate model and a smaller detail model than just to increase the grid size of the latter. As the computer costs are - roughly speaking - inversily proportional to the third power of the grid size, this can be a feasible alternative. Given the dimensions of the detail model, can one generate the topography by some kind of interpolation scheme from the coarse grid information or should one start from the original data? Conservation of cross-section can be accomplished by (bi-) linear interpolation, but even for gentle slopes the topography soon becomes distorted. This leads us to the conclusion that the topography always must be based on the original data.
The schematization of local topography near in- and outflow boundaries might differ appreciably between the two grids. A section of one or two coarse grid intervals at these boundaries, in which the bottom profile is gradually adjusted, smoothens the transition between the two grids. This bottom adjustment zone decreases certain boundary condition problems.

Boundary conditions for detail models

In 1975 Van Der Kuur and Verboom studied the calibration of a small coastal area model with three open boundaries as a function of the type of boundary conditions used. The best combination of boundary conditions from a calibration point of view

was found to be waterlevel at one of the main in- or outflow boundaries and velocities at the other boundaries. Prescribing the waterlevel at an almost flow parallel boundary makes the flow field in the entire model rather sensitive to small changes in this boundary condition. In recent applications most of the detail models are substancially smaller than this coastal area model, but a similar combination of boundary conditions is often used to start with. However, the final combination not seldomly is different from the initial one.
In Figure 1A-D four detail models are sketched together with the main flow directions.

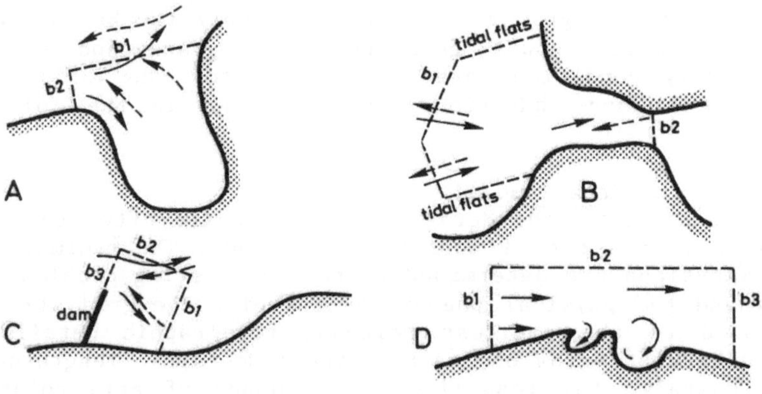

Figure 1 Four applications of nested models. (Only the detail models are shown)

Model 1A is an estuary-type with a scaling factor (coarse - fine grid) of four; the bottom is rather flat and the boundary conditions used are waterlevel at b1 and waterlevel or velocity at b2. The results didn't depend on which combination was used and only minor adjustments were required to trim the model; no bottom adjustment zone was applied. Model 1B pertains to the flow in an estuary with deep channels and large tidal flats, the scaling factor is about 3.5. The channels are crossed normal to their axis by the open boundaries and a bottom adjustment zone of one coarse grid interval is applied. Boundary b1 is waterlevel and b2 is velocity controlled. No adjustments were required at all to trim the model. Model 1C represents a coastal area with a scaling factor of five. The bottom slope is rather mild and no bottom adjustment zone was applied. We started with a ζ-ζ-u combination for b1, b2, and b3, respectively, but major adjustments were required to represent the flow across b1 correctly. With an u-ζ-u combination, however, no adjustments were required at all. Models 1B and 1C are discussed in more detail in the following section. The fourth and last model, 1D, pertains to a very small coastal area in an estuary entrance. The grid size is only 45 m and the boundary data is taken from an 400 m grid model in two

steps. The boundary conditions used are u-u-ζ and only minor adjustments were necessary to trim the model. Normal to the coast there is a steep bottom slope that covers about half the width; no bottom adjustment zone was applied.
From these applications we concluded that a combination of waterlevel and velocity at the main in- and outflow boundaries is a good first choice. However, for some cases the flow near the waterlevel controlled boundary is very sensitive to small changes in this boundary condition or on the bottom profile at this boundary: velocities at all main in- and outflow boundaries are then to be preferred. To prevent uncontrolled waterlevel run-away, this quantity must be prescribed at some part of one of the open boundaries. A bottom adjustment zone was applied with good result in one rather extreme case, but more applications are required before a definite statement can be made. We have met a similar kind of problems - though not that extreme - as reported by Boulot 1980. Instead of a velocity correction factor we solved the problems by changing the type of boundary conditions used.
A final remark concerns the use of weakly reflective boundary conditions to get rid of short wave components. In detail models bottom friction is rather inactive to dissipate unwanted wave components, such as eigenfrequencies and disturbances from the initial condition. Following a suggestion of Stelling 1983 we perturbed the boundary conditions in such a way that they become weakly reflective for short wave components only. A velocity is now prescribed as

$$u + \alpha \, Ri_t = f(t),$$

where Ri_t is the time derivative of the ingoing Riemann-invariant and α is a constant. For a linear 1-D problem with u-ζ boundary conditions the reflection coefficient can be shown to be given by

$$R = 1/\sqrt{1 + (4\pi\delta/T)^2}$$

where T is the wave period. The reflection coefficient is about zero for short waves and about one for long waves. Short wave components indeed disappear within a few transition times. To accomplish this for tidal waves, weakly reflective boundary conditions are now being implemented along the lines as proposed by Verboom, this conference.

Morphological computations

When a practical problem in an estuary or a coastal area involves morphological aspects, morphology is often included in the detail model only. This introduces specific questions, in addition to those mentioned before.
In the simplest case, the detail flow is based on the same computer program as the overall model, the sediment transport is evaluated using a local transport formula and only the initial rate of accretion/erosion of the bottom is determined. Then there are no additional nesting problems and the only point of attention is, that the flow and sediment transport computations are accurate enough to calculate the accretion/

erosion rate of the bottom.
Especially in coastal applications, short waves and wave-induced currents can play an important part in (relatively small scale) morphological processes. For practical reasons introduction of wave effects into the tidal flow model WAQUA is not quite feasible yet. Therefore, steady flow models based on FRIMO or ODYSSEE have to be applied in this case, which introduces a number of nesting problems, such as:

- working with other programs and essentially different computational grids than in the overall model requires complicated interface-programs;
- schematization of the tidal cycle into a series of steady state situations;
- correction of the boundary conditions of the tidal flow to account for the lack of storage in the detail flow domain;
- if the flow induced by the vertical motion of the water surface in the detail domain is important: derivation of the vertical velocity of the water surface from the overall model results and formulation of the boundary conditions for the corresponding flow components;
- combination with a third model, describing the propagation and dissipation of short waves in the detail model area;
- formulation of boundary conditions for the wave-driven currents (cannot be derived from the overall tidal model);
- interaction between wave effects in the detail model area and the flow pattern in the overall model (via an enhanced bottom friction, for instance).

Except for the last one, these problems have been solved succesfully for the combination WAQUA-FRIMO. The combination WAQUA-ODYSSEE, however is more complicated and needs further development in this respect.
If the sediment concentration is determined not only by local quantities, but also by horizontal interaction processes like convection and diffusion, a convection-diffusion equation has to be solved to determine the sediment transport rate. The solution of this equation requires boundary conditions for the sediment concentration. If the overall model concerns only the flow, these conditions will have to be derived from field measurements or from assumptions on the transport of sediment in the overall model.
If the long-term aggregation and degradation of the bottom have to be described, the morphological changes have to be described as a time-dependent process (though usually on a time scale that is substantially longer than the tidal period). In that case, also the bed level computation requires boundary conditions, which have to be formulated with caution in order to keep spurious boundary disturbances from penetrating into the area of interest. On the other hand, bottom disturbances leaving the detail model area can give rise to discontinuities in the bottom level at the boundaries with the overall model and hence to special nesting problems. Experience with time-dependent morphological models in two horizon-

tal dimensions, either nested or not, is still very limited and effective guidelines to deal with this kind of problems cannot be given yet.

APPLICATIONS OF NESTED MODELS

An approach channel study

The Asan Bay Approach Channel Study concerns the influence of an approach channel to a LNG-terminal on the flow and sediment transport in the bay and particularly the sedimentation of the particularly channel itself. Asan Bay, Korea, and its approaches form an average sized estuary with a length of about 50 km. It strongly converges up to a narrow passage with a width of about 2 km. The terminal site is located just southeastwards of this passage. The bathymetry shows two relatively deep channels separated by a partly drying bank and is bordered by large tidal flats. The partly drying bank is to be intersected by the approach channel. The overall model, Δx=750 m and Δt = 300 s (Cr = 8) is shown in Figure 2. The tide in Asan Bay is mainly semi-diurnal with a small daily inequality; the main spring tidal range is about 8 m with a maximum current velocity of about 1.2 m/s. The spring tide maximum flood current pattern for the overall model is shown in Fig. 3.

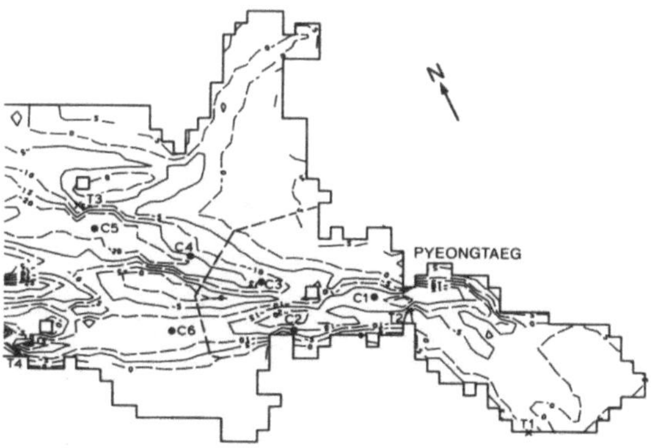

Figure 2 Overall model Asan Bay

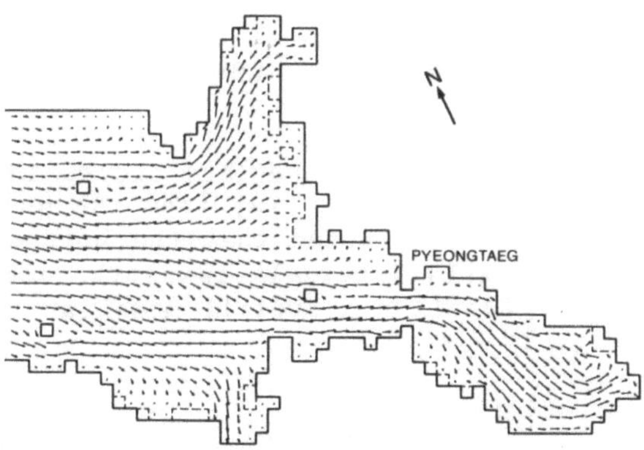

Figure 3 Current pattern spring tide maximum flood.

The detail model covers an area of about 13 x 13 km^2, Fig. 2; this relatively large extent was necessary in view with a proper modelling of the confluence of the two deep channels. The buckled seaward boundary is waterlevel controlled and the east boundary is velocity controlled. Bottom schematization has been derived from available bathymetric charts.
A bottom adjustment zone of one coarse grid interval is applied along the boundaries of the detail model. For the detail model a grid size of 200 m and a time step of 150 s (Cr = 12) is used. The overall model was calibrated by adapting the boundary conditions (field data) and the bottom roughness. The boundary conditions for the detail model were taken from the coarse grid model by bi-linear interpolation; no further calibration was necessary. The current pattern of the detail model for slack and maximum velocities are given in Figs. 4A-D.

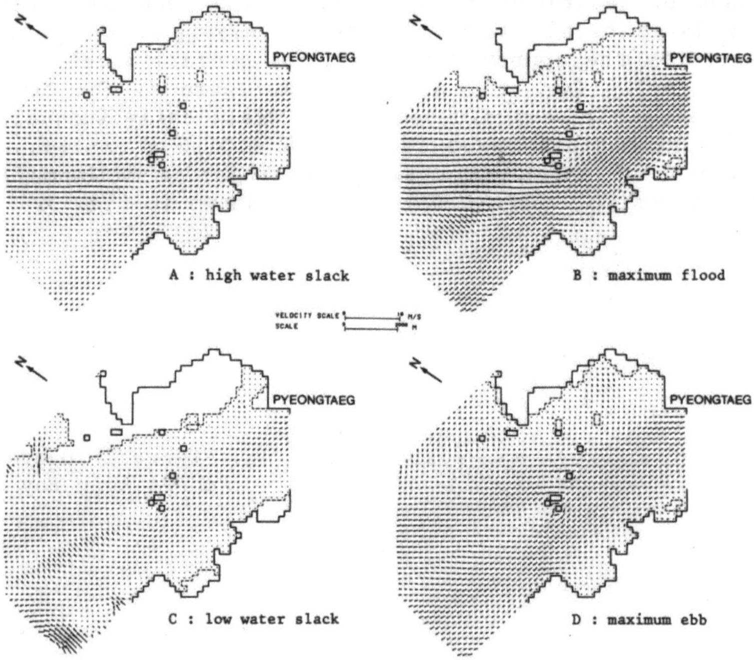

Figure 4 Detail model spring tide current pattern.
Velocity scale: 1 Δx = 1 m/s

At low water slack an unrealistic flow pattern occurs in the west corner, but it is sufficiently far away from the problem area to be irrelevant. It is to be noted that the bottom adjustment zone of three fine grid intervals is still included in Fig. 4. A comparison of detail and overall model at three locations is given in Fig. 5.

A cooling water in- and outlet system

For a new steam power and desalination complex at the west coast of the United Arab Emirates, a cooling water intake

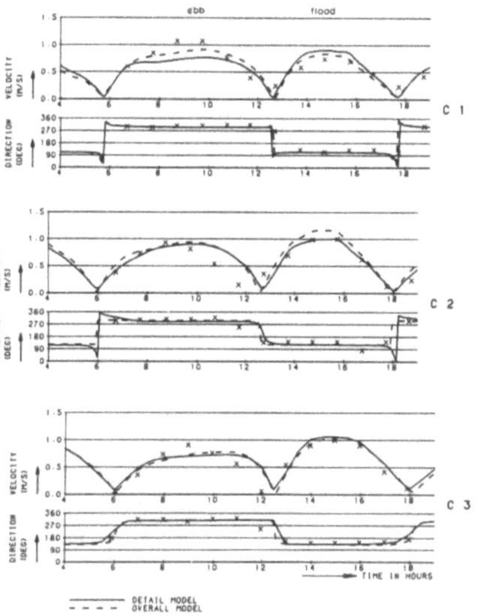

Figure 5 Comparison detail and overall model at three velocity locations.

channel must be designed. The study involved recirculation and sedimentation aspects. The bathymetry is rather flat with a mild slope normal to the shore line. The tide is mainly semi-diurnal with an important diurnal component; the tidal range is just over 1 m and the maximum velocity is about 0.5 m/s. The overall model, Δx = 500 m and Δt = 150 s (Cr = 4.5), covers an area of about 17 x 8 km^2, Fig. 6.

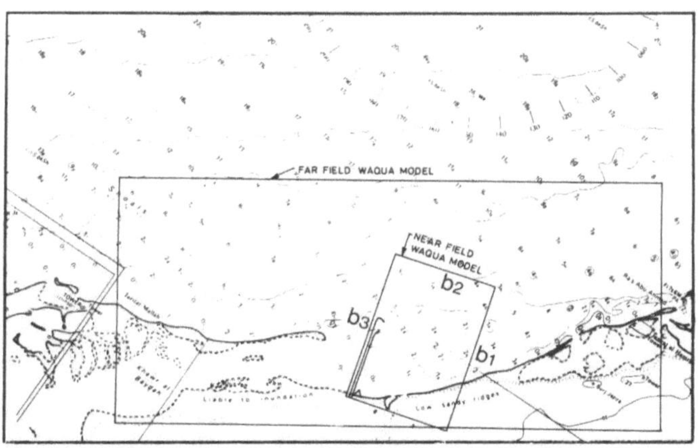

Figure 6 Lay-out overall and detail model

The current pattern in the overall model at about maximum ebb is given in Fig. 7A. The flow in the detail model with a boundary condition combination of ζ-ζ-u is given in Fig. 7B.

Because of the mild bottom slope no bottom adjustment zone is applied. As can be seen from Fig. 7B the flow at b1, water-level controlled, hardly resembles the flow in the overall

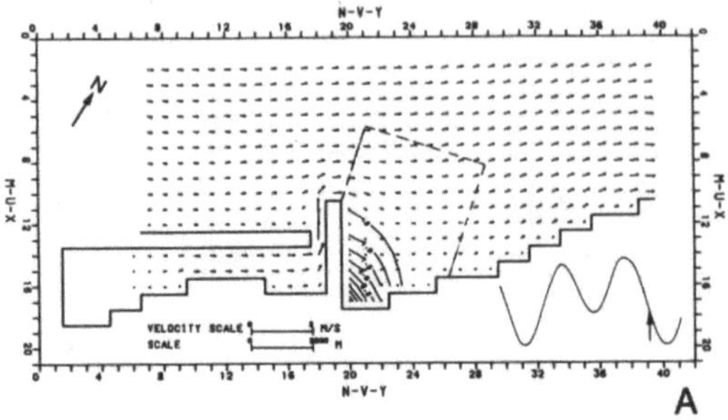

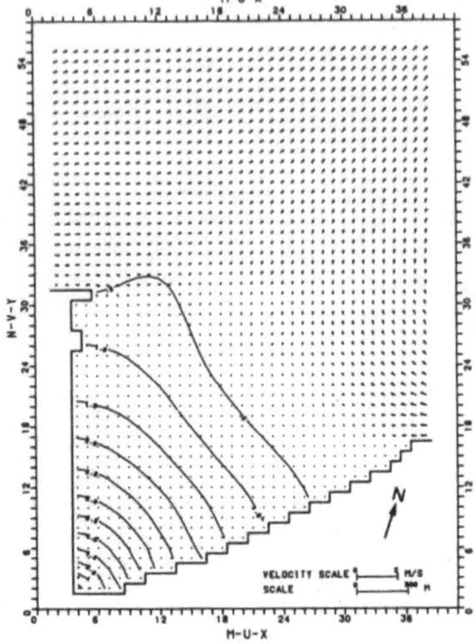

Figure 7 Maximum ebb current pattern in overall model (A) and in detail model (B) with ζ-ζ-u boundary conditions

B

model. If however, the boundary condition at bl is changed to velocity no further calibration is required at all, Fig. 8A. In Fig. 8B the time histories are compared at two velocity check points.

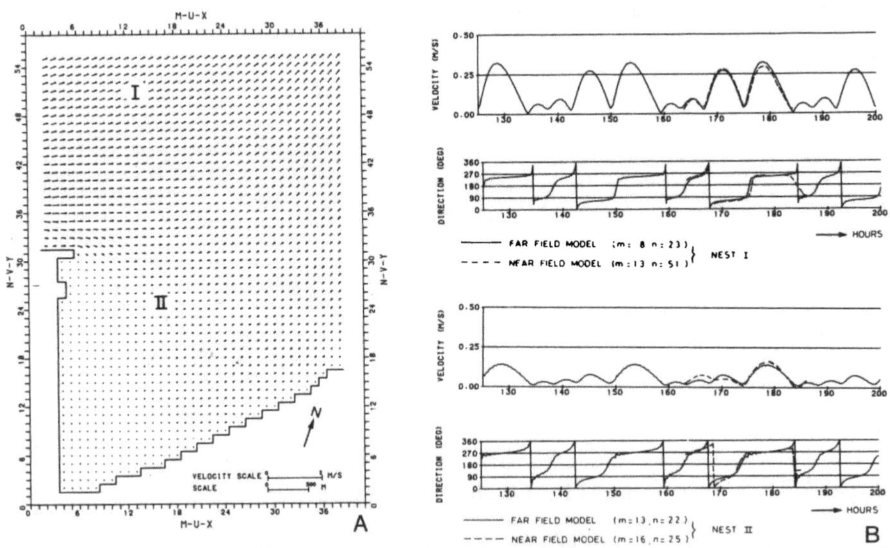

Figure 8 Maximum ebb current pattern in detail model with u-ζ-u boundary condition (A) and comparison between overall and detail model at two velocity checkpoints (B).

CONCLUSION

Nested models are applied succesfully to solve detailed flow problems in hydraulic engineering. The amount of work required to trim the detail model (strongly) depends on the combination of boundary conditions used. A bottom adjustment zone at the open boundaries of the detail model decreases this dependence. Short wave components and eigenfrequencies are succesfully suppressed by using a special kind of weakly reflective boundary condition. Several aspects are discussed related to nested morphological computations.

REFERENCES

Alfrink, B.J., Officier, M.J., Vreugdenhil, C.B., Wind, H.G. (1983) Applications in hydraulics of a curvilinear finite-difference method for the Navier-Stokes equations. Proc. Int. Conf. Num. Methods in Laminar and Turbulent Flow, Seattle.

Boulot, F. (1980) Modeling of heated water discharges on the French coast of the Englisch Channel. In "Transport Models for Inland and Coastal Waters". Proc.Symp. on Predictive Ability, (Ed. H.B. Fisher), Berkeley.

van der Kuur, P. and Verboom, G.K. (1975) Computational analysis for optinal boundary control of twodimensional tidal models. Proc. 16th IAHR Congress, Sao Paulo.

Stelling, G.S. (1983) On the construction of computational methods for shallow water flow problems. Ph. D. Thesis Delft University of Technology.

Verboom, G.K. (1984) Weakly-reflective boundary conditions for two dimensional shallow water flow problems. This conference.

A MODIFIED WAVE EQUATION MODEL FOR 3D FLOW IN SHALLOW BODIES OF WATER

J.P. LAIBLE, Department of Civil/Mechanical Engineering, University of Vermont, Burlington, Vermont, USA

INTRODUCTION

This paper presents a numerical model of a set of equations which represent the depth varying nearly horizontal flow conditions observed in shallow bodies of water. Flows of this type may be due to wind, various seiches of adjacent bodies, tidal action, or river inflows. Since shallow regions are commonly the initial recipient of pollutants and often the most severly affected, a realistic depiction of the physical condition, is an important component of an environmental study. Numerical modeling of three dimensional flows can be used to assess exchange rates and resident times between shallow bays and adjacent lakes or seas, or serve as a basis for direct modeling of pollutant transport.

The key features of the model presented here are: 1) a wave equation type formulation is used to control numerical noise; 2) the vertical variation of flow over the depth is directly obtained using Hermitian polynomial approximations in the vertical direction; 3) full advantage of the "integral lumping " integration technique is used to avoid the formation of large system matrices; 4) isoparametric quadralaterals are used for the horizontal discretization to facilitate the modeling of curved boundaries; 5) a combination of finite element based equations on the boundary and finite difference equations on an irregular grid in the interior can be naturally generated using the nodal integration technique and the quadralateral elements. The major objective of this paper is to describe the aspects of the model pertinent to the wave equation scheme in the context of a 3D description of the flow field. Details of features 1,3 and 4 as originally developed by Gray and Lynch for the shallow water equations can be found in Gray(1976), Gray and Lynch (1977), and Lynch and Gray (1979).

THE GOVERNING EQUATIONS AND VERTICAL INTEGRATION

The flow in a shallow body of water is assumed to be governed by two horizontal momentum equations and a continuity equation.

$$\frac{\partial u}{\partial t} + u\frac{\partial u}{\partial x} + v\frac{\partial u}{\partial y} - fv + g\frac{\partial \zeta}{\partial x} - \frac{\partial}{\partial z}\left(\varepsilon_z \frac{\partial u}{\partial z}\right) = 0 \qquad (1)$$

$$\frac{\partial v}{\partial t} + u\frac{\partial v}{\partial x} + v\frac{\partial v}{\partial y} + fu + g\frac{\partial \zeta}{\partial y} - \frac{\partial}{\partial z}\left(\varepsilon_z \frac{\partial v}{\partial z}\right) = 0 \qquad (2)$$

$$\frac{\partial \zeta}{\partial t} + \frac{\partial}{\partial x}\int_{-h}^{\zeta} u \, dz + \frac{\partial}{\partial y}\int_{-h}^{\zeta} v \, dz = 0 \qquad (3)$$

in which f is the Coriolis parameter, $\varepsilon(z)$ the variable eddy viscosity coefficient, g is gravity and all other notations are as defined in figure 1. These equations serve as the basis for deriving the shallow water equations. In the model presented here these equations are not analytically integrated over the depth but are first treated by the Galerkin method in the vertical direction to obtain a set of z independent equations. The variables in the solution however define the variation of flow over the depth. The variables for the x direction, X_1 and X_2, are the surface and bottom velocities while X_3 and X_4 are the gradients multiplied by the depth. Using variables of the type X_3 and X_4 , it is possible to formulate the required matrices and tensor products in a nondimensional form, reducing storeage requirements for programming. With the analogous set of variables for the y direction and the surface elevation there are a total of nine unknowns at each node. The description of the flow field over the depth is defined by the nondimensional Hermitian polynomials.

$$\tilde{V} \simeq \bar{N}_i \tilde{Y}_i \quad \bar{N}_2 = 2s^3 + 3s^2 \quad \bar{N}_4 = s^3 + 2s^2 + s \qquad (4)$$

$$\tilde{U} \simeq \bar{N}_i \tilde{X}_i \quad \bar{N}_1 = -2s^3 - 3s^2 + 1 \quad \bar{N}_3 = s^3 + s^2$$

Applying the Galerkin method over the vertical dimension with the boundary conditions

$$\varepsilon_z(o)\left(\frac{\partial\tilde{u}}{\partial z}\right)_o = C_w W_x^2 \qquad \varepsilon_z(o)\left(\frac{\partial\tilde{v}}{\partial z}\right)_o = C_w W_y^2$$

$$\varepsilon_z(-h)\left(\frac{\partial\tilde{u}}{\partial z}\right)_{-h} = C_b\tilde{U}_b \qquad \varepsilon_z(-h)\left(\frac{\partial\tilde{v}}{\partial z}\right) = C_b\tilde{V}_b \tag{5}$$

one obtains (using integration by parts for the eddy viscosity term)

$$[\bar{\alpha}]\{\frac{\partial X}{\partial t}\} + \frac{\varepsilon}{h^2}[\overline{\kappa\lambda}]\{\tilde{X}\} - f[\bar{\alpha}]\{\tilde{Y}\} - \frac{1}{h}\{\tau_{wx}\} + \frac{1}{h}\{C_x\} + g\{\bar{\beta}\}\frac{\partial\tilde{\zeta}}{\partial x} = 0 \tag{6}$$

$$[\bar{\alpha}]\{\frac{\partial Y}{\partial t}\} + \frac{\varepsilon}{h^2}[\overline{\kappa\gamma}]\{\tilde{Y}\} + f[\bar{\alpha}]\tilde{X} - \frac{1}{h}\{\tau_{wy}\} + \frac{1}{h}\{C_y\} + g\{\bar{\beta}\}\frac{\partial\tilde{\zeta}}{\partial y} = 0 \tag{7}$$

$$\frac{\partial\tilde{\zeta}}{\partial t} + \frac{\partial}{\partial x}(h\llcorner\bar{\beta}\lrcorner\{\tilde{X}\}) + \frac{\partial}{\partial y}(h\llcorner\bar{\beta}\lrcorner\{\tilde{Y}\}) = 0 \tag{8}$$

in which C_w is the wind shear coefficient, C_b is the bottom friction parameter, W_x and W_y the x and y components of the wind velocity. The remaining vectors and tensor products are defined in the appendix. The bottom friction term is incorporated into the dissipation matrix $[\overline{\kappa\gamma}]$ via the term $\gamma = C_b h/\varepsilon$, in which ε is some base eddy viscosity used in the term $\varepsilon(s)=\varepsilon P_\varepsilon(s)$. The $[\overline{\kappa\gamma}]$ matrix is

$$[\overline{\kappa\gamma}] = [\bar{\kappa}] + \begin{bmatrix} o & & \\ & \gamma & \\ & & o_o \end{bmatrix} \tag{9}$$

The matrix $[\bar{\kappa}]$ is symmetric but singular. If γ equals zero, a stationary steady state solution to 5 and 6 will not exist. System damping therefore depends strictly on non zero values of γ which causes $[\overline{\kappa\gamma}]$ to be nonsingular. For $\gamma = 0$ a periodic nondamping time dependent solution can be obtained.

HORIZONTAL INTEGRATION

Discretization in the horizontal domain is accomplished by applying the Galerkin method, using the 9 node isoparametric quadratic element, and integral lumping. The advantage of using the integral lumping technique is that the equations are uncoupled. For any node j in the mesh, we eventually obtain , for the x momentum equation

$$[\bar{\alpha}]\{\frac{\partial X}{\partial t}\}_j + \frac{\varepsilon}{h_j^2}[\overline{\kappa\gamma}]\{X\}_j - f[\bar{\alpha}]\{Y\}_j - \frac{1}{h_j}\{\tau_{wx}\}_j + \frac{1}{h}\{C_x\}_j + g\{\bar{\beta}\}\left(\frac{\partial\zeta}{\partial x}\right)_j = 0 \tag{10}$$

The treatment of curvilinear coordinates and expressions for the convective terms and the surface gradients can be found in Laible (1984). The evaluation of the surface gradient term via this approach is similar to that employed by Cheng (1977) and Forristal (1980). The approach used here allows for the use of either a finite difference or finite element treatment. Equation 10, its y direction analogy and a similar treatment of the continuity equation lead to the three dimensional version of the **primitive equations model** (PEM). Because of difficulties with numerical noise these equations have a limited range of applicability. Consequently, a modified version of the wave equation formulation (WEQ) has been developed to investigate the potential benifits for a 3D model. The modified model is hereafter referred to as the MWEQ model.

WAVE EQUATION MODELS

Lynch and Gray (1979) developed a wave equation formulation for the shallow water equations. This approach controls the spurious $2\Delta X$ waves which inevitably enter the solution, ultimately leading to instability. Experience with the PEM model for depth varying flow has also revealed the presence of numerical waves. For this reason a similar wave equation model has been develped, starting with equations 1-3. Here we will consider the problem of wind driven circulation in a shallow region for which h can be taken as independent of time since the surface variation ζ are much smaller than the depth. To develop the MWEQ model we begin by differentiating the continuity equation (8) with respect to time

$$\frac{\partial^2\zeta}{\partial t^2} + \frac{\partial}{\partial x}(h\,\underline{\bar{\beta}}\,\{\frac{\partial \tilde{X}}{\partial t}\}) + \frac{\partial}{\partial y}(h\,\underline{\bar{\beta}}\,\{\frac{\partial \tilde{Y}}{\partial t}\}) = 0 \tag{11}$$

Applying the the Galerkin method and integrating the last two terms by parts yields an equation for each node j

$$\int \phi_j \frac{\partial^2\tilde{\zeta}}{\partial t^2}\, dA - \int \frac{\partial \phi_j}{\partial x}(h\,\underline{\bar{\beta}}\,\{\frac{\partial \tilde{X}}{\partial t}\}) + \frac{\partial \phi_j}{\partial y}(h\,\underline{\bar{\beta}}\,\{\frac{\partial \tilde{Y}}{\partial t}\})\, dA$$
$$+ \int \ell_x(\phi_j \dot{\tilde{q}}_x) + \ell_y(\phi_j \dot{\tilde{q}}_y) ds = 0 \tag{12}$$

Substituting the two momentum equations in the form of equation 10 , and the basis function expansions for the flux and surface elevation terms leads to

$$\int \phi_j \phi_i dA \left(\frac{\partial^2 \zeta}{\partial t^2}\right)_i + \left(\int \frac{\partial \phi_j}{\partial x} \phi_i dA\right) (\dot{q}_{xm})_i + \left(\int \frac{\partial \phi_j}{\partial y} \phi_i dA\right) (\dot{q}_{ym})_i$$

$$+ g\,\overline{(\beta^T \alpha^{-1} \beta)} \left(\int \left(\frac{\partial \phi_j}{\partial x} h \frac{\partial \phi_i}{\partial x} + \frac{\partial \phi_j}{\partial y} h \frac{\partial \phi_i}{\partial y}\right) dA \right) \zeta_i$$

$$+ \int \ell x (\phi_j \phi_i) (\dot{q}_x)_i + \ell y (\phi_j \phi_i) (\dot{q}_y)_i \; ds = 0 \qquad (13)$$

$$\dot{q}_{ym} = h \underline{\beta} \left[\frac{\varepsilon}{h^2} [\overline{\alpha^{-1} \kappa \gamma}] \{\tilde{Y}\} - f\{X\} - \frac{1}{h} \{\overline{\alpha^{-1}} \tau_{xy}\} + \frac{1}{h} \{\overline{\alpha^{-1} \tilde{C}_y}\} \right] \qquad (13a)$$

$$\dot{q}_{xm} = h \underline{\overline{\beta}} \left[\frac{\varepsilon}{h^2} [\overline{\alpha^{-1} \kappa \gamma}] \{\tilde{X}\} + f\{Y\} - \frac{1}{h} \{\overline{\alpha^{-1}} \tau_{wx}\} + \frac{1}{h} \{\overline{\alpha^{-1} \tilde{C}_x}\} \right] \qquad (13b)$$

$$\dot{\tilde{q}}_y = h \underline{\overline{\beta}} \left\{\frac{\partial \tilde{Y}}{\partial t}\right\} \qquad \dot{\tilde{q}}_x = h \underline{\overline{\beta}} \left\{\frac{\partial \tilde{X}}{\partial t}\right\} \qquad (13c)$$

$$\phi_i = \text{quadratic basis functions} \qquad (13d)$$

These equations can now be integrated by the integral lumping technique to yield the uncoupled equations at each node in the grid. It is therfore possible to solve for second order time derivative of the surface elevation term at any node j from each of the node equations.

TIME INTEGRATION SCHEME

For the wave equation the second order time derivative is approximated by

$$\left(\frac{\partial^2 \zeta}{\partial t^2}\right)_j = \frac{(\zeta^{k+1} - 2\zeta^k + \zeta^{k-1})_j}{\Delta t^2} \qquad (14)$$

For the momentum equations the time derivative is approximated by

$$\left\{\frac{\partial X}{\partial t}\right\}_j^k = \delta\left\{\frac{X^{k+1} - X^k}{\Delta t}\right\} + (1-\delta)\left\{\frac{X^k - X^{k-1}}{\Delta t}\right\} \qquad (15)$$

An implicit but time centered treatment of the dissipation term can be accomplished by using

$$\{X\}^* = \delta\{X\}^{k+1} + (1-2\delta)\{X\}^k + \delta\{X\}^{k-1} \qquad (16)$$

The surface elevation term is also expressed by a time centered approximation

$$\zeta^* = \frac{\theta}{2}(\zeta^{k+1} + \zeta^{k-1}) + (1-\theta)\zeta^k \tag{17}$$

The x momentum equation (10) and its y analogy are now solved for $\{\partial X/\partial Dt\}$ and $\{\partial Y/\partial t\}$. Likewise equation 13 is solved for the second order time derivative $\partial^2\zeta/\partial t^2$. Substituting the difference approximations and the centered evaluations of $\{X\}^*$, $\{Y\}^*$ and $\{\zeta\}^*$ the following three equations are obtained

$$[\,\delta[I] + \delta\frac{\Delta t\varepsilon}{h_j}[\overline{\alpha^{-1}\kappa\gamma}]\,]\{X^{k+1} - X^k\}_j = \Delta t(f_x)_j - (1-\delta)\{X^k - X^{k-1}\}_j \tag{18}$$

$$[\,\delta[I] + \delta\frac{\Delta t\varepsilon}{h_j}[\overline{\alpha^{-1}\kappa\gamma}]\,]\{Y^{k+1} - Y^k\}_j = \Delta t(f_y)_j - (1-\delta)\{Y^k - Y^{k-1}\}_j \tag{19}$$

$$\zeta_j^{K+1} = -\Delta t^2(f_\zeta)_j^k + 2\zeta_j^k - \zeta_j^{k-1} \tag{20}$$

The expressions $(f_x)_j$, $(f_y)_j$, and $(f_\zeta)_j$ are rather lengthy and space limitations prohibits inclusion here. Details may be found in Laible (1984). The terms $(f_x)_j$ and $(f_y)_j$ include values of $\{X\}$ and $\{Y\}$ at time level k and k-1 and values of ζ at time level k+1 . The term $(f_\zeta)_j$ is based on values of $\{X\}$, $\{Y\}$ and ζ at time level k . Consequently $(f_\zeta)_j$ is computed first and used in equation 20 to compute ζ at time level k+1 . The new ζ is then used to evaluate $(f_x)_j$ and $(f_y)_j$ and subsequently the values of $\{X\}$ and $\{Y\}$ at k+1 via equations 18 and 19.

MODEL ANALYSIS

Based on a von Neumann stability analysis the most accurate solution would be obtained with $\delta=1/2$. Gray and Kinnmark (1982) have shown that a value of δ less than 1 will produce oscillations in the **velocity** from time step to time step for the shallow water equations. The same type of behavior was observed for the 3D, verifying that temporal differencing involving a first order derivative should only include values at time levels k and k+1. Gray and Lynch (1977, 1979) have also performed extensive Fourier analyses of several numerical schemes and have demonstrated , via the wave propogation factor, that the wave equation model (WEQ) posesses the ability to dampen and progogate the $2\Delta X$ waves. This approach controls numerical noise automatically without resorting to various "adjustments" such as increasing bottom friction ,the use of extra viscous terms, numerical smoothing (averaging of nodal values) or the use of time integration schemes which contain excessive numerical damping.

While the modified wave equation effectively controls the spurious spacial waves, the solution to a wave equation does not automatically guarantee a solution to the contiuity equation. The wave equation is derived by a time differentiation of the contiuity equation. The numerical counterpart of equation 3 (the continuity equation) could therfore be equal to a constant rather than zero, and still satisfy the wave equation. Continuity could therfore be violated. To demonstrate this formally consider a Fourier analysis of the linearize one dimensional vertically integrated equations for each of the models PEM, WEQ and MWEQ. Table 1 compares the basic equations and the roots of the system. These values were obtained by expressing u and ζ in terms of the Fourier components $u=u_o \exp(i\beta t+i\sigma x)$ $\zeta=\zeta_o \exp(i\beta t+i\sigma x)$. β= time frequency, σ= spacial wave frequency, $i=\sqrt{-1}$. For the primitive equation there are two roots as expected. For the wave models an artificial analytic root exists due to the time differentiation of the continuity equation. Substituting these extra roots into the Fourier components and the result into the continuity equation we find that the artificial components of the continuity equation C are of the form

$$C=F(x,y)\exp(-kt) \text{ for WEQ} \qquad C=F(x,y)\exp(0) \text{ for MWEQ}$$

These components are not zero unless F(x,y) is zero. This can be achieved by assuring that the initial conditions of the flow field satisfy continuity. The WEQ model has the added advantage that if continuity is violated the term exp(-kt) will cause this mode of response to decay , at least analytically.

	EQUATION	MATRIX	ROOTS
PEM	$\frac{\partial\zeta}{\partial t} + h\frac{\partial u}{\partial x} = 0$ $\frac{\partial u}{\partial t} + g\frac{\partial\zeta}{\partial x} + \kappa u = 0$	$\begin{bmatrix} i\beta & i\sigma h \\ (i\sigma)\, g & i\beta + \kappa \end{bmatrix}$	$i\beta = \alpha\{-\frac{\kappa}{2\sigma} + i\sqrt{gh-(k/2\sigma)^2}\}$ $i\beta = \alpha\ -\frac{\kappa}{2\sigma} - i\sqrt{gh-(\kappa/2\sigma)^2}\}$
WEQ	$\frac{\partial^2\zeta}{\partial t^2} + \kappa\frac{\partial\zeta}{\partial t} - gh\frac{\partial^2\zeta}{\partial x^2} = 0$ $\frac{\partial u}{\partial t} + g\frac{\partial\zeta}{\partial x} = \kappa u = 0$	$\begin{bmatrix} (i\beta)^2+\kappa(i\beta)-(i\sigma)^2 gh & o \\ (i\sigma)\, g & i\beta+\kappa \end{bmatrix}$	$i\beta = -\kappa$ $i\beta = \sigma\{\frac{-\kappa}{2\sigma} + i\sqrt{gh-(\kappa/2\sigma)^2}\}$ $i\beta = \sigma\{\frac{-\kappa}{2\sigma} - i\sqrt{gh-(\kappa/2\sigma)^2}\}$
MWEQ	$\frac{\partial^2\zeta}{\partial t^2} - h\frac{\partial u}{\partial x} - gh\frac{\partial^2\zeta}{\partial x^2} = 0$ $\frac{\partial u}{\partial t} + g\frac{\partial\zeta}{\partial x} + \kappa u = 0$	$\begin{bmatrix} (1\beta)^2-(i\sigma)^2 gh & -h\kappa(i\sigma) \\ (i\sigma)g & i\beta+\kappa \end{bmatrix}$	$i\beta = 0$ $i\beta = \sigma\{\frac{-\kappa}{2\sigma} + i\sqrt{gh-(\kappa/2\sigma)^2}\}$ $i\beta = \sigma\{\frac{-\kappa}{2\sigma} - i\sqrt{gh-(\kappa/2\sigma)^2}$

κ = linearized bottom friction term

TABLE 1

EXAMPLE ANALYSES and FUTURE WORK

Using five nine node isoparametric elements a model of a 1000 m long straight channel was analyzed with the following parameter values: $\varepsilon = 0.01\ m^2/s$ (constant over the depth), $g = 10\ m/s^2$, $C_b = 0.005$ m/s, h=10.0 m, $Wx^2 = 10.0 m^2/s^2$, $C_w = 0.001$. The wind direction was along the major axis of the channel. With the boundary conditions that the flux at the ends of the channel be zero an analytic solution at steady state is obtained:

$$u = -g\frac{\partial\zeta}{\partial x}\left(\frac{h}{C_b} + \frac{h^2-z^2}{2\varepsilon}\right) + C_w W_x^2\left(\frac{1}{C_b} + \frac{h-z}{\varepsilon}\right)$$

$$g\frac{\partial\zeta}{\partial x} = \frac{C_w W_x^2}{h}\left(\frac{1+\gamma/2}{1+\gamma/2}\right)$$

This solution was matched exactly by the MWEQ solution because the form of the exact solution is quadratic while the hermitian polynomials are cubic.

As a test of the time dependent solution without any bottom friction, a similar model of length 200 m, h=4m $C_w W^2 = 0.00981 m^2/s^2$, $g = 9.81\ m/s^2$, $\varepsilon = 0.001\ m^2/s$, t=7.5 sec., $\theta = 1.0$, $\delta = 1.0$, was analyzed with the boundary conditions $\zeta = 0$ at x=0 and U = vertically integrated velocity=0 at x=L. The results of the 3D model were then integrated over the depth to compare with the exact periodic solution as described by Wang (1975). The wave form consists of a series of straight lines (not a sine wave). The exact solution is a superposition of sine waves which converges rather slowly. A minimum of 50 terms were necessary. Figure 2 illustrates the time history response of the analytic and finite element solution. Both phase and amplitude are in good agreement. The exact velocity response has an abrupt change at the peak which is extremely difficult to model by any numerical solution unless the time steps are significantly smaller than the 7.5 sec. step used here. In spite of the overshoot at the peak the response always recovers:

A shallow bay adjacent to a larger lake (St. Albans Bay on Lake Champlain, VT.) was analyzed using the MWEQ (figure 3). The depths were very irregular ranging from 1m on the boundaries to 8m in the channel section down the middle of the Bay. The values of γ ranged from 1.0 to 20.0. As the Bay has a major axis which is oriented on a NE bearing, a southwest(prevailing) wind analysis was carried out to assess the transport of water in and out of the Bay. The boundary condition between the Bay and the Lake for this analysis was a fixed head ($\zeta = 0$). No node to node oscillations were observed and the solution demonstrated the classical two cell circulation pattern in the shallow regions. The strong reverse currents, which have been measured using drogues, are apparent in the numerical solution. Work is proceeding to collect more reliable flow data which is in the range of 15 cm/s on the surface and 5-10 cm/sec in the region of maximum reverse flow at about the 3/4 depth under moderate (10mph) wind loads. Both the measured flows and the simulated flows are being used to estimate the exchange rates between the Bay and the Lake (under various wind loadings) for subsequent studies of phosphorous transport. A version of the MWEQ is also being implemented on a Compupro microcomputer.

REFERENCES

Cheng, R.T., `Transient three-dimensional circulation of lakes', Journal of the Engineering Mechanics Division, Vol. 103, No. EM1, (1977) 17.

Forristall, G.Z., `A two-layered model for hurricaine driven currents on an irregular grid', J. Physical Oceanography 10 (1980) 1417-1438.

Gray, W.G., `An efficient finite element scheme for two-dimensional surface water computation', Proceedings of the First International Conference on

Finite Elements in Water Resources, Princeton University, Pentech Press, 1976.

Gray, W.G. and Lynch, D.R. `Time stepping schemes for finite element tidal computations', Adv. in Water Resources 1,2, 83-95. 1977.

Gray, W.G., Kinnmark, P.E., `Time-weighting of the momentum equation in explicit wave models of surface water flow'. Proceedings of the 4th International Conference on Finite Elements in Water Resources, Hannover, Germany, Published by Springer-Verlag Berlin, A Computational Mechanics Centre Publication. 1982.

Laible, `A Finite Element/Finite Difference Wave Model for Depth Varying Nearly Horizontal Flow', accepted for publication in Advances in Water Resources, 1984.

Lynch, D.R. and Gray, W. G., `A wave equation model for finite element tidal computations', Computers and Fluids 7 (1979) 207-208.

Wang, J.D. and Connor, J.J., `Mathematical modeling of near coastal circulations', Ralph M. Parsons Laboratory for Water Resources and Hydromechanics, MIT, Department of Civil Engineering, Report No. 200,1975.

APPENDIX

Matrices and tensor products for equations (6),(7),(8).

$$\bar{\alpha}_{ij} = \int_{-1}^{0} \bar{N}_i \bar{N}_j \, ds$$

$$\bar{\kappa}_{ij} = \int_{-1}^{0} \frac{\partial \bar{N}_i}{\partial s} \bar{P}_\varepsilon \frac{\partial \bar{N}_j}{\partial s} \, ds$$

$$\beta_i = \int_{-1}^{0} \bar{N}_i \, ds$$

$$\psi_{irs} = \int_{-1}^{0} \bar{N}_i \bar{N}_r \frac{\partial \bar{N}_s}{\partial h} \, ds$$

$$\bar{\Lambda}_{irs} = \int_{-1}^{0} \bar{N}_i \bar{N}_r \bar{N}_s \, ds$$

$$\overline{\kappa\gamma}_{ij} = \bar{\kappa}_{ij} + \bar{\gamma}_{ij}$$

$$\bar{\gamma}_{ij} = \begin{cases} \gamma & \text{for } i = j = z \\ 0 & \text{for } i \neq j \ \& \ i = j \neq z \end{cases}$$

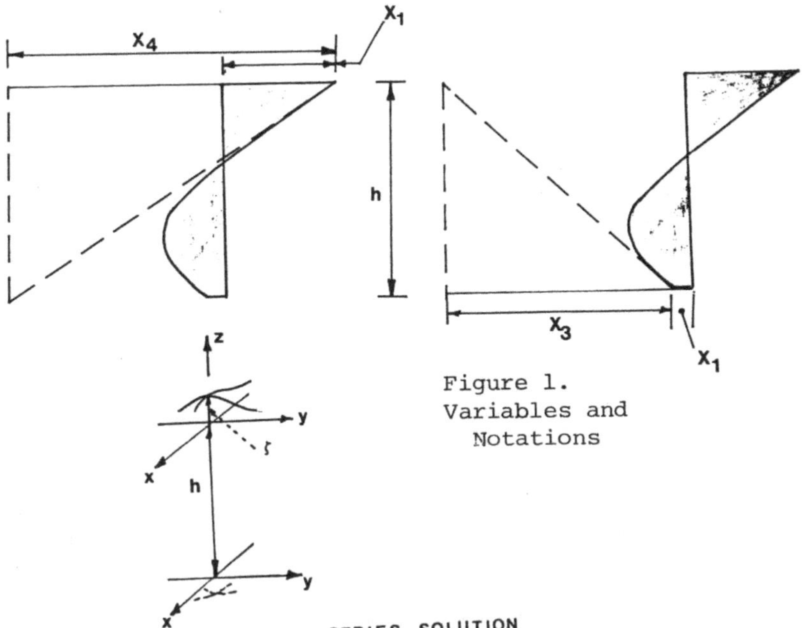

Figure 1.
Variables and
Notations

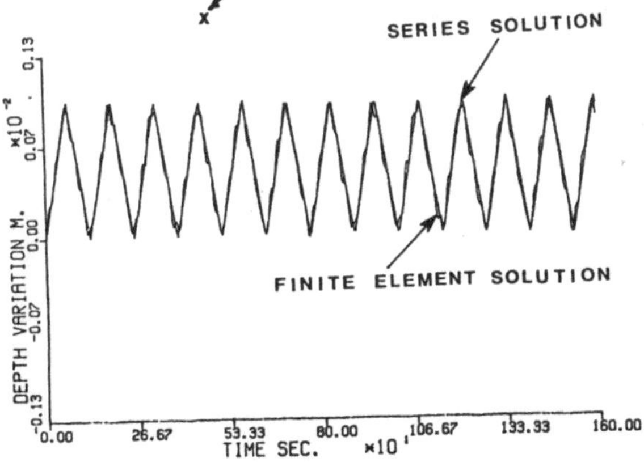

Figure 2a
Surface elevation
response at end
of channel

Figure 2b
Vertically
integreated
velocity at the
center of channel

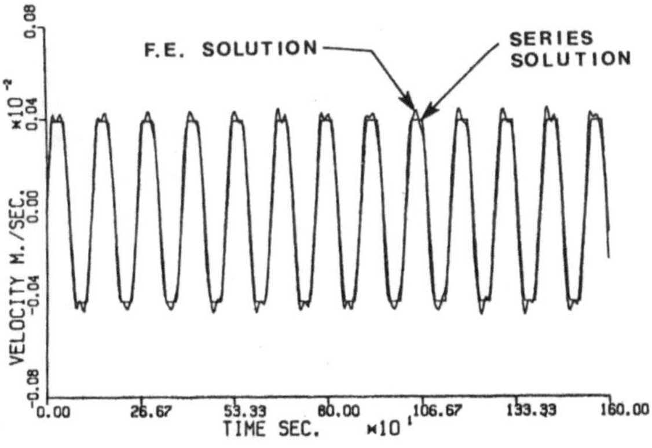

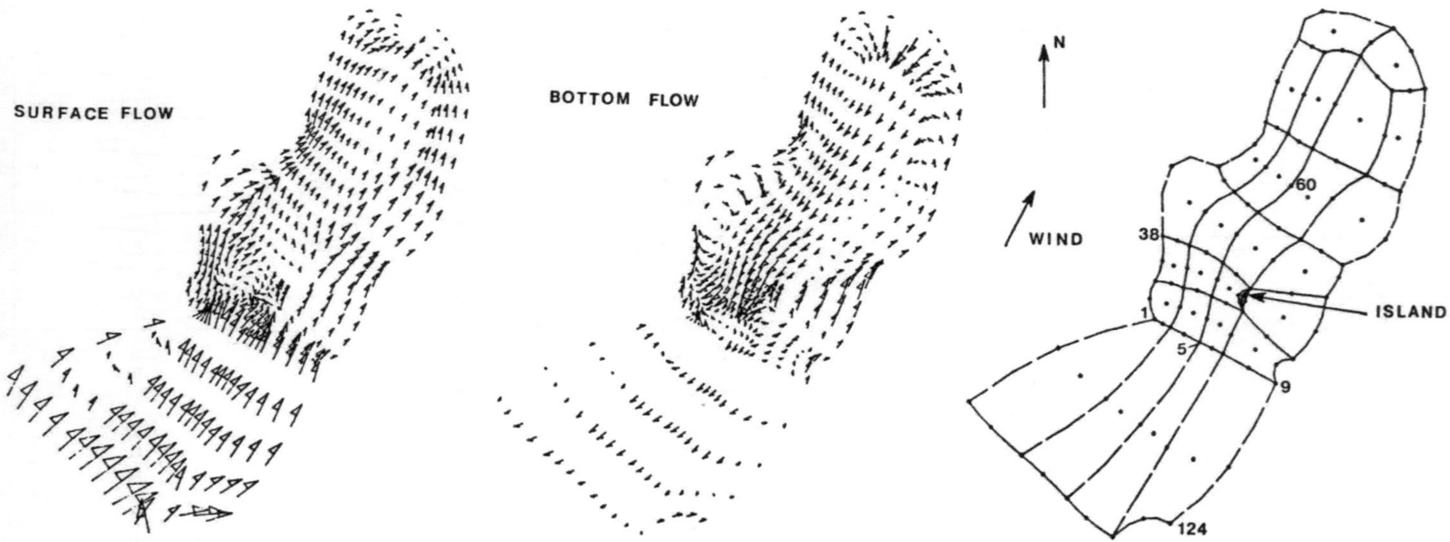

Figure 3. St. Albans Bay VT. Flows 10-15 m.p.h. S.W. wind

WEAKLY-REFLECTIVE BOUNDARY CONDITIONS FOR TWO-DIMENSIONAL SHALLOW WATER FLOW PROBLEMS

Verboom, G.K., Slob, A.
Delft Hydraulics Laboratory, Delft, the Netherlands

SUMMARY

In this paper weakly-reflective boundary conditions are derived for the two-dimensional shallow water equations, including bottom friction and Coriolis force. The essential aspects of the derivation are given. Zeroth and first order approximations are applied to the test problem of an intially Gaussian shaped free surface elevation. For the numerical solution a finite element program is used and various aspect of the numerical implementation are discussed. For small scale practical problems a rather simple (one parameter) formulation might be sufficient. The influence of this parameter is discussed on the weakly-reflectiveness of the boundary condition.

INTRODUCTION

In the numerical solution of many hydraulic engineering problems waves play a dominant role. As the area of interest generally is just a small part of a much larger system artificial boundaries are introduced to obtain a limited domain. In nature waves can cross these artificial boundaries unhampered in both directions, but in our numerical model we must include this property explicitly. In literature so-called non- and weakly-reflective boundary conditions are derived for hyperbolic equations: Taylor 1975; Engquist and Majda 1977. For the quasi-linear one-dimensional shallow water equations Verboom 1982 derived upto second order weakly-reflective boundary conditions. For tidal wave problems reflection coefficients of only a few percent were realized with the Preissmann-scheme. For more general schemes numerical reflections limit the effect of higher order conditions.
In this paper we derive zeroth and first-order weakly-reflective boundary conditions for the two-dimensional shallow water equations including bottom friction and Coriolis force. The essential aspects of the derivation are given. Next the conditions are applied to the test problem of an initially Gaussian

free surface elevation. For the numerical solution we use a finite element program with bi-linear base functions and an explicit time integration scheme. For many practical problems in which the boundary conditions are derived from a larger area (coarser grid) model a much simpler formulation proposed by Stelling 1983 will do, Verboer et al. 1984.
This condition is discussed and results are reported for three values of the parameter included in that formulation.

THEORETICAL ASPECTS

Formulation of the problem

The two-dimensional shallow water equations in primitive variables read

$$\vec{w}_t + A_1\vec{w}_x + A_2\vec{w}_y + A_3\vec{w} + \vec{F} = \vec{0} \ , \qquad (1)$$

with $\vec{w} = (u,v,\zeta)^T$ and

$$A_1 = \begin{pmatrix} u & o & g \\ o & u & o \\ h+\zeta & o & u \end{pmatrix} \quad ; \quad A_2 = \begin{pmatrix} v & o & o \\ o & v & g \\ o & h+\zeta & v \end{pmatrix} ,$$

$$A_3 = \begin{pmatrix} \lambda & -f & o \\ f & \lambda & o \\ o & o & o \end{pmatrix} \quad ; \quad \vec{F} = (-gh_x,-gh_y,o)^T \ .$$

The notation is explained at the end of the paper. The bottom friction parameter, λ, is generally given by

$$\lambda = \frac{g\sqrt{u^2+v^2}}{C^2(h+\zeta)} \ .$$

External forces, such as wind can be accounted for in $\vec{F}$. For most practical computations we solve System 1 with an ADI-finite difference program, but for specific relatively small scale problems we can use an explicit finite element program that solves the corresponding system written in conservative form. However, for the derivation of weakly-reflective boundary conditions it is advantageous to symmetrize and/or diagonalize matrix A_1 or A_2. As System 1 is a quasi-linear strictly hyperbolic system, A_1 and A_2 can be symmetrized simultaneously, but only one can be diagonalized at a time. Matrices A_1 and A_2 are symmetrized by the transformation

$$\psi = 2\sqrt{g(h+\zeta)} \ . \qquad (2)$$

For the analysis in the following section we diagonalize matrix A_1 with the additional transformation

$\vec{v} = V.(u,v,\psi)^T$

with

$$V = \begin{pmatrix} 1/\sqrt{2} & o & 1/\sqrt{2} \\ o & 1 & o \\ 1/\sqrt{2} & o & -1/\sqrt{2} \end{pmatrix}. \tag{3}$$

The system for $\vec{v}$ reads

$$\vec{v}_t + A\,\vec{v}_x + B\,\vec{v}_y + C\,\vec{v} + V\,\vec{F} = \vec{0}\ , \tag{4}$$

with $\vec{v} = (\frac{1}{\sqrt{2}}(u+\psi),\ v,\ \frac{1}{\sqrt{2}}(u-\psi))^T$ and

$$A = V\ A_1 V^{-1} = \begin{pmatrix} u+\frac{1}{2}\psi & o & o \\ o & u & o \\ o & o & u-\frac{1}{2}\psi \end{pmatrix}$$

$$B = V\ A_2 V^{-1} = \begin{pmatrix} v & \psi/2\sqrt{2} & o \\ \psi/2\sqrt{2} & v & -\psi/2\sqrt{2} \\ o & -\psi/2\sqrt{2} & v \end{pmatrix}$$

$$C = V\ A_3 V^{-1} = \begin{pmatrix} \lambda/2 & -f/\sqrt{2} & \lambda/2 \\ f/\sqrt{2} & \lambda & f/\sqrt{2} \\ \lambda/2 & -f/\sqrt{2} & \lambda/2 \end{pmatrix}.$$

Derivation of weakly-reflective boundary conditions

The general solution of System 4 contains progressive waves which even in the limit of vanishing C and $\vec{F}$ are coupled (because A and B cannot be diagonalized simultaneously). Therefore, it is impossible to derive truly non-reflective (local) boundary conditions, i.e. a complete decoupling of ingoing and out-going waves. As a result, the boundary conditions to be derived are only weakly-reflective in some kind of approximation. For the one-dimensional shallow water equations, Verboom 1982, the approximation parameter is (λ/ω), where ω is the wave frequency. For the two-dimensional problem two additional parameters exist: (f/ω) and the angle of incidence of waves at the boundary.

To derive weakly-reflective boundary conditions for the half-space problem $x < 0$ one can proceed along the following lines:

i) freeze the coefficients in System 4 in order to arrive at a linear system,

ii) perform a Fourier transform in time and in y-direction,

with dual variables ω and η, respectively, and write the transformed system as

$$\hat{v}_x = G\,\hat{v}\ , \tag{5}$$

with

$$G = -i\omega\, A_f^{-1}\left(I + \frac{\eta}{\omega} B_f + \frac{1}{i\omega} C_f\right)\ , \tag{6}$$

where the index f refers to "frozen", and I is the identity matrix. In Equation 5 we neglected $\hat{F}$, but we will return to this matter later on.
For A_f^{-1} to exist we must exclude the cases where $u_f = 0$, no flow across the boundary, and $u_f^2 = \frac{1}{2}\psi$, i.e. no critical flow.
iii) perform a transformation $\hat{w} = W\,\hat{v}$, such that $D = W\,GW^{-1}$, is a diagonal matrix. The components of $\hat{w}$ are decoupled and these quantities or more precisely an approximation to these quantities must be prescribed at the boundaries. The number of conditions to be prescribed at $x = 0$ is given by the number of positive eigen-values of D; similarly the number of boundary conditions for the right half-space problem, $x > 0$, equals the number of negative eigen-values of D.
iv) write the transformation W as a polynomial in (η/ω) and $(1/i\omega)$, and truncate this polynomial. A local weakly-reflective boundary condition is obtained by an inverse Fourier-transform.

The advantage of this formal derivation is that $\hat{w}$ still is an exact result: approximations are introduced in the final stage of the derivation only. Unfortunately, it is not possible to write the eigen-values and eigen-vectors of G, Equation 6, in an explicit form if f and λ are non-zero simultaneously, as they are in the general case. To solve this problem we can proceed along two lines.
Firstly, following Engquist and Majda 1977 instead of diagonalizing G exactly and then approximating W one can diagonalize G approximately with a much simpler transformation. Above that, it is sufficient to decouple the ingoing waves from the outgoing waves upto the desired order of approximation, whereas the outgoing waves still might depend more strongly on the ingoing waves. For the matrix $W\,GW^{-1}$ this means that only some of the off-diagonal terms need to be zero (in the order of approximation). The transformation matrix W can now be written as

$$W \simeq \sum_{\substack{p=0\\ q=0}}^{\infty} \left(\frac{\eta}{\omega}\right)^p \left(\frac{1}{i\omega}\right)^q W_{pq} \tag{7}$$

For System 4 W_{pq} will turn out to be rather simple for (p and q) ≤ 1.
Secondly, one can by-pass all matrix manipulations and substitute a solution of the form

$$\vec{v} = \hat{v} \exp(i\omega t + ik_x x + ik_y y) \ , \tag{8}$$

in the frozen coefficient problem, and look for boundary conditions of the form

$$\alpha \hat{u} + \beta \hat{v} + \gamma \hat{\psi} = \hat{f} \ ,$$

where α, β, γ and $\hat{f}$ are functions of ω, k_x and k_y, such that only ingoing waves are prescribed. Weakly-reflective boundary conditions are obtained by approximating the coefficients α, β and γ in terms of ω, k_x, and k_y. The number of boundary conditions required equals the number of ingoing waves and is found from the analysis. However, additional information is required on what kind of combinations of $\hat{u}$, $\hat{v}$, and ψ might be physically relevant.
A drawback of both methods is that the boundary conditions derived do not guarantee a stable computation, see also Verboom et al. 1982. If, as a rule of thump, the boundary condition contains a term like $y_t + by$, then b must be negative irrespective of the other terms.

We proceed with a derivation along the lines indicated by Engquist and Majda 1977: apply the transformation given by Equation 7 to Equations 5 and 6. In a rather straightforward analysis weakly-reflective boundary conditions can be derived for inflow and outflow boundaries. The results for the left half space problem, $x \leq 0$, subcritical flow, and (p and q) ≤ 1 are summarized in Table 1.
The boundary conditions given in Table 1 apply (as well) for a boundary at x=L for a problem defined at x=[0,L], L C R, of course.
For p=1 and q=1 two conditions instead of one are derived without introducing a second order derivative in space or time. Set B differs from set A in that it is a higher order approximation in $(\lambda/i\omega)$-terms; for $(f/i\omega)$-terms this was found to be not possible. A similar result was found by Pakvis 1983, for the one-dimensional shallow water equations. The one-dimensional weakly-reflective boundary condition, equivalent to set B, with p=1 and q=1, reads

$$(u-\psi)_t + \frac{\lambda_f}{4} \left\{(3+\beta+ \frac{2u_f}{\psi_f})u-(1+\beta- \frac{2u_f}{\psi_f})\psi\right\} = g_1 . \tag{9}$$

The parameter β can be chosen arbitrarily, but Equation 9 is a higher order approximation in $(\lambda/i\omega)$-terms, if $\beta = 0$. For the linear equations Pakvis proved that the reflection coefficient depends rather strongly on β. The optimal value is about zero for small values of (λ/ω) and approaches -0.8 if (λ/ω) is of the order of 100. For these extremely high values of (λ/ω), which can easily be encountered in, for instance, flood wave problems, the reflection coefficient with the optimal value of β was found to be less than half the reflection coefficient with $\beta = 0$.

Subcritical inflow ($u_f < 0$ and $u_f < \sqrt{\frac{1}{2}\psi_f}$)

$$p=0,\ q=0: \begin{pmatrix} u-\psi \\ v \end{pmatrix} = \vec{g}$$

$$p=1,\ q=1: \begin{pmatrix} (u-\psi)_t \\ v_t+\frac{1}{2}(u_f+\frac{1}{2}\psi_f)(u+\psi)_y \end{pmatrix} = \vec{g}$$

$$p=1,\ q=1\quad A: \begin{pmatrix} (u-\psi)_t + \frac{\lambda_f}{4}(1+\frac{2u_f}{\psi_f})(u+\psi) \\ v_t + \frac{1}{2}(u_f+\frac{1}{2}\psi_f)(u+\psi)_y + \frac{f}{2}(1+\frac{2u_f}{\psi_f}) \end{pmatrix} = \vec{g}$$

$$B: \begin{pmatrix} (u-\psi)_t + \frac{\lambda_f}{4}\{(3+\frac{2u_f}{\psi_f})u-(1-\frac{2u_f}{\psi_f})\psi\} - \frac{1}{4}(u_f+\frac{1}{2}\psi_f)v_y \\ v_t + \frac{1}{2}(u_f+\frac{1}{2}\psi_f)(u+\psi)_y + v_f v_y \end{pmatrix} = \vec{g}$$

Subcritical outflow ($0 < u_f < \sqrt{\frac{1}{2}\psi_f}$)

$$p=0,q=0 \qquad u-\psi = g_1$$

$$p=1,q=0 \qquad (u-\psi)_t - u_f v_y = g_1$$

$$p=1,q=1 \quad A:\ (u-\psi)_t + \frac{\lambda_f}{4}(1+\frac{2u_f}{\psi_f})(u+\psi) - \frac{2u_f}{\psi_f} fv - u_f v_y = g_1$$

$$B:\ (u-\psi)_t + \frac{\lambda_f}{4}\{(3+\frac{2u_f}{\psi_f})u-(1-\frac{2u_f}{\psi_f})\psi\} - \frac{2u_f}{\psi_f} fv - u_f v_y = g_1$$

Table 1 Weakly-reflective boundary conditions for subcritical in- and outflow boundaries for the left half-space problem, $x < 0$.

For the two-dimensional linear shallow water equations without friction and Coriolis force Wagatha 1983 derived a set of parameters to perturbe the boundary conditions. He could decrease the reflection by optimizing these parameters, though the influence was much less compared to the factor reported by Pakvis. As Wagatha obtained his results from numerical computations it might be that spurious reflections due to the numerical scheme blurred the results. From these results it is expected that the boundary conditions summarized in Table 1 can be optimized by introducing certain parameters. However, this is left for the future.

The external influences, driving forces, are included in the right hand sides of the boundary conditions. As the components of $\vec{g}$ are unknown in general and cannot be taken from field measurements one must construct $\vec{g}$ from previous knowledge on u, v, ψ and their derivatives if these occur in the boundary conditions. For practical problems this information can be obtained from a larger area (coarser grid) model only. To get rid of short wave disturbances due to the initial condition Stelling 1983 proposed a much simpler formulation. For a velocity controlled inflow boundary at x=0 this formulation is given by

$$u + \alpha(u+\psi)_t = g \ , \tag{10}$$

where α is a constant, $\alpha > 0$. The influence of the second term readily follows from an analysis of a linear one-dimensional problem given by

$$u_t + g\zeta_x = 0$$

$$\zeta_t + hu_x = 0 \ .$$

The reflection coefficient at x=0 is given by

$$Re = 1/\sqrt{1+(4\pi\alpha/T)^2} \ , \tag{11}$$

where T is the wave period. For short waves Re goes to zero, but Re $\simeq$ 1 for long (tidal) waves. It is to be noted that Equation 10 has some resemblance with Equation 9 for $\beta = -1$ and $\alpha \sim 2/\lambda_f$; the main difference is, of course, the right hand side. Equation 10 has been applied successfully in many so-called nested model applications, Verboom et al. 1984.

Steady-state solution and external forces

In the derivation of the preceeding sections we neglected not only $\vec{F}$ (and external forces) but also the steady-state solution of System 4. The latter can be accounted for by substituting $\vec{v}' = \vec{v} - \vec{v}_o$ in the frozen coefficient formulation, where $\vec{v}_o$ is the steady-state solution. In the final result $\vec{v}$ is replaced by $\vec{v}-\vec{v}_o$. In actual computations the (approximate) steady-state of the original problem is used with good results. The influence of steady external forces such as wind can be accounted for in the same way, though it may require a larger area model to get a reliable quess for $\vec{v}_o$. Time dependent winds must be accounted for in the boundary forcing function g. Apart of nested models no general procedure seems to be available to solve this problem.

NUMERICAL RESULTS

The first example concerns the application of a finite element program with weakly-reflective boundary conditions to the evolution of a Gaussian-shaped free surface elevation, Figure 1.

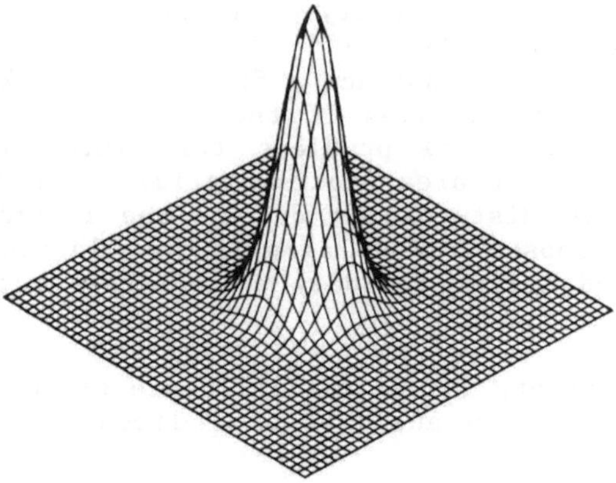

Figure 1 Initially Gaussian-shaped free surface elevation

The relevant parameters are: h = 10 m, $\zeta(0,x,y) = \exp(-(x^2+y^2)/L^2$, L = 200 m, bottom friction and Coriolis force are neglected. System 2 but expressed in conservative form with variables $(h+\zeta)u$, $(h+\zeta)v$, and $(h+\zeta)$ instead of u, v and ζ, respectively, and is solved with a Galerkin finite element method. Square elements with bi-linear base functions are used; the time discretization is basically explicit and characterized by

$$
\begin{aligned}
U^{n+1} &= f_1(U^n,V^n,H^n) \\
V^{n+1} &= f_2(U^{n+1},V^n,H^n) \\
H^{n+1} &= f_3(U^{n+1},V^{n+1},H^n) \ ,
\end{aligned}
\tag{12}
$$

with $U = u(h+\zeta)$, $V = v(h+\zeta)$, and $H = h+\zeta$.
The numerical parameters are Δx = 50 m and Δt = 5 s. The results at four points in time are given in Figures 2A-D. At the boundaries zeroth order conditions are prescribed, i.e. p=0, q=0.

To study the reflective properties in more detail these results are subtracted from a larger area solution and a reflection coefficient is defined as

$$Re = \frac{\max_t(\Delta\zeta)}{\max_t \zeta} \ .$$

Though Re is not a proper reflection coefficient in that its value is bounded to the domain [-1,1], it provides an indication of the degree of reflections.

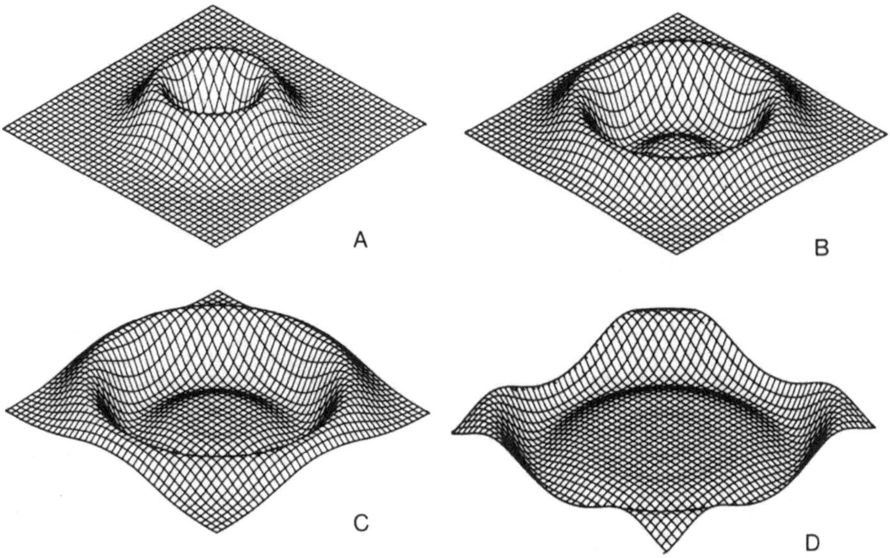

Figure 2 Time evolution of an initially Gaussian shaped free surface elevation after 30, 60, 80 and 110 s.

Figure 3 shows the results for a zeroth order, p=0 and q=0, and a first order, p=1 and q=0, boundary conditions, respectively. The reflection coefficient is given as a function of the angle of incidence of the wave, which is proportional to the coordinate along a boundary.

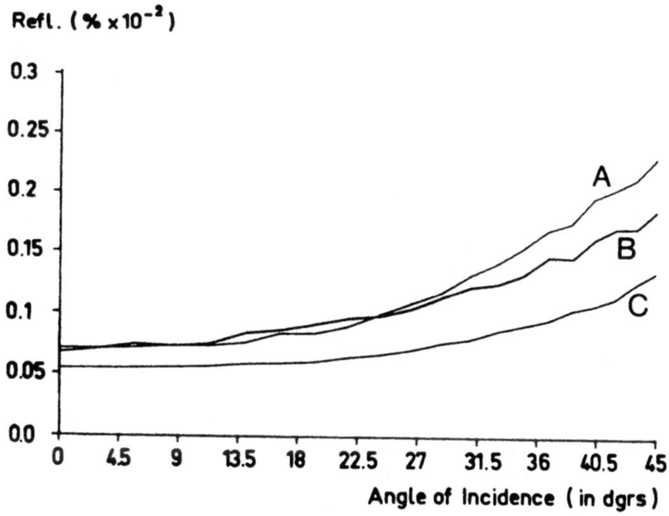

Figure 3 Reflection coefficient as a function of the angle of incidence and the boundary conditions. A: zeroth order, B: first order, C: first order, Δx and Δt halved.

The reflection coefficient depends rather strongly on the angle of incidence and increases about a factor of 3 the range (0-45)°. The reflections at normal incidence are solely due to numerical reflections; they especially confuse the results of the first order condition.
Concerning the numerical implementation there is a lot of freedom in a finite element program, because the program itself even does not demand a specific number of boundary conditions. For the zeroth order condition we used the following implementation at outflow.

$$\frac{\bar{u}^{n+1}+u^{n}}{2} - \psi^{n} = u_o - \psi_o$$

where $\bar{u}^{n+1}$ is a temporary value of u^{n+1} and is used to find $\bar{y}^{n+1}$ and $\bar{\psi}^{n+1}$, and $\bar{U}^{n+1} = \bar{u}^{n+1}H^{n}$. The definite values of u^{n+1} and ψ^{n+1} are found from

$$u^{n+1} - \psi^{n+\frac{1}{2}} = u_o - \psi_o$$

$$u^{n+1} + \psi^{n+\frac{1}{2}} = \bar{u}^{n+1} + \bar{\psi}^{n+\frac{1}{2}}$$

$$\psi^{n+\frac{1}{2}} = 2\sqrt{g(H^{n+1}+H^{n})/2}$$

Finally, the new value of U^{n+1} is found from $U^{n+1} = u^{n+1}H^{n+1}$. Several other formulations were tested, but the best results were obtained if u and ψ were shifted one half time step, as follows from Equation 12 and $u+\psi$ was kept constant.

The second example concerns the influence of the parameter α in Equation 10. This boundary condition is used in an ADI-finite difference program with a space staggered grid. For a schematized river section (2 km long and 5 m depth) the water-level is prescribed and kept constant at the downstream end and the velocity is abruptly prescribed at 1 m/s at the upstream end. The numerical parameters used are: Δx = 100 m and Δt = 100 s.
Figure 4A shows the velocity at the downstream boundary as a function of time for α = 100, 500 and 1.10^4. For α = 100, Figure 4A-II, strong eigen-oscillations are generated which are damped in time by bottom friction. The initial period is about 1.5 hours, whereas the wave transition time is only 300 s. For α = 500, Figure 4A-I, the boundary condition is also transparant for the eigen-frequencies and hardly any eigen-oscillation is generated. The initial period is about halved. If α is increased beyond about 1000 the initial period starts to increase again, not because of oscillations but because the variables now approach the final solution in the limit of large time (like a super-critically damped resonator), Figure 4 B-I.

As the transition time is only about 300 s the initial period should drop well below one half hour if a weakly-reflective boundary condition of Table 1 is used.

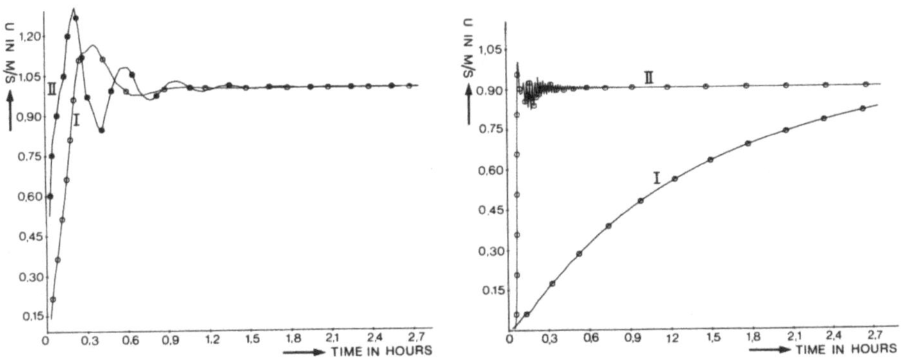

Figure 4 Velocity as a function of time. Influence of α, A-I, A-II and B-I and of a zeroth-order boundary conditions, B-II on the initial period.

Figure 4B shows the result if the velocity boundary is replaced by a zeroth order condition. Indeed, no eigen-oscillations are generated and the initial period is about 0.3 hour. The short ($2\Delta t$) waves are due to the abrupt starting condition only: they do not occur if the boundary condition is increased to its final value in one or two wave transition times. This boundary condition, with α about 100, is succesfully used in many practical applications, Verboom et.al. 1984.
A final remark concerns the influence of the parameter α on the eigen-frequencies of a problem area. If $\alpha > 0$ the eigen-frequencies are lowered: for the example discussed above the oscillation period was increased by about 27% for $\alpha = 100$ and by about a factor of 3 for $\alpha = 500$. These figures are confirmed by an analysis for a simple one-dimensional formulation.

CONCLUSION

Weakly-reflective boundary conditions have been derived for the two-dimensional shallow water equations, including bottom friction and Coriolis-force. With a finite element formulation reflection coefficients of only a few percent were obtained. The reflection coefficient increases by about a factor of three if the angle of incidence increases from zero to 45°. The effect of higher order conditions is blurred by reflections generated by the numerical scheme itself. For many practical applications a much simpler formulation proposed by Stelling 1983, suppresses effectively short wave disturbances and eigen-oscillations of the problem area. Initial periods of only a few wave transition times can be realized by using the weakly-reflective boundary conditions derived in this paper.

REFERENCES

Engquist, B. and Majda, A., 1977, Absorbing boundary conditions for the numerical simulation of waves. Math. Comp. 31, 629-651.

Pakvis, J., 1983, Weakly-reflective boundary conditions for the shallow water equations (in Dutch). Ms. Sc. thesis, Technical University Delft, Dep. Num. Math., Delft Hydraulics Laboratory, Rep. S 545-I.

Slob, A., 1983, Weakly-reflective boundary conditions for the two-dimensional shallow water equations (in Dutch).
Ms. Sc. thesis, Technical University Delft, Dep. Num. Math. Delft Hydraulics Laboratory, Rep. S545-II.

Slob, A., Verboom, G.K. and Segal, G., 1984, Weakly-reflective boundary conditions for the two-dimensional shallow water equation. To appear.

Stelling, G.S., 1983, On the construction of computational methods for shallow water flow prolems. Ph.D Thesis Technical University Delft.

Taylor, M.E., 1975, Reflections of singularities of solutions to systems of differential equations. Comm. Pure Applied Math., 28, 457-478.

Verboom, G.K., 1982, Weakly-reflective boundary conditions for the shallow water equations. Presented at 4th International Conference on Finite Elements in Water Resources, Hannover, June 21-25. Not included in the proceedings. Available as Delft Hydraulics Publication No. 266.

Verboom, G.K., Stelling, G.S., and Officier, M.J., 1982, Boundary conditions for the shallow water equations. In Engineering Applications of Computational Hydraulics; Homage to Alexandre Preissmann (Ed. Abbott, M.B. and Cunge, J.A.) Pitman, London.

Verboom, G.K., de Vriend H.J., Akkerman, G.J., Thabet, R.A.H., and Winterwerp, J.C., 1984, Nested models, applications to practical problems. This conference.

Wagatha, L., 1983, Approximation of pseudo-differential operators in absorbing boundary conditions for hyperbolic equations. Num. Math. 42, 51-64.

NOTATION

$\vec{w}$	:	$(u,v,\zeta)^T$
u,v	:	velocity components in x and y direction, respectively
ζ	:	free surface elevation above a reference plan
h	:	bottom below a reference plane
λ	:	bottom friction parameter
C	:	de Chézy-coefficient
f	:	Coriolis parameter
g	:	gravitational acceleration
x,y,t	:	space and time coordinates
$\vec{F}$	:	vector, includes bottom gradients and external forces
A_j	:	coefficient matrix
ψ	:	$\psi = 2\sqrt{g(h+\zeta)}$
$\vec{v}$	:	$(1/\sqrt{2}\ (u+\psi),\ v,\ 1/\sqrt{2}\ (u-\psi))^T$
A,B,C	:	coefficient matrices
V	:	transformation defined in Equation 3
ω	:	$2\pi/T$
T	:	wave period
$\hat{v}$	:	Fourier transform of v in t and y
G	:	matrix defined in Equation 6
g_j	:	boundary forcing function
Re	:	reflection coefficient

SOLUTION OF THE NAVIER STOKES EQUATIONS

TURBULENCE MODELLING FOR MARINE CURRENT AND THE RELATED FINITE ELEMENT SYSTEM OF EQUATIONS

An-Sheng Cheng

Numerical Analysis Research Department
Shanghai Institute of Computing Technology
Shanghai, China

INTRODUCTION

The Reynolds equations for marine current are obtained by averaging the Navier-Stokes equations in a suitable period of time [5]. We often consider the steady flow of sea water. This supposition will help us to conveniently analyse the three-dimensional distribution of current in the sea. In fact,the numerical modelling established in this case may give the valuable computation results for the investigation of the nature of the three-dimensional marine current.

In this paper, the property of turbulence is considered, thus our modelling will be able to correspond with the canonical Reynolds equations for marine current. Therefore, the symmetrical finite element system of equations is also derived by discretization.

We first start from the turbulent energy equation and suppose the convective acceleration terms vanish as a steady flow. If first level large-scale of marine current is taken, i.e., a high Reynolds number is used, thus the viscous diffusion terms can be dropped. According to the Helmhotz-Korteweg principle, a functional with constraint conditions is formed by using the dissipation terms of the equation and the continuity equation, where the viscous dissipation terms are expressed as the production terms and the convective diffusion terms of turbulent energy.

It is found that the first variation of the functional gives the Euler-Lagrange equation, which is just the Reynolds equations and the natural boundary conditions. This generalized variational can be discretized by the finite element technique. There is a symmetric matrix in the discretized system, thus it is favourable to compute and solve its large linear sparse system of equations. In addition, an error estimation can be made by using a way similar to that in treating the Stokes problem.

TURBULENT ENERGY EQUATION

The common three-dimensional turbulent energy equation [4] is represented by

$$\frac{D}{Dt}\frac{\overline{q^2}}{2} = -\frac{\partial}{\partial x_i}\overline{u_i\left(\frac{p'}{\rho}+\frac{q^2}{2}\right)} - \overline{u_iu_j}\frac{\partial U_j}{\partial x_i}$$

$$+\nu\frac{\partial}{\partial x_i}\overline{u_j\left(\frac{\partial u_i}{\partial x_j}+\frac{\partial u_j}{\partial x_i}\right)} - \nu\overline{\left(\frac{\partial u_i}{\partial x_j}+\frac{\partial u_j}{\partial x_i}\right)\frac{\partial u_j}{\partial x_i}} \qquad (1)$$

Where x_i is the Cartesian coordinate in the ith direction. U_j is the average velocity component. u_i, u_j and p' are the instantaneous fluctuation velocity and pressure about its average value respectively. ρ is a constant, the average value of sea water density. The turbulent energy $\frac{1}{2}\overline{q^2}=\frac{1}{2}\overline{u_iu_i}$. ν is the kinetic viscosity. The convective acceleration operator

$$\frac{D}{Dt} = \frac{\partial}{\partial t} + U_i\frac{\partial}{\partial x_i}$$

The Einstein summation convention is performed on a repeated subscript.

Equation (1) may be expressed as

$$\underset{I}{\frac{D}{Dt}\frac{\overline{q^2}}{2}} = \underset{II}{-\frac{\partial}{\partial x_i}\overline{u_i\left(\frac{p'}{\rho}+\frac{q^2}{2}\right)}} \underset{III}{-\overline{u_iu_j}\frac{\partial U_j}{\partial x_i}} \underset{IV}{+\nu\overline{u_i\nabla^2 u_i}} \qquad (2)$$

Where I is the change of turbulent kinetic energy including convective transport, II is the convective diffusion of turbulent energy, III is the production term of turbulence, IV is the viscosity term of turbulence.

Under the supposition of equilibrium, we drop the convective acceleration term $\frac{D}{Dt}\frac{\overline{q^2}}{2}$. The viscous diffusion term does not remain in this equation because of the high Reynolds number ($Re = UL/\nu$) which is due to the large-scale of marine turbulence [1] . Therefore, the viscous term $\nu\overline{u_i \nabla^2 u_i} \approx -\nu\overline{(\frac{\partial u_i}{\partial x_j})^2}$, where the right is the viscous dissipation terms.

According to above result, (2) may be rewritten as

$$\nu\overline{\left(\frac{\partial u_i}{\partial x_j}\right)^2} = -\overline{u_i u_j}\frac{\partial U_j}{\partial x_i} - \frac{\partial}{\partial x_i}\overline{u_i\left(\frac{p'}{\rho} + \frac{q^2}{2}\right)} \tag{3}$$

This expression is considered as an equilibrium between first level scale for the fluctuation produced by the production term and the heat dissipated by the molecule viscosity, in the successive transmissions under the turbulent diffusion.

Now we examine the diffusion term $-\frac{\partial}{\partial x_i}\overline{u_i(\frac{p'}{\rho} + \frac{q^2}{2})}$ in(3). An useful form of $-\frac{\partial}{\partial x_i}(u_i\frac{p'}{\rho})$ to include a mean strain is $-Cq^2(\frac{\partial U_i}{\partial x_j} + \frac{\partial U_j}{\partial x_i})$ [2] , where C =constant. At the same time, because $-\frac{\partial}{\partial x_i}(u_i\frac{q^2}{2})$and $-\frac{\partial}{\partial x_i}(u_i\frac{p'}{\rho})$ are all used for redistributing or transfering the energy without any net production or loss in the space, the form of $-\frac{\partial}{\partial x_i}\overline{(u_i\frac{q^2}{2})}$ may be same as that of $-\frac{\partial}{\partial x_i}(u_i\frac{p'}{\rho})$. We use the common form with the normal strain, then the form of the convective diffusion term is taken as $\frac{1}{4}q^2\delta_{ij}(\frac{\partial U_i}{\partial x_j} + \frac{\partial U_j}{\partial x_i})$. The production term can be written as $-\frac{1}{4}(\overline{u_i u_j} + \overline{u_j u_i})(\frac{\partial U_i}{\partial x_j} + \frac{\partial U_j}{\partial x_i})$. Therefore, (3) is represented by

$$\mu\overline{\left(\frac{\partial u_i}{\partial x_j}\right)^2} = -\frac{1}{4}(\rho\overline{u_i u_j} + \rho\overline{u_j u_i})\left(\frac{\partial U_i}{\partial x_j} + \frac{\partial U_j}{\partial x_i}\right) + \frac{1}{4}\rho q^2\delta_{ij}\left(\frac{\partial U_i}{\partial x_j} + \frac{\partial U_j}{\partial x_i}\right) \tag{4}$$

where the dynamic density $\mu = \rho\nu$.

VARIATIONAL PROCEDURE

From the expression (4), we have

$$\mu\overline{\left(\frac{\partial u_i}{\partial x_j}\right)^2} = R_{ij}\, d_{ij} \tag{5}$$

where

$$d_{ij} = \frac{1}{2}\left(\frac{\partial U_i}{\partial x_j} + \frac{\partial U_j}{\partial x_i}\right)$$

and

$$R_{ij}=-\frac{1}{2}(\rho\overline{u_i u_j}+\rho\overline{u_j u_i}-\rho\,\overline{u_k u_k}\,\delta_{ij})$$

When introducing the eddy viscosity coefficients $\mu(j)$ $(j=1,2,3)$, (5) becomes the following form

$$R_{ij}=\frac{1}{2}(\mu(j)\frac{\partial U_i}{\partial x_j}+\mu(i)\frac{\partial U_j}{\partial x_i}-\mu(k)\frac{\partial U_k}{\partial x_\ell}\delta_{k\ell}\,\delta_{ij}) \tag{6}$$

For sea water, generally it is true that

$$\mu(j)=\begin{cases}\mu_\ell & j=1,\ 2\\ \mu_z & j=3\end{cases} \tag{7}$$

and

$$\mu_z << \mu_\ell$$

We can express the total stress σ_{ij} by using R_{ij}, i.e.,

$$\sigma_{ij}=-p\,\delta_{ij}+2R_{ij} \tag{8}$$

where p is the average value of pressure.
We consider a bounded domain D, and the boundary conditions on the bound Bd D of the domain are

$$U_i=f_i \quad \text{on } Bd^{(u)}D \tag{9}$$

and

$$\mu(j)\frac{\partial U_i}{\partial x_j}n_j=g_i \quad \text{on } Bd^{(t)}D \tag{10}$$

where $Bd\,D=Bd^{(u)}D\bigcup Bd^{(t)}D$, and $\mathbf{n}=\{n_1,n_2,n_3\}$ is the unit outward vector normal to $Bd^{(t)}D$. Usually the boundary conditions for the foundational equations of marine current are as follows: there is the Dirichlet boundary condition of side surfaces on $Bd^{(u)}D$, and the Neumann condition of top or bottom surfaces on $Bd^{(t)}D$.

Thus we consider the functional

$$J=\int_D R_{ij}\, d_{ij}\, dv+\int_D \Lambda \frac{\partial U_i}{\partial x_i} dv-\int_{Bd^{(t)}D} g_i U_i\, ds \tag{11}$$

where Λ is the Lagrange multiplier function. There exists a function vector $(U_1, U_2, U_3) \in C^2$ satisfying the boundary condition on $Bd^{(u)}D$ so that the functional J is stationary, thus we arrive at the variation

$$\delta J=\int_D \left(\mu(j)\frac{\partial^2 U_i}{\partial x_j^2}+\frac{\partial \Lambda}{\partial x_i}\right)\delta U_i\, dv+\int_D \frac{\partial U_i}{\partial x_i}\delta\Lambda\, dV$$

$$+\int_{Bd^{(t)}D}\left(\mu(j)\frac{\partial U_i}{\partial x_j}n_j+\Lambda n_i-g_i\right)\delta U_i\, ds \tag{12}$$

Let x_3 indicate the gravity direction, and we have

$$n=(0,\ 0,\ \pm 1)$$

on the top or bottom surface. Now we take

$$\frac{\partial \Lambda}{\partial x_i}=-\frac{\partial p}{\partial x_i}-2\rho\,\varepsilon_{ijk}\Omega_j U_k+\rho\frac{\partial \Phi}{\partial x_i} \tag{13}$$

where $2\rho\,\varepsilon_{ijk}\Omega_j U_k$ is the coriolis force, and Φ is the gravity potential. Therefore, we have

$$\frac{\partial \Phi}{\partial x_1}=\frac{\partial \Phi}{\partial x_2}=0\,, \qquad \frac{\partial \Phi}{\partial x_3}=g$$

and g is the gravity acceleration.

From the above expressions, we can determine the Reynolds equation

$$\mu(j)\frac{\partial^2 U_i}{\partial x_j^2}-\frac{\partial p}{\partial x_i}-2\rho\,\varepsilon_{ijk}\Omega_j U_k+\rho\frac{\partial \Phi}{\partial x_i}=0 \tag{14}$$

the continuity equation

$$\frac{\partial U_i}{\partial x_i}=0 \tag{15}$$

and the Neumann condition

$$\mu(3)\frac{\partial U_i}{\partial x_3}=g_i\ \ (\text{or}-g_i),\ i=1,2,3 \tag{16}$$

on Bd $^{(t)}$ D, where g_i is given by wind, topography, and other special conditions.

FINITE ELEMENT SYSTEM OF EQUATIONS

Let U_i , P , and Λ in each finite element e be approximated to the following form

$$\begin{cases} U_i^{(e)} = \psi_N^{(e)}(X)\, U_{Ni}^{(e)} \\ P^{(e)} = \psi_N^{(e)}(X)\, P_N^{(e)} \\ \Lambda^{(e)} = \psi_N^{(e)}(X)\, \Lambda_N^{(e)} \end{cases} \tag{17}$$

where $U_{Ni}^{(e)}$, $P_N^{(e)}$, $\Lambda_N^{(e)}$ are the nodal values of U_i , P and Λ respectively, $\psi_N^{(e)}(X)$ is the local interpolating function. There are Ne nodes in the element e.

Substituting (17) for the corresponding functions in (11), one can arrive at

$$\begin{aligned} J^{(e)} = \frac{1}{4}\int_{De} \Big(& \mu(j)\frac{\partial \psi_N^{(e)}}{\partial x_j}\frac{\partial \psi_M^{(e)}}{\partial x_j} U_{Ni}^{(e)} U_{Mi}^{(e)} + (\mu(i)+\mu(j))\frac{\partial \psi_N^{(e)}}{\partial x_j}\frac{\partial \psi_M^{(e)}}{\partial x_i} U_{Ni}^{(e)} U_{Mj}^{(e)} \\ & + \mu(i)\frac{\partial \psi_N^{(e)}}{\partial x_i}\frac{\partial \psi_M^{(e)}}{\partial x_i} U_{Nj}^{(e)} U_{Mj}^{(e)} - \mu(K)\frac{\partial \psi_N^{(e)}}{\partial x_\ell}\frac{\partial \psi_M^{(e)}}{\partial x_j} U_{NK}^{(e)} U_{Mi}^{(e)} \delta_{K\ell}\delta_{ij} \\ & - \mu(K)\frac{\partial \psi_N^{(e)}}{\partial x_\ell}\frac{\partial \psi_M^{(e)}}{\partial x_i} U_{NK}^{(e)} U_{Mj}^{(e)} \delta_{K\ell}\delta_{ij}\Big) dV + \int_{De}\frac{\partial \psi_N^{(e)}}{\partial x_\ell}\psi_M^{(e)} U_{N\ell}^{(e)} \Lambda_M^{(e)} dV \\ & - \int_{Sp} g_i^{*} \psi_N^{(e)} U_{Ni}^{(e)} ds \end{aligned} \tag{18}$$

in the bounded domain of the element e, where Sp is the intersection between the element e and the bound Bd$^{(t)}$D. Obviously, if an element is not the boundary element on Bd$^{(t)}$D, Sp =empty, and g_i^* is determined by the values of g_i at the nodes on Sp .

Differentiating $J^{(e)}$ with respect to U_{Ni}, we can find the finite element system of equations without including the Dirichlet boundary condition,

$$\begin{bmatrix} A_{NiMj}^{(e)} & B_{Ni}^{(e)} \\ B_{Mj}^{(e)} & 0 \end{bmatrix} \begin{bmatrix} U_{Mi}^{(e)} \\ \Lambda_M^{(e)} \end{bmatrix} = \begin{bmatrix} F_i^{(e)} \\ 0 \end{bmatrix} \tag{19}$$

where

$$A_{NiMi} = \int_{De} \mu(K)\frac{\partial \psi_N^{(e)}}{\partial x_K}\frac{\partial \psi_M^{(e)}}{\partial x_K} dV \tag{20}$$

$$A^{(e)}_{NiMj}=\frac{1}{2}(\mu(i)+\mu(j))\int_{De}\left(\frac{\partial\psi^{(e)}_N}{\partial x_j}\frac{\partial\psi^{(e)}_M}{\partial x_i}-\frac{\partial\psi^{(e)}_N}{\partial x_i}\frac{\partial\psi^{(e)}_M}{\partial x_j}\right)dV\,,\ i\neq j \tag{21}$$

$$B^{(e)}_{Ni}=\int_{De}\psi^{(e)}_M\frac{\partial\psi^{(e)}_N}{\partial x_i}dV \tag{22}$$

$$B^{(e)}_{Mj}=\int_{De}\psi^{(e)}_N\frac{\partial\psi^{(e)}_M}{\partial x_j}dv \tag{23}$$

$$F^{(e)}_i=\int_{S_p}g^*_i\,\psi^{(e)}_N ds \tag{24}$$

herein i,j=1,2,3, N,M=1,2,...Ne. Substituting (17) in (13), we have

$$\frac{\partial\psi^{(e)}_N}{\partial x_i}p^{(e)}_N=-\frac{\partial\psi^{(e)}_N}{\partial x_i}\Lambda^{(e)}_N-2\rho\,\varepsilon_{ijk}\Omega^{(e)}_j\psi^{(e)}_N U^{(e)}_{NK}-\rho\frac{\partial\Phi}{\partial x_i} \tag{25}$$

Applying a Boolean matrix connecting the local node labels with the global ones, we may form the global system of equations by using the collection of (19) and(25)for all elements [6]. The representation of global system, with the relations given by Dirichlet boundary condition,is obtained, i.e.,

$$\begin{bmatrix}A & B^T\\ B & 0\end{bmatrix}\begin{bmatrix}U\\ \Lambda\end{bmatrix}=\begin{bmatrix}F\\ 0\end{bmatrix} \tag{26}$$

This is a matrix form for the saddle point method.

ON THE ERROR ESTIMATE AND THE SOLUTION

Resembling the general Stokes problem, the system of equations may be rewritten by

$$\begin{cases} -\mu(j)\dfrac{\partial^2 U_i}{\partial x_j^2}=\dfrac{\partial\Lambda}{\partial x_i} & i=1,2,3\\[2ex] \dfrac{\partial U_i}{\partial x_i}=0 & \\[2ex] U_i=f_i \quad \text{on } Bd^{(u)}D & i=1,2,3\\[2ex] \mu(j)\dfrac{\partial U_i}{\partial x_j}n_j=g_i \quad \text{on } Bd^{(t)}D & i=1,2,3 \end{cases} \tag{27}$$

Now we express the weak formulation of (27). By

utilizing Green formula, we may derive the following form:

$$\int_D \mu(j) \frac{\partial W_i}{\partial x_j} \frac{\partial V_i}{\partial x_j} dv + \int_D \Lambda \frac{\partial V_i}{\partial x_i} dV = \int_{Bd^{(t)}D} g_i V_i dV$$

$$- \int_D \mu(j) \frac{\partial U_{0i}}{\partial x_j} \frac{\partial V_i}{\partial x_j} dV \tag{28}$$

where

$$\frac{\partial V_i}{\partial x_i} = \frac{\partial U_{0i}}{\partial x_i} = 0$$

$$V_i = 0 \ , \ U_{0i} = f_i \qquad \text{on } Bd^{(u)}D$$

and

$$\frac{\partial W_i}{\partial x_i} = 0$$

thus the solution

$$U_i = W_i + U_{0i}$$

Under certain conditions, we can make an error estimate for this discretization by using the results for the Stokes problem [3].

An equivalent form of the system is

$$\Lambda = (BA^{-1}B^T)^{-1}BA^{-1}F \tag{29}$$

$$U = A^{-1}F - A^{-1}B^T\Lambda \tag{30}$$

If we adopt the penalty method and introduce a diagonal matrix C [7], then the matrix expression for the system is

$$\begin{bmatrix} A & B^{-1} \\ B & -\varepsilon C \end{bmatrix} \begin{bmatrix} U \\ \Lambda \end{bmatrix} = \begin{bmatrix} F \\ 0 \end{bmatrix} \tag{31}$$

After eliminating Λ, we have

$$(A + \varepsilon^{-1} B^T C^{-1} B)U = F \tag{32}$$

where ε is a little value. When it is estimated in whole error,(32)is a replacement of the original system.

CONCLUSIONS

In foregoing paragraphs, we started from the fundational turbulent energy equation, and then established the variational formulation corresponding to Reynolds equations of steady flow for marine current, by utilizing the property of turbulence. In fact,this modelling possesses a wide significance. Just as François Thomasset said[7], solving Navier-Stokes equations is an iteration procedure in which a number of Stokes problems are solved succesively. Therefore, when considering Reynolds equations of unsteady flow for marine current, we can construct the schemes refering to this modelling of steady flow.

REFERENCES

[1]Bradshaw, P. (1964) An Introduction to Turbulence and its Measurement. Pergaman, 186-193

[2]Frost, W. and Moulden, T. H. (1977) Handbook of turbulence. Plenum, N. Y. & Lond., 240-242.

[3]Girault, V. and Raviart, P. A. (1979) Finite element approximation of the Navier-Stokes equations. Springer-Verlag, Berlin Heidelberg, 58-85.

[4]Hinze, J. O. (1975) Turbulence. McGraw-Hill, 72.

[5]Jing, Z. -H. (1966) Principle of Marine Current. Science Press of China. 11-14. (in Chinese).

[6]Oden, J. T. and Reddy, J. N. (1976) An Introduction to the Mathematical Theory of Finite Element. John Wiley & Sons, 199-235.

[7]Thomasset, F. (1981) Implementation of Finite Element Methods for Navier-Stokes Equations. Springer-Verlag, N. Y., 72-100.

FINITE ELEMENT MODELING OF DENSITY-DRIVEN RECIRCULATING TURBULENT FLOW

B.A. DeVantier, Southern Illinois University, Carbondale
B.E. Larock, University of California, Davis

INTRODUCTION

For many fluid flows in both natural and man-made settings, density gradients influence the nature, direction, and magnitude of the flow. Differences in fluid density may be the result of differences in fluid temperature, salinity or the presence of suspended solids. In all of these flows the thermal energy or contaminant transport is intimately linked to momentum transport in the determination of the flow pattern. Usually these flows are turbulent, and as a consequence a proper representation of the transport phenomena must correctly characterize the interplay between turbulent fluctuations and density variations in the flow. A two-dimensional steady model of the mean flow then consists of a mass conservation equation, two component momentum conservation equations and one transport equation for conservation of a scalar.

Recent efforts in modeling turbulent flows have demonstrated the value of second order closure models which employ transport equations to describe the changes in turbulence parameters (Rodi, 1980). A complete flow model then consists of an even larger set of coupled nonlinear partial differential equations, which can only be solved numerically. For a predominantly two-dimensional recirculating flow, it is difficult to construct a numerical solution scheme which can adequately resolve the spatial variations in all of the model variables and yet remain both computationally manageable and convergent. This paper presents a model for turbulent recirculating flow in sedimentation basins in which the local sediment concentration significantly influences the determination of the flow field. The model equations are solved approximately by the use of the Galerkin weighted residual technique.

MODEL DESCRIPTION AND SOLUTION

The development of the turbulent continuum model equations from the instantaneous conservation equations for fluid and sediment is relatively lengthy and has been presented recently by DeVantier and

Larock (1983). The Galerkin finite element representation of the element level (element domain Ω_e and element boundary Γ_e) residuals F_q for the six governing equations for steady two-dimensional radia flow in a vertical plane (r, z are the radial and vertical coordinates are now presented:

$$F_P = \int_{\Omega_e} M_i \left(\frac{\partial}{\partial r} (rU) + r \frac{\partial W}{\partial z} \right) d\Omega \tag{1}$$

$$F_U = \int_{\Omega_e} N_i \left[r \left(U \frac{\partial U}{\partial r} + W \frac{\partial U}{\partial z} + \frac{\partial P}{\partial x} \right) - \overline{vv} \right] d\Omega$$

$$- \int_{\Omega_e} r \left(\overline{uu} \frac{\partial N_i}{\partial r} + \overline{uw} \frac{\partial N_i}{\partial z} \right) d\Omega$$

$$+ \int_{\Gamma_e} r N_i \left(\overline{uu}\, \ell_r + \overline{uw}\, \ell_z \right) d\Gamma \tag{2}$$

$$F_W = \int_{\Omega_e} N_i \; r \left(U \frac{\partial W}{\partial r} + W \frac{\partial W}{\partial z} + \frac{\partial P}{\partial z} + \eta \Lambda_{AVE} g \right) d\Omega$$

$$- \int_{\Omega_e} r \left(\overline{uw} \frac{\partial N_i}{\partial r} + \overline{ww} \frac{\partial N_i}{\partial z} \right) d\Omega$$

$$+ \int_{\Gamma_e} r N_i \left(\overline{uw}\, \ell_r + \overline{ww}\, \ell_z \right) d\Gamma \tag{3}$$

$$F_k = \int_{\Omega_e} N_i \; r \left(U \frac{\partial k}{\partial r} + W \frac{\partial k}{\partial z} - Pr + \varepsilon - B \right) d\Omega$$

$$+ \int_{\Omega_e} r \frac{\nu_t}{\sigma_k} \left(\frac{\partial k}{\partial r} \frac{\partial N_i}{\partial r} + \frac{\partial k}{\partial z} \frac{\partial N_i}{\partial z} \right) d\Omega$$

$$- \int_{\Gamma_e} r N_i \frac{\nu_t}{\sigma_k} \left(\frac{\partial k}{\partial r} \ell_r + \frac{\partial k}{\partial z} \ell_z \right) d\Gamma \tag{4}$$

$$F_\varepsilon = \int_{\Omega_e} N_i \, r \left(U \frac{\partial \varepsilon}{\partial r} + W \frac{\partial \varepsilon}{\partial z} - c_{\varepsilon 1} \frac{\varepsilon}{k} \left[Pr + (1 - c_{\varepsilon 3}) \, B \right] + c_{\varepsilon 2} \frac{\varepsilon^2}{k} \right) d\Omega$$

$$+ \int_{\Omega_e} r \frac{\nu_t}{\sigma_\varepsilon} \left(\frac{\partial \varepsilon}{\partial r} \frac{\partial N_i}{\partial r} + \frac{\partial \varepsilon}{\partial z} \frac{\partial N_i}{\partial z} \right) d\Omega$$

$$- \int_{\Gamma_e} N_i \, r \frac{\nu_t}{\sigma_\varepsilon} \left(\frac{\partial \varepsilon}{\partial r} \ell_r + \frac{\partial \varepsilon}{\partial z} \ell_z \right) d\Gamma \qquad (5)$$

$$F_\Lambda = \int_{\Omega_e} N_i \, r \left(U \frac{\partial \Lambda}{\partial r} + W \frac{\partial \Lambda}{\partial z} \right) d\Omega + \int_{\Omega_e} r \Lambda (1 - \Lambda) V_s \frac{\partial N_i}{\partial z} d\Omega$$

$$+ \int_{\Omega_e} r \gamma \nu_t \left(\frac{\partial \Lambda}{\partial r} \frac{\partial N_i}{\partial r} + \frac{\partial \Lambda}{\partial z} \frac{\partial N_i}{\partial z} \right) d\Omega$$

$$- \int_{\Gamma_e} r \, N_i \left[\gamma \nu_t \left(\frac{\partial \Lambda}{\partial r} \ell_r + \frac{\partial \Lambda}{\partial z} \ell_z \right) + \Lambda (1 - \Lambda) V_s \ell_z \right] d\Gamma \qquad (6)$$

The vector of nodal variables at node j is $q_j = [P, U, W, k, \varepsilon, \Lambda]_j^T$, where P = difference from clear water hydrostatic pressure, U,W = horizontal and vertical mean velocity components, k = turbulence kinetic energy per unit mass, ε = dissipation rate of k, and Λ = mean sediment volume concentration. All variables in these equations are non-dimensionalized on U_o (a characteristic velocity for idealized flow through a basin), h (mean depth of fluid in the basin) and ρ (clear fluid density). The pressure P is interpolated linearly over 4-node quadrilateral elements by the linear functions M_i; all other unknowns in q_j are interpolated quadratically over 8-node elements by functions N_i. The factors ℓ_r, ℓ_z are the direction cosines of the unit outer normal to the element boundary Γ_e. The element residuals F_q are assembled into global sets of equations for subsequent solution; the element boundary integrals need only be evaluated when element boundaries coincide with the global domain boundary.

Owing to the definition of P, the body force term $\eta\Lambda_{AVE}g$ represents only the contribution of the sediment to the body force, where g = gravity. In this term a single value of excess density ($\eta\Lambda_{AVE} = \Delta\rho/\rho$, where ρ = bulk fluid density) has been used within each element by calculating Λ_{AVE} as the arithmetic mean of the element nodal values of Λ. Treating the body force in this fashion is consistent with the linear approximation of pressure and is found to alleviate certain spurious numerical results in buoyancy-driven flows, as noted by King (1982).

The current turbulence model is the k-ε turbulence model for density-affected flows proposed by Rodi (1980). In it the overbarred terms in Eqs. 2 and 3, which are turbulent velocity correlation terms, are modeled by relations similar to the constitutive relations of laminar flow, viz.

$$\overline{uu} = \frac{2}{3}k - 2\nu_t \frac{\partial U}{\partial r} \quad , \quad \overline{ww} = \frac{2}{3}k - 2\nu_t \frac{\partial W}{\partial z}$$
$$\overline{vv} = \frac{2}{3}k - 2\nu_t \frac{U}{r} \quad , \quad \overline{uw} = -\nu_t\left(\frac{\partial U}{\partial z} - \frac{\partial W}{\partial x}\right) \tag{7}$$

Here the kinematic eddy, or turbulent, viscosity is expressed as

$$\nu_t = c_\mu k^2/\varepsilon \tag{8}$$

with $c_\mu = 0.09$. The empirical determination of this and the other k-ε model constants σ_k, σ_ε, $c_{\varepsilon1}$, $c_{\varepsilon2}$, and $c_{\varepsilon3}$ is described by Rodi (1980). The transport equations that describe the distribution of k and ε complete the turbulence model; these relations are contained in Eqs. 4 and 5. These equations are driven by two kinds of source terms, Pr which represents the production of k due to shearing of the mean flow, and B which is the buoyancy-induced production of k.

Equation 6 is the sediment transport equation. The turbulent transport coefficient for sediment is represented as $\gamma\nu_t$, where γ is an experimentally determined coefficient of order one. In the current application $\gamma = 1.2$ is assumed. The motion of the continuum representation of the sediment is determined by the superposition of the mean fluid flow (U, W) and the settling velocity V_s. This settling velocity is not simply that of a single particle but is best interpreted as the subsidence velocity for a suspension undergoing hindered settling; it is in general an experimentally determined function of the local sediment volume concentration Λ.

When one notes that the velocities U and W appear in all six conservation equations and that the turbulent viscosity ν_t causes the combination k^2/ε to appear in all of these equations except Eq. 1, the strong coupling and nonlinearity of the model equations are clearly evident. The solution of these equations was based upon the Newton iteration scheme

$$\frac{\partial f_i}{\partial q_j} \Delta q_j = f_i \ , \ q_j^{new} = q_j^{old} + \alpha \, \Delta q_j \tag{9}$$

In some early solution attempts f_i was naively chosen to be the assembled global nodal residual array corresponding to the entire set of six equations; the system of equations was unmanageably large and difficult to initialize and did not yield a convergent solution for a problem of any appreciable size. In Eq. 9 q_j now represents the global number of nodal unknowns in the set of equations. Also $\alpha \leqslant 1$ is a relaxation parameter which is relatively small during the first few iterations.

Convergent solutions were successfully obtained by use of the partitioned matrix used earlier by Schamber and Larock (1981). However, the successful choice for the partitioning of the variables was not immediately obvious. Partitioning of the variables into the subsets (U_j, W_j, P_j) and $(k_j, \varepsilon_j, \Lambda_j)$ leads to two subsets of equations f_i of roughly equal size, and this grouping was attempted first. The very strong dependence of P upon Λ caused this solution scheme to diverge; these two variables must therefore appear in the same subset. The successful partition of variables is thus $(U_j, W_j, P_j, \Lambda_j)$ and (k_j, ε_j). The solution scheme alternately iteratively solves first one subset of equations while holding the unknowns from the other subset constant, and then iteratively solves the second subset while the first set of unknowns is kept constant. By this procedure the quadratic convergence rate of a pure Newton method is sacrificed, but the solution is eventually obtained when both equation subsets have converged.

MODEL APPLICATION

Sedimentation basins are used to remove suspended sediment from a carrier fluid (most often water). Frequently the concentration of sediment is sufficiently high that density currents are created. Among others, Larsen (1977) has noted such currents in secondary sedimentation basins at sewage treatment plants. Figure 1 is a schematic diagram of a typical radial flow basin shown in vertical cross-section. Fluid flows radially outward from a center feed well and exits over the effluent weir. A baffle ring is placed near the center well to prevent short circuiting to the weir. Sedimented solids are removed from the floor in a concentrated sludge by rotating pipes with suction devices.

The mathematical model is not applied to the entire physical domain of the basin cross-section because the model is only applicable in turbulent flow regions. Thus the computational region ends a small

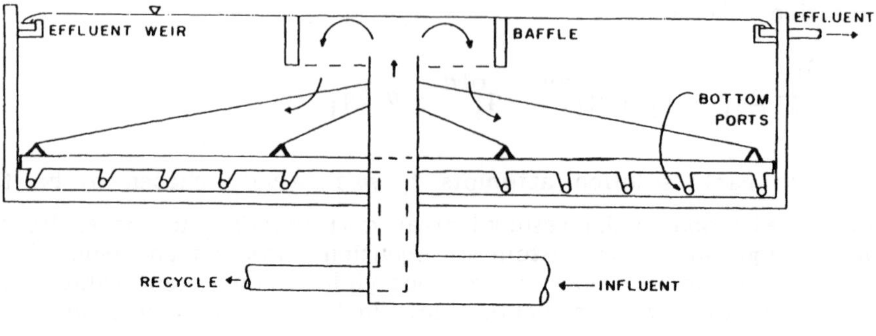

Figure 1. Schematic Diagram of a Center Feed Radial Flow Sedimentation Basin.

distance away from the physical locations of the walls, and a slip boundary condition on velocity is applied there. The entrance and exit regions are also excluded from the computational modeling region, and simple velocity distributions are prescribed at these boundaries. The details of flow in these localized regions are believed not to be significant in the determination of transport properties over the majority of the domain, and an important reduction in overall computational effort is thereby achieved.

Figures 2 and 3 present model predictions of the sediment concentration and velocity distribution for steady flow conditions that are typical for secondary settling basins. The rectangular border defines three physical surfaces and one axis of symmetry. The bottom and right borders are near walls, and the upper boundary is a liquid free surface. The left border is the central axis of symmetry (z axis). The recirculation of the flow is clearly apparent in the velocity vector plot of Figure 3. Velocity vectors near the center are not shown, because these vectors would be too large at this scale. The sediment concentration gradients which create this density current are evident in Figure 2. The steep drop in sediment concentration near the center corresponds to a fluid "dive" region. The left edge of the

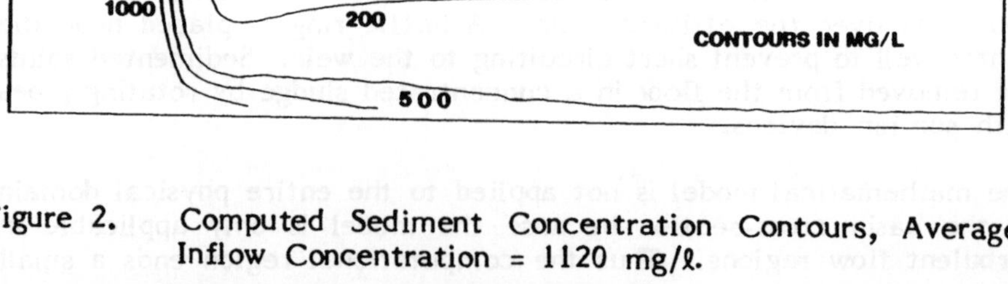

Figure 2. Computed Sediment Concentration Contours, Average Inflow Concentration = 1120 mg/ℓ.

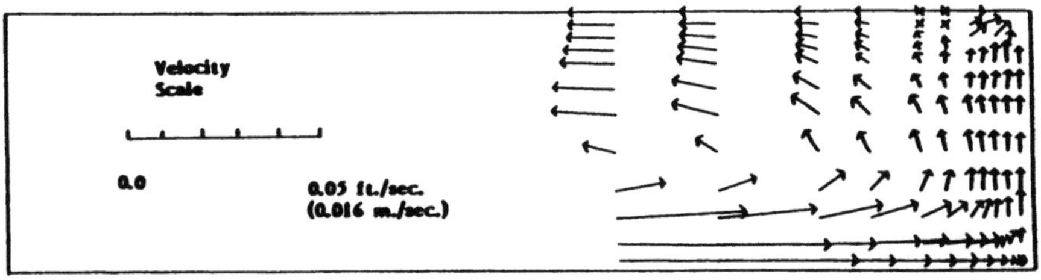

Figure 3. Velocity Vector Plot in Outer Portion of Basin.

contours is the edge of the computational region where a linear concentration profile is specified. This edge is also the location of the baffle. A uniform velocity profile is specified at the inflow section, but this is quickly swept downward by the diving, recirculatory flow. The density current is thus the primary feature of the flow field. This is consistent with the measurements of Larsen (1977). The presence of the suction apparatus is ignored in the computations, except for the choice of boundary conditions along the floor.

Figures 4, 5, and 6 present computed turbulence parameters for this flow. The higher turbulence levels associated with the higher velocities near the center of the basin can be noted in all of the figures. It is significant that the turbulent viscosity ν_t varies by nearly an order of magnitude in the radial direction. This variation directly affects the sediment concentration distribution via turbulent diffusion and indirectly through the determination of the velocity distribution. The downward flow near the entrance region affects each of these parameters, as is apparent in the figures, and it also affects the selection of a finite element grid that is proportioned so that a convergent solution procedure can be achieved. Refinement of the grid near the free surface in the dive region was required to keep the turbulence parameters from becoming locally negative during the solution process. Negative values of these quantities have no physical meaning and quickly lead to divergent results, as noted by Schamber et al. (1982) and others. Since a single mesh was used for all unknowns, the addition of a few extra elements in the dive region increased computational time significantly.

The finite element grid used in obtaining a solution contained 260 elements and 853 node points. Use of this grid creates 2662 U-W-P-Λ equations and 1530 k-ε equations. The initial solution was obtained after 17 U-W-P-Λ iterations and 35 k-ε iterations with a total CPU time of 1700 seconds on a CDC 7600 computer. Subsequent solutions with different conditions required significantly less time because in each case a previous solution was used to provide a relatively good initial estimate of the distribution of each variable. To illustrate this point, the solution shown in Figs. 2-6 is not the original solution but is the result of a subsequent computer run with

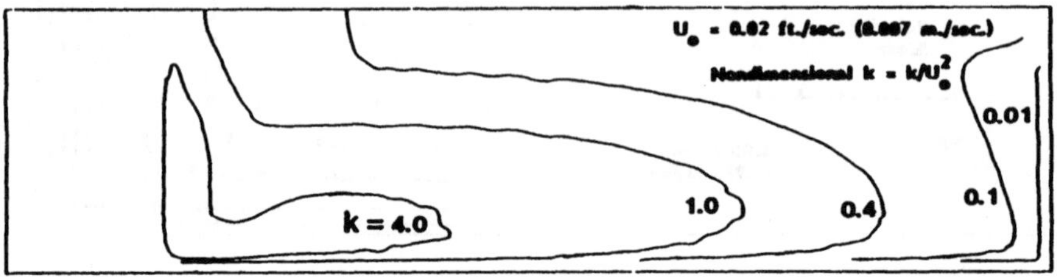

Figure 4. Contours of Computed Nondimensional Turbulence Kinetic Energy k.

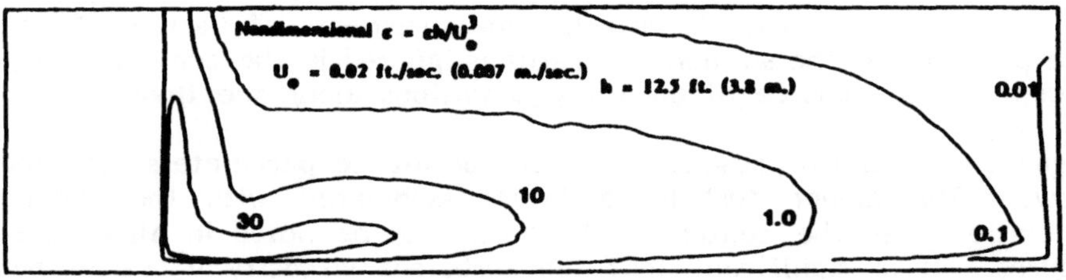

Figure 5. Contours of Computed Nondimensional Turbulence Kinetic Energy Dissipation Rate ε.

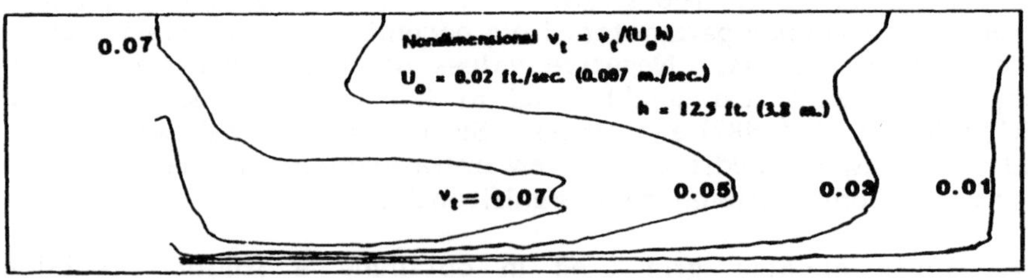

Figure 6. Contours of Computed Nondimensional Turbulent Viscosity ν_t.

a sediment settling velocity distribution having a range of settling velocities approximately twice as large as the distribution used in the initial solution. CPU time was 325 seconds which is less than 1/5 of the original solution time.

SUMMARY

This article describes a finite element model which can be used to predict density-driven recirculating flow in sedimentation basins. The model uses a second order closure for turbulence which makes it more generally applicable than a constant eddy viscosity formulation but also greatly increases the required computer solution time. The model equations are strongly nonlinear and coupled, and converged solutions are difficult to obtain if the initialization of variables is far from the final solution. The initialization of variables from a similar problem largely overcomes this difficulty and leads to significant computational economies as well.

ACKNOWLEDGEMENT

Support of this work by the National Science Foundation Grant CME-7914762 is gratefully acknowledged.

REFERENCES

DeVantier, B.A. and B.E. Larock (1982) Computation of Coupled Turbulent Flow and Sediment Transport. Finite Element Flow Analysis, T. Kawai, ed., Univ. of Tokyo Press, Tokyo, Japan, 211-218.

DeVantier, B.A. and B.E. Larock (1983) Sediment Transport in Stratified Turbulent Flow. J. Hyd. Engineering, ASCE, 109, HY12, 1622-1635.

King, I.P. (1982) A Finite Element Model for Three Dimensional Flow. Report 9150 to U.S. Army Corps of Engineers, Vicksburg, Mississippi; RMA Inc., Lafayette, California.

Larsen, P. (1977) On the Hydraulics of Rectangular Sedimentation Basins: Experimental and Theoretical Studies. Report No. 1001, Dept. Water Res. Engr., Lund Inst. Tech., Univ. of Lund, Sweden.

Rodi, W. (1980) Turbulence Models and Their Application in Hydraulics, IAHR Publication, Delft, Netherlands.

Schamber, D.R. and B.E. Larock (1981) Numerical Analysis of Flow in Sedimentation Basins. J. Hyd. Div., ASCE, 107, HY5, 575-591.

Schamber, D.R., Larock, B.E. and B.A. DeVantier (1982) Continuation Methods for the Finite Element Solution of Turbulent Flows. Finite Elements in Water Resources, Holz, K.P. et al., eds., Proc. Fourth Int. Conf. Finite Elements in Water Resources, Hannover, Germany, Springer, Berlin, 3.23-3.32.

RECENT DEVELOPMENTS IN THE APPLICATION OF THE F.E.M. TO FULLY THREE DIMENSIONAL TURBULENT FLOW

C. Taylor*, J. Rance* and J.O. Medwell**

*Department of Civil Engineering, University College, Singleton Park, Swansea SA2 8PP, U.K.

**Department of Mechanical Engineering, University College, Singleton Park, Swansea SA2 8PP, U.K.

SUMMARY

A one equation model of turbulence, the k model, is used to predict heat transfer from a duct rotating about an axis normal to its longitudinal centre-line. Such ducts form part of the cooling mechanism of rotating turbine blades.

1. INTRODUCTION

The transfer of heat to a fluid flowing in a tube rotating about an axis perpendicular to its longitudinal centre line is markedly affected by secondary transverse flows initiated and maintained by Coriolis forces. Such secondary flows may alter the efficiency of cooling ducts in gas turbine blades and an effective robust numerical model of this phenomenon would be a useful additional design tool.

Experimentation, excluding heat transfer, aimed at quantifying flow in cylindrical rotating ducts with a fully developed longitudinal velocity profile was undertaken by Trefethen [1,2]. Ito and Nanbu [3] performed similar experiments over a much wider range of Reynolds numbers and produced associated predictive formulae for friction factors.

Laminar flow with heat transfer was the subject of a study by Mori and Nakayama [4]. Their analytically predicted friction factors did, however, differ from those measured by Trefethen [2]. An experimental investigation by Mori, Fukada and Nakayama [5] resulted in predictions for both heat and mass transfer coefficients.

Numerically predicted results, obtained using a finite difference scheme, by Skiadaressis and Spalding [6] were compared with the measured values published by Ito and Nanbu [3].

The predicted results compared quite favourably with the measured values for isothermal conditions.

The present paper reports on recent developments of a finite element based approach to predict secondary flow patterns and the associated heat transfer in the cooling passages of turbine blades.

2. GENERAL EQUATIONS

A one equation turbulence model concept is used to define the equations used to predict the flow characteristics of the fluid within the rotating cylinder. Figure 1, which illustrates a cylinder rotating about an axis normal to its longitudinal centre-line.

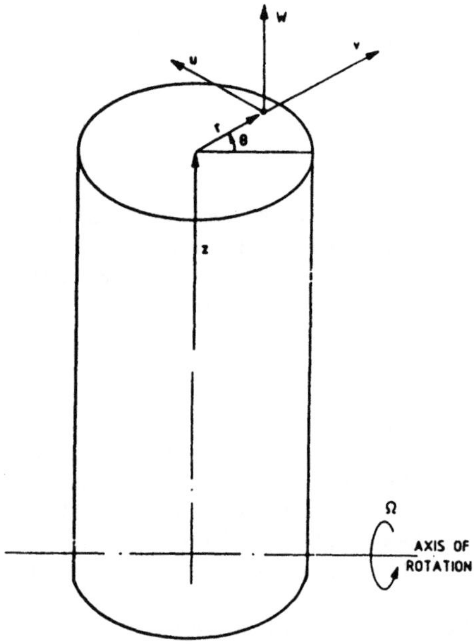

Fig.1.
Cylindrical duct rotating about an axis that is perpendicular to the duct axis.

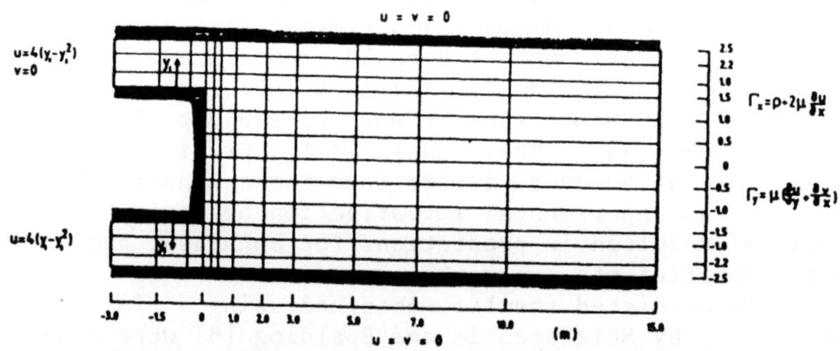

Fig.2. Element mesh and boundary conditions

Ignoring any coupling, via body force terms, between the energy and momentum equations and assuming that the dynamic viscosity is temperature independent, a generalised set of governing equations used for present purposes is,

$$\rho\left[w\frac{\partial v}{\partial z} + v\frac{\partial v}{\partial r} + \frac{u}{r}\frac{\partial v}{\partial \theta} - \frac{u^2}{r}\right] = -\frac{\partial p}{\partial r} + \left[\frac{1}{r}\frac{\partial}{\partial r}(r\,\tau_{rr}) + \frac{1}{r}\frac{\partial}{\partial \theta}(\tau_{r\theta}) - \frac{\tau_{\theta\theta}}{r} + \frac{\partial \tau_{rz}}{\partial z}\right] + 2\rho\Omega\omega\,\mathrm{Cos}\theta \tag{1}$$

depicting momentum transfer in the radial r-direction

$$\rho\left[w\frac{\partial u}{\partial z} + v\frac{\partial u}{\partial r} + \frac{u}{r}\frac{\partial u}{\partial \theta} + \frac{uv}{r}\right] = -\frac{1}{r}\frac{\partial p}{\partial \theta} - \left[\frac{1}{r^2}\frac{\partial}{\partial r}(r^2\,\tau_{r\theta}) + \frac{1}{r}\frac{\partial}{\partial \theta}(\tau_{\theta\theta}) + \frac{\partial}{\partial z}(\tau_{\theta z})\right] - 2\rho\Omega\omega\,\mathrm{Sin}\theta \tag{2}$$

corresponding to momentum in the θ-direction

$$\rho\left[w\frac{\partial w}{\partial z} + \frac{u}{r}\frac{\partial w}{\partial \theta} + v\frac{\partial w}{\partial r}\right] = -\frac{\partial p}{\partial z} + \left[\frac{1}{r}\frac{\partial}{\partial r}(r\,\tau_{rz}) + \frac{1}{r}\frac{\partial}{\partial z}(\tau_{zz})\right] + \rho\Omega^2 z \tag{3}$$

depicting momentum in the longitudinal, z, direction. Assuming the fluid to be incompressible the continuity equation is

$$\frac{\partial w}{\partial z} + \frac{1}{r}\frac{\partial u}{\partial \theta} + \frac{\partial v}{\partial r} + \frac{v}{r} = 0 \tag{4}$$

The transfer of turbulence kinetic energy, k, is assumed to be represented by

$$\rho\left[w\frac{\partial k}{\partial z} + v\frac{\partial k}{\partial r} + \frac{u}{r}\frac{\partial k}{\partial \theta}\right] = -\frac{1}{r}\frac{\partial}{\partial r}\left[r\frac{\mu_t}{\sigma_k}\frac{\partial k}{\partial r}\right] + \frac{1}{r}\left[\frac{\mu_t}{\sigma_k}\frac{\partial k}{\partial \theta}\right] + G_k - \frac{\rho k^{3/2}}{\ell} \tag{5}$$

finally, the energy equation is written in the form

$$\rho\left[v\frac{\partial T}{\partial r} + \frac{u}{r}\frac{\partial T}{\partial \theta} + w\frac{\partial T}{\partial z}\right] = \frac{1}{r}\frac{\partial}{\partial r}\left[r\frac{\mu_t}{\sigma_h}\frac{\partial T}{\partial r}\right] + \frac{1}{r}\frac{\partial}{\partial \theta}\left[\frac{\mu_t}{\sigma_h}\frac{\partial T}{\partial \theta}\right] \tag{6}$$

in which u, v and w are the velocities corresponding to the tangential radial and axial directions, respectively, ρ, p and T are the local values of fluid density, pressure and temperature, and Ω the Coriolis coefficient. The turbulence viscosity, μ_t, is related to the turbulence kinetic energy, k, and a characteristic length 'ℓ', using an expression of the form,

$$\mu_t = C_\mu \rho k^{1/2} \ ,$$

and C_u, C_1, C_2, σ_k and σ_h are all assumed to be constants.

Following standard 'law of the wall' procedure a Van Driest [7] model is used in the near wall region and the finite element domain is started some distance away from the wall.

3. METHOD OF APPROACH

A conventional Galerkin based finite element formulation of the governing equation is used [8] and twenty noded isoparametric 'brick' elements effect a spatial discretisation.

The method of solution, once the matrix equivalent of the governing equation had been established, followed precisely the same procedure as that adopted by the authors whilst solving other turbulent flow problems [8]. The only additional refinement was the development of a technique where the downstream boundary conditions could be evaluated as the solution proceeds. Typically, for the current class of problems boundary conditions of the following form can be imposed:-

i) Dirichlet

ii) Neumann, and

iii) 'Traction'

The first two are self explanatory and have been used extensively throughout the literature. However, the third has not been used quite so extensively and its full potential for solving turbulent flow problems has not been realised. A simple technique has been developed [9] where initially arbitrary values of traction on downstream boundaries are imposed which are successively updated to eventually comply with the upstream flow regime.

The above procedure can be demonstrated by a closer inspection of typical traction type boundary conditions and the procedure employed for solving the generalised turbulent form of the Navier-Stokes equations.

Consider, for simplicity, a two dimensional flow situation where the orthogonal axes are represented by x' and y' and corresponding velocities by u' and v'. The traction condition referred to the momentum equations can be written,

$$\left.\begin{aligned} \tau_x &= -p' + 2\mu_e \frac{\partial u'}{\partial x} \\ \tau_y &= \mu_e \left(\frac{\partial v'}{\partial x} + \frac{\partial u'}{\partial y}\right) \end{aligned}\right\} \qquad (7)$$

On a downstream boundary, where no disturbances are propagated through the boundary in an upstream direction, the values of τ_x and τ_y are set to arbitrary, usually zero, values. The primitive variables p, u and v are then evaluated. Obviously, the initial, arbitrary, values of traction will not now comply with (7) if the updated primitive variables and μ_e are incorporated. These 'updated' values are then used, with an appropriate relaxation factor, to calculate more accurate values of the boundary traction terms. This process is repeated until both the tractions associated with the governing equations and the imposed boundary tractions are compatible. This technique proves particularly useful when downstream boundaries are located in regions where fully developed conditions cannot be assumed. Indeed, such a technique has been successfully employed where the downstream boundary is located within a recirculation zone [9].

A comparative study of the velocities within the recirculation zone of a backward facing step, Figure 2, is shown on Figure 3. The step height is 1.5 m and the Reynolds number, based on step height, is 40. Although the illustrative example is for a low Reynolds number the technique is equally applicable for turbulent flow and was used during the execution of the example printed in the current text.

Example

The illustrative example chosen for demonstrating the applicability of the finite element technique to solving three dimensional flow problems with both centrapetal acceleration and Coriolis forces is that of flow through a rotating cylindrical duct.

Since the secondary flow, in the r,θ plane, is usually small compared with that in the longitudinal, z, direction, the following model could be used for evaluating the characteristic length scale, ℓ, [10]

$$\ell = \left(1 - e^{-y^+/26}\right) \ell_m \tag{8}$$

where

$$\ell_m = (0.14-0.08\,(y/R)^2 - 0.06\,(y/R)^4)\,R \times 2.5$$

$$y^+ = \sqrt{\frac{\tau_w}{\rho}}\cdot\frac{y}{\nu}$$

in which τ_w is the value of the shear stress on the boundary, y is measured normal to the boundary, ν the kinematic viscosity of the fluid and R the duct radius.

The physical dimensions of the duct, boundary conditions imposed and finite element mesh used are shown on Figure 4. The properties of the fluid and values of coefficients are as follows,

$\rho = 1.77\ Kg/m^3$, $\mu = 1.983 \times 10^{-5}\ Kg/m.s$

$\sigma = 0.7$, $\sigma_k = 1.53$, $\sigma_h = 0.9$, $C_\mu = 0.22$, $C_1 = 1.45$ and $C_2 = 0.18$.

The non-dimensional rotating speed, ω, imposed is 0.04 where

$$\omega = \frac{\Omega\ d}{\bar{\omega}}$$

where d is the diameter of the duct and $\bar{w}$ the mean velocity in the z direction. The Reynolds number based on duct diameter corresponding to the above is 25,000.

The calculated profile of longitudinal velocity is as shown on Figure 5 and the associated streamline plot on Figure 6. When utilising the energy equation the temperature profiles, Figure 7, were obtained.

An interesting feature of the current developments is the plot of Nusselt number variation with increasing rotation, Figure 8 . These are compared with both experimental results and those predicted numerically by Spalding [6]. It is apparent that the values determined using the finite element method compare more favourably with experiment.

The authors are of the opinion that upwinding, not used in the finite element method, may be a factor to be investigated since spurious damping may affect the secondary flow pattern and give rise to the discrepancies in the Nusselt number values. This is currently being investigated by the authors using a numerical model where 3-D streamwise upwinding has been introduced.

4. CONCLUSIONS

The finite element model is now becoming a viable technique in solving practical three dimensional turbulent flow problems. With the advent of the updated traction boundary condition approach, matrix symmetrising techniques [11] and the E.B.E. technique [12] more complex problems can now be analysed within the restrictions imposed by core and c.p.u. time availability. The updated traction boundary condition technique enables the analyst to solve problems without the restriction of a priori, Neumann, mixed Neumann and traction or mixed Dirichlet and traction conditions imposed on downstream boundaries. In addition, matrix symmetrising techniques obviate much of the core requirement restriction.

5. ACKNOWLEDGEMENTS

The authors wish to acknowledge the financial assistance of the S.E.R.C. without which the current work would not have been possible.

6. REFERENCES

1. Trefethen, L. (1957) Fluid flows in radial rotating tubes. Actes IXe Congrés International de Mécanique Appliquée, Vol. 2., Université de Bruxelles, pp. 341-350.

2. Trefethen, L. (1957) Flow in rotating radial ducts: Report R55GL350 on laminar flow in rotating heated horizontal and bent tubes, extended into transition and turbulent regions, General Electric Company, pp. 7-27.

3. Ito, H. and Nanbu, K. (1971) Flow in rotating straight pipes of circular cross-section, A.S.M.E. Transactions, J. Basic Engineering, pp. 383-394.

4. Mori,Y. and Nakayama, W. (1968) Convective heat transfer in rotating radial circular pipes (1st Report Laminar Region), Int. J. Heat and Mass Transfer, Vol. 11, pp. 1027-1040.

5. Mori, Y., Fukada, T. and Nakayama, W. (1971) Convective heat transfer in a rotating radial circular pipe, Int. J. Heat and Mass Transfer, Vol. 14, pp. 1807-1824.

6. Skiadaressis, D. and Spalding, D.B. (1977) Heat transfer in a pipe rotating around a perpendicular axis, Internal Report, Heat Transfer Section, Imperial College of Science & Technology, HTS/77/3.

7. Van Driest, E.R. (1956) On turbulent flow near a wall. Journal of the Aeronautical Sciences, November.

8. Taylor, C., Thomas, C.E. and Morgan, K. (1983) Turbulent heat transfer via the F.E.M. : Heat transfer in rod bundles, I.J.N.M.F., 3, 4, pp. 363-375.

9. Taylor, C., Rance, J. and Medwell, J.O. (1984) A note on the imposition of traction boundary conditions, submitted J. Comp. Science.

10. Nikuradse, J. (1932) Gesetzmassigkeit der turbulenten stromung in glatten rohren, Ver. Deutsch. Ing. Forschungsheft, 356, 21.

11. Garg, N.K. (1984) A matrix symmetrising technique in the finite element context, Internal Report, University College of Swansea.

12. Hughes, T.J.R., Raefsky, A., Muller, A., Winger, J.H. and Levit, I. (1984) A progress report on E.B.E. 2nd Int. Conf. Non-Linear Problems, Barcelona, Pineridge Press.

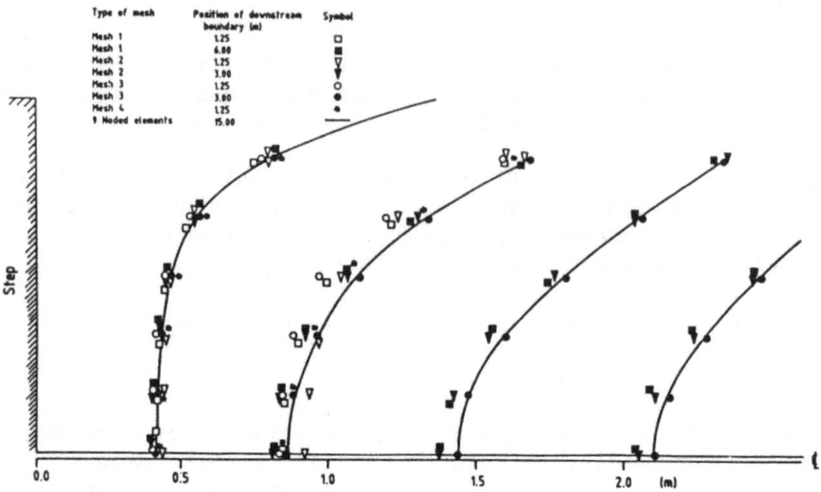

Fig.3. Comparison of velocities within the recirculation zone with mesh refinement at the downstream boundary

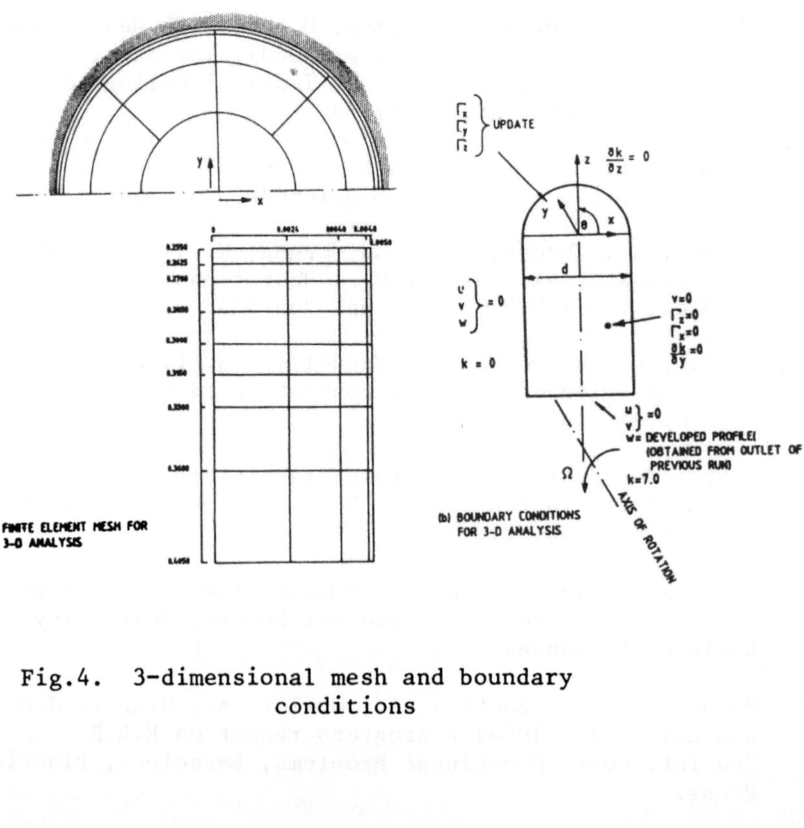

Fig.4. 3-dimensional mesh and boundary conditions

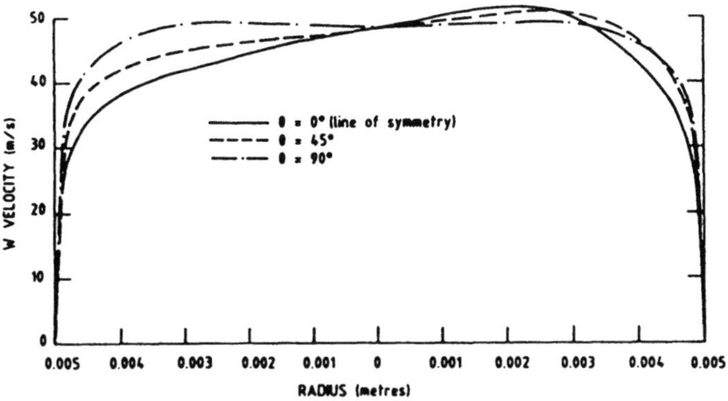

Fig. 5. Fully developed longitudinal velocity profiles at various angular positions to the line of symmetry (Re=25,000, ω=0.04)

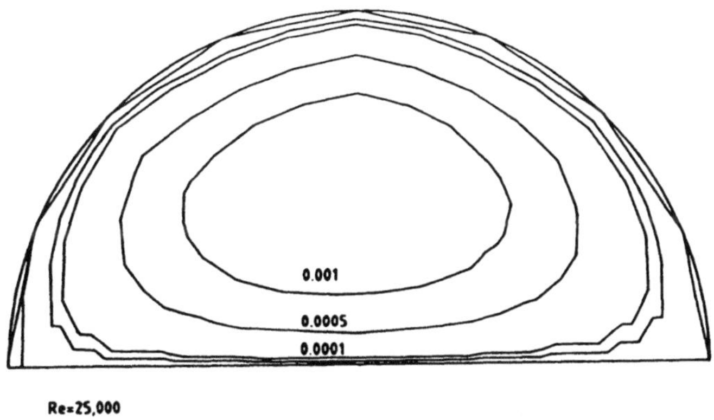

Fig. 6. Streamline plot for Coriolis effect (Re=25,000, ω=0.04)

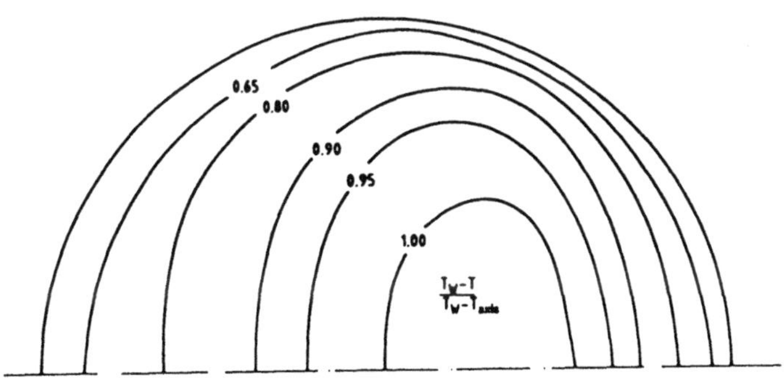

Fig. 7. Temperature contours for fully developed turbulent flow (Re=25,000, ω=0.04)

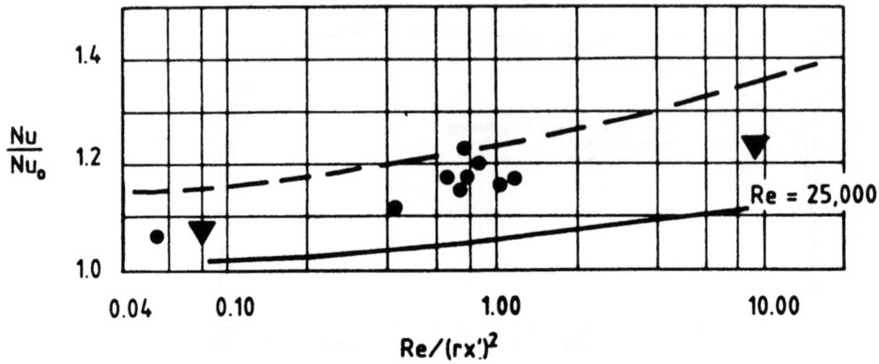

Fig. 8. Nusselt numbers for fully developed turbulent flows

——— Spalding predictions

— — Predictions by Mori, Fukada and Nakayama

● ● Experimental data by Mori, Fukada and Nakayama

$r = \bar{w}/(R\,\Omega)$ $x' = (1+5.136(1/r^2))^{1/2} - 2.27(1/r)$

▼ F.E.M. predictions

ANOTHER ATTEMPT TO OVERCOME THE BENT ELEMENT BLUES

P. M. Gresho and J. M. Leone, Jr.
Lawrence Livermore National Laboratory
P. O. Box 808
Livermore, CA 94550

1. INTRODUCTION

This paper is the second of a pair in which we attempt, mostly unsuccessfully, to patch a perceived defect in the Galerkin finite element approximations to the pressure gradient. The first was presented earlier this year (Gresho, et al. (1984a)).

Specifically, we address the problems associated with the bilinear velocity, piecewise-constant pressure element for solving the incompressible Navier-Stokes equations. When the discrete approximation to ∇P is examined on a mesh of distorted elements (via the bilinear isoparametric mapping), it appears to be not very accurate; from a Taylor series viewpoint, this element is not meant to be bent. After exposing these alleged defects, we proceed to patch the gradient approximation. In the limited context of our techniques used to integrate the time-dependent equations, we discuss and demonstrate the advantages of the new approximation. But we also discuss the numerous disadvantages uncovered during our investigation and, in the end, admit failure — and we continue to sing the "Bent Element Blues."

2. NAVIER-STOKES EQUATIONS AND DISCRETE APPROXIMATIONS

The governing equations of interest herein are the incompressible Navier-Stokes equations:

$$\partial \underline{u}/\partial t + \underline{u}\cdot\nabla\underline{u} = -\nabla P + \nu\nabla^2\underline{u} + \underline{a}, \quad \nabla\cdot\underline{u} = 0, \qquad (1a,b)$$

where $\underline{a}$ represents a body "force" term, such as buoyancy when non-isothermal flow is considered. The application of a (modified - see Gresho et al. 1984b) GFEM scheme to (1a) and a time-differentiated form of (1b) leads to

$$M\dot{u} + CP = f, \quad C^T\dot{u} = 0\ , \qquad (2a,b)$$

where $u(0) = u_o$ and $C^T u_o = g$. Here u is a vector of all nodal velocities, P is a vector of discrete pressures, and f is a vector function of velocity and temperature that includes all physical processes except the pressure gradient (advection, diffusion, buoyancy, boundary conditions). The vector g, assumed independent of time for simplicity, reflects the imposition of imposed velocity boundary conditions.

The matrix of interest herein is C, which approximates the pressure gradient. For the bilinear velocity (ϕ_i), piecewise-constant pressure (ψ_i) basis functions, it is given by

$$C^x_{ij} = -\int_\Omega \phi_{i,x}\psi_j\,, \quad C^y_{ij} = -\int_\Omega \phi_{i,y}\,\psi_j\,. \tag{3a,b}$$

An alternative but equivalent form of the semi-discrete equations, is (2a) and

$$(C^T M^{-1} C)\,P = C^T M^{-1}\,f\,, \tag{4}$$

the discrete Poisson equation for pressure, which is easily derived from (2). Since we are primarily interested in explicit time integration techniques, we will use (2a) and (4) in the lumped mass approximation, $M_{ij} = \delta_{ij} \int_\Omega \phi_i$, to compute P which is then inserted into (2a) to permit time integration. Using the explicit Euler method, the following algorithm results: given u_n satisfying $C^T u_n = g$ at time t_n, (i) evaluate $f_n = f(u_n)$ and solve (4) for P_n, (ii) update the velocity from a discrete approximation to (2a): $u_{n+1} = u_n + \Delta t\, M^{-1}(f_n - CP_n)$.

Additional and important details are given in Gresho et al. (1984b), especially regarding the process of subcycling in which the pressure update is not performed at each time step.

3. THE PERCEIVED PROBLEM

If one examines the discrete approximation to the pressure gradient at a particular node in a patch (typically 4) of distorted (isoparametric) elements, e.g. $M^{-1}C^x \sim \partial/\partial x$, he will be disappointed in that the locations (x, y-coordinates) of the 4 "corner" nodes in the 4-patch are not recognized by the discrete approximation; they can be anywhere and the stencil is unchanged; see also Gresho et al. (1984a). This leads to a Taylor-series (or finite difference) interpretation that the approximation is generally zero order accurate in h (the generic measure of mesh size). While this fact does not preclude convergence as $h \to 0$, since Taylor series consistency is sufficient, but not necessary for convergence, it does suggest that the ∇P approximation may be quite inaccurate on distorted meshes.

Figure 1 shows the simplest possible 4-patch of this type, with only one (simply) distorted element. We will examine the discrete approximation to $\partial P/\partial x$ at the "central" node of this patch, first using the C-matrix and, later, with more accurate discrete approximations; see also Cullen (1983). From (3a) and the lumped mass approximation ($M = 4h^2(1 + (1/12)\tan\theta)$), we get

$$M^{-1}C^x P = (1/2)[(P_1-P_2)/h + (P_4-P_3)/h]/[1 + (1/2)\tan\theta], \tag{5}$$

which, for $h \to 0$ and θ fixed, gives

$$M^{-1}C^{x}P = [\partial P/\partial x - (1/8) \tan\theta \; \partial P/\partial y]/(1 + (1/12) \tan\theta) + 0(h \tan\theta) + 0(h^2), \qquad (6)$$

which is Taylor-series inconsistent (it does not yield simply $\partial P/\partial x$ as $h \to 0$). The spurious "cross-gradient" term can generate significant errors if $\partial P/\partial y >> \partial P/\partial x$, a situation that can arise, for example, when buoyancy forces are significant and gravity acts in the y-direction.

In practice, this type of inaccuracy has caused significant errors (in the more important variable - velocity) on meshes which might otherwise appear to be adequate. For example, in Gresho et al. (1981), we presented vortex shedding results for which the error in ∇P produced spurious results. This example is also treated in Gresho et al. (1984a), in which a higher-order pressure approximation (bilinear - i.e. equal-order interpolation) generated good results (velocity only) on the same mesh.

4. SOME AD HOC PATCH JOBS

Based on the observation that the gradient approximation is poor, yet the divergence approximation (C^T, an element-level approximation) is good (it is generally one order more accurate in h (Gresho et al. (1984b)), we next present some new approximations which retain C^T in the continuity equation, but replace $M^{-1}C$ by an ostensibly more accurate approximation to the gradient, which we call G. The goal is to obtain a gradient approximation which, using the same (centroid) locations for P as for the C-matrix, is never worse than first order accurate.

4.1 Internal nodes. Referring to the general 4-patch shown in Fig. 2, we present 3 discrete gradient approximations which achieve the above goal.

(1) A geometric construction. This technique, due to M. Cullen (private communication), is as follows: First connect the 4 centroids with straight line segments, then locate the points where these line segments intersect the x- and y-axes (A,B,C,D), and finally, assuming linear variation of P along these line segments, compute

$$G_x P|_0 = (P_D - P_B)/\ell_{BD} \qquad (7)$$

to approximate $\partial P/\partial x$ and similarly for the y-component. In (7), the interpolated pressures are given by $P_D = (\ell_{4D} P_1 + \ell_{D1} P_4)/\ell_{41}$ and $P_B = (\ell_{3B} P_2 + \ell_{B2} P_3)/\ell_{32}$, where the ℓ's are the distances between the points indicated in the figure; these are "easily" obtained from the given geometry (i.e., from the x, y-coordinates of the 9 nodes comprising the 4-patch).

This technique, however, was abandoned after brief testing, for the following reasons: (i) it can lead to strange (and probably quite inaccurate) results on the boundary for certain meshes, (ii) the logic involved for the general case (arbitrary geometry) became quite complicated, (iii) the prospects for a 3D extension did not look encouraging, and (iv) the technique below is much simpler.

(2), (3) Bilinear interpolation. With the concurrence of several of our numerical analyst consultants at LLNL, we then generated a gradient approximation by using bilinear interpolation on the quadrilateral connecting the centroid points 1,2,3,4 after mapping it to the unit square; the general result is

$$\frac{\partial P}{\partial x}(\xi,\eta) = \left(\frac{\partial y}{\partial \eta}\frac{\partial P}{\partial \xi} - \frac{\partial y}{\partial \xi}\frac{\partial P}{\partial \eta}\right) / \left(\frac{\partial y}{\partial \eta}\frac{\partial x}{\partial \xi} - \frac{\partial y}{\partial \xi}\frac{\partial x}{\partial \eta}\right), \tag{8}$$

and similarly for $\partial P/\partial y$, where x, y, and P are expressed by the bilinear functions in (ξ,η); e.g. $P = \Sigma P_j\phi_j(\xi,\eta)$. (Note that this approach is less general than the geometric construction, which is not limited to 4-patches.)

Although this technique generates a first order gradient approximation anywhere inside the quadrilateral, we tested only two locations: (i) the simplest is to evaluate ∇P at the centroid of the quadrilateral; i.e. at $\xi = \eta = 0$. In this case, we obtain a special case (a 4-patch) of the general gradient operater presented by Noh (1964). Unfortunately, and surprisingly, this simple scheme was found to generate an inconsistent linear system (via the analog of (4)) in certain cases. E.g. for a thermally-driven and tilted (gravity not aligned with x or y) cavity ($\underline{u} = \underline{0}$ on $\partial\Omega$) with distorted elements inside, the code generated two zero pivots during the LU decomposition of the matrix (C^TG); one corresponds to the familiar hydrostatic pressure mode, but the other corresponds to a spurious eigenvector which, unfortunately, was not in the range of C^TG. This technique too was abandoned. (ii) The final method consists, logically one might add, of evaluating ∇P at the point where the momentum equation is needed (node 0 in Fig. 2); requiring, of course, the evaluation of the inverse mapping. While this technique has the drawback that node 0 is now constrained to lie inside the quadrilateral 1-2-3-4, it was by-and-large the most successful. (Brief experiments with bilinear <u>extrapolation</u> in cases where node 0 is outside the quadrilateral proved near-fatal; not only is it often quite inaccurate, but the mapping does not even exist for certain regions.)

In spite of the fact that only one of the three approximations survived, it is interesting to compare them for two special cases:

(i) For the patch of 4 different rectangles shown in Fig. 3, the following discrete approximations to $\partial P/\partial x$ are obtained.

<u>C-matrix:</u>

$$M^{-1}C^xP|_o = 2\,[h_2\,(P_1-P_2) + h_1\,(P_4-P_3)]/(\ell_1+\ell_2)(h_1+h_2), \tag{9a}$$

which is still first-order accurate (since the elements are not distorted).

<u>Geometric construction:</u>

$$G_x\,P|_o = 2[h_1(P_1-P_2)+h_2(P_4-P_3)]/(\ell_1+\ell_2)\,(h_1+h_2), \tag{9b}$$

which is also first-order accurate. It is interesting that the weighting factors, h_1 and h_2, are reversed from their appearance in the Galerkin result ($M^{-1}C_xP$).

Bilinear interpolation with centroid evaluation (the triangle in the figure:

$$\partial P/\partial x|_{\Delta} = 2\,[(P_1-P_2) + (P_4-P_3)]/(\ell_1+\ell_2), \tag{9c}$$

which is also first-order accurate at node 0, although it is second-order accurate at the centroid (where it is not needed).

Bilinear interpolation evaluated at node 0: Same as the geometric construction, (9b); these two schemes are identical for rectangular meshes. In addition, they are also then exact for the bilinear term, xy. Finally we remark that all four schemes are the same when the rectangles are the same size — and the approximation is second-order accurate.

(ii) For the simple 4-patch shown in Fig. 1, we obtain three first-order accurate approximations (vis-a-vis (5) from the C-matrix):

Geometric construction:

$$G_xP = [(P_1-P_2)+(P_4-P_3)+(1/4)\tan\theta(2P_1-P_2-P_3)]/2h[1+(1/4)\tan\theta], \tag{10a}$$

Bilinear interpolation with centroid evaluation:

$$\partial P/\partial x|_o = [(P_1-P_2)+(P_4-P_3)+(1/4)\tan\theta(P_1-P_3)]/2h[1+(1/8)\tan\theta], \tag{10b}$$

Bilinear interpolation evaluated at the "central" node:

$$\partial P/\partial x|_{node} = [32(P_1-P_2+P_4-P_3)+8\tan\theta(2P_1-P_2-P_3)+\tan^2\theta(P_1-P_2)] \tag{10c}$$
$$/h(8+\tan\theta)^2.$$

The truncation error in each of these is $O(h\tan\theta) + O(h^2)$.

4.2 Boundary nodes. When boundary conditions other than Dirichlet (specified velocity) are employed, it may be necessary to also approximate one or both components of ∇P at boundary nodes. After studying several potential methods, the one finally selected is based on simple linear extrapolation from the nearest internal node. Figure 4 shows two nodes on the boundary (A and B) at which an approximation to ∇P is required. Given that ∇P is already available at node 0 as discussed above, the approximations used are:

$$\nabla P|_A \approx \nabla P|_{A'} = 2\nabla P|_{\bar{P}} - \nabla P|_o, \tag{11a}$$

$$\nabla P|_B \approx \nabla P|_{B'} = 2\nabla P|_1 - \nabla P|_o, \tag{11b}$$

where $\nabla P|_{\bar{P}}$ is obtained from the bilinear interpolant evaluated at $\xi = 0$, $\eta = 1$ and $\nabla P|_1$ is evaluated at $\xi=\eta=1$. These (first-order?) approximations are obviously more accurate when the elements are all rectangular, since then A'=A and B'=B.

Earlier, and subsequently abandoned approximations included: (i) Retain the C-matrix approximation on the boundary. While this technique may ultimately be the only one that can work for flow-through boundaries (for reasons discussed later), it was found to be too inaccurate for certain thermal convection simulations involving a (fixed) free surface and a shear stress boundary condition. (ii) Zero-order extrapolation: Simply use the computed gradient at node 0 (Fig. 4) at the boundary nodes. This too was, not surprisingly, found to be too crude. (iii) Bilinear extrapolation. For reasons discussed earlier, this is too dangerous; bilinear extrapolation (in the mapped plane) is not recommended for a distorted quadrilateral (it is, of course, more acceptable for rectangles, for which the approximation remains bilinear in the x,y-coordinates).

5. MODIFICATIONS TO THE TIME INTEGRATION SCHEME

Rather than (2a) and (4), the semi-discrete differential-algebraic system is now

$$\dot{u} + GP = M^{-1}f, \tag{12a}$$

and (12b) (below), where G is the matrix approximation (bilinear) to the gradient operator.

Using (12a) and (2b), P is first obtained from the new Poisson equation (replacing (4))

$$(C^TG)\,P = C^TM^{-1}f \equiv b, \tag{12b}$$

where the (global) matrix C^TG is an unsymmetric approximation to the Laplacian operator — and this causes (at least) two new problems, one obvious (a more expensive solution procedure) and one not (a null space of possibly larger dimension than that of $C^TM^{-1}C$). We address these, and other, related, issues below. Note that, while G approximates the grad operator, G^T does not approximate the div operator.

5.1 The Null Space of C^TG and its Effects. To set the stage, we first briefly review some relevant theory from linear algebra. The linear system Ax=b, where A is a square, but unsymmetric matrix with real elements, has a solution if and only if the given vector b is orthogonal to every vector in the null space of A^T. I.e., if the dimension of the null space of A^T (and A, since they have the same eigenvalues) is n, there are n nontrivial vectors y_i such that $A^Ty_i=0$, i=1,2...n and it is required that $b^Ty_i=0$, i=1,2,...n in order that Ax=b have a solution. If these conditions are satisfied, the solution is given by

$x = x_p + \sum_{j=1}^{n} a_jz_j$, where x_p is a particular solution, the a_j are arbitrary scalars, and the z_j are the vectors in the null space of A, $Az_i=0$, i=1,2...n. Thus, when n>0 we have two "problems": (i) the solution may not exist, and (ii) when it does exist, it is never unique.

Application of the above to the Poisson equation of (12b) leads to the requirement, with $A=C^TG$, that $f^TM^{-1}C\tilde{P}_i=0$ where $\tilde{P}_i$ satisfies $G^TC\tilde{P}_i=0$, $i=1,2,...n$ in order for a solution to exist. If a solution exists, it is given by

$P=P_P + \sum_{j=1}^{n} a_j P_{*j}$, where P_{*i} solves $C^TG\, P_{*i}=0$, $i=1,2,...n$, and P_P is a particular solution of (12b).

Consider first the pure pressure modes of C (see Sani et al. 1981) which are also in the null space of A^T above: These are the two vectors, Pm_i, satisfying $CPm_i=0$. The first of these is the hydrostatic mode, $Pm_1=P_H$, a constant vector; and the second is the checkerboard (CB) mode, $Pm_2=P_{CB}$, where $(P_{CB})_i = \pm 1/A_i$ for rectangular elements with area A_i (the plus sign applies to "black" elements and the minus sign to "red" ones). If u_o and g (see (2b)) are "well-posed" in the sense described by Sani et al. (1981), then these modes cause no more problem using G than C; i.e. if the system is well-posed with C, it is also well-posed with G.

Next, an examination of the G-matrix reveals that it too has the hydrostatic pressure mode as a null vector and a new version of the CB mode (at least on a rectangular grid). The former is of course not surprising and, considering the "inverted weights" in the gradient stencil (cf (9a), (9b)), neither is the latter. For the G-matrix, the CB mode is $(P_{CB})_i = \pm A_i$ for rectangular elements and since it is this mode that can show up in the pressure solution, a new CB filter is required. Rather than the C-matrix filter,

$P_o = \sum_{j=1}^{4} (P_j A_j) / \sum_{j=1}^{4} A_j$ where A_i is the area of element i (e.g. $A_1 = \ell_2 h_2$ in Fig. 3), P_j is the (polluted) centroid pressure, and P_o is the filtered pressure at node 0, the CB mode for G is filtered by

$$P_o = \sum_{j=1}^{4} (P_j/A_j) / \sum_{j=1}^{4} (1/A_j).$$

Finally we discuss the case that leads to insurmountable difficulty for the G-matrix: if outflow boundary conditions are employed and gradient extrapolation (ostensibly of any type) is used at the outflow nodes, we have the case where $GP_H = 0$ (P_H is the hydrostatic pressure mode), yet $CP_H \neq 0$ — there is no hydrostatic mode for the C-matrix owing to the natural boundary condition obtained from integration by parts. Here P_H is a null space vector of the A matrix ($C^TG\, P_H = 0$), but not of the A^T matrix ($G^TC\, P_H \neq 0$). Yet there must exist some vector in the null space of G^TC corresponding to this zero eigenvalue, since A and A^T have the same eigenvalues. In numerical experiments corresponding to this case, we have observed that this elusive vector, call it P_e, satisfying $G^TC\, P_e = 0$ but not $CP_e = 0$ exists and, unfortunately, seems to preclude the existence of a solution (i.e., apparently $f^TM^{-1}CP_e \neq 0$). This is the reason, mentioned earlier, that "flow through" domains may require that the C-matrix be retained at outflow boundaries (or perhaps at just one node there?) where $\underline{u}$ cannot (or should not) be specified.

5.2 Direct Solution of the Unsymmetric Problem. As mentioned in Section 2, we are using explicit Euler to integrate the semi-discretized equations and the algorithm is now (given u_n at time t_n satisfying $C^T u_n = g$): (i) evaluate $f_n = f(u_n)$ and solve (12b) for P_n, (ii) advance the velocity via $u_{n+1} = u_n + \Delta t\ (M^{-1} f_n - GP_n)$. While straightforward, this is an expensive scheme, relative to the symmetric system associated with the C matrix. In order to at least test the new scheme, however, we simply replaced our efficient symmetric skyline solver with an available unsymmetric banded solver from our library; in addition to the obviously increased storage and execution times, the routine uses partial pivoting which further increases the cost. The approximately three-fold increase can be sharply reduced when the ideas presented in the next section prove successful (which is only sometimes).

5.3 A Potentially Cost-Effective Scheme: Dragged Pressure. Since we have seen that "G=C" on a uniform mesh of rectangles, we shall assume that G is never very far from C and use a perturbation argument to recover two of the major advantages of C: (i) an efficient symmetric matrix, and (ii) the implementation of subcycling, a trick wherein the pressure update is not required at every time step (see Gresho et al. (1984b) for details).

Define $\delta C \equiv MG - C$ so that $G = M^{-1}(C+\delta C)$ and (12) becomes

$$M\dot{u} + (C+\delta C)P = f, \quad C^T M^{-1}(C+\delta C)P = b. \tag{13a,b}$$

The trick is to "drag" the small part of the pressure gradient; i.e. when (13) is evaluated at $t=t_n$, evaluate δCP at t_{n-1} instead to give

$$M\dot{u}_n = f_n - CP_n - \delta CP_{n-1}, \quad (C^T M^{-1} C)P_n = b_n - C^T M^{-1} \delta C\, P_{n-1}, \tag{14a,b}$$

where (14b) is solved first and P_{n-1} is known. Inserting the definition of δC leads to the equivalent pair

$$\dot{u}_n = M^{-1} f_n - GP_{n-1} - M^{-1} C\, \delta P_n, \quad (C^T M^{-1} C)\delta P_n = b_n - C^T G\, P_{n-1}, \tag{15a,b}$$

where $\delta P_n \equiv P_n - P_{n-1}$, and the algorithm is: (i) solve (15b) for δP_n, (ii) integrate (15a) using explicit Euler, via $u_{n+1} = u_n + \Delta t \dot{u}_n$. Finally, (15) can be shown to be equivalent to

$$\dot{u}_n = (M^{-1} f_n - GP_n) + M^{-1} \delta C\, \delta P_n, \quad (C^T G)P_n = b_n + C^T M^{-1} \delta C \delta P_n; \tag{16a,b}$$

i.e. the "corrections" to the desired equations ($\delta C \delta P_n = 0$) are the products of two (hopefully) small quantities, δC and δP_n.

Remarks: (i) The first step must be done honestly, since no old pressure is available; i.e. the unsymmetric system (12b) is solved for P_o. (ii) The remaining steps are much more efficient in that only the symmetric linear system need be solved and subcycling can be invoked (Gresho et al., 1984b), the principle extra cost being that both C and G (but not $C^T G$) must be stored. (iii) The stability of the (2-step) scheme is not guaranteed.

We have in fact encountered situations in which (15) is unconditionally unstable. But for many other cases, including those to be discussed below, it was stable and performed very well (as determined by comparing detailed results from (15), with subcycling, to those from the unsymmetric system, (12), without subcycling).

6. NUMERICAL RESULTS

After verifying that the G matrix is indeed more accurate than the C-matrix on a series of simple test cases which verified the Taylor-series accuracy (viz: both are second-order accurate on a mesh of equal rectangles, C is first-order on unequal rectangles whereas G is first-order plus bilinear (the xy term), and C is zero-order while G is first-order on a mesh of arbitrary quadrilaterals), we compared them on some cavity simulations.

6.1 Thermal cavity. Since there is now a good deal of accurate information available (DeVahl Davis, (1983)) regarding the side wall-heated (cooled) thermal cavity, we used this problem to compare G with C. Specifically, we compared the two schemes against the benchmark solution (at $Ra = 10^6$, $Pr = .71$) for two graded meshes of rectangular elements (16x16 and 24x24) and one mesh (16x16) containing a number of distorted quadrilaterals. Unfortunately, this exercise proved nearly futile as it revealed no significant advantages for either: comparing maxima in streamfunction, and horizontal and vertical (centerline) velocities, indicated that C is as accurate as G.

6.2 Tilted Thermal Cavity. In order to try a more difficult simulation, we tilted the thermal cavity (60° clockwise) so that gravity no longer acts along a coordinate direction. For this case, we used our own previously-obtained results (Upson et al. (1980)) from a higher-order element and on a mesh which proved to be quite accurate on the benchmark problem discussed above as a first measure of error. In addition to the meshes discussed above for the untilted cavity, we also used a 48x48 graded rectangular mesh. Since the results on this mesh agreed with those from the 9-node element to within 1-2%, for both G and C, and the contour plots looked virtually identical, we used one of these (G) to make the comparisons shown below.

Figure 5 shows the velocity field (vectors) on the 48x48 mesh (for C, G, or the 9-node element) and the corresponding temperature field, showing the stable stratification that exists over most of the fluid. Fig. 6 shows the stream function (ψ) for both G and C and suggests a nearly "grid-converged" result, at least regarding the ∇P approximation. The ψ curves here, from the G results, will be used to assess the accuracy of G and C on coarser meshes.

Fig. 7 shows the C result on the 24x24 mesh, Fig. 8 shows it on the 16x16 mesh, and Fig. 9 shows the G result on the 16x16 mesh (the G result on the 24x24 mesh is very close to "exact" and is not shown) from which we observe that the G result on a 16x16 mesh is no worse than the C result on a 24x24 mesh.

Moving away from rectangles, Fig. 10 shows a 16x16 distorted mesh (a perturbation of the rectangular mesh), and the G result on this mesh. Fig. 11 is the C result on this mesh. Whereas the G results are not too different from those on a rectangular mesh, the C results are quite poor; distortion of the elements on an already coarse mesh is harmful to C.

6.3 Lid-Driven Cavity. As a final test, we return to the lid-driven cavity at a rather large Reynolds number (10^4), for which the C-matrix displayed the wrong size and shape for the two lower corner eddies. As reported by Gresho et al. (1984b), the graded 50x50 mesh was not quite adequate at this Re. In Fig. 12 is the solution at the lower right corner from the G-matrix; it is, unfortunately, virtually identical to that from C. We seem to (again) see the result that C is about as good as G on reasonably fine graded meshes of rectangular elements. The errors here are apparently caused by other terms than the pressure gradient.

7. DISCUSSION

We begin by summarizing the advantages and disadvantages of G relative to C. Advantages: (i) It is never worse than first-order accurate (Taylor series sense). (ii) It can be more accurate, especially on coarse grids with distorted elements. Disadvantages: (i) The unsymmetric approximation to the Laplacian operator is more costly in storage and execution. (ii) Outflow boundaries generate an ill-posed algebraic system; apparently the C-matrix must be retained at these boundaries. (iii) The loss of the subcycling technique incurs large additional costs. (iv) The combination of dragged pressure and subcycling, while accurate and cost-effective when stable, can be unconditionally unstable. (v) The additional complications for the 3D case appear to be formidable.

On balance, we have decided that the gain in accuracy, which is not consistently high, is not sufficient to justify the extra cost — at least using our current technique of explicit time integration.

In closing, we should point out that our opinions may well be tainted by our solution technique. If one uses implicit time integration, or is solving the steady equations, both of which (usually) already involve the solution of unsymmetric linear systems, the G-matrix may lead to a gain in cost-effectiveness.

Acknowledgments. We wish to thank Drs. S. T. Chan and A. C. Hindmarsh (LLNL) and Prof. R. L. Sani (University of Colorado) for their assistance.

This work was performed under the auspices of the U. S. Department of Energy by the Lawrence Livermore National Laboratory under contract No. W-7405-Eng-48.

References

Cullen, M. J. P. (1983) Analysis of Some Low Order Finite Element Schemes for the Navier-Stokes Equations. J. Comp. Phys., 51, 2:273.

DeVahl Davis, G. (1983) Natural Convection of Air in a Square Cavity: A Benchmark Solution. Int. J. Num. Meth. Fluid, 3, pp 249-264.

Gresho, P. M., R. L. Lee and C. D. Upson (1981) FEM Solution of the Navier-Stokes Equations for Vortex Shedding Behind a Cylinder: Experiments with the Four-Node Element. Adv. Water Resources, 4, p. 175.

Gresho, P. M., R. L. Lee, and R. L. Sani (1984a) Further Studies on Equal-Order Interpolation for Navier-Stokes. In Proc. of Fifth Int. Conf. on Finite Elements in Flow Problems, TICOM, Ed. G. Carey & J. Oden.

Gresho, P. M., S. T. Chan, R. L. Lee and C. D. Upson (1984b) A Modified Finite Element Method for Solving the Time-Dependent, Incompressible Navier-Stokes Equation. Int. J. Num. Meth. Fluids, in press.

Noh, W. F. (1964) CEL: A Time-Dependent, Two-Space-Dimensional, Coupled Eulerian-Lagrange Code. In Methods in Computational Physics, Vol 3, Fundamental Methods in Hydrodynamics, Academic Press, NY.

Sani, R. L., P. Gresho, R. Lee, D. Griffiths (1981) The Cause and Cure(?) of the Spurious Pressures Generated by Certain FEM Solutions of the Incompressible Navier-Stokes Equations: Part 1. Int. J. Num. Meth. Fluids, 1, p. 17.

Upson, C., P. Gresho, and R. Lee (1980) Finite Element Simulations of Thermally-Induced Convection in an Enclosed Cavity. Lawrence Livermore National Laboratory report UCID-18602.

0085F

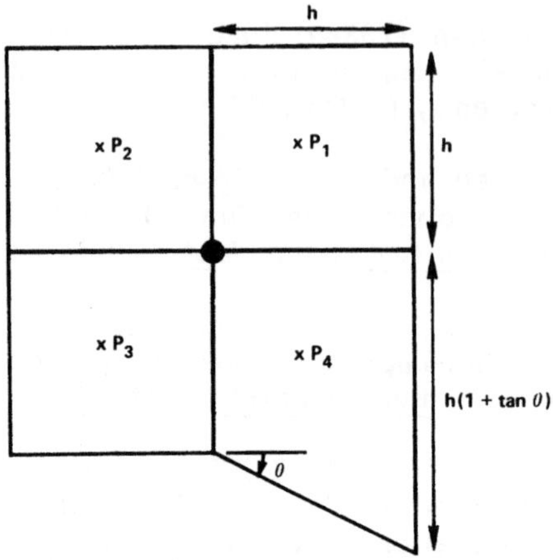

Fig. 1. A 4-patch with one bent element.

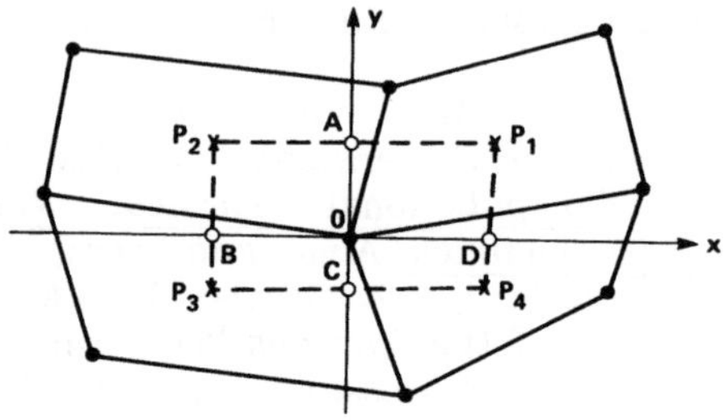

Fig. 2. A general 4-patch.

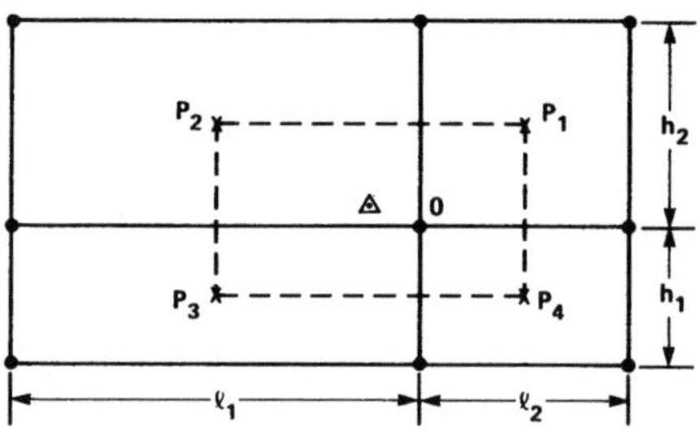

Fig. 3. A 4-patch of unequal rectangles.

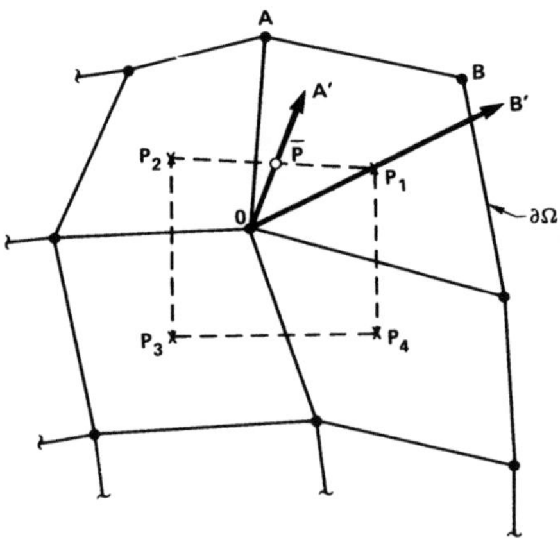

Fig. 4. A boundary 4-patch.

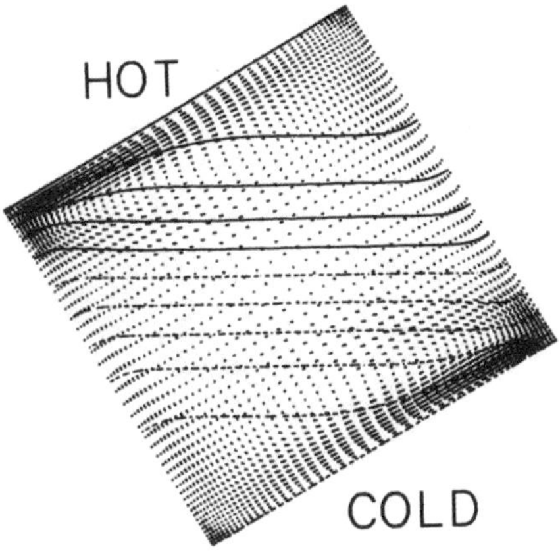

Fig. 5. Vectors and isotherms on 48x48 mesh; Ra = 10^6, Pr = 0.71.

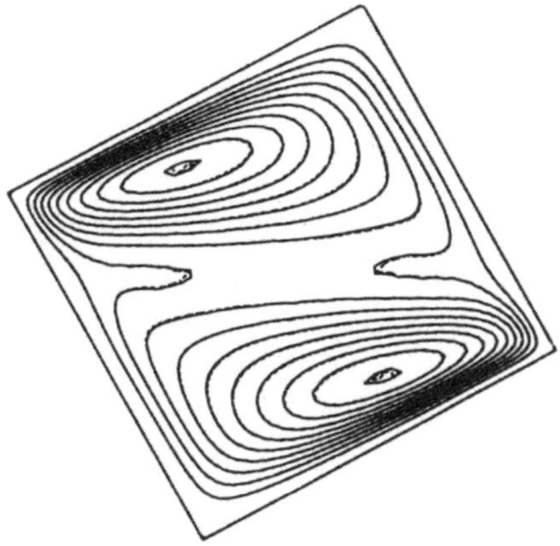

Fig. 6. Streamlines on same mesh. Solid: G-matrix; dashed: C-matrix.

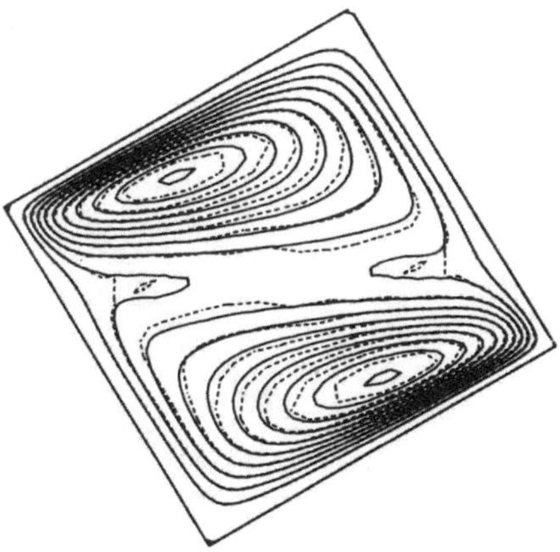

Fig. 7. Streamlines. Solid: G on 48x48; dashed: C on 24x24.

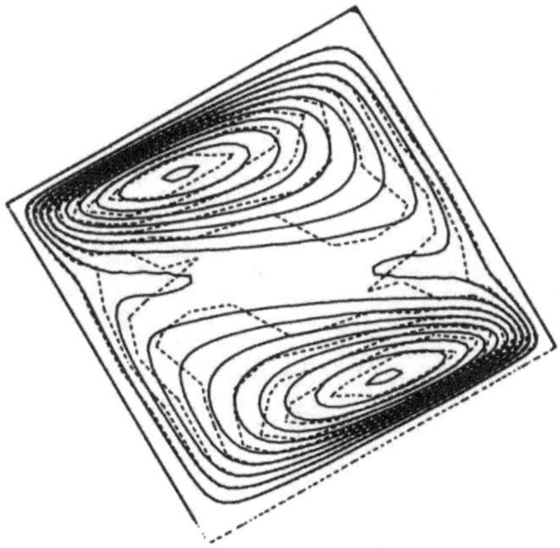

Fig. 8. Streamlines. Solid: G on 48x48; dashed: C on 16x16.

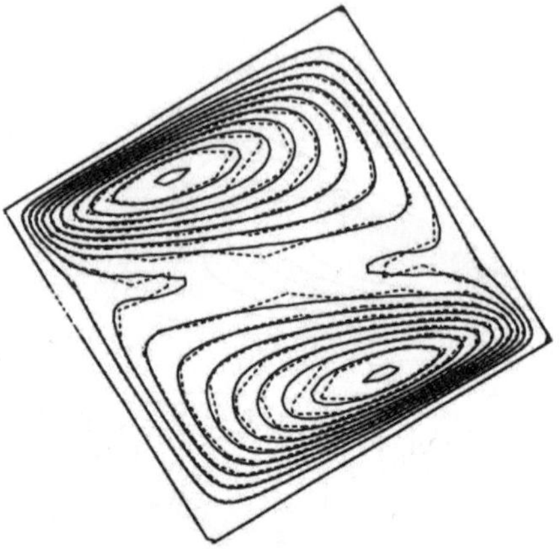

Fig. 9. Streamlines. Solid: G on 48x48; dashed: G on 16x16.

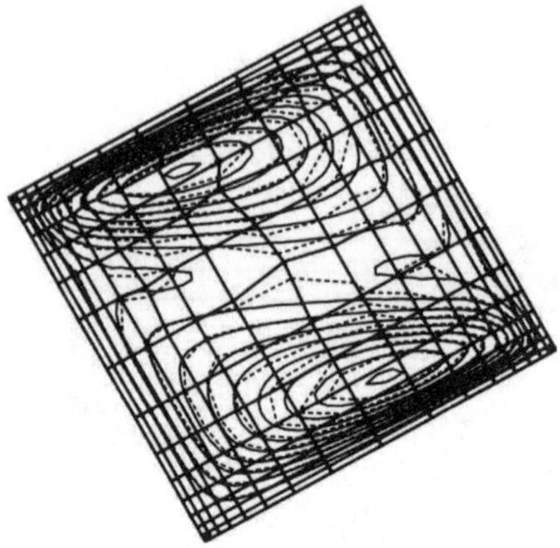

Fig. 10. Distorted 16x16 mesh and streamlines. Solid: G on 48x48 rectangular mesh; dashed: G on 16x16 distorted mesh.

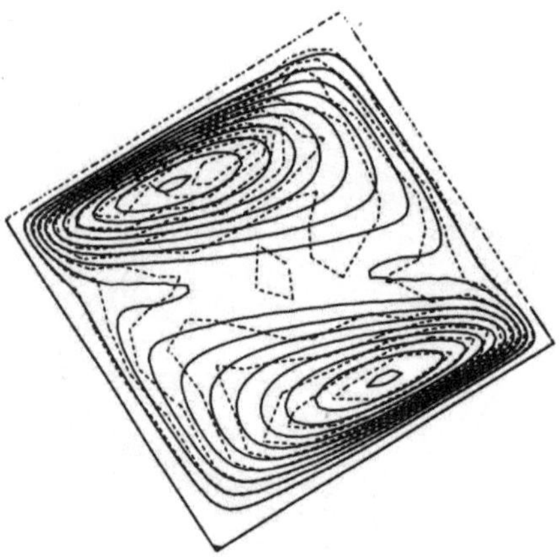

Fig. 11. Streamlines. Solid: G on 48x48; dashed; C on 16x16 distorted mesh.

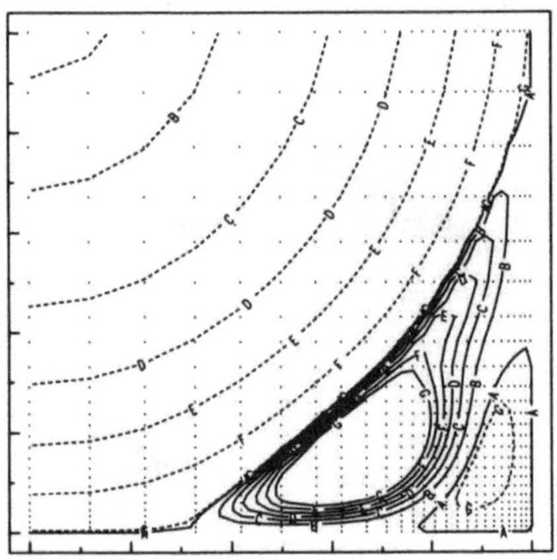

Fig. 12. Streamlines in lower right corner: G-matrix.

SEDIMENTATION PROCESSES

COASTAL SEDIMENTATION PROCESSES MODELING

S. Y. Wang

Center for Computational Hydroscience and Engineering
The University of Mississippi, University, MS 38677 USA

INTRODUCTION

If one observes closely and carefully, one finds that even topography of beach and seabed in the coastal environment are highly dynamic one. Beach erosion and migration are occuring constantly, no matter how they are caused, either storm surges or just the ordinary tides and currents. Similarly, the scouring and deposition of materials on the bed under coastal waters are also happening continuously, especially around dredged channels and near man-made structures. All of these phenomena are associated with sediment movement in the coastal zone. The sedimentation processes in the coastal zone have profound effects on marine environmental quality as well as survivability, on the cost of coastal engineering projects such as the design and maintenance of dredged shipping channels, construction of harbor and beach protection structures, among many others. Therefore, the enhancement of basic understanding of sedimentation processes in coastal region is of great importance.

Water motions in coastal zones are more complicated than those in inland waterways. The major complication is caused by the existance of a variety of waves and currents including tides, tsunamis, storm surges, thermohaline circulations, ocean currents, wind-driven currents, etc. Not only these wave motions intensify the turbulent flows of coastal waters, but, the water particles' obital motions are undergoing a series of deformation and subsequent destruction when approaching to the beaches. Besides, the boundary conditions of diffractions and reflections of these waves further complicate the coastal flow problems. One has to obtain, at least, an approximate solution adequately before the modeling of sedimentation processes become possible. Therefore, a brief discussion on current methods for modeling of coastal flows is to be presented before the discussion of sediment transport simulation.

The basic physical mechanism of all phases of sedimentation processes including scouring, entrainment, diffusion, suspension, transport, and deposition, is also too complex to allow pure analytic modeling. Therefore only the most effective models based on numerical-empirical methodology are to be presented in this paper after the review of the coastal flow modeling.

COASTAL FLOW MODELING

Three dimensional viscous flows subject to a free-surface were modeled using the Finite Difference and its related methods by Nichols and Hirt (1971) and Harlow and Welch (1972). Although the solutions obtained by these complex and sophisticated schemes are quite adequate, it is somewhat difficult to realistically apply them for modeling flows in a highly irregular environment such as natural streams, estuaries, and coastal regions. The versatility of the FEM in handling complex boundary geometry, heterogeneous physical properties of media in the solution domain, as well as the discontinuities in the solution has raised its popularity in the field of hydraulics and fluid dynamics. A summary of its recent applications to free-surface flows, shallow water circulation, and two-dimensional viscous flows had been presented by Connor and Brebbia (1976), and Norrie and de Vries (1978). Readers can find detailed information on the development of FE Modeling of Fluid Flows in general from these books and other similar papers.

A two-dimensional quasi-linear FE Model for Open Channel Flow near Critical Conditions was reported recently by Katopodes (1980). It successfully demonstrated the capability of FEM to simulate a supercritical flood-wave. The truly three-dimensional finite element modeling of viscous flows in an open channel with and without the existence of obstructions was carried out by Wang and Alonso (1978). Three-dimensional linear hexahedral, isoparametric elements were used to obtain solutions of very slow viscous flows in open channels of varying cross-section and around an isolated obstruction. Although results obtained were physically reasonable, the requirement of computer storage and computing time were prohibitive.

One of the most effective alternatives is the Depth-Averaging Scheme. It has been used primarily in the simulation of currents and water waves in lakes, estuaries and shallow coastal waters using both FDM and FEM. Some typical contributions may be found in the work of Leendertse (1967) and Niemeyer (1977). More recently the utilization of depth-averaging models in the finite element simulation of flows in open channels, rivers, lakes, and coastal areas were reported by Thienpont and Berlamong (1980), Wang, Su, and Alonso (1980),

Tanaka (1980), and Kawahara (1980). And, interestingly enough, similar successes have also been reported using the Finite Difference Methods (FDM) to solve the Depth-Averaging Models of the free surface flows including the work of Leendertse (1977) and Simons (1979). Therefore, one can choose either FDM or FEM to establish the hydrodynamic model to be used for sediment transport simulation. But, it is quite obvious that the FEM does have the advantage of better-suited for highly irregular boundaries common to the flows in coastal seas as demonstrated in Figure 1, for the flow in English Channel.

Presently, numerous Finite Element Models capable of simulating coastal flows have been developed. They may be categorized into the following groups: (1) Utilization of primitive variables including Kawahara et. al. (1976), Gartling (1980), Gresho, et. al. (1980), and Wang, et. al. (1980), (2) Application of penalty function including Hughes, et. al. (1979), Berccouier-Englemen (1979), and Sani (1982); (3) Upwind method including Fortin-Thomasset (1979), Hughes, et. al. (1982), and Kikuchi-Ushijima (1982), among many others.

There are several Finite Element Models which are currently being developed including McAnally (1984), Wang, et. al. (1980) using three-dimensional linear and quadratic hexahedra elements and isoparametric interpolation functions; Wang, et. al. (1984), incorporating a combination of interpolation functions into the depth-averaged model to account for the separation and vortices in vertical planes; and Kawahara and Wang (1984) who are collaborating in the development of a multilevel depth-averaged model to simulate the estuarine and coastal flows capable of handling the vortices in verticle planes. This last approach has the potential of being cost-effective, because it can produce a three-dimensional flowfield approximately using a multilevel two-dimensional model. Obviously, that the computer storage and computing time required for a two-dimensional model is much less than those required for a three-dimensional one.

The basic approach of the multilevel depth-averaged model is briefly presented below using a two-layered tidal flow case for simplicity. The variables and coordinates are defined in Figure 2. The continuity and momentum equations are:

$$\dot{u}_i + u_j u_{i,j} + gS_{,i} - A_h(u_{i,j} + u_{j,i}) + \alpha(u_i - v_i) = 0 \qquad (1)$$

$$\dot{S} - \dot{d} + ((S - d)u_i)_{,i} = 0 \qquad (2)$$

$$\dot{v}_i + v_j v_{i,j} + g(S_{,i} + \Delta d_{,i}) - A_{hl}(v_{i,j} + v_{j,i}) - \beta(u_i - v_i) + \gamma v_i = 0 \qquad (3)$$

$$\dot{d} + ((d-b)v_i),_i = 0 \tag{4}$$

where: u_i, S: upper velocity and water elevation

v_i, d: lower velocity and water elevation

$$\Delta = \frac{\rho_1 - \rho}{\rho_1}: \text{ density difference ratio}$$

$$\alpha = \frac{fm}{(S-d)\rho} \cdot \sqrt{(u_1 - v_1)^2 + (u_2 - v_2)^2}$$

friction coefficients

$$\beta = \frac{fm}{(d-b)\rho_1} \cdot \sqrt{(y_1 - \nu_1)^2 + (u_2 - \nu_2)^2}$$

$$\gamma = \frac{f_b}{(d-b)\rho_1} \cdot \sqrt{\nu_1^2 + \nu_2^2}$$

The boundary conditions for the upper layer are:

$$u_i = \hat{u}_i \quad \text{on } S_1^u \tag{5}$$

$$q_i = (-gS\,\delta_{ij} + A_h(u_{i,j} + u_{j,i}))n_n = \hat{q}_i \quad \text{on } S_2^u \tag{6}$$

$$S = \hat{S} \quad \text{on } S_3^u \tag{7}$$

And, for lower layer they are:

$$v_i = \hat{v}_i \quad \text{on } S_1^\ell \tag{8}$$

$$r_i = (-g(S + \Delta d)\,\delta_{ij} + A_{hl}(v_{i,j} + v_{j,i}))n_j = \hat{r}_i \quad \text{on } S_2^\ell \tag{9}$$

$$d = \hat{d} \quad \text{on } S_3^\ell \tag{10}$$

Introducing the finite element interpolating functions into the Galerkin's conditions, one obtains the following set of governing differential equations:

$$A_{\lambda\mu}(\dot{S}_\mu - \dot{d}_\mu) + D_{\lambda\mu\beta}(S_\mu - d_\mu)u_\beta = 0 \tag{11}$$

$$M_{\alpha\beta}\dot{u}_\beta + K_{\alpha\beta\delta}u_\beta u_\delta + H_{\alpha\lambda}S_\lambda + A_h T_{\alpha\beta}u_\beta + \alpha M_{\alpha\beta}(u_\beta - \nu_\beta) = 0 \tag{12}$$

$$A_{\lambda\mu}\dot{d}_{\mu} + D_{\lambda\mu\beta}(d_{\mu} - b_{\mu})v_{\beta} = 0 \qquad (13)$$

$$M_{\alpha\beta}\dot{v}_{\beta} + K_{\alpha\beta\delta}v_{\beta}v_{\delta} + H_{\alpha\lambda}(S_{\lambda} + \Delta d_{\lambda}) + A_{hl}T_{\alpha\beta}v_{\beta} - \beta M_{\alpha\beta}(u_{\beta} - v_{\beta}) + \delta M_{\alpha\beta}v_{\beta} = 0 \qquad (14)$$

This set of equations can be solved iteratively. A typical set of results of a two-layered tidal flow model is shown in Figure 3. It is seen clearly that the characteristics of flowfields of the upper and lower levels are indeed different as expected. The results of a more involved multileveled model is presented in Figure 4, which shows that the flow in lower levels is sometimes in the opposite direction to those of the upper layers. These facts have significant effects on the accuracy of sediment transport prediction. Therefore, the depth-averaged models based on linear and/or quadratic interpolating functions may not be adequate for coastal sedimentation modeling. The multilevel, multi-interpolation functions, or three-dimensional models should be used.

SEDIMENTATION PROCESSES MODELING

The sedimentation processes of scouring, entrainment, suspension, transportation, and deposition, are quite involved. Physically the detachment of particles in the erosion or scouring process occurs due to the shear force and/or turbulent momentum transfer generated by flowing water. Once a particle has been detached, it must be entrained into the water before it is transported away. Both entrainment and transportation depend on the shape, size, and specific weight of the particle and forces exerted on the particle by the flow. These forces depend highly on the properties as well as the flowing conditions of the fluid. When these forces are diminished to the extent that the transport rate is reduced, sediment depositions will occur. If one takes a closer look at the motion of sediment particles, he shall find that some of the particles roll and slide intermittently along the bed; with increasing flow speed they may make a few short jumps, leaving the bed for short instants of time and returning either to come to rest or to continue in motion on the bed or by executing further jumps; and some other particles move with the fluid in suspension for appreciable lengths of time. These phenomena become even more difficult to describe due to the fact that the motions mentioned above are usually of stochastic nature.

Tremendous efforts have been, as well as are being, made all over the world trying to provide theoretically well-founded answers to the sediment movement problems arising in engineering practice. The volume of knowledge has been accumulated and expanded at a rapidly accelerating rate in the past few decades. The inclusion of all efforts and advances in the area of sediment transport research would be practically impossible in this brief review. Therefore, only a few typical examples will be presented.

Early contributions were primarily from physical modeling. DuBoys' tractive-force formula (1879), and modified by Kramer (1982), Lane (1938), et. al. Also developed are classical impact theory concerning the critical bottom velocity by Einstein (1943), Indri (1941) and many others. Based upon these theories, numerous bed load empirical formula were developed including Chang (1947), Meyer-Peter and Muller (1948), Colby (1964), Engelund-Hansen (1966), Toffaleti (1969), among others. The dimensional analysis have also been applied in attempting to develop universal bed load transport formulas in terms of several physically relevant non-dimensional parameters. Additional empirical relations have been determined using new testings with these non-dimensional parameters as well as data previously obtained. Some typical contributors are Barr and Herbertson (1966,68), Garde and Albertson (1961) and Bogardi (1964).

Parallel to the bed load studies, the sediment transport capacity in suspension has also been studied. Contrary to the bed load which is defined as bed material moving on or near the bed in the form of rolling, sliding, and jumping (saltating), the suspended load covers only the solid particles being surrounded by fluid during their entire motion above a thin layer over the bed. Physically the movement of suspended sediment is just as complicated as that of bed load. Therefore, empirical approaches have been adopted to predict the suspended load for some time. Rouse (1939), Vanoni and Brooks (1946), Frankl (1955), Bogardi (1949), Sanoyan (1968), and many others have made significant contributions in the past few decades. In addition, total load formulas have also been developed. Several examples are presented in the book of Vanoni (1979).

Although many existing sediment transport functions imply that they simply represent local correlation between the fluid flow characteristics and sediment particles' properties and distribution, so that they may be used in various conditions; recent studies have shown some special empirical functions to predict sediment transport capacity in coastal waters should be used in wave-dominated coastal flows. Again, one may choose one of the three approaches such as suspended load, bed load, and total load to model the coastal sediment movement. The solution procedure of coastal sediment transport in suspension is not much different from those for sediment transport modeling in inland waterways, have been reported extensively by Jobson (1970), Cunge (1973), Smith and O'Connor (1977), and Kerssens, et. al. (1979).

Applications of bed load and total load empirical functions to the simulation of onshore-offshore as well as along-shore sediment movement were studied by Inman and Bagnold (1963), Komar (1969), Dean (1973), van de Graaff and van Overeem (1979), Thornton (1979), and Hallermeier (1982), among others. Thornton (1979) proposed the integration in time of local oscillatory bed load rate due to near shore wave action to obtain a mean bed load empirical function. Hallermeier (1982) reported such a functional relationship resulted from laboratory physical experimentation. It has the following form:

$$q_B = \omega U^3 D^{0.5}/(r'g)^{1.5} \tag{15}$$

where ω is the flow frequency, U is the peak velocity near the bed, D is the sediment mean grain diameter, $r' = (\rho_s - \rho)/\rho$, ρ_s and ρ are densities of sediment and water respectively, g is the gravitational acceleration, and q_B is the bed load sediment transport rate having the units of volume of sediments per unit time per unit bed width normal to the flow.

A typical total load sediment transport function as given by Price, Tomlinson, and Willis (1972) is presented below:

$$q_s = \frac{0.70\ E_b\ (nC)_b \mathrm{Sin}\alpha_b \cos\alpha_b}{\gamma\ (1 - p)\ (S - 1)} \tag{16}$$

in which E represents the wave energy density, nC is the group velocity, α is the angle between the breaking wave front and the shoreline, γ is the specific weight of water, p is the sediment porosity, and S is the specific gravity of the sediment. The subscript b represents values of those variables at breaking. This total load function and Komars' (1969) equation produced reasonable agreement with physical model studies on sediment transport simulation in the vicinity of coastal structures.

To simulate the bed elevation and beach migration, one may combine the bed load or total load sediment transport rate equations such as those given in this paper to the sediment continuity equation as shown below:

$$(1 - p)\ \frac{\partial \zeta}{\partial t} + \frac{\partial q_x}{\partial x} + \frac{\partial q_y}{\partial y} = 0 \tag{17}$$

in which p is the porosity, ζ is the bed elevation, q_x and q_y are either bed load or total load sediment transport rate in x- and y-direction respectively. This equation can be conveniently solved by finite element or finite difference method to obtain the bed elevation changes as a time dependent phenomenon. Typical results of coastal sediment transport simulations are given in Figures 5-7. It is seen that the numerical-empirical models can indeed produce reasonable results. It is worthwhile to mention that all of these results can be generated in a short time and a several order of magnitude less than the cost required by pure physical modeling.

More detailed review of the recent development in sediment transport modeling were presented by Wang (1982, 1983). It is clear that the most successful methodology for coastal flow and sedimentation modeling is the numerical-empirical model, which utilizes the most accurate numerical solutions of fluid flow equations and sediment continuity equation as well as the most reliable empirical sediment transport functions available to date for investigating the coastal sedimentation processes. Before the detailed interactions between fluid and sediment particles can be well-described theoretically, the numerical-empirical modeling is going to be the most cost-effective methodology for the study of sediment transport phenomena in the foreseeable future.

ACKNOWLEDGEMENT

This work is a result of research sponsored in part by NOAA Office of Sea Grant, Department of Commerce, under Grant #NA81AA-D-00050, the Mississippi-Alabama Sea Grant Consortium, and The University of Mississippi.

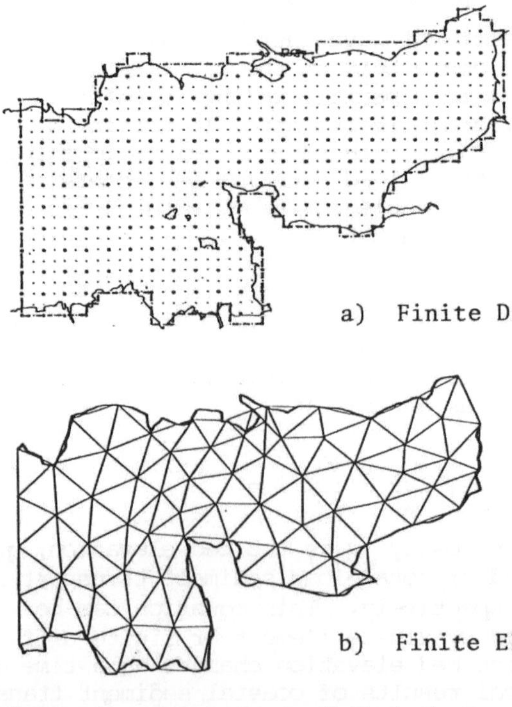

a) Finite Difference

b) Finite Element

Figure 1. Comparison of Mesh Systems

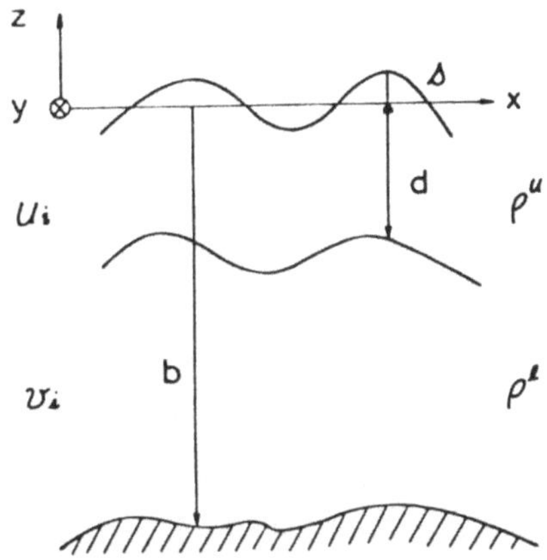

Figure 2. Two-layered tidal flow model

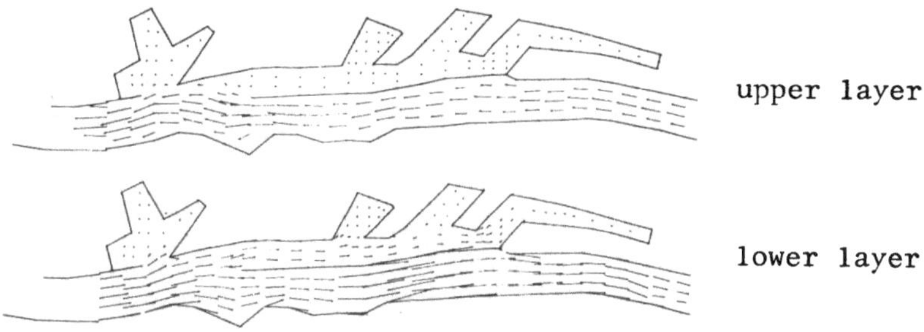

Figure 3. Numerical results of two-layered tidal flow.

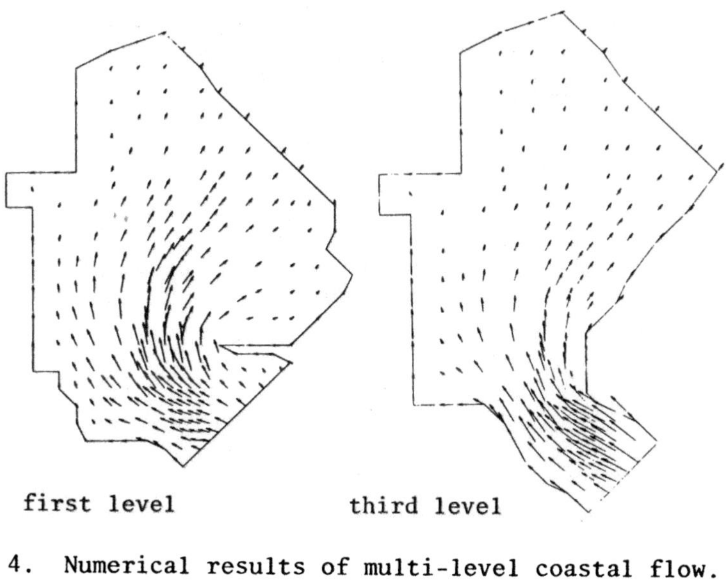

Figure 4. Numerical results of multi-level coastal flow.

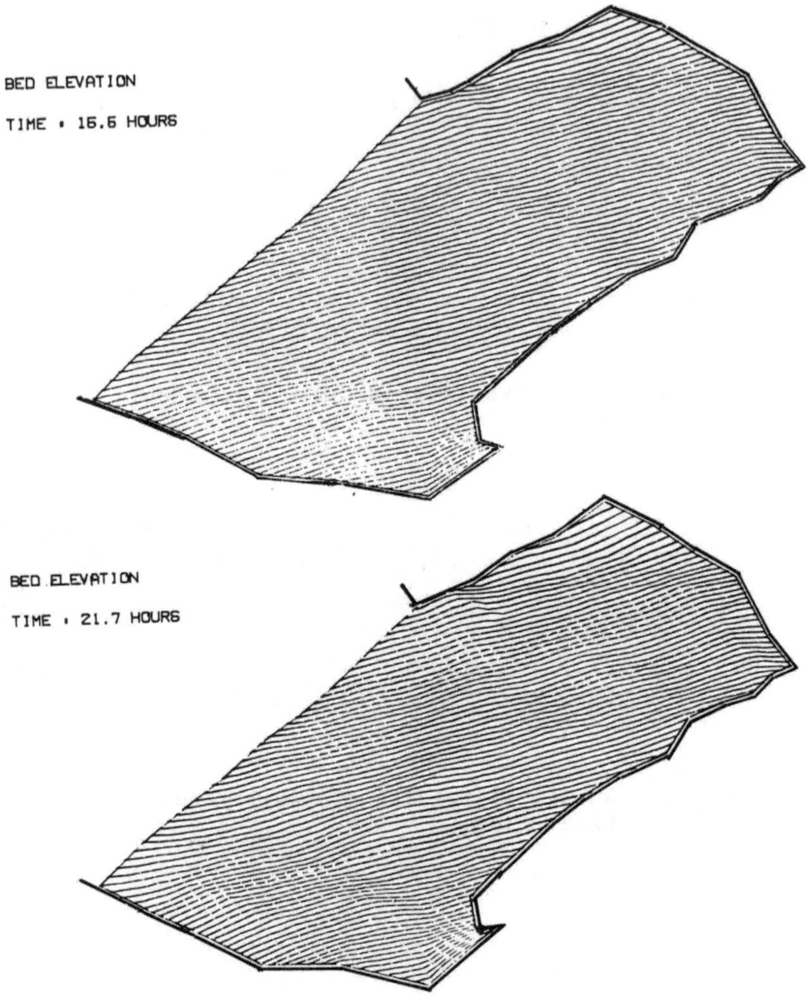

Figure 5. Bed deformation in a bay due to tidal currents.

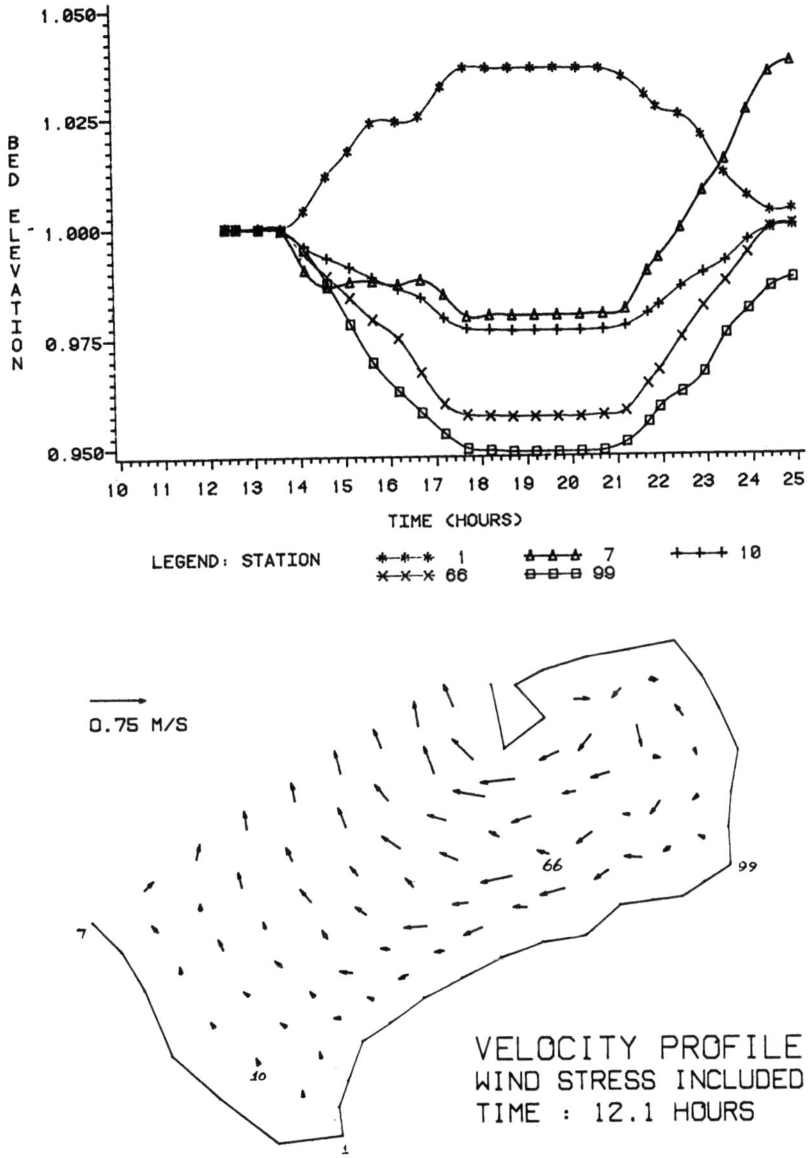

Figure 6. Bed elevation variation with time at selected location in a bay.

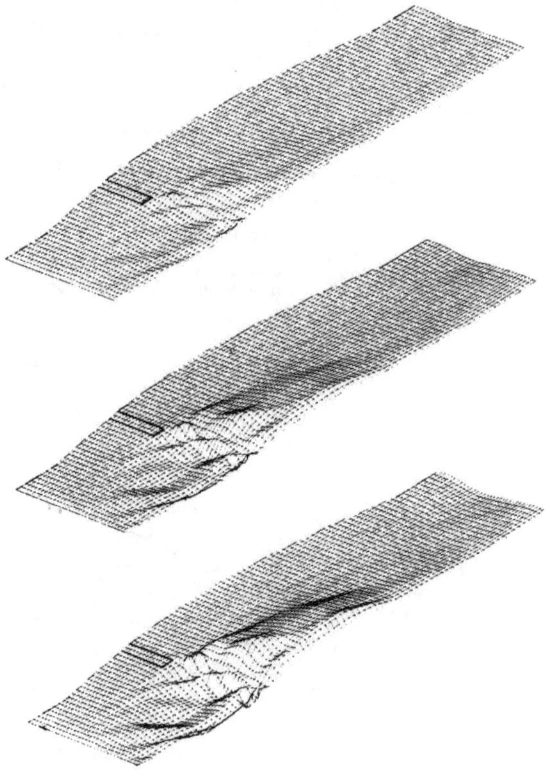

Figure 7. Finite element modeling of scouring phenomenon in the vicinity of a beach protection dike due to along-shore current.

REFERENCES

Alonso, C. V. and Wang, S. Y. (1978) Finite Element Modeling of Flow Around an Open Channel Obstruction, Proceedings Second International Conference of Finite Elements in Water Resources, Imperial College, London, England.

Alonso, C. V. and S. Y. Wang, Simulation of Local Scour and Fill in Sand Bed Streams in Finite Elements in Water Resources, Vol. III, 6.57-6.68, University of Mississippi.

Barr, D. I. H. and J. G. Herbertson (1966) Discussion of Sediment Transportation Mechanics: Initiation of Motion, by the Task Committee on Preparation of Sedimentation Manual, Committee on Sedimentation of the Hydraulics Division, Vito A. Vanoni, Chmn., Journal of Hydraulics Division, ASCE, Volume 92, No. HY. 6, Proceeding Paper 4959, pp. 248-251.

Barr, D. I. H. and J. G. Herbertson (1968) Similitude Theory Applied to Correlation of Flume Sediment Transport Data, Water Resources Research, Volume 4, No. 2, pp. 307-316.

Bercovier, M. and Engleman, M. (1979) A Finite Element for the Numerical Solution of Viscous Incompressible Flows, J. Comp. Physics, Volume 29, pp. 181-201.

Bogardi, J. (1949) The Movement of Suspended Sediment Between Zahony and Razompuszta in the Tisza River, Vizugyi Kozlemenyek, No. 3-4.

Bogardi, J. (1964) Dimensional Analysis and Similarity, Lectures Delivered in Utah State University C.E., 260.

Colby, B. R. (1964) Discharge of Sands and Mean Velocity and Relationships in Sands-Bed Streams, Professional Paper 462-A, U. S. Geological Survey, Washington, DC.

Connor, J. J. and Brebbia, C. A. (1976) Finite Element Techniques for Fluid Flow, Newnes-Butterworth, London.

Cunge, J. A. and Perdreau, N. (1973) Mobile Bed Fluvial Mathematical Models, La Houille Blanche, No. 7.

Dean, R. G. (1973) Heuristic Models of Sand Transport in the Surf Zone, in the Proceedings of a Conference on Engineering Dynamics in the Surf Zone, pp. 208-214. The Institution of Engineers, Australia.

DuBoys, P. (1879) Le Rhone et les Rivieres a Lit Affouillable, Annales des Pontes et Chaussees.

Einstein, H. A. (1943) Flow on Movable Bed, Proceedings of the Second Hydraulic Conference, University of Iowa Studies in Engineering, Bulletin 27.

Engelund, F. and Hansen, E. (1966) Investigations of Flow in Alluvail Streams, Acta Polytech. Scand., Ci-35.

Fortin, M. and Thomasset, J. (1979) Mixed Finite-Element Methods for Incompressible Flow Problems, J. Comp. Physics, Volume 31, pp. 113-145.

Frankl, F. I. (1955) Energy Equations of Suspended-Sediment Transportation, DANSSR, Volume 192.

Garde, R. J. and M. A. Albertson (1961) Bed-Load Transport in Alluvial Channels, La Huille Blanche, N. 3.

Gartling, D. K. (1980) Finite Element Analysis of Thermal Convection in Deep Ocean Sediments, Finite Elements in Water Resources, Volume 3, pp. 7.30-7.44.

Gresho, P. M. and Upson, C. D. (1982) Current Progress in Solving the Time-Dependent, Incompressible Navier-Stokes Equations in Three-Dimensions by (Almost) the FEM, Finite Elements in Water Resources, Volume 4, pp. 3-43 to 3-58.

Hallermeier, R. J. (1982) Oscillatory bedload transport: Data Review and Simple Formulation, submitted to Cont. Shelf Res.

Harlow, F. H. and Welch, J. D. (1972) Numerical Calculation of Time-Dependent Viscous Incompressible Flows of Fluid with Free Surface, Physics of Fluids, 10-2, 324-340.

Hughes, T. J. R., Taylor, R. L. and Levy, J-f. (1979) High Reynolds Number, Steady, Incompressible Flow by a Finite Element Method, Finite Elements in Fluids, Volume 3, pp. 55-72.

Hughes, T. J. R., Tezduyar, T. E. and Brooks, A. N. (1982) Streamline Upwind Formulations for Advection-Diffusion, Navier-Stokes, and First-Order Hyperbolic Equations, Finite Element Flow Analysis Volume 4, pp. 97-104.

Indri, E. (1941) Problem del Transporto Solido ed i Risultati di Recenti Richerche, Societa Editrice Riviste Elettriche.

Inman, D. L., and R. A. Bagnold (1963) Littoral Processes, in The Sea, Vol. 3, edited by M. N. Hill, pp. 523-549, Interscience, New York.

Katopodes, N. D. (1980) Finite Element Model for Open Channel Flow Near Critical Conditions, Third International Conference on Finite Elements in Water Resources, Volume 2, 5.37-5.46.

Kawahara, M. (1980) Finite Element Methods in Two Layer and Multileveled Flow Analysis, Presented at Third International Conference on Finite Elements in Water Resources, University of Mississippi.

Kawahara, M. and Wang, S. Y. (1984) Multi-level Modeling of Coastal Flows, under development in a cooperative research project.

Kawahara, J., Yoshimura, N., Nakagawa, K. and Ohsaka, K. (1976) Steady and Unsteady Finite Element Analysis of Incompressible Viscous Fluid, Int. J. Num. Meth. Eng., 10, pp. 437-456.

Kerssens, P. J. M., van Rijn, L. C. and Van Wijngaarden, N. J. (1979) Model for Non-Steady Suspended Sediment Transport IAHR Modelling of Sediment Transport pp. 113-120.

Kikuchi, N. (1982) On a Smoothing Scheme of Penalty/Finite Element Methods for Stokes Flow Analysis, Finite Element Flow Analysis, Volume 4, pp. 45-50.

Komar, P. D. (1969) The Longshore Transport of Sand on Beaches, Ph.D. Thesis, 143 pp. Univ. of Calif., San Diego.

Kramer, H. (1932) Modellgeschiebe und Schleppkraft, Berlin.

Lane, W. (1938) Collection of Data of the Solids Load of Flowing Streams, Journal of the Association of Chinese and American Engineers, XIX, No. 3

Leendertse, J. J. (1967) Aspects of a Computational Model Long-Period Water-Wave Propogation, Rand Corp. Memorandum. RM-5294-PR.

Leendertse, J. J. and S. K. Liu (1977) A Three-Dimensional Model for Estuaries and Coastal Seas: Colume IV, Turbulent Energy Computation, prepared for the Office of Water Research and Technology, Department of the Interior R-2187-OWRT.

Lopez, S. J. L. (1978) Mathematical Modeling of Sediment Deposition in Reserviors, Hydrology Papers, No. 95 CSU, Fort Collins, Colorado.

McAnally, W. H., and W. A. Thomas (1980) Finite Element Models in an Hybrid Model Study of Estuarine Sedimentation, in Finite Elements in Water Resources, Vol. III, 6.47-6.56, University of Mississippi.

McAnally, W. H. (1984) FEM of 3-D Estuary Flows (under development, private communication.)

Meyer-Peter, E. and R. Muller (1948) Formulas for Bed-Load Transport, Report on Second Meeting of International Association for Hydraulic Research, Stockholm, Sweden, pp. 39-64.

Nichols, B. D. and Hirt, C. W. (1971) Improved Free Surface Boundary Conditions for Numerical Incompressible-Flow Calculations, Journal of Computational Physics, 8-3, 434-49.

Niemeyer, G. L. (1977) Numerical Method for the Simulation of Hydrodynamic and Ecological Processes, with application to Kaneohe Bay, Oahu, Hawaii.

Norrie, D. H. and DeVries, G. (1978) A Survey of the Finite Element Applications in Fluid Mechanics, Finite Elements in Fluids, Volume 3, 363-384.

Price, W. A., Tomlinson, K. W., and Willis, D. H. (1972) "Predicting Changes in the Plan Shape of Beaches," Proceedings of the 13th Conference on Coastal Engineering, American Society of Civil Engineers, pp. 1321-1329.

Rouse, H. (1939) Experiments on the Mechanics of Sediment Suspension, Proceedings Fifth International Congress of Applied Mechanics.

Sani, R. L. (1982) On the Solution of the Incompressible Navier-Stokes and Boussinesq Equations via a Penalty Galerkin Finite Element Method, Finite Element Flow Analysis Volume 4, pp. 179-184.

Sanoyan, V. G. (1968) Theoretical Determination of the Sediment Transporting Capacity of Watercourses..., Symposium on Current Problems in River Training and Sediment Movement, 3, Budapest.

Simons, D. B., Y. H. Chen, and V. M. Ponce (1979) Development of 2-D Water and Sediment Routing Model and its Application to Study Lower Pool and in the Upper Miss. River Sys., Vol. I, Engineering Research Center, Colorado State University, Fort Collins, Colorado, 80523.

Smith, T. J. and O'Connor, B. A. (1977) A Two Dimensional Model for Suspended Sediment Transport, IAHR Proceedings Seventeenth Congress, Volume 1, pp. 79-87.

Su, T. Y. (1980) Computer Simulation of Sediment Transport, Ph.D. Dissertation, University of Mississippi.

Su, T. Y. and Wang, S. Y. (1980) Depth-Averaging Models of River Flows, presented at Third International Conference on Finite Elements in Water Resources, University of Mississippi.

Tanaka, T. Y., Ono and T. Ishise (1980) The Open Boundary Value Problems in Ocean Dynamics by Finite Elements, presented at Third International Conference on Finite Elements in Water Resources, University of Mississippi.

Thienpont, M. and Berlamont, J. (1980) A Finite Element Solution for Depth-Averaged Two-Dimensional Navier-Stokes Equations for Flows in Rivers and Channels, Third International Conference on Finite Elements in Water Resources, Volume 2, 5.28-6.36.

Thornton, E. B. (1979) Energetics of Breaking Waves within the Surf Zone, J. Geophys. Res., 84, 4931-4938.

Toffaleti, F. B. (1969) Definitive Computations of Sand Discharge in Rivers, Journal of the Hydraulic Division, ASCE, Volume 95, No. HY1, Proceedings Paper 6350, pp. 225-248.

Ushijima, T. and Kanayama, H. (1982) A Mathematical Analysis of a Finite Element Scheme for a Linearized Viscous Shallow-Water System, Finite Element Flow Analysis, Volume 4, pp. 51-58.

van de Graaff, J., and J. van Overeem (1979) Evaluation of Sediment Transport Forulae in Coastal Engineering Practice, Coastal Eng., 3, 1-32.

Vanoni, V. A. (1946) Transportation of Suspended Sediment by Water, Transactions ASCE iii.

Vanoni, V. A., Editor (1975) Sedimentation Engineering, ASCE-Manuals and Reports on Engineering Practice No. 54.

Wang, S. Y. (1982) Computer Simulation of Sedimentation Processes, invited lecture presented at and paper published in the Proceedings of Finite Elements in Water Resources, Vol. IV, (16.35-16.47) Springer-Verlag, Hannover, West Germany.

Wang, S. Y. (1982) "Finite Element Modeling of Sediment Transport -- The State of the Art," invited paper published in the Proceedings of the 4th International Symposium on Finite Element Methods in Flow Problems, Tokyo, Japan.

Wang, S. Y. (1982) Recent Developments on Numerical Modeling of Sediment Transport, invited paper published in the Proceedings of the International Conference on Computational Methods and Experimental Measurements, pp. 595, Springer-Verlag, Berlin.

Wang, S. Y. and Alonso, C. V. (1978) Numerical Modeling of Three Dimensional Viscous Flows in an Open Channel, 2:IV-2, 1-2, 13, Applied Numerical Modeling, Pentech Press, England.

Wang, S. Y., Hu, K. K., and Su, T. Y. (1984) Multi-interpolation-function Approach in Free Surface Flow Simulation, in preparation.

APPLICATION OF FINITE ELEMENT TECHNIQUES TO HEAT FLOW INVESTIGATIONS WITHIN SEDIMENTARY BASINS

Allan D. Woodbury and Leslie Smith

Department of Geological Sciences, University of British Columbia, Vancouver, B.C., Canada V6T 2B4

INTRODUCTION

A proper understanding of the thermal regime of the earth's crust is essential in many Geo-Science studies. Such studies include global tectonics, investigations of low and high temperature geothermal environments, oil maturation and earthquake research. The basic approach to investigate thermal conditions at depth is to measure temperature gradients in shallow boreholes, which when combined with measurements of the thermal conductivity, provide estimates of surface heat flow. In areas of low relief with low permeability rocks at the surface, temperature fields are essentially conductive. However, there is ample evidence to show that the thermal regime of basins with high relief, infilled with permeable sediments, can be highly influenced by circulating groundwater (see review by Smith and Chapman, 1983). In these cases, interpretation of surface heat flow values in terms of thermal conditions at depth cannot be carried out without a complete understanding of the coupled thermal-hydrogeologic system.

Mathematical modeling provides a framework for dealing with data collected in an advectively-disturbed thermal regime. Because the water table commonly exhibits subdued relief orthogonal to the regional slope, a three-dimensional approach is necessary. The purpose of this paper is to outline the appropriate boundary value problem describing the flow of heat and groundwater, to develop a finite element solution, and to present an application of the modeling approach to illustrate how heat redistribution by groundwater influences surface heat flow.

GOVERNING EQUATIONS

The equations governing heat and fluid transfer are the continuity equation for the fluid, continuity equation for thermal energy, momentum transfer, and the constitutive equations relating fluid density and viscosity to temperature. Bear

(1972) and Frind (1980) proposed the use of a pseudo-potential h_0, which Frind (1980) described as the "equivalent freshwater head." The freshwater head is defined as:

$$h_0 = (P/\rho_0 g) + Z \tag{1}$$

where P is the fluid pressure $\{M/Lt^2\}$, ρ_0 is some reference density $\{M/L^3\}$ at a reference temperature, g is the acceleration of gravity $\{L/t^2\}$ and Z is the elevation above datum $\{L\}$. A generalized form of Darcy's law for density and viscosity-dependent flow is (Bear, 1972):

$$\bar{q}_r = -\frac{k}{\mu}(\nabla P + \rho g \nabla Z) \tag{2}$$

where $\bar{q}_r$ is the relative specific discharge vector $\{L/t\}$, k is the permeability tensor $\{L^2\}$, μ is the dynamic viscosity $\{M/Lt\}$, and ρ is the fluid density $\{M/L^3\}$. Substituting (1) into (2) and simplifying yields:

$$\bar{q}_r = -\frac{k\rho_0 g}{\mu}(\nabla h_0 + \rho_r \nabla Z) \tag{3}$$

where ρ_r is the relative density defined as: $\rho_r = (\rho/\rho_0) - 1$. In the treatment of many problems in regional groundwater flow, the approximation of steady state fluid and temperature fields can be used. Substituting (3) into a mass conservation equation results in the fluid continuity equation:

$$\nabla \cdot \frac{\rho\rho_0 g}{\mu} k(\nabla h_0 + \rho_r \nabla Z) = 0 \tag{4}$$

This equation describes steady-state, temperature-dependent flow in a saturated porous medium with no sources or sinks. The thermal energy equation written for a steady-state system is:

$$\nabla \cdot \lambda_m \cdot \nabla T - \rho C_{vf} \bar{q}_r \cdot \nabla T = 0 \tag{5}$$

where C_{vf} is the specific heat capacity of the fluid $\{L^2/Tt^2\}$, and λ_m is the coefficient of thermal conductivity of the porous medium $\{ML/t^3T\}$. Written in this form, the effects of thermal dispersion have been neglected (for justification, see Smith and Chapman, 1983).

Boundary conditions

The boundary conditions needed to solve (4) are usually of the first (Dirichlet) or second (Neumann) type. The first type boundary can be written as:

$$h = c \quad \text{or} \quad h = Z \text{ on } \Gamma_1 \tag{4a}$$

where c is a constant. The second type boundary can be written in terms of a flux q perpendicular to the boundary, given as:

$$q_n = q_0 \text{ on } \Gamma_2 \tag{4b}$$

where q_0 is a specified mass fluid flux $\{M/tL^2\}$ and

$$\Gamma_1 + \Gamma_2 = \Gamma$$

is the boundary of the porous medium. q_0 is equal to zero on an impermeable boundary.

Boundary conditions for equation (5) are usually of the first or second type, and sometimes of the third (Cauchy) type. The first type is of the form:

$$T = c \quad \text{on} \quad \Gamma_3 \tag{5a}$$

where c is a constant or a function of the surface elevation, i.e., c = c(Z). The second type is of the form:

$$q_n = q_1 \quad \text{on} \quad \Gamma_4 \tag{5b}$$

where q_1 is some specified heat flux $\{M/t^3\}$. If q_1 is zero, then the boundary is perfectly insulated. Boundary conditions along a surface in contact with the atmosphere or a fluid continuum are complicated, and the heat flux across such a boundary is the result of combinations of heat transfer due to: evaporation, transient atmospheric temperatures, radiative heat transfer, convection from flowing air, transpiration, and possibly, thermal springs. When the depths to which these effects penetrate are small with respect to the depth of the basin, simple first or second type boundary conditions are usually sufficient. Should it be necessary to include approximations to these effects, a third type boundary condition can be invoked. A third type <u>exit</u> boundary has the form:

$$-\lambda\, \partial T/\partial\eta + \rho C_{vf} T(\bar{q}_r \cdot \eta) = H(T - T_\infty) \text{ on } \Gamma_5 \tag{5c}$$

The first two terms represent heat conduction and advection, and the term on the right hand side represents an approximation of radiative heat transfer. Other effects can be considered. H is the coefficient of surface heat transfer $\{M/t^3 T\}$, T_∞ is the outside ambient air temperature, η is the unit outward normal vector, $\partial T/\partial\eta = \nabla T \cdot \eta$, and T is the temperature on the boundary surface. For a third type <u>entry</u> condition, such as when the external medium is a liquid continuum (Bear, 1972, p.621), the boundary condition becomes:

$$\lambda \partial T/\partial\eta = \bar{q}_r \cdot \eta (T - T_\infty)\rho C_{vf} \quad \text{on} \quad \Gamma_6$$

and

$$\Gamma_3 + \Gamma_4 + \Gamma_5 + \Gamma_6 = \Gamma \tag{5d}$$

Note that heat transfer through an unsaturated zone is neglected in this formulation.

GALERKIN FINITE ELEMENT FORMULATION

The Galerkin finite element technique has been used to solve the coupled equations (4) and (5), with boundary conditions (4a,b) and (5a,b,c,d). We will show that the Galerkin technique combined with simple three dimensional elements can handle non-linear problems of reasonable size. A major saving is achieved through the use of linear elements, which eliminates the need for repetitive numerical integration of higher order element matrices.

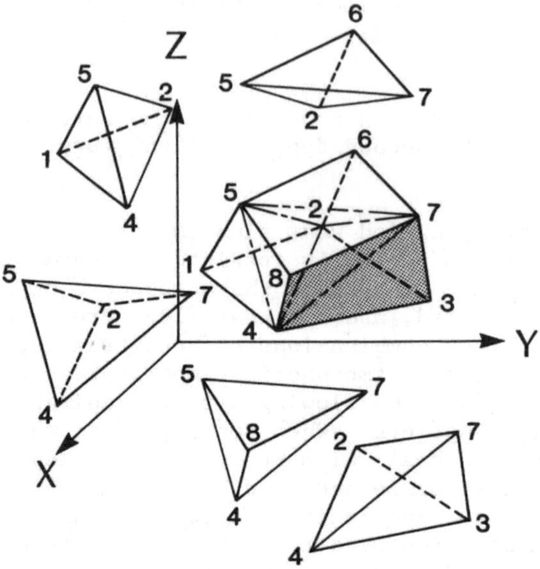

Figure 1. Basic tetrahedral element and assemblage of elements (adapted from Zienkiewicz, 1977).

Basis functions

The basic element, a simplex tetrahedron, is shown on Figure 1. We can define an approximate solution to dependent variables h_0 and T as:

$$h_0 \triangleq \hat{h}_0 = \sum_{i=1}^{n} h_i N_i(x,y,z) \tag{6}$$

and

$$T \triangleq \hat{T} = \sum_{i=1}^{n} T_i N_i(x,y,z) \tag{7}$$

Functional coefficients of the fluid properties ρ, μ, and ρ_r are adopted, such that fluid properties are defined at the nodes and vary linearly across the elements. The values $N_i(x,y,z)$ are defined as:

$$\begin{aligned} N_i &= (a_i x + b_i y + c_i z + d_i) \\ N_j &= (a_j x + b_j y + c_j z + d_j) \\ N_k &= (a_k x + b_k y + c_k z + d_k) \\ N_l &= (a_l x + b_l y + c_l z + d_l) \end{aligned} \tag{8}$$

where the a,b,c, and d coefficients are linear functions of x, y, and z. These functions are developed and presented in Woodbury (1983).

Fluid continuity equation

In the Galerkin formulation a differential operator L can be defined as:

$$L(h_0) = \nabla \cdot \left[\frac{\rho\rho_0 g}{\mu} k\,(\nabla h_0 + \rho_r \nabla Z)\right] = 0 \tag{9}$$

Substituting equation (6) into (9), integrating the resulting residual and the n basis functions over the domain, and applying Green's first identity results in a matrix equation written as:

$$A_{nm} \cdot h_m + B_n = Q_n \tag{10}$$

A_{nm} is the global stiffness matrix defined as (assuming that the principal directions of anisotropy are co-incident with the coordinate axes):

$$A_{nm} = \sum_e \frac{\rho_0 g}{144V} \left(\sum_{j=1}^{4} \left\{\frac{\rho_j}{\mu_j}\right\}\right) \left[k_{xx} a_n a_m + k_{yy} b_n b_m + k_{zz} c_n c_m\right] \tag{11}$$

B_n is the buoyancy term equal to:

$$B_n = \sum_e \frac{\rho_0 g}{24} \left(\sum_{j=1}^{4} \left\{\frac{\rho_{rj}}{\mu_j}\right\}\right) k_{zz} c_n \tag{12}$$

Type 1 and 2 boundary conditions can be imposed through the Q_n vector:

$$Q_n = \sum_e q_n \frac{A^e}{3} \tag{13}$$

where A^e is the area of the element face on which flux q_n is imposed. Specific discharge components q_x, q_y, q_z can be calculated for each element through the use of the following expressions:

$$q_x = -\sum_{m=1}^{4} \frac{k_{xx} \rho_0 g}{24V} \left(\sum_{j=1}^{4} \frac{1}{\mu_j}\right) a_m h_m \tag{14}$$

$$q_y = -\sum_{m=1}^{4} \frac{k_{yy} \rho_0 g}{24V} \left(\sum_{j=1}^{4} \frac{1}{\mu_j}\right) b_m h_m$$

$$q_z = -\sum_{m=1}^{4} \frac{k_{zz} \rho_0 g}{24V} \left(\sum_{j=1}^{4} \frac{1}{\mu_j}\right) \left[c_m h_m + \frac{V}{2.67} \left\{\sum_{\ell=1}^{4} \rho_{r\ell}\right\}\right]$$

V is the volume of the tetrahedral element (L^3).

Heat transfer

For heat transfer we again adopt the same trial solution, and another differential operator M, such that

$$M(T) = \nabla \cdot \lambda_m \cdot \nabla T - \rho C_{vf} \bar{q}_r \cdot \nabla T = 0 \tag{15}$$

In an identical manner to that used in dealing with the fluid flow equation, the Galerkin procedure is followed and another matrix equation is formed:

$$S_{nm} \cdot T_m = F_n \tag{16}$$

where S_{nm} is a global matrix and F_n is a vector of size n of applied fluxes. The global matrix can be expressed as:

$$S_{nm} = \sum_e \left[\frac{1}{36V} (\lambda_{xx} a_n a_m + \lambda_{yy} b_n b_m + \lambda_{zz} c_n c_m) + \frac{C_{vf}}{96} \left(\sum_{j=1}^{4} \rho_j \right) \left\{ q_x a_m + q_y b_m + q_z c_m \right\} \right] \tag{17}$$

The flux vector F_n is expressed as:

$$F_n = - \sum_e \frac{J_T A^e}{3} \tag{18}$$

The normal flux is J_T and A^e is the area of the element face on the boundary of the domain. For type 3 boundaries additional terms must be added to elements within the global stiffness matrix and to the F_n flux vector.

Solution strategy

The two systems of equations (10) and (16) define a coupled heat and fluid transfer problem. An iterative technique is used to solve the system of equations. The first step in the procedure is to solve the system assuming a conductive heat flow regime and then assign densities and viscosities to each node. The fluid flow equation is then solved and elemental specific discharges computed. The heat transfer equation is solved next to determine nodal temperatures. Then, densities and viscosities are updated using this new estimate of the temperature field. An iterative sequence is then employed until the maximum temperature change between iterations is less than a specified tolerance (0.1° C).

Programming considerations

A separate mesh generation code was developed to rapidly construct a large, simple three-dimensional domain in which element prisms consisting of eight nodes are first formed and then subdivided into five elements per prism (Figure 1). Element faces must be rotated during assembly to ensure that faces across prism boundaries match perfectly. This procedure is followed in order to guarantee element compatibility (Zienckewicz, 1977).

We have used Cholesky decomposition to solve (10) and Gaussian elimination with pivoting to solve (16). Both of these routines require the full or one half bands of the system to be stored. A second version of the program was developed that employed subroutines to reorder and compress matrices, which resulted in reducing the number of operations required to factor the global matrix and reduced core storage requirements (UBC SPARSPAK - Nicol, 1982). However, savings in storage costs were offset by increased charges for execution time. The primary value in using this second version of the program was in the denser meshes which could be accommodated. Typical meshes consisted of 19 nodal spacings in the X direction, 10 in Z and 8 in Y for a total of 1980 nodes and 7600 elements. For advectively disturbed cases, nonlinearities in density and viscosity required the program to iterate about 11 times to converge, requiring about 12 minutes of CPU time on an Amdahl 470/V8.

APPLICATIONS

To demonstrate the application of this modeling approach in simulating a hydrologic disturbance, consider the hypothetical basin shown schematically in Figure 2. The upper boundary of the flow domain is the water table. A linear relief of 40 m is imposed in the direction orthogonal to the regional slope, which has a linear slope of 500 m. A line of symmetry exists at Y equals 10 k. We shall refer to the section along this line of symmetry as the ridge line. A temperature of 10°C is imposed on the water table. A conductive heat flux of $60 mW/m^2$ is prescribed on the basal boundary which is assumed to be impervious to fluid flow. At the basin divides along the sides of the model, a condition of zero lateral heat and fluid flux is set, a result of symmetry with adjacent basins. Other relevant material properties are: a specific heat of the fluid of 4186 J/Kg°C, a thermal conductivity of the fluid of 0.586 W/m°C and the rock of 2.51 W/m°C. Porosity is assumed to decrease with depth in a piecewise manner from 0.20 at the surface to 0.06 at the base of the model. Surface heat flow is calculated as the vertical component of the conductive heat flux across the water table. Nodal positions in the finite element grid

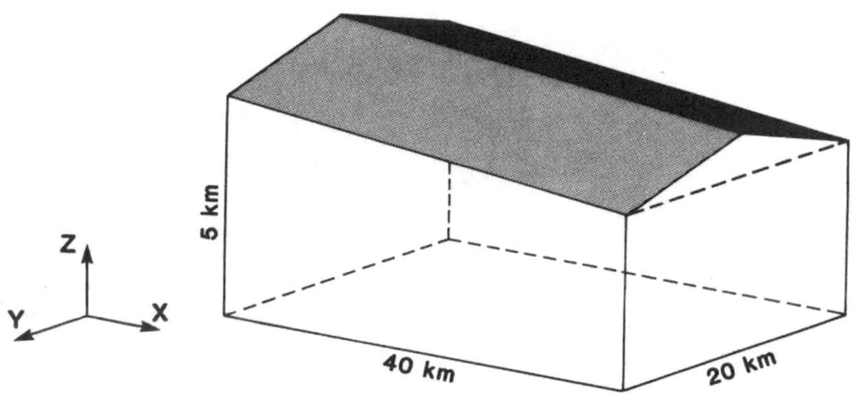

Figure 2. Schematic of hypothetical basin - linear water table configuration.

are designed so that the temperature gradient is computed over a 100 m depth interval immediately below the water table. The results should therefore be compared with field measurements obtained over the same interval.

Figure 3 shows surface heat flow for two different values of permeability. For a permeability of $1 \times 10^{-16} m^2$, the regime is dominated by heat conduction, with variations in heat flow due primarily to topographic effects. For illustrative purposes we have shaded the lower plot in that region where surface heat flow is within 16% of the basal heat flux, or between 50 and 70 mW/m^2. At $k = 5 \times 10^{-16} m^2$, the system is advectively-dominated. Surface heat flow in the upper reaches of the basin and along the ridge line is reduced due to downward groundwater flow. Along the edge of the basin, elevated heat flow exists

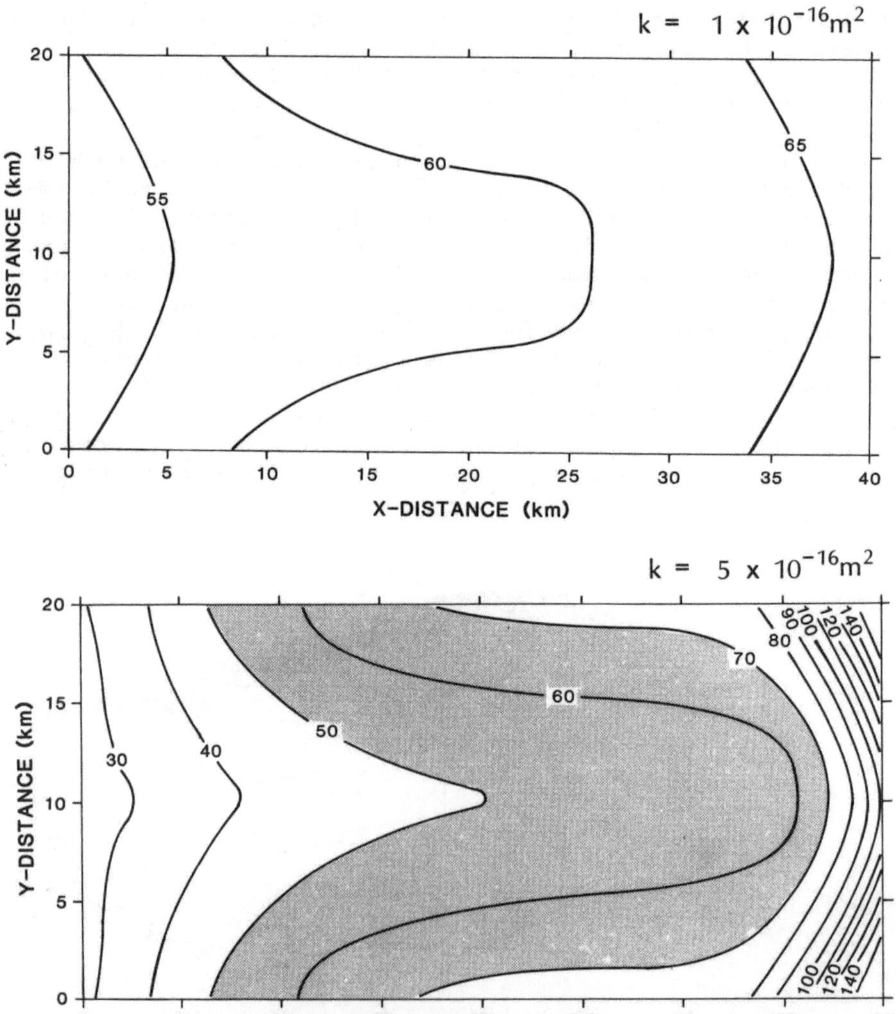

Figure 3. Surface heat flow for two levels of groundwater disturbance.

due to upward groundwater flow. The highest heat flow is observed along the lower edge of the basin, commensurate with groundwater discharge from the deep regional flow system. In a more complex environment, the character of the surface heat flow variation will reflect differences in groundwater recharge/discharge caused by the influence of geologic layering, anisotropy in permeability, the depth of flow, and variations in the configuration of the water table.

Temperature/gradient — depth plots for the advectively-disturbed system are shown on Figure 4. At location (a), the thermal disturbance is not strongly developed; hence, the near-linear slope of the temperature — depth plot. Surface heat flow in the upper 100 m is about 50 mW/m^2, close to the imposed basal value of 60 mW/m^2. Also plotted as a dashed line is the geothermal gradient for a purely conductive case. The gradient varies slightly because of the reduction in porosity with depth. At location (b), the disturbance is marked; surface heat flow in the upper 100 m is about 120 mW/m^2. The convex temperature-depth plot is indicative of upward groundwater flow. If one were to assume that the gradient in the bottom portion of the hole is the most representative, then we have a situation where the gradient, and hence heat flow depends on the depth of hole.

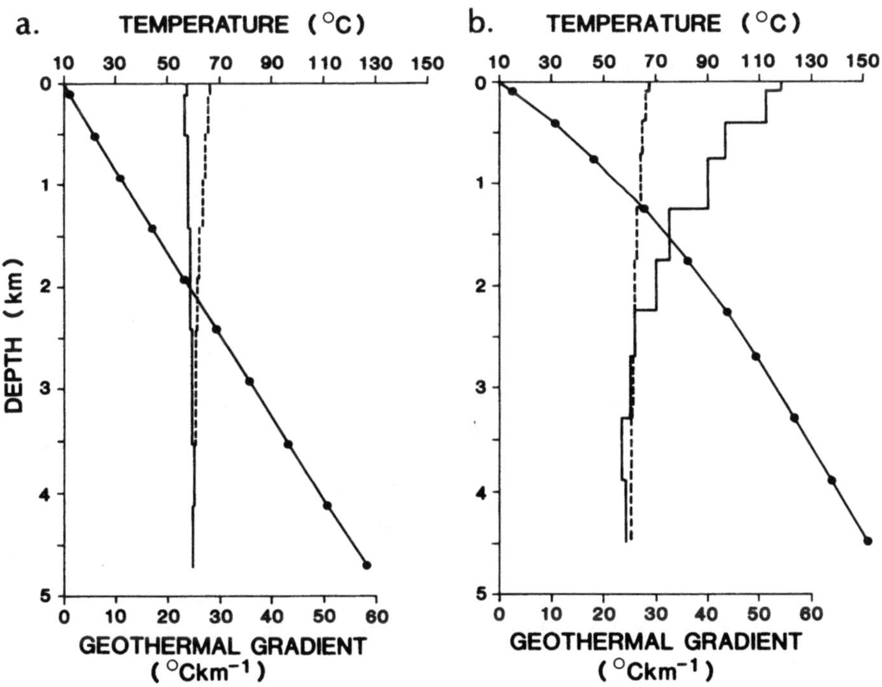

Figure 4. (a) Temperature/gradient depth plots for coordinate position X = 20 km, Y = 10 km, and (b) X = 37.5 km, Y = 2.5 km.

A deeper hole in this case would lead to lower estimated surface heat flow, and as can be seen, only gradients and thermal conductivities measured beyond 2.5 km in depth would yield heat flow values close to the basal heat flux.

CONCLUSIONS

Analyses of the kind discussed in the last section provide insight into how a topographically-driven flow system can perturb the thermal region of a basin. A model study carried out during the initial phases of a field study can be used as a guide in locating heat flow measurement sites, in selecting borehole depths and in locating sites where hydrologic data should be collected. If groundwater is redistributing heat within the basin, with the result that surface heat flow measurements may bear no direct relationship to thermal conditions at depth, then combined modeling of the thermal and hydrologic regimes provides an approach to unravel the dependence of surface heat flow on groundwater flow. Work is continuing on a project to develop techniques for estimating directly basal heat flux values, given data on the temperature-depth profile and hydraulic head distribution at selected locations within a basin.

ACKNOWLEDGEMENTS

This research was supported by funds provided by the Natural Sciences and Engineering Research Council of Canada.

REFERENCES

Bear, J. (1972) Dynamics of Fluids in Porous Media, Elsevier, The Netherlands

Frind, E.O. (1980) Seawater intrusion in continuous coastal aquifer-aquitard systems, Proc. Third Inter. Conf. on Finite Elements in Water Resources, Univ. Miss., Oxford, U.S.A.

Nicol, T. (1982) UBC SPARSPAK User's Guide, University of British Columbia Computing Centre.

Smith, L. and D.S. Chapman (1983) On the thermal effects of groundwater flow; 1, Regional scale systems, J. Geophys. Res., 88, 593-608.

Woodbury, A.D. (1983) The thermal effects of three dimensional groundwater flow, M.Sc. Thesis, The University of British Columbia.

Zienkiewicz, O.C. (1977) The Finite Element Method, 3rd ed., McGraw-Hill, New York.

BOUNDARY ELEMENT METHOD

SIMULATION OF STEADY-STATE FLOW IN THREE-DIMENSIONAL FRACTURE NETWORKS USING THE BOUNDARY ELEMENT METHOD

Allen M. Shapiro

Technion - Israel Institute of Technology
Faculty of Civil Engineering
Technion City, Haifa 32000, Israel

Johan Andersson

Royal Institute of Technology
Department of Water Resources Engineering
S-100 44 Stockholm, Sweden

INTRODUCTION

The modelling of transport phenomena in fractured rock has been a topic of increasing interest over the past several years. In studies which have been undertaken to date, the means by which transport phenomena are mathematically conceptualized have taken two distinct routes. The necessity for alternative idealizations of fractured rock has arisen from the fact that the length scale of a given problem in relation to the density of fracturing is not consistent from one rock formation to another. This has led to conceptualizations of fractured rock as either a system of individual and possibly interconnected fractures in a permeable or impermeable host rock, or as one or more overlapping continua, in a similar manner to the mathematical treatment of granular porous media.

In applying the continuum conceptualization to a fractured rock, it is assumed that a sufficient number of interconnected fractures exist to define statistically meaningful properties associated with the fracture network at all points in the investigated domain. The physical structure of the individual fractures is no longer considered in the modelling procedure. Instead, average characteristics and responses are hypothesized for both the rock and the fluid phases within the fractures. It is only in this context that rock properties such as porosity and conductivity have meaning.

Clearly, the continuum conceptualization is applicable to those formations where there is a high degree of fracturing in relation to the problem scale. For rock formations where there are few fractures, it is necessary to consider the processes of transport in the network of individual fractures. Therefore, the explicit geometry of the fracture network plays an important role in the responses which are predicted. In this paper we concern ourselves with the discrete fracture conceptualization, and address the problem of modelling steady-state fluid movement in the fractures of an impervious host rock.

In modelling subsurface hydrologic phenomena, reducing the spatial dimensionality of a given problem is desirable for computational reasons. However, when considering discrete fracture networks, it is necessary to retain the three-dimensional geometric characteristics. Only in the case of specific fracture configurations, subject to specific fluid responses, can the three-dimensional problem of flow in a fractured rock be reduced to an equivalent two-dimensional problem. Such fracture geometries, however, are not often physically realistic.

Although it is necessary to consider the three-dimensional nature of a fracture network, it is possible to attain certain efficiencies in the solution procedure by making assumptions with regard to the fluid movement within the individual fractures. The fractures may be of any orientation within the rock, however, we assume flow in the plane of the fracture to be essentially two-dimensional. Such an assumption is appropriate if the fracture aperture is small in relation to its overall extent. The lines along which fractures intersect are also considered as (one-dimensional) fluid conduits. Under these assumptions it is possible to construct the three-dimensional geometric characteristics of flow in a fractured formation by considering a series of coupled, one- and two-dimensional equations.

GOVERNING EQUATIONS

Fractures

The movement of fluid at any point in a fracture, under the assumptions of steady-state flow and constant mass density, is described by

$$\nabla \cdot \underset{\sim}{V} = 0 \tag{1}$$

where $\underset{\sim}{V}$ is the fluid velocity. Since an impervious host rock is being considered, we impose the condition of no flow at the fracture walls, i.e.,

$$\underset{\sim}{V} \cdot \underset{\sim}{n} = 0 \tag{2}$$

where $\underset{\sim}{n}$ is a unit normal vector outwardly directed from the fracture.

The fractures are assumed to be sufficiently narrow in relation to their overall extent. Therefore, flow is essentially two-dimensional in the surface defining the fracture axis. An equation of fluid movement which is appropriate under such conditions is generated by integrating (1) over the thickness of the fracture in the direction orthogonal to the fracture axis (Shapiro and Andersson, 1983). With the use of (2), the resulting two-dimensional equation takes the form

$$\underset{\sim}{\nabla}'\cdot(b\underset{\sim}{V}') = 0 \tag{3}$$

where $\underset{\sim}{V}'$ is the average fluid velocity defined over the fracture thickness, b, and $\underset{\sim}{\nabla}'$ is the gradient operator in the orthogonal, two-dimensional coordinate system which defines the fracture axis.

The product $b\underset{\sim}{V}'$ represents the fluid discharge over the thickness of the fracture at a given point. We shall assume that the fluid discharge is proportional to the gradient of an average hydraulic potential, h, defined over the fracture thickness, i.e.,

$$b\underset{\sim}{V}' = -K\,\underset{\sim}{\nabla}'h \tag{4}$$

where the fracture conductance, K (L^2/T), is assumed to be a scalar quantity.

Substituting (4) into (3) yields

$$\underset{\sim}{\nabla}'\cdot(K\,\underset{\sim}{\nabla}'h) = 0 \tag{5}$$

This expression is valid at all points in the two-dimensional surface defining the extent of a given fracture. The areal boundaries of a fracture are taken to be its abrupt ending in the impervious host rock, or at the lines of intersection with other fractures.

<u>Fracture intersections</u>

Lines of fracture intersections are treated as separate fluid conduits since the conductance along the axis of the intersection need not be the same as that of the fractures which form it. We shall assume that flow in the intersection is essentially in the direction of its axis. However, at points along the intersection fluid can enter from (or exit into) the adjacent fractures. From (1), which is valid at all points in the fracture intersection, an appropriate one-dimensional equation of fluid movement can be generated in a manner similar to the development of (3). The form of this equation is

$$\frac{d}{d\xi}(aV_\xi) - \sum_{\ell=1}^{N} q^\ell = 0 \tag{6}$$

where V_ξ is the average fluid velocity in the direction of the intersection axis (defined by the coordinate ξ), $a(\xi)$ is the cross-sectional area of the fracture intersection and q^{ℓ} denotes sources of fluid from the (N) fractures which form the intersection.

Assuming that the fluid discharge through the fracture intersection can be defined by

$$aV_{\xi} = -\kappa \ (dh/d\xi) \tag{7}$$

where κ (L^3/T) is the conductance of the intersection, (6) takes the form

$$\frac{d}{d\xi}(\kappa \frac{dh}{d\xi}) + \sum_{\ell=1}^{N} q^{\ell} = 0 \tag{8}$$

In this analysis, we shall assume that the hydraulic head at a given point in the fracture intersection is equal to the hydraulic head at the boundary of the fractures which form it. In addition, we shall limit our discussion by considering only planar fractures. Consequently, fracture intersections are linear.

BOUNDARY ELEMENT FORMULATION

The three-dimensional geometry of the fracture network is retained in the solution of (5) and (8). These equations, however, cannot be solved independently since the fluid entering (or being discharged from) the fracture intersection depends on the responses in the adjacent fractures. The responses in the fractures, in turn, cannot be evaluated without the fluxes from the fracture intersections acting as boundary conditions. These equations are linear, however, and may be coupled to obtain a simultaneous evaluation of both flow regimes.

In practice, most any numerical technique for solving equations of the type given by (5) and (8) can be employed in this analysis. Here, however, we choose to apply the boundary element method in order to avoid internally discretizing the areal domain of a given fracture. The boundary element method computes the steady-state fluid responses by considering the fluid potential and fluid flux at the boundaries of the fracture planes. Since the flux at the boundary of the fracture plane appears as a solution variable, further numerical manipulation in unnecessary in order to obtain the flux as an input to (8).

<u>Fractures</u>

A boundary element solution to (5) is developed by first multiplying this expression by a function ψ and integrating over the investigated domain, in this case the area of a given fracture plane, A, i.e.,

$$\int_{(A)} \psi \nabla^2 h \, da = 0 \tag{9}$$

Here we have assume the conductance of the fracture to be

constant over the fracture plane. In addition, we have dropped the superscript associated with the gradient operator since it is understood that we are considering a two-dimensional domain. Successive use of the divergence theorem in the above expression yields

$$\int_{(A)} h \nabla^2\psi \, da + \int_{(S)} (\psi \frac{\partial h}{\partial n} - h \frac{\partial \psi}{\partial n}) \, ds = 0 \qquad (10)$$

where S is the closed boundary defining the areal domain and $\partial(\ )/\partial n$ is the outwardly directed normal derivative to the boundary S.

The function ψ is chosen to be a solution to $\nabla^2\psi + \delta(x-x_m)\,\delta(y-y_m) = 0$ where $\delta(\ )$ is the Dirac delta function. In two-dimensions, ψ is the logarithmic potential function, i.e.,

$$\psi(x,y;x_m,y_m) = -\frac{1}{2\pi} \ln(\{(x-x_m)^2+(y-y_m)^2\}^{\frac{1}{2}}) \qquad (11)$$

With the definition of ψ, (8) can be written for an arbitrary point (x_m,y_m) on the boundary as (Brebbia, 1978)

$$\frac{1}{2} h(x_m,y_m) = -\int_{(S)} (\psi \frac{q}{K} + h \frac{\partial \psi}{\partial n}) \, ds \qquad (12)$$

where in place of $\partial h/\partial n$ we have introduced the outwardly directed fluid flux vector, $q=(b\underset{\sim}{V}'\cdot\underset{\sim}{n})$, on the boundary of the fracture plane.

Approximate solutions for h and q on the boundary are obtained by discretizing the previous equation. As an assumption, over each boundary element on S, h and q are taken to be constant. Consequently, the discretized form of (12) is

$$\frac{1}{2} h_m = -\sum_{j=1}^{E} \{ \frac{q_j}{K} \int_{\Delta s_j} \psi_{jm} \, ds + h_j \int_{\Delta s_j} \frac{\partial \psi}{\partial n}_{jm} \, ds \} \qquad (13)$$

where E is the number of elements comprising S with each element denoted by Δs_j, h_j and q_j are the element values of the hydraulic head and fluid flux, respectively, and the subscripts j and m denote nodal locations on the boundary.

At boundary elements, other than those at the fracture intersections, either h or q is prescribed as a boundary condition. At those elements which lie along a fracture intersection, both h and q are unknown, and must be solved in conjunction with a discretized form of (8).

<u>Fracture intersections</u>

Along fracture intersections, the discretization is chosen to be consistent with that of the fracture planes. At the midpoint of each element we write the following discretized form of (8),

$$\frac{\kappa}{(\Delta\xi)^2}(h_{j-1} - 2h_j + h_{j+1}) + \sum_{\ell=1}^{N} q_j^{\ell} = 0 \tag{14}$$

where the subscript j denotes the element under consideration and $\Delta\xi$ is the length of the boundary elements (which are assumed to be constant) along a given fracture intersection.

At points where fracture intersections intersect, we apply a condition of continuity with regard to the fluid fluxes along the intersection axes, i.e.,

$$\sum_{i=1}^{I}(aV_{\xi})_i = -\sum_{i=1}^{I}(\kappa \frac{dh}{d\xi})_i = 0 \tag{15}$$

where I denotes the number of lines of intersection at the point in question. At the end points of fracture intersections, it is possible to prescribe the hydraulic head (e.g., if the fracture intersection ends at a boundary of the rock domain) or a fluid flux (e.g., if the intersection ends at a point internal to the rock domain, the flux is zero).

A determinant system of equations is provided by writing (13) for all fracture planes, and (14), (15) and the appropriate boundary conditions for all fracture intersections. The solution variables are the hydraulic head and the fluid flux at the boundaries of the fracture planes.

<u>Treatment of wells</u>

In our analysis, we additionally consider the possibility of boreholes from which fluid can be extracted. For those fractures which are intersected by a borehole, the borehole is treated as an internal boundary. The shape of the borehole in the fracture plane is discretized by a series of boundary elements. This is entirely consistent with the form of (13) where E can be considered as the total number of boundary elements, including those elements which define the boreholes in the fracture plane.

Since a given borehole may intersect several fractures, the fluid production from each of the fractures is only a portion of the total discharge. Furthermore, the amount of fluid discharged from each fracture, and the distribution with respect to the elements which define the borehole is unknown. Over the length of the borehole we shall assume that the hydraulic head is constant. Although this acts as a boundary condition applied to the elements defining the borehole, the value of the hydraulic head which allows the correct discharge from all fractures is unknown. Therefore, for boreholes which intersect more than one fracture plane, it becomes necessary to iteratively obtain a solution, i.e., to prescribe a value of the head in the borehole and check if the solution for the total discharge from all fractures satisfies the prescribed discharge from the well.

NUMERICAL RESULTS

The accuracy of the boundary element method as applied to problems similar to this is demonstrated in Shapiro and Andersson (1983). Here, we consider a hypothetical example of steady-state fluid movement in a three-dimensional fracture network in order to demonstrate the boundary element formulation presented above. We consider the three-dimensional section of a fractured rock with the dimensions shown in Fig. 1. In this domain there are three planar fractures (F1, F2 and F3) forming two fracture intersections. In addition, we consider the presence of a borehole of radius 0.1 m, which is vertically oriented, and located at a position which intersects the three fractures. Fracture planes F1 and F2 are assigned a conductance of 0.75 m^2/day, while the conductance of F3 is given a value of 10 m^2/day. The conductance of the fracture intersections are each assigned a value of 20 m^3/day. At the rock faces in the x-direction ($x=0$ and $x=-40$), we impose the condition of a prescribed hydraulic head, $h=0$. For all other rock faces an impervious boundary is assumed.

For these physical conditions, and for the parameter values given, we examine the fluid responses in the fracture network as a result of injecting fluid in the borehole at a rate of 100 m^3/day. Results for this sample problem are given in Fig. 2 where lines of equipotential are plotted for those segments of the fractures which are intersected by the borehole. The values of the hydraulic head at the interior points of the fracture planes are obtained from the solutions for h and q on the boundaries (see, Brebbia, 1978).

The results indicate large hydraulic gradients in the vicinity of the borehole in fractures F1 and F2. This arises due to the lower conductance assigned to these fractures. Although there is a larger hydraulic gradient in the vicinity of the borehole in F1 and F2, the percent of the total injected fluid which enters these fractures is only 11.5 and 10.5 percent, respectively. The majority of the injected fluid enters the fracture network through F3.

SUMMARY AND CONCLUSIONS

An efficient method of predicting steady-state fluid responses in three-dimensional fracture networks is formulated with the use of the boundary element method. The three-dimensional geometric characteristics of the fracture geomentry are retained while solving coupled sets of one- and two-dimensional equations. In this analysis, fractures may have any orientation in an impervious host rock. However, the fractures are treated as surfaces where fluid movement is assumed to be essentially two-dimensional. In using the boundary element method, the steady-state fluid responses in the fractures are evaluated by considering the hydraulic head and fluid flux at the boundaries of the fracture planes. These boundaries are depicted as either the abrupt ending of the fracture in the impervious host rock, or as lines of intersection with other fractures.

Fracture intersections, which are also treated separate fluid conduits, are described by one-dimensional equations written in terms of the hydraulic head and the fluid flux from those fractures which form the intersection. These are the same solution variables which define the fluid responses in the fracture surfaces. Consequently, a direct coupling of the discretized equations for each flow regime is easily attained.

REFERENCES

Brebbia, C. (1978) The Boundary Element Method for Engineers, Pentech, London, 1978.

Shapiro, A. and J. Andersson (1983) Steady-state fluid response in fractured rock - A boundary element solution for a coupled, discrete fracture-continuum model, Water Resour. Res., 19, 959-969.

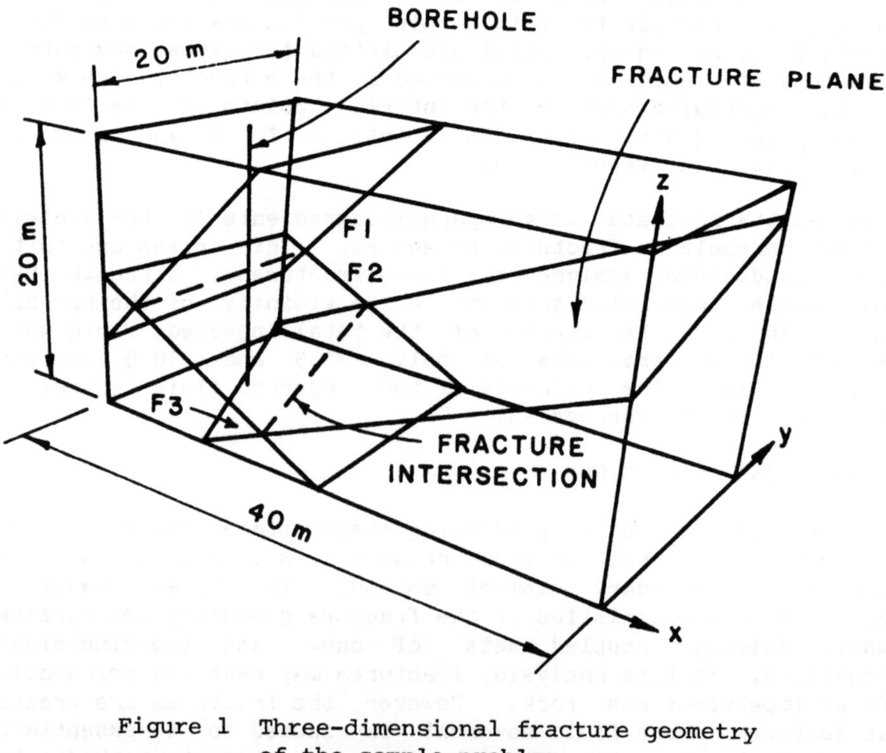

Figure 1 Three-dimensional fracture geometry of the sample problem.

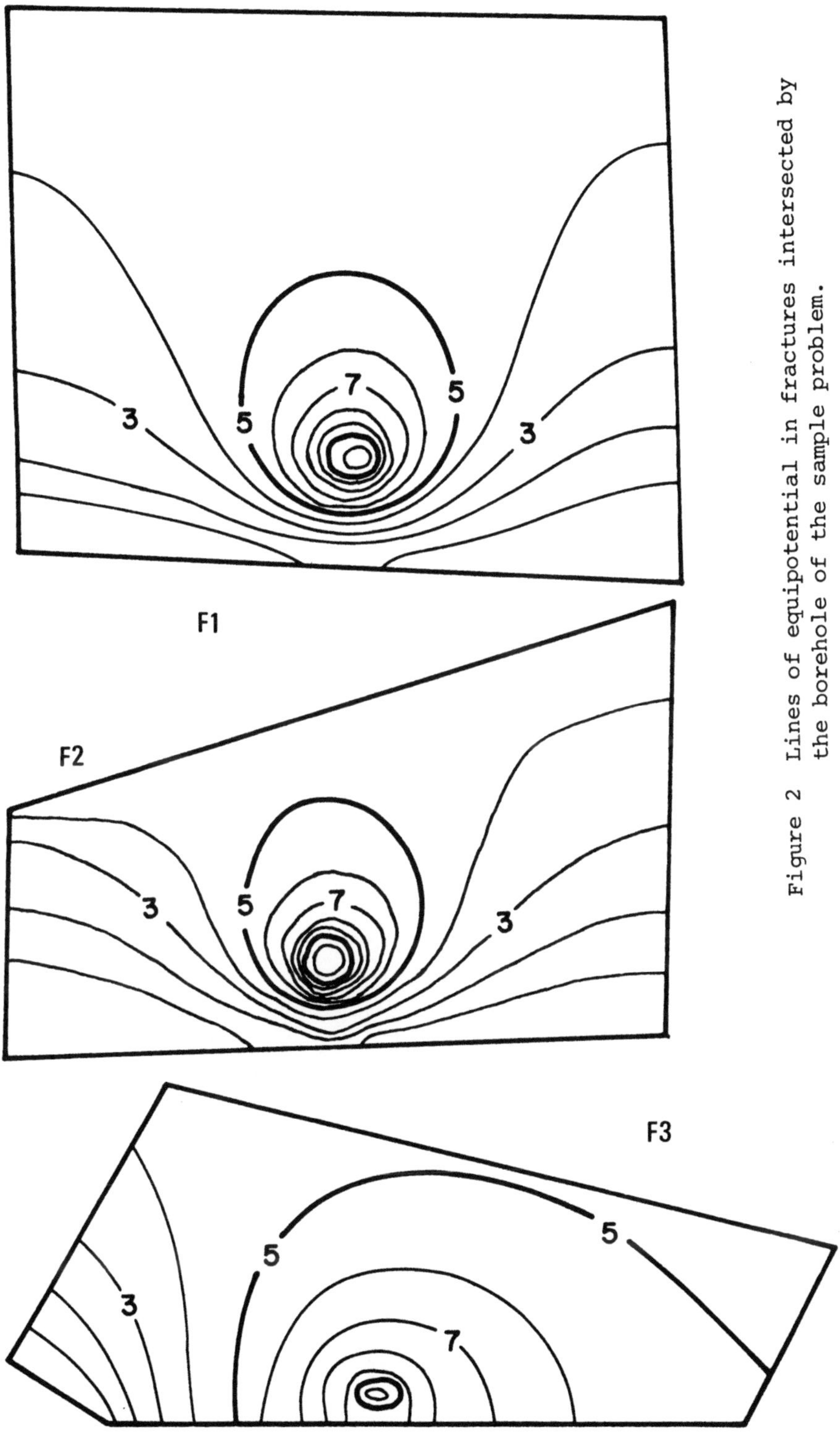

Figure 2 Lines of equipotential in fractures intersected by the borehole of the sample problem.

BOUNDARY ELEMENT ANALYSIS OF FLOW IN AQUIFERS

Miodrag Radojković
Josip Pečarić

INTRODUCTION

Finite differences and finite element techniques were almost exclusively used to solve the equations governing the flow in aquifers. Recently it was shown that Boundary Element Method (BEM) can also be applied sucessefully [1].

The BEM model designed to aquifer flow analysis has to fulfill two basic requirements:

- modelling of wells
- modelling of couplied subregions with constant transmissivity.

In the paper presented are some contributions to meet these requirements:

- drawdown in a well can be computed for given discharge and viceversa (only the first possibility exists in [1])
- double node approach is avoided for all nodes on external and internal boundaries with mixed element (linear for potential, constant for flux) retaining the accuracy of linear elements

The presentation includes the basic theory of these contributions and some results of practical application to the design of artificial recharge system for one aquifer in Yugoslavia).

2. BASIC THEORY

Let Ω be a two-dimensional domain and Γ, $S_1, S_2, \ldots, S_n$ its boundaries (see Fig. 1), where S_j $(j = 1,\ldots,n)$ are circles with radius r_{oj} $(j = 1,\ldots,n)$ representing system of wells.

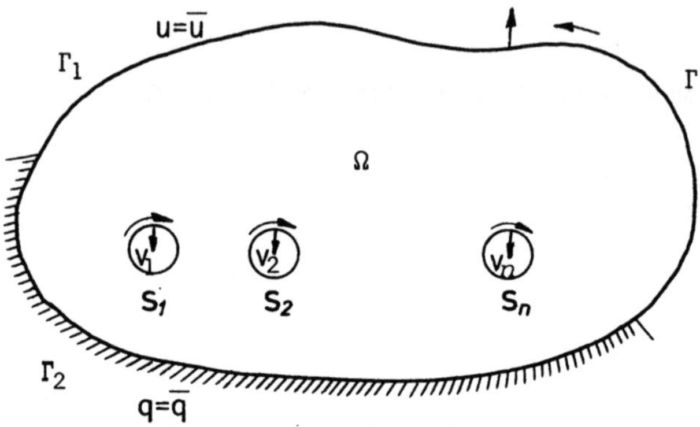

Fig. 1. Explanation of symbols

Consider the Laplace equation:

$$\nabla^2 u = 0 \qquad \text{on} \quad \Omega \tag{1}$$

with boundary conditions:

$$u = \bar{u} \quad \text{on} \quad \Gamma_1$$

$$q = \frac{\partial u}{\partial n} = \bar{q} \quad \text{on} \quad \Gamma_2 \qquad (\Gamma = \Gamma_1 + \Gamma_2)$$

and:

$$u = \bar{u}_j \quad \text{on} \quad S_j, \quad j \in I$$

$$Q = \int_{S_j} q \, d\,S_j = \bar{Q}_j \quad \text{on} \quad S_j,\ j \in I_n \setminus I$$

where $I_n = \{1,\ldots,n\}$.

The boundary integral formulation of the above problem can be derived starting from Green's second identity which in this case reads:

$$\int_\Omega (u^*\nabla^2 u - u\nabla^2 u^*)d\Omega = \int_\Gamma (u^*q - uq^*)d\Gamma$$

$$+ \sum_{j=1}^{n} S \int_{S_j} (u^*q - uq^*)dS_j \qquad (q^* = \frac{\partial u^*}{\partial n}) \tag{2}$$

where u satisfies (1) and u^* is the fundamental solution of the Laplace equation (solution for concentrated potential acting at a point "i")

$$u^* = \frac{1}{2\pi} \ln \frac{1}{r} \tag{3}$$

where r is the distance from the point "i" to any point. Then (2) becomes:

$$C^i u_i = \int_\Gamma (u^*q-uq^*)d\Gamma + \sum_{j=1}^{n} \int_{S_j} (u^*q-uq^*)dS_j \tag{4}$$

where C^i is a constant from $[0,1]$ ($C^i = 1$ for an internal point, $C^i = 0$ for an external point, and $C^i = 1/2$ for a point on the smooth boundary).

As the integrals along Γ cannot be evaluated analytically, the integration has to be done numerically by subdivision of the boundary Γ to N boundary elements. Assuming the case when $u_{N+j} = \text{Const}_j$ along all boundaries S_j ($j = 1,\ldots,n$), eq.(4) becomes:

$$C^i u_i + \sum_{K=1}^{N} \int_{\Gamma_K} uq^*d\Gamma + \sum_{j=N+1}^{N+n} u_{N+j} \int_{S_{j-N}} q^*dS_{j-N}$$
$$= \sum_{K=1}^{N} \int_{\Gamma_K} qu^*d\Gamma_K + \sum_{j=N+1}^{N+n} \int_{S_{j-N}} u^*qdS_{j-N} , \tag{5}$$

where u_{N+j} ($j = 1,\ldots,n$) are the potentials on S_j ($j = 1,\ldots,n$).

Let V_j be an interior domain with circle S as its boundary. Then:

$$\int_{S_j} q^*dS_j = 0 \tag{6}$$

if "i" is an external point, and:

$$\int_{S_i} q^*dS_i = 1 \tag{7}$$

if "i" is an internal point.

The proof of these statements follows from Green's second identity:

$$\int_{V_j} (u^*\nabla^2u-u\nabla^2u^*)dV_j = \int_{S_j}(u^* \frac{\partial u}{\partial n} - u \frac{\partial u^*}{\partial n})dS_j \tag{8}$$

applied to the case when u = 1.

Then:

$$\int_{S_j} \frac{\partial u^*}{\partial n} dS_j = \int_{V_j} \nabla^2u^*dV_j \tag{9}$$

i.e.

$$\int_{S_j} \frac{\partial u^*}{\partial n} dS_j = - \int_{V_j} \delta(\vec{x} - \vec{x}_i)dV_j \tag{10}$$

where δ is the Dirac delta function. Now, from (10) one can obtain (6) in the case when "i" is an external point because in this case $\delta(\vec{x} - \vec{x}_i) = 0$ for $\bar{x} \in V_j$, and (7) if "i" is an internal point since the normal to S_j is directed to the interior od V_j (see Fig.1).

Consider at first the case when "i" is a point from the boundary Γ. Then, due to (6), eq.(5) becomes:

$$C^i u_i + \sum_{K=1}^{N} \int_{\Gamma_K} u q^* d\Gamma_K = \sum_{K=1}^{N} \int_{\Gamma_K} q u^* d\Gamma_K + \sum_{j=N+1}^{N+n} \int_{S_{j-N}} u^* q dS_{j-N} \tag{11}$$

Let r_j be the distance from the point "i" to the centre of circle S_j. If the case when $r_j >> r_{oj}$ ($j = 1,\ldots,n$) is assumed, one can put:

$$u^*_{N+j} \approx \frac{1}{2\pi} \ln \frac{1}{r_j}\,, \tag{12}$$

and (11) becomes:

$$C^i u_i + \sum_{K=1}^{N} \int_{\Gamma_K} u q^* d\Gamma_K = \sum_{K=1}^{N} \int_{\Gamma_K} q u^* d\Gamma_K + \sum_{j=N+1}^{N+n} u^*_j Q_j \tag{13}$$

Now, consider the case when points "i" are in the centre of circles S_i ($i = 1,\ldots,n$). In this case $C^i = 0$, and due to (7) eq. (5) becomes :

$$u_{N+i} + \sum_{K=1}^{N} \int_{\Gamma_K} u q^* d\Gamma_K = \sum_{K=1}^{N} \int_{\Gamma_K} q u^* d\Gamma_K + \sum_{j=N+1}^{N+n} \int_{S_{j-N}} u^* q dS_{j-N} \tag{14}$$

For circle S_i one has:

$$u^* = u^*_{N+i} = \frac{1}{2\pi} \ln \frac{1}{r_{oi}} \tag{15}$$

In the case when $r_j >> r_{oi} + r_{oj}$ (r_j is also the distance from the point "i" to the centre of circle S_j) one can use (12), so that (14) becomes:

$$u_{N+i} + \sum_{K=1}^{N} \int_{\Gamma_K} u q^* d\Gamma_K = \sum_{K=1}^{N} \int_{\Gamma_K} q u^* d\Gamma_K + \sum_{j=N+1}^{N+n} u^*_j Q_j \tag{16}$$

Equation (16), after selection of the boundary element type (constant, linear, quadratic, etc), can be solved for all unknown u's and q's on the boundary Γ, and all unknown u's and Q's on the boundaries S_j ($j = 1,\ldots,n$). Once this is done one can similarly calculate the values of u's and q's at any interior point $M(x_i, y_i)$ using equations:

$$u_i = \sum_{K=1}^{N} \int_{\Gamma_K} (q u^* - u q^*) d\Gamma_K + \sum_{j=N+1}^{N+n} \int_{S_{j-N}} u^* q dS_{j-N} \tag{17}$$

and

$$q_{ix} = \frac{\partial u_i}{\partial x_i} = \sum_{K=1}^{N} \int_{\Gamma_K} (q \frac{\partial u^*}{\partial x_i} - u \frac{\partial q^*}{\partial x_i}) d\Gamma_K + \sum_{j=N+1}^{N+n} \int_{S_{j-N}} q \frac{\partial u^*}{\partial x_i} dS_{j-N} \quad (18)$$

$$q_{iy} = \frac{\partial u_i}{\partial y_i} = \sum_{K=1}^{N} \int_{\Gamma_K} (q \frac{\partial u^*}{\partial y_i} - u \frac{\partial q^*}{\partial y_i}) d\Gamma_K + \sum_{j=N+1}^{N+n} \int_{S_{j-N}} q \frac{\partial u^*}{\partial y_i} dS_{j-N} \quad (19)$$

3. MIXED BOUNDARY ELEMENTS

In this work the mixed elements with a linear variations for the potential and constant flux along the elements are applied. Therefore, variation of potential reads:

$$u(\xi) = N_1(\xi)u_K + N_2(\xi)u_{K+1} \quad (20)$$

where N_1 and N_2 are linear shape functions:

$$N_1(\xi) = \frac{1}{2}(1 - \xi), \quad N_2(\xi) = \frac{1}{2}(1 + \xi) \quad (-1 \leq \xi \leq 1) \quad (21)$$

Note that the boundary Γ is discretized by straight segments.

Using the above aproximation eq.(13) becomes:

$$C^i u_i + \sum_{K=1}^{N} \hat{H}_{iK} u_K = \sum_{K=1}^{N} G_{iK} q_K + \sum_{j=K+1}^{N+n} u_j^* Q_j \quad (22)$$

where:

$$\hat{H}_{iK} = A_{K-1,2} + A_{K,1} \; ; \; G_{iK} = \int_{\Gamma} u^* d\Gamma_K$$

$$A_{K1}^i = \int_{\Gamma_K} N_1 q^* d\Gamma_K, \quad A_{K2}^i = \int_{\Gamma_K} N_2 q^* d\Gamma_K \quad (23)$$

As it is seen the equation (22) can be applied to the extreme points or to the center of an element. In [3] it was shown that the best combination is one from [2], i.e. for elements from Γ_2 (u unknown, q known) the integration is done from the extreme points of the element, and for elements from Γ_1 (u known, n unknown) the integration is done from the center of the elements (see Fig. 2).

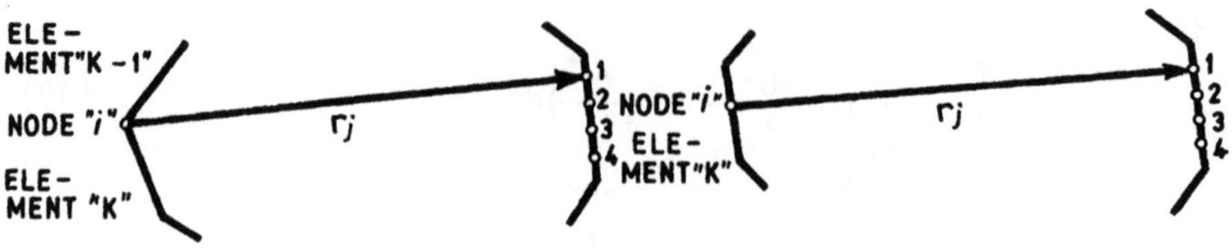

a. Unknown potential at "i" b. Unknown flux at "i"

1,2,3,4 - points for Gaussian quadrature

Fig 2. Integration rules for mixed elements

For element (i) from Γ_2 $A^i_{i,1} = A^i_{i,2} = A^i_{i-1,1} = A^i_{i-1,2} = 0$ because $q^* = 0$ for elements "i" and "$i-1$". If one puts:

$$H_{iK} = \begin{cases} \hat{H}_{iK}, & K \neq i \\ C^i, & K = i \end{cases} \tag{24}$$

(22) becomes:

$$\sum_{K=1}^{N} H_{iK} u_K = \sum_{K=1}^{N} G_{iK} q_K + \sum_{j=N+1}^{N+n} u_j^* Q_j \tag{25}$$

When integrating from the extreme point to an element (K) not contiguous to it, the coefficients are calculated numerically (four point Gaussian quadrature formula is applied). The coefficient G_{ii} is calculated analytically, and the coefficients H_{ii} are found by application of a constant potential u. Then from (25) it follows:

$$C^i = - \sum_{\substack{K=1 \\ K \neq 1}}^{N} H_{iK} \tag{26}$$

For element (i) from Γ_1 one has three values for potential in equation (22) because the node is a center of element. This limits direct use of mixed element from [2] in problems with subregions of different material properties where both u and q are unknown along internal boundary. Since the linear variation for the potential function u along the element is used, one has (see [3])

$$u_i \rightarrow \frac{u_i + u_{i+1}}{2} \quad \text{and} \quad C^i = 1/2$$

and equation (20) becomes:

$$\frac{1}{4}(u_i + u_{i+1}) + \sum_{K=1}^{N} \hat{H}_{iK} u_i = \sum_{K=1}^{N} G_{iK} q_K + \sum_{j=N+1}^{N+n} u_j^* Q_j \tag{28}$$

If one puts:

$$H_{iK} = \begin{cases} \hat{H}_{iK} & , \quad K \neq i,i+1 \\ \hat{H}_{iK} + \frac{1}{4} & , \quad K = i,i+1 \end{cases} \tag{29}$$

then (25) is obtained from (28).

The coefficients A^i_{K1}, A^i_{K2} ($K \neq i$) are also calculated numerically, the coefficient G analytically, while $A^i_{i1} = A^i_{i2} = 0$ because $q^* = 0$ on element "i".

For points from S_j ($j = 1,\ldots,n$), due to (14), one has:

$$u_{N+j} + \sum_{K=1}^{N} \hat{H}_{iK}\, u_K = \sum_{K=1}^{N} G_{iK}\, q_K + \sum_{j=N+1}^{N+n} u^*_j\, Q_j \ , \tag{30}$$

where $\hat{H}_{iK}, A^i_{K1}, A^i_{K2}$ and G_{iK} are given by (23). All these coefficients are calculated numerically.

Mixed elements were also used for computing the values of u's and q's at any interior point. Coefficients obtained in this way using (17), (18) and (19), are also calculated numerically.

Note that in many problems of aquifer flow, instead of the Laplace equation, one has to solve the Poisson equation:

$$\nabla^2 u = w \tag{31}$$

This can be done by adding an extra term to eq.(4) in form of the area integral $\int_\Omega w u^* d\Omega$ which is easily computed in the way described in [1].

4. GUIDELINES FOR PRACTICAL APPLICATION

When applying the theory developed in the previous chapters, one of the major problems is appropriate discretization of the external boundary and internal boundaries (between zones of different permeabilities) for accurate modelling of wells.

The general solution of the problem is difficult to find out theoretically, thus an attempt was done to do it empirically. Consider the flow to an array of wells as given in Fig.3.

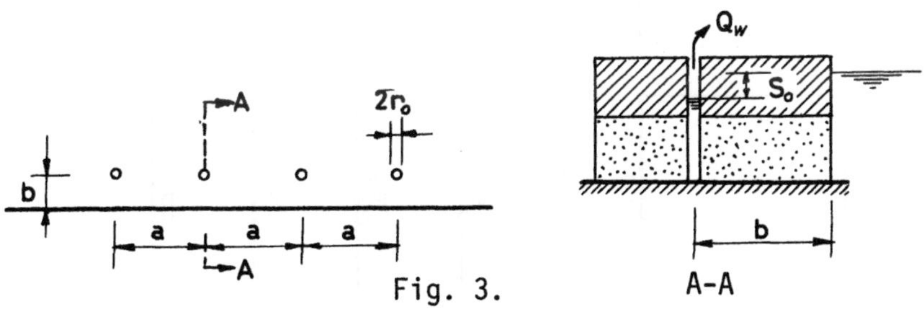

Fig. 3.

Theoretical relation between drawdown and discharge in a well from the array is:

$$\frac{Q_w}{TS_o} = \frac{2\pi}{\ln \frac{a}{\pi r_o} \, sh \, \frac{2\pi b}{a}} \tag{32}$$

The accuracy of the BEM solution for the same problem was investigated by changing the number of nodes along external potential boundary and internal boundaries (dividing the flow domain to a number of subdomains). The results of comparison are given in Fig.4 (note that all subdomains were with the same permeability).

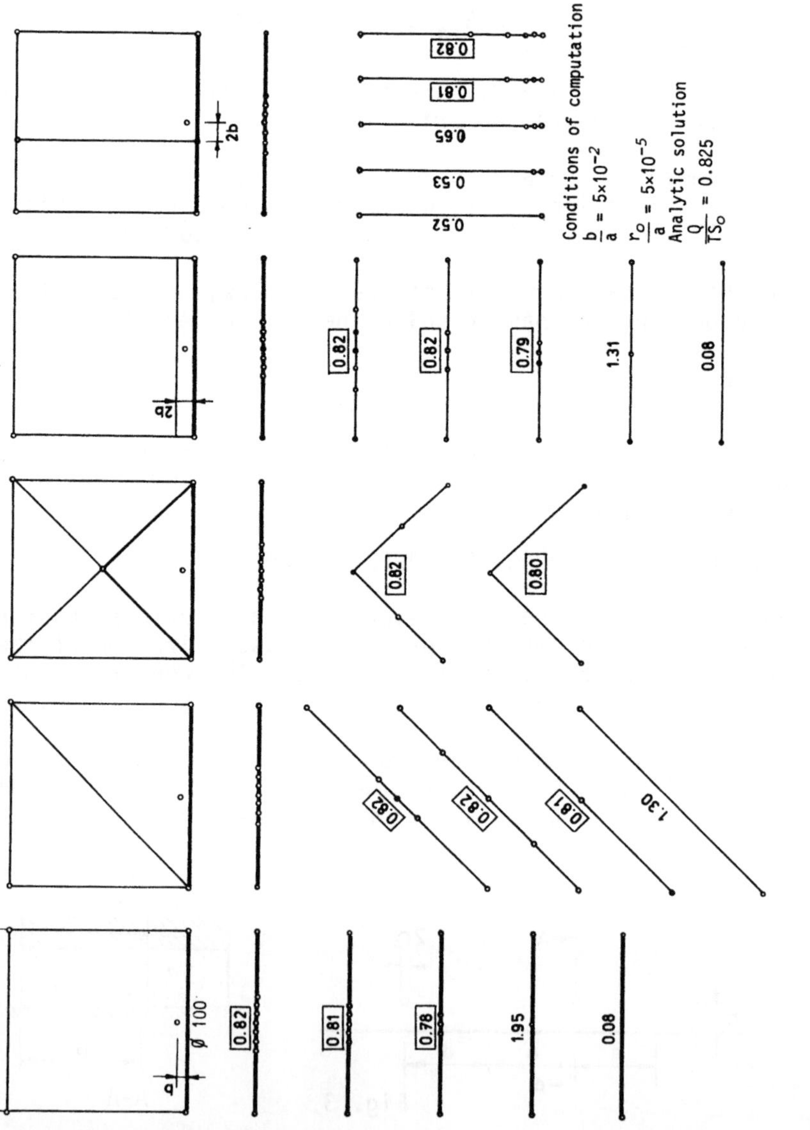

Fig. 4. Influence of discretization of external and internal boundaries on the accuracy for the computed discharge in well
(values in rectangles are with error less than 5%)

It is seen that the accuracy of BEM solution for a well discharge (when unit drawdown is given as boundary condition in the well) depends strongly on both external and internal boundary discretization. For reasonable accuracy one needs at least three nodes along the nearest boundaries in the vicinity of a well. These nodes should be placed at distance that is nearly equal to the distance between the well and this boundary.

It is interesting to note that similar "rule of thumb" was derived by Liggett and Liu in [1] but for internal potential when evaluated by (17).

5. EXAMPLE OF APPLICATION

On the base of the foregoing theory the boundary element code has been developed at the Institute for Hydraulic Engineering of the Belgrade University. It was successfully applied for different groundwater flow problems concerning both aquifer flow and seepage. One of these studies is described in the sequel.

5.1. Problem description

The problem studied was the aquifer located adjacent to the Danube river a few kilometers upstream from Novi Sad, the center of the northeastern region of Yugoslavia. The aquifer is currently exploited by the system of vertical wells. A drastic reduction of discharge capacity of wells was observed (about 100% within only two years). The reason is probably high content of iron in groundwater (considerably higher than in the Danube).

If such a trend of decrease in well capacity continues the economy of further exploitation of water catchment will be under question. So, it was decided to investigate the possibility of increasing the well capacity by artificial recharge through the system of infiltration basins. As the recharging water will be taken directly from the Danube, there is also a possibility that in these way the process of well capacity reduction will be retarded.

5.2. Model calibration

In order to investigate the parameters necessary for design of infiltration basins the experimental study was carried out on the small infiltration basin with dimensions 30×8 m. The basin was located near W-11 at distance of 30 m (see Fig.5).

Field study included the following hydrodynamic observations:

- level in Danube
- discharges in wells
- drawdowns in the wells
- levels in piezometers
- level in infiltration basin
- discharge to infiltration basin

The study also included traser injection and detection but the results are not directly relevent to the topic of the paper and will not be reported here.

The task of the mathematical model was to determine those values of aquifer transmissivities that, for given discharges in wells and basin and level in the Danube, predict drawdowns in wells

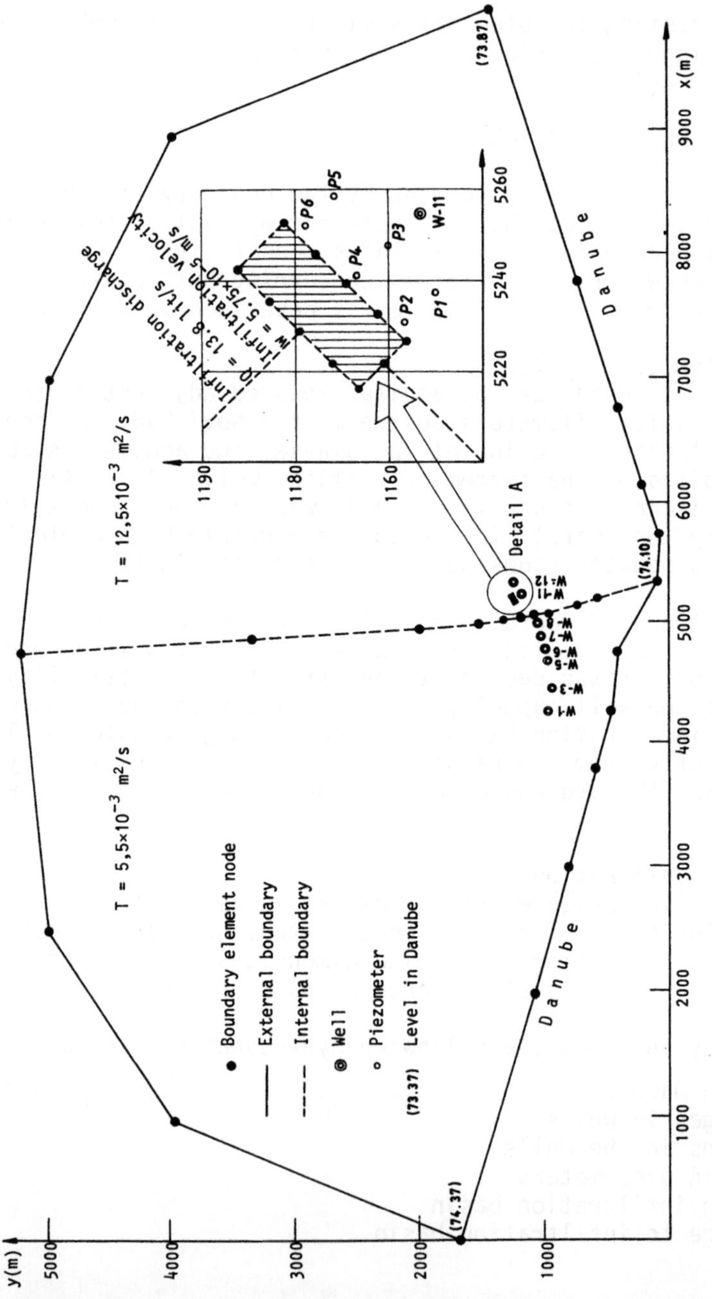

Fig. 5. Boundary element mesh used for identification of model piezometers (39 nodes, 8 wells, 4 macroelements)

and piezometers with reasonable accuracy. The boundary element mesh used for model calibration is given in Fig. 5.

The part of external aquifer boundary with unknown potential was located in such a way that its effect on the flow to wells was the same as these boundaries were infinite. This location was estimated on auxiliary BEM model results of which were compared to analytical solution for the flow to the system of wells in homogeneous aquifer with straight equipotential boundary. Position of the boundary was determined so that the difference between computed and analytical drawdowns in wells were less than 1%.

Note that the infiltration basin was modeled as separate macroelement for which Poisson eq.(31) applies (term w is the ratio of basin discharge and bottom area).

It was shown that by subdivision of the aquifer to only two subdomains with different transmissivities a fairly good agreement of observed and measured drawdowns was obtained in piezometers between the basin and W-11 (table 1). These values of transmissivities are in close agreement with ones obtain from pumping tests carried out at the begining of aquifer exploitation.

Table 1
Comparison of observed and computed drawdowns

No	Wells		Piezometers	
	No	Discharge	Observed	Computed
P1	1	12.80	70.81	70.88
P2	3	11.90	70.94	70.95
P3	5	12.50	70.50	70.57
P4	6	15.60	70.98	70.95
P5	7	13.00	70.99	70.70
P6	8	9.80	71.01	70.92
P7	11	22.80	71.11	71.06
Danube		25.70	74.10	

Note that the observed drawdown in the wells include also the quadratic losses at the well screen that are not encountered in the BEM model. For that reason, in absence of more piezometers away from wells, there were no practical reasons for calibration that will result in "better" agreement of observed and computed drawdowns although it can be achivied in principe by subdivision of the aquifer to more than two macroelements.

5.3. Predictions

Calibrated BEM model was used to predict the hydrodynamic effects of five infiltration basins with dimensions 150×20 m located at distance of 100 m far from the well. The boundary conditions for basins was such infiltration rate that the maximum level below the basin should not exceed the level of basin bottom. Of course, the infiltration rate also should not exceed the value observed in experimental basin.

The reason for the above limitation was the system must have some initial reserve to compensate the resistance due to subse-

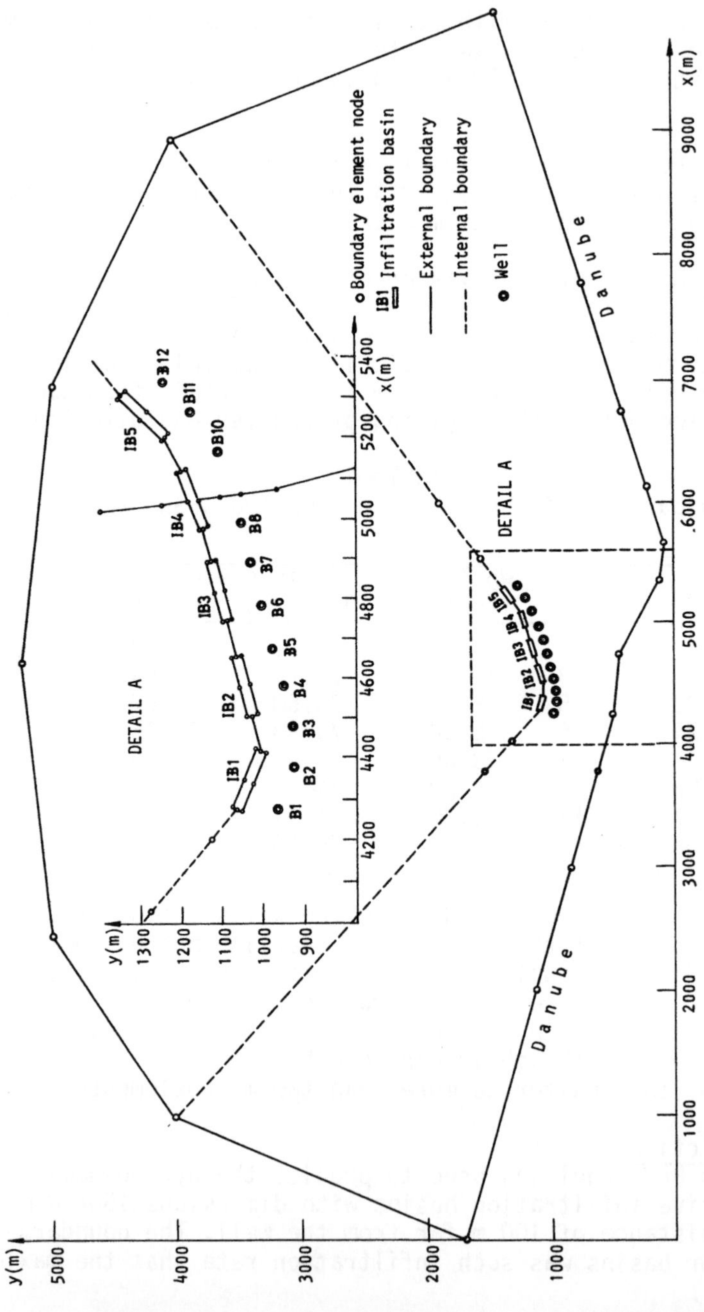

Fig. 6. Boundary element mesh used for prediction of the effect of infiltration basins on discharge in wells (75 nodes, 11 wells, 9 macroelements)

quent clogging of the basin bottom before the bottom cleaning at the end of one exploitation cycle.

The mesh of BEM model is given in Fig. 6.

The model was used to predict different operating conditions (normal operation with 5 basins, operation when one of the basins is cleaned etc.).

6. CONCLUSIONS

The authors have presented the theory of BEM model that is well suited for the study of aquifer flow problems. Due to the simplicity of well modelling, a potential user of the model can easily discretize the flow domain properly which is not the case in the finite element method.

The model can be readily extended to nonlinear free-surface problems and also to transient problems and than it will be applicable to most of the real-life problems in aquifer flow analysis.

ACKNOWLEDGEMENTS

This study was supported by the Civil Engineering Faculty of the Belgrade University (Grant No 36/22-9-82). The authors are also grateful to Vladimir Janković from the Institute for Hydraulic Engineering for careful typing the manuscript.

LYST OF SYMBOLS

A_{iK} - integration constants
C^i - constant
G_{iK} - elements of G matrix
H_{iK} - elements of H matrix
N - number of nodes on external boundary
N_1, N_2 - linear shape functions
Q - integral of flux along interior circle
Q_w - discharge in well
S_j - boundary of interior circle
S_o - drawdown in well
T - transmissivity of aquifer
V_j - interior domain bounded by circle S_j
b - distance between well and equipotential boundary
a - distance between wells in an array
n - number of interior circles
r_i - distance from point "i" to any point in Ω
r_{oj} - radius of interior circle
q - flux
q* - normal derivative of flux
u - potential
u* - normal derivative of potential
w - infiltration rate

Γ - exterior boundary of domain Ω
Ω - domain
ξ - dimensionless coordinate

REFERENCES

1. Liggett, J.A. and Liu, P.L.F. (1983) The Boundary Integral Equation Method for Porous Media Flow. London-Boston-Sydney.
2. Martin, A., Rodrigez, I. and Alarcon, E. (1980): Mixed Element in Boundary theory. In: Recent Advances in Boundary Element Methods. Edited by C.A.Brebbia. Southampton, pp.34--42.
3. Pečarić, J.E. and Radojković, M. Mixed Boundary Element for Potential Problems. To appear.

BOUNDARY INTEGRAL SOLUTION OF DARCY'S FLOW WITH VARIABLE PERMEABILITY

Alexander H-D. Cheng

Department of Civil Engineering and Engineering Mechanics
Columbia University

INTRODUCTION

For flow of groundwater that obeys Darcy's law the governing equation can be written as

$$\nabla \cdot (k \nabla H) = 0 \tag{1}$$

where k = hydraulic conductivity; H = piezometric head = p/γ + z; p = pore-water pressure; γ = specific weight of water; and z = elevation above datum. If k is constant throughout the medium Equation (1) reduces to the well-known Laplace equation. But in general k is not a constant and can vary spatially due to variation of soil type, previous inelastic deformation, etc. In the present note we assume a continuous distribution of hydraulic conductivity, $k = k(x,y,z)$, being known a priori. Numerical solutions of Equation (1) are sought for given geometries and boundary conditions.

The variable permeability assumption does not pose fundamental solution difficulty for the Finite Element Method. But we have in mind the Boundary Integral Equation Method (BIEM) and its ability in incorporating the material inhomogeneity into its formulation. The boundary method, when applicable, is a more efficient solution method in that it requires only boundary discretization in the numerical implementation, as compared to the domain discretization of the Finite Element and Finite Difference Methods. Recently there has been continuing effort in using BIEM for solution of a variety of groundwater flow problems (see, e.g., Liggett and Liu, 1983). The current work intends to contribute to that effort by enhancing the BIEM solution capability in dealing with Darcy's flow with variable permeability.

BIEM

For homogeneous medium Equation (1) becomes the Laplace equation and the boundary integral equation is the Green's second identity (Liggett and Liu, 1983)

$$\alpha H = \int_S \left(G^* \frac{\partial H}{\partial n} - H \frac{\partial G^*}{\partial n}\right) dS \tag{2}$$

where G^* = free-space Green's function of the Laplace equation; n = outward normal of the solution boundary, S; and α = a constant determined from the Cauchy's principal value integration of the Green's function singularity. For inhomogeneous soil, the piecewise homogeneous assumption may be invoked. That is, the soil mass is divided into zones each with constant permeability. Equation (2) is then applicable for each zone. The resultant is combined with interzonal compatibility relations of head and flux to form a sufficient solution system (Lafe, et al., 1981). Numerical discretization in this case is needed for both the external boundary and the interfaces between zones. This approach is efficient if the number of zones involved is not very large.

For continuously varied permeability, $k = k(x,y,z)$, the following integral equation is similar to that given in Banerjee (1979)

$$\alpha H = \int_S \left(kG^* \frac{\partial H}{\partial n} - kH \frac{\partial G^*}{\partial n}\right) dS + \int_V (H\nabla G^* \cdot \nabla k)\, dV \tag{3}$$

Equation (3) utilizes the Green's function of the Laplace equation but contains a volume integral which may pose numerical difficulty in a boundary solution scheme. Banerjee (1979) suggested an iterative procedure which treats the volume integral as integration of known quantities using trial values of H. The trials are then successively updated until a convergence is achieved. The success of this scheme is tied to the effectiveness of the domain integration and the rate of solution convergence. These essential aspects, however, are not properly addressed in the literature.

The third and current approach uses the Green's function defined by

$$\nabla \cdot (k \nabla G) = -2\pi\delta(\bar{x} - \bar{x}') \tag{4}$$

where $\bar{x}$ and $\bar{x}'$ = general points in space; and $\delta(\bar{x} - \bar{x}')$ = Dirac delta function located at $\bar{x}'$. The boundary integral equation now becomes

$$\alpha H = \int_S \left(kG \frac{\partial H}{\partial n} - kH \frac{\partial G}{\partial n}\right) dS \tag{5}$$

Equation (5), similar to Equation (2), does not contain volume integral. In fact, the two formulae differ only in the definition of the kernel functions, kG and $k \frac{\partial G}{\partial n}$ in

Equation (5), as compared to G^* and $\frac{\partial G^*}{\partial n}$ in Equation (2). Therefore one expects identical computer codes for Equations (2) and (5), except for appropriate function statements defining kernel functions be used in each case. One can also incorporate Equation (5) with a multizone program based on Equation (2). The variable permeability assumption within each zone can be switched on and off by calling different function statements in the program. This capacity should reduce the number of zones required to simulate an aquifer.

Equation (5), however, is not without its drawback. The coding of Equation (5) requires knowledge of the Green's function which must be solved from Equation (4). For a general k this itself may be a difficult task. While we are aware that Equation (4) can be attacked by successive approximation or solved numerically, we are only interested in Green's functions in explicit expressions, and preferably in closed forms. This is because the numerical implementation of Equations (2) and (5) involves with repeated evaluation of the kernel functions. Numerical efficiency will be heavily taxed if large amount of effort should go to their evaluation. In the ensuing section we present a few cases in which the closed form Green's functions are available.

GREEN'S FUNCTION

For soils in layered formation or exhibiting a predominant direction of permeability variation, we may assume $k = k(y)$. For two-dimensional geometries we find the Green's function for the singular equation

$$\frac{\partial}{\partial x} k(y) \frac{\partial G}{\partial x} + \frac{\partial}{\partial y} k(y) \frac{\partial G}{\partial y} = - 2\pi\delta(x - x';y - y') \quad (6)$$

for a few permeability distributions. The three-dimensional cases can be similarly treated, but are not discussed here for brevity.

A popular permeability variation in the literature seems to be (Schiffman and Gibson, 1964; Mahmoud and Deresiewicz, 1980)

$$\frac{k}{k_o} = (1 + \beta \frac{y}{d})^{\eta} \quad (7)$$

where k_o = reference hydraulic conductivity at $y = 0$; d = reference thickness in the direction of permeability variation; β and η = parameters of curve fitting. The Green's function for this case is given by Clements (1980) in a double series. The series truncates at a finite number of terms for $\eta = 2N$ where $N = 0, \pm 1, \pm 2$, etc. We adopt herein only $\eta = 2$ in Equation (7). Some of the typical permeability profiles based on this function are sketched in Fig. 1. The Green's function is

$$G(x,x';y,y') = \frac{-\ln r}{[k(y)k(y')]^{1/2}} \tag{8}$$

where $r = [(x - x')^2 + (y - y')^2]^{1/2}$

The permeability profiles in Fig. 1 are rather rigid because arbitrary selection of η is not allowed if closed form Green's function is desired. We therefore work two more cases in which the profiles are more flexible and closed form Green's functions can be found. Georghitza (1969, as cited in Bear, 1972) suggested a transformation which transforms Equation (1) into Helmholtz and modified Helmholtz equations, respectively, if the permeabilities themselves satisfy those equations. In the current context this implies the following permeability functions

$$\frac{k}{k_o} = (\cos \frac{\beta y}{d} + \eta \sin \frac{\beta y}{d})^2 \tag{9}$$

$$\frac{k}{k_o} = (e^{\beta y/d} + \eta \sinh \frac{\beta y}{d})^2 \tag{10}$$

And we find the Green's functions, respectively for Equation (9) and Equation (10), as

$$G(x,x';y,y') = -\frac{\pi}{2} \frac{Y_o(|\beta| r)}{[k(y)k(y')]^{1/2}} \tag{11}$$

$$G(x,x';y,y') = \frac{K_o(|\beta| r)}{[k(y)k(y')]^{1/2}} \tag{12}$$

where Y_o and K_o, respectively, = Bessel function and modified Bessel function of second kind of order zero. It is noted that Equations (9) and (10) each contains three parameters of fit, k_o, β and η which is one more than Equation (7) with η = 2. Typical profiles based on Equations (9) and (10) are plotted as Figs. 2 and 3. The profiles seem to be flexible enough for general use.

EXAMPLES

A two-dimensional BIEM program utilizing some of the advanced numerical techniques in FEM is written. It features the following: geometrical and functional interpolations are done by quadratic isoparametric elements; solution singularities are handled by quarter-point singular element; Gauss and logarithmic weighted quadrature rules are used for numerical integration; Bessel functions are evaluated using polynomial approximation; coefficient α is calculated by the "rigid body motion analogy"; and discontinuity in flux is treated exactly at corner nodes. Details of some of the above features can be found in Cheng and Liggett (1983). Two numerical examples are investigated as follows.

One-Dimensional Seepage

Solution of one-dimensional problem does not require sophisticated numerical method. Here it is solved as a full two-dimensional problem by BIEM for the purpose of verification with available exact solution. The solution geometry is a one by one square box bounded by the limits $0 < x < 1$ and $-1 < y < 0$. The following boundary conditions are prescribed

$$\begin{aligned} \frac{\partial H}{\partial n} &= 0 \quad \text{along} \quad x = 0, \quad -1 < y < 0 \\ &\quad\;\; \text{and} \quad x = 1, \quad -1 < y < 0; \\ H &= 0 \quad \text{along} \quad y = 0, \quad 0 < x < 1; \\ H &= 1 \quad \text{along} \quad y = -1, \quad 0 < x < 1 \end{aligned} \tag{13}$$

Two quadratic elements are used for each side of the box. The number of elements are purposely kept small to demonstrate the accuracy of the numerical approximation. Two permeability profiles, profile 5 in Fig. 2 and profile 4 in Fig. 3 are chosen for their maximum variability. A comparison of BIEM solution with exact solution along the impermeable wall is shown in Table 1 for the two cases. The agreement is within a fraction of a percentage point which is excellent for this coarse mesh.

Seepage Below Dam

The seepage under an impervious foundation is depicted as Fig. 4. For simplicity we use $\ell = 1$, $d = 1$, and $k_o = 1$. The boundary conditions are

$$\begin{aligned} \frac{\partial H}{\partial n} &= 0 \quad \text{along AB and CE} \\ H &= 1 \quad \text{along BD} \\ H &= 0 \quad \text{along AC} \end{aligned} \tag{14}$$

For convenience of numerical solution we truncate the left side of the domain at a distance of $x = -10$. The no flux boundary condition is prescribed there although infinite element can be used for better approximation. The solution boundary is discretized into twenty-five quadratic elements; two of them are quarter-point elements with tip located at B. Three permeability distributions are investigated.

In case 1 constant permeability is assumed. This is the special case of $\beta = 0$ in Equation (7) or $\beta = \eta = 0$ in Equations (9) and (10). But Equations (11) and (12) are not defined for $\beta = 0$. Instead $\beta = 0.01$ is actually used for Equations (9) and (10). Solutions for all three runs turn out to be practically identical. They are shown as single data point and are designated as case 1 in Figs. 5 and 6. Case 2 adopts permeability profile 5 in Fig. 2, and case 3 follows profile 4 in Fig. 3. Results of those cases are also plotted in Figs. 5 and 6.

The right half of Fig. 5 shows the piezometric head distribution along the foundation base, AB. The left half gives the normal gradient along BD. Shown as solid lines in Fig. 5 are the exact solutions by Polubarinova-Kochina (1962) for constant permeability. The agreement between the exact solution and case 1 of BIEM solution is excellent and the singularity at B is accurately represented. The left of Fig. 6 gives the normal flux $\frac{\partial H}{\partial n}$ along AC and the right depicts the specific discharge $q = - k \frac{\partial H}{\partial n}$ at the same location. Again, close agreement between case 1 and the exact solution is observed.

CONCLUSION

A BIEM which incorporates the variable permeability assumption in its formulation is developed. For three permeability functions, given as Equations (7), (9) and (10), closed form Green's functions are found. The permeability profiles sketched as Figs. 1, 2 and 3 can be used for some simple fits of field measured permeabilities. The current computer code can be easily combined with a zoned BIEM program to reduce the number of zones required in representing an aquifer.

References

Banerjee, P.K. (1979) "Non-Linear Problems of Potential Flow," in Developments in Boundary Element Methods-1, P.K. Banerjee and R. Butterfield, eds., Applied Science, London.

Bear, J. (1972) Dynamics of Fluid in Porous Media, American Elsevier, New York.

Cheng, A.H-D. and Liggett, J.A. (1983) "Boundary Integral Equation Method for Linear Porous-Elasticity with Applications to Soil Consolidations," International Journal for Numerical Methods in Engineering, in press.

Clements, D.L. (1980) "A Boundary Integral Equation Method for the Numerical Solution of a Second Order Elliptic Equation with Variable Coefficients," J. Austral. Math. Soc., Vol. 22, Series B, pp. 218-228.

Georghitza, St. I. (1969) "On the Plane Steady Flow of Water Through Inhomogeneous Porous Media," 1st Symp. The Fundamentals of Transport Phenomena in Porous Media, IAHR, Haifa.

Lafe, O.E., Liggett, J.A. and Liu, P.L-F. (1981) "BIEM Solutions to Combinations of Leaky, Layered, Confined, Unconfined, Nonisotropic Aquifers," Water Resources Research, Vol. 17, No. 5, pp. 1431-1444.

Liggett, J.A. and Liu, P. L-F. (1983) The Boundary Integral Equation Method for Porous Media Flow, Allen and Unwin, London.

Mahmoud, M.S. and Deresiewicz, H. (1980) "Settlement of Inhomogeneous Consolidating Soils - I: The Single-Drained Layer Under Confined Compression," International Journal for Numerical and Analytical Methods in Geomechanics, Vol. 4, pp. 57-72.

Polubarinova-Kochina, P. Ya. (1962) The Theory of Ground Water Movement, translated by J.M.R. DeWiest, Princeton Univ. Press, Princeton, N.J.

Schiffman, R.L. and Gibson, R.E. (1964) "Consolidation of Nonhomogeneous Clay Layers," Journal of the Soil Mechanics and Foundations Division, ASCE, Vol. 90, No. SM5, pp. 1-30.

Table 1

y	k^*(eq. 9)$^+$		k^*(eq. 10)$^{++}$	
	H(exact)	H(BIEM)	H(exact)	H(BIEM)
0.	0.	0.	0.	0
-0.25	0.245	0.245	0.271	0.271
-0.5	0.403	0.402	0.583	0.583
-0.75	0.584	0.583	0.841	0.841
-1.	1.	1.	1.	1.

+ Eq. 9 with $\beta = 2.0$ and $\eta = -1.2$

++ Eq. 10 with $\beta = 1.7$ and $\eta = -0.46$

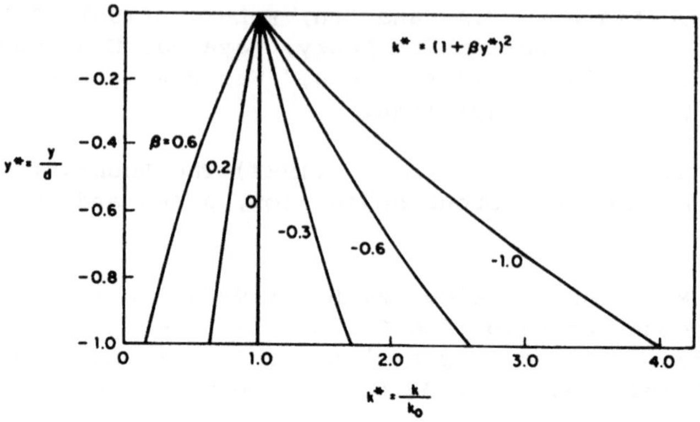

Figure 1: Permeability profile based on eq. 7

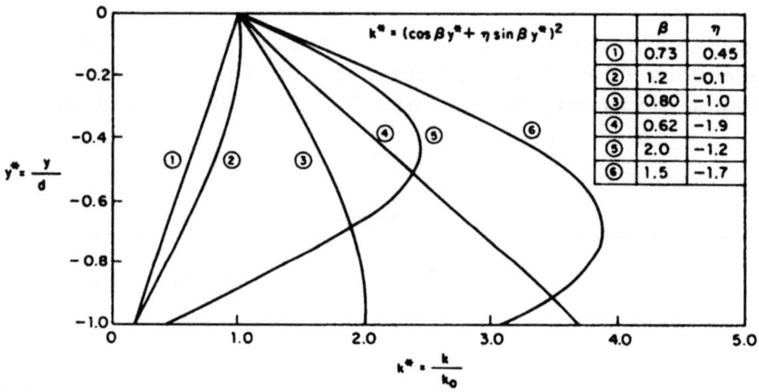

Figure 2: Permeability profile based on eq. 9

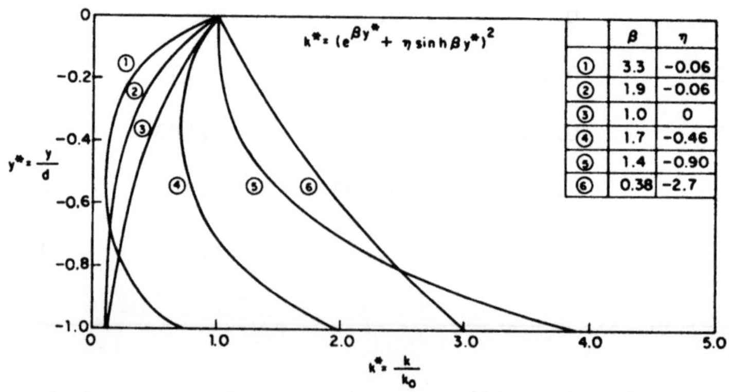

Figure 3: Permeability profile based on eq. 10

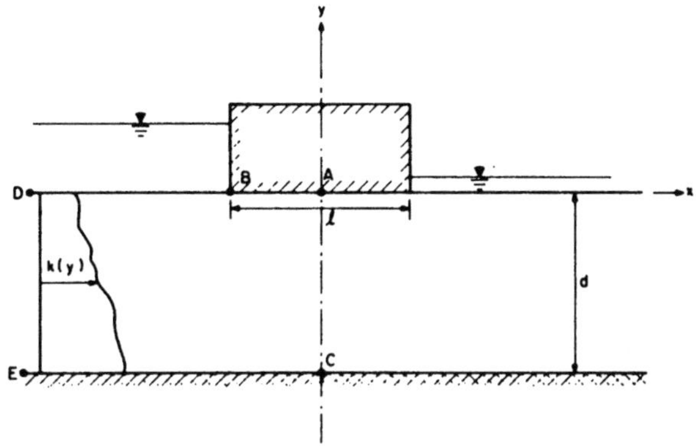

Figure 4: Seepage below dam

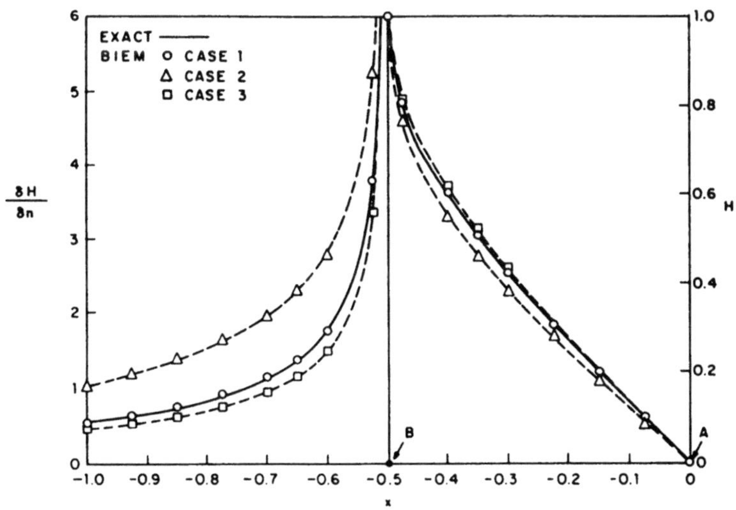

Figure 5: Solution along the upper boundary

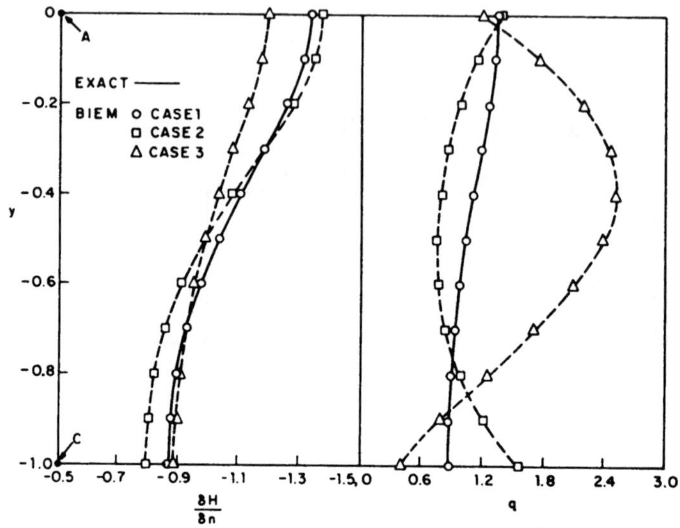

Figure 6: Solution along the centerline

DIFFUSION-CONVECTION PROBLEMS USING BOUNDARY ELEMENTS

C.A. Brebbia and P. Skerget

Computational Mechanics Institute, England, & University of Maribor, Yugoslavia.

1. INTRODUCTION

In this paper the boundary element technique is applied to solve steady and time dependent transport problems governed by the diffusion-convection equation following previous work presented by the authors in references [1] to [4].

For time independent transport problems the classical fundamental solution used in Laplace's type problem is applied. Time dependent parabolic type problems are solved instead using a fundamental solution function of time and space. This effectively reduces the dimensionality of the problem transforming it into a boundary problem. However, due to the presence of the convective terms domain integrals need also to be computed. These integrals increase the amount of computation required and special care should be taken as the manner in which the integrations are carried out.

The method of solution is extended to problems with non-linear constitutive laws by applying the Kirchhoff's transform. This technique used by the authors in reference [2] reduces the problem to a linear one.

Some examples are solved using higher order elements in space and linear time interpolation. Results are compared with those presented by other authors or against known analytical solutions. The agreement of numerical results versus known solutions demonstrates the validity of the present approach. In general very accurate results are obtained when using boundary element methods in diffusion-convection problems.

2. GOVERNING EQUATIONS

Let us first consider the problem governed by the steady state diffusion convection equation for homogeneous isotropic medium, i.e.

$$\alpha\nabla^2 u - \left(v_x \frac{\partial u}{\partial x} + v_y \frac{\partial u}{\partial y} + v_z \frac{\partial u}{\partial z}\right) = 0 \qquad (1)$$

This is an inhomogeneous elliptic partial differential equation in the scalar function u(s). The coefficient α is the diffusivity, $\alpha = k_o/c\rho$, where k_o is the constant conductivity c the specific heat and ρ the density. The velocity vector has three components, i.e. $\vec{v} = \{v_x, v_y, v_z\}$.

Two sets of linear boundary conditions can be associated with equation (1), i.e.

$$\begin{aligned} &\text{essential conditions,} \quad u = \bar{u} \quad \text{on } \Gamma_1 \\ &\text{natural conditions,} \quad k_o \frac{\partial u}{\partial n} = -\bar{q} \quad \text{on } \Gamma_2 \end{aligned} \qquad (2)$$

If the conductivity coefficient is function of the potential, i.e. k = k(u), the governing equation (1) becomes nonlinear, i.e.

$$(\vec{\nabla} k).(\vec{\nabla} u) - c\rho\left(v_x \frac{\partial u}{\partial x} + v_y \frac{\partial u}{\partial y} + v_y \frac{\partial u}{\partial z}\right) = 0 \qquad (3)$$

where $\vec{\nabla}$ represents the gradient vectors, i.e. $\vec{\nabla}(\ ) = \left\{\frac{\partial(\ )}{\partial x}, \frac{\partial(\ )}{\partial y}, \frac{\partial(\ )}{\partial z}\right\}$.

Time dependent diffusion-convection problems with constant conductivity are represented by the following parabolic partial differential equation in the scalar function u(s,t).

$$\frac{\partial u}{\partial t} = \alpha\nabla^2 u - \left(v_x \frac{\partial u}{\partial x} + v_y \frac{\partial u}{\partial y} + v_z \frac{\partial u}{\partial z}\right) \qquad (4)$$

The linear boundary conditions corresponding to this equation are like those given in formulae (2). In addition a set of initial conditions is needed, i.e.

$$u = u_i \quad \text{in } \Omega \quad \text{at } t = t_i \qquad (5)$$

Sometimes, when higher order time-interpolation functions are introduced during the integration, fluxes on the boundary at $t = t_i$ also have to be prescribed, i.e.

$$q = q_i \quad \text{on } \Gamma \quad \text{at } t = t_i \qquad (6)$$

If the conductivity k is function of the potential equation (4) becomes non-linear, i.e.

$$c\rho \frac{\partial u}{\partial t} = (\vec{\nabla}k)(\vec{\nabla}u) - c\rho\left(v_x \frac{\partial u}{\partial x} + v_y \frac{\partial u}{\partial y} + v_z \frac{\partial u}{\partial z}\right) \tag{7}$$

Kirchhoff's transform

One way of solving the non-linear formulations resulting when the conductivity is function of the potential is to linearize them using Kirchhoff's transform, [1-4]. This operation results in a linear system when the problem presents only natural and essential boundary conditions of the type described by equations (2) and conductivity dependent on u, i.e. k(u).

The Kirchhoff's transform is defined as,

$$\psi = K[u] = \int_{u_o}^{u} k(u)du \tag{8}$$

where u_o is an arbitrary reference value. Equations (3) and (7) can be transformed into a linear system in terms of the new variable $\psi(u)$. For steady state problems (equation (3)) application of Kirchhoff's transform produces the following governing equation,

$$\alpha\nabla^2\psi - \left(v_x \frac{\partial\psi}{\partial x} + v_y \frac{\partial\psi}{\partial y} + v_z \frac{\partial\psi}{\partial z}\right) = 0 \tag{9}$$

For time dependent problems (equation (7)) one obtains

$$\frac{\partial\psi}{\partial t} = \alpha\nabla^2\psi - \left(v_x \frac{\partial\psi}{\partial x} + v_y \frac{\partial\psi}{\partial y} + v_z \frac{\partial\psi}{\partial z}\right) \tag{10}$$

Thus the linear form of equations (1) and (4) is preserved. The transformed boundary conditions are also linear, i.e.

$$\begin{aligned} &\text{essential} \quad \psi = \bar{\psi} = K[\bar{u}] \quad &\text{on } \Gamma_1 \\ &\text{natural} \quad \frac{\partial\psi}{\partial n} = -\bar{q} \quad &\text{on } \Gamma_2 \end{aligned} \tag{11}$$

The initial conditions are as follows,

$$\psi = \psi_i = K[u_i] \quad \text{in } \Omega \quad \text{at } t = t_i \tag{12}$$

and the initial fluxes (if needed) are

$$\frac{\partial \psi}{\partial n} = q_i \qquad \text{on } \Gamma \quad \text{at } t = t_i \tag{13}$$

3. STEADY STATE PROBLEM

Let us first describe problems governed by the following inhomogeneous elliptic equation,

$$\nabla^2 u(s) + b = 0 \qquad \text{in } \Omega \tag{14}$$

Weighting this formula and the corresponding boundary conditions by the fundamental solution of the elliptic equation $u^*(\xi,s)$ as indicated in reference [5] one obtains a boundary integral equation relating boundary values for potentials and normal derivatives, i.e.

$$c(\xi)u(\xi) + \int_\Gamma u(S)q^*(\xi,S)d\Gamma(S) = \int_\Gamma q(S)u^*(\xi,S)d\Gamma(S) + \int_\Omega bu^*(\xi,s)d\Omega(s) \tag{15}$$

where ξ represents the source point, S and s are reference points on the boundary Γ and in the domain Ω, respectively. The coefficient $c(\xi)$ is given by;

$c(\xi) = 1$ if ξ is in the domain

$c(\xi) = \frac{1}{2}$ if ξ lies on a smooth boundary

$c(\xi) = \frac{(\pi+\alpha_1-\alpha_2)}{2}$ if ξ lies on a non-smooth boundary Γ - α_1 and α_2 angles are defined in figure 1.

For the type of two dimensional problems to be discussed here the fundamental solution and its normal derivative are given by,

$$u^*(\xi,S) = \frac{1}{2\pi} \ln \frac{1}{r(\xi,S)}$$
$$q^*(\xi,S) = \frac{1}{2\pi} \frac{d(\xi,S)}{r^2(\xi,S)} \tag{16}$$

$r(\xi\ S)$ is the distance from the point of application of the source to the reference point and $d(\xi,S) = (x(\xi)-x(S)).n_x(S) + (y(\xi)-y(S)).n_y(S)$, where $n_x(S)$ and $n_y(S)$ are the direction cosines of the normal at the point S.

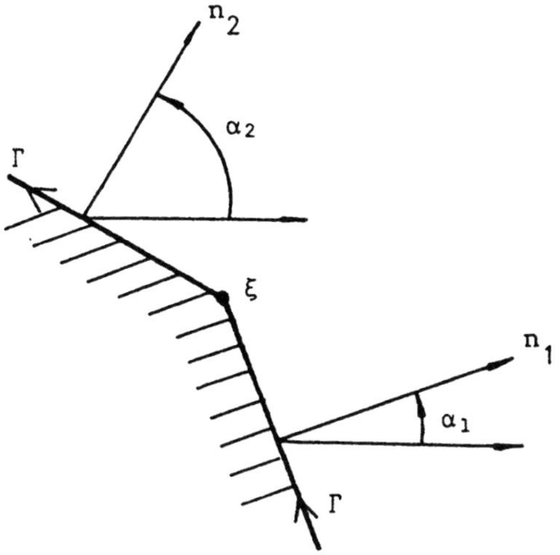

Figure 1 Definitions for the Computation of Coefficient $c(\xi)$

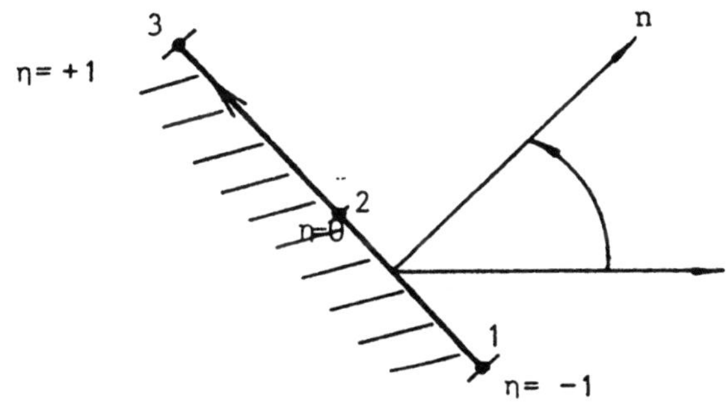

Figure 2 Quadratic boundary element

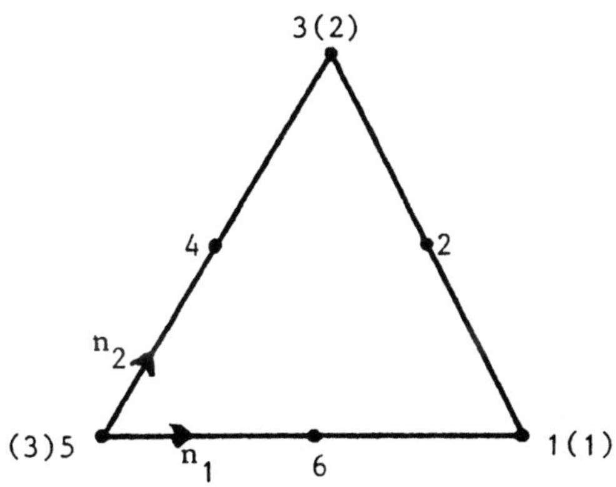

Figure .3 Quadratic domain cell

Convective terms

One can easily extend the boundary integral formulation (15) to the diffusion-convection case simply by taking the body force term, b, as equivalent to the convective terms in formula (1). After some mathematical manipulations this produces,

$$c(\xi)u(\xi) + \int_{\Gamma} u(S)q^*(\xi,S)d\Gamma(S) = \int_{\Gamma} q(S)u^*(\xi,S)d\Gamma(S)$$
$$- \frac{1}{\alpha}\int_{\Gamma} u(S)\{v_x(S).n_x(S) + v_y(S).n_y(S)\}u^*(\xi,S)d\Gamma(S)$$
$$+ \frac{1}{\alpha}\int_{\Omega} u(S)\left\{v_x(s)\frac{\partial u^*(\xi,s)}{\partial x(s)} + v_y(s)\frac{\partial u^*(\xi,s)}{\partial y(s)}\right\} d\Omega(s) \qquad (17)$$

Notice that equation (17) describes the transport process in integral equation form. The first two integrals represent the diffusion from the boundary and the third describes the convection from the boundary. The last-domain-integral is due to the convective effect.

4. BOUNDARY ELEMENT DISCRETIZATION OF THE ELLIPTIC EQUATION

Let us consider a series of boundary elements as given in figure 2, for the boundary integrals on Γ and cells within the region - figure 3 - for the Ω integrations.

The values of the u,q variables and the v_x, v_y velocity components are assumed to vary within each element or cell according to a space interpolation function (see also section 7).

$$u(\eta) = \underset{\sim}{\phi}^T \underset{\sim}{U}^n \quad ; \quad v_x(\eta_1,\eta_2) = \underset{\sim}{\varphi}^T \underset{\sim}{V}_x^n$$
$$q(\eta) = \underset{\sim}{\phi}^T \underset{\sim}{Q}^n \quad ; \quad v_y(\eta_1,\eta_2) = \underset{\sim}{\varphi}^T \underset{\sim}{V}_y^n \qquad (18)$$

The coordinates η and η_1,η_2 are homogeneous coordinates defined in figures 2 and 3. The index n refers to the number of the nodes within each element or cell which are associated with the nodal values of u, q, v_x, v_y. Assuming that the boundary Γ is divided into E elements with N_E boundary nodes and the domain Ω is discretized into C cells with N_I internal points, one can write the discretized form of (17) as,

$$c(\xi)u(\xi) + \sum_{e=1}^{E}\left[\int_{\Gamma_e} \underset{\sim}{\phi}^T q^* d\Gamma\right]\underset{\sim}{U}^n = \sum_{e=1}^{E}\left[\int_{\Gamma_e} \underset{\sim}{\phi}^T u^* d\Gamma\right]\underset{\sim}{Q}^n$$

$$- \frac{1}{\alpha} \sum_{e=1}^{E} \left\{ \int_{\Gamma_e} \underset{\sim}{\phi}^T [\underset{\sim}{\phi}^T \underset{\sim}{V}_x^n . n_x + \underset{\sim}{\phi}^T \underset{\sim}{V}_y^n . n_y] u^* d\Gamma \right\} \underset{\sim}{U}^n \tag{19}$$

$$+ \frac{1}{\alpha} \sum_{c=1}^{C} \left\{ \int_{\Omega_c} \underset{\sim}{\varphi}^T [\underset{\sim}{\varphi}^T \underset{\sim}{V}_x^n q_x^* + \underset{\sim}{\varphi}^T \underset{\sim}{V}_y^n q_y^*] d\Omega \right\} \underset{\sim}{U}^n$$

Notice that we have considered that the interpolation functions φ for the velocities become ϕ when they are referred to the boundary, $q_x^* = \partial u^*(\xi,s)/\partial x(s)$ and $q_y^* = \partial u^*(\xi,s)/\partial y(s)$.
One can also express the coordinates in terms of the local coordinates and their nodal values, i.e.

$$x(\eta_1 \eta_2) = \underset{\sim}{\varphi}^T \underset{\sim}{x}^n \quad ; \quad y(\eta_1 \eta_2) = \underset{\sim}{\varphi}^T \underset{\sim}{y}^n \tag{20}$$

in the domain and,

$$x(\eta) = \underset{\sim}{\phi}^T \underset{\sim}{x}^n \quad ; \quad y(\eta) = \underset{\sim}{\phi}^T \underset{\sim}{y}^n \tag{21}$$

on the boundary.

Since the interpolation functions are expressed in terms of local coordinates, it is necessary to transform the elements of the surface differentials $d\Gamma$ and the domain differential $d\Omega$, from the global cartesian system to the local system of coordinates, i.e.

$$d\Gamma = |J| \, d\eta \quad ; \quad d\Omega = |G| \, d\eta_1 d\eta_2 \tag{22}$$

Applying equations (19) to all boundary nodes the following matrix equations can be obtained,

$$[\underset{\sim}{H} + \underset{\sim}{C} - \underset{\sim}{D}] \underset{\sim}{U} = \underset{\sim}{G} \underset{\sim}{Q} \tag{23}$$

$\underset{\sim}{H}$ and $\underset{\sim}{G}$ are standard boundary element matrices. $\underset{\sim}{C}$ and $\underset{\sim}{D}$ are due to the effect of boundary and domain convective terms. Adding all the left hand side matrices together we obtain,

$$\underset{\sim}{E} \underset{\sim}{U} = \underset{\sim}{G} \underset{\sim}{Q} \tag{24}$$

After rearranging columns due to the application of boundary conditions the above equation can be rewritten as the following system,

$$\underset{\sim}{A} \underset{\sim}{X} = \underset{\sim}{F} \tag{25}$$

where $\underset{\sim}{A}$ is the linear system matrix and $\underset{\sim}{X}$ is the vector of unknowns.

5. TIME DEPENDENT PROBLEMS

Let us now consider those problems described by the following scalar inhomogeneous parabolic equation,

$$\frac{\partial u(s,t)}{\partial t} - \alpha\nabla^2 u(s,t) = b \qquad \text{in } \Omega \tag{26}$$

A boundary integral equation relating boundary values to potentials and normal derivatives can be obtained by weighting equation (26) and the corresponding boundary conditions with respect to a space and time dependent fundamental solution of the parabolic equation, i.e. $u^*(\xi,t_F;s,t)$. After certain manipulations this weighting gives

$$c(\xi)u(\xi,t_f) + \int_{t_{f-1}}^{t_f}\int_\Gamma \alpha u(S,t)q^*(\xi,t_F;S,t)d\Gamma(S)st =$$

$$= \int_{t_{f-1}}^{t_f}\int_\Gamma \alpha q(S,t)u^*(\xi,t_f;S,t)d\Gamma(S)dt \tag{27}$$

$$+ \int_\Omega u_{f-1}(s,t=t_{f-1})u^*(\xi,t_f;s,t=t_{f-1})d\Omega(s)$$

$$+ \int_{t_{f-1}}^{t_f}\int_\Omega bu^*(\xi,t_f;s,t)d\Omega(s)dt$$

For the type of two dimensional problem discussed here the fundamental solution and its normal derivative are given by

$$u^*(\xi,t_f;S,t) = \frac{1}{4\pi\alpha\tau}\exp\left(-\frac{r^2(\xi,S)}{4\alpha\tau}\right)$$

$$q^*(\xi,t_f;S,t) = \frac{d(\xi,S)}{8\pi\alpha^2\tau^2}\exp\left(-\frac{r^2(\xi,S)}{4\alpha\tau}\right) \tag{28}$$

where the parameter τ is given by $\tau = t_f - t$.

Convective terms

Following the same mathematical procedure as for the steady state, one can introduce the convective terms into the time dependent problems and after some manipulation write the following integral expression [3,4]

$$c(\xi)u(\xi,t_f) + \int_{t_{f-1}}^{t_f}\int_{\Gamma} \alpha u(S,t)q^*(\xi,t_f;S,t)d\Gamma(S)dt =$$

$$\int_{t_{f-1}}^{t_f}\int_{\Gamma} q(S,t)u^*(\xi,t_f;S,t)d\Gamma(S)dt \tag{29}$$

$$- \int_{t_{f-1}}^{t_f}\int_{\Gamma} u(S,t)\ \{v_x(S,t)n_x(S)+v_y(S,t)n_y(S)\}\ u^*(\xi,t_f;S,t)d\Gamma(S)dt$$

$$+ \int_{\Omega} u_{f-1}(s,t=t_{f-1})u^*(\xi,t_f;s,t=t_{f-1})d\Omega(s)$$

$$+ \int_{t_{f-1}}^{t_f}\int_{\Omega} u(s,t)\left\{v_x(s,t)\frac{\partial u^*(\xi,t_f;s,t)}{\partial x(s)} +\right.$$

$$\left. + v_y(s,t)\frac{\partial u^*(\xi,t_f;s,t)}{\partial y(s)}\right\} d\Omega(s)dt$$

6. BOUNDARY ELEMENT DISCRETIZATION OF THE PARABOLIC EQUATION

The values of the variables u, q, v_x and v_y will now be assumed to vary not only within each element but also within each time step according to the space ϕ, and time ψ interpolation functions, i.e. on the boundary.

$$u(\eta,t) = \underset{\sim}{\phi}^T\ \underset{\sim}{\psi}\ \underset{\sim}{U}_m^n \quad ; \quad v_x(\eta,t) = \underset{\sim}{\phi}^T\ \underset{\sim}{\psi}\ \underset{\sim}{V}_{xm}^n$$

$$q(\eta,t) = \underset{\sim}{\phi}^T\ \underset{\sim}{\psi}\ \underset{\sim}{Q}_m^n \quad ; \quad v_y(\eta,t) = \underset{\sim}{\phi}^T\ \underset{\sim}{\psi}\ \underset{\sim}{V}_{ym}^n \tag{30}$$

(and similarly for the v_x v_y in the domain).

The index m refers to the degree of variation of the function $\underset{\sim}{\psi}$, i.e. m = 1 if $\underset{\sim}{\psi}$ is constant ; m = 1,2 if $\underset{\sim}{\psi}$ is linear on time, etc.

The discretized form of equation (29) can be written as,

$$c(\xi)u(\xi,t_f) + \alpha \sum_{e=1}^{E} \left[\int_{\Gamma_e} \underset{\sim}{\phi}^T \int_{t_{f-1}}^{t_f} q^* \underset{\sim}{\psi}\, dt d\Gamma\right] \underset{\sim}{U}_m^n =$$

$$\alpha \sum_{e=1}^{E} \left[\int_{\Gamma_e} \underset{\sim}{\phi}^T \int_{t_{f-1}}^{t_f} u^* \underset{\sim}{\psi}\, dt\, d\Gamma\right] \underset{\sim}{Q}_m^n \tag{31}$$

$$- \sum_{e=1}^{E} \left\{\int_{\Gamma_e} \underset{\sim}{\phi}^T \int_{t_{f-1}}^{t_f} [\underset{\sim}{\phi}^T \underset{\sim}{\psi}\, \underset{\sim}{V}_{xm}^n . n_x + \underset{\sim}{\phi}^T \underset{\sim}{\psi}\, \underset{\sim}{V}_{ym}^n\, n_y] u^* \underset{\sim}{\psi}\, dt d\Gamma\right\} \underset{\sim}{U}_m^n$$

$$+ \sum_{c=1}^{C} \left[\int_{\Omega_c} \underset{\sim}{\varphi}^T u^*\, d\Omega\right] \underset{\sim}{U}_{f-1}^n + \sum_{c=1}^{C} \left\{\int_{\Omega_c} \underset{\sim}{\varphi}^T \int_{t_{f-1}}^{t_f} [\underset{\sim}{\varphi}^T \underset{\sim}{\psi}\, \underset{\sim}{V}_{xm}^n\, q_x^*\right.$$

$$\left. + \underset{\sim}{\varphi}^T \underset{\sim}{\psi}\, \underset{\sim}{V}_{ym}^n\, q_y^*\,]\, \underset{\sim}{\psi}\, dt d\Omega\right\} \underset{\sim}{U}_m^n$$

where $q_x^* = \partial u^*(\xi,t_f;s,t)/\partial x(s)$ and $q_y^* = \partial u(\xi,t_f;s,t)/\partial y(s)$.

Let us assume a linear variation on time within each time step for the functions u, q, v_x, v_y, according to the following interpolation functions,

$$\psi_1 = \frac{t_f - t}{\Delta t} \quad ; \quad \psi_2 = \frac{t - t_{f-1}}{\Delta t} \tag{32}$$

where $\Delta t = t_f - t_{f-1}$. Applying equation (31) to all boundary nodes, the following matrix equations can be obtained;

$$[\underset{\sim}{H}_2 + \underset{\sim}{C}_2 - \underset{\sim}{D}_2]\underset{\sim}{U}_f - \underset{\sim}{G}_2\, \underset{\sim}{Q}_f = [-\underset{\sim}{H}_1 - \underset{\sim}{C}_1 + \underset{\sim}{D}_1 + \underset{\sim}{B}_1]\underset{\sim}{U}_{f-1} + \underset{\sim}{G}_1\, \underset{\sim}{Q}_{f-1} \tag{33}$$

Notice that the new type of term $\underset{\sim}{B}$ is due to the initial conditions. Adding together all the terms in the brackets one can write,

$$\underset{\sim}{E}_2 \underset{\sim}{U}_f - \underset{\sim}{G}_2 \underset{\sim}{Q}_f = \underset{\sim}{E}_1 \underset{\sim}{U}_{f-1} + \underset{\sim}{G}_1 \underset{\sim}{Q}_{f-1} \tag{34}$$

After applying the corresponding boundary conditions, equation (34) can be rewritten as (25) i.e.,

$$\underset{\sim}{A} \underset{\sim}{X} = \underset{\sim}{F} \tag{35}$$

7. HIGHER ORDER ELEMENTS

Let us assume a quadratic space variation over each element for the functions u, q, v_x and v_y, while the element is still represented by straight lines (figure 2).

The interpolation functions, $\underset{\sim}{\phi}$ and the vectors $\underset{\sim}{U}^n$, $\underset{\sim}{Q}^n$, $\underset{\sim}{V}_x^n$, $\underset{\sim}{V}_y^n$ are given by,

$$\underset{\sim}{\phi}^n(\eta) = \begin{Bmatrix} \frac{1}{2}\eta(\eta-1) \\ 1-\eta^2 \\ \frac{1}{2}\eta(\eta+1) \end{Bmatrix} \; ; \quad \underset{\sim}{U}^n \begin{Bmatrix} u^1 \\ u^2 \\ u^3 \end{Bmatrix} \; ; \quad \underset{\sim}{Q}^n = \begin{Bmatrix} q^1 \\ q^2 \\ q^3 \end{Bmatrix} \; ;$$

$$\underset{\sim}{V}_x^n = \begin{Bmatrix} v_x^1 \\ v_x^2 \\ v_x^3 \end{Bmatrix} \; ; \quad \underset{\sim}{V}_y^n = \begin{Bmatrix} v_y^1 \\ v_y^2 \\ v_y^3 \end{Bmatrix} \tag{36}$$

With the above assumptions the geometrical shape functions in local coordinates are linear, i.e.

$$\underset{\sim}{\hat{\phi}}^n(\eta) = \begin{Bmatrix} \frac{1}{2}(1-\eta) \\ \frac{1}{2}(1+\eta) \end{Bmatrix} \; ; \quad \underset{\sim}{x}^n = \begin{Bmatrix} x^1 \\ x^2 \end{Bmatrix} \; ; \quad \underset{\sim}{y}^n \begin{Bmatrix} y^1 \\ y^2 \end{Bmatrix} \tag{37}$$

(The superscript ^ is to differentiate the linear functions given by (37) from the quadratic equations (36)). The Jacobian for a straight element - equation (22) - is simply half the boundary element length.

The domain cells are quadratic triangles , figure 3. The $\underset{\sim}{\varphi}$ and $\underset{\sim}{U}^n$, $\underset{\sim}{V}_x^n$, $\underset{\sim}{V}_y^n$ vectors are given by

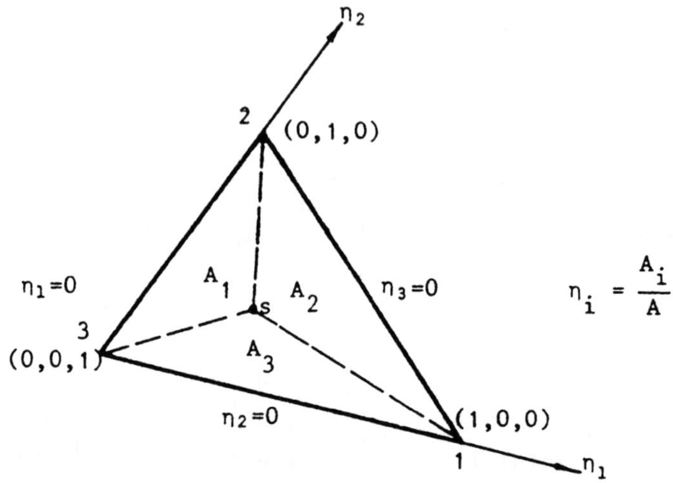

Figure 4 Triangular cell and definition of area coordinate system (η_1, η_2, η_3)

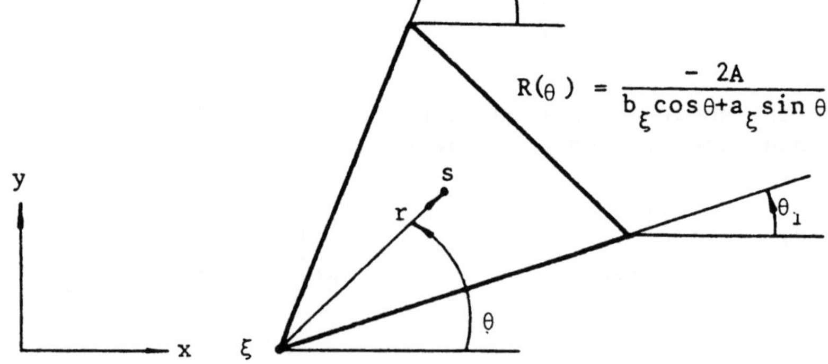

Fkgure 5 Cylindrical coordinate system based at the singular point ξ

$$\varphi^n(\eta_1\eta_2) = \begin{Bmatrix} \eta_1(2\eta_1-1) \\ 4\eta_1\eta_2 \\ \eta_2(2\eta_2-1) \\ 4\eta_2\eta_3 \\ \eta_3(2\eta_3-1) \\ 4\eta_3\eta_1 \end{Bmatrix} ; \quad \underset{\sim}{U}^n = \begin{Bmatrix} u^1 \\ u^2 \\ u^3 \\ u^4 \\ u^5 \\ u^6 \end{Bmatrix} ; \quad \underset{\sim x}{V}^n = \begin{Bmatrix} v_x^1 \\ v_x^2 \\ v_x^3 \\ v_x^4 \\ v_x^5 \\ v_x^6 \end{Bmatrix} ; \quad \underset{\sim y}{V}^n = \begin{Bmatrix} v_y^1 \\ v_y^2 \\ v_y^3 \\ v_y^4 \\ v_y^5 \\ v_y^6 \end{Bmatrix}$$

(η_3 is a redundant coordinate used for symmetry $\eta_3 = 1-\eta_1-\eta_2$. η's are area coordinates as described in figure 4).

The cartesian coordinates at points within each cell are expressed by the following equations [6],

$$\begin{aligned} x &= x_1\eta_1 + x_2\eta_2 + x_3\eta_3 \\ y &= y_1\eta_1 + y_2\eta_2 + y_3\eta_3 \end{aligned} \tag{39}$$

The Jacobian in this case is equal to twice the area of the triangle shown in figure 4.

When the source point does not lie within the element of integration the integrals can be computed by standard Gauss formulae. If the source point lies on the element under consideration the boundary integrals contain singularities and special care has to be taken in their evaluation [2,4,7].

Domain integrals necessary for the cells, can be computed using Hammer's rule [7], for the case when the source point does not lie on the cell under consideration. If the source point lies on the cell the integrals contain a singularity. In order to perform the integration of the singular part, it is more convenient to define a cylindrical system of coordinates (r, θ) with the origin at the singular point (figure 5). Local area coordinates can be now written in function of polar coordinates as follows,

$$\eta_i = \frac{1}{2A}\,(2A_i^\circ + b_i x_\xi + a_i y_\xi) + \frac{r}{2A}(b_i \cos\theta + a_i \sin\theta) \tag{40}$$

where $a_i = x_k - x_j$; $b_i = y_j - y_k$; $2A_i^\circ = x_j y_k - x_k y_j$ for $i = 1,2,3$; $j = 2,3,1$ and $k = 3,2,1$. A is the area of the triangle.

Analytical integration with respect to r does not present now any difficulties because the singularity of the integral has been removed by introducing the polar coordinate system. The integration with respect to the angle θ, can be carried out numerically using a standard Gaussian quadrature formula.

8. APPLICATIONS

Several applications will now be described to illustrate the applicability of the method. They are

i) Plane-Poiseuille Flow (Steady State)
ii) Couette type flow problem (Steady State)
iii) Prescribed concentration in a channel (Time Dependent)

i) Plane Poiseuille Flow
The first example is a steady state thermal problem as shown in figure 6. The fluid is assumed to enter the channel with a temperature $u = 0$ at $x = 0$ and is heated by contact with the solid wall where a temperature value $u = 1$ is prescribed. For fully developed laminar flow the velocity field is given by

$$v_x = 2(1-4y^2) \quad ; \quad v_y = 0 \tag{a}$$

The problem was studied for different values of diffusivity ($\alpha = 0.5$; 0.05 and 0.0167) with the boundary element mesh of figure 7. Taking symmetry into consideration the domain was discretized using only 26 quadratic elements (with 52 nodes) and 44 quadratic internal cells (with 63 internal points). The boundary element results are shown in figure 8.

ii) Couette Type Flow Problem
The steady thermal problem defined in figure 9 was studied next. This case represents Couette type flow between two parallel plates. Temperature boundary conditions on the two plates are prescribed as shown in the figure. The problem was studied for different values of diffusivity ($\alpha = 14.3 \times 10^{-6}$; $\alpha = 14.3 \times 10^{-7}$ and $\alpha = 14.3 \times 10^{-8}$) which can also be expressed as different types of Peclet's numbers (Pe = 35; Pe = 350 ; Pe = 3500).

The boundary element mesh shown in figure 10 represents 24 quadratic boundary elements, with 48 nodes and 64 quadratic internal cells with 105 internal points. The boundary element results for temperatures at $x = 0.1$ are shown in figure 11. The numerical solution is in excellent agreement with the analytical results. (Notice that the results are practically the same for Pe = 35 and 350).

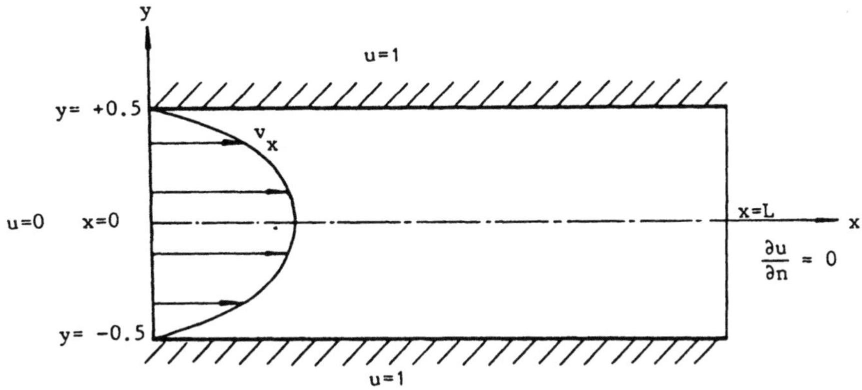

Figure 6 Thermal entry problem

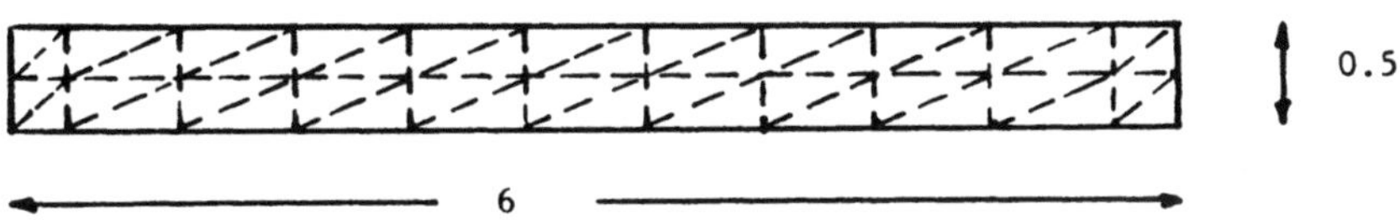

Figure 7 Discretization of the plane Poiseuille-problem

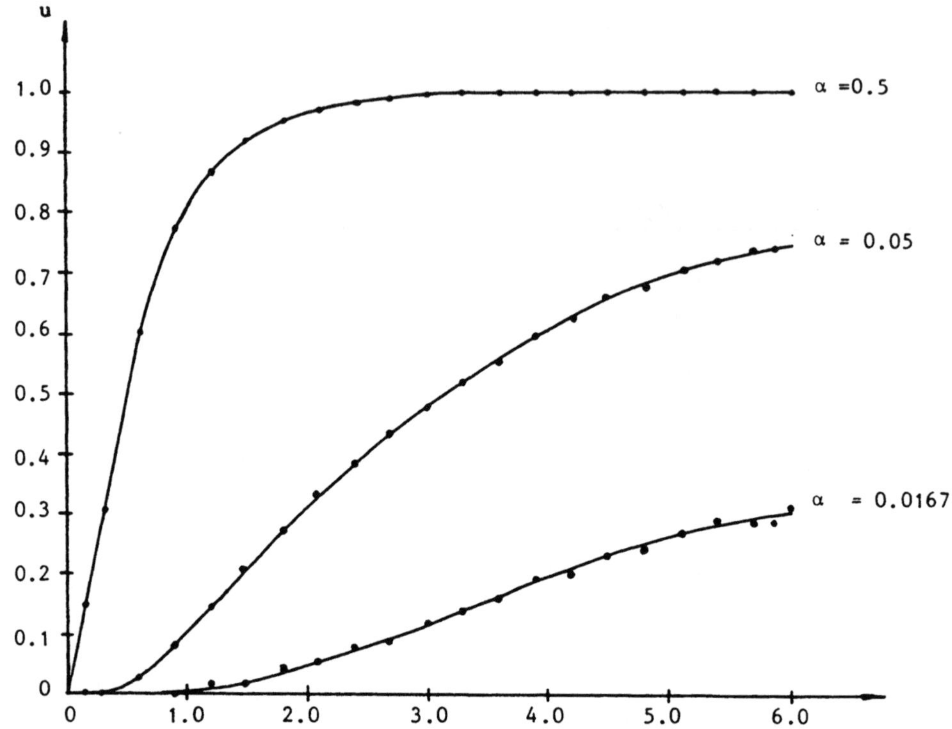

Figure 8 Centreline temperature for steady plane - Poiseuille flow

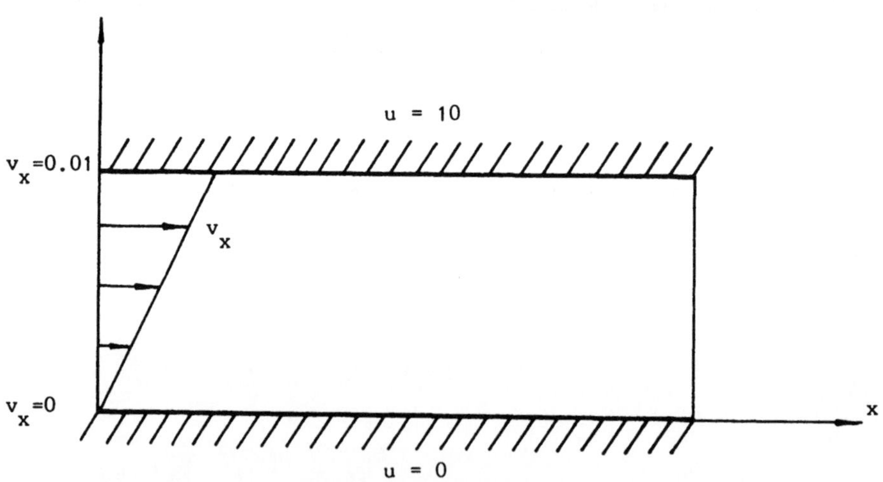

Figure 9 Couette flow thermal problem

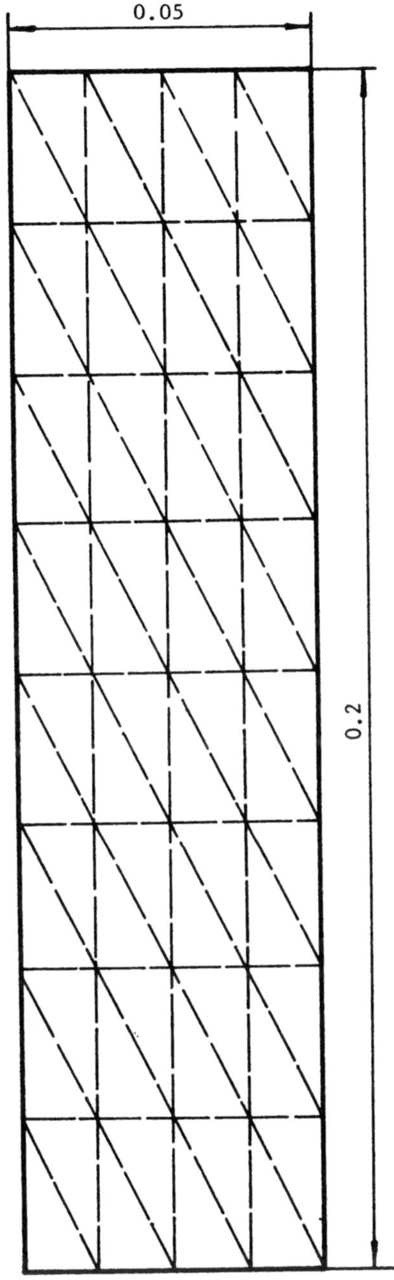

Figure 10 Discretization for the Couette flow problem

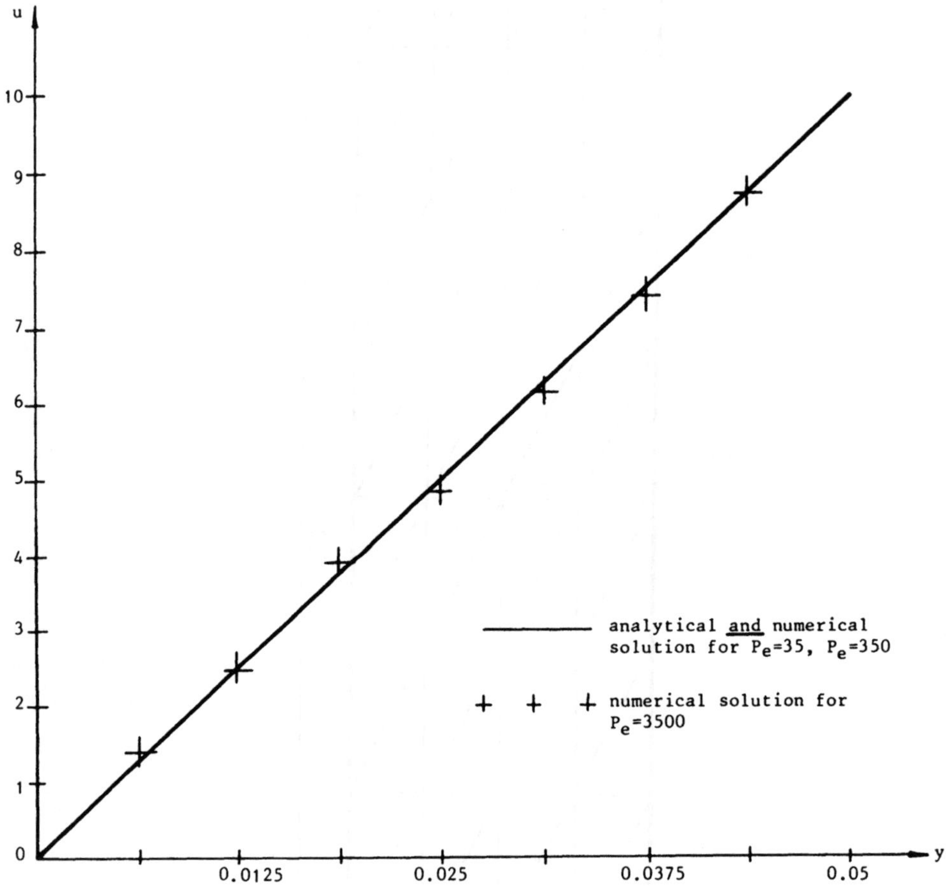

Figure 11 Temperature results at x = 0.1 for Couette flow problem

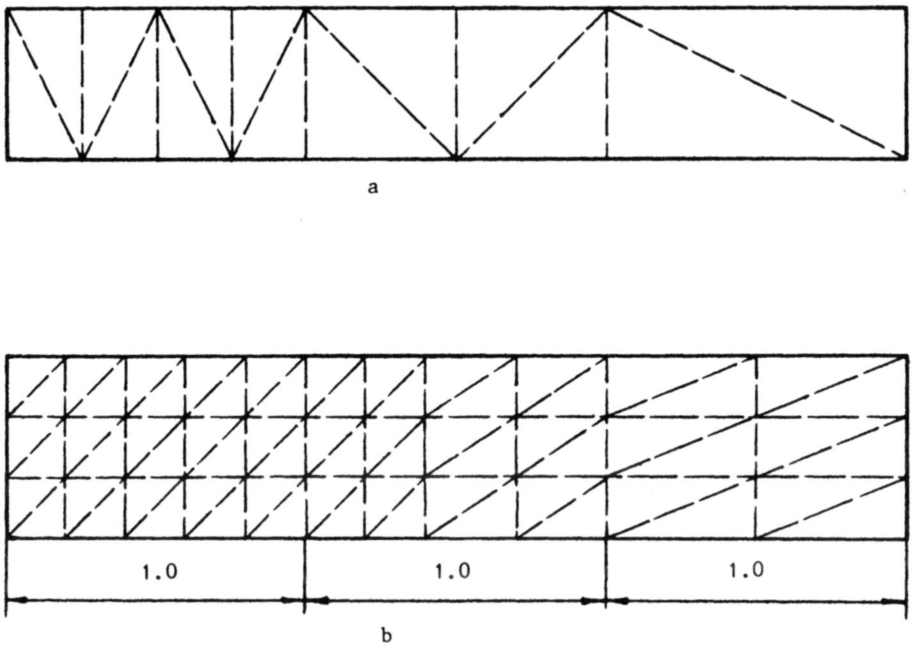

Figure 12 Discretization of a channel into boundary elements and internal cells
(a - coarse mesh b - fine mesh)

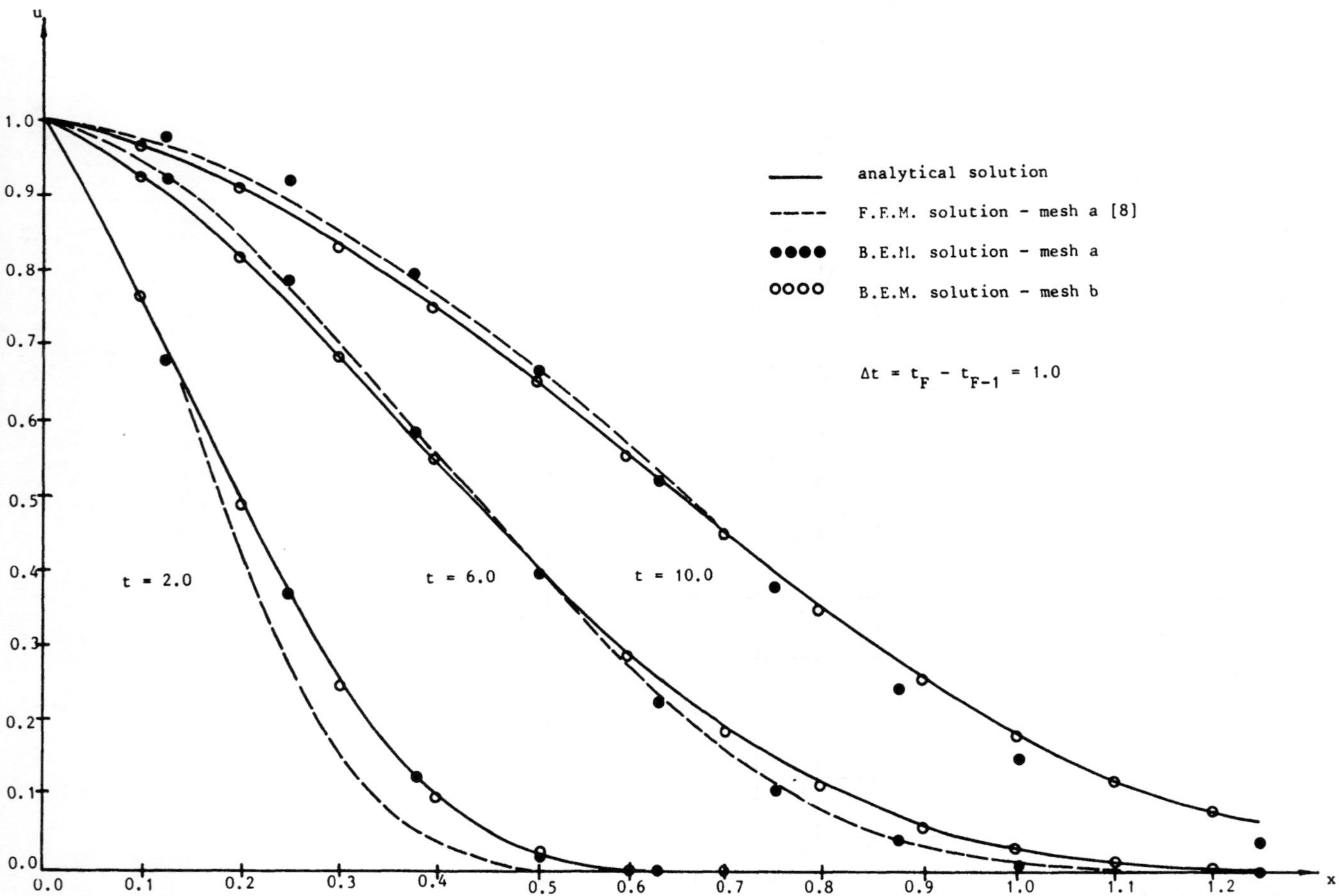

Figure 13 One dimensional diffusion-convection

iii) Prescribed Concentration in a Channel

A simple time dependent diffusion-convection problem was solved using elements quadratic in space and linear on time. The domain shown in figure 12 represents the discretization of a channel in which a constant velocity field $v_x = 0.05$ is assumed to be acting. A discharge producing a concentration $u = 1$ is applied as a step function at $x = 0$ and $t = 0$. The material properties are such that $c = \rho = 1$ and $k_o = 0.01$.

The coarse boundary element mesh (a) employed 16 elements with 32 nodes and 14 internal cells with 13 internal points. The quadratic finite element mesh used for comparison purposes consisted of 14 elements and 45 nodes [8]. The boundary and finite elements results in figure 13 were obtained using $\Delta t=1.0$.

The fine boundary element mesh (fig. 12b) was tried next as it was suspected that the bad agreement of B.E. for $t = 6$ and $t = 10$ was due to the coarseness of the cell discretization The new mesh employed 28 boundary elements with 56 nodes and 66 internal cells with 105 internal nodes. The boundary element results shown in figure 13 were obtained using $\Delta t = 1.0$. Notice that the accuracy of the solution is now very high and the results are practically identical to the analytical solution.

9. CONCLUSIONS

The boundary element has been applied to solve steady state and time dependent transport problems governed by the diffusion-convection equation. The starting integral equations for the problems are obtained using weighted residual techniques in time and space. A space dependent fundamental solution is applied for steady state problems and a time and space dependent one is used to solve time dependent problems. These fundamental solutions introduce singularities in the problem which can be dealt with using numerical and analytical techniques described in the chapter. The formulation is implemented for higher order elements, i.e. quadratic in space and linear on time.

Results are compared against those presented by other authors or against known analytical solutions and their agreement demonstrates the validity of the present approach. Accurate results can be obtained using boundary elements in diffusion-convection problems when the convective terms are dealt with in the manner described in this chapter.

The method can be extended to solve a range of non-linear material problems which can be linearized using the Kirchhoff' transform as shown in section 2. The interested reader can also consult references [1-3,9] on this topic.

REFERENCES

1. SKERGET, P. and BREBBIA, C.A. Non linear potential problems Chapter 1 in Progress in Boundary Element Methods, Vol.2 Pentech Press, London, Springer-Verlag, New York (1983).

2. SKERGET, P. and BREBBIA, C.A. Time dependent non-linear potential problems, Chapter 1 in Progress in Boundary Element Methods, Vol.3. Springer-Verlag, Berlin, NY, 1983.

3. BREBBIA, C.A. and SKERGET, P. Time dependent diffusion convection problems using boundary elements. Third International Conference on Numerical Methods in Laminar and Turbulent Flow, the University of Washington, Seattle, U.S.A. Pineridge Press (1983).

4. SKERGET, P. and BREBBIA, C.A. The solution of convective problems in laminar flow. 5th International Conference on Boundary Elements, Hiroshima, Japan. Springer-Verlag, Berlin, N.Y. 1983.

5. BREBBIA, C.A. "The Boundary Element Method for Engineers" Pentech Press, London, Halstead Press, NY, 1978.

6. BREBBIA, C.A. and WALKER, S. "Boundary element techniques in engineering." Butterworths, London, 1980.

7. BREBBIA, C.A., TELLES, J. and WROBEL, L. "Boundary Element Methods - Theory and Applications in Engineering" Springer-Verlag, Berlin, NY, 1983.

8. ADEY, R.A. Numerical prediction of transient water quality and tidal motion in estuaries and coastal waters. Ph.D. Thesis, University of Southampton, 1974.

9. BIALECKI, R. and NOVAK, A.J. Boundary value problems in heat conduction with non-linear material and non-linear boundary conditions. Appl. Math. Modelling, 5, p.417-421, 1981.

BOUNDARY ELEMENT SOLUTIONS TO AN INVERSE GROUNDWATER PROBLEM

Philip L-F. Liu

School of Civil and Environmental Engineering
Cornell University

INTRODUCTION

Recently, a rather unique inverse groundwater problem has been suggested to the author: If the groundwater table in a confined aquifer basin (piezometric heads) is determined from the records of many observation wells, is it possible to estimate the amount of groundwater being pumped out of the aquifer at certain wells? The reason for posing this question is because some hotels and factories within the groundwater basin have been suspected of pumping groundwater illegally, thus causing a rather serious decrease in the groundwater table and associated land subsidence problems. It would be interesting to develop a scientific tool for identifying the guilty parties. In this paper a numerical scheme is developed to answer the question. For simplicity, only the steady state flow in a confined homogeneous aquifer is discussed. The numerical scheme is then checked with an idealized example. It shows that the method can calculate the pumping rate accurately. Applications of this scheme to field problems will be reported in the near future.

THEORETICAL FORMULATION

Consider steady state groundwater flows in a confined homogenous aquifer. The governing equation for the piezometric head, $\Phi = P/\rho g + z$, can be written as

$$\nabla^2 \Phi = \sum_{k=1}^{NW} \frac{Q_k}{T} \delta(x - x_k)\, \delta(y - y_k) \qquad \text{in } D \tag{1}$$

where T is the transmissivity of the aquifer, Q_k is the pumping rate for the k-th well located at (x_k, y_k), and the corresponding depth-integrated velocity vector is defined as

$$\vec{q} = T \nabla \Phi \tag{2}$$

The general solution to the Poisson equation, equation (1), can be expressed as the sum of the homogeneous solution Φ and the particular solution so that

$$\Phi = \tilde{\Phi} + \sum_{k=1}^{NW} \frac{Q_k}{2\pi T} \ln r_k \tag{3}$$

where

$$r_k = [(x - x_k)^2 + (y - y_k)^2]^{1/2} \tag{4}$$

The homogeneous solution $\tilde{\Phi}$ satisfies the Laplace equation

$$\nabla^2 \tilde{\Phi} = 0 \qquad \text{in } D \tag{5}$$

For the present problem, the Dirichlet boundary condition is prescribed along the boundary Γ, i.e.

$$\Phi = \Phi_b \qquad \text{on } \Gamma \tag{6}$$

Then in terms of $\tilde{\Phi}$

$$\tilde{\Phi} = \Phi_b - \sum_{k=1}^{NW} \frac{Q_k}{2\pi T} \ln r_k \qquad \text{on } \Gamma \tag{7}$$

Furthermore, there are M observation wells within the domain D at which the piezometric heads are given

$$\Phi = \Phi_j(x_j, y_j) \quad , \quad j = 1, 2, 3, \ldots, M \tag{8}$$

In terms of $\tilde{\Phi}$,

$$\tilde{\Phi}_j = \Phi_j(x_j, y_j) - \sum_{k=1}^{NW} \frac{Q_k}{2\pi T} \ln r_k^j; \tag{9}$$

with

$$r_k^j = [(x_j - x_k)^2 + (y_j - y_k)^2]^{1/2} \tag{10}$$

The pumping rates, Q_k, and the locations of the wells are to be determined from the given boundary conditions, equation (6), and the information given at observation wells, equation (8).

NUMERICAL FORMULATION

To investigate the inverse problem presented in the previous section, the boundary integral equation method (BIEM) is chosen (Liggett and Liu, 1983). The reason for the particular choice is simply because the pumping wells (point singularities) can cause considerable difficulty when using domain methods (finite elements or finite differences) (e.g. Charbeneau and Street, 1979). Since the potential function $\tilde{\Phi}$ satisfies the Laplace equation, the corresponding integral equation can be expressed as (Liggett and Liu, 1983)

$$\alpha \tilde{\Phi} (P) = \int_\Gamma [\tilde{\Phi} (Q) \frac{\partial}{\partial n} \ln r - \ln r \frac{\partial \tilde{\Phi}}{\partial n} (Q)] \, ds \qquad (11)$$

where r is the distance from a base point P to a boundary point Q. If the point P is an interior point, $\alpha = 2\pi$ and the above equation becomes

$$\tilde{\Phi} (P) = \frac{1}{2\pi} \int_\Gamma [\tilde{\Phi} (Q) \frac{\partial}{\partial n} \ln r - \ln r \frac{\partial \tilde{\Phi}}{\partial n} (Q)] \, ds \qquad (12)$$

when the point P is a boundary point, the coefficient α becomes the interior angle of the boundary at point P; for a smooth boundary $\alpha = \pi$.

The numerical procedure can be described as follows: (1) The boundary Γ is first divided into N linear segments. Both Φ and its normal derivatives $\partial\Phi/\partial n$ are assumed to be linearly varying functions over each element. (2) Integrating equation (11) with the point P being located at each nodal point successively yields the following matrix equation

$$\underset{N\times N}{[RN]} \; \underset{1\times N}{\{\tilde{\Phi}\}} = \underset{N\times N}{[RLN]} \; \underset{1\times N}{\{\frac{\partial \tilde{\Phi}}{\partial n}\}} \qquad (13)$$

where $\{\tilde{\Phi}\}$ and $\{\partial\tilde{\Phi}/\partial n\}$ are column vectors for $\tilde{\Phi}$ and $\partial\tilde{\Phi}/\partial n$ at each nodal point, and coefficient matrices [RN] and [RLN] depend only on the geometry of the boundary. (3) Applying the base point P at the location of observation wells and integrating equation (12) give the following set of equations

$$\underset{N\times M}{[A]} \; \underset{1\times N}{\{\tilde{\Phi}\}} = \underset{N\times M}{[B]} \; \underset{1\times N}{\{\frac{\partial \tilde{\Phi}}{\partial n}\}} + \underset{1\times M}{\{\tilde{\Phi}_j\}} \qquad (14)$$

where $\{\tilde{\Phi}_j\}$ is the value of $\tilde{\Phi}$ at the j-th observation well.

One can combine equations (13) and (14) into one equation, thus

$$\underset{A}{[\overset{RN}{\cdots}]}\ \{\tilde{\Phi}\} = \underset{B}{[\overset{RLN}{\cdots}]}\ \{\frac{\partial\tilde{\Phi}}{\partial n}\} + \{\cdot\underset{\Phi_j}{\overset{0}{\vdots}}\cdot\} \tag{15}$$

$$N\times(N+M) \quad 1\times N \quad N\times(N+M) \quad 1\times N \quad 1\times(N+M)$$

(4) Substitutions of the boundary conditions, equation (7), and the observation well data, equation (9), into equation (15) yield a system of nonlinear algebraic equations for unknown $(\partial\Phi/\partial n)$, on the boundary, pumping rates Q_k and the location of pumping wells. Thus

$$F_i(Q_1,\ldots,Q_{NW},x_1,\ \ldots x_{NW},y_1,\ \ldots,y_{NW},(\frac{\partial\tilde{\Phi}}{\partial n})_1,\ldots(\frac{\partial\tilde{\Phi}}{\partial n})_N) = 0$$

$$i = 1,\ 2,\ \ldots,\ N+M \tag{16}$$

There are (N+M) equations for (3NW+N) unknowns. If M is greater or equal to 3NW, equation (16) can be solved for the unknown quantities.

Linear Problem

If the locations of pumping wells are known, equation (16) becomes linear and can be written as

$$\sum_{j=1}^{N+NW} H_{ij}\ X_j - f_i = 0, \qquad i = 1,\ 2,\ \ldots,\ N+M \tag{17}$$

where $\{X\}$ is the column vector containing unknowns Q_k and $(\partial\Phi/\partial n)$, on the boundary, the coefficient matrix $[H]$ and the column vector $\{f\}$ are known and calculated from numerical integration and boundary conditions. When the number of observation wells M is greater than that of the pumping wells, equation (17) represents an overly determined system. One can find the solutions by the least square approximation. Assuming the residuals of equation (17) are

$$R_i = \sum_{j=1}^{N+NW} H_{ij}\ X_j - f_i \quad , \qquad i = 1,\ 2,\ \ldots,\ N+M \tag{18}$$

One minimizes the residuals in a least square sense, i.e.

$$\frac{\partial}{\partial x_j}\left[\sum_{i=1}^{N+M} R_i^{\,2}\right] = 0 \quad , \qquad j = 1,\ 2,\ \ldots,\ N+NW \tag{19}$$

which leads to a system of equation

$$[H]^T [H] \{x\} = [H]^T \{f\} \tag{20}$$

Equation (20) can be used to solve for pumping rates.

NUMERICAL EXAMPLE

To verify the numerical procedure presented here for the linear problem, a simple example is created. The dimensions of the groundwater basin are 20 × 20. There are two pumping wells located at (1,0) and (-1,0), respectively. The piezometric heads, Φ, along the boundary are given according to the following formula

$$\Phi = x + 1 - \frac{1}{2\pi} [\ln \sqrt{(x-1)^2 + y^2} + \ln \sqrt{(x+1)^2 + y^2}] \tag{21}$$

In other words, the groundwater flows are the superposition of a uniform flow in the x-direction and two pumping wells with the same pumping rate Q = 1.0. The transmissivity of the aquifer is assumed to be one.

Table 1. The location and the piezometric head of observation wells.

x	y	Φ
4.0	5.0	4.4080723
-5.0	5.0	-4.6226496
-2.0	5.0	-1.5398897
1.0	5.0	1.4758891
3.0	-5.0	3.4365225
-6.0	5.0	-5.6538155
3.0	-1.0	3.6464651
3.0	9.0	3.2824211

In computations the boundary is divided into forty equal size elements (N = 40). Eight interior points are arbitrarily chosen as observation wells (M = 8); their locations and piezometric heads are listed in Table 1. The results for the pumping rates are almost exact; Q_1 = 0.9998 (x_1 = 1.0, y_1 = 0.0) and Q_2 = 1.0002 (x_2 = 1.0, y_2 = 0.0). To test further the capability of the numerical procedure, two additional wells are added as unknown; their locations are Q_3 (x_3 = 5.0, y_3 = -3.5) and Q_4 (x_4 = -6.5, y_4 = -2.5). The same boundary conditions and interior solutions as those in the previous example are used. If the procedure could give exact solutions, Q_3 and Q_4 should be zero. The actual numerical solutions are $Q_1 = 1.0001$, $Q_2 = 0.9995$, $Q_3 = -8.1311 \times 10^{-6}$ and $Q_4 = 1.1246 \times 10^{-3}$.

CONCLUDING REMARKS

This paper presented an efficient and accurate numerical scheme to find solutions for an inverse groundwater problem. Although the paper only addresses homogeneous aquifer cases, it can be extended to zoned inhomogeneous problems without too much difficulty (Liggett and Liu, 1983). There are two topics which deserve further research: (1) When the locations of the pumping wells are also not known, the matrix equation for unknowns becomes nonlinear. Numerical procedure must be developed to solve a set of overly determined nonlinear equations. (2) When the readings in the observation wells and the values of boundary condition contain "noises", the accuracy of the numerical results must be influenced. A systemetic study in the direction is needed.

Acknowledgement

This research is, in part, supported by a research grant to Cornell University (Grant No. CEE 7902803A03) and a research grant to Humboldt State University (Grant No. INT-8115382) from National Science Foundation. Professor Y-S. Tsao, of National Taiwan University, who introduced me to the problem, is also acknowledged.

References

1. Liggett, J. A. and Liu, P. L-F., 1983, The Boundary Integral Equation Method for Porous Media Flow, George Allen & Unwin, London.

2. Charbeneau, R. and Street, R., 1979, Modeling groundwater flow fields containing point singularities: A technique for singularity removal, Water Resour. Res., 15(3), p. 583-594.

ADDITIONAL PAPERS

AN UPSTREAM WEIGHT FINITE ELEMENT METHOD FOR SOLVING THE 3-DIMENSIONAL CONVECTION-DISPERSION EQUATIONS

Ne-Zheng Sun, Department of Mathematics, Shandong University (now at UCLA), and
William W-G. Yeh and Chuching Wang, Civil Engineering Department, University of California, Los Angeles

A simple upstream weight finite element method is presented to simulate the 3-dimensional transport of contaminant in aquifer. The principal procedure of this method is the same as the general Galerkin finite element method, except that six invented nodes are added to each triangular prism element. The concentrations of the invented nodes are defined as the weighted combinations of those of the real nodes. The weighting coefficients are determined by the local Peclet number. When convective transport dominates the dispersive transport, this technique can eliminate the oscillation of the numerical solutions efficiently with only a nominal increase of computation effort over that of the general linear Galerkin finite element method. A classical example is used to illustrate the accuracy of the proposed numerical model.

INTRODUCTION

In recent years, numerical models have been developed for solving the solute transport problem in groundwater, such as the Finite Difference Methods, the Method of Characteristics (Konikow and Bredehoeft, 1978), the Finite Element Methods (Pinder and Gray, 1977). Anderson (1979) and Bachmat, et al. (1980) summarized recent developments in this area.

Most of the established numerical models have been limited to the 2-dimensional case. However, in many field problems, a 3-dimensional model would be more appropriate to define the shape of a contaminant plume. A few 3-dimensional models have been formulated and used in field problems (Gupta, 1975; Khaleel and Redell, 1977; Segol, 1977). Generally, 3-dimensional models require more input data, storage space and computation time than the 2-dimensional models. When convective transport dominates the dispersive transport, overshoot and numerical dispersion will occur in numerical solutions if precautions are not taken (Gray and Pinder, 1977). The numerical solutions of the dispersion equation are frequently characterized by oscillation and by numerical errors around a sharp concentration front. In order to eliminate the oscillation of numerical solutions, various improved models have been presented, such as high-order Finite Difference Methods (Morton, 1980), the Method of Characteristics (Konikow and Bredhoeft, 1978), Finite Element Method (FEM) with higher-order basis functions (Van Genuchten, 1977), FEM with nonsymmetric weighting functions (Heinrich et al., 1977; Huyakorn and Nilkuha, 1979; Hughes, 1978; Payre, et al., 1982), Orthogonal Collocation FEM (Pinder and Shapiro, 1979; Herbst, 1981), moving and deforming coordinate method (O'Neill, 1981).

Most of these techniques are difficult to apply to the 3-dimensional problems because of the complexity in programming and the expense of computation effort. Sun and Yeh (1983) presented a simple upstream weight numerical method for solving the 2-dimensional dispersion equations. In this paper, we extend it to the 3-dimensional case. The advantages of this technique are as follows: 1. it can eliminate the numerical oscillation effectively, 2. the weighting coefficients are adaptively added based upon the local Peclet number, and 3. the computational procedure is essentially the same as the general Galerkin FEM with only a nominal increase in computational effort.

GOVERNING EQUATION AND GALERKIN FEM

The governing equation for a 3-dimensional solute-transport process in a saturated porous medium is (Bear, 1979)

$$\frac{\partial C}{\partial t} = \text{div}\ (\underline{D}\ \text{grad}\ C) - \text{div}\ (\bar{V}C) - M \tag{1}$$

where

C = concentration of solute (M/L^3);
$\underline{\underline{D}}$ = hydrodynamic dispersion coefficient tensor (L^2/T);
$\bar{V}$ = average velocity vector of fluid in porous medium (L/T);
M = source or sink (M/L^3T);

In Cartesian coordinates with summation convention assumed, Equation (1) can be written as

$$L(C) \equiv \frac{\partial C}{\partial t} - \frac{\partial}{\partial x_\alpha}\left(D_{\alpha\beta}\frac{\partial C}{\partial x_\alpha}\right) + V_\alpha \frac{\partial C}{\partial x_\alpha} + M = 0, \tag{2}$$

$$(\alpha, \beta = 1, 2, 3)$$

where x_1, x_2, x_3 are Cartesian coordinates; V_1, V_2, V_3 are the components of velocity $\bar{V}$ in the x_1, x_2, x_3 directions respectively.

The Galerkin FEM is used to solve Equation (2). Let the trial function be

$$\hat{C}(x_\alpha,t) = \sum_{\ell=1}^{N} C_\ell(t)\phi_\ell(x_\alpha) \quad (3)$$

To minimize the residues of L(C), we obtain N-orthogonality conditions

$$\iiint_{(\Omega)} L(\hat{C})\phi_\ell(x_\alpha)d\Omega = 0 \quad (4)$$

$$(\ell=1,\ 2,\ \ldots,\ N)$$

where the set of basis functions $\{\phi_\ell;(\ell=1,2,\ldots,N)\}$ is the first N-functions of an orthogonal and complete system in the region (Ω). The N-equations, Equation (4), are served to determine the N-undetermined coefficients C_ℓ in Equation (3). Following the general Galerkin's method, one can translate Equation (4) into a system of ordinary differential equations:

$$[A]\{C\} + [B]\ \{\frac{dC}{dt}\} = \{F\} \quad (5)$$

where [A] and [B] are $(N-N_1) \times (N-N_1)$ matrices, N_1 is the number of nodes on boundary (S_1). {C} and $\{\frac{dC}{dt}\}$ are vectors which contain the undetermined coefficients C_ℓ and their time derivatives. Typical elements of [A], [B] and {F} are

$$A_{ij} = \iiint_{(\Omega)} [D_{\alpha\beta} \frac{\partial\phi_i}{\partial x_\alpha} \frac{\partial\phi_j}{\partial x_\beta} + V_\alpha\phi_i \frac{\partial\phi_j}{\partial x_\beta}]\ d\Omega \quad (6)$$

$$B_{ij} = \iiint_{(\Omega)} \phi_i\ \ \phi_j\ d\Omega \quad (7)$$

$$F_i = \iint_{(S_2)} g\phi_i dS - \iiint_{(\Omega)} M\phi_i d\Omega \quad (8)$$

In order to calculate [A], [B] and {F}, one should divide the flow region (Ω) into elements, select the location of nodes and define the basis functions. Dirichlet and Neumann conditions are prescribed on (S_1) and (S_2). In the following analysis, we replace x_1, x_2, x_3, by x, y, z.

The flow region is divided into triangular prism elements. The top and bottom of these triangular-prism are parallel to the x,y-plane. The other sides are parallel to the z axis. The six vertices of each element are taken as nodes (Figure 1). They are

$$P_1\ (x_i,\ y_i,\ z_0),\ P_2\ (x_j,\ y_j,\ z_0),\ P_3\ (x_k,\ y_k,\ z_0)$$

$$P_4\ (x_i,\ y_i,\ z_1),P_5\ (x_j,\ y_j,\ z_1),\ P_6\ (x_k,\ y_k,\ z_1)$$

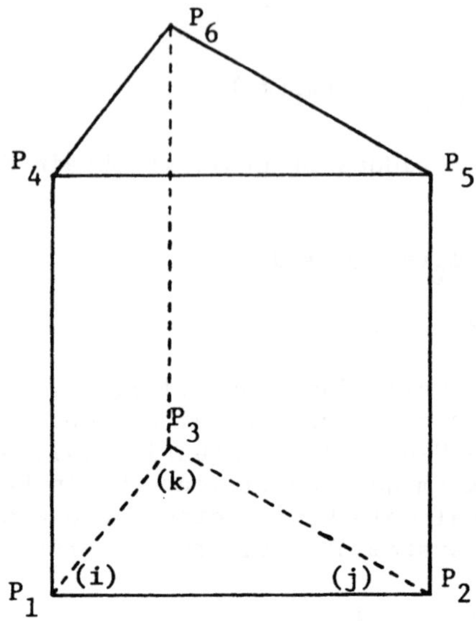

Figure 1. 3-D Finite Element

For each element, the basis functions corresponding to the six nodes are:

$$
\begin{aligned}
\phi_1(x,y,z) &= z^-\phi_i(x,y)\\
\phi_2(x,y,z) &= z^-\phi_j(x,y)\\
\phi_3(x,y,z) &= z^-\phi_k(x,y)\\
\phi_4(x,y,z) &= z^+\phi_i(x,y)\\
\phi_5(x,y,z) &= z^+\phi_j(x,y)\\
\phi_6(x,y,z) &= z^+\phi_k(x,y)
\end{aligned}
\qquad (9)
$$

where

$$z^- = \frac{z_1 - z}{\Delta z}\ ;\ z^+ = \frac{z - z_o}{\Delta z};\ \Delta z = z_1 - z_o\ ;$$

ϕ_i, ϕ_j and ϕ_k are two-dimensional linear basis functions for the triangle element (Δ) which is the bottom of (e), i.e.,

$$\phi_\ell(x,y) = \frac{1}{2\Delta}(a_\ell + b_\ell x + c_\ell y) \qquad (\ell = i,\ j,\ k) \qquad (10)$$

$$(x,y)\varepsilon(\Delta),$$

where

$$
\begin{aligned}
a_i &= x_j y_k - x_k y_j & b_i &= y_j - y_k & c_i &= x_k - x_j\\
a_j &= x_k y_i - x_i y_k & b_j &= y_k - y_i & c_j &= x_i - x_k\\
a_k &= x_i y_j - x_j y_i & b_k &= y_i - y_j & c_k &= x_j - x_i
\end{aligned}
\qquad (11)
$$

Δ is the area of triangle (Δ).

If one uses the basic element and basis functions mentioned above, less input data will be required. Also, one can compute the coefficients of Equation (5) directly without having to use numerical integration.

For each element, the 6x6 element matrix which relates matrix [A] and element (e) is noted by $[A]^e$. For example, we have

$$\begin{aligned} A^e_{11} &= \frac{(\Delta z)}{12\Delta}[D_{xx}b_ib_i + D_{xy}b_ic_i] - \frac{1}{12} D_{xz}b_i \\ &+ \frac{(\Delta z)}{12\Delta}[D_{yx}c_ib_i + D_{yy}c_ic_i] - \frac{1}{12} D_{yz}c_i \\ &- \frac{1}{12} [D_{zx}b_i + D_{zy}c_i] + \frac{1}{6} D_{zz} \frac{\Delta}{(\Delta z)} \\ &+ \frac{(\Delta z)}{18} [V_xb_i + V_yc_i] - \frac{1}{12} V_z \Delta; \end{aligned} \tag{12}$$

Similar expressions are developed for the remaining elements of matrix $[A]^e$, Obviously $[A]^e$ is a nonsymmetric matrix. The 6x6 matrix $[B]^e$ which relates matrix [B] and element (e) is a symmetric matrix. We have

$$\begin{aligned} B^e_{ij} &= \frac{\Delta(\Delta z)}{18}, && \text{when } i = j; \\ B^e_{ij} &= \frac{1}{2}\,\frac{\Delta(\Delta z)}{18}, && \text{when nodes i and j have same (x,y) coordinates or have same z coordinate.} \\ B^e_{ij} &= \frac{1}{4}\,\frac{\Delta(\Delta z)}{18}, && \text{other cases.} \end{aligned} \tag{13}$$

By combining all the element matrices, one can obtain the global matrices [A] and [B]. Once [A] and [B] are determined, one can solve Equation (5) by applying any ODE solver to obtain the concentration distribution.

UPSTREAM WEIGHT FINITE ELEMENT METHOD

For a general convection-dispersion problem, the governing equation can be solved satisfactorily by the Galerkin FEM as previously described. However, numerical oscillation will occur if convective transport dominates the dispersive transport. In order to eliminate the numerical oscillation, a simple upstream weight FEM is presented. In each element (e), six invented nodes are added. They are:

$$P_7(x_i,y_i,\frac{z_1+z_o}{2}),\ P_8(x_j,y_j,\frac{z_1+z_o}{2}),\ P_9(x_k,y_k,\frac{z_1+z_o}{2})$$

$$P_{10}(x_m,y_m,z_o),\ P_{11}(x_m,y_m,\frac{z_1+z_o}{2}),\ P_{12}(x_m,y_m,z_1)$$

where

$$x_m = \frac{1}{3}(x_i+x_j+x_k), \quad y_m = \frac{1}{3}(y_i+y_j+y_k)$$

The concentrations of the six invented nodes are presented as weighted combinations of the concentrations of the real nodes. The weighted combinations are as follows:

$$\begin{aligned} C_7 &= \lambda_1 C_1 + \lambda_4 C_4 & C_{10} &= \omega_1 C_1 + \omega_2 C_2 + \omega_3 C_3 \\ C_8 &= \lambda_2 C_2 + \lambda_5 C_5 & C_{12} &= \omega_4 C_4 + \omega_5 C_5 + \omega_6 C_6 \\ C_9 &= \lambda_3 C_3 + \lambda_6 C_6 & C_{11} &= \sigma_1 C_{10} + \sigma_2 C_{12} \end{aligned} \tag{14}$$

where all ω, λ, and σ are weighting coefficients which satisfy

$$\begin{aligned} \omega_1 + \omega_2 + \omega_3 &= 1 & \omega_4 + \omega_5 + \omega_6 &= 1 \\ \lambda_1 + \lambda_4 &= 1 & \sigma_1 &= \frac{1}{3}(\lambda_1+\lambda_2+\lambda_3) \\ \lambda_2 + \lambda_5 &= 1 & \sigma_2 &= \frac{1}{3}(\lambda_4+\lambda_5+\lambda_6) \\ \lambda_3 + \lambda_6 &= 1 & \omega_i &> o, \ \lambda_i > o \ (i=1,2,\ldots,6) \end{aligned} \tag{15}$$

These coefficients will be determined later.

By connecting invented nodes and real nodes, the original element is divided into six subelements (e_s').

In each subelement (e'), the unknown concentration function C is expressed as a combination of the concentrations of its six vertices. For example, for the subelement (e_1') with vertices: P_1, P_2, P_{10}, P_7, P_8 and P_{11}, the concentration function is

$$\begin{aligned} C(x,y,z,t) = {} & \phi_1' C_1 + \phi_2' C_2 + \phi_3' C_{10} \\ & + \phi_4' C_7 + \phi_5' C_8 + \phi_6' C_{11}; \ (x,y,z) \ (e_1') \end{aligned} \tag{16}$$

where ϕ_i $(i=1,2,\ldots,6)$ are basis functions for element (e_1'). According to Equation (13), we have

$$\begin{aligned} \phi_1'(x,y,z) &= (z^- - z^+)\ \phi_{ki}(x,y) \\ \phi_2'(x,y,z) &= (z^- - z^+)\ \phi_{kj}(x,y) \\ \phi_3'(x,y,z) &= (z^- - z^+)\ \phi_{km}(x,y) \\ \phi_4'(x,y,z) &= 2z^+ \phi_{ki}(x,y) \end{aligned} \tag{17}$$

$$\phi_5'(x,y,z) = 2z^+ \phi_{kj}(x,y)$$

$$\phi_6'(x,y,z) = 2z^+ \phi_{km}(x,y)$$

where ϕ_{ki} ϕ_{kj} ϕ_{km} are the two-dimensional basis functions for triangular $\Delta(ijm)$, which is the bottom of subelement (e_1'), (Figure 2).

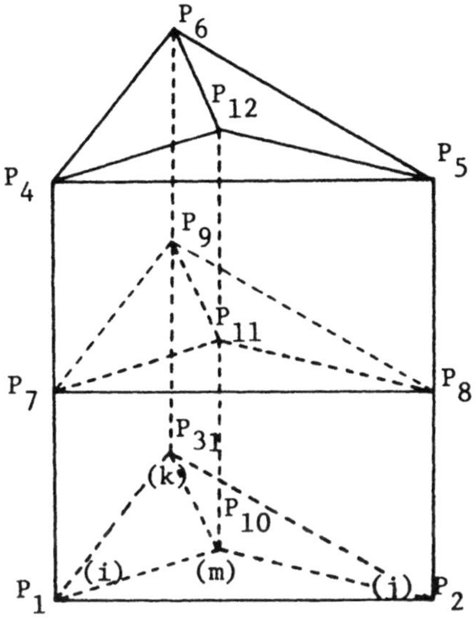

Figure 2. Invented Nodes and Subelements

$$\phi_{k\ell} = \frac{3}{2\Delta}(a_{k\ell} + b_{k\ell}x + C_{k\ell}y), \quad (\ell = i,j,m) \tag{18}$$

where

$$\begin{aligned} a_{ki} &= x_j y_m - x_m y_j & b_{ki} &= y_j - y_m & C_{ki} &= x_m - x_j \\ a_{kj} &= x_m y_i - x_i y_m & b_{kj} &= y_m - y_i & C_{kj} &= x_i - x_m \\ a_{km} &= x_i y_j - x_j y_i & b_{km} &= y_i - y_j & C_{km} &= x_j - x_i \end{aligned} \tag{19}$$

The concentrations of the invented nodes can be eliminated by utilizing the relationship among real nodes and invented nodes, which is defined by Equation (14). By substituting Equations (14) and (17) into Equation (16), we have

$$C(x,y,z,t) = \phi_1^{(1)}C_1+\phi_2^{(1)}C_2+\phi_3^{(1)}C_3+\phi_4^{(1)}C_4+\phi_5^{(1)}C_5+\phi_6^{(1)}C_6 \tag{20}$$

where

$$\phi_1^{(1)}=[z^-+(2\lambda_1-1)z^+]\phi_{ki}+\omega_1[z^-+(2\sigma_1-1)z^+]\phi_{km}$$

$$\phi_2^{(1)}=[z^-+(2\lambda_2-1)z^+]\phi_{kj}+\omega_2[z^-+(2\sigma_1-1)z^+]\phi_{km}$$

$$\phi_3^{(1)} = \omega_3[z^- + (2\sigma_1 - 1)z^+]\phi_{km} \qquad (21)$$

$$\phi_4^{(1)} = 2\lambda_4 z^+ \phi_{ki} + 2\omega_4\sigma_2 z^+ \phi_{km}$$

$$\phi_5^{(1)} = 2\lambda_5 z^+ \phi_{kj} + 2\omega_5\sigma_2 z^+ \phi_{km}$$

$$\phi_6^{(1)} = 2\omega_6\sigma_2 z^+ \phi_{km}$$

which are the basis functions with weighting coefficients included. If all $\omega = 1/3$, and all $\lambda = \frac{1}{2}$, then according to

$$\phi_{ki} + \frac{1}{3}\phi_{km} = \phi_i$$

$$\phi_{kj} + \frac{1}{3}\phi_{km} = \phi_j \qquad (22)$$

$$\frac{1}{3}\phi_{km} = \phi_k$$

Equation (21) can be reduced to the same as original basis functions of element (e) without weight coefficients.

The same procedure can be applied to all other subelements, thus the unknown concentration C has six different expressions in the six subelements of element (e) respectively, i.e.,

$$\hat{C}(x,y,z,t) = \sum_{\ell=1}^{6} C_\ell(t)\ \phi_\ell^{(r)}(x,y,z)$$
$$(x,y,z)\ \varepsilon(e'_r),\ r = 1,2,\ldots,6 \qquad (23)$$

Beside Equation (21), other weighted basis function would be

$$\phi_1^{(2)} = [z^- + (2\lambda_1 - 1)z^+]\phi_{ji} + \omega_1[z^- + (2\sigma_1 - 1)z^+]\phi_{jm}$$

$$\phi_2^{(2)} = \omega_2[z^- + (2\sigma_1 - 1)z^+]\phi_{jm}$$

\- - - - - - - - - - - - - - (24)

$$\phi_5^{(6)} = [z^+ + (2\lambda_5 - 1)z^-]\phi_{ij} + \omega_5[z^+ + (2\sigma_2 - 1)z^-]\phi_{im}$$

$$\phi_6^{(6)} = [z^+ + (2\lambda_6 - 1)z^-]\phi_{ik} + \omega_6[z^+ + (2\sigma_2 - 1)z^-]\phi_{im}$$

If and only if all $\omega = \frac{1}{3}$ and all $\lambda = \frac{1}{2}$, these weighted basis functions reduce to original basis functions, Equation (9), and the $\hat{C}$ in Equation (23) returns to its unity expression, Equation (3).

The element A_{ij}^e of matrix $[A]^e$ now becomes

$$\bar{A}_{ij}^e = \iiint_{(e)} [\cdot]\, d\Omega = \sum_{r=1}^{6} \iiint_{(e_r')} [\cdot]\, d\Omega \tag{25}$$

where $[\cdot]$ is the integration function in Equation (6). Substituting Equations (21) and (24) into Equation (25), and calculating term by term, we can obtain the expression of $\bar{A}_{ij}^e$. For example, assume

$$\begin{aligned} \lambda_1 &= \lambda_2 = \lambda_3 = \lambda \\ \lambda_4 &= \lambda_5 = \lambda_6 = 1 - \lambda \end{aligned} \tag{26}$$

then we have

$$\begin{aligned} \bar{A}_{11}^e = {} & \frac{(\Delta z)A(\lambda)}{8\Delta} [D_{xx}\bar{b}_i(\omega_1)\cdot\bar{b}_i(\omega_1) + D_{xy}\bar{b}_i(\omega_1)\cdot\bar{c}_i(\omega_1)] \\ & - \frac{1}{12} D_{xz}\bar{b}_i(\omega_1)\cdot\bar{I}_i(\omega_1) \\ & + \frac{(\Delta z)A(\lambda)}{8\Delta} [D_{yx}\bar{b}_i(\omega_1)\cdot\bar{c}_i(\omega_1) + D_{yy}\bar{c}_i(\omega_1)\cdot\bar{c}_i(\omega_1)] \\ & - \frac{1}{12} D_{yz}\bar{c}_i(\omega_1)\cdot\bar{I}_i(\omega_1) \\ & - \frac{1}{12}[D_{zx}\bar{b}_i(\omega_1) + D_{zy}\bar{c}_i(\omega_1)]\cdot\bar{I}_i(\omega_1) + \frac{\Delta}{9(\Delta z)} B(\lambda)E_1(\omega_1,\omega_1) \\ & + \frac{(\Delta z)A(\lambda)}{36}[V_x\bar{b}_i(\omega_1) + V_y\bar{c}_i(\omega_1)]\cdot\bar{I}_i(\omega_1) - \frac{\Delta}{36} E_1(\omega_1,\omega_1) \end{aligned} \tag{27}$$

where

$$\begin{aligned} \bar{b}_i(\omega) &= (\omega b_{im},\ b_{ji} + \omega b_{jm},\ b_{ki} + \omega b_{km}) \\ \bar{c}_i(\omega) &= (\omega c_{im},\ c_{ji} + \omega c_{jm},\ c_{ki} + \omega c_{km}) \\ \bar{I}_i(\omega) &= (\omega,\ 1+\omega,\ 1+\omega) \\ E_1(\omega_1,\omega_1) &= 2 + 2\omega_1 + 3\omega_1^2 \\ A(\lambda) &= 2\lambda^2 + \lambda + 1 \\ B(\lambda) &= 2\lambda^2 - 2\lambda + 1 \end{aligned} \tag{28}$$

If all $\omega = \frac{1}{3}$, $\lambda = \frac{1}{2}$, then Equations (27) reduce to Equation (12). Similar expressions are developed for the remaining $\bar{A}_{ij}^e$.

Using the same procedures, the matrix $[B]^e$ can be translated into a weighted matrix $[\bar{B}]^e$. With Equation (26) we have

$$\bar{B}^e_{11} = \frac{(\Delta z)\Delta}{18}\frac{1}{6} A(\lambda)E_1\ (\omega_1,\omega_1)$$

Similar expressions are developed for the remaining $\bar{B}^e_{ij}$.

Without Equation (26), we can also obtain similar expressions for $\bar{A}^e_{ij}$ and $\bar{B}^e_{ij}$ which include more weighting coefficients.

By combining all the element matrices $[\bar{A}]^e$ and $[\bar{B}]^e$ as before, one can obtain the global matrices $[\bar{A}]$ and $[\bar{B}]$. Thus, instead of the system of equations, Equation (5), we have

$$[\bar{A}]\{C\} + [\bar{B}]\ \{\frac{dC}{dt}\} = \{\bar{F}\} \tag{29}$$

where $\{\bar{F}\}$ is given by Equation (8) except using the weighted basis function.

The backward finite-difference scheme is used to replace the time derivatives of concentration and then Equation (29) becomes a system of linear equations in each time step

$$([\bar{A}]+[\bar{B}]/\Delta t)\ \{C\}_{t+\Delta t} = ([\bar{B}]/\Delta t)\ \{C\}_t + \{\bar{F}\} \tag{30}$$

A block-iteration technique is applied to solve Equation (30). The nodes which locate at the same level are regarded as a "block." The iteration procedure is level by level. In each level, the solutions are obtained by a direct method.

UPSTREAM WEIGHT MULTIPLE-CELL BALANCE METHOD

Based upon the multiple-cell balance method (Sun and Yeh, 1983) a system of weighted equations for 3-dimensional problems can be derived.

First, each element (e) is divided into six subelements (d'_S) by the six invented nodes introduced in the previous section, but in a different way (Figure 3). For example, (d'_1) is the prism with vertices P_1, Q_1, P_{10}, Q_3, P_7, Q_2, P_{11}, Q_4. The exclusive subdomain (d_i) of a node i is then defined as a multiangular prism that consists of all subelements with node i as one of their corners.

The mass balance equation of subdomain (d_i) can be described as

$$\iiint_{(d_i)} (\frac{\partial C}{\partial t} + M)\, d\Omega = \iint_{(s_i)} D_{\alpha\beta}\frac{\partial C}{\partial x_\beta} n_\alpha ds - \iiint_{(d_i)} V_\alpha \frac{\partial C}{\partial x_\alpha}\, d\Omega \tag{31}$$

The partial derivatives $\frac{\partial C}{\partial x}$, $\frac{\partial C}{\partial y}$, $\frac{\partial C}{\partial z}$, and $\frac{\partial C}{\partial t}$ can be expressed as a combination of the nodal values C_i, C_j, $\frac{dC_i}{dt}$ and $\frac{dC_j}{dt}$ by utilizing equation (23), where j represents all the neighboring nodes of i. Therefore, the three terms involved in Equation (31) can be integrated directly. For example, by taking node i

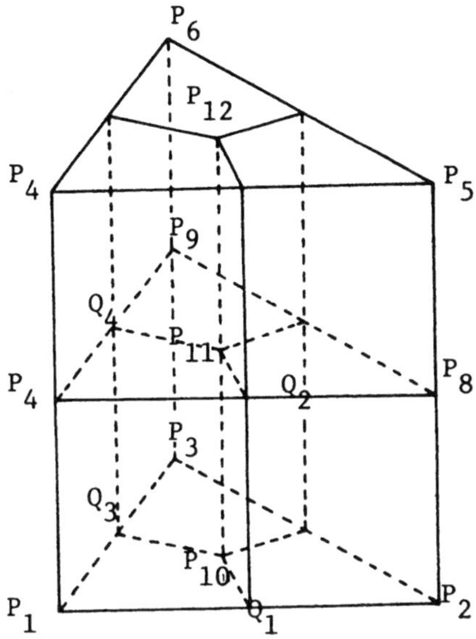

Figure 3. Invented Nodes and Subelements

; P_1 in Figure 3, the integrations related the subelement
1{)^1 can be obtained as

$$\iint_{(s_1')} D_{\alpha\beta}\frac{\partial C}{\partial x_\beta}\, n_\alpha ds = \sum_{\ell=1} \alpha^e_{\ell,1} C_\ell \tag{32}$$

iere (s_1') consists of plans $Q_1Q_2P_{11}P_{10}$, $P_{11}P_{10}Q_3Q_4$ and
,$Q_2P_{11}Q_4$.

$$\begin{aligned}
\alpha^e_{1,1} = \frac{(\Delta z)}{16\Delta}\,(1+\lambda)\ &\{D_{xx}[\bar{b}_{ki}(\omega_1)(b_j-b_i)+\bar{b}_{ji}(\omega_1)(b_k-b_i)] \\
&+D_{xy}[\bar{c}_{ki}(\omega_1)(b_j-b_i)+\bar{c}_{ji}(\omega_1)(b_k-b_i)]\} \\
&-\frac{Dxz}{48\Delta}\,(1-\lambda)(1+2\omega_1)(b_ic_j-c_ib_j)(b_j-b_k) \\
+\frac{(\Delta z)}{16\Delta}\,(1+\lambda)\{&D_{xy}[\bar{b}_{ki}(\omega_1)(c_j-c_i)+\bar{b}_{ji}(\omega_1)(c_k-c_i)] \\
&+D_{yy}[\bar{c}_{ki}(c_j-c_i)+\bar{c}_{ji}(c_k-c_i)]\} \\
&-\frac{D_{yz}}{48}\,(1-\lambda)(1+2\omega_1)(b_ic_j-c_ib_j)(c_k-c_j) \\
+\frac{\lambda}{4}\,\{D_{zx}[\bar{b}_{ki}(\omega_1)+\bar{b}_{ji}(\omega_1)] &+ D_{zy}[\bar{c}_{ki}(\omega_1)+\bar{c}_{ji}(\omega_1)]\} \\
&-\frac{\Delta(1-\lambda)}{3(\Delta z)}\,D_{zz}(1+\frac{2}{3}\omega_1)
\end{aligned} \tag{33}$$

$$\iiint_{(d_1')} V_\alpha \frac{\partial C}{\partial x_\alpha} d\Omega = \sum_{\ell=1}^{6} \beta_{\ell,1} C_\ell \tag{34}$$

where

$$\beta^e_{1,1} = \frac{1}{16}(\Delta z)(1+\lambda)\{V_x[\bar{b}_{ki}(\omega_1)+\bar{b}_{ji}(\omega_1)]+V_y[\bar{c}_{ki}(\omega_1)+\bar{c}_{ji}(\omega_1)]\}$$
$$- \frac{\Delta}{6}(1-\lambda)\, V_z \,(1+\frac{2}{3}\omega_1) \tag{35}$$

In Equations (33) and (35),

$$\bar{b}_{ki}(\omega_1) = b_{ki} + \omega_1 b_{km} \qquad \bar{c}_{ki}(\omega_1) = c_{ki} + \omega_1 c_{km}$$
$$\bar{b}_{ji}(\omega_1) = b_{ji} + \omega_1 b_{jm} \qquad \bar{c}_{ji}(\omega_1) = c_{ji} + \omega_1 c_{jm} \tag{36}$$

Similar expressions are developed for the remaining $\alpha^e_{\ell,1}$ and $\beta^e_{\ell,1}$.

$$\iiint_{(d_1')} \frac{\partial C}{\partial t} d\Omega = \frac{(\Delta z)\Delta}{24}\{(1+\lambda)[(1+\frac{2}{3}\omega_1)\frac{\partial C_1}{\partial t} + (\frac{1}{6}+\frac{2}{3}\omega_2)\frac{\partial C_2}{\partial t}$$
$$+ (\frac{1}{6}+\frac{2}{3}\omega_3)\frac{\partial C_3}{\partial t}$$
$$+(1-\lambda)[(1+\frac{2}{3}\omega_4)\frac{\partial C_4}{\partial t} + (\frac{1}{6}+\frac{2}{3}\omega_5)\frac{\partial C_5}{\partial t} + (\frac{1}{6}+\frac{2}{3}\omega_6)\frac{\partial C_6}{\partial t}]\} \tag{37}$$

The integrations over the subdomain (d_i) can be obtained by summing up the integrations of its subelements. Equation (31) can then be expressed as an ordinary differential equation (ODE).

If the same procedure is applied to all nodes with unknown concentrations, a system of ODE, which is of the same form as Equation (29), would be obtained.

DETERMINATION OF WEIGHTING COEFFICIENTS

The determination of weighting coefficients is described below. First, one can solve the groundwater flow problem by the 3-dimensional Galerkin FEM to obtain the head of each node. Then the Darcy's Law is used to calculate the average flow velocity within each element. Let V_L denote the average horizontal velocity between element (e) and the element below (e). Similar to the 2-dimensional case (Sun and Yeh, 1983), we define

$$\omega_1 = \frac{1}{3} + \eta \frac{V_L^{ij}|ij| - V_L^{ki}|ik|}{\alpha_L V_L}$$

$$\omega_2 = \frac{1}{3} + \eta \frac{V_L^{jk}|jk| - V_L^{ij}|ij|}{\alpha_L V_L} \tag{38}$$

$$\omega_3 = \frac{1}{3} + \eta \frac{V_L^{ki}|ik| - V_L^{jk}|jk|}{\alpha_L V_L}$$

where

α_L is the longitudinal dispersivity;

$|ij|$, $|jk|$, $|ki|$ are the lengths of sides of the triangle (Δ) respectively;

V_L is the magnitude of $\bar{V}_L$;

η is a positive undetermined coefficient.

The weighting coefficients ω_4, ω_5, and ω_6 can be obtained in the same manner. The only difference is that $\bar{V}_L$ now becomes the average horizontal velocity between element (e) and the element above (e).

Denoting V_z^i as the average vertical velocity in the z -direction among those elements which contain both node P_1 and node P_4, we define

$$\lambda_1 = \frac{1}{2} + \eta \frac{V_z^i(\Delta z)}{\alpha_L|V_z^i|} \tag{39}$$

$$\lambda_2 = \frac{1}{2} + \eta \frac{V_z^i(\Delta z)}{\alpha_L|V_z^i|} \tag{40}$$

where η is a positive undetermined coefficient. Similarily, one can obtain λ_2, λ_3, λ_5 and λ_6.

The weighting coefficients which are defined as above obviously satisfy the requirement of Equation (15).

The value of η can be determined by a trial and error procedure. However, it has been found that a η value of 0.001 gives good results under a variety of conditions.

NUMERICAL EXAMPLE

A classical 1-dimensional convection-dispersion problem in a semi-infinite column is used to verify the proposed 3-dimensional scheme.

That is

$$\frac{\partial C}{\partial t} = D \frac{\partial^2 C}{\partial x^2} - V \frac{\partial C}{\partial x}$$

$$C(x,o) = o$$

$$C(o,t) = C_o \qquad (41)$$

$$C(\infty,t) = o$$

where D is the longitudinal hydrodynamic coefficient and V is the average velocity. The analytical solution of this problem is given by (e.g., Fried and Combornous, 1971).

$$C(x,t) = \frac{C_o}{z}\{\mathrm{erfc}[\frac{x-vt}{2(Dt)^{\frac{1}{2}}}] + \exp(\frac{Vx}{D})\ \mathrm{erfc}\ [\frac{x+vt}{2(Dt)^{\frac{1}{2}}}]\} \qquad (42)$$

$C_o = 10$, $V = 1, \Delta x = 5$, $D = 10$, then the mesh Peclet number $P_e = 0.5$. In this case, the numerical solution obtained by using the Galerkin FEM mentioned above fits the analytical solution exactly. For the case of large Peclet number, for example, $C_o = 10$, $V = 1$, $\Delta x = 5$, $D = 0.05$, then $P_e = 100$. The numerical solution still compares favorably with analytical solution. However, slight oscillations occur behind the front of numerical solution. When we use the upstream weight MCB method to solve this problem, the oscillations of numerical solutions are damped.

CONCLUSION

Based upon the concept of upstream weighting and Galerkin FEM, a new numerical method has been developed for simulating the 3-dimensional transport process in groundwater. Triangular-prism elements are used, and six invented nodes are added to each element. With the aid of these invented nodes, upstream weights that depend on the local Peclet number are directly added to the linear basis functions. The proposed method has the following advantages. It can be used to simulate the transport problems with small as well as large Peclet numbers. For a nonuniform flow field, upstream weights are increased automatically at parts with large P_e, and are automatically decreased at parts with small P_e. The adaptive scheme makes it possible not only to avoid the oscillations of solutions but to minimize the additional numerical dispersion with a nominal increase in computer time when compared to the general linear Galerkin FEM.

Also, in this scheme, there are six invented nodes in each element, so the weighting coefficients can be added with a great flexibility by taking into account the flows in any direction.

The numerical experiments and the application of this method to practical 3-dimensional problems and the problems of multi-layered aquifer will be discussed in another paper.

ACKNOWLEDGMENT

This research was supported by a National Science Foundation grant CEE-8301230. The authors wish to thank L. Becker who reviewed the original manuscript.

REFERENCES

Anderson, M.P. (1979) Using Models to Simulate the Movement of Contaminants Through Groundwater Flow Systems. CRC Crit. Rev. Environ. Control, 9(2), 97-156.

Bachmat, Y., J. Bredehoeft, B. Andrews, D. Holtz, and S. Sebastian, (1980) Groundwater Management: The Use of Numerical Models. Water Resour. Monogr. Ser., 5, AGU, Washington, D.C.

Bear, J. (1979) Hydraulics of Groundwater. McGraw-Hill, New York.

Fried, J.J. and M.A. Combernous (1971) Dispersion in Porous Media, Advan. Hydrosci. 7, 170 pp.

Gupta, S.K., K.K. Tanji and J.N. Luthin (1975) A Three-dimensional Finite Element Groundwater Model, California Water Research Center, Contribution, 152.

Heinrich, J.C., P.S. Huyakorn, O.C. Zienkiewicz and A.R. Mitchell, (1977) An "Upwind" Finite Element Scheme for Two-Dimensional Convective-Transport Equation. Int. J. Numer. Method. Eng., 11, 131-143.

Herbst, B.M. (1981) Collocation Method and the Solution of Conduction-Convection Problems. Int. J. Numer. Method. Eng., 17, 1093-1102.

Hughes, T.J.R. (1978) A Simple Scheme for Developing "Upwind" Finite Elements. Int. J. Numer. Method. Eng., 12, 12, 1359-1365.

Huyakorn, P.S. and K. Nilkuha (1979) Solution of Transient Transport Equation Using an Upstream Finite Element Scheme. Appl. Math. Model., 3, 7-17.

Khaleel, R. and D.L. Reddell (1977) Simulation of Pollutant Movement in Groundwater Aquifer. Water Resources Institute Technical Report 81, Texas A&M University.

Konikow, L.F. and J.D. Bredehoeft (1978) Computer Model of Two-Dimensional Solute Transport and Dispersion in Groundwater. U.S. Geological Survey Techniques of Water Resources Investigations, Book 7, C2.

Morton, K.W. (1980) Stability of Finite Difference Approximations to a Diffusion-Convection Equation. Int. J. Numer.

Method Eng., 15, 677-683.

O'Neil, K. (1981) Highly Efficient, Oscillation Free Solution of the Transport Equation over Long Times and Spaces. Water Resour. Res., 17(6), 1665-1675.

Payre, G., M. deBroissia and J. Bazinet (1982) An "Upwind" Finite Element Method via Numerical Integration. Int. J. Numer. Method. Eng., 18, 381-396.

Pinder, G.F. and G. Gray (1977) Finite Element Simulation in Surface and Subsurface Hydrology. Academic Press, New York.

Pinder, G.F. and A. Shapiro (1979) A New Collocation Method for the Solution of the Convection-Dominated Transport Equation. Water Resour. Res., 15(5).

Segol, G. (1977) A Three-Dimensional Galerkin-Finite Element Model for the Analysis of Contaminant Transport in Saturated-Unsaturated Porous Media in Finite Elements in Water Resources, Gray, W.G., G.F. Pinder, and C.A. Brebbia, Pentech Press, London, 2, 123.

Sun, N.-Z. and W. W-G. Yeh (1983) A Proposed Upstream Weight Numerical Method for Simulating Pollutant Transport in Groundwater. Water Resour. Res., 19(6), 1489-1500.

Van Genuchten, M.T. (1977) On the Accuracy and Efficiency of Several Numerical Schemes for Solving the Convective Dispersive Equation, in Finite Elements in Water Resources, Gray, W.G., G.F. Pinder and C.A. Brebbia, Eds., Pentech Press, London, 1, 71.

APPLICATION OF FINITE-ELEMENT MODELS IN THE EVALUATION OF ENGINEERED BARRIERS FOR A MINED GEOLOGIC REPOSITORY

R. G. Baca, J. C. Sonnichsen, M. K. Altenhofen
Rockwell Hanford Operations, Richland, Washington 99352

D. W. Langford
BCS Richland, Inc., Richland, Washington 99352

INTRODUCTION

Considerable research is being conducted in the United States and other countries to develop the technology for safe disposal of nuclear wastes in mined geologic repositories (Klingsberg and Duguid, 1980). The basic concept of geologic disposal is to construct a large facility deep beneath the land surface in stable rock formations that will contain the nuclear waste for 10,000 yr. Over this period of time, the process of radioactive decay would significantly reduce the radiotoxicity of the nuclear waste and assure that risks to future generations would be no greater than risks from an equivalent amount of unmined uranium ore (EPA, 1982). The capability of a repository to isolate the nuclear waste from man and the accessible environment depends on the performance of the natural barriers, i.e., geologic strata, and engineered barriers. In the United States, the federal regulatory agencies have established numerical criteria and standards that will be used to assess the suitability of candidate repository systems.

One of the technical challenges facing scientists and engineers is the problem of developing mathematical models capable of predicting the long-term performance of the repository and its subsystems. Mathematical models will play a dual role in quantifying the isolation capability of the repository system and evaluating alternate designs for the engineered barriers. In this paper, we describe a finite-element simulation approach that is being used to evaluate and compare waste emplacement geometries for a repository in a basalt geology. The approach is based on the use of: (1) a deterministic fluid flow model to predict groundwater movement as a function of hydraulic properties and the thermal regime and (2) a stochastic transport model to predict long-term migration of radionuclides. The simulation results summarized herein represent a preliminary evaluation of relative performance for two proposed barrier designs.

BARRIER DESIGN PROBLEM

Background

The conceptual design for a mined geologic repository in a basalt geology consists of an underground facility of drifts, access tunnels, shafts, and emplacement holes (RKE/PB, 1983). The proposed facility would be constructed at a depth of about 950 m in a dense rock layer about 70 m thick. The subsurface facility would have an areal extent of about 5 km^2 with a total of about 40 km of tunneling. This facility will be designed to accommodate several thousand high-integrity canisters containing nuclear waste with a total inventory equivalent to 70,000 metric tonnes of heavy metal.

During the preclosure phase of the repository, each waste canister would be placed in the repository and then surrounded with a packing material composed of a bentonite clay-crushed basalt mixture (Smith et al., 1983). The emplaced canister with packing material is commonly referred to as the waste package. The performance of this engineered barrier subsystem is a complex function of many factors (i.e., radiation intensity, stress and temperature fields, changes in the geochemical environment) that may affect the canister corrosion rate and the physical properties of the packing material. Another important design factor is the disturbed rock zone around the emplacement hole. The radial extent and hydraulic properties of this zone depend on the method of excavation, i.e., blasting or boring, and the size of the excavation (Kelsall et al., 1982). After placement of all the canisters in the underground facility, the repository would be completely backfilled and sealed.

Analysis Objectives

The basic objective of this preliminary analysis is to characterize the fluid flow environment and study postclosure performance of the waste package as a function of emplacement geometry. Performance indicators relevant to the waste package are the maximum radionuclide release rate and the cumulative radionuclide release over a 10,000-yr period. In this paper, we calculate the performance indicators for two emplacement geometries: (1) long horizontal borehole storage and (2) in-tunnel placement. The main differences between the two geometries are: packing thickness and extent of the disturbed rock zone. These two factors have a significant influence on the flow regime and waste package performance.

ANALYSIS METHODOLOGY

In the vicinity of the repository facility, the isolation performance will be determined by the hydraulic properties of the packing and adjacent host rock, canister lifetime, and the geochemical properties of the waste package. Physical and chemical processes considered to be of major importance are: (1) groundwater flow by natural hydraulic head gradients and thermally induced hydraulic gradients and (2) mass transport resulting from the combined effects of advection, dispersion/diffusion, sorption, and decay. Two separate finite-element analyses were performed to evaluate groundwater flow and mass transfer processes for two candidate waste package design concepts.

Governing Equations

In formulating the mathematical model for this particular analysis, the rock mass is treated as a continuum, i.e., an equivalent porous medium. The numerical models, however, are applicable to fractured-porous media (Baca et al., 1984). For this application, the following set of assumptions has been made:

- Groundwater flow is single-phase, laminar, and steady-state.
- Fluid flow in porous regions obeys Darcy's law.
- Fluid density variations are important, as they induce buoyancy effects.
- Mass transport occurs slowly relative to the rate of chemical interaction (i.e., sorption).

The rock mass is assumed to possess average hydraulic and transport properties. With these assumptions, the governing equations are written in indicial notation as:

Flow law:

$$q_i = - K_{ii}\left(\frac{\partial h}{\partial x_i} + \delta_b \delta_{i3}\right) \tag{1}$$

Continuity:

$$\hat{\varepsilon} = \frac{\partial}{\partial x_i} K_{ii} \left(\frac{\partial h}{\partial x_i} + \delta_b \delta_{i3}\right) \tag{2}$$

Conservation of Mass:

$$\varepsilon = R_d \phi \frac{\partial C}{\partial t} + q_i \frac{\partial C}{\partial x_i} - \frac{\partial}{\partial x_i} (D_{ii} \phi \frac{\partial C}{\partial x_i}) + \lambda R_d \phi C \qquad (3)$$

The quantities ε and $\hat{\varepsilon}$ are referred to as the residuals. In the flow law equation, q_i are the Darcian velocity components, K_{ii} are the principal hydraulic conductivity components, δ_b is the density disparity (buoyancy driving force), and δ_{i3} is the Kronecker delta. In the equation for mass transport, R_d is the radionuclide retardation factor, ϕ is the effective porosity, C is the radionuclide concentration, D_{ii} are the principal components of the hydrodynamic dispersion tensor, and λ is the decay constant. The quantities x_i and t represent the independent variables of space and time, respectively.

Simulation Approach

For a two-dimensional analysis, the physical system is conceptualized in terms of a vertical cross-section through the waste package. In this plane, the physical system consists of the four regions: canister, packing, disturbed rock zone, and host rock. Boundaries of the system are defined by two vertical lines of symmetry and two horizontal boundaries (10 m above and below canister centerline). For the first design, horizontal emplacement, the packing thickness is 0.15 m with a disturbed rock zone thickness of 0.2 m. In the second design, in-tunnel placement, the packing thickness is 1.0 m with a disturbed rock zone thickness of 1.0 m. The performance of the engineered barrier is quantified in terms of the two indicators, calculated at the boundary between the packing material and the damaged rock zone.

The fluid flow process is modeled using a deterministic finite-element model for nonisothermal flow. The physical system is approximated by a two-dimensional cartesian system. Groundwater movement around the canister is caused by the buoyancy driving forces because of the large temperature gradients in the vicinity of the canister. For this analysis, the groundwater flow regime is calculated for the waste package and near-field temperatures expected to occur at approximately 100 yr after waste emplacement (Altenhofen, 1981). Assuming the packing is saturated with groundwater at 100 yr, the results of this analysis would correspond to the peak bouyant driving force in the vicinity of the canister. However, the peak bouyancy condition is not expected to persist beyond a few thousand years. The boundary conditions consist of specified hydraulic heads on the top and bottom boundaries. The lateral boundaries are no-flow boundaries.

For the mass transport analysis, the physical system was analyzed using radial coordinates. Mass transport is modeled as a stochastic process to account for data uncertainty. The first random parameter is the distribution coefficient, K_d, which enters directly into the calculation of the radionuclide retardation factor, R_d, given by:

$$R_d = 1 + \frac{\rho K_d}{\phi} \tag{4}$$

where ρ is the bulk density of the medium. The second random parameter is the canister lifetime, which determines the initial time of the release. The final random parameter is the boundary condition at the canister boundary; this boundary condition is assumed to be equal to the radionuclide solubility, which is represented by a statistical distribution. For conservatism, a zero concentration boundary condition is imposed at the disturbed zone/host rock interface.

A Monte Carlo sampling technique is used in conjunction with a finite-element solution of the mass transport equation. This stochastic modeling technique consists of a four-step procedure: (1) select random values for the random parameters according to the applicable statistical distributions, (2) execute the finite-element simulation model, (3) compute the appropriate performance indicators, and (4) tabulate results and compute the statistical moments. This procedure is repeated until convergence in the statistical moments for each performance indicator is achieved. The basic outputs of the Monte Carlo sampling procedure are cumulative and complementary probability curves for the two performance indicators: maximum radionuclide release rate and cumulative release over 10,000 yr.

Finite-Element Method

Two finite-element codes have been developed for two-dimensional analysis of groundwater flow and mass transport. The numerical solution techniques are based on the general method of weighted residuals with the Galerkin criteria. For brevity, the discussion of the numerical method is limited to the finite-element approach developed for the mass transport equation. Details on the finite-element technique developed for solution of the flow equation may be found in Baca et al., (1984).

To obtain a finite-element formulation in terms of incremental changes in concentration, ΔC, a first-order Taylor series approximation is introduced into the standard Galerkin functional. For the mass transport equation, this functional equation is of the form:

$$\chi = \iint_{R'} \omega_j \varepsilon dR + \iint_{R'} \omega_j \frac{\partial \varepsilon}{\partial C} \Delta C dR \tag{5}$$

where R' is the continuum domain and w_j represents a set of weighting functions.

The first integral will yield the load vector of the finite-element equations. Similarly, the second integral will produce the coefficient matrix. Substituting Equation 3 into the functional equation, the load vector is computed from:

$$\iint_{R'} \omega_j \varepsilon dR = R_d \phi \iint_{R'} \omega_j \left(\frac{\partial C}{\partial t} + \lambda C\right) dR + \iint_{R'} \omega_j \left(u \frac{\partial C}{\partial x} + v \frac{\partial C}{\partial y}\right) dR$$
$$+ D_{11} \phi \iint_{R'} \frac{\partial \omega_j}{\partial x} \frac{\partial}{\partial C} \left(\frac{\partial C}{\partial x}\right) dR + D_{22} \phi \iint_{R'} \frac{\partial \omega_j}{\partial y} \frac{\partial}{\partial C} \left(\frac{\partial C}{\partial y}\right) dR \tag{6}$$

Similarly, the coefficient matrix is computed from:

$$\iint_{R'} \omega_j \frac{\partial \varepsilon}{\partial C} dR = R_d \phi \iint_{R'} \omega_j \left\{ \frac{\partial}{\partial C}\left(\frac{\partial C}{\partial t}\right) + \lambda \right\} dR$$
$$+ \iint_{R'} \omega_j \left\{ u \frac{\partial}{\partial C}\left(\frac{\partial C}{\partial x}\right) + v \frac{\partial}{\partial C}\left(\frac{\partial C}{\partial y}\right) \right\} dR + D_{11} \phi \iint_{R'} \frac{\partial \omega_j}{\partial x} \frac{\partial}{\partial C}\left(\frac{\partial C}{\partial x}\right) dR$$
$$+ D_{22} \phi \iint_{R'} \frac{\partial \omega_j}{\partial y} \frac{\partial}{\partial C}\left(\frac{\partial C}{\partial y}\right) dR \tag{7}$$

Time derivatives in the above equations are replaced and/or approximated using the formula:

$$\frac{\partial C^{m+1}}{\partial t} = \frac{\alpha}{\Delta t}(C^{m+1} - C^{m}) + (1 - \alpha)\frac{\partial C^{m}}{\partial t} \tag{8}$$

The coefficient, α, is an integrating factor whose value may range from 1 to 2; Δt is the time step size, and m is the time plane index. The concentration, C, is approximated by quadratic shape functions for quadrilaterals, triangles, and line elements. One of the advantages of the Taylor series approach is that it eliminates the need to modify the finite-element matrices to incorporate Dirichlet boundary conditions, i.e., ΔC for nodes with specific values is zero.

EVALUATION OF BARRIER DESIGNS

The finite-element simulations presented herein are examples from recent theoretical analyses The simulations are based on various simplifying assumptions and data estimates, and therefore, they do not represent absolute or final performance predictions. The simulations are intended to provide insight into controlling phenomena and significance of data uncertainty. Fluid flow simulations were generated using the deterministic version of the MAGNUM 2-D code. Mass transport simulations were made using the stochastic version of the CHAINT code.

Fluid Flow

Deterministic simulations of groundwater flow fields around the waste package were performed using a nonisothermal fluid flow model. A three-step procedure was used to produce the flow simulations for each waste placement design. First, temperature boundary conditions were assigned to the periphery of the finite-element network and the heat equation was then solved to obtain the temperature field around the waste package; for conservatism, the heat transfer was assumed to occur by conduction only. Second, the temperature field was input to flow simulation along with appropriate hydraulic boundary conditions. The hydraulic head field around the waste package was calculated as a direct function of the in situ hydraulic gradient and buoyancy driving force. Third, the computed hydraulic head distribution was input to a computer code that calculated the groundwater streamlines and velocity vectors.

The model geometry for each design was represented with a finite-element grid composed of quadrilateral isoparametric elements. For the horizontal borehole design, the finite-element grid consisted of 132 elements and 453 nodes, the in-tunnel placement design was represented with a 354-element grid system with 1,151 nodes. For both cases, the finite-element mesh was refined in the area near the package and coarse in other areas. The hydraulic properties for each material type of the finite-element mesh were as follows. The hydraulic conductivity and effective porosity of the packing material were 10^{-9} m/s and 0.3, respectively. The host rock was assigned a hydraulic conductivity of 10^{-13} m/s and an effective porosity of 10^{-4}.

The hydraulic conductivity of the disturbed zone was estimated using the approach developed by Kelsall et al. (1982). For the horizontal borehole, a single zone representing a stress-induced transition between the edge of the borehole and the host rock was assumed. This stress-induced zone was assumed to extend a distance of 2.0R from the canister centerline into the host rock, where R is the radius of the borehole. The hydraulic conductivity of the stress-induced zone was assumed to be 10^{-12} m/s. For the in-tunnel emplacement, two zones consisting of an inner blast damage zone and an outer stress-induced zone were assumed to exist. The blast damage zone was assumed to extend a distance of 1.3R from the canister centerline into the host rock. The hydraulic conductivity of the blast damage zone was assumed to be 10^{-10} m/s, and the hydraulic conductivity of the stress-induced zone was assumed to be 10^{-12} m/s. The porosity of the rock in the damage zone was related to the porosity of the host rock using an empirical relation.

The groundwater flow simulations for the two designs are presented in Figure 1. These results indicate that the flow patterns are characterized by distinct regimes. Both cases exhibit a convection cell in the vicinity of the waste package, whereas in the host rock the flow is, for the most part, rectilinear. In the first design, the convection cell is almost entirely confined in the packing material. This is primarily due to the large relative difference in hydraulic conductivity between the packing material and the stress-induced zone, i.e., 3 orders of magnitude. In contrast, the flow patterns for the second design show that the convection cell extends into the damaged rock zone. The circulation in this localized region is again due to the contrast in hydraulic properties between the damaged and disturbed rock zones.

The calculated water traveltimes along the streamlines indicate that the average residence times in the packing material are significantly longer than in the host rock. Moreover, the groundwater velocities in the packing material are small relative to mass transfer velocities. The large

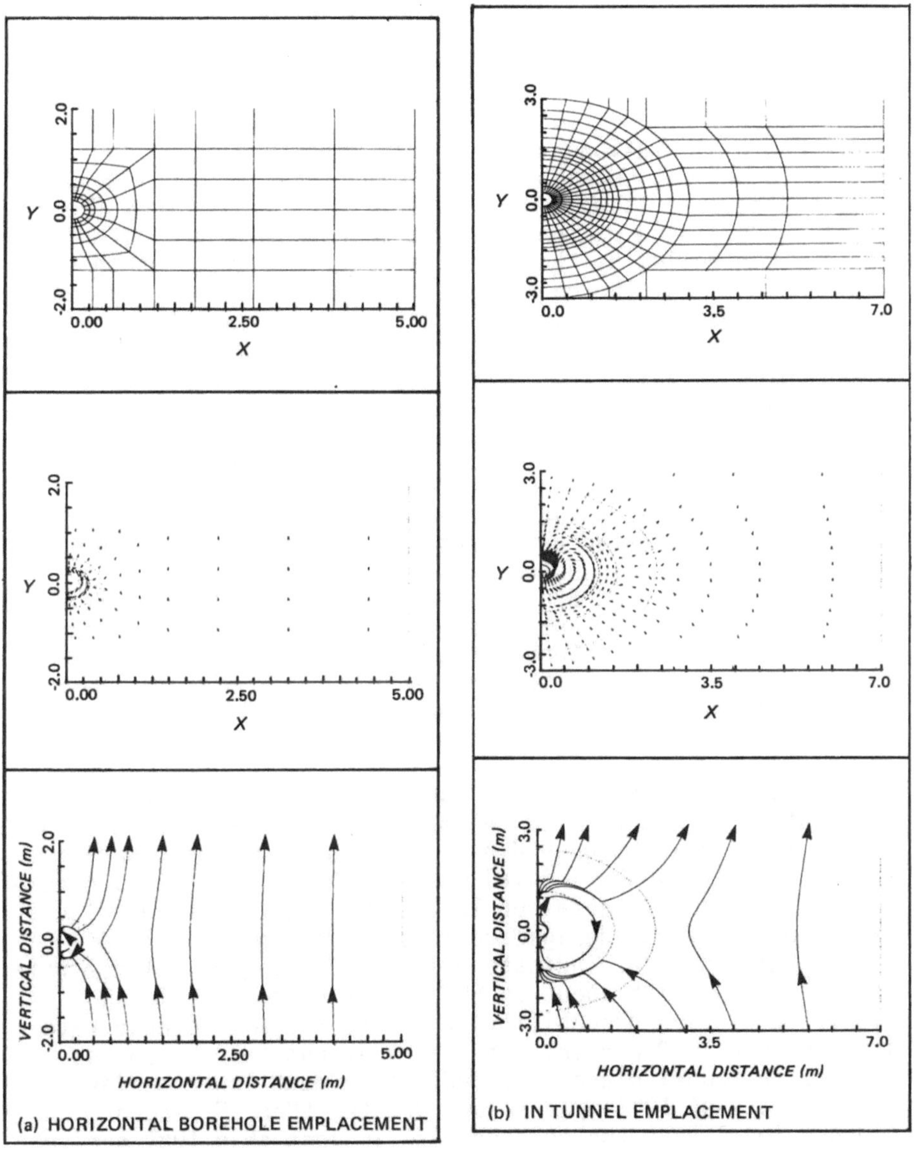

Figure 1. Flow Field Geometry and Simulated Groundwater Flow Characteristics for Two Emplacement Configurations.

contrast in residence times and low water velocities suggest that the mass transport from the canister to the boundary of the disturbed rock zone will be diffusion-controlled, i.e., the dominant transport mechanism is molecular diffusion. This effect becomes increasingly pronounced as the radionuclide sorption becomes large. This result motivated the axisymmetric approximation used in the mass transport analysis.

Mass Transport

Monte Carlo simulations of mass transport were conducted for one radionuclide, selenium-79 (^{79}Se). This radionuclide was selected because it was identified in a previous study (Wood et al., 1983) as a radionuclide whose releases were not adequately controlled by the waste package. Movement of this radionuclide from the canister to the host rock was modeled as a diffusion-controlled process, attenuated by sorption and decay. The molecular diffusion coefficient, D, was assigned a fixed value of 3.14×10^{-4} m^2/yr (Torstenfelt, et al., 1983). The random model parameters consisted of the distribution coefficient, K_d, and solubility, C_s, which were assumed to be log-normally distributed. Sample means and standard deviations were derived from available laboratory data and theoretical estimates (Salter and Jacobs, 1983). Canister lifetime, also treated as a random parameter, was assumed to be log-normally distributed with a median of 1,000 yr and standard deviation of 630 yr. In the Monte Carlo simulation, each trial required a numerical solution of the mass transport equation for a time interval from the canister failure time (a random parameter) to 10,000 yr. A total of 600 trials were required to obtain convergence on the statistical moments of the output.

The radionuclide release rate for each trial was obtained by calculating and tabulating the maximum fractional release rate (i.e., mass flux integrated over the waste package boundary and normalized by the initial mass of the radionuclide) occurring in the time interval of 10,000 yr. Release rate calculations for the first design, canister placement in a long horizontal borehole, indicate a median fractional release rate of 10^{-8} per yr with about an 80 percent confidence level (i.e., cumulative probability) that the maximum release rate is less than 10^{-5} per yr. The calculations for the second design, in-tunnel placement, indicate a much lower median release rate of 10^{-11} per yr and greater than 95 percent confidence level that the release rate is less than 10^{-5} per yr. The lower median release rate for the second design reflect the impact of sorptive capacity in a thicker barrier, i.e., the radionuclide traveltime to the waste package boundary is much longer. The numerical results for both designs are compared in Figure 2.

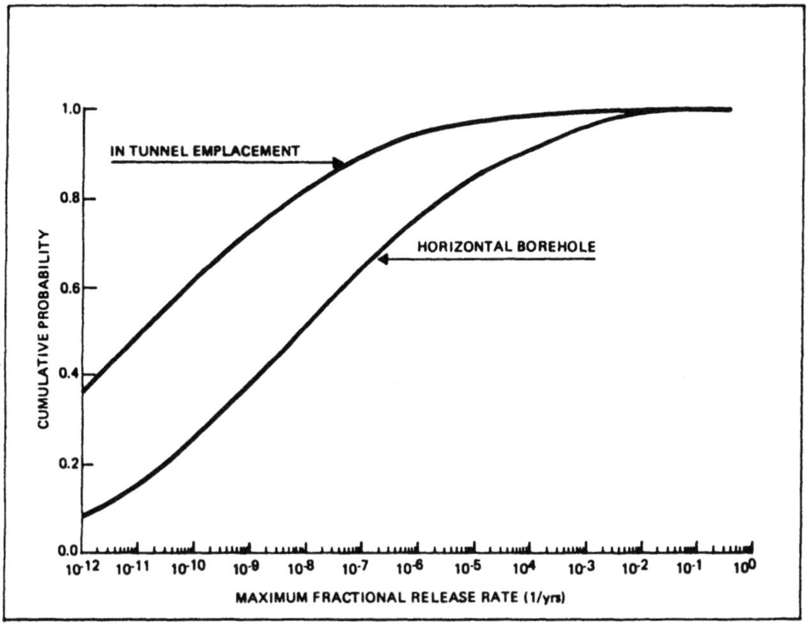

(a) RADIONUCLIDE RELEASE RATE (1/yrs)

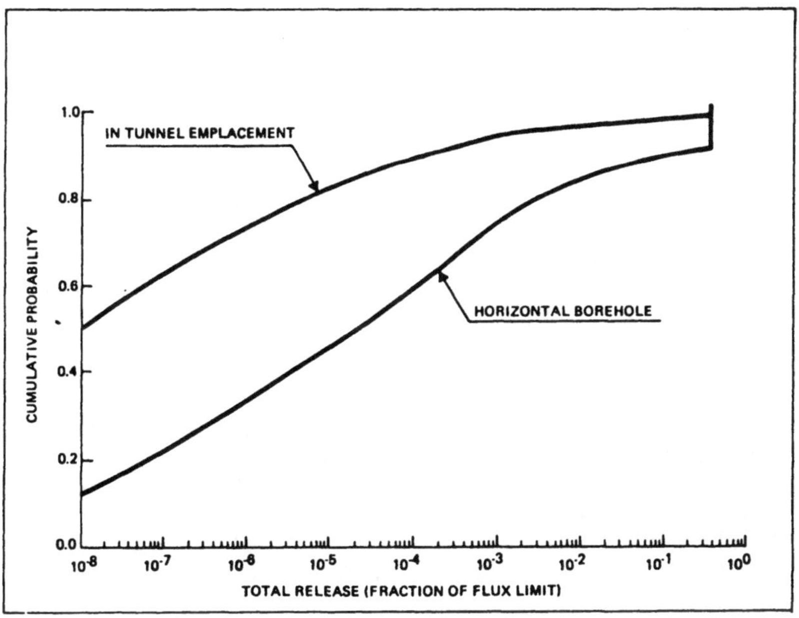

(b) CUMULATIVE RADIONUCLIDE RELEASES

Figure 2. Probability Curves for Maximum Fractional Release Rate And Total Release of ^{79}Se for Two Emplacement Configurations.

The cumulative radionuclide release for each trial was obtained by tabulating the total mass released from the waste package over the time interval of 10,000 yr. The simulations for the first design case indicate a median cumulative release of ^{79}Se of about 10^{-5} flux limits, i.e., total curie release normalized by the proposed EPA curie limit, with a 95 percent confidence that the cumulative release is less than 0.35 flux limit. For the second design, the calculations show a median cumulative release of 10^{-8} flux limits with a greater than 95 percent confidence that the total release is less than 0.35 flux limits. The cumulative probability curves for both designs are shown in Figure 2. These preliminary results suggest that uncertainty in the predicted waste package performance can be compensated for in a systematic way by adjusting design factors such as packing thickness to meet waste package performance goals, i.e., desired level of confidence that the releases do not exceed a specified amount. This approach can also be applied to other design factors of the engineered barriers.

CONCLUDING REMARKS

The numerical models developed for repository performance evaluations employ finite-element solution techniques within a deterministic and stochastic modeling framework. The model applications clearly demonstrate the versatility of the finite-element technique to complex engineering analysis problems. This unique versatility permits the analyst to tailor the predictive methodology to meet the requirements of the engineering analysis problem.

A practical application of finite-element models to the problem of evaluating alternative designs has been presented. The computer simulations described in this paper illustrate how the finite-element models can provide increased understanding of engineered barrier performance regarding its dependence on hydrologic parameters and geotechnical factors.

ACKNOWLEDGMENTS

The authors wish to thank various colleagues for their critical review of the manuscript and helpful suggestions. The authors thank Messrs. R. C. Arnett, P. M. Clifton, and B. Sagar. The work summarized in this paper was conducted for the Basalt Waste Isolation Project, Rockwell Hanford Operations, under contract to the U.S. Department of Energy.

REFERENCES

Altenhofen, M. K.
1981. Waste Package Heat-Transfer Analysis: Model Development and Temperature Estimates for Waste Packages in a Repository Located in Basalt. RHO-BWI-ST-18, Rockwell Hanford Operations, Richland, Washington.

Baca, R. G., Arnett, R. C., and Langford, D. W.
1984. Modeling Fluid Flow in Fractured-Porous Rock Masses by Finite-Element Techniques. International Journal for Numerical Methods in Fluids (in press), also RHO-BW-SA-297 P, Rockwell Hanford Operations, Richland, Washington.

EPA.
1982. Environmental Standards for the Management and Disposal of Spent Nuclear Fuel, High-Level and Transuranic Radioactive Wastes. Environmental Protection Agency, 40 CFR 191. Federal Register, Vol. 47, No. 250, December 29, 1982, Proposed Rule.

Kelsall, P. C., Case, J. B., and Chabannes, C. R.
1982. A Preliminary Evaluation of the Rock Mass Disturbance Resulting From Shaft, Tunnel, or Borehole Excavation. ONWI-411, Prepared for Battelle Memorial Institute by D'Appolonia Consulting Engineers, Inc., Albuquerque, New Mexico.

Klingsberg, C. and Duguid, J.
1980. Status of Technology for Isolating High-Level Radioactive Wastes in Geologic Repositories. DOE/TIC 11207, U.S. Department of Energy, Washington, D.C.

RKE/PB.
1983. Conceptual System Design Description, Nuclear Waste Repository in Basalt. Raymond Kaiser Engineers/Parsons Brinkerhoff Quade & Douglas, Inc., Oakland, California, SD-BWI-TI-005, Rockwell Hanford Operations, Richland, Washington.

Salter, P. F. and Jacobs, G. K.
1983. BWIP Data Package for Reference Solubility and K_d Values SD-BWI-DP-001, Rockwell Hanford Operations, Richland, Washington.

Smith, M. J., Barney, G. S., and Moore, E. L.
1983. Waste Packages for a Repository Located in Basalt. In the proceedings of the Civilian Radioactive Waste Management Information Meeting, Washington, D.C., also RHO-BW-SA-320 P, Rockwell Hanford Operations, Richland, Washington.

Torstenfelt, B., Allard, B., Andersson, K., Kipsati, H. Eliasson, L., Olofsson, U., and Persson, H.
1983. Radionuclide Diffusion and Mobilities in Compacted Bentonite. KBS TR-83-34, Chalmers University of Technology, Goteborg, Sweden.

Wood, M. I., Relyea, J. F., Myers, J., and Apted, M. J.
1983. The Near-Field Waste Package Environment in Basalt and Its Effect on Waste Form Releases. In the proceedings of the Civilian Radioactive Waste Management Information Meeting, Washington, D.C., also RHO-BW-SA-331 P, Rockwell Hanford Operations, Richland, Washington.

NUMERICAL SOLUTION OF FOURTH ORDER EQUATIONS ARISING IN INCOMPRESSIBLE FLOW PROBLEMS

George J. Fix

Carnegie-Mellon University, Pittsburgh, PA 15213, USA

1. INTRODUCTION

The use of stream function-vorticity formulations for two dimensional flow problems, and in addition, vector potentials for three dimensional problems, has been widespread. Such formulations reduce the number of variables, and offer potentially greater resolution than formulations in primative variables.

The major drawback is that these methods are intrinsically of the mixed type and hence subject to instabilities not seen in standard formulations (Fix, Gunzburger, Nicolaides 1981; Babuska, Oden, Lee 1978, Brezzi 1975). Moreover, both finite difference and finite element discretizations share these problems.

The goal of this paper is to discuss these instabilities in the context of the Brezzi canonical form for mixed methods. Results from selected numerical experiments will also be discussed.

2. BASIC FORMULATION

In the interest of brevity we shall confine attention to stream function-vorticity formulations using finite element discretizations. In particular, let φ denote the stream function, u the vorticity, and g a body force. Then in the flow region Ω we have

$$u - \Delta\varphi = 0 \quad \text{in} \quad \Omega \tag{2.1}$$

$$\Delta u - J(u,\varphi) = g \quad \text{in} \quad \Omega \tag{2.2}$$

$$\varphi = \frac{\partial\varphi}{\varphi n} = 0 \quad \text{on} \quad \Gamma, \tag{2.3}$$

where Γ is the boundary of Ω, n is the outer normal, and

$$J(u,\varphi) = u \frac{\partial u}{\partial x}\frac{\partial \varphi}{\partial y} - \frac{\partial \varphi}{\partial x}\frac{\partial u}{\partial y}. \tag{2.4}$$

Let H_1 denote the space of functions v for which

$$\int_\Omega |\nabla v|^2 < \infty,$$

and let $\mathring{H}_1$ denote those functions in H_1 which vanish on the boundary Γ of Ω. Then the standard variational formulation (Fix 1975_1; Fix 1975_2) of (2.1)-(2.3) is to find a u in H_1 (no boundary conditions) and a φ in $\mathring{H}_1$ (zero boundary data) for which

$$\int_\Omega uv + \int \nabla v \cdot \nabla\varphi = 0 \quad \text{all} \quad v \quad \text{in} \quad H_1 \tag{2.5}$$

$$\int_\Omega \nabla u \cdot \nabla\psi + \int_\Omega J(u,\varphi)\psi = \int g\psi \quad \text{all} \quad \psi \quad \text{in} \quad \mathring{H}_1 \tag{2.6}$$

The boundary condition $\frac{\partial\varphi}{\partial n} = 0$ on Γ is natural or implicit in this formulation.

Approximations are obtained by replacing H_1 with a finite dimension subspace

$$\mathcal{S}_h \subset H_1,$$

parameterized by the generic mesh spacing $h > 0$. Letting

$$\mathring{\mathcal{S}}_h \subset \mathcal{S}_h$$

denote the subspace of functions v_h in $\mathcal{S}_h$ vanishing on the boundary, the approximate system is to seek u_h in $\mathcal{S}_h$ and a φ_h in $\mathring{\mathcal{S}}_h$ such that (2.5)-(2.6) holds (with u_h replacing u and φ_h replacing φ) for all v in $\mathcal{S}_h$ and ψ in $\mathring{\mathcal{S}}_h$.

Once a basis is chosen for $\mathcal{S}_h$ and $\mathring{\mathcal{S}}_h$, (2.5)-(2.6) reduces to a system of nonlinear algebraic equations (Strang and Fix 1973).

3. BREZZI CANONICAL FORM

Omitting the nonlinear term for the moment the system has the general form

$$Au + B\varphi = f \tag{3.1}$$

$$B^* u = g, \tag{3.2}$$

which is commonly called the Brezzi Canonical Form (Brezzi 1975).

Variational formulations of the mixed type can generally be reduced to either (3.1)-(3.2) or a minor perturbation of it.

There are two cases of interest. One is the so-called "smoothing system" where A dominates B in a sense that can be made precise. An example of such a system is provided by the Stoke's equations at a small Reynolds number:

$$(\nu\Delta+\underline{v})u - \text{grad}\ \varphi = f \quad \text{in} \quad \Omega \tag{3.3}$$

$$\text{div}\ u = 0 \quad \text{in} \quad \Omega \tag{3.4}$$

$$u = 0 \quad \text{on} \quad \Gamma. \tag{3.5}$$

Here u is the velocity, φ the pressure, and the viscosity $\nu > 0$ is sufficiently large for

$$A = \nu\Delta + v \tag{3.6}$$

to dominate

$$B = -\text{grad}. \tag{3.7}$$

An example of a "nonsmoothing system" is provided by the stream function-vorticity formulation in Section 2, i.e.,

$$A = I, \qquad B = \Delta. \tag{3.8}$$

The Poisson equation

$$u - \text{grad}\ \varphi = 0 \quad \text{in} \quad \Omega \tag{3.9}$$

$$\text{div}\ u = 0 \quad \text{in} \quad \Omega \tag{3.10}$$

$$\varphi = 0 \quad \text{on} \quad \Gamma$$

is another example.

Both smoothing and nonsmoothing systems share intrinsic instabilities due to so-called divergence constraint

$$B^*u = g \quad \text{in} \quad \Omega. \tag{3.11}$$

In the Stokes and Poisson equations B^* is indeed the divergence operator div, however in the (2.1)-(2.3), B^* is the laplacian Δ. These instabilities have been widely studied (Girault and Raviart 1979; Gunzburger and Peterson; Fix, Gunzburger, and Nicolaides 1979; Fix, Gunzburger, and Nicolaides 1981; Boland 1983), and necessary and sufficient conditions on $\mathcal{S}_h$ have been established for when they will not be present. They are called Babuska-Brezzi angle conditions.

In smoothing systems these conditions are also necessary and sufficient for the overall stability of the approximation. Moreover, a large number of elements have been developed which satisfy these conditions. These include so-called bubble elements (Girault and Raviart 1979), the union jack element (Babuska and Aziz 1972), and box elements (Boland 1983) to cite only a few.

In nonsmoothing systems these conditions are also necessary but not sufficient. That is, there are elements which work very well for smoothing systems like low Reynolds number incompressible flows (Stokes), but display erratic behavior in the compressible case (Poisson).

Suppose for the moment we take the Poisson equation (3.9)-(3.10), where u denotes the mass flow. Observe that the function is in the space

$$\mathfrak{n} = \{v \in H_1 \ : \ \operatorname{div} v = 0\} \qquad (3.12)$$

of divergence free functions (which is an expression of mass conservation). The approximation $u_h \in \mathcal{S}_h$ is only in the space

$$\mathfrak{n}_h = \{v_h \in \mathcal{S}_h \ : \ \int \operatorname{div} v_h \psi_h = 0 \ \text{ all } \ \psi_h \in \mathring{\mathcal{S}}_h\}, \qquad (3.13)$$

and these functions may not be divergence free. In fact, the union jack element is the only one of the ones cited above that satisfies the <u>inclusion property</u>

$$\mathfrak{n}_h \subset \mathfrak{n}, \qquad (3.14)$$

which is equivalent to exact mass conservation.

It has been shown in (Fix, Gunzburger, Nicolaides 1981), that (3.14) is (along with the Babuska-Brezzi angle condition) sufficient for overall stability for nonsmoothing systems. Moreover, various new elements satisfying (3.14) have been recently introduced (Fix and Suri).

The inclusion property (3.14) for the (2.1)-(2.3) introduces a far more serious restriction on $\mathcal{S}_h$. Indeed, in this case

$$\mathfrak{n} = \{v \in H_1 \ : \ \Delta v = 0\}$$

is the space of harmonic functions, while

$$\mathfrak{n}_h = \{v \in \mathcal{S}_h \ : \ \int \nabla v \cdot \nabla \psi = 0 \ \text{ all } \ \psi \in \mathring{\mathcal{S}}_h\}$$

consists of discrete harmonic functions. Thus, (3.14) boils down to using appropriate C^1 elements.

4. RESULTS

A large number of numerical experiments have been conducted with the stream function-vorticity equations using C^1 elements (Fix, Gunzburger, Nicolaides and Peterson 1984), and they have confirmed the optimal accuracy and stability of these formulations. On the other hand, one of the major reasons for going to a mixed formulation like (2.5)-(2.6) is that it, unlike standard fourth order formulations, permits the use of linear elements. The above suggests that there will be problems with such formula tions, but because of the practicality of lower order elements some attention has been given to how much accuracy is actually lost in these approaches.

As far as the stream function φ is concerned, and in particular the flow velocities $\nabla\varphi$, the general theory shows that optimal accuracy is achieved in the sense that

$$\{\int_\Omega |\nabla\varphi-\nabla\varphi_h|^2\}^{1/2} = O(h^{k-1}) \tag{4.1}$$

if piecewise polynomials of degree $k-1$ are used. Thus with linear elements ($k=2$) one gets first order accuracy in the velocities.

The same is not true for the vorticity. The general theory gives an upper bound of the form

$$\{\int_\Omega |u-u_h|^2\}^{1/2} = O(h^{k-3/2}) \tag{4.2}$$

This predicts $O(h^{3/2})$ for noninclusive quadratic elements ($k=3$), and numerical experiments indicate that this is sharp.

The bound (4.2) predicts $O(h^{1/2})$ convergence for linear elements ($k=2$), however selected numerical experiments indicate that this may not be sharp. Indeed, $O(h^2)$ convergence was seen for homogeneous boundary data, and this dropped to $O(h)$ with inhomogeneous data.

Interestingly, a closer examination of the algebraic equations arising from (2.5)-(2.6) with linear and quadratic noninclusive elements shows that in effect these formations force the partial differential equation (2.1) to hold on the boundary Γ of Ω. This yields boundary values for the vorticity u and with this the equation (2.6) reduces to a standard finite element formulation of a second order elliptic problem. Such approaches have been used in finite difference formulations with similar results (Roache 1972).

The inclusive elements, on the other hand, do the exact opposite. Here the partial differential equation (2.1) is not imposed on the boundary, but rather the boundary conditions (2.3) are extended slightly into the interior. That such a formulation with inclusive quadratic elements gives $O(h^3)$ accuracy to the vorticity and $O(h^2)$ accuracy to the velocities may be a key to the development of more accurate finite difference formulations.

ACKNOWLEDGMENTS

This work was supported in part by ONR under contract N00014-81-K-0306 and ARO under contract DAAG29-83-K-0084.

REFERENCES

Babuska, I., Oden, J. T., Lee, J. K. (1978) Mixed Hybrid Finite Element Approximation of Second Order Elliptic Boundary Value Problems. Comput. Methods Appl. Mech. Engrg. 14, 1-23.

Boland, James M. (1983) Finite Elements and The Divergence Constraint for Viscous Flows. Ph.D. Thesis, Dept. of Math., Carnegie-Mellon University.

Brezzi, F. (1975) On the Existence, Uniqueness and Application of Saddle Point Problems Arising from Lagrange Multipliers. R.A.I.R.O. 8, 129-150.

Fix, G. J. (1975) Finite Element Models for Ocean Circulation Problems. SIAM J. Appl. Math. 29, 371-387.

Fix, G. J. (1975) Finite Elements and Fluid Dynamics. ICASE Report 75-1, NASA Langley Research Center, Hampton, Virginia.

Fix, G. J., Gunzburger, M. D., Nicolaides, R. A. (1979) Theory and Applications of Mixed Finite Element Methods. Proceedings of a symposium in honor of R. J. Duffin, Academic Press, 275-393.

Fix, G. J., Gunzburger, M. D., Nicolaides, R. A. (1981) On Mixed Finite Element Methods for First Order Elliptic Systems. Numerische Mathematik 37, 29-48.

Fix, G. J., Gunzburger, M. D., Nicolaides, R. A., Peterson, J. S. Peterson (1984) Mixed Finite Element Approximations for the Biharmonic Equation. Proceedings of the Fifth International Conference on Finite Elements and Flow Problems, Austin, Texas, to appear.

Fix, G. J., Suri, M., Three-Dimensional Mass Conserving Elements for Compressible Flows, in preparation.

Girault, V., Raviart. P.-A. (1979) Finite Element Approximation of the Navier-Stokes equations. Lecture Notes in Mathematics (A. Dold and B. Eckmann, eds.), New York: Springer-Verlag.

Gunzburger, M. D., Peterson, J. S., On Conforming Finite Element Methods for the Inhomogeneous Stationary Navier-Stokes Equations. Accepted for publication, Numerische Mathematik.

Pitkaranta, J., Stenberg (1982) Analysis of some mixed finite element methods for plane elasticity equations. Report HTKK-MAT-A194, Helsinki University of Technology.

Roache, P. (1972) Computational Fluid Dynamics. Hermosa Publishers.

Strang, G., Fix, G. J. (1973) An Analysis of the Finite Element Method. New York: Prentice Hall.

AUTHOR INDEX

J. Carrera	Department of Hydrology and Water Resources, University of Arizona, USA	199-211
F. Cavalieri	Tecnosistem S.p.A, Napoli Italy	339-351
A. H-D. Cheng	Department of Civil Engineering and Engineering Mechanics Columbia University, Canada	737-746
A.S. Cheng	Numerical Analysis Research Department, Shanghai Institute of Computing Technology Shanghai, China	637-645
X. Chunhong	Nanjing University, China	89-102
B.A. Devantier	Southern Illinois University Carbondale, USA	647-655
T.J. Durbin	US Geological Survey, USA	265-282
P. Esposito	Laboratoire National d'Hydraulique, Electricite de France France	477-494
M.C.G. Foreman	Institute of Ocean Sciences, Sidney, Canada	447-456
D.G. Fredlund	Department of Civil Engineering University of Saskatchewan, Canada	113-122
E.O. Frind	Department of Earth Sciences University of Waterloo Ontario, Canada	363-381
G. Gambolati	Istituto di Matematica Applicata Italy	25-36
G. Abi-Ghanem	Environmental and Water Resources, Minneapolis, USA	299-307
W.G. Gray	Department of Civil Engineering University of Notre Dame, USA	533-543
P.M. Gresho	Lawrence Livermore National Laboratory, Livermore, USA	667-680
A.M. Harte	Department of Civil Engineering University College Galway, Ireland	427-436

M.A. Noor	Mathematics Department King Saud University Saudi Arabia	231-240
K. O'Neill	US Army Cold Regions Research Hanover, USA	231-229
A. Pandit	Florida Institute of Technology USA	319-328
J. Pecaric	Faculty of Civil Engineering University of Belgrade, Yugoslavia	723-736
A.M. Perdon	Istituto di Matematica Appli- cata, Italy	25-36
G.F. Pinder	Department of Civil Engineering Princeton University, USA	165-174
P. Prinos	Department of Civil Engineering University of Ottawa, Canada	509-520
C. Le Provost	Institute of Mechanics Grenoble, France	567-576
M. Radojkovic	Faculty of Civil Engineering University of Belgrade, Yugoslavia	723-736
J. Rance	Department of Civil Engineering University College, Swansea, Wales	657-66
R. Ratke	Hannover University, Germany	521-532
H.W. Reeves	University of Notre Dame, USA	309-318
G. Ricceri	Istituto di Costruzioni Marit- time e Geotecnica, University of Padua, Italy	25-36
W. Rottmann-Sode	Institute of Fluid Mechanics University of Hannover, West Germany	405-415
G. Schmitz	The Technical University of Munich, Germany	495-508 555-566
G.J. Seus	The Technical University of Munich, Germany	495-508 555-566

MIX
Papier aus verantwortungsvollen Quellen
Paper from responsible sources
FSC® C105338
FSC
www.fsc.org

If you have any concerns about our products,
you can contact us on
ProductSafety@springernature.com

In case Publisher is established outside the EU,
the EU authorized representative is:
Springer Nature Customer Service Center GmbH
Europaplatz 3, 69115 Heidelberg, Germany

Printed by Libri Plureos GmbH
in Hamburg, Germany